HETERO-AROMATIC NITROGEN COMPOUNDS

HETERO-AROMATIC NITROGEN COMPOUNDS

Pyrroles and Pyridines

K. SCHOFIELD, Ph.D., D.Sc., F.R.I.C.

University of Exeter

Springer Science+Business Media, LLC

1967

©

Springer Science+Business Media New York

1967

Originally published by Butterworth & Co. (Publishers) Ltd in 1967.

Softcover reprint of the hardcover 1st edition 1967

ISBN 978-1-4899-5894-5   ISBN 978-1-4899-5892-1 (eBook)

DOI 10.1007/978-1-4899-5892-1

Suggested U.D.C. *No.* 547·52–17

Suggested additional No. 547·82

Library of Congress Catalog Card Number 67-26477

CONTENTS

PREFACE

It is the author's aim to provide a comprehensive account of the chemistry of the major mono- and bi-cyclic nitrogen heteroaromatic systems. The most characteristic feature of recent work on heterocyclic chemistry has been the application of the methods of physical organic chemistry. It is fair to say that current textbooks and monographs on heterocyclic chemistry do not give these developments adequate treatment. An attempt is made here to redress these shortcomings.

In this volume the two families, the pyrroles and pyridines, the prototypes of these systems, are discussed. All the characteristics of nitrogen heteroaromatic systems are to be found in the behaviour of pyrroles and pyridines, and thus both systems are of the greatest theoretical interest. In addition, the pyridines are of great medical, technical and practical importance. Pyrrole chemistry is dominated by electrophilic substitutions, whereas electrophilic, nucleophilic and radical substitutions abound in pyridine chemistry. This contrast, combined with the practical importance of the pyridines, has created a very big difference between our knowledge of the pyridines and of the pyrroles.

The major theme is the aromatic reactivity of the pyrrole and pyridine nuclei. Consequently, an attempt is made to set their chemistry into the historical development of the theory of aromatic structure and reactivity, and to present that theory as it has been applied to heterocyclic systems. A recurring theme is the poverty of the experimental evidence available for comparison with theory.

Second only to the vast amount of material to be surveyed, there is among the problems facing the author of a work of this kind that of deciding how to arrange it. The only proper basis for discussing the majority of chemical reactions reported is clearly their division into electrophilic, nucleophilic and radical substitutions or additions. This course has been followed. The blurring of the division between substitutions and additions, particularly when the primary attack is by a nucleophile, makes it desirable to include some reactions which in their outcome are additions, in the discussion of substitutions. Inside each of the three great divisions mentioned, substitutions have been, for want of any fundamental basis, arranged alphabetically. The idea of substitution is used in its generalized sense, and with each type, after examples resulting in the replacement of hydrogen, those resulting in the replacement of other groups are presented in alphabetic sequence of groups replaced. Further, the replacement of each type of substitution is reported in the order of substrates: pyridines, pyridine oxides and quaternary pyridinium salts.

A number of other reactions, including many additions to aromatic nuclei, are not so easily dealt with, often because their mechanisms are uncertain. These have been grouped under the general heading of Reactions which modify the Nucleus.

These simple ring systems present no difficult problems of nomenclature.

Arguments about the naming of tautomeric systems, such as that exemplified by the hydroxypyridine–pyridone pair, seem to the author to be sterile. Such tautomerism is carefully discussed at appropriate points, but in other places names are used without implications about the tautomerism.

With the Appendices to Chapters 4, 5 and 6, the literature is covered up to the end of 1965. I am indebted to my colleague Dr. K. G. Orrell for the Appendix on the nuclear magnetic resonance spectroscopy of pyridines.

A. INTRODUCTION

1

THE HISTORICAL DEVELOPMENT OF
AROMATIC STRUCTURES

(1) INTRODUCTION

No problem in organic chemistry has attracted interest so unfailingly as that of accounting for the properties of aromatic compounds by means of chemical structures. Many accounts have been given of the protracted feelings towards the true solution, but only in the last thirty years have the grounds for that solution been established.

Three phases are discernible in the history of aromatic theory. From Kekulé's perception of the cyclic nature of benzene until 1922, classical valency theory was exploited to its limits. The decade 1920–30 saw the emergence of electronic theories, and the confusion of electronic and classical ideas presented by the chemical literature of that time is difficult to resolve in retrospect. From about 1930 onwards, the significant ideas of the previous decade assumed clearer outlines and crystallized in a quantum theory of valency which to a great extent has solved the problem of aromatic stability and reactivity.

Every gain in our understanding of benzene has enriched our insight into the nature of heteroaromatic compounds such as pyridine and pyrrole, prototypical of all those to be discussed in this book. It is interesting, therefore, to consider the development of structures for pyridine and pyrrole against the background of the benzene problem in the first phase of its growth and as it was affected by some of the ideas from the second phase.

(2) EARLY BENZENE STRUCTURES

A satisfactory benzene structure must account for the symmetry of the molecule as revealed by the equivalence of the six hydrogen atoms[1, 2] and the observed numbers of isomers of substituted benzenes. It must also account for the stability of the benzene nucleus, as shown qualitatively by its resistance to degradation and to addition, and by its ability to undergo and survive substitution reactions. We shall not at first stress the demands of stability, for it is now clear that classical organic chemistry could not meet them from its own resources.

Of all the suggestions drawing on these resources, that due to Kekulé[3a], in its modified oscillatory form[3b], has proved to be of the most abiding interest and value. It accounted for the symmetry of benzene, and proved to be a convenient symbol by which to express the characteristically 'aromatic' substitution reactions.

Other early benzene structures[4] fell short of Kekulé's in one or other of these two respects, and only certain of them retain interest relevant to hetero-aromatic chemistry. Dewar's bridged structure[5], inadequate with regard to

3

symmetry in its original form, assumed new interest[6] when regarded as one phase in a dynamic conception of benzene, occupying 'a position intermediate between the two double-bonded (Kekulé) forms'. Such a conception went as far as was possible in non-electronic terms, accounting for symmetry and transformations (especially the relatedness in substituted benzenes of o- and p-positions), but again leaving unconsidered the problem of stability.

Other structures, sufficiently symmetrical, attempted to remove the double bonds which appeared to endow Kekulé's structure with a fatally high degree of unsaturation, by adopting ad hoc postulates divorced from the classical concepts of organic chemistry as a whole. Thus, Armstrong*[7] in his centric formula (1) consigned the fourth valencies of the six carbon atoms to the hexagonal internuclear limbo, and supposed the excess 'affinity' to be there 'uniformly and symmetrically distributed'. Von Baeyer[8] used the idea, and it retains interest as the source of Bamberger's views (discussed below). Superficially similar but firmly based on studies linking open-chain unsaturated compounds with aromatic substances, was Thiele's structure[9]. Recognizing the profound influence of the 'conjugation' of saturated and unsaturated centres and the resulting suppression in reactivity of the central parts of such systems, Thiele conceived a theory of 'partial valencies' which throughout a chain might neutralize each other, leaving only the ends of the chain reactive (2). In benzene, such terminal reactivities were not possible, conjugation achieved its apotheosis, and a system (3) resulted which lacked additive tendencies to a marked degree.

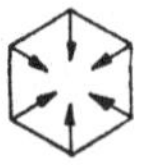

(1)    (2)    (3) 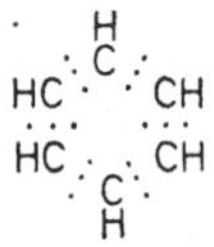   (4)

Early electronic formulations of benzene could not *explain* aromatic stability but did nevertheless produce new systematizations. The classical paper of Kermack and Robinson[10] in which benzene was represented by the symbol (4), essentially a restatement of Thiele's structure, led Robinson[11] to the idea of the aromatic sextet. Electronic theory at the time recognized the importance of stable associations of electrons in pairs and in octets but provided no basis for the stability attributed to a group of six electrons.

(3) PYRIDINE AND PYRROLE

Pyridine was first isolated in a pure state from bone oil by Anderson[12] who had earlier obtained picoline from coal tar. He established the molecular

* Armstrong's intentions are best conveyed in his own words: 'The only difference between this symbol and those employed hitherto arises from the fact that I do not consider that, apart from its connection with the other carbon atoms owing to their association in a ring, any one carbon atom is directly connected with any other atom not contiguous to it in the ring; my opinion being that each individual carbon atom exercises an influence upon each and every other carbon atom.'

formula of pyridine and showed it to be a tertiary base, capable of forming quaternary salts. In trying to account for the isomerism of aniline and picoline, Körner[13] proposed, in a letter to Cannizzaro, a structure for pyridine of the Kekulé type (5). The suggestion escaped general notice and was stated independently by Dewar[14]. Priority probably lay with Dewar who based his suggestion on the stability of pyridine, and on the analogy between the pyridine-dicarboxylic acids and the phthalic acids on the one hand, and the relationship of pyridine to quinoline and of benzene to naphthalene on the other. The virtues and shortcomings of (5) as a structure for pyridine are those of the Kekulé structure for benzene: the stability of the base was not *explained* but implicitly related to the same characteristic of benzene. On the other hand, (5) did represent the symmetry of pyridine, as evidenced, for example, by the existence of three isomeric monosubstituted derivatives. The proposed *skeleton* was confirmed by the reduction of pyridine to piperidine, by the reverse oxidation and by the synthesis of piperidine[15]. The equivalence of the α- and α'-positions was demonstrated by Scholtz and Wiedemann's[16] synthesis of the compound (8) from (6) and (7).

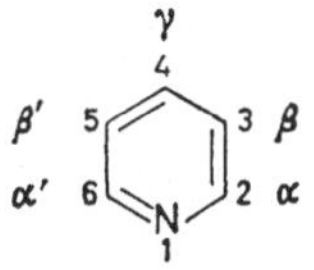

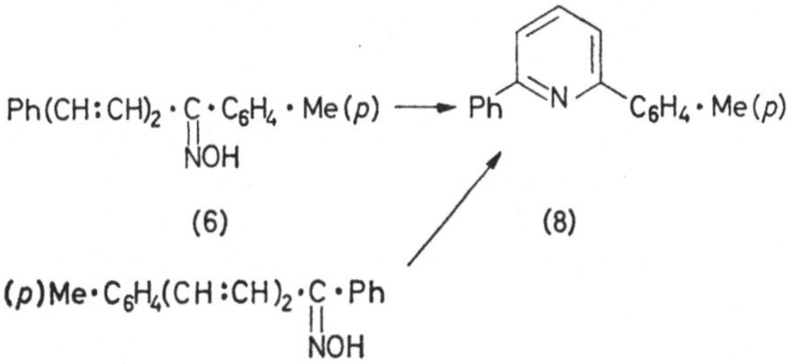

Later suggestions concerning pyridine followed closely the pattern of the history of benzene structures. Consideration of the conversion of pyrrole to 3-chloropyridine (see p. 88) led Ciamician and Dennstedt[17] to propose a prism formula. Riedel[18], from the degradation of acridine to quinoline, and thence to pyridine derivatives, supposed the diagonal structure which he had proposed for acridine (9) to persist in pyridine (10), while Bamberger[19] (see below) suggested a centric structure (11), and a Thiele formula was also considered.

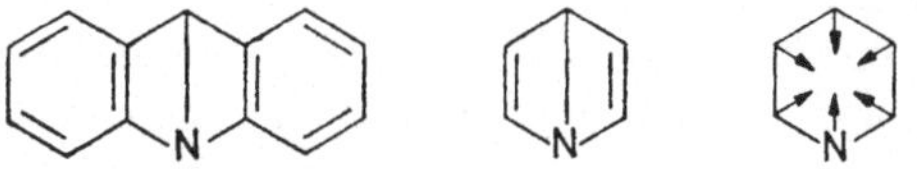

Pyrrole was isolated from bone oil by Runge[20] and purified and analysed some years later by Anderson[21]. Following work in the indole series, v. Baeyer and Emmerling[22] proposed for pyrrole the structure (12), the nearest analogy to the Kekulé structures of the six-membered ring compounds. This

5

structure accounted[23] for the reduction of pyrrole to pyrrolidine (13) by the addition of four hydrogen atoms, and for the fact that the hydrogen atom attached to nitrogen was replaceable by metals. It did not conflict with evidence from synthetic methods which were developed for pyrrolic compounds. However, the degree of unsaturation required by (12) was not found in pyrrole, which possesses much of the stability and tendency to undergo substitution characteristic of aromatic compounds. Bamberger[19] related pyrrole

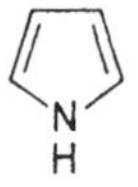

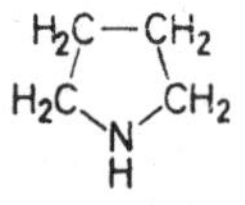

to the latter by the centric representation (14), whilst Ciamician[24] attempted to apply Thiele's theory in the expression (15).

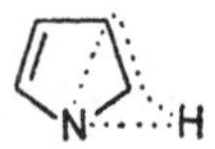

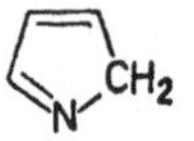

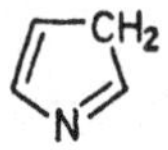

The peculiar reactivity of pyrrole (see p. 89) prompted efforts to devise new structures up to very recent times, but the results were of doubtful physical significance. Thus, in (16) and (17) Oddo[25a, b] tried to express the ambiguity in the position of the hydrogen atom [attached to nitrogen in (12)] which was thought to be revealed by the reactions of pyrrole. In this respect, they are elaborations of the suggestion[26] that pyrrole is a tautomeric substance capable of reacting in either the pyrrole (12) or the pyrrolenine (18 and 19) forms.

Armstrong's centric structure for benzene and its extensions to pyridine and pyrrole by Bamberger[19] have been mentioned. Bamberger's ideas[19, 27] retain great historical interest, containing intimation of later developments. Pyrrole is only very weakly basic (p. 60); it does not form stable salts nor react directly with alkyl halides to give quaternary derivatives. Combining this fact with the axiom of hexacentric valency equilibrium in aromatic systems, Bamberger concluded that in pyrrole nitrogen was pentavalent, and therefore saturated, because the nucleus required from it a contribution of two valencies. Reduction to pyrroline and pyrrolidine removed the hexacentric requirement, and in these compounds the nitrogen was tervalent and basic. In pyridine, maintenance of the hexacentric system required only one valency from the nitrogen atom, leaving it still capable of forming salts. In this way Bamberger elegantly accounted for one of the most striking differences between the pyrrole and pyridine series. The idea was extended to many other systems[19], such as imidazole (20) which contains nitrogen atoms in both states. With bicyclic systems, similar representation was possible only if ordinary linkages between the bridge atoms were omitted, as in the expressions for naphthalene (21), indole (22) and quinoline (23).

6

If Kermack and Robinson's[10] representation of benzene (4) restated Thiele's structure in electronic terms, it also re-interpreted Armstrong and Bamberger's[11] ideas which were now transformed into the view that aromatic stability was associated with a sextet of electrons[28]. Two of nitrogen's five valency electrons were used to link the nitrogen atom to the two adjacent carbon atoms in pyridine, one contributed to the aromatic sextet, and two,

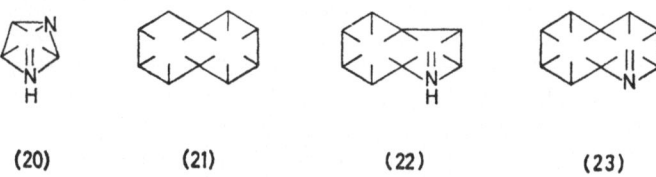

(20)	(21)	(22)	(23)

the 'lone pair', remained for salt formation. In pyrrole, three of the electrons linked the nitrogen atom to the two carbon atoms and a hydrogen atom, and two were absorbed in the sextet, leaving none for salt formation. In bicyclic systems, such as naphthalene or quinoline, sextets could be maintained only by dissolving the bridging valency, or by assuming that one electron could be shared between two groups of five.

This was the limit to which pre-quantum theories could reach. The Lewis–Langmuir concept of the covalent bond, formed between atoms by the sharing of two electrons, could replace directly the straight lines of classical valency structures, but superimposed on this framework there was, in aromatic compounds, a sextet of electrons[11] identified as the seat of aromatic stability. Contemporary electronic theory could not in 1925 provide a physical basis for the stability attributed to such a group of electrons. At this stage, an advance in the theory of aromatic stability could only follow the discovery of such a physical basis for the evident difference in state between certain electrons in aromatic compounds from the state of those in saturated compounds. At the same time, great success attended the use of Kekulé's structure in representing aromatic transformations (Fieser[4]). It was this success which led to overstressing of Kekulé forms in polynuclear compounds and of bond fixation of such compounds in those forms.

These considerations lead to a final point. Kekulé structures proved useful in representing the transformations of aromatic substances. Allowing further that these structures *implicitly* represented the stability of such compounds, nineteenth-century chemists could assume the replacement of —CH= by —N= and of —CH=CH— by —NH— to generate structures (pyridine and pyrrole) also endowed with this stability. They could not from such relationships deduce the differing reactivities of benzene, pyridine and pyrrole towards the common chemical reagents. Further, the linking of Kekulé- and Dewar-type structures indicated a relationship between o- and p-positions in benzene derivatives[6] which could have been extended to the 2- and 4-position of pyridine. However, although the pyrrolenine forms, (18) and (19), historically held a position similar to that of the Dewar benzene structure[26], they did not permit deductions about the relative reactivities of α- and β-positions in pyrrole. In 1904, Ciamician[24] commented upon these problems, remarking that although pyridine was regarded as an aromatic compound, it showed only a few of the less characteristic reactions of benzene. It was, on the other hand, one of the future tasks of chemical theory to explain why pyrrole resembled phenol in its reactivity.

At this point in its history, the theory of aromatic structures, and of valency in general, entered the confused period of transition from classical to quantum concepts. Ideas proliferated and were discarded with disturbing rapidity. It remains profitable only to indicate those contributions which emerged as the sound basis for further development. In 1916, the electronic theory of the covalent bond was formulated[29] and in 1927 given quantum theoretical treatment by Heitler and London[30]. From 1926 to 1929, Ingold[31] developed the concept of mesomerism, and in 1931–33, Hückel[32] applied quantum-mechanical theory to benzene while Pauling[33] discussed simple molecules in terms of resonance and Ingold[34a] examined the problem of aromatic structures. Finally, in 1933 Pauling and Wheland[35] developed Hückel's work on benzene, and in 1935 applied their methods to pyridine and pyrrole. Thus was the theory of aromatic stability firmly established, and it has since been continuously developed.

Lewis's account[29] (1916) of the inductive effect and Lowry's[36] idea of electromeric shifts in unsaturated bonds (1923) were combined by Lucas[37] in 1924–5, and developed by Robinson[38] (1926) and Ingold[31, 34b] (1926) into a comprehensive qualitative theory of reactions.

Table 1.1

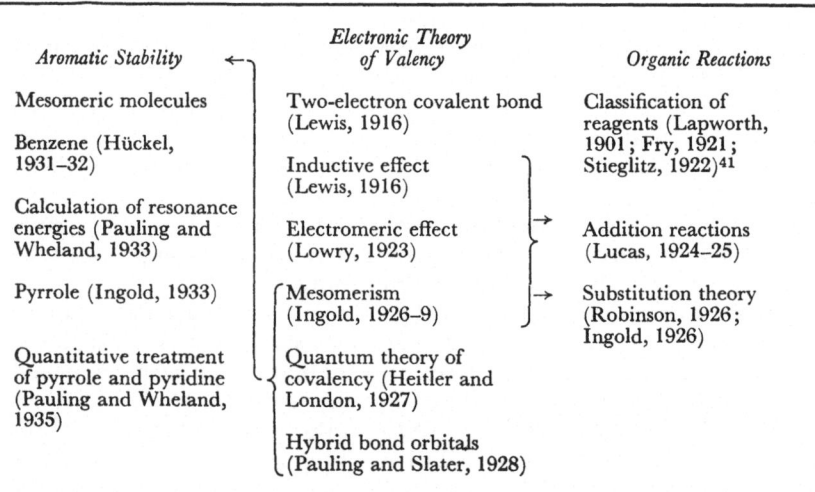

Aromatic Stability		Electronic Theory of Valency	Organic Reactions
Mesomeric molecules		Two-electron covalent bond (Lewis, 1916)	Classification of reagents (Lapworth, 1901; Fry, 1921;
Benzene (Hückel, 1931–32)		Inductive effect (Lewis, 1916)	Stieglitz, 1922)[41]
Calculation of resonance energies (Pauling and Wheland, 1933)		Electromeric effect (Lowry, 1923)	Addition reactions (Lucas, 1924–25)
Pyrrole (Ingold, 1933)		Mesomerism (Ingold, 1926–9)	Substitution theory (Robinson, 1926; Ingold, 1926)
Quantitative treatment of pyrrole and pyridine (Pauling and Wheland, 1935)		Quantum theory of covalency (Heitler and London, 1927)	
		Hybrid bond orbitals (Pauling and Slater, 1928)	

The interconnections of these developments are outlined in *Table 1.1*, and their consequences are considered in Chapter 2. Any such summary of the efforts of a period of such intense activity necessarily does injustice to many workers. Two essays by Robinson[39] and Ingold[34c], seminal in their influence, are useful in supplying perspective (for discussion of early theories, see also[4, 34b, 40]).

REFERENCES

[1] Ladenburg, A. *Ber. dtsch. chem. Ges.* **2** (1869) 140; *Justus Liebigs Annln Chem.* **172** (1874) 331

[2] Wroblevsky, E. *Ber. dtsch. chem. Ges.* **5** (1872) 30; *Justus Liebigs Annln Chem.* **168** (1873) 147; **192** (1878) 196; Hübner, M. and Petermann, A. *ibid.* **149** (1869) 129; Hübner, H. **195** (1879) 1; **222** (1884) 67, 166

[3] Kekulé, F. A. (*a*) *Ber. dtsch. chem. Ges.* **2** (1869) 362; (*b*) *Justus Liebigs Annln Chem.* **162** (1872) 77

[4] Excellent accounts in Ingold, C. K. *Structure and Mechanism in Organic Chemistry*, London (Bell) 1953 and in *The Chemistry of Carbon Compounds*, Vol. III A Ch. 1, ed. Rodd, E. H., London (Elsevier) 1954; also Badger, G. M. *The Structure and Reactions of the Aromatic Compounds*, Cambridge University Press, 1954, and Fieser, L. F. in *Organic Chemistry. An Advanced Treatise*, 2nd Ed., Vol. I, Ch. 3, ed. Gilman, H., New York (Wiley) 1943

[5] Dewar, J. *Proc. R. Soc. Edinb.* (1866) 84

[6] Ingold, C. K. *J. Chem. Soc.* **121** (1922) 1133

[7] Armstrong, H. E. *J. chem. Soc.* **51** (1887) 258

[8] Baeyer, A. v. *Justus Liebigs Annln Chem.* **245** (1888) 103

[9] Thiele, J. *Justus Liebigs Annln Chem.* **306** (1899) 87

[10] Kermack, W. O. and Robinson, R. *J. chem. Soc.* **121** (1922) 427

[11] Armit, J. W. and Robinson, R. *J. chem. Soc.* **127** (1925) 1604

[12] Anderson, T. *Trans. R. Soc. Edinb.* **16** (1849) 123, 463

[13] Körner, G. *G. sci. nat. econ. Palermo* **5** (1869); reprinted in Calm, A. and Buchka, K. v. *Die Chemie des Pyridins und seiner Derivate*, Braunschweig, 1889–1891, which contains an excellent account of the early history of these bases.

[14] The tangled history of the pyridine structure is interestingly recounted by Dobbin, L. *J. chem. Educ.* **11** (1934) 596

[15] Ladenburg, A. *Justus Liebigs Annln Chem.* **247** (1888) 1; *Ber. dtsch. chem. Ges.* **18** (1885) 2956; **19** (1886) 780; Königs, W. *ibid.* **12** (1879) 2341; Lellmann, E. and Geller, W. *ibid.* **21** (1888) 1921; Tafel, J. *ibid.* **25** (1892) 1619

[16] Scholtz, M. and Wiedemann, A. *Ber. dtsch. chem. Ges.* **36** (1903) 845

[17] Ciamician, G. L. and Dennstedt, M. *Ber. dtsch. chem. Ges.* **14** (1881) 1153

[18] Riedel, C. *Ber. dtsch. chem. Ges.* **16** (1883) 1609

[19] Bamberger, E. *Ber. dtsch. chem. Ges.* **24** (1891) 1758; cf. Pechmann, H. V. and Baltzer, O. *ibid.* 3144

[20] Runge, E. *Annln Phys.* **31** (1834) 65

[21] Anderson, T. *Justus Liebigs Annln Chem.* **105** (1858) 335

[22] Baeyer, A. v. and Emmerling, A. *Ber. dtsch. chem. Ges.* **3** (1870) 514

[23] Ciamician, G. and Magnachi, P. *Gazz. chim. ital.* **15** (1885) 481

[24] Ciamician, G. L. *Ber. dtsch. chem. Ges.* **37** (1904) 4200

[25] Oddo, B. *Gazz. chim. ital.* (*a*) **37** (1907) 83; **52** (1922) 42, 56; **55** (1925) 174; **61** (1931) 699; (*b*) **64** (1934) 584

[26] König, W. *J. prakt. Chem.* **84** (1) (1911/2) 194

[27] Bamberger, E. *Justus Liebigs Annln Chem.* **257** (1890) 1; **273** (1893) 373

[28] Robinson, R. *Tetrahedron* **3** (1958) 323

[29] Lewis, G. N. *J. Am. chem. Soc.* **38** (1916) 762

[30] Heitler, W. and London, F. *Z. Phys.* **44** (1927) 455

[31] Ingold, C. K. and E. H. *J. chem. Soc.* (1926) 1310

[32] Hückel, E. *Z. Phys.* **70** (1931) 204; **72** (1931) 310; **76** (1932) 626

[33] Pauling, L. *J. Am. chem. Soc.* **54** (1932) 3570

[34] Ingold, C. K. (*a*) *J. chem. Soc.* (1933) 1120; (*b*) *A. Rep. Chem. Soc.* **23** (1926) 129; (*c*) *Chem. Rev.* **15** (1934) 225

[35] Pauling, L. and Wheland, G. W. *J. chem. Phys.* **1** (1933) 362; Wheland, G. W. and Pauling, L. *J. Am. chem. Soc.* **57** (1935) 2086

[36] Lowry, T. M. *J. chem. Soc.* **123** (1923) 822, 1866; *Nature, Lond.* **115** (1925) 276

[37] Lucas, H. J. and Jameson, A. Y. *J. Am. chem. Soc.* **46** (1924) 2475; Lucas, H. J. and Moyse, H. W. *ibid.* **47** (1925) 1462

[38] Allan, J., Oxford, A. E., Robinson, R. and Smith, J. C. *J. chem. Soc.* (1926) 401; Ing, H. R. and Robinson, R. *ibid.* 1655

[39] Robinson, R. *Outline of an Electrochemical (Electronic) Theory of the Course of Organic Reactions*, London, R. Institute of Chemistry, 1932

[40] Remick, A. E. *Electronic Interpretations of Organic Chemistry*, 2nd Ed., New York (Wiley) 1949

[41] Lapworth, A. *J. chem. Soc.* **79** (1901) 1273; *Mem. Proc. Manch. lit. phil. Soc.* **64** (1920) No. 3; *J. chem. Soc.* **121** (1922) 416; *Nature, Lond.* **115** (1925) 625; Fry, H. S. *The Electronic Conception of Valence and the Constitution of Benzene*, London and New York (Longmans, Green) 1921; Stieglitz, J. *J. Am. chem. Soc.* **44** (1922) 1293

2

AROMATIC STRUCTURE

(1) MESOMERISM AND RESONANCE

(a) Mesomerism

Lowry's recognition[1] of the process (1) related originally to the polarization of unsaturated linkages at the demand of a chemical reagent, in preparation for chemical reaction; that is, it referred to the *polarizability* of unsaturated linkages. This mechanism of electronic redeployment, within the limit set by the duplet and octet principles, is called the *conjugative mechanism**, and its particular embodiment in polarizability, the *electromeric effect*[3a]. Ingold perceived[3a, b] that the situation in (1) might also represent a *polarization*, or permanent state of unsaturated linkages. In illustration he discussed structures such as (2), which were represented by symbols as in (3), with the implication that in its resting state the unit (2) is actually removed some degree towards the state (4). Such *polarization* was named[3c] the *mesomeric effect*. A

(1)

$$R_2N-C=C-C=O$$

(2)

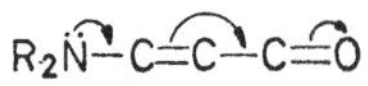

$$R_2\overset{\frown}{N}-C\overset{\frown}{=}C-C\overset{\frown}{=}O$$

(3)

$$R_2\overset{+}{N}=C-C=C-\overset{-}{O}$$

(4)

(5)

$$X \longleftarrow Y$$

(6)

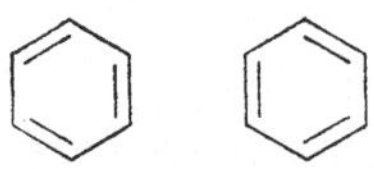

(7)

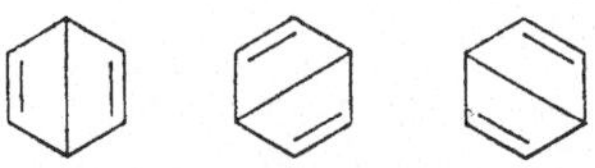

(8)

mesomeric molecule is, in its resting condition, more stable than any classical structure which might be used to represent it. Benzene (5) is a mesomeric molecule, and consequently more stable than the Kekulé structures were expected to be. It remained to explain why electrons behave in this

---

* For historical reasons it was known as the 'tautomeric mechanism', but the later description is preferable[2].

way; if a source of energy could be found for the mesomeric effect, aromatic stability in particular would be explained.

(b) Resonance. Valency States. Hybridization. Double Bonds

This source of energy was discovered in quantum theory, and the principle, as it affects this discussion, can be stated roughly as follows: if the electrons in a classical valency structure can be delocalized whilst obeying the Pauli exclusion principle, then the increased indeterminacy in their positions results in the molecule having a lower energy in the ground state than would be expected from the classical structure.

The phenomenon is not, of course, restricted to aromatic structures, but is at the very basis of covalent bond formation, as can be seen by Heitler and London's[4] discussion of the hydrogen molecule. That discussion underlines the distinction between electrostatic forces, which account for the attraction or repulsion of ions and for the unequal sharing of valency electrons in a saturated link between unlike atoms or groups (6)—the *inductive effect*[5]—and those quantum forces responsible for holding atoms together by covalent bonds. The new principle originated in Heisenberg's work on the quantum mechanical phenomenon of resonance and in his Uncertainty Principle[6], and it has given rise to a body of ideas known as 'Resonance Theory' which, though using a different terminology, is equivalent to Ingold's theory of mesomerism[3c, d]. The idea of resonance of molecules among several valence-bond structures was developed by Slater[7] and Pauling[8a], and applied especially to benzene by Hückel[9a].

The resonance theory can be better appreciated in the light of modern views on valency. In wave-mechanical theory, the state of an electron is described by means of its wave function, ψ, a function of the coordinates of the system, corresponding to the amplitude in classical theory of wave motion. The wave function of a single electron, moving in the field of one atom, is called an 'atomic orbital'[10a]. Since the orbitals describe the space in which a particular electron is most probably to be found, the electron is said to 'occupy' the orbital. The extra-nuclear electrons of the normal carbon atom are, on spectroscopic grounds[11], known to be in the state $1s^2 2s^2 2p^2$. The description indicates that the K shell, the one of lowest energy, is complete, having two electrons in the $1s$ orbital which differ only in their spins, these being anti-parallel or coupled. The essential fact about an s orbital is its spherical symmetry, the electrons in such an orbital occupying a spherical shell of space surrounding the nucleus. The quantity $\psi^2 dv$ measures the probability of finding the electron in the element of volume dv, and this quantity varies in such a way that the chance of finding an s electron outside the spherical shell is very small. In the L shell the carbon atom has four electrons. The two $2s$ electrons, with coupled spins, occupy an orbital which is again spherically symmetrical but which has a greater radius of maximum density than the $1s$ orbital, and an internal spherical node. Again, since the two electrons differ only in spin, this sub-group is complete. Whilst the $2p$ orbitals differ but little from the $2s$ orbitals in their radial distribution, they do, in contrast, depend on the angular distribution. For any given energy level, three p orbitals are possible, identical in energy but mutually at right angles and having dumb-bell shapes with a node through the nucleus[7, 12].

11

B

In building atoms, any one orbital in a sub-group cannot take two electrons until all the sub-group orbitals contain at least one each. Accordingly, if the p orbitals are assigned arbitrary x, y and z coordinates, the L shell of the carbon atom is more accurately described as $2s^2 2p_x 2p_y$.

With such a picture of the carbon atom it was possible, following Heitler and London's theory of the covalent bond in hydrogen[4], to develop the theory of covalent compounds in general, and carbon compounds in particular. Slater[7] and Pauling[8a] first showed that directed valencies could be accounted for by using the p orbitals to form covalent bonds. The water molecule is a simple illustration. If the two covalent bonds in water were formed by the fusion of the two hydrogen s orbitals with the two oxygen p orbitals, the angle between the bonds would be $90°$. It is actually[13, 14] $104°27'$, the deviation being explained below.

The main points in this theory of covalent compound formation are[8a]: (1) a single bond arises from two unpaired electrons; (2) the bond is formed in the direction in which the orbitals of the two electrons overlap most (for then the exchange energy of the two electrons is greatest); (3) of two orbitals in one atom which might be used for bond formation, that which could overlap most with the orbital of the electron from the second atom will form the stronger bond.

In the carbon atom, only the two unpaired $2p_x$ and $2p_y$ electrons could form bonds. If they did, carbon would be divalent, combining with, say, two hydrogen atoms to form bonds at right angles. But, of course, carbon is not divalent but commonly tetravalent. To be able to form four bonds, the carbon atoms must be excited to the state $2s 2p_x 2p_y 2p_z$. Such an atom would be tetravalent, with three valencies directed mutually at right angles and the fourth having no favoured direction. But again, this cannot be the usual state of combined carbon, long known from stereochemical studies to be of tetrahedral configuration. The accepted explanation was given by Pauling in the idea of hybridization[8]. The real 'valency state'* (a term due to Van Vleck[15]) is a 'hybridized' sp^3 carbon atom; besides the states described above, any linear combination of the $2s$ and $2p$ orbitals will provide a solution of the wave equation for the carbon atom. The $2s$ orbital and the $2p_x$, $2p_y$, and $2p_z$ orbitals can be blended to produce four new orbitals allowing maximum overlap when combined in bond formation with, say, four hydrogen atoms[17]. It turns out that these orbitals[19] are oriented tetrahedrally. Whilst such sp^3 hybridization produces regular tetrahedral symmetry in, say, methane, many degrees of hybridization are possible, and in saturated compounds less symmetrical than methane, bond angles vary[13, 14, 20] from the regular tetrahedral value of $109°28'$. For water there must be some mixing of the s with the p orbitals.

Nitrogen is of special interest here. In the state $2s^2 2p_x 2p_y 2p_z$, nitrogen would be trivalent, the three valencies being mutually at right angles. In fact, the bond angles in ammonia[13, 14] are $106°47'$, presumably because of repulsion among the hydrogen atoms, and the orbitals must again utilize some degree of s mixing.

---

* Free atoms in the valency states are not spectroscopically observable. The energetics of the processes of excitation are still not entirely clear[15-18].

Double bonds can now be considered. In the classical tetrahedral theory of carbon stereochemistry, a double bond is represented by the fitting together of edges of the valency tetrahedran, and originally Pauling[8a] and Slater[7] attempted to describe double bonds in terms of the overlapping of tetrahedrally directed orbitals. This description, insofar as it is embodied in bent bonds, has received recent consideration[21-2], but a more frequently used one was indicated by Hückel[9b] and by Mulliken[10b] and worked out more fully by Penney[23a]. Instead of sp^3 hybridization, this uses sp^2 or plane-trigonal hybridization. One of the carbon p orbitals, arbitrarily chosen as p_z, remains unchanged in its dumb-bell form, and the $2s$, $2p_x$ and $2p_y$ orbitals are hybridized to give three new orbitals at right angles to the p_z orbital and at $120°$ to each other. Ethylene then arises from the overlapping of two of these trigonal orbitals, one from each carbon atom, to form a bond ('σ bond') between them, and by the attachment of hydrogen atoms through their s orbitals to the remaining four trigonal orbitals. The distinctive point is that the remainder of the 'double bond' results from the overlapping of the two p_z orbitals, to produce a different type of bond—a π bond—having a nodal plane in the plane of the nuclei. According to this description, it is in this exposed component that the familiar unsaturation of the double bond resides. The existence of the π bond depends on the overlapping of the p_z orbitals, and thus a restriction is laid upon the ability of the two ends of the molecule to rotate about the C–C σ bond. Maximum overlapping of the p_z orbitals requires as small a distance as possible between the carbon nuclei, and the C–C bond length in ethylene is shorter than that in ethane[13, 14]. Plane-trigonal hybridization requires HCC or HCH angles of $120°$. For tetrahedral hybridization, the figure is $109°28'$, and varying degrees of hybridization between tetrahedral and plane-trigonal would require intermediate values. Experimentally the HCH angle in ethylene has been found to agree much better with the trigonal requirement[13, 14]. The degree to which the geometry of ethylene is to be ascribed to π-electron effects or to the consequences of the type of hybridization present in the σ bonds is a matter of controversy[24].

(c) The Valence-Bond Picture of Benzene, Pyridine and Pyrrole. Geometrical Consequences

This picture of ethylene leads to a formal description of the construction of benzene. If six plane-trigonally hybridized carbon atoms are fitted together by means of σ bonds, six trigonal orbitals remain outside the hexagonal framework, whereon to attach hydrogen atoms. The six p_z orbitals stand parallel to each other at right angles to the plane of the nuclei. When it is attempted to form these into spin-coupled pairs, producing three π bonds, it is seen that there is no unique way of doing this, and the two structures produced are entirely equivalent; each clearly represents a classical Kekulé structure. In other words, there is exchange degeneracy, with a resulting gain in energy. Each Kekulé structure represents an approximate wave function for benzene, and a better representation of the molecule is given by a linear combination of the two. A truer picture results if the forms equivalent to the Dewar structures for benzene are included, and the wave function

13

is represented by a linear combination of the wave-functions of the two Kekulé and three Dewar forms

$$\psi = 2a\psi_K + 3b\psi_D$$

where a and b are numerical coefficients representing the different impor-tances of the two forms, their 'weights', in the true structure. This true structure, called by Ingold a 'mesomeric' molecule, finds then an equivalent description as a 'resonance hybrid' of the Kekulé and Dewar structures. The five Kekulé and Dewar forms, (7) and (8), are called a 'canonical set' and, as will be seen later, the Kekulé structure, though having no real existence, serves as an arbitrary reference point wherefrom to estimate the amount by which benzene is 'stabilized by resonance'. As should be clear from the formal construction of benzene just described, 'resonance' is not a process through which molecules go but a state in which they exist[3b, 13].

Some consequences of this description of benzene are obvious and accord with long-established facts. The desirable condition in benzene would be such as to provide maximum overlapping of the plane-trigonally hybridized orbitals, and also of the p_z orbitals; this condition would be fulfilled by a planar, hexagonal structure. Further, since resonance exists in the molecule, the six nuclear bonds must all have the same character: rather than three single and three double bonds, there are six hybrid bonds which are expected to be shorter than single but longer than double bonds, and which result in the hexagon being regular. Single bonds and isolated double bonds are found experimentally to be of remarkably constant lengths, and deviation from these values has been taken as a guide to the presence of hybridized bonds arising from resonance[13, 14]. The planarity of benzene is accurately attested by spectroscopic data[25], and the bond lengths obtained[13, 14, 26] are indicated in (9).

The delocalization of π bonds represented by resonance in benzene re-sults in a gain in stability according to the principle stated above. It should be noticed, however, that by this description benzene owes its *special* posi-tion as the most stable of the cyclic hydrocarbons, C_nH_n, particularly to the fact that its six-membered ring exactly accommodates the 120° angle, re-sulting in maximum overlap in the σ bonds. In any carbon ring other than the six-membered this is not the case[23b].

The chemical stability of benzene, its lack of unsaturation and tendency to undergo substitution rather than addition reactions follow from the existence of resonance, which would have to be destroyed to permit addition.

In representing benzene by the canonical set (7) and (8), dipolar struc-tures such as (10) are neglected. The 'weight' of such structures would, since they are of higher energy than (7) and (8), be small in the expression for ψ given above. It is a feature of the resonance or valence bond method that in its simplest approximations such dipolar structures are neglected and homo-polar forms overstressed. The shortcoming is not inherent, however, and later workers[27] have improved the theory in its quantitative form by taking account of dipolar forms.

The qualitative resonance theory yields interesting results when extended to polynuclear compounds such as naphthalene and anthracene. In the first case, besides three Kekulé-type structures (11), thirty-nine other non-

14

dipolar forms are possible containing trans-annular bonds, making a canonical set of forty-two[28]. Discussion of the complete set is impossible, but a rough picture of the state of affairs in naphthalene is gained even if all but the Kekulé forms (11) are neglected. Considering these, the 1:2 bond in naphthalene would be expected to differ from the 2:3 bond; if the bonds in benzene are described as possessing $\frac{1}{2}$ double bond character[13, 29a], then the

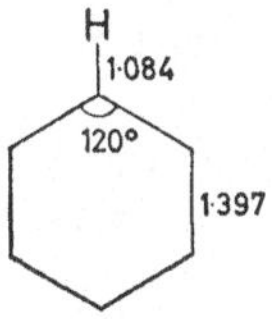

(9) 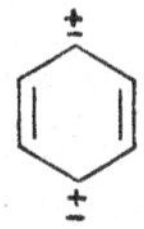   (10) 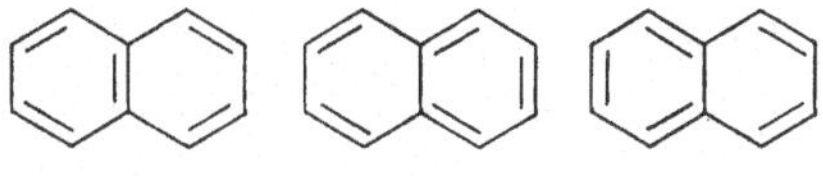   (11)

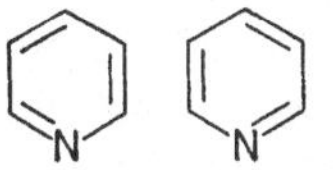

(12) 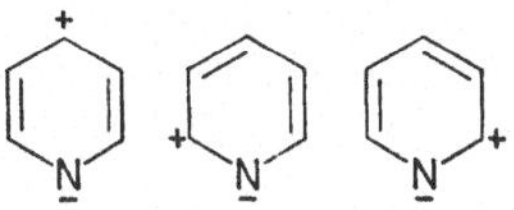   (13)

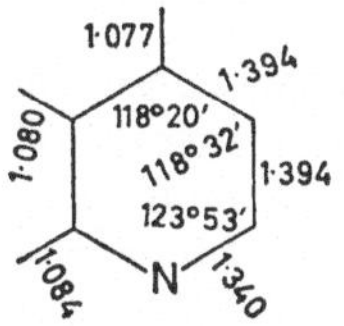

(14)

1:2 bond in naphthalene could be said to possess $\frac{2}{3}$, and the 2:3 bond $\frac{1}{3}$ double bond character. Similar remarks apply to anthracene.

This qualitative resonance picture of benzene can now be extended to pyridine and pyrrole, the two substances prototypical of all heteroaromatic nitrogen compounds.

Clearly, the formal construction of the pyridine molecule could be carried through by extracting a –CH– group from benzene and replacing it by nitrogen. If the valency state of the nitrogen atom were such as to produce four orbitals resembling those of the plane-trigonally hybridized carbon atom, the exchange would but slightly disturb the general architecture of the original structure. The state of the nitrogen atom mentioned above, K $2s^2 2p_x 2p_y 2p_z$, would not do, but hybridization produces a different state similar to the plane-trigonal form of carbon. The $2p_z$ orbital remains unchanged and the four electrons (originally $2s^2 2p_x 2p_y$) enter plane-trigonal-type orbitals, with the important difference that two spin-coupled electrons are to be found in one of the streamers. Fitted into the residue of the benzene skeleton, this produces a structure closely resembling the Kekulé form of benzene, except that one of the nuclear atoms, the nitrogen atom, cannot carry a hydrogen atom since the appropriate orbital is now filled. For the same reason as in benzene, exchange degeneracy exists, and pyridine can be described as a resonance hybrid of the structures (12). Again as with benzene, a more complete description must include other structures representing approximate wave functions, such as (13). Nitrogen is more electronegative

15

than carbon[29b], and consequently dipolar forms (13) are of much greater significance in the canonical set of pyridine than are those (10) in the set for benzene. The resulting concentration of negative charge on the nitrogen atom, and loss of it from the 2- and 4-positions, has important consequences for the chemical and physical properties of pyridine. It should be noticed that pyridine, in contrast to benzene, would be expected to possess a finite dipole moment (p. 123). The geometry of pyridine, based on electron diffraction[26] and microwave spectroscopic studies[30], is in accord with a planar molecule closely resembling benzene (14).

If the Kekulé-type structures (12) above are considered, as was done in discussing the double-bond character of the linkages in the carbocyclic compounds, it can be seen that the C–C bond lengths in pyridine would be expected to approach closely those in benzene, as indeed they do. Discussion of the C–N bond lengths is more difficult, as will be seen later (p. 27). The bond angles have been fairly satisfactorily accounted for[31].

The trigonal orbital of the nitrogen atom in pyridine, projecting out from the ring, holds two spin-coupled electrons and so cannot link a hydrogen atom to the nucleus. It can, however, accommodate a proton. In other words, pyridine is basic (see p. 145).

If pyridine closely resembles benzene, pyrrole is interesting because of the very different situation which it reveals. If hexagonal structures are particularly stable because they can achieve both maximum overlap of plane-trigonal and parallelism of p_z orbitals, then any change of ring size must produce a less favoured structure. Nevertheless, a compromise can be reached, for example in a five-membered ring. Suppose that the nitrogen atom is in the plane-trigonally hybridized form and that three of the five

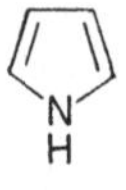

(15) 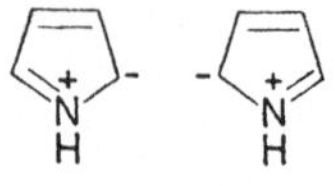   (16) 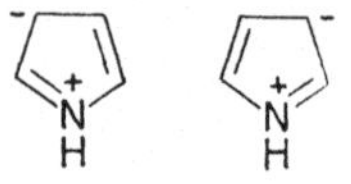   (17)

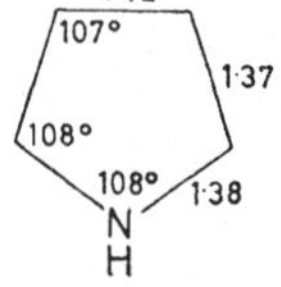

(18)

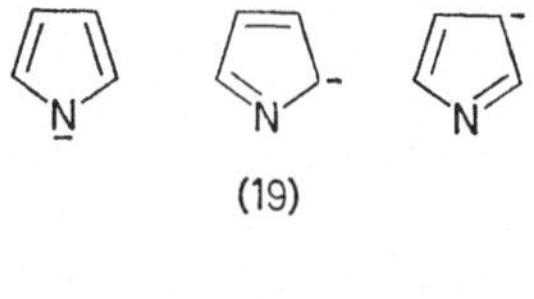

(19)

valency electrons are placed one in each of the trigonal orbitals. These orbitals can now be used to unite the nitrogen atom to a hydrogen atom and to the $-C_4H_4-$ framework. The remaining two nitrogen valency electrons could now go only into the p_z orbital, and a structure (15) would result. Not until one of the electrons in the nitrogen p_z orbital had been given up to a carbon atom could the remaining p_z electrons form a π bond with the p_z electron of an adjacent carbon atom. As in previous examples, exchange degeneracy would exist between alternative forms, but in both cases the nitrogen atom would be positively charged, and a carbon atom negatively. The canonical set for pyrrole would thus include (16) and (17) as well as (15), and the important difference is that here these dipolar forms are fundamental to the existence of resonance in the molecule[3c]. Again, the consequences for the properties of pyrrole are far-reaching.

It is understandable that pyrrole should possess a finite dipole moment (p. 51). Further, the resonance description requires the molecule to be planar. The classical structure (15) could accommodate a pyramidal nitrogen atom, but although dipole-moment studies[32] allow the extra-annular nitrogen valency an out-of-plane latitude of 7°, and Raman and infra-red spectroscopic studies favour the planar form but do not rigorously exclude a non-planar form[33], the microwave rotational spectrum[34] confirms the truly planar structure (18).

This description of pyrrole, taken in conjunction with that developed for pyridine and benzene, illuminates the earlier theories of Bamberger (p. 6) and the idea of the aromatic sextet (p. 7). Pyrrole is only very weakly basic (p. 60) because, to function as a base, it must lose its resonance stabilization. On the other hand, two factors contribute to the acidity of the hydrogen atom attached to nitrogen; first, there is the character of the nitrogen atom [cf. (17)] which results in the binding of the bonding electrons nearer to the nitrogen nucleus, and second, and perhaps more important, there is the fact that resonance in the pyrrole anion (19) must have a more stabilizing influence than in pyrrole itself, since in the anion it does not involve charge separation.

In all the heterocyclic compounds considered in this volume, the nitrogen atoms resemble that in pyrrole or that in pyridine; in some substances both conditions are found. The broad consequences for the molecular geometry of the various systems will be obvious from what has been said above, and the experimental data relating to this are reported at appropriate points in later Chapters.

It remains now to consider the consequences as observed in the thermodynamic stabilities of these various systems, and later as seen in their reactivities.

(d) Empirical Resonance Energies

As was mentioned above, the hypothetical classical structure considered intuitively to be the most stable of the canonical set presents an arbitrary norm from which to estimate the stabilization occasioned by resonance in the molecule. More precisely, in the case of, say, benzene, the stabilization would be obtained by calculating the energy of an approximate valence-bond wave function such as that represented by a Kekulé structure and also the energy of the real molecule from the linear combination of the totality of approximate wave functions; the difference between these two quantities is the vertical resonance energy (see p. 25).

17

Such calculations are not possible, and the stabilization is usually taken to be represented by the empirical resonance energy. This quantity is the difference between the experimentally determined energy of a compound and the energy calculated for the most stable member of the canonical set of structures representing that compound, by comparison with related real compounds.

Empirical resonance energies are obtained[13a] from heats of either hydrogenation or combustion. For compounds in which resonance involves no more than adjacent atoms, heats of hydrogenation are roughly additive[13a, 35]. Thus, from the heat of hydrogenation of ethylene that for, say, 1,4-pentadiene can be predicted. In the same way, the heat of hydrogenation of cyclohexene can be used to estimate that of a cyclic structure having three double and three single bonds. Experiment shows benzene to be more stable than such a structure by 36·0 kcal mole^{-1}. This is the empirical resonance energy of benzene. If both cyclohexene and 1,3-cyclohexadiene are used for the comparison[36], the value obtained is 34·2 kcal mole^{-1}. The empirical resonance energies for aromatic compounds are always at least an order of magnitude greater than those for aliphatic compounds such as 1,3-butadiene[13a]. Unfortunately, the study of heats of hydrogenation, which provides results of considerable accuracy, has not been carried on extensively. In the heterocyclic series direct hydrogenations have not been used, though some heats of hydrogenation, and thence empirical resonance energies, have been deduced more or less indirectly[37, 38] (*Table 2.1*).

Heats of combustion can be used to estimate empirical resonance energies, but this method suffers from relatively large experimental errors. Approximate values can be arrived at by using the empirical rule that the heat of combustion of a substance with a unique classical structure is expressible as the sum of approximately constant contributions from the various bonds present. The method can be improved by including terms for other structural features, but such corrections necessarily lead to complications. The inaccuracy of the method is least disabling in the case of aromatic substances which possess the largest empirical resonance energies.

It is not always appreciated how very complicated is the process of evaluating the bond-energy contributions to be used in calculating heats of combustion. The experimental combustion data to be used must first be selected and corrected to refer to the substance in a standard state (usually as a gas at 25 °C and 1 atm pressure); both of these selections leave room for differing opinions. Nevertheless, with carbocyclic compounds, for which very extensive combustion data are available, consistent and reliable values have been derived. The most important compilations are due to Klages[41] and Wheland[13a]. The difficulties met with in applying the method are well described by the latter author. He has also used heats of combustion for the purpose of calculating empirical resonance energies, in rather a different way. This variation stems from the work of Franklin[42] who used heats of formation. Instead of summing contributions from bonds, Franklin summed contributions from larger structural units. The method permits the calculation of heats of combustion with great accuracy, for substances which are not resonance hybrids. It is, however, complicated and offers no improvement with other classes of compounds.

Table 2.1. Empirical Resonance Energies (kcal mole^{-1})

	Kistiakowsky[39]	Pauling[29a]	Wheland[40]	Wheland[13a]	Klages[41]	Others
Butadiene	3·5					
Benzene	36·0	37	41	36·0* 36·4†	35·9	36[44]
Naphthalene		75	77	61·0 61·2	61·0	
Anthracene		105	116	83·5 83·7	85·8	
Phenanthrene		110	130	91·3 92·4	99·2	
Pyrrole		31	24	21·2 24·5	21·6	27[42]
Pyrazole		28·5† 41·5†			27·6**	22·1§ 17·4‖ 30·9*** 26·8§ 29·3[45]
						32·7‖
Imidazole		32·1‡			22·3**	12·2§ 14·2[45] 17·7‖
1,2,4-Triazole		49·2‡				20·0§ 36·2‖
Tetrazole						55·2§ 63·1‖
Indole		54 57·6§	63	47 49	51·0**	52·0§
Indazole						41·8‖ 53·8***
Benzimidazole		73·9†			52·3**	59·0§ 59·5‖
Benztriazole		68·8†				48·7§ 48·8‖
						82·9§ 74·7‖
Carbazole		91 100·7†	97	74 75	87·4**	94·9§
Pyridine		43	43	23·0	27·9	79·2‖ 89·7***
Quinoline		69	75	47·3 —	48·4	27·2 or 37·3[44] 21 or 31[46]
Acridine						22[46] 35[46] 32[38]
Phenazine						106[47] 84·0[37] 105[47]

Calculated by *Klages' method[41] or by †Franklin's method[42]; from new thermochemical data[43] using ‡ Pauling's;[29a] ** bond energy terms; *** Klages'[41] bond contributions to heats of combustion; §Cottrell's[17] and ‖Coates and Sutton's[44] bond energy terms; *** Franklin's[44] bond contributions to heats of combustion.

19

B*

Heats of combustion have been used to calculate empirical resonance energies in another way, of theoretical interest. The method is illustrated by reference to a compound $C_aH_bN_c$. Consideration of the equations set out below leads to the relationship shown between ΔH_c, the heat of combustion, and ΔHf_a, the atomic heat of formation.

$$C_aH_bN_c \text{ (gas)} \rightarrow aC\text{(gas)} + bH \text{ (gas)} + cN \text{ (gas)} \qquad \Delta H = \Delta Hf_a$$

$$C_aH_bN_c \text{ (gas)} + (a + b/4)O_2 \rightarrow aCO_2 \text{ (gas)} + b/2.H_2O \text{ (liquid)}$$
$$+ c/2.N_2 \text{ (gas)} \qquad \Delta H = \Delta H_c$$

$$CO_2 \text{ (gas)} \rightarrow C \text{ (graphite)} + O_2 \text{ (gas)} \qquad \Delta H = x$$

$$C \text{ (graphite)} \rightarrow C \text{ (gas)} \qquad \Delta H = y$$

$$H_2 \text{ (gas)} + \tfrac{1}{2}.O_2 \text{ (gas)} \rightarrow H_2O \text{ (liquid)} \qquad \Delta H = l$$

$$\tfrac{1}{2}.H_2 \text{ (gas)} \rightarrow H \text{ (gas)} \qquad \Delta H = m$$

$$\tfrac{1}{2}.N_2 \text{ (gas)} \rightarrow N \text{ (gas)} \qquad \Delta H = n$$

$$\Delta Hf_a = -\Delta H_c - a(x + y) - b(m - l/2) - cn$$

where x = heat of combustion of graphite

$\quad y$ = heat of atomization of graphite

$\quad l$ = heat of combustion of molecular hydrogen

$\quad m$ = heat of atomization of hydrogen

$\quad n$ = heat of atomization of nitrogen

The heat of formation of $C_aH_bN_c$ (suppose it to be ethylamine) can also be expressed as the sum of contributions from each bond, the so-called bond energy terms[44], assumed to be approximately constant in each case:

$$\Delta Hf_a = -E(C\text{—}C) - 5E(C\text{—}H) - E(C\text{—}N) - 2E(N\text{—}H)$$

Now $E(C\text{—}H)$ and $E(C\text{—}C)$ can be calculated from combustion data for the paraffins. Thus for CH_4, $\Delta Hf_a = -4E(C\text{—}H) = -\Delta H_c - (x + y) - 4(l + m)$ whence $E(C\text{—}H)$ follows.

Similarly, $E(C\text{—}C)$ is derived from data for ethane and $E(N\text{—}H)$ from ammonia. Clearly, then, a value for $E(C\text{—}N)$ can be obtained. Values for this particular bond energy term, derived from a group of aliphatic amines by Coates and Sutton[44], vary from 47·8 to 56·4 kcal mole^{-1}. They used an average value of 52·5 kcal mole^{-1} or, with a different value for the heat of atomization for graphite, 73·1 kcal mole^{-1}. The last value is comparable with that of 74·7 kcal mole^{-1} given by later workers[45].

With a resonance hybrid such as pyridine, the atomic heat of formation is given by

$$\Delta Hf_a = -H_c - 5(x + y) - 5(m - l/2) - n$$

and for a single Kekulé-type structure, the calculated heat of formation would be

$$\Delta Hf'_a = -2E(\text{C—C}) - 2E(\text{C═C}) - 5E(\text{C—H}) - E(\text{C═N}) - E(\text{C—N}$$

so that the empirical resonance energy $(-\Delta Hf'_a - \Delta Hf_a)$ requires for its evaluation only $E(\text{C═N})$ and $E(\text{C═C})$, in addition to the quantities already discussed. As with $E(\text{C—C})$, so with $E(\text{C═C})$ a range of accurate thermochemical data is available for its evaluation, but with $E(\text{C═N})$ the situation is not so satisfactory. This point is commented upon below.

At this stage, other points require stressing. Bond energy terms such as $E(\text{C—C})$ and $E(\text{C═C})$, based on a considerable range of experimental data, can be made more nearly additive by structural corrections. Thus, in discussing benzene, a more appropriate value of $E(\text{C═C})$ is obtained from *cis*-hex-3-ene, and so on[44]. Even so, the theoretical significance of the bond energy terms is not simple; they certainly do not represent the true heats required to rupture the particular type of bond to which they refer[18, 20, 26, 44] but are merely average values. For this reason these quantities have been called[44] bond energy terms rather than bond energies. In deriving bond energy terms from thermochemical data, the heats of atomizing elements from their standard states are used; the accepted values of these constants have been frequently revised, and so different authors have assigned different values to bond energy terms. Any author using one set of combustion data would, of course, deduce the same empirical resonance energies directly from these data or from manipulating them in the form of bond energy terms.

With these points in mind it is useful to consider values for the empirical resonance energies of representative aromatic compounds deduced from thermochemical data by various authors. Some of these are given in *Table 2.1*.

The values given by Pauling[29a] were evolved from earlier values[26, 48], and they differed from Kistiakowsky's values, derived from heats of hydrogenation, mainly because of inadequate correction for structural effects[44]. The early estimates also suffer in some cases from the use of old and unreliable combustion data, for example in the case of pyrrole[26, 49]. It is fair to say that with homoaromatic compounds the more recent estimates of resonance energies based on combustion data are in good agreement with those from heats of hydrogenation.

With the heteroaromatic substances the situation is not so satisfactory. The difference is due partly to the more extensive data available in the carbocyclic series, but there are also other reasons. The major difficulty is that of assigning a value to the term $E(\text{C═N})$. Apart from the relatively small range of experimental data available in general, no values for the C═N group present in a six-membered ring are known[46]. In any case, with bonds formed between elements of different electronegativities it is probable that energy terms vary from compound to compound[44]. This is true for $E(\text{C—N})$, as mentioned above, and almost certainly for the value for the more polarizable carbon–nitrogen double bond. By using a value for $E(\text{C═N})$ of 96 kcal mole^{-1} instead of the 106 kcal mole^{-1} resulting from their experimental data, Coates and Sutton[44] obtained a resonance energy for pyridine roughly equal to that accepted for benzene (see *Table 2.1*). Since, on general grounds, this would be expected, these authors conclude that the value of $E(C{=}N)$ varies from 96 to 106 kcal mole.

Further comments on the significance of empirical resonance energies will be made later (p. 25).

(2) THE MOLECULAR ORBITAL METHOD

(a) General Considerations

Besides the valence-bond theory, a second type of theoretical treatment, widely used, has found particularly fruitful application in aromatic chemistry. This, the molecular orbital method*, was devised by Mulliken, Hund and Lennard-Jones[52b] and extended by Hückel[9] to unsaturated organic molecules. In the valence-bond treatment of, for example, benzene, the p_z electrons are coupled in pairs to form three bonds which are, however, mobile. The molecular orbital method proceeds in the case of aromatic systems by assuming the nuclear framework, cemented together by localized σ bonds. The π electrons are then supplied to occupy not atomic orbitals but molecular orbitals[10a] extending over the whole molecule. Each π electron is treated as moving in the field due to the nuclei and the average charge distribution of the other electrons. Pauli's principle enters when, these molecular orbitals having been defined, the π electrons are assigned in pairs to those of lowest energy.

The virtues and shortcomings of these two most commonly used theoretical methods have frequently been discussed[19, 53, 54]*. A few general points are appropriately mentioned here. The valence-bond method, using as it does structures familiar to organic chemists, is especially valuable in qualitative discussion. On the other hand, whilst it leads to a picture of an aromatic molecule such as benzene which in no way recalls, say, the Kekulé structure, there is no doubt that in semi-quantitative applications the molecular orbital method is very much the simpler. Further, it has proved more readily adaptable to the handling of heteroaromatic molecules. In the present context, both treatments are useful in different ways, but the molecular orbital method especially so. Both methods are, in their semi-quantitative forms, gross approximations, but they have nevertheless been used to compute quantities of great interest to the organic chemist. The particular relevance of the molecular orbital method to problems of heteroaromatic chemistry demands that it be given more extended discussion here.

To obtain the molecular orbitals, we should need to solve the one-electron Schrödinger equation

$$-\frac{h^2}{8\pi^2 m}\left(\frac{\partial^2\psi}{\partial x^2} + \frac{\partial^2\psi}{\partial y^2} + \frac{\partial^2\psi}{\partial z^2}\right) + v\psi = E\psi$$

written in the Hamiltonian form as

$$\mathbf{H}\psi = E\psi$$

Almost universally the initial approximation is made[52b] of expressing the wave function of each of the molecular orbitals as a linear combination of

---

* Excellent accounts of the theory will be found in[19, 50-52a].

atomic orbitals, i.e. the atomic orbitals, generally taken as the $2p_z$ ones, of the carbon atoms constituting the aromatic framework (the L.C.A.O.–M.O. approximation[55]). Thus, in the case of benzene, the six atomic orbitals would result in six molecular orbitals, each of the form

$$\psi = c_1\phi_1 + c_2\phi_2 + c_3\phi_3 + c_4\phi_4 + c_5\phi_5 + c_6\phi_6$$

If this wave function were an accurate solution of the Schrödinger equation above, we could write the energy of the jth molecular orbital as

$$E_j = \int \psi_j \mathbf{H}\psi_j dv / \int \psi_j^2 dv$$

$$= \frac{(\Sigma\limits_r C_{rj}^2 H_{rr} + 2\Sigma\Sigma\limits_{r<s} C_{rj} C_{sj} H_{rs})}{(\Sigma\limits_r C_{rj}^2 + 2\Sigma\Sigma\limits_{r<s} C_{rj} C_{sj} S_{rs})} \quad (r = 1, ..6)$$

where

$$H_{rs} = \int \phi_r \mathbf{H}\phi_s dv \qquad S_{rs} = \int \phi_r \phi_s dv$$

To make ψ_s a good solution, and the above expression a better approximation to the energy of the orbitals, the conditions

$$\frac{\partial E_j}{\partial C_{rj}} = 0$$

are imposed. These partial differentiations lead to a set of equations (the secular equations), six in the case of benzene:

$$(H_{rr} - ES_{rr})c_r + \Sigma_s(H_{rs} - S_{rs}E)c_s = 0$$

[that is, $c_1(H_{11} - ES_{11}) + c_2(H_{12} - S_{12}E) + c_3(H_{13} - S_{13}E) +$

$$\ldots + C_6(H_{16} - S_{16}E) = 0, \text{ etc.]}$$

Elimination of the coefficients leads to the secular determinantal equation

$$\begin{vmatrix} H_{11}-E & H_{12}-ES_{12} & H_{13}-ES_{13}\ldots\ldots H_{16}-ES_{16} \\ H_{12}-ES_{12} & H_{22}-E & H_{23}-ES_{23}\ldots\ldots H_{26}-ES_{26} \\ \cdot\quad\cdot\quad\cdot & \cdot\quad\cdot\quad\cdot\quad\cdot & \cdot\quad\cdot\quad\cdot\quad\cdot\quad\cdot\quad\cdot \\ H_{16}-ES_{16} & H_{26}-ES_{26} & H_{36}-ES_{36}\ldots\ldots H_{66}-E \end{vmatrix} = 0$$

The integrals H_{rr} are the coulombic integrals, representing the interaction of the electron of atom r in the orbital ϕ_r with all the other electrons and nuclei. They reflect the electronegativity of the atom r, and in the case of benzene are all set equal to α. Quantities of the type H_{rs}, the resonance integrals, measure the exchange energy of one electron between atoms r and s. As a first approximation, the resonance integrals for non-adjacent atoms ar sete equal to zero, and all those for neighbouring atoms equal to β. The overlap integrals S_{rs} are also set equal to zero, whilst the particular cases $S_{rr} = \int \phi_r^2 dv$ become equal to unity by normalization. The secular determinantal equation then becomes

$$\begin{vmatrix} \alpha-E & \beta & 0 & 0 & 0 & \beta \\ \beta & \alpha-E & \beta & 0 & 0 & 0 \\ 0 & \beta & \alpha-E & \beta & 0 & 0 \\ 0 & 0 & \beta & \alpha-E & \beta & 0 \\ 0 & 0 & 0 & \beta & \alpha-E & \beta \\ \beta & 0 & 0 & 0 & \beta & \alpha-E \end{vmatrix} = 0$$

with the six roots (the energies of the six molecular orbitals) $E_1 = \alpha + 2\beta$, $E_2 = E_3 = \alpha + \beta$, $E_4 = E_5 = \alpha - \beta$, and $E_6 = \alpha - 2\beta$. Substitution back into the secular equations yields the ratios of the coefficients, and their absolute magnitudes follow from the normalization condition

$$c_1^2 + c_2^2 + \ldots c_6^2 = 1$$

This method of treating hydrocarbons, with the salient features that all resonance integrals for adjacent atoms are assumed equal and that overlap is neglected, is known as the Hückel approximation[9a]. It has been the one most used.

In a modified treatment, Wheland[56] allowed for overlap between adjacent atoms. The secular determinantal equation then becomes

$$\begin{vmatrix} \alpha-E & \beta-ES & 0 & 0 & 0 & 0 \\ \beta-ES & \alpha-E & \beta-ES & 0 & 0 & 0 \\ 0 & \beta-ES & \alpha-E & \beta-ES & 0 & 0 \\ \text{-} & \text{-} & \text{-} & \text{-} & \text{-} & \text{-} \\ \beta-ES & 0 & 0 & 0 & \beta-ES & \alpha-E \end{vmatrix} = 0$$

Some results will be noted below.

Of greater consequence for the theory of aromatic reactivity (Chapter 3), particularly regarding heterocyclic compounds, is the modification introduced by Wheland and Pauling[57]. By using different Coulomb integrals for different atoms, corrections are made for the differing electronegativities of atoms in the aromatic nucleus. The method has been criticized[58], but also arguments have been made justifying it[59].

Numerous other forms of molecular orbital theory have been introduced[60a, 61], of greater sophistication than the Hückel method. However, as regards problems of aromatic stability and reactivity, the latter is of great value and has achieved successes which are remarkable in view of its simplicity[61]. In other problems, such as those of spectroscopy, this is not nearly so true.

(b) Resonance or Delocalization Energies

It was seen above that the Hückel approximation, applied to benzene, led to the energies of the molecular orbitals

$$E_1 = a + 2\beta, \qquad E_2 = E_3 = a + \beta, \qquad E_4 = E_5 = a - \beta,$$

$$E_6 = a - 2\beta$$

The energies of the first three of these are, since β is negative, lower than that of the isolated $2p_z$ atomic orbital, a. Accordingly, the first three are bonding orbitals, and the remainder are antibonding. In the unexcited benzene molecule, each of the first three orbitals contains two electrons, the antibonding orbitals remaining empty. The total π-electron energy is then $(6a + 8\beta)$. But the π-electron energy of an isolated double bond is $(2a + 2\beta)$, and of the unperturbed Kekulé form, $(6a + 6\beta)$. The delocalization energy for benzene is, therefore, 2β.

In general, for such structures the orbital energies are of the form

$$E_j = a + m_j\beta$$

bonding when m_j is positive, antibonding when it is negative. If each bonding orbital contains two electrons, the delocalization energy (summed over occupied orbitals only) becomes

$$D.E. = \sum_j 2(m_j - 1)\beta$$

The Hückel approximation has been used to derive delocalization energies for numerous hydrocarbons[50b], some of which are given in *Table 2.2*. The delocalization energy corresponds to the vertical resonance energy of valence bond theory (p. 17). Comparison of columns 3 and 4 of *Table 2.2* reveals a good correlation between these quantities. There is also a surprisingly good correlation between delocalization energies and empirical resonance energies. Depending on the way in which the latter have been evaluated, this can lead to a value for β of between 16 and 18 kcal mole^{-1}, much lower than those obtained from other data. There is no justification at all for using the value of β derived from empirical resonance energies for any purpose other than that of predicting empirical resonance energies, for there is no sound reason to expect any simple, direct relationship between delocalization energies (or vertical resonance energies) and empirical resonance energies.

The truth of this last remark can be seen by a closer consideration of the nature of the empirical resonance energy[66]. The reference state used in evaluating it (e.g. the Kekulé structure of benzene) is really a structure composed or ordinary double and single bonds. Therefore, in trying to evaluate the stabilization to be attributed to delocalization, some allowance must be made for the energy needed to change all the bonds in the reference structure to bonds of the same length as those in benzene. Thus we must add to the empirical resonance energy this compression energy to give the true delocalization or vertical resonance energy. For the compression energy, a value of 27 kcal mole^{-1} has been estimated[66], giving a vertical resonance,

Table 2.2. Delocalization and Resonance Energies

	Empirical resonance energies* kcal mole⁻¹	Vertical resonance energies† (valence-bond method)	Delocalization energies† M.O. method (Hückel approximation)	(with overlap)	Calculated resonance energies‡ kcal mole⁻¹ valence-bond	M.O.	(with overlap)
Benzene	36	1·11J	2·00β	1·07γ	67·3	66·2	63·2
Naphthalene	75	2·04J	3·68β	1·86γ	97·4	95·8	88·7
Anthracene	105	2·95J	5·31β	2·61γ	99·7	98·1	93·2
Phenanthrene	110	3·02J	5·45β	2·74γ	24·7	34·2	34·6
Pyrrole	31		1·9 β	1·03γ	37	35·8	
Pyridine	37		1·99β		38·1	37·8	
			2·1 β				
Pyridazine**	—		1·94β	0·96γ	22	35·0	32·3
§	—		2·07β	1·08γ	—	37·2	36·2
Pyrimidine	—		1·98β	0·99γ	33	35·7	33·4
			2·2 β			39·6	
Pyrazine	—		1·95β	0·96γ	33	35·1	32·3
			2·2 β		40	39·6	
s-Triazine	—		1·98β	0·97γ	29	35·6	32·1

* Table 2.1.
† For carbocyclic compounds, cf.[56] for heterocyclic compounds,[61-66]
‡ Obtained by using the value 36 kcal mole⁻¹ for benzene and the derived values of J, β and γ.
‡ Pyridazine: with reference to the structure containing =N.N=** and to that containing —N:N—§

* Table 2.1.
† For carbocyclic compounds, cf.[56] for heterocyclic compounds,[61-66]
‡ Obtained by using the value 36 kcal mole⁻¹ for benzene and the derived values of J, β and γ.

or delocalization, energy of $(27 + 36) = 63$ kcal mole^{-1}. This and other[67], even higher, estimates lead to values of β more nearly comparable to those obtained in other ways. The close correlation mentioned between delocalization energies and empirical resonance energies implies a close parallelism between compression energies and empirical resonance energies.

One further point needs to be raised concerning the significance of empirical resonance energies and the values of vertical resonance energies derived from them in the way described. The point is this: in deriving empirical resonance energies, it is always assumed that σ bonds have the same bond energy, no matter what the character of the carbon atoms which they join. However, powerful arguments have been made for thinking that σ bonds between sp^2 hybridized carbon atoms are considerably more stable than those between sp^3 hybridized carbon atoms. That being the case, the stabilization of benzene represented by the empirical resonance energy cannot all be attributed to delocalization. Indeed, when allowances were made for hybridization changes, a value for the vertical resonance energy of benzene as low as 13 kcal mole^{-1} was estimated[24].

Clearly, empirical resonance energies are quantities with no simple theoretical significance[69]. They, and parameters derived from them, should be used cautiously.

To return to the contents of *Table 2.2*: Wheland[56] has shown that a simple relationship exists between the expression given above for delocalization energy, and that derived for calculations including overlap. In this case, the delocalization energy (summed over occupied orbitals only) becomes

$$D.E. = \Sigma\, 2\left(k_j - \frac{1}{1+S} \right) \gamma$$

where $\gamma = \beta - Sa$, $k_j = m_j/1 + Sm_j$. Wheland gave the overlap integral the value 0·25. Then

$$D.E. = \underset{j}{\Sigma}\, 2\,(k_j - 0\cdot 8)\,\gamma$$

Some of the delocalization energies from calculations including overlap are given in *Table 2·2*.

Some calculations of delocalization energies of heterocyclic compounds have been reported, notably by Davies[62] who included overlap for adjacent orbitals and used the value of S also used by Wheland[56]. Orgel, Cottrell, Dick and Sutton[65] and Dewar[60b] also reported some values. The assumptions used by these authors are mentioned later (pp. 35 and 36).

Maccoll[63] and Simonetta[64] applied valence bond theory (neglecting all but Kekulé structures) to some heterocycles. A number of these results are collected in *Table 2.2*.

(c) Bond Orders and Bond Lengths

Modern theories of molecular structure have been applied with considerable success to the study of bond lengths in aromatic substances[51, 52a, 70-1]. The idea of a connection between bond length and fractional double bond character was first suggested by Pauling, Brockway and Beach[72],

27

and their approach, based on the valence bond method, was indicated briefly above (p. 15). Thus, in the case of naphthalene, consideration of the three structures (11) leads to a description of the $C_{(1)}$—$C_{(2)}$ bond, say, as possessing $\frac{2}{3}$ double-bond character. The important idea was that this bond character was uniquely related to the bond length. Pauling, Brockway and Beach[72] drew the first correlation curve between doublebond character and experimentally determined bond lengths. Up to a point, the method can be refined by allowing for more contributing structures and by attributing to these the correct weights[28, 72, 73]. However, it is impossible to consider more than a fraction of the large number of possible contributing structures, and in any case such a process would give increasing weight to structures indicating non-bonding or antibonding at particular points[51]. In practice, the simple Pauling method, using only Kekulé structures to which are ascribed equal weights, is surprisingly successful[51, 74].

A different approach to the problem, designed to overcome the difficulties in the Pauling method, was described by Penney[75], and Lennard-Jones[76] used molecular orbital methods in an attempt to calculate bond lengths directly. Of more immediate interest, because of attempts to extend it to heterocyclic compounds, is the semi-empirical method due to Coulson[71]. This uses the coefficients of the expansion of molecular orbitals as the sum of atomic orbitals. The quantity c_{jr}^2 indicates the contribution made by an electron in the jth orbital to atom r, and consequently the π-electron density at this atom is

$$q_r = \sum_j n_j c_{jr}^2$$

n_j being the number of electrons in the jth orbital[57]. Similarly, Coulson expressed the sum of the contributions made by the π electrons to the π bond between atoms r and s as

$$p_{rs} = \sum_j n_j c_{jr} c_{js}$$

p_{rs} being the π- or mobile-bond order of the bond r–s. For pure single, double and triple bonds, $p_{rs} = 0$, 1 and 2. The order of a σ bond is regarded as constant at unity, and so the *total* bond order is $(1 + p_{rs})$. As in Pauling's method, a correlation curve established between bond orders so defined and bond lengths as measured, provides a means of predicting unknown bond lengths. The success of the method is striking: it is capable of predicting aromatic bond lengths to within 0·02 Å, a figure inside the accuracy of most x-ray determinations[51, 70-1]. Both the theoretical and experimental studies clearly indicate that the bond lengths in a polynuclear compound vary with their location, and the different varieties of theoretical treatment agree fairly well[74].

The common assumption made in most treatments of bond orders is that, as already mentioned, the π electrons can be treated separately from the σ bonds, and that the latter can be regarded as constant, shortening being caused by π delocalization. As in other connections, this assumption has been challenged, and an attempt has been made to explain changes in bond lengths as being due to changes in σ-bond hybridization[24]. Consequently, the precise theoretical significance of π-bond orders is not clear.

When it is attempted to extend bond order–bond length theories to heterocyclic compounds, a number of difficulties arise. In the first place, there is no sound theory of the Coulombic terms which must now be used. Thus, allowance should be made for the change of valency angle (and consequently of hybridization ratios) caused by heteroatoms. The molecular orbital approach does not lead to the value $p_{rs} = 1$ for a double bond between heteroatoms. The actual difference from unity will depend on the values used for the adjustable parameters, and with the carbon–nitrogen double bond will probably be negligible compared with the experimental uncertainty[77]. There is also the fact that no direct experimental value is known for the length of the 'pure' carbon–nitrogen double bond. Ionic character in bonds also presents difficulties.

Cox and Jeffrey[77] first attempted to establish a length–order correlation curve for carbon–nitrogen linkages. Data for a variety of aliphatic compounds suggested the point 1/1·475 Å, and the covalent radii of Schomaker and Stevenson[78], with a small adjustment, gave 2/1·28 Å. Theoretical bond orders and experimental bond lengths available for three heteroaromatic compounds were also used. Experimental data on melamine were taken to indicate that all the carbon–nitrogen links in this molecule were equivalent. Pauling and Sturdivant[79] had deduced for this compound a delocalization energy of 5·83 β. The total π energy would then be 11·83 β, and the π-bond order[71], p ($= E/2n\beta$), 0·658; thus melamine provided the point 1·66/1·345 Å. For pyridine and pyrrole, bond orders had been calculated

Table 2.3. Bond Lengths and Bond Orders[82]

	a_N	p_{C-N}	Bond length, Å
Melamine	$a + 0.5\beta$	0·563	1·343 ± 0·05
	$a + \beta$	0·520	1·353 ± 0·05
s-Triazine	$a + 0.5\beta$	0·654[62]	1·319 ± 0·005
Phenazine	$a + 0.6\beta$	0·603	1·345 ± 0·009
Pyrimidine	$a + 0.6\beta$	0·656[83]	1·33
Pyridine	$a + 0.5\beta$	0·654	1·340[30]

Bond	1–2	2–3	3–4	5–6	6–7	7–8	References
Pyrrole	0·555	0·725	0·603				84
	0·661	0·566	0·686				84
	0·50	0·74	0·62				65
Pyridine	0·652	0·671	0·664				62
	0·656	0·665	0·666				62
	0·651	0·611	0·663				83*
	0·273	0·539	0·406				85†
Quinoline	0·524	0·678	0·669	0·718	0·615	0·701	80
	0·283	0·500	0·432	0·518	0·392	0·473	85†
	0·306	0·219	0·616	0·505	0·367	0·568	85†

* Using $a_N = a + 0.6\beta$ for parameters; used in the other references, cf. *Table 3.1*, page 36.
† Valence-bond method.

by Longuet-Higgins and Coulson[80]. Combined with the electron-diffraction data of Schomaker and Pauling[26], the value for pyridine gave the point 1·534/1·37 Å, and the correlation curve was drawn through this, the melamine point, and the end points. This curve was almost rectilinear, but the pyrrole point stood off it.

More recently the problem has been reconsidered by Goodwin, who studied[81] the polynuclear heteroaromatic compound flavanthrone. Goodwin and Porte[82] made the detailed examination of pteridine the occasion for attempting to construct a new correlation curve, based on more extensive experimental data than were available to Cox and Jeffrey[77]. The parameters used by Goodwin and Porte are indicated in *Table 2.3*, as are the points used in devising the correlation curve. To include melamine, the bond orders for this compound were recalculated, and the experimental (x-ray) data were taken to show different lengths for the two types of carbon–nitrogen bonds present. Cox and Jeffrey's value for the length of the 'pure' carbon–nitrogen single bond was modified to 1·435 Å to allow for sp^2 hybridization. Leaving aside pyridine, the tabulated data suggested a point 0·600/1·336 Å which was used as the datum point. The carbon–carbon correlation curve developed by Goodwin and Vand[86] was then displaced and proportionally compressed to fit the datum point and 1/1·435 Å. The resulting curve suggested the length 1·267 Å for the pure carbon–nitrogen double bond, which agrees well with the value derived from covalent radii[78]. With regard to pteridine, the correlation curve gives good agreement with the experimental values except in the case of the bridging bond. For this, the measured length is remarkably short (1·35 Å).

It is obvious that in the heterocyclic series the situation is not so advanced as in the carbocyclic series. The main requirement is for more accurate experimental data on bond lengths in nitrogen heteroaromatic compounds*.

Addendum on work reported during 1965

Correlation of the available data on bond angles in nitrogen heterocycles suggests that the valence angle of nitrogen which carries a hydrogen atom is significantly larger ($125 \pm 0·2°$) than that ($115·7 \pm 0·2°$) with no attachment[88].

There is a simple additivity relationship for bond angles in azabenzenes. Replacement of annular .CH: by .N: can be treated as a perturbation changing all ring angles by a definite amount[89].

REFERENCES

[1] Lowry, T. M. *J. chem. Soc.* **123** (1923) 822, 1866; *Nature, Lond.* **115** (1925) 376

[2] Ingold, C. K. *Structure and Mechanism in Organic Chemistry*, p. 62, London (Bell) 1953

[3] (a) Ingold, C. K. and E. H. *J. chem. Soc.* (1926) 1310; (b) Ingold, C. K. *Chem. Rev.* **15** (1934) 225; (c) idem, *J. chem. Soc.* (1933) 1120; (d) idem, *Nature, Lond.* **133** (1934) 946

[4] Heitler, W. and London, F. *Z. Phys.* **44** (1927) 455

[5] Lewis, G. N. *J. Am. chem. Soc.* **38** (1916) 762

[6] See Heisenberg, W. *The Physical Principles of the Quantum Theory*, University of Chicago Press, 1930

[7] Slater, J. C. *Phys. Rev.* **37** (1931) 481; **38** (1931) 1109

[8] Pauling, L. (a) *J. Am. chem. Soc.* **53** (1931) 1367, 3225; **54** (1932) 988, 3570; *Proc. natn. Acad. Sci. U.S.A.* **18** (1932) 293; (b) *ibid.* **14** (1928) 359

[9] Hückel, E. *Z. Phys.* (a) **70** (1931) 204; (b) **60** (1930) 423; (c) **72** (1931) 310; **76** (1932) 628

* For examples of the use of bond length–bond order relationships in discussing the geometry of pyridine, see [30, 87].

REFERENCES

[10] Mulliken, R. S. *Phys. Rev.* (a) **41** (1932) 49; (b) **43** (1933) 279

[11] Herzberg, G. *Atomic Spectra and Atomic Structure*, 2nd Ed., New York (Dover Publications) 1944

[12] White, H. E. *Phys. Rev.* **37** (1931) 1416

[13] (a) Wheland, G. W. *Resonance in Organic Chemistry*, New York (Wiley), 1955; (b) p. 129

[14] *Tables of Interatomic Distances and Configuration in Molecules and Ions*, Chem. Soc. *Spec. Publ. No. 11* 1958;

[15] Van Vleck, J. H. and Sherman, A. *Rev. mod. Phys.* **7** (1935) 167

[16] Cottrell, T. L. *The Strengths of Chemical Bonds*, 2nd Ed., London (Butterworths) 1958

[17] Coulson, C. A. *Valence*, 2nd Ed., Oxford University Press, 1961

[18] Long, L. H. *Experientia* **7** (1951) 195

[19] For accurate shapes of these orbitals see Moffitt, W. E. and Coulson, C. A. *Phil. Mag.* **38** (1947) 634

[20] Van Vleck, J. H. *J. chem. Phys.* **1** (1933) 219

[21] Pople, J. A. *Qu. Rev. chem. Soc.* **11** (1957) 272

[22] Pauling, L. *Theoretical Organic Chemistry*, London (Butterworths) 1958

[23] Penney, W. G. *Proc. R. Soc.* (a) **A144** (1934) 166; (b) **A146** (1934) 223

[24] Dewar, M. J. S. and Schmeising, H. N. *Tetrahedron* **5** (1959) 166; **11** (1960) 196; Mulliken, R. S. *ibid.* **6** (1959) 68; Dewar, M. J. S. *Hyperconjugation*, New York (Ronald Press) 1962

[25] Ingold, C. K. *et al.*, *J. chem. Soc.* (1936) 971; (1946) 316

[26] Schomaker, V. and Pauling, L. *J. Am. chem. Soc.* **61** (1939) 1769

[27] Sklar, A. *J. chem. Phys.* **5** (1937) 669; Craig, D. P. *Proc. R. Soc.* **A200** (1950) 401

[28] Pauling, L. and Wheland, G. W. *J. chem. Phys.* **1** (1933) 362

[29] (a) Pauling, L. *The Nature of The Chemical Bond*, 3rd Ed., New York (Cornell University Press) 1960; (b) *ibid.* p. 89

[30] De More, B. B., Wilson, N. S. and Goldstein, J. H. *J. chem. Phys.* **22** (1954) 876; Bak, B., Hansen, L. and Rastrup-Andersen, J. *ibid.* 2013; *J. molec. Spectrosc.* **2** (1958) 361; Bak, B. *Acta chem. scand.* **9** (1955) 1355

[31] Kim, H. and Hameka, H. F. *J. Am. chem. Soc.* **85** (1963) 1398; Coulson, C. A. *J. chem. Soc.* (1963) 5893

[32] Kofod, H., Sutton, L. E. and Jackson, J. *J. chem. Soc.* (1952) 1467

[33] Lord, R. C., Jr. and Miller, F. A. *J. chem. Phys.* **10** (1942) 328

[34] Wilcox, W. S. and Goldstein, J. H. *J. chem. Phys.* **20** (1952) 1656; Bak. B., Christensen, D., Hansen, L. and Rastrup-Andersen, J. *ibid.* **24** (1956) 720

[35] Baker, J. W. *Hyperconjugation*, p. 39, London (Oxford University Press) 1952

[36] Turner, R. B., Meador, W. R., Doering, W. v. E., Knox, L. H., Mayer, J. R. and Wiley, D. W. *J. Am. chem. Soc.* **79** (1957) 4127

[37] Jackman, L. M. and Packham, D. I. *Proc. chem. Soc.* (1957) 349

[38] Bedford, A. F., Beezer, A. E. and Mortimer, C. J. *J. chem. Soc.* (1963) 2039

[39] Kistiakowsky, G. B., Rukoff, J. R., Smith, H. A. and Vaughan, W. E. *J. Am. chem. Soc.* **58** (1936) 146, and other papers in this series

[40] Wheland, G. W. *The Theory of Resonance and its Application to Organic Chemistry*, New York (Wiley) 1944

[41] Klages, F. *Chem. Ber.* **82** (1949) 358

[42] Franklin, J. L. *Ind. Engng Chem. analyt Edn* **41** (1949) 1070; *J. Am. chem. Soc.* **72** (1950) 4278

[43] Zimmermann, H. and Geisenfelder, H. *Z. Elektrochem.* **65** (1961) 368

[44] Coates, G. E. and Sutton, L. E. *J. chem. Soc.* (1948) 1187

[45] Bedford, A. F., Edmondson, P. B. and Mortimer, C. T. *J. chem. Soc.* (1962) 2927

[46] Cox, J. D., Challoner, A. R. and Meetham, A. R. *J. chem. Soc.* (1954) 265

[47] Albert, A. and Willis, J. B. *Nature, Lond.* **157** (1946) 341

[48] Pauling, L. and Sherman, J. *J. chem. Phys.* **1** (1933) 606

[49] Elderfield, R. C. *Heterocyclic Compounds*, Vol. I, p. 282, New York (Wiley) 1950

[50] (a) Pullman, B. and A. *Les Theories Electroniques de la Chimie Organique*, Paris (Masson) 1956; (b) p. 224

[51] Coulson, C. A. In *Determination of Organic Structures by Physical Methods*, Ed. Braude, E. A. and Nachod, F. C. Chapter 16, New York (Academic Press) 1955

[52] Lennard-Jones, J. E. (a) *Proc. R. Soc.* **207A** (1951) 75; (b) *Trans. Faraday Soc.* **25** (1929) 668

[53] Partington, J. R. *An Advanced Treatise on Physical Chemistry*, Vol. V. p. 189, London (Longman) 1954

[54] Wheland, G. W. *J. chem. Phys.* **2** (1934) 474

[55] Mulliken, R. S. *J. chem. Phys.* **3** (1935) 375

[56] Wheland *J. Am. chem. Soc.* **63** (1941) 2025

[57] Wheland, G. W. and Pauling, L. *J. Am. chem. Soc.* **77** (1935) 2086

[58] McWeeny, R. *Proc. R. Soc.* **A237** (1956) 355

[59] Brown, R. D., Coller, B. A. W. and Heffernan, M. L. *Tetrahedron* **18** (1962) 343

[60] Dewar, M. J. S. (a) *A. Rep. chem. Soc.* **53** (1956) 126; (b) *Trans. Faraday Soc.* **42** (1946) 764

[61] Streitwieser, A. *Molecular Orbital Theory for Organic Chemists*, New York (Wiley) 1961

31

[62] Davies, D. W. *Trans. Faraday Soc.* **51** (1955) 449
[63] Maccoll, A. *J. chem. Soc.* (1946) 670
[64] Simonetta, M. *J. Chim. phys.* **49** (1952) 68
[65] Orgel, L. E., Cottrell, T. L., Dick, W. and Sutton, L. E. *Trans. Faraday Soc.* **47** (1951) 113
[66] Coulson, C. A. and Altmann, S. L. *Trans. Faraday Soc.* **48** (1952) 293
[67] Mulliken, R. S. and Parr, R. G. *J. chem. Phys.* **19** (1951) 1271
[68] Glockler, G. *J. chem. Phys.* **21** (1953) 1249
[69] Mortimer, C. T. *Reaction Heats and Bond Strengths,* London (Pergamon) 1962
[70] Coulson, C. A. *Proc. R. Soc.* **A207** (1952) 91
[71] Coulson, C. A., *Proc. R. Soc.* **A169** (1939) 413
[72] Pauling, L., Brockway, L. O. and Beach, J. Y. *J. Am. chem. Soc.* **57** (1935) 2705
[73] Sherman, J. *J. chem. Phys.* **2** (1934) 488
[74] Coulson, C. A., Daudel, R. and Robertson, J. M. *Proc. R. Soc.* **A207** (1951) 306
[75] Penney, W. S. *Proc. R. Soc.* **A158** (1937) 306
[76] Lennard-Jones, J. E. *Proc. R. Soc.* **A158** (1937) 280
[77] Cox, E. G. and Jeffrey, G. A. *Proc. R. Soc.* **A207** (1951) 110
[78] Schomaker, V. and Stevenson, D. P. *J. Am. chem. Soc.* **63** (1941) 37
[79] Pauling, L. and Sturdivant, J. H. *Proc. natn. Acad. Sci. U.S.A.* **23** (1937) 615
[80] Longuet-Higgins, H. C. and Coulson, C. A. *Trans. Faraday Soc.* **43** (1947) 87
[81] Goodwin, T. H. *J. chem. Soc.* (1955) 1689
[82] Goodwin, T. H. and Porte, A. L. *J. chem. Soc.* (1956) 3595
[83] Chalvet, O. and Sandorfy, C. *C. r. hebd. Séanc. Acad. Sci., Paris* **228** (1949) 566
[84] Brown, R. D. *Aust. J. Chem.* **8** (1955) 100
[85] Pullman, B. (a) *Bull. Soc. chim. Fr.* **15** (1948) 533; (b) Daudel, R. and Martin, M. *ibid.* 559
[86] Goodwin, T. H. and Vand, V. *J. Chem. Soc.* (1955) 1683
[87] Liquori, A. M. and Vaciago, A. *Ricerca scient.* **26** (1956) 1848; Somayajulu, G. R. *J. chem. Phys.* **28** (1958) 822
[88] Singh, C. *Acta crystallogr.* **19** (1965) 861
[89] Coulson, C. A. and Looyenga, H. *J. chem. Soc.* (1965) 6592

3

AROMATIC REACTIVITY

(a) INTRODUCTION

The theory developed mainly by Lapworth, Robinson, and Ingold gives a qualitative account of the influence of substituents upon the state of activation of the aromatic nucleus, and of the orienting powers of substituents towards electrophilic and nucleophilic substituting reagents, which is now widely familiar[1]. Broadly, it assumes that reagents of these types attack an aromatic nucleus preferentially at centres which are enriched with, or denuded of, electronic charge. At such centres the rates of the respective types of substitution will be higher. The immediate relevance of the theory is epitomized in expressions (1) and (2), drawing the parallel between nitrobenzene and pyridine (pp. 215, 271), and (3) and (4), illustrating the similarity between phenol and pyrrole (pp. 8 and 63). The first two expressions lead us to expect that pyridine, like nitrobenzene, will, with respect to benzene, be deactivated towards electrophilic substitution (which will occur preferentially at $C_{(3)}$), and activated towards nucleophilic substitution (particularly at $C_{(2)}$ and $C_{(4)}$). Similarly, pyrrole will be activated towards

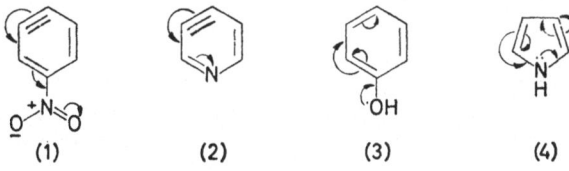

(1)   (2)   (3)   (4)

electrophilic substitution. These broad conclusions are borne out by experiment (cf. Chapters 4–6).

The theory thus most often centres attention upon the electronic situation in the ground state, as being of controlling importance. However, the stress which it lays upon substitution as a developing process[1, 2], and upon the importance of polarizability, allows it to recognize that the perturbing influence of the approaching reagent may generate a state of affairs different from that indicated by the ground state.

The strength of this theory is the generality of its concepts. Its weaknesses are its inability to yield more than qualitative information and its incapacity for dealing with free-radical substitution. The polar requirements of the latter are small, and their attacks upon aromatic nuclei are not to be understood in terms of the gross electronic distribution in the ground state, or of the redistribution permitted by polarizability.

33

The picture of aromatic substitution given by qualitative resonance theory[3] is roughly equivalent to the above treatment, but, as usually presented, less flexible in failing to stress the influence of the approaching reagent[2].

It is characteristic of these qualitative theories that they concentrate attention upon the aromatic molecule being attacked, when the reaction system is at the very beginning of its path, or at the most, slightly modified towards the condition of the transition state[4]. A complete theory must take into account the difference in energy between the transition state and the reactants in their initial conditions, i.e. of the activation energy*. In fact, theories of aromatic reactivity have developed at two levels of approximation, distinguished, respectively, by a concentration upon the condition of the aromatic substance near the beginning of the reaction, and by a preoccupation with the transition state. It is necessary to consider now how the first sort of approach has been developed beyond the stage represented above. The first step in this direction was taken by Wheland and Pauling using molecular orbital theory[3a].

(b) MOLECULAR ORBITAL TREATMENT OF THE ISOLATED MOLECULE METHOD

The work of Wheland and Pauling[3a] may be outlined as follows. The π-electron density at atom r in the aromatic structure is given by

$$q_r = \sum_j n_j c_{jr}^2$$

n being the number of electrons in the jth molecular orbital (p. 28). To evaluate q_r it is necessary, in general, to obtain the roots of the appropriate secular determinantal equation (that is, the energies of the molecular orbitals), then to substitute these into the secular equations and apply the normalization condition (p. 24). In general, the Coulomb integrals of the various carbon atoms and of the hetero-atoms will not be equal. Instead of a we may write for these quantities $a + h\beta$, where a is the Coulomb integral for carbon in benzene, h being a constant for a particular atom. If the atom has a greater electron affinity than carbon, $h > 0$, if a smaller, $h < 0$. In the case of pyridine (with the usual numbering of the positions) h_1, the constant for the nitrogen atom, will be positive. The inductive effect of the nitrogen atom upon the adjacent carbon atoms makes it necessary to assign positive values ($h_2 = h_6$)—auxiliary inductive parameters; h_4 in general—of the constant to them also. Beyond the neighbouring carbon atoms, the inductive effect is neglected. The secular determinantal equation for pyridine would then differ from that for benzene (p. 24) and would assume the form shown, overlap being neglected:

---

* It is generally assumed for comparisons of theory with experiment that entropies of activation are constant[4]. For discussion of this point see [1](e).

$$\begin{vmatrix} x+h_1 & 1 & 0 & 0 & 0 & 1 \\ 1 & x+h_2 & 1 & 0 & 0 & 0 \\ 0 & 1 & x & 1 & 0 & 0 \\ 0 & 0 & 1 & x & 1 & 0 \\ 0 & 0 & 0 & 1 & x & 1 \\ 1 & 0 & 0 & 0 & 1 & x+h_6 \end{vmatrix} = 0$$

where $x = (a - E)/\beta$. This equation cannot be solved without ascribing numerical values to h_1 and h_2. Wheland and Pauling found that, with $h_1/28 < h_2 < h_1/4$, a reasonable picture was obtained of pyridine as a molecule generally deactivated towards electrophilic attack, but least so at the β position. The calculated π-electron densities obtained by using $h_1 = 2$, $h_2 = h_6 = 0.2$, are given in *Table 3.1*. Wheland and Pauling carried through a similar treatment of pyrrole and found that neglect of $h_2(= h_5)$ (with the obvious notation) showed the 3- and 4-positions to be more strongly activated towards electrophilic substitution, whilst with $h_2(=h_5) > h_1/25$ the 2- and 5-positions were more strongly activated. It will be seen later (Chapter 4) that experiment usually indicates the 2- and 5-positions to be most reactive in this sense. As mentioned already (p. 24), the use of the auxiliary inductive parameter, h_t, has been both criticized[11] and justified[16] by later workers.

This treatment was extended by Longuet-Higgins and Coulson[5a] to a number of heteroaromatic substances. They used the values $h_N(= 2)$ and $h_t(= 0.25)$. All resonance integrals were equated to β, the value for the carbon–carbon bond. Some of their results are included in *Table 3.1*. Apart from all the other approximations in this treatment (p. 24) and the lack of self-consistency in deducing different values of q for any atoms to which are attributed equal values of the coulombic integral[4], its great weakness is the absence of a reliable method for deciding the value of h for the hetero-atom. A large number of approaches to this problem have been used[17]. The general conclusion to be drawn is that Wheland and Pauling's original value for $h_N(= 2)$ was too big. In calculating the dipole moments of a number of heterocyclic substances, Löwdin[18] found the value $h_N = 0.6$ to be satisfactory, and a similar value (0.59) was favoured by Jaffé[19] from studies of Hammett substitution constants (see below, Section h). Orgel, Cottrell, Dick and Sutton[20], who were also interested mainly in dipole moments, found it necessary to use two values for h_N, one ($h_N = 2$) for pyrrolic compounds and another ($h_N = 1$) for pyridine-like substances. The experimental value for the dipole moment of pyridine[21] supports a value for h_N near to unity in this case. Orgel, Cottrell, Dick and Sutton also used a modified value for the resonance integral, β_{CN}, writing it as 1.2β. Brown[22] derived values for h_N of 0.5 and of $\beta_{CN} = \beta$ from a correlation of molecular orbital theory and experimental data on the phenylation of pyridine (p. 252) and pointed out that there is no reason why the best values for use in dipole-moment calculations should also be the best for consideration of chemical reactivity. The same author has given a more sophisticated theoretical treatment of the

Table 3.1. *π-Electron Densities in Some Heterocyclic Compounds; Selected Examples**

| Compound | Parameters | | π-Electron densities, q_r | | | | | | | | Ref. |
	l_N	h_i	1	2	3	4	5	6	7	8	
Pyridine	1	0	1·37	0·85	1·01	0·91					3a
	2	0	1·62	0·76	1·01	0·84					3a
	3	0	1·79	0·71	1·01	0·79					3a
	2	0·2	1·59	0·83	0·96	0·81					3a
	2	0·25	1·586	0·849	0·947	0·822					5a
	0·9	0·1	1·327	0·901	0·982	0·910					6
	0·58	$h_N/8$	1·204	0·940	0·986	0·943					7
	with overlap ($S = 0·25$)		1·202	0·939	0·988	0·945					7
	0·5	0		0·923	1·004	0·950					8
	0·6	0	1·234	0·909	1·006	0·940					9
			1·399	0·867	1·000	0·867					10
	valence bond method		1·100	0·951	1·010	0·979					11a†
Pyrrole	1	0·125	1·497	1·122	1·129						12
	2	0·2	1·707	1·081	1·065						12
	2	0·25	1·692	1·096	1·058						5
	2	0·0	1·598	1·071	1·129						9
	0·6	0·0	1·406	1·124	1·173						9
	0·6	0·03	1·399	1·133	1·166						13
	1	0	1·518	1·085	1·156						13
	−1	0	0·813	1·353	1·238						
Quinoline	2	0	1·886	0·573	1·035	0·721	0·953	1·015	0·933	1·053	14a
	2	0·25	1·810	0·676	0·996	0·727	0·956	0·994	0·935	1·023	14a
	2	0·25	1·633	0·789	0·978	0·772	0·958	0·989	0·947	1·003	5a
	1	0		0·806	1·013	0·873	0·980	1·005	0·970	1·024	15
	‡		−0·405	0·162	0·002	0·137	0·022	0·003	0·033	−0·012	5b
	valence bond method		1·463	0·866	1·000	0·882	0·938	1·000	0·933	1·000	10
Isoquinoline	2	0·25	0·767	1·594	0·942	0·938	0·996	0·946	0·984	0·948	5a
	‡		0·156	−0·365	0·057	0·027	−0·001	0·032	0·008	0·029	5b
	valence bond method		0·866	1·419	9·912	1·000	1·000	0·938	1·000	0·933	10

* A large number of calculations on many heterocyclic systems have been reported; other examples are quoted at the appropriate points. † Self-consistent field molecular orbital theory.
‡ Perturbation method (p. 38)

nitrogen molecular orbital parameters[23]. In comparing the conclusions drawn by different workers it is important to ascertain that their results are based on the use of the same values for these parameters.

It is useful to digress slightly at this point to notice some relationships[14b] which justify the importance attached to the quantities q_r and the bond order, p_{rs}, already seen to be of great value in discussing aromatic structure. From the expression for E_j (p. 23), the energy of the jth molecular orbital, we have, by normalizing and neglecting overlap:

$$E_j = \Sigma_r c_{jr}^2 a_r + 2\Sigma\Sigma_{r<s} c_{rj}c_{sj}\beta_{rs}$$

whence the total π-electron energy

$$E = \Sigma n_j E_j = \Sigma\Sigma_{j\ r} n_j c_{jr}^2 a_r + 2\Sigma\Sigma\Sigma_{j\ r<s} c_{rj}c_{sj}\beta_{rs}$$

$\therefore$
$$E = \Sigma_r q_r a_r + 2\Sigma\Sigma_{r<s} p_{rs}\beta_{rs}$$

and
$$\frac{\partial E}{\partial a_r} = q_r \qquad \frac{\partial E}{\partial \beta_{rs}} = 2p_{rs}$$

These expressions show that q_r and p_{rs} are indeed basically important to the properties of the system.

Evaluation of the coefficients c_{jr} makes possible a more complete description of the isolated aromatic molecule than is provided merely by the π-electron densities and the bond orders, p_{rs}. A quantity related to p_{rs}, the free valency at the rth atom, F_r, is useful in considerations of reactivity. Coulson[24] defined this quantity as

$$F_r = N_{max} - N_r$$

where
$$N_r = 3 + \Sigma p_{rs}$$

The numerical factor in this equation refers to the σ bonds in which atom r is engaged. N_{max} is the maximum degree of bonding of which atom r is capable. Since N_r measures the extent to which the atom is already committed to bond formation, it is not unreasonable to regard F_r as a measure of its residual valency, and thus, of its reactivity towards attack by free radicals[24]. Various values have been used for N_{max}[25, 26a]; that used by a particular author is indicated with the related value of F_r in *Table 3.2*. The introduction of the idea of free valency provided the 'Isolated Molecule' theory of aromatic reactivity with a means of dealing with free radical substitution.

Another attempt to meet this shortcoming of the theory, and also to provide a unified approach to electrophilic, nucleophilic and radical substitution, is found in the idea of 'frontier electron densities'[28]. Instead of referring ease of electrophilic and nucleophilic substitutions to high and low total π-electron density, respectively, this approach refers the former to the π-electron density due to the electron in the highest molecular orbital in the ground state, and the latter to the density in the lowest unoccupied orbital. Radical attack is similarly related to the density of charge at a particular position due to a pair of electrons, one occupying the highest occupied, and the other the lowest unoccupied orbital of the ground state. Although these 'frontier electron densities' have been calculated for different kinds of attack upon several aromatic and heteroaromatic molecules[7, 28, 29], the physical basis of the concept has been questioned[26b, 17].

Table 3.2. Free Valence

Compound	N_{max}	1	2	3	4	5	6	7	8	Ref.
						Position					
Benzene					0·398						27
Pyrrole	4·732	0·623	0·453	0·404	—	—	—	—	—	13
		0·409	0·505	0·481	—	—	—	—	—	13
Pyridine	4·414	0·110	0·091	0·079	0·086	—	—	—	—	7
	4·414	0·102	0·092	0·083	0·081	—	—	—	—	7
		—	0·399	0·398	0·402	—	—	—	—	8
		—	0·515	0·488	0·438	—	—	—	—	8
Naphthalene	4·732	—	0·387	0·345	0·360	—	—	—	—	15
		0·448	0·400	—	—	—	—	—	—	27
Quinoline		—	0·344	0·290	0·356	0·338	0·290	0·296	0·331	15

It was mentioned above (p. 24) that Wheland[30] had developed a molecular orbital treatment which allowed for overlap between adjacent atomic orbitals. He found that inclusion of overlap made little difference in the calculated resonance energies of hydrocarbons, and this is generally to be expected[31] for bond orders and charge densities with this class of compounds. However, with heterocyclic molecules some differences do arise. Davies[7] applied the method to a number of heterocyclic compounds and derived values for charge densities, bond orders and free valencies. The coulombic integral for nitrogen is now written* $a_N = a_c + h_1'\gamma$, where $\gamma = \beta - ha$ (p. 34). Other integrals are similarly modified. Davies used the same value for the overlap integral S ($= 0·25$) as was used by Wheland. A value for h_1' ($= 0·535$) was chosen to give approximately the same dipole moments for the compounds considered from calculations with overlap as were obtained without overlap. Though the π-electron densities obtained by the two methods are almost identical (*Table 3.1*), differences are found in the values of the free valencies (*Table 3.2*).

(c) THE MOLECULAR ORBITAL ACCOUNT OF POLARIZABILITY

In their original application of molecular orbital theory to the problem of aromatic reactivity[3a], Wheland and Pauling considered the effect, upon the ground state of an aromatic compound, of an approaching electrophilic reagent. The tendency of the latter to increase the electron affinity of the atom which it is approaching was allowed for by modifying the coulombic integral of the carbon atom through a subsidiary parameter. The general conclusion was that for molecules which, like pyridine, show marked differ-

---

* This notation is used to conform to what has gone before; it is not exactly that used by Davies.

ences between the various positions due to the permanent polarization, the polarizability factor was not of great importance. On the other hand, for molecules with an even distribution of charge in the ground state, polarizability might become of great importance.

If the self-polarizability of atom r is defined[14b] as

$$\pi_{r,\,r} = \frac{\partial q_r}{\partial a_r}$$

being a measure of the change of π-electronic charge at atom r caused by a change in the coulombic integral such as might be produced by an approaching reagent, then, from the relationships given above

$$\pi_{r,\,r} = \frac{\partial^2 E}{\partial a_r^2}$$

As a first approximation, all energy changes produced by the approaching reagent other than changes in the π-electron energy, E, are neglected. Then, since E is a function only of the resonance integrals of bonds and coulombic terms for atoms, the change in E will come mainly from the change in a_r produced by the polarizability effect, and might be expressed[14a] as a Taylor series in δa_r, giving

$$\delta E = \frac{\delta E}{\delta a_r} \cdot \delta a_r + \tfrac{1}{2} \frac{\delta^2 E}{\delta a_r^2} \cdot \delta a_r^2 + \dots$$

$$= q_r \cdot \delta a_r + \frac{\pi_{r,\,r}}{2} \delta a_r^2 + \dots$$

The first term is the most important, and since for an electrophilic agent δa_r is negative, the position favoured for attack will be that at which q_r is greatest. The opposite is true for nucleophilic attack. Here we see a justification for the significance attached to q_r above. When, however, the initial charge distribution is uniform, as e.g. in naphthalene and other members of the important group of alternant hydrocarbons (see below), the second term in the equation becomes important. Depending upon δa_r^2, that is, upon the magnitude and not the sign of δa_r, this indicates that in such cases substitution will occur preferentially at the most polarizable site by reagents of any kind (self-polarizabilities being negative quantities).

It is convenient to notice here a similar argument which justifies the significance attributed to the quantity F_r. When a free radical approaches the aromatic system, it will not change the coulombic integrals significantly, and its influence in terms of the above description will be small. If, however, we consider the radical as it is forming a bond with atom r, and so removing that atom from the conjugated system, since the total π-electron energy is

$$E = \sum_{r<s} 2\beta p_{rs}$$

then

$$\delta E = 2\beta \Sigma_s p_{rs} = 2\beta(N_r - 3) = 2\beta(N_{\max} - 3) - 2\beta F_r$$

39

(assuming as a first approximation that, during the isolation of r, the other bonds in the system are unchanged). The first term in the above equation is a constant, and the ease of radical substitution will depend[4, 25] on the magnitude of F_r.

Some quantities related to the self-polarizability, $\pi_{r, r}$, are of practical utility in connection with heteroaromatic molecules. They are the mutual polarizabilities of atoms r and s and the polarizability of atom r by the bond s–t, defined, respectively, as

$$\pi_{r,\,s} = \frac{\partial q_r}{\partial a_s} \quad \text{and} \quad \pi_{r,\,st} = \frac{\partial q_r}{\partial \beta_{st}}$$

These quantities measure the influence upon the π-electron density at atom r of a change in the coulombic integral of atom s and in the resonance integral of bond s–t. If we regard the formal change of, say, benzene to pyridine, or of naphthalene to quinoline, as a perturbation of the original hydrocarbon structure, a simplified means of calculating the charge distributions in the derived heterocyclic compounds becomes available. Thus[5b], if the π-electron density at atom r in the aromatic hydrocarbon is q_r^0, the new value produced by replacing =CH– by =N– will be

$$q_r = q_r^0 + \Sigma \pi_{r,s}^0 (a_s - a_s^0) + \Sigma \pi_{r,tu}^0 (\beta_{tu} - \beta_{tu}^0)$$
$$\phantom{q_r = q_r^0 + {}}{}_s \phantom{\pi_{r,s}^0 (a_s - a_s^0) + {}} {}_{t,u}$$

The point of practical importance is that, in certain important cases, namely when the parent hydrocarbon is an alternant hydrocarbon (see below), this equation can be greatly simplified. Then, q_r^0 is always unity (i.e. the π-electron distribution is always uniform), and thus $\pi_{r,tu}$ is zero[14a]. Therefore, for heteroaromatic systems which are *isoconjugate* with alternant hydrocarbons,

$$q_r = 1 + \Sigma \pi_{r,s}^0 a_s'$$
$$\phantom{q_r = 1 + {}}{}_s$$

(a_s' being the coulombic integral referred now to that for carbon in an aromatic system as zero, i.e. $a_s' = h\beta$). Clearly, if the values of $\pi_{r,s}^0$ for the parent hydrocarbon are evaluated, they can be used to deduce q_r in any derived aza-hydrocarbon. This is useful because the symmetry of the hydrocarbon often makes the calculations relatively easy, and once made they can be used for a number of aza-derivatives—a much more rapid process than that of solving the secular equations for each aza-hydrocarbon separately[5b]. There is, as always, the difficulty of selecting the value of h to be used for the nitrogen atom. Longuet-Higgins and Coulson[5b] selected the familiar value $h_N = 2$, and for atoms adjacent to nitrogen, $h_i = 0.25$. To successive degrees of approximation we then have

$$q_r = 1 + \pi_{r,N} a_N' \qquad \text{(neglecting the inductive effect)}$$

and, using quinoline as an example,

$$q_r = 1 + \pi_{r,N} a_N' + \pi_{r,2} a_2' + \pi_{r,9} a_9' \quad \text{(including the inductive effect)}$$

Calculations for quinoline to these degrees of approximation have been compared with values obtained by direct solution of the secular equations[14a]. Longuet-Higgins and Coulson[5b] have used the method to deduce charge distributions for mono- and di-aza-derivatives of naphthalene, anthracene and phenanthrene. Some of the values are included in *Table 3.1*, others are referred to elsewhere.

(d) TRANSITION STATE THEORIES: WHELAND'S METHOD

The method of discussing chemical reactivity represented by theories of the type discussed above cannot in general be justified. The rate of a chemical reaction depends on the difference in energy between the reactants and the transition state for the reaction[32]. In principle, then, it is essential to consider the nature of this transition state rather than merely to concentrate on the nature of the reactants.

The importance of the transition state in aromatic substitution has long been recognized. Thus Ingold and his co-workers[33] frequently stressed the change in character, from the sp^2 to the tetrahedral configuration, of the

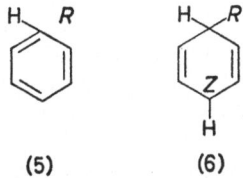

(5)   (6)

carbon atom being attacked, as the attacking reagent became bonded to it. However, the first major contribution to a transition state theory of aromatic substitution was made by Wheland[34]. He regarded the activated complex in a substitution reaction as a resonance hybrid resulting from contributing structures such as those of 'Type I' (5) in which the reagent R is merely near to the carbon atom being attacked, and those of 'Type II' (6) in which the reagent has become bonded to the carbon atom and the hydrogen atom is not yet expelled. In (6), z is a positive charge if R is an electrophilic reagent, a negative charge if R is a nucleophilic reagent, and an odd electron in the case of radical substitution. If Type I structures are so important that their stabilities determine the stability of the transition state or, more generally, if all other structures are merely neglected, we have the Isolated Molecule theory of reactivity. The English School's stress on polarizability, and Wheland and Pauling's treatment of the same factor (see above) represent a bridge between theories concentrating on Type I structures and those admitting the importance of others, such as those of Type II. Wheland[34], in 1942, explored the view that only Type II structures need to be considered. In other words, he assumed a model for the transition state in aromatic substitution in which the attacked carbon atom is isolated from conjugation with the rest of the system by being converted to the sp^3 condition. The assumption was made that, in electrophilic substitution, reaction would occur most readily at the position which most easily supplied a pair of electrons, nucleophilic substitution most readily where an open sextet was forthcoming, and radical substitution most readily at the carbon atom which most willingly provided an odd electron. The theory has thus the advantage over the Isolated Molecule approach of treating uniformly all three types of substitution. Molecular orbital theory was then applied to calculate the π-electronic energy of the aromatic substrate, and of the conjugated residue remaining in the transition state after isolation of the attacked carbon atom, overlap being included. The method has been outlined already in Chapter 2. With the overlap integral set at 0·25 in both structures, E_π for benzene is $6a + 5 \cdot 8667\ \gamma_0$, where γ_0 is the value of γ appropriate to benzene and was taken as -38 kcal. For each of the three kinds of substitution, the corresponding value for the transition state is $6a + 4 \cdot 0174\ \gamma_0$, giving $\Delta E_\pi = -1 \cdot 8493\ \gamma_0$, or about 70 kcal/mole. ΔE_π is not, of course, the activation energy for the reaction, for no account has been taken of bonds made or broken, of Type I structures, of σ electrons, etc.; nevertheless, it is suggested that the smaller the value of ΔE_π, the more rapid the reaction concerned. With substituted benzenes or heterocyclic compounds, the usual difficulty arises of assigning to atoms in the substituent, or to the nitrogen atom, electron affinities different from that of carbon. Numerical values must be assigned to a number of adjustable parameters. However, whilst the quantitative significance of the results is thus made uncertain, the qualitative conclusions often depend only on the sign and not on the magnitude of the parameters, and are therefore more reliable. Examples of Wheland's results are given in *Table 3.3*. The conclusions follow that pyridine will resist

Table 3.3. Wheland's Localization Energies[34]

| Parameters | | Position | ΔE_π (in units of $-\gamma_0 = 38$ kcal mole^{-1}) | | |
			Electrophilic	Nucleophilic	Radical
Benzene	—	—	1·849	1·849	1·849
Pyridine	*	α	1·849 + 0·239h_1 + 0·012h_2	1·849 − 0·427h_1 + 0·012h_2	1·849 − 0·084h_1 + 0·012h_2
		β	1·849 + 0·007h_1 + 0·530h_2	1·849 + 0·007h_1 − 0·804h_2	1·849 + 0·007h_1 − 0·137h_2
		γ	1·849 + 0·290h_1 + 0·013h_2	1·849 − 0·376h_1 + 0·013h_2	1·849 − 0·043h_1 + 0·013h_2
Nitro- benzene† $\begin{cases} h_b = 0 \cdot 8 \\ h_a = 0 \cdot 4 \\ \rho_{a1} = 0 \cdot 6 \end{cases}$		o	1·886	1·783	1·834
		m	1·852	1·853	1·852
		p	1·861	1·757	1·809

* For pyridine, agreement with experiment follows with any positive value of h_1 and positive values of $h_2 < h_1$
† h_b refers to oxygen atoms, h_a to the nitrogen atom, and ρ_{a1} is an adjustment of the resonance integral of the C—N bond.

electrophilic attack more than does benzene, and that $C_{(3)}$ will be the most reactive. Pyridine will be more susceptible to nucleophilic attack, the order of reactivity being $C_{(2)}$, $C_{(4)} >$ benzene, $C_{(3)}$. In radical attack, the order will be $C_{(2)}$, $C_{(4)} >$ benzene, $C_{(3)}$. These conclusions are broadly correct, as will be seen later (Chapter 5).

This sort of approach can be criticized in several ways. First, of course, it shares the general shortcomings of all theories of reactivity in neglecting steric factors and solvation factors. More particularly, the application of molecular orbital methods to ions, using the same parameters that are used for neutral molecules, is not strictly justifiable. Also, in selecting the Type II structures as reasonable models for the transition state, Wheland's treatment excludes all distinctions between *different* electrophilic, nucleophilic or radical reagents.

Calculations of atom localization energies by Wheland's method have been made by a number of authors and will be considered at appropriate points in this work.

(e) THE NON-BONDING MOLECULAR ORBITAL METHOD

As a general treatment of the problem of substitution, Wheland's method has the disadvantage of not being very easy to apply. For one class of aromatic substances, namely the alternant hydrocarbons and heterocyclic compounds isoaromatic with them, a comprehensive and rapid treatment has been developed. Alternant hydrocarbons have no odd-numbered rings. They have the property that the constituent carbon atoms fall into two sets distinguished as starred and unstarred, such that no atom of one set is adjacent to another of the same set. Odd alternant hydrocarbons are radicals or ions. As an

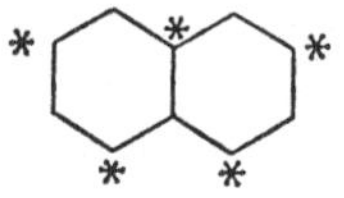

(7) 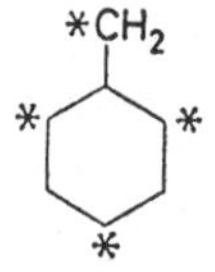   (8) 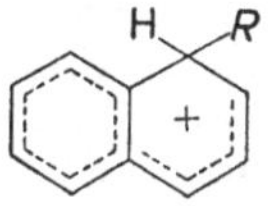   (9)

example of an alternant hydrocarbon, we might consider naphthalene (7), and as an odd alternant structure, the benzyl radical (8). Further examples of odd alternant structures are the residual molecules in the transition states for substitution into alternant hydrocarbons, e.g. that (9) for electrophilic substitution at the α-position in naphthalene. A number of important theorems relating to such structures have been deduced by Coulson and Rushbrooke[35], Coulson and Longuet-Higgins[14a], and Longuet-Higgins[36]. Thus, in alternant hydrocarbons, the π-electron distribution is uniform. In a neutral odd alternant hydrocarbon (radical), the odd electron inhabits a molecular orbital which to the degree of approximation of this treatment has the same energy as that in a carbon $2p$ orbital; further, this *non-bonding molecular orbital* is confined to the starred atoms (the larger of the two sets). Since ions corresponding to neutral odd alternant hydrocarbons differ from them only by one electron more or less in this non-bonding molecular orbital, the formal charge in such an ion is spread over the starred centres only. A rule due to Longuet-Higgins[36] allows rapid evaluation of the coefficients of the non-bonding molecular orbital. The rule is that the sum of the coefficients at the starred atoms adjacent to any unstarred atom vanishes. The ratios of the coefficients follow, and their absolute values are given by normalization. The process is illustrated for electrophilic substitution in benzene (10), and nucleophilic substitution at the α-position of naphthalene (11).

43

C

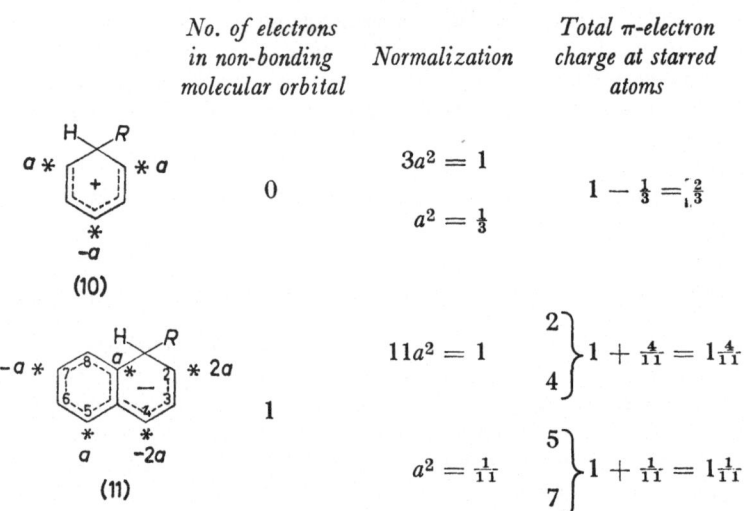

	No. of electrons in non-bonding molecular orbital	Normalization	Total π-electron charge at starred atoms

For (10):

$$3a^2 = 1$$
$$a^2 = \tfrac{1}{3}$$

$$0$$

$$1 - \tfrac{1}{3} = \tfrac{2}{3}$$

For (11):

$$11a^2 = 1$$
$$a^2 = \tfrac{1}{11}$$

$$1$$

$$\left.\begin{matrix} 2 \\ 4 \end{matrix}\right\} 1 + \tfrac{4}{11} = 1\tfrac{4}{11}$$

$$\left.\begin{matrix} 5 \\ 7 \end{matrix}\right\} 1 + \tfrac{1}{11} = 1\tfrac{1}{11}$$

Longuet-Higgins[36] has shown how such considerations can be used to illuminate a number of problems in aromatic chemistry. His results will be referred to later at appropriate points, but the method is usefully illustrated here by reference to aromatic substitution, as for example in the case of nucleophilic substitution in benzene and pyridine. The essential point is that not only benzene and pyridine themselves, but also the transition states (12), (13), (14) and (15) form isoaromatic series. Qualitatively, it is obvious that, since in (12) negative charge resides on the starred atoms,

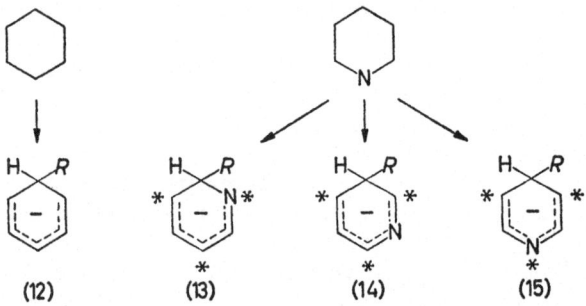

replacement of =CH– by –N= to give (13), (14) or (15) will produce the greatest effect if the replacement is made at a starred position. There the electron-attracting nitrogen atom would have a facilitating influence; thus nucleophilic substitution will be relatively easy at the α- and γ-pyridine positions [(13) and (15)]. To a fair approximation, however, the β-position (14) will not be affected.

The method is capable of more exactitude[36, 37]. One way of seeing this is by using the equation $\partial E/\partial a_r = q_r$ (p. 37). From the isoaromatic pair represented by an alternant hydrocarbon and its aza-derivative, we have $E_{\pi_{aza}} - E_{\pi_{A.H.}} = \Sigma q_r \cdot a_r$, where a_r is the difference between the coulombic term at the rth atom in the aza-compound and the coulombic term for carbon. Similarly, for the two transition states in a substitution, $E_{\pi'_{aza}} - E_{\pi'_{A.H.}} = \Sigma q'_r a_r$. Subtraction and substitution of the non-bonding molecular orbital coefficient (a_{0r}) for the hydrocarbon transition state in place of q'_r gives

$$\Delta E_{\pi_{aza}} = \Delta E_{\pi_{A.H.}} - A \Sigma a_{0r}^2 a_r$$

where $\Delta E_{\pi_{aza}}$ is the difference in π-electronic energy between the aza-compound and its transition state in substitutions, and $\Delta E_{\pi_{A.H.}}$ has the same significance for the alternant hydrocarbon. A is a numerical factor, having the values -1, 0 or $+1$ for electrophilic, radical, or nucleophilic attack[38]. $\Sigma a_{0r}^2 a_r$ is made up of terms expressing the effect of replacing $-CH=$ by $-N=$, and others for the influence of the nitrogen atom upon the coulombic integrals of the remaining ring carbon atoms. To a first approximation, we might neglect all of these except the nitrogen coulombic term and write for, say, nucleophilic substitution

$$\Delta E_{\pi_{aza}} = \Delta E_{\pi_{A.H.}} - a_{0r}^2 a_N$$

Applying this to a-substitution in naphthalene and into aza-derivatives it is obvious [see (7)] that, *with this degree of approximation*, we must expect no facilitation of nucleophilic substitution when the nitrogen atom is placed at position 3, 6, or 8, slight activation from positions 5 and 7, and the greatest activation from positions 2 and 4. That is, nucleophilic substitution will proceed most readily at the 1-position in isoquinoline and at the 4-position in quinoline; the 5-position in quinoline and the 8-position in isoquinoline would be very slightly activated, and the 4- and 5-position in isoquinoline and the 8-position in quinoline would be unactivated as compared with positions in naphthalene. The lowering of activation energies would be roughly in the ratios $4 : 4 : 1 : 1 : 0 : 0 : 0$. The deactivation towards electrophilic substitution, and the lack of influence (*at this degree of approximation*) upon radical substitution of aza-replacements can be deduced similarly. The method will be illustrated again later.

This treatment reproduces exactly the predictions of simple resonance theory[36], but with a degree of semi-quantitative differentiation. However, a full utilization of the more exact equation for $\Delta E_{\pi_{aza}}$ given above demands some means of evaluating $\Delta E_{\pi_{A.H.}}$, the difference in π-electronic energy between two systems, the alternant hydrocarbon and a transition state derived from it, which are not isoaromatic. In general, to predict the effect of substituents on the relative rates of reactions, this ability to evaluate relative energies of non-isoaromatic systems is essential. This necessary extension of the method was provided by Dewar[39]. So far as aromatic substitution is concerned, the central result obtained by Dewar, using molecular orbital theory in its simplest unrefined form, is expressed in the equation

$$\Delta E_{\pi_{A.H.}} = 2\beta(a_{0s} + a_{0t})$$

where a_{0s} and a_{0t} are the non-bonding molecular orbital coefficients, in the transition state, at the atoms s and t, adjacent to the atom at which substitution is occurring. A simple application of this equation is to the problem of α- and β-substitution in naphthalene. We have

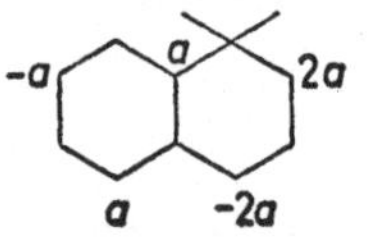

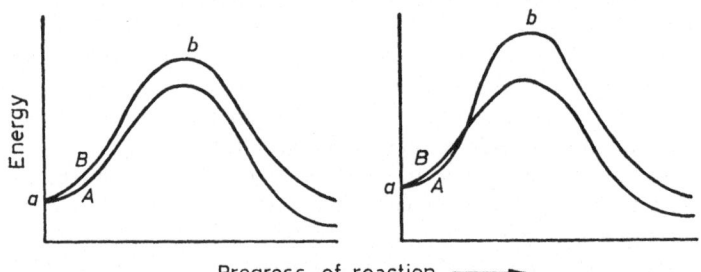

(16) 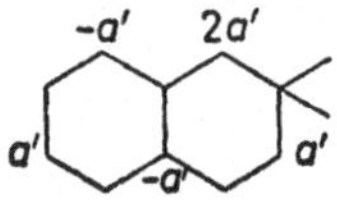   (17)

Clearly, the α-position, $\Delta E_{\pi_\alpha} = 2\beta(2a + a) = 1\cdot81\beta$ [see (16)], should be more readily substituted than the β-position, $\Delta E_{\pi_\beta} = 2\beta(2a' + a') = 2\cdot12\beta$ [see (17)], as it is. The use of the above expression for $\Delta E_{\pi_{A.H.}}$ and $\Delta E_{\pi_{aza}}$ has made possible a semi-quantitative treatment of aromatic and heteroaromatic substitution, which will be illustrated later.

(f) THE RANGE OF VALIDITY OF THE ISOLATED MOLECULE METHOD. GENERAL REMARKS

It is important to enquire as to the conditions under which the Isolated Molecule treatment can be expected to agree with the Transition State method in predicting correct sequences of reactivity. Leaving aside the problem of using the Transition State method correctly, it can be seen that it and the Isolated Molecule method can give conflicting answers in cases such as the following. In each of the two diagrams shown, the curves represent the changes in π-electron energy during the course of, say, substitution

Progress of reaction ⟶

at positions A and B in an aromatic system from the reactants at a to the transition state at b. In the first case, both theories would predict greater reactivity at position A. In the second, the Isolated Molecule theory would predict greater reactivity at A, the Transition State method at B. These are simple illustrations of a complicated position discussed by Brown[4] in terms of a chemical 'non-crossing rule'.

In general, the approximate nature of the theory in its present state cannot be too much stressed. All the methods used are based on the approximation of the linear combination of atomic orbitals, and usually there is of necessity the further gross approximation of the Hückel method. Conse-

46

quently, the quantitative results can be allowed only relative significance. With this stricture constantly in mind, it can hardly be denied that the theory has achieved great successes, some of which will be exemplified in what follows.

(g) THE VALENCE-BOND METHOD

The discussion of aromatic reactivity has proceeded in terms of molecular orbital theory. The valence-bond method has also been used[40a, 41] to evaluate quantities which are parallel to those discussed above. The methods have in common the charge densities, and the free valency and bond order of molecular orbital theory have counterparts of similar but not identical significance in the valence-bond method. The latter has not been used widely in dealing with heterocyclic compounds, and so will not be much referred to here. In *Table 3.1* are included data obtained by the valence-bond method for pyridine and quinoline. The use of molecular diagrams originated with studies in the valence-bond method[40b].

(h) THE HAMMETT EQUATION

A different, and basically empirical method of correlating aromatic reactivity with aromatic structures is represented by Hammett's equation[42a]

$$\log (k/k^0) = \sigma\rho$$

Here, k and k^0 are rate or equilibria constants referring to reactions of the substituted and unsubstituted compounds, respectively; σ is the 'substituent constant' which depends on the position and character of the substituent to which it refers, and ρ is the 'reaction constant' characteristic of the equilibrium or rate process appropriate to it. Since $\log (k/k^0)$ is proportional to the difference in the free energies of the reactions of the substituted and unsubstituted species in the case of equilibria, or to differences in activation energies for kinetic processes, Hammett's equation expresses a linear free-energy relationship. Attempts to remove deficiencies in the original Hammett equation have been numerous, and modified treatments can now be used to correlate experimental observations on a large range of equilibria and reactions[43]. Progress has been made[42] in correlating the empirical substituent constant with changes in electron density caused by substitution, use being made of molecular orbital theory. The Hammett equation has been employed to some extent in the heterocyclic series, and it will be mentioned again at the appropriate point.

REFERENCES

[1] (a) Ingold, C. K. *A. Rep. Chem. Soc.* **23** (1926) 129; (b) Robinson, R. *Two Lectures on an Outline of an Electrochemical (Electronic) Theory of the Course of Organic Reactions*, London (Institute of Chemistry) 1932; (c) Ingold, C. K. *Chem. Rev.* **15** (1934) 225; (d) idem, *Structure and Mechanism in Organic Chemistry*, London (Bell) 1953; (e) ibid. p. 255

[2] Waters, W. A. *J. chem. Soc.* (1948) 727

[3] (a) Wheland, G. W. and Pauling, L. *J. Am. chem. Soc.* **57** (1935) 2086; (b) Pauling, L. *Nature of the Chemical Bond*, 3rd Ed., New York (Cornell Univ. Press) 1960

[4] Brown, R. D. *Qu. Rev. chem. Soc.* **6** (1952) 63

[5] Longuet-Higgins, H. C. and Coulson, C. A. (*a*) *Trans. Faraday Soc.* **43** (1947) 87; (*b*) *J. chem. Soc.* (1949) 971

[6] Ploquin, J. *C.r. hebd. Séanc. Acad. Sci., Paris* **226** (1948) 245

[7] Davies, D. W. *Trans. Faraday Soc.* **51** (1955) 449

[8] Brown, R. D. and Heffernan, M. L. *Aust. J. chem.* **9** (1956) 83

[9] Gyoerffy, E. *C.r. hebd. Séanc. Acad. Sci., Paris* **232** (1951) 515

[10] Pullman, B. *Bull. Soc. chim. Fr.* **15** (1948) 533; Daudel, R. and Martin, M. *ibid.* 559

[11] (*a*) McWeeny, R. and Peacock, T. E. *Proc. phys. Soc.* **70** (1957) 41; (*b*) McWeeny, R. *Proc. R. Soc.* **A237** (1956) 355

[12] Pickett, L. W., Corning, M. E., Wieder, G. M., Semenow, D. A. and Buckley, J. M. *J. Am. chem. Soc.* **75** (1953) 1618

[13] Brown, R. D. *Aust. J. Chem.* **8** (1955) 100

[14] Coulson, C. A. and Longuet-Higgins, H. C. *Proc. R. Soc.* (a) **A192** (1947) 16; (*b*) **A191** (1947) 39

[15] Sandorfy, C. and Yvan, P. *C.r. hebd. Séanc. Acad. Sci., Paris* **229** (1949) 715; *Bull. Soc. chim. Fr.* **17** (1950) 131

[16] Brown, R. D., Coller, B. A. W. and Heffernan, M. L. *Tetrahedron* **18** (1962) 343

[17] Streitwieser, A. *Molecular Orbital Theory for Organic Chemists*, New York (Wiley) 1961

[18] Löwdin, P. *J. chem. Phys.* **19** (1951) 1323

[19] Jaffé, H. H. *J. chem. Phys.* **20** (1952) 1554

[20] Orgel, L. E., Cottrell, T. L., Dick, W. and Sutton, L. E. *Trans. Faraday Soc.* **47** (1951) 113

[21] De More, B. B., Wilcox, W. S. and Goldstein, J. H. *J. chem. Phys.* **22** (1954) 876

[22] Brown, R. D. *J. chem. Soc.* (1956) 272

[23] Brown, R. D. and Penfold, A. *Trans. Faraday Soc.* **53** (1957) 397

[24] Coulson, C. A. *Discuss. Faraday Soc.* **2** (1947) 16

[25] Burkitt, F. H., Coulson, C. A. and Longuet-Higgins, H. C. *Trans. Faraday Soc.* **47** (1951) 553

[26] Greenwood, H. H. (*a*) *Trans. Faraday Soc.* **48** (1952) 677; (*b*) *J. Am. chem. Soc.* **77** (1955) 2055

[27] Hey, D. H. and Williams, G. H. *Discuss. Faraday Soc.* **14** (1953) 216

[28] Fukui, K., Yonezawa, T., Nagata, C. and Shingu, H. *J. chem. Phys.* **22** (1954) 1433

[29] Mason, S. F. *The Chemistry and Biology of Purines* (Ciba Foundation Symposium) 1957, p. 72

[30] Wheland, G. W. *J. Amer. chem. Soc.* **63** (1941) 2025

[31] Chirgwin, B. H. and Coulson, C. A. *Proc. R. Soc.* **A 201** (1950) 196; Löwdin, P. *J. chem. Phys.* **18** (1950) 365

[32] Eyring, H. *Chem. Rev.* **10** (1932) 103; Glasstone, S., Laidler, K. J. and Eyring, H. *The Theory of Rate Processes*, New York (McGraw-Hill) 1941

[33] Cowdrey, W. A., Hughes, E. D., Ingold, C. K., Masterman, S. and Scott, A. D. *J. chem. Soc.* (1937) 1257; Hughes, E. D. and Ingold, C. K. *ibid.* (1941) 608

[34] Wheland, G. W. *J. Am. chem. Soc.* **64** (1942) 900; *Resonance in Organic Chemistry*, Chapter 8, New York (Wiley) 1955

[35] Coulson, C. A. and Rushbrooke, S. *Proc. Camb. phil. Soc. math. phys. Sci.* **36** (1940) 193

[36] Longuet-Higgins, H. C. *J. chem. Phys.* **18** (1950) 265, 275, 283

[37] Brown, R. D. *J. chem. Soc.* (1951) 1955

[38] Dewar, M. J. S. and Maitlis, P. M. *J. chem. Soc.* (1957) 2521

[39] Dewar, M. J. S. *J. Am. chem. Soc.* **74** (1952) 3341, 3345, 3350, 3353, 3355, 3357

[40] Pullman, B. and A. (*a*) *Les Théories Electroniques de la Chimie Organique*; (*b*) *ibid.* p. 143, Paris (Masson) 1952

[41] Pullman, B. and A. 'Free Valence in Conjugated Organic Molecules', in *Progress in Organic Chemistry*, Vol. 4 (ed. Cook, J. W.) London (Butterworths) 1958

[42] (*a*) Hammett, L. P. *J. Am. chem. Soc.* **59** (1937) 96; *Trans. Faraday Soc.* **34** (1938) 156; *Physical Organic Chemistry*, New York (McGraw-Hill) 1940; Jaffé, H. H. *Chem. Rev.* **53** (1953) 191; (*b*) *idem, J. chem. Phys.* **20** (1952) 279, 778, 1554

[43] Leffler, J. E. and Grunwald, E. *Rates and Equilibria of Organic Reactions*, New York (Wiley) 1963; Stock, L. M. and Brown, H. C. in *Advances in Physical Organic Chemistry*, Vol. I (ed. Gold, V.), London (Academic Press) 1963

B. PROTOTYPICAL SYSTEMS

4

PYRROLE AND ITS DERIVATIVES

(1) PHYSICAL PROPERTIES

(a) General. Dipole Moments

Some of the physical constants of pyrrole and of a selection of its derivatives are collected in *Table 4.1*. The boiling point of pyrrole is higher than might have been expected, and the closer similarity in this property of 1-methylpyrrole to, say, toluene suggests that in pyrrole the imino group is responsible for some sort of association (see below). The characterization of pyrrole and simple alkylpyrroles through the formation of crystalline derivatives is not always easy. Picrates are usually unstable. In cases where picric acid causes dimerization, the dimer picrate is often a satisfactory derivative[13, 233]. The reaction with phenyl isocyanate (p. 66) is useful.

From the general discussion of its geometry and stability already given (p. 17), it has been seen that in these respects pyrrole shows a fairly highly developed aromatic character.

Although dipole-moment measurements did not permit a conclusive decision regarding the planarity of pyrrole (p. 17), such studies[1] do provide interesting information in other ways. Assuming standard values for the σ-bond moments and a regular pentagonal structure of side length 1.4 Å, the σ moment can be calculated to be $+0.4$ D. If then the dipole moment of pyrrole $(1.80$ D$)$ is positive (i.e. if the positive pole is towards the nitrogen atom and the negative towards the ring), the π moment must be $+1.40$ D. This agrees fairly well with the value calculated from the π-electron distribution deduced from Hückel molecular orbital theory (p. 35). More advanced molecular orbital treatments attributed almost the whole of the dipole moment to π-electron delocalization[167b, 326, 327]. That the moment of pyrrole and N-methylpyrrole is indeed positive is confirmed by examination of the influence of substituents such as methyl. This fact of a positive dipole moment accords with the description of the aromatic character of pyrrole already given (p. 16).

(b) Spectra

The absorption spectrum of pyrrole in the vacuum ultra-violet region (*Table 4.2*) has been studied by a number of workers[2]. Three band systems are found, and consideration[2a] of a series of cyclic dienes (cyclopentadiene, furan, thiophen and pyrrole) has been taken to indicate that, in pyrrole, homocyclic conjugation is not extensive. It has been argued[2c] that the appearance in N-methylpyrrole of a region (2,000–2,250 Å) not found with pyrrole[2b] points to fuller conjugation of the nitrogen lone pair in N-methylpyrrole with other π electrons, made possible by the electron-releasing properties of the methyl group. Recently the opposite view has been supported[2f] that the similarities between the spectra of cyclopentadiene, pyrrole,

C*

Table 4.1. Pyrrole and its Alkyl and Aryl Derivatives[232, 233, 253, 117a]

Substituents	M.p., °C	B.p., °C	Derivatives
—		132 (d_4^{20} 0·9691; n_D^{20} 1·5085)	picrate, m.p. 69° (dec.)*; 1-acetyl, b.p. 180–1°; 1-benzoyl, b.p. 276°/715 mm
1-Me		116	
2-Me†		146–9	1-p-nitrobenzoyl, m.p. 198°[135]
3-Me		143–143·5	'picrate'‡, m.p. 103–103·5
1-Et		129–130	
2-Et[57]		68/14 mm	
3-Et[295]		65/14 mm	
2-Pr		92/33 mm	
3-Pr		90/30 mm	
2-iPr†		80/25 mm	
3-iPr		90/23 mm	
1-Allyl[61]		50/22 mm	
2-Allyl[61]		83–4/24 mm	
1-CH₂Ph[290]		67–9/0·20–0·25 mm	
2-CH₂Ph[290]		85–9/0·12–0·15 mm	
1,2-Me₂		139–141	
2,3-Me₂†		97/65 mm	
2,4-Me₂**§		166	picrate, m.p. 91–2°
2,5-Me₂**	33	164–166/755 mm	1-p-nitrobenzoyl, m.p. 189°[135]
3,4-Me₂[296]		66/14 mm	1-p-nitrobenzoyl, m.p. 195°[135]
1,2,5-Me₃[246]		168–170/755 mm	
2,3,4-Me₃§	36–8	72/10 mm	picrate, m.p. 140°; hydrochloride, m.p. 137–8° (dec.)[297]
2,3,5-Me₃			hydrochloride, m.p. 147–9[14a]
2-Et-3-Me†[22]		76·5/16 mm	picrate, m.p. 83°
2-Et-4-Me[68c]			
2-Et-5-Me[120c]		93/40 mm	picrate, m.p. 86°
3-Et-2-Me†		74·5/11 mm	
3-Et-4-Me		75·5–76·5/15 mm	
3-Et-4-Me (opsopyrrole)		73–5/13 mm	

Table 4.1. (continued)

Substituents	M.p., °C	B.p., °C	Derivatives
4-Et-2-Me[298]	112	86/20 mm	picrate, m.p. 109°
2,3,4,5-Me4			picrate, m.p. 127°; hydrochloride, mp. 116–119°[14a]
4-Et-2,3-Me2 (haemopyrolle)		113/16 mm	picrate, m.p. 124°
2-Et-3,5-Me2		112–4/63 mm	
3-Et-2,4-Me2‖§		85·5–7/12·5 mm	picrate, m.p. 138°; hydrochloride, m.p. 109–117° (dec.)[12c]; 1-p-nitrobenzoyl, m.p. 162°[135]
3-Et-2,5-Me2	69	121/63 mm	picrate, m.p. 104°
2-Et-3,4,5-Me3	70	114–5/27 mm	picrate, m.p. 104–5°
3-Et-2,4,5-Me***	60–1		
1,2,3,4,5-Me5[233]	129		
1-Ph††	179		
2-Ph††	143		
2,4-Ph2[299]	140–1		
2,5-Ph2††	211–2		
2,3,5-Ph3[125b]	282		
2,3,4,5-Ph4[300]			
1,2,3,4,5-Ph5[301]			

*The picrate and the trinitrobenzene complex are unstable. For these compounds and general comments on picrates of the pyrrole series, see Treibs and Dieter[13]
†With ethereal hydrogen chloride a hydrochloride, and with ethereal picric acid a picrate of the dimer is formed (p. 83)
‡This is not a true picrate of 3-methylpyrrole***; this compound gives amorphous polymeric material with ethereal hydrogen chloride or ethereal picric acid (p. 83)
**2,4- and 2,5-Dimethylpyrrole give unstable monomer hydrochlorides which form dimers when kept[13c, 14a]
§For reactions in which the monomeric hydrochloride is used to produce a dimer, see p. 84
‖Cryptopyrrole; for possible dimerization by picric acid see p. 84
***Phyllopyrrole; for possible dimerization by picric acid see p. 84
††See Allen et al. in[14c].

Table 4.2. Ultra-violet Absorption Spectroscopy of Pyrroles

Substituent	Solvent	Free base		Conjugate acid		References
		λmax, mμ	εmax	λmax, mμ	εmax	
A. Pyrrole and its homologues						
—	Hexane	217	7,900	—	—	233*, 234a
—	EtOH	208	7,300	—	—	
—		205	6,700	—	—	
1-Me	H₂O	210	5,800	241	3,900	233*
2-Me		208	7,100	247	4,100	
3-Me		208	5,900	233	4,500	
1,2-Me₂		210	7,200	258	4,800	
2,3-Me₂		208	5,600	240	3,800	
2,4-Me₂		209	5,800	246	5,200	
2,4-Me₂	EtOH	218	4,700	249		4b
2,5-Me₂	H₂O	209	7,700	237	2,900	233*
				275†	740†	
2,5-Me₂	Iso-octane	217	6,300	—	—	234b
3,4-Me₂		205	4,400	271	5,800	
1,2,5-Me₃	H₂O	211	8,300	243‡	3,100‡	233*
2,3,4-Me₃		208	5,100**	262	5,000**	
2,3,5-Me₃		212	6,600	252	5,700	
1,2,3,5-Me₄		216	7,500	259	5,600	
2,3,4,5-Me₄		216	5,800	265	5,200	
Me₅		216	7,000	269	5,600	
3-Et-4-Me	EtOH	~203	5,670	—	—	4a
3,4-Et₂		~208	5,900	—	—	
3-Et-2,4-Me₂		214	7,000	—	—	4b
3-Et-2,4-Me₂	EtOH + HCl	~200	7,450	261	4,000	

Table 4.2. (continued)

Substituent	Solvent	Free base				Conjugate acid		References
		λmax, mμ	εmax	λmax, mμ	εmax	λmax, mμ	εmax	
B. Acylpyrroles								
2-CHO	Hexane			246	5,370	279	18,600	235
2-CHO				251	3,100	287	13,300	4b, 236
1-MeCO				238·5	1,000	288	760	236
2-MeCO				251	4,100	290	16,400	4b, 236
2-MeCO-1-Me	EtOH			245(sh)	4,900	287	10,050	236
2-MeCO.CH:CH						352	16,900	4b
3-MeCO-2,4-Me₂		209	10,300	250	10,400	282	4,900	4
2-MeCO-4-Et-3,5-Me₂				266(sh)	4,700	308	19,600	4a
2,5-(MeCO)₂		229	12,800	—	—	305	16,400	4b
C. Alkoxycarbonylpyrroles§								
2-EtO₂C-3,4-Me₂	EtOH			240(sh)	5,000	273	14,500	4a
2-EtO₂C-3,5-Me₂				253	6,800	276	19,300	4a
2-EtO₂C-3,4,5-Me₃				250(sh)	5,900	285	15,800	4b
2-EtO₂C-4Et-3,5-Me₂				258	5,300	283·5	18,900	4a
3-EtO₂C-2,4-Me₂		232	9,000	263·5	3,740	—	—	4b
3-EtO₂C-4,5-Me₂		231·5	9,980	270	4,700	—	—	} 4a
3-EtO₂C-2,4,5-Me₅		232·5	9,400	243	5,640	—	—	
2,3-(EtO₂C)₂-5-Me		210	12,400			273	16,100	4b
2,4-(EtO₂C)₂-3,5-Me₂		221	25,300			282	22,000	4a
2,5-(EtO₂C)₂-3,4-Me₂		220	16,700			—	—	4b
3,4-(EtO₂C)₂C-2,5-Me₂		215	11,200	267·5	8,450	—	—	4a
2,4,5-(EtO₂C)₃-3-Me		224	24,000			276	17,500	4b

Table 4.2. (continued)

Substituent	Solvent	Free base λ_{max}, mμ	Free base ϵ_{max}	Conjugate acid λ_{max}, mμ	Conjugate acid ϵ_{max}	References
D. *Arylpyrroles*						
1-Ph	EtOH	253	13,500			215
2-Ph	EtOH	287	~11,000			237
2,5-Ph_2	EtOH	230	11,900	325	26,400	238
	EtOH			330	26,700	238
1-Me-2,5-Ph_2	EtOH	231	8,400	307	20,500	238
1,2,5-Ph_3	EtOH			301	19,500	238
E. *Cyanopyrroles*						
3-CN-2,5-Me_2	EtOH	246	5,500			4*b*
3,4-$(CN)_2$-2,5-Me_2	EtOH	252	8,100			4*b*
F. *Nitropyrroles*						
2-NO_2	H_2O–pH 2			370	3,980	20*c*
	H_2O–pH 13			390	15,900	20*c*
	AcOH			330	7,900	20*c*
1-Me-2-NO_2	AcOH			330	12,600	20*c*
1-Me-3-NO_2	AcOH			281	5,500	20*c*
2,4-$(NO_2)_2$	H_2O–pH 2			320	12,600	20*c*
	H_2O–pH 13			395	20,000	20*c*
	AcOH	230	7,940	319	13,500	20*c*
2,5-$(NO_2)_2$	H_2O–pH 2			348	19,950	20*c*
	H_2O–pH 13	300	6,310	400	25,100	20*c*
	AcOH	240	5,010	330	19,950	20*c*
1-Me-2,4-$(NO_2)_2$	AcOH	230	7,940	310	12,700	20*c*
1-Me-3,4-$(NO_2)_2$	AcOH	292	5,010			20*c*

*Data on conjugate acids refer to solutions in sulphuric acid.
†Assigned to the β-protonated form.
‡Absorption tailing to 300 mμ due to β-protonated form.
**Inaccurate because of impure specimen.
§Ref. 354 records the longest wavelength absorption maximum of C–Me and C–OMe derivatives of methyl pyrrole-2- and pyrrole-3-carboxylate in methanol.

furan and thiophen indicate that homocyclic conjugation is present in cyclopentadiene rather than that it is absent in the heterocycles. The early ultra-violet absorption data for hexane solutions of pyrrole are inaccurate[233] and have led to some incorrect conclusions (see below). The transition energies observed for pyrrole in the vacuum ultra-violet have been compared[2d] with those derived from Hückel molecular orbital calculations of the orbital energies (*Table 4.3*). The calculations give only average values for the singlet and triplet levels. The variable electronegativity and self-consistent field methods have been used successfully to calculate the energy levels of pyrrole[167b, 327].

Table 4.3. Transition Energies of Pyrrole[2d]

cm^{-1} or β*		Calculated (in units of β) for	
		$h_N = 1, h_i = 0.125$	$h_N = 2, h_i = 0.2$
46,250	1.85	1.934	1.743
54,560	2.18	2.224	2.241
57,400	2.30	2.295	2.333
		2.585	2.831

* Putting $\beta = 3.1$ eV.

The absorption spectra of pyrrole and its simple alkyl derivatives in the ultra-violet region more familiar to organic chemists are singularly unexciting. Extensive data are now available (a small selection is given in *Table 4.2*) for functional derivatives of pyrrole. Early, inaccurate determinations of the absorption spectrum of pyrrole led to some incorrect deductions concerning the effect of methyl substituents upon this spectrum[4, 233]. In fact, methyl groups cause small bathochromic shifts of the pyrrole maximum, and these shifts are larger for the conjugate acids than for the free bases[233]. The effects of other substituents are generally similar to those met with in the benzene series; a methyl group or a bromine atom produces a red shift of 5–10 mμ, whilst replacement of carbethoxyl by formyl or acetyl causes a bathochromic shift of about 20 mμ. A chromophoric –M group conjugated with the pyrrole nucleus produces a new absorption band (a *K* band)[3, 4]; λ_{max} for this band moves to longer wavelengths with increasing conjugative power in the substituent. Since 3-substituted pyrroles absorb at wavelengths 15–20 mμ shorter than the 2-derivatives, conjugation in the excited state of the 2-derivative must be more extensive. Comparison of ethyl 3,4-dimethylpyrrole-2-carboxylate with ethyl 2,4-dimethylpyrrole-3-carboxylate indicates a difference in conjugation energy in the excited states of about 6 kcal/mole. As in the benzene series, steric hindrance between the conjugated substituent and, say, an alkyl group lowers the *K*-band intensity. Examination of λ_{max} for *K* bands in the series of compounds Ar.CH:CH:COMe, where Ar is 2-pyrryl (λ_{max} 352 mμ), 2-furyl (λ_{max} 310 mμ) and phenyl (λ_{max} 285 mμ) shows conjugative abilities in these aromatic nuclei decreasing in that order.

The vibrational spectra of pyrrole and some deutero-pyrroles have been analysed for C$_{2v}$ symmetry and assignments made for the twenty-four normal vibrational modes (Chapter 2, p. 17). The occurrence of a weaker band at a slightly longer wavelength than the strong second harmonic of the N–H band is due to vibrational excitement[5, 6]. Infra-red data for a considerable number of pyrrole derivatives are now available[10a, b]. Numerous studies have been made of the NH-stretching region of the pyrrole spectrum[235, 239, 240]. The phenomena uncovered by studying this stretching frequency over a wide range of concentration and solvents have been attributed[7, 12b] to hydrogen bonding of a kind not clearly understood. Other workers speak of other types of association[8]. The frequency of the free NH band is 3,500 cm^{-1} in hexane, and 3,219 cm^{-1} in pyridine[7]. Shifts of this band with change of solvent are due to specific interactions and not to a general dielectric effect[241]. α-Carbonyl derivatives of pyrrole[242] and of dipyrrylmethanes[240] show hydrogen bonding of related types:

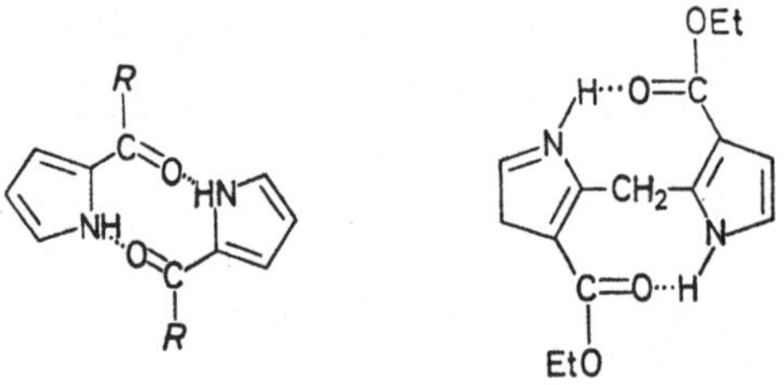

There is no support from spectroscopic studies for the occurrence of pyrrole–pyrrolenine tautomerism[9].

The infra-red stretching band of a carbonyl group, substituted at either the 2- or the 3-position of the pyrrole nucleus, is moved to lower frequencies than usual[10a, 369]. This effect is attributed to contributing forms such as (1) in the hybrid structure. However, carbonyl stretching frequencies in both cases are very similar and not useful in indicating the position of the carbonyl substituent. N–H absorption produces a sharp band in the range 3,435–3,400 cm^{-1}, and its frequency is not much changed by a carbonyl substituent at the 2-position. Sorocco and Nicolaus[11a] observed a linear relationship between the frequency difference of the NH vibration in the pure compound and in carbon tetrachloride and the acid strengths (*Table 4.5*, p. 62) of pyrrole derivatives.

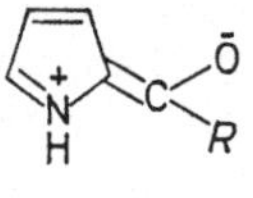

(1)

The proton magnetic resonance spectrum of pyrrole[11b] shows a very broad line at low field due to the proton of the imino group, and a spectrum of 8 lines due to the other protons. Temperature effects show the broad peak to arise from quadrupole-induced relaxation of ^{14}N. In pyrrole, the spin couplings of the imino proton with the protons at $C_{(2)}$ and $C_{(3)}$ are nearly equal.

Relative chemical shifts of ring protons are strongly dependent on solvent and concentration[11b, 243]. The dilution of pyrrole with cyclohexane breaks up interactions between imino protons and the π-electrons of another pyrrole molecule, whilst the addition of pyridine introduces interactions between the imino proton and the pyridine lone pair[243]. The evidence agrees with that from infra-red studies about the nature of the self-association of pyrrole. Chemical shifts have been recorded for a number of pyrroles[12b, 233, 244–45], and coupling constants have been studied[286]. Proton magnetic resonance spectroscopy was important in proving the structure of 2-thiocyanatopyrrole[283] (p. 81) and of some halogenopyrroles[372].

Proton magnetic resonance studies have played a central part in the elucidation of the sites of protonation of pyrroles by acids. Thus, in hydrochloric acid[12b], the following situations arise:

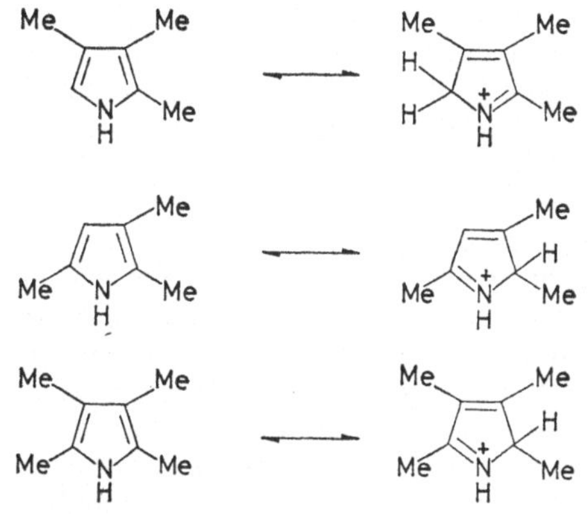

Experiments using sulphuric acid solutions show that pyrroles and methyl-pyrroles form stable α-protonated salts. Polymerization (p. 83) is avoided if the solutions are dilute and sufficiently acid to ensure complete protonation—observations consistent with the mechanism proposed[247] for pyrrole trimerization (p. 86). Altogether, the data[12b, 246, 247] show that α-protonation is preferred; an α-methyl group directs protonation to the α'-position, a β-methyl group to the α-position, and a methyl group lowers the basicity of the nuclear carbon atom carrying it. These rules add interest to the cases of 2,5-di- and 1,2,5-tri-methylpyrrole. Proton resonance spectra show that, in 12 M sulphuric acid solutions of these compounds, both α- and β-protonated forms are present in roughly equal amounts[246]. Ultra-violet spectroscopy supports this conclusion (*Table 4.2*); in addition to a maximum at 237 mμ, due to the α-protonated form[12c], solutions of 2,5-dimethylpyrrole show a broad shoulder at $\sim$ 275 mμ, due to the β-protonated form. This feature is also shown by 1,2,5-trimethyl- and 2,5-dimethyl-1-phenylpyrrole, but not by pyrrole, 1-methyl- and 2-methyl-pyrrole which undergo only α-protonation[246]. No N-protonation occurs in solution. This may not be true of crystalline salts[12c].

The general preference for α-protonation is understandable in terms of the two following groups of structures:

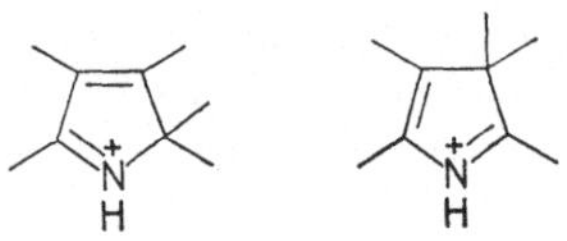

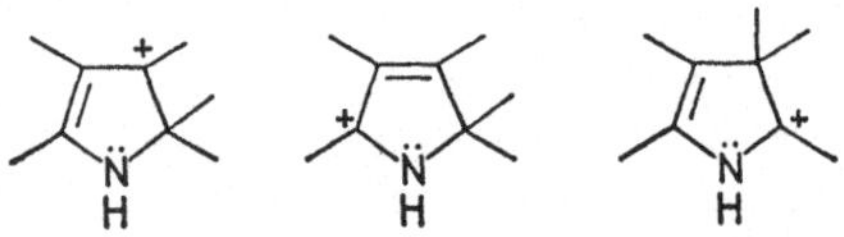

Since α-predominance depends upon its being associated with two excited structures, it should be less marked than, say, β-predominance in indole which depends on a first-order effect. The effects of the methyl group are understandable intuitively[246]. The preference for α-protonation is also to be expected from consideration of π-electron densities (p. 60) and localization energies (pp. 90–91).

(c) Basic and Acidic Properties

The base strengths of pyrrole and some methylpyrroles in aqueous sulphuric acid have been determined by a method which amounts to the definition of an acidity function specifically for these compounds[246–47]. These and other results are collected in *Table 4.4*. An earlier value[12a] for pyrrole was much too high[247]. Values of pK_a obtained by use of buffer solutions are not strictly comparable with those for sulphuric acid solutions[247]. Pyrrole is clearly a very weak base, and alkyl groups are markedly base-strengthening. The effect of methyl substitution at a particular site upon protonation at another site is roughly constant[247]. The sites of protonation have already been discussed.

*Table 4.4. Basic Strengths of Pyrroles**

In aqueous sulphuric acid[246–47]		In buffer solutions[12b]	
Substituents	pK_a	Substituents	pK_a
—	−3·80		
1-Me	−2·90		
2-Me	−0·21		
3-Me	−1·00		
2,4-Me₂	−2·55†	2,4-Me₂	1·9
2,5-Me₂	− 0·71, − 1·07‡		
3,4-Me₂	0·66		
1,2,5-Me₃	−0·24, −0·49‡	2,3,4-Me₃	3·9
2,3,5-Me₃	2·00	3-Et-2,4-Me₂	3·5
2,5-Me₂-1-Ph	−2·01, −2·73‡	2,3,4,5-Me₄	3·7

* The values refer to α-protonation unless otherwise indicated.
† Also in perchloric acid.
‡ The first value is for α-, the second for β-protonation. For the three compounds concerned, the composite pK_a values were −0·55, −0·05 and −1·93, and the ratios of β- to α-protonation[246] 0·42, 0·56 and 0·191, respectively.

Fischer and Treibs[13] used the differing basicities of substituted pyrroles to effect separations by means of hydriodic acid. 2,3,4-Trimethylpyrrole gave a crystalline hydrochloride, and with hydrobromic in acetic acid, crypto-pyrrolecarboxylic (2,4-dimethylpyrrole-3-β-propionic) and related acids gave hydrobromides which could be recrystallized from acetic acid[13]. Cookson[4a] showed that addition of hydrochloric acid to an alcoholic solution of cryptopyrrole (3-ethyl-2,4-dimethylpyrrole) removed the ultra-violet absorption band at 200 mμ and produced a new maximum at 261 mμ (ϵ, 4,000). The acid solution was stable for at least 10 min and the spectrum reverted to its original form upon addition of alkali. Other pyrroles with electron-attracting substituents did not show these changes which are clearly due to salt formation.

Treibs and Kolm[14a] treated a number of pyrroles in chloroform or carbon tetrachloride with hydrogen chloride and observed in many cases the formation of crystalline hydrochlorides (see p. 83 and *Table 4.1*). As would be expected, alkyl groups facilitate salt formation. Their failure to obtain a crystalline salt from 2,5-dimethylpyrrole may be due to the occurrence of both α- and β-protonation[246] (see above). Pyrrole aldehydes also form stable derivatives under these conditions, as do 2,4-diaryl-5-nitrosopyrroles[14b], but ester groupings inhibit salt formation. Ultra-violet spectroscopic data, recalling the observations of Cookson, were also reported by Treibs and Kolm. In pyrrole aldehydes, the oxygen atom of the substituent may be involved in salt formation. Treibs and Kolm[14a] were not able to isolate salts of pyrrole itself. If air were rigorously excluded, pyrrole could be recovered unchanged from solution in hydrochloric acid after several hours in the dark. The reactivity of the cation makes it difficult to avoid the formation of pyrrole trimer, unless protonation is complete (see above).

The electronic structure of pyrrole allows us to understand not only the very weak basic properties of pyrrole but also its acidic character (p. 17). Pyrrole reacts with fused potassium hydroxide, or with potassium, to give potassium pyrrole, C_4H_4NK. The reaction with potassium proceeds smoothly, with evolution of hydrogen, in ligroin solution[15]. Sodium pyrrole has not been obtained under such conditions, but pyrrole reacts with sodamide in liquid ammonia to give a compound C_4H_4NNa,NH_3. This loses ammonia at 20°. Calcium and silver pyrrole can be prepared in the same way[16].

Potassium reacts readily with pyrrole and alkylpyrroles in boiling toluene; sodium is much less reactive, and lithium unreactive. Pyrrole salts of all three metals are readily formed from solutions of the metals in liquid ammonia[257].

Lithium derivatives have been obtained by treating pyrrole with phenyl or n-butyl lithium in ether[17] or in benzene[257]. Hydrogen–lithium exchange can also accur between N-methyl- or N-phenyl-pyrrole and n-butyl lithium, replacement proceeding at the α-position[17b]. Alkyl Grignard reagents react with pyrroles, an alkane being liberated and a 'pyrrole Grignard reagent' being formed[18]; 1-methylpyrrole does not form a Grignard reagent[19] (p. 105).

The alkali metallic derivatives are white, crystalline compounds which are decomposed by water to give back pyrrole and which have some value in synthesis. From colorimetric and spectroscopic studies, McEwen[20a] derived the value $pK_a = 16·5$ for pyrrole (as an acid). It may be noted that the value for methanol is $pK_a = 16$. As would be expected, pyrroles containing electron-attracting substituents are stronger acids than pyrrole (*Table 4.5*). The aldehydes (2) and (3) are soluble in alkali, and the nitrile (4)

Table 4.5. Acid Strengths of Pyrrole Derivatives

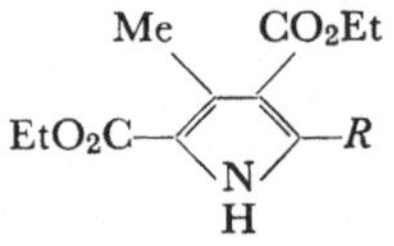
Me
CO_2Et
EtO_2C
R
N
H

		Nitropyrroles[20c]*	
R[11a]	pK_a	*Substituent*	pK_a
H	13·1	2-NO_2	10·6
Me	13·4	2,5-$(NO_2)_2$	3·6
Br	10·5	2,4-$(NO_2)_2$	6·15
Cl	10·2		

* Determined spectrophotometrically in aqueous solution.

can be titrated as a monobasic acid with phenolphthalein as indicator[20b]. Treibs and Kolm[14a] compared the abilities of a variety of pyrroles to form sodium salts when treated with sodium ethoxide in ether. Pyrrole itself does not react under these conditions. Electron-attracting substituents have the expected effect.

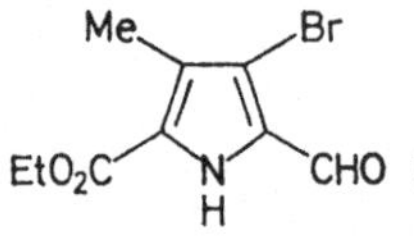
Me
Br
EtO_2C
N
H
CHO

(2)

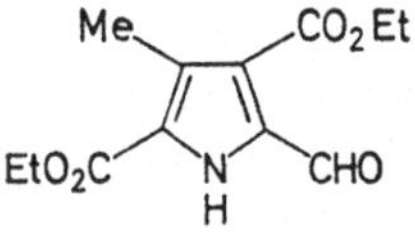
Me
CO_2Et
EtO_2C
N
H
CHO

(3)

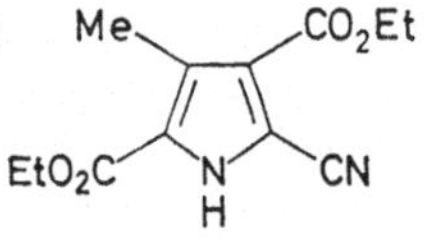
Me
CO_2Et
EtO_2C
N
H
CN

(4)

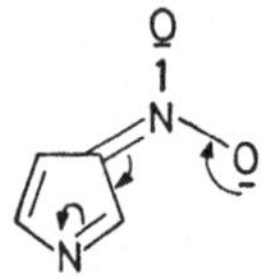
O
N
O
N

(5)

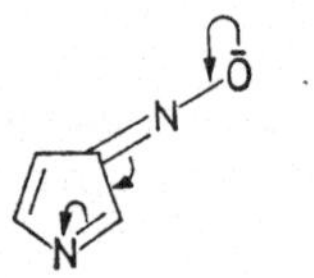
N
O
N

(6)

Halogenopyrroles with a carbonyl substituent at $C_{(2)}$ dissolve in aqueous alkali, and such di- or tri-halogeno compounds have pK_a values similar to those of simple phenols. This acidity is the cause of the bathochromic shifts shown by the spectra of these compounds in alkali[372].

Nitro- and nitroso-pyrroles are markedly acidic. The former are pale yellow compounds which give yellow solutions in alkali. The simple nitroso compounds, lacking aryl or alkoxycarbonyl substituents, are only stable as their yellow salts. The sodium salts of nitro- and nitroso-pyrroles result from nitrations and nitrosations effected, respectively, by means of ethyl nitrate or amyl nitrite in presence of sodium or sodium ethoxide (see below). These salts must involve mesomeric anions such as (5) and (6). The acidity of the nitroso compounds is one of the factors which has led, perhaps mistakenly, to their formulation as isonitrosopyrrolenines (p. 107).

(2) BEHAVIOUR IN SUBSTITUTION REACTIONS
ELECTROPHILIC SUBSTITUTION

The chemistry of pyrrole is rich in substitution reactions which can readily be recognized to involve attack by an electrophilic reagent. On the other hand, little is known of the reactions of pyrroles with radicals or with nucleophilic reagents. With regard to the electrophilic substitutions it should be realized that whilst their general character is usually immediately apparent, in hardly any case is anything known about the detailed mechanism.

(a) Acylation and Carboxylation

These can be effected by a number of methods familiar from their application to phenols. Pyrrole-2-aldehyde was first prepared, in poor yield, by treating pyrrole with chloroform in aqueous potassium hydroxide solution[21]. The Reimer–Tiemann reaction, although it has been improved since its original application[22], is not, however, of much practical value in the pyrrole series. In this reaction, 2,5-dimethylpyrrole gives[23] not only the aldehyde but also 2-dichloromethyl-2,5-dimethylpyrrolenine (7), a product analogous to those formed in the 'abnormal' Reimer–Tiemann reaction of phenols[24]. This pyrrolenine and its analogues are not intermediates in the conversion of pyrroles into 3-chloropyridines during the Reimer–Tiemann reaction (p. 88).

For preparative purposes the Gattermann reaction is much more useful. It was first used[25] to prepare ethyl 2-formyl-3,5-dimethyl- and ethyl 3-formyl-2,5-dimethyl-pyrrole-4-carboxylate, by means of hydrogen cyanide and hydrogen chloride in ether. Later applications sometimes involved minor changes such as the use of chloroform as solvent, or of Adams' modification of the reaction[26]. Formylation generally occurs at an α-position, but if none is open there is usually no difficulty in β-substitution. Pyrrole-2-aldehyde cannot be made this way, for it reacts further to form dyestuffs, but several 1-alkylpyrroles have been satisfactorily formylated[27]. The initial

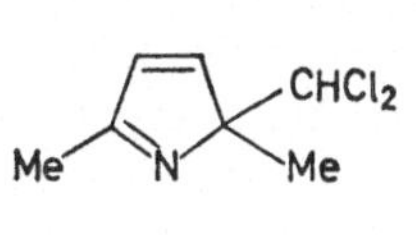

(7) 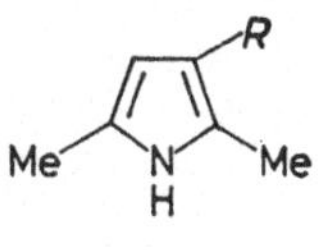   (8 ; R = Me, $(CH_2)_2CO_2H$, CN or CO_2Et)

product of the reaction is a salt of the aldehyde imine, and with crypto-pyrrole and haemopyrrole, substances of the type [pyrrole nucleus]$_2$CHNH$_2$ have also been isolated[28]. Among the few examples of failures may be noted 2-acetylpyrrole[27] and compounds of the type (8)[29]. Curiously, the reaction is said to fail with 1,2,5-trimethylpyrrole in ether, but to succeed in chloroform[30].

Some eliminations have been reported, as of the carboxyl group in the preparation of 4-bromo-2-formyl-3,5-dimethylpyrrole[31]. Replacement of a bromine atom by chlorine or by formyl has also been observed[26].

The most satisfactory formylation of pyrrole, giving pyrrole-2-aldehyde in 83 per cent yield, occurs with excess of dimethylformamide and one equivalent of phosphoryl chloride[32]. The reaction works with 1-methylpyrrole. 3-Methylpyrrole gives[233] 2-formyl-3-methyl- and 2-formyl-4-methylpyrrole in the ratio 4:1. Ester groupings do not prevent formylation, and ethyl pyrrole-2-carboxylate undergoes α'- and β'-formylation[248] in the ratio 3:1. Many examples of the reaction are now known[249-51], including some of diformylation[331, 337], and by the use of N,N-dimethylamides other than dimethylformamide, it has been extended to the synthesis of ketones[249, 252]. These substitutions may be represented by the scheme

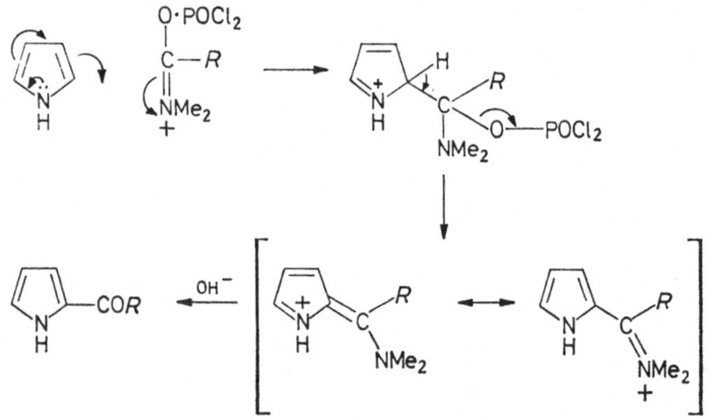

The Hoesch ketone synthesis also functions well in the pyrrole series[22, 28, 33], ketones resulting from reaction with a variety of reagents $R.CN$ ($R = $ Me, $\cdot CH_2Cl$, $\cdot CO_2Et$, CN, $\cdot CH_2CN$) in the presence of ethereal hydrogen chloride. Orientation follows the same rules as in the Gattermann reaction. 2,3,4-Trimethylpyrrole is said not to react with acetonitrile in the Hoesch synthesis.

When pyrrole is boiled with acetic anhydride, alone or containing sodium acetate, a mixture of 1- and 2-acetylpyrrole results[34]. At higher temperatures (240–250°), 2,5-diacetylpyrrole is formed[35]. Similar reactions have been carried out with benzoic and propionic anhydride[36a], and with trifluoroacetic anhydride at 0°, pyrrole gives 2-trifluoroacetylpyrrole[36b]. Substituted pyrroles have been acylated[37] by this method. With cryptopyrrole and 2,3,4-trimethylpyrrole, short boiling with acetic acid suffices to cause C acetylation[38]. 1-Methyl- or 1-benzyl-pyrrole with acetic anhydride at 250° provide 2,5-diacetyl derivatives[39]. In some cases, as with 3-chloro- or 3-hydroxy-1,2,4-triphenylpyrrole, C acetylation is brought about by acetic anhydride containing a drop of sulphuric acid[40]. With hydroxypyrroles in general, both C- and O-acetylation are observed (p. 105). 1-Methylpyrrole reacts with acetyl chloride in ether, giving 2-acetyl-1-methylpyrrole[41]. Thioacetic acid converts 2,4-dimethylpyrrole into 2-acetyl-3,5-dimethylpyrrole[48].

The great ease of these C-acylations makes it unlikely that in general they proceed by preliminary N-acylation (p. 109).

A particular case of this type of acylation which deserves mention occurs when pyrrole is heated with phthalic anhydride, the product being 'pyrrolene phthalide'. The structure of this substance, long debated, is now established[42] as (9).

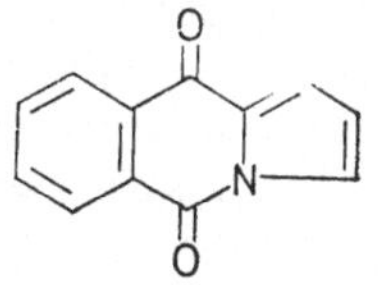

(9)

The above examples make it not surprising that pyrroles containing de-activating groups such as carbethoxyl can be acylated under Friedel–Crafts conditions[43, 250]. The usual reagents are an acyl chloride and aluminium chloride in carbon disulphide, but sometimes the solvent is omitted, and stannic chloride with acid anhydrides is also useful[44]. Both α- and β-positions have been acylated by this method. Acetic anhydride in the presence of zinc chloride or boron trifluoride etherate acetylates 1-methylpyrrole, and 1-methyl-2- and -3-nitropyrrole (giving 4-acetyl-1-methyl-2- and 2-acetyl-1-methyl-4-nitropyrrole, respectively)[45]. An interesting claim is that pyrrole reacts with acetic anhydride in the presence of perchloric acid at 70–80° to give 1- and 2-acetylpyrrole in a ratio determined by the time of heating, the relative amount of the 2-isomer increasing with time[46].

A little explored reaction, which causes acylation of the pyrrole nucleus, is that with diazoketones in the presence of copper powder:

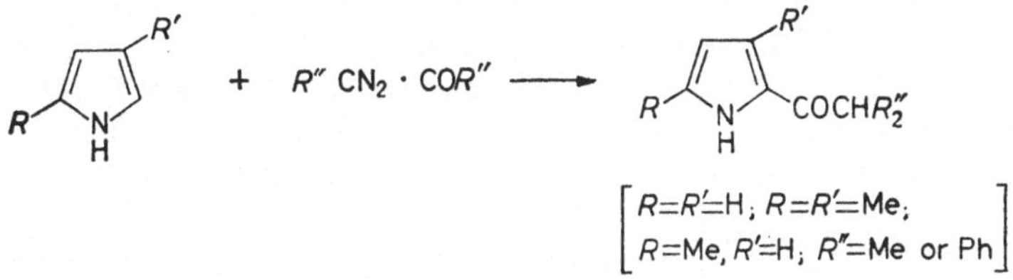

Pyrrole itself undergoes α,α'-disubstitution[47]. Diazo-acetoacetic ester can be used in this reaction. The mechanism of this reaction, which may be a radical substitution or involve the intermediate formation of a ketene, is uncertain, but it is interesting to note that a number of pyrroles react readily with diketene, giving C-acetoacetyl derivatives, and that ketene converts 2,4-dimethylpyrrole into 2-acetyl-3,5-dimethylpyrrole[48]. In view of these results, the earlier report[49] that diphenylketene converted pyrrole into 1-diphenylacetylpyrrole must be questioned.

Besides being readily acylated, pyrroles are easily carboxylated. Thus, with aqueous ammonium carbonate at 130–140°, pyrrole gives* pyrrole-2-carboxylic acid[37], and potassium carbonate and carbon dioxide give the

---

* The early literature relating to this and other reactions causing carboxylation is confused because of the dimorphism of pyrrole-2-carboxylic acid[50].

same result[338]. The reaction indicates a degree of reactivity about the same as that shown by resorcinol[51]. Carbonyl chloride[52, 29] and carbamyl chloride[53] react with several pyrroles to give carboxylic acid chlorides and amides, respectively. Pyrrole-2-carboxylic acid is also formed from pyrrole by the action of carbon tetrachloride and alcoholic potassium hydroxide[54] and, in a reaction analogous to the Kolbe synthesis in the phenol series, from potassium pyrrole and carbon dioxide[54].

Pyrrole reacts with aryl isocyanates to give anilides of pyrrole-2-carboxylic acid. Pyrrole homologues behave similarly: 2,4-dimethylpyrrole gives derivatives of 3,5-dimethylpyrrole-2-carboxylic acid, and 2,5-dimethylpyrrole reacts at the β-position. The reactions are useful for characterizing simple pyrroles[253, 302].

C-Acylation and -carboxylation also occur with pyrrole Grignard reagents (p. 106). Whilst lithium pyrrole and lithium 2,4-dimethylpyrrole react with ethoxycarbonyl chloride to give ethyl pyrrole-2- and 2,4-dimethylpyrrole-5-carboxylate, respectively[257], the corresponding potassium salts give the N-substituted pyrroles (p. 81).

(b) Alkylation, substituted Alkylation and Alkenylation

Early workers[55] converted pyrrole into its α-homologues by reaction with an alcohol and zinc chloride at 270–280°, and the complex products formed from pyrrole and aldehydes or ketones in the presence of zinc chloride give C-alkylpyrroles when distilled[37b] (see carbonyl reactions, p. 71). More clear-cut are the substituted alkylations brought about by some carbinols formed when pyrroles react with carbonyl compounds (p. 72).

The reactions of 1-methyl- or 2,3,4,5-tetramethyl-pyrrole with methyl iodide and potassium carbonate result in C-methylation and then in further substitution, giving polymethyl pyrrolenines. Methyl iodide and magnesium oxide in ether produce similar results[56].

The structures of the products other than the simple 1- or 2-alkylpyrroles formed in these reactions are not proved in details. They are formulated as pyrrolenines by analogy with reactions in the indole series. The reactions have been used[51] as an argument that pyrroles react as resonance hybrids and not as the tautomeric pyrrolenines.

The reactions of alkali metal salts of pyrroles with alkyl halides have generally been reported to give 1-alkylpyrroles. This is certainly the case with potassium salts and methyl iodide[41, 132, 122] or dimethyl sulphate[257], but with higher alkyl halides the situation is confused. In some cases, as with potassium pyrrole and n-butyl bromide[287] and n-hexyl iodide[150], a considerable degree of N-alkylation certainly occurs, but ethyl iodide produces some 2-ethylpyrrole* as well as 1-ethyl- and 1,2-diethyl-pyrrole, though 1-ethylation is dominant[58, 60]. Propyl iodide behaves similarly[59], and higher halides may give relatively more C-alkylation[62]. Alkali metal salts of pyrroles are said to give 1-alkyl derivatives with alkyl halides in liquid ammonia[288], and 2-alkoxycarbonyl-4-nitropyrroles behave similarly with alkyl bromides and sodium ethoxide in ethanol[289].

This confused situation, and observations such as those of the different

---

* The early literature is confused about the identity of 2-ethylpyrrole. It is almost certain that in every case the 2- and not the 3-isomer was obtained[57].

behaviour of lithium and potassium pyrroles with ethoxycarbonyl chloride (pp. 66, 81), suggest that a systematic study of the effect upon the occurrence of N- and C-alkylation of varying the metal cation, the alkyl halide, the medium and the temperature would be profitable.

Such a study has been reported for alkenylation with allylic halides[61, 290]. In this work, lithium, sodium and potassium pyrrole were treated with allyl, crotyl and benzyl halides in the presence of a solvent. According to the combination of reagents used, the reactions were either homogeneous or heterogeneous. They produced mixtures of 1- and 2-substituted pyrroles (usually in high overall yield) with varying amounts of dialkylated products. Traces of 3-substituted pyrroles were also formed in the reactions which produced relatively large proportions of 2-substituted pyrroles. In all cases an increase in reaction temperature caused a slight increase in the relative amount of the 2-substituted pyrrole and favoured further substitution. Disubstituted products arose from 2- and not from 1-substituted pyrroles. In both heterogeneous and homogeneous conditions, the nature of the medium and of the cation greatly influenced the relative proportions of 1- and 2-substitution. The following general conclusions were drawn:

1. With a particular cation, under both heterogeneous and homogeneous conditions, the most polar medium gave the greatest, and the least polar the lowest, proportion of N-substitution.

2. For a given medium, with both heterogeneous and homogeneous reactions, the proportion of N-substitution depends on the cation in the order $Li^+ < Na^+ < K^+ < Me_3\overset{+}{N}Ph$.

3. With potassium pyrrole, the proportion of N-substitution increases with increasing solubility of the salt in the medium used.

4. For a series of salts in one medium, heterogeneous or homogeneous reaction produces a proportion of N-substitution increasing with decreasing solubility of the salt in the medium. For homogeneous reactions, the same is true for decreasing concentrations of the pyrrole salt.

5. Addition of $Bu_4\overset{+}{N}Br^-$ to any reaction increases the proportion of 1-alkylation.

To account for these observations, it was suggested that dissociation of the salts favours N-substitution and ionic association favours C-substitution in homogeneous reactions:

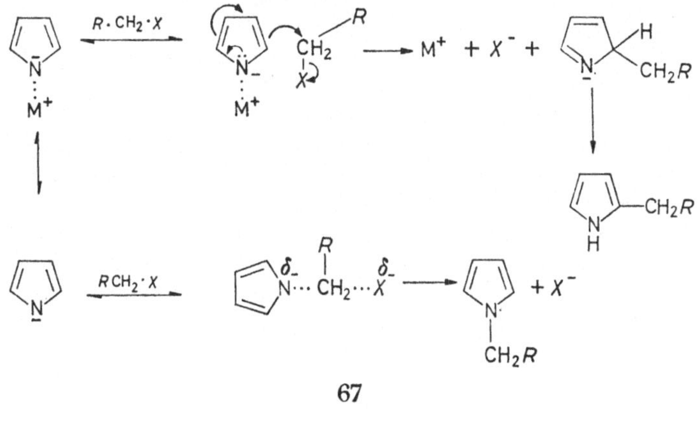

Dialkylation could proceed by similar processes from the 2-substituted pyrrole but not from the 1-substituted compound. The case of heterogeneous reaction is, of course, difficult to unravel because of uncertainty as to whether the reactions are truly heterogeneous or occur in very dilute solution. The transition state proposed for 2-substitution suggests why this preponderates over 3-substitution, though the difference may also be accounted for electronically (p. 90). It is also suggested[290] that the readiness with which the polarizable allyl halides would fit into such a transition state explains why allyl halides produce 2-substituted pyrroles so much more markedly than do alkyl halides. In view of the marked lack of evidence about the behaviour of alkyl halides (see above), this idea may be premature.

The competition between N- and C-substitution in the alkali metal derivatives of pyrrole is, of course, only one example of the general problem of substitution into ambident ions[61, 216, 225-26].

It is unfortunate that the pyrryl magnesium halides were not included in the systematic study of alkenylation described above. Pyrryl magnesium halides give C-alkylpyrroles with alkylating reagents (p. 106). This behaviour and other substitution reactions undergone by these reagents is discussed separately, but the evidence about their structure suggests that they are probably ionic without being much dissociated (p. 107) and that C-substitutions occurring with them follow a path similar to that shown above for C-substitution in alkali metal derivatives.

Reactions which almost certainly involve 2-substitution by a carbonium ion are those between pyrrole and a triaryl carbinol, or with xanthydrol in acetic acid. Triaryl chlorides also cause 2-substitution by a carbonium ion mechanism[63].

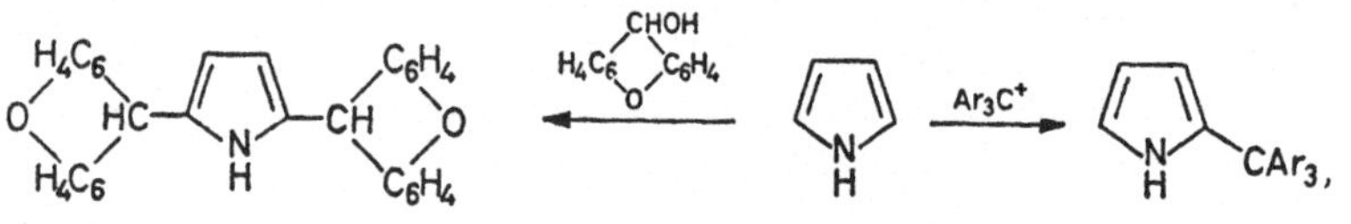

Methyl pyrrole-2-carboxylate with isopropyl chloride and aluminium chloride gives 4-, 5-, and 4,5-di-isopropyl derivatives. Primary attack is probably at $C_{(4)}$, and most of the 5-isopropyl compound is likely to arise by rearrangement, which has been shown to occur[371].

An interesting alkylation reaction occurs when a pyrrole is heated at about 220° with a solution of a sodium alkoxide in the corresponding alcohol[64]. The method is also applicable to sufficiently reactive phenols[65] and to indoles. The reaction has mostly been used with polyalkylpyrroles and proceeds at either an α- or a β-position, though the β-position is said to be more resistant. In the process, acyl, hydrazone, ketazine, and ester groups can be replaced by alkyl substituents[66]. It cannot be pretended that the mechanism of this type of substitution is understood, but it has been argued that the alkoxide is not itself the alkylating reagent and that methoxide, for example, may be oxidized to 'activated formaldehyde' which then behaves as an electrophilic substituting reagent. The last step would then be a reduction[65].

The reactions of pyrrole with propiolactone[67], and of various pyrroles with ethyl methoxymethylmalonate in the presence of acids (ethanolic hydrochloric acid, or stannic chloride in carbon disulphide)[22, 29, 68], closely resemble each other:

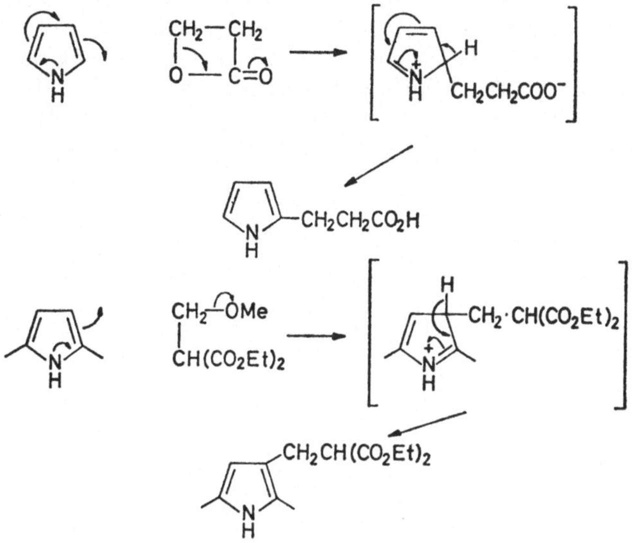

Reported examples of the second reaction occur at the β-position in an a,a'-disubstituted pyrrole. Whilst for the second reaction there is no evidence to indicate the intermediate formation of ethyl methylenemalonate, followed by Michael addition, pyrroles can participate in such additions. Examples are the ready reaction with ethynyl ketones[69], with quinones[70] and with acrylic esters (in the presence of boron trifluoride etherate)[71]:

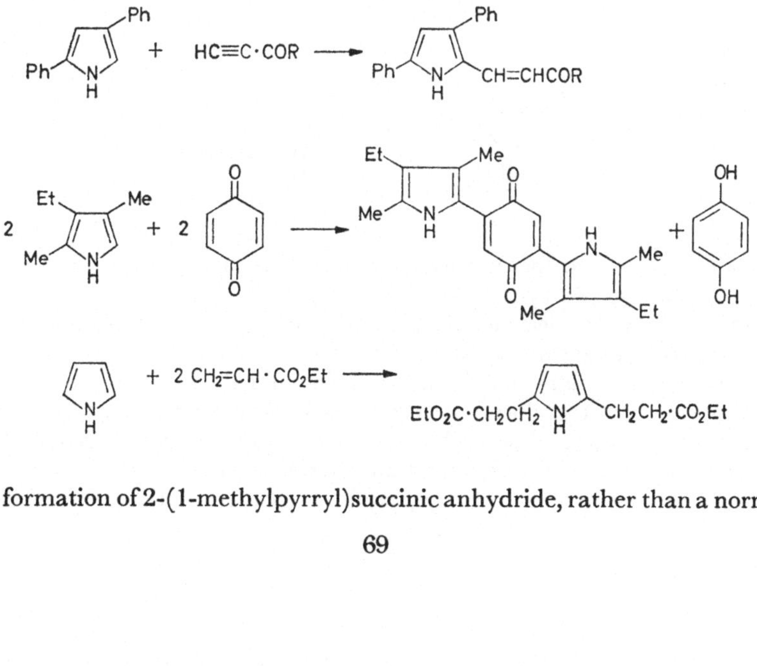

The formation of 2-(1-methylpyrryl)succinic anhydride, rather than a normal

Diels–Alder addition product, from 1-methylpyrrole and maleic anhydride[72] is a further example of this sort of reaction. 2-Methylpyrrole behaves similarly, but the initial product (10) from pyrrole is obtained in small yield because it reacts further (p. 88). Pyrroles such as 2-methyl-, 1,2-dimethyl- and 2,4-dimethyl-pyrrole undergo similar Michael additions with methyl acetylenedicarboxylate[72, 254], in the third instance two molecules of the pyrrole reacting to form (11). 1-Methoxycarbonylpyrrole reacts at the 2-position with acetylene dicarboxylic acid[254] as well as giving a Diels–Alder type of product (p. 82). With the same acid, ethyl 3,5-dimethylpyrrole-2-carboxylate undergoes Michael addition at the vacant β-position[72], giving (12). Other examples of reactions with dienophiles are discussed below (p. 82).

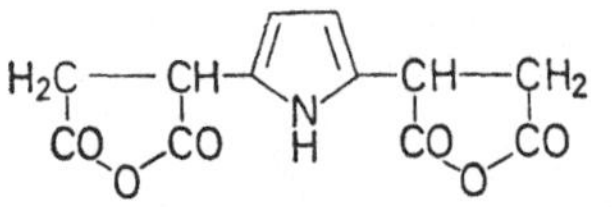

(10) 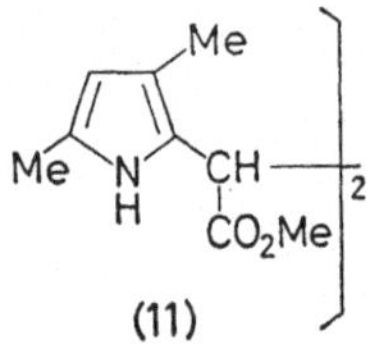   (11)

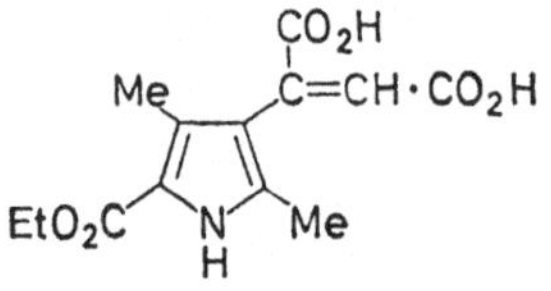

(12)

A reaction of uncertain character which deserves mention is that between pyrroles and ethyl diazoacetate in the presence of copper powder[47, 253]. This carbethoxymethylation might be a radical substitution or it might proceed through an initial adduct

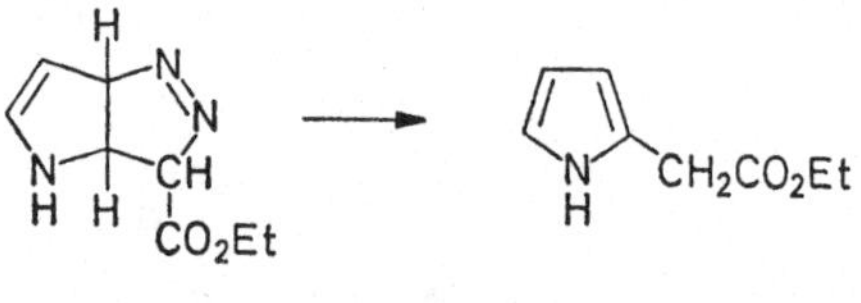

In any case, it is clear that in this reaction pyrrole, unlike some related structures, does not give a cyclopropane derivative[73].

Whatever the detailed mechanism of the reaction between formaldehyde, secondary or primary amines and a suitably reactive aromatic compound, it is clear that essentially the Mannich aminomethylation is an electrophilic substitution. It is therefore not surprising that pyrrole reacts very readily with formaldehyde and secondary amines to give 'Mannich bases'. If an excess of pyrrole is not maintained during the reaction, more highly substituted products result[74]. 1-Methylpyrrole reacts, if anything, more readily than pyrrole, and 2,5-bis-aminomethyl derivatives are readily obtained[75]. On the other hand, 1-phenylpyrrole undergoes only monosubstitution. Substituted pyrroles, including 2,5-disubstituted examples which react at the β-positions, have been used in the Mannich reaction[76], and whilst acetaldehyde is of little value[77], primary amines are satisfactory[78]. Substitution at the pyrrole nitrogen atom does not occur in the Mannich reaction.

An interesting substitution which gives a product (13) related to the Mannich bases, by means of a modified Vilsmeier reaction, occurs between pyrrole, pyrrolidone and phosphoryl chloride[255]. Other examples are known[271, 313].

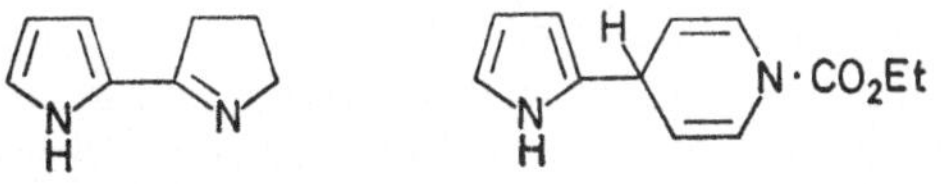

(13)   (14)

The acid polymerizations of pyrroles (p. 83) are essentially Mannich reactions.

(c) Arylation

Sufficiently electrophilic aromatic compounds can effect substitution into pyrroles. Thus, acylpyridinium compounds give products formulated in the manner of (14), though the evidence for attachment at $C_{(4)}$ in the pyridine nucleus, rather than at $C_{(2)}$, is not conclusive[256-58]. In contrast, the product from 2-ethoxy-3,4-dimethylpyrrole, ethyl chloroformate and pyridine is formulated[259] as the fully aromatized 3,4-dimethyl-2,5-di(4-pyridyl)pyrrole. Pyrroles react with acridine, giving either complexes of 9-(2-pyrrolyl)acridan with acridine, or 9-(2-pyrrolyl)acridine[258]. 1-Methylpyrrole does not react with acridine, but 2,5-dimethylpyrrole reacts at both β-positions.

2-Phenylpyrrole is formed from pyrrole and benzyne[308], a reaction which again shows the reluctance of pyrrole to undergo Diels–Alder addition (pp. 70 and 82).

(d) Carbonyl Reactions

An especially important group of reactions is that with carbonyl compounds. Pyrrole reacts vigorously with formalin giving condensed products, presumbly of the type (15), which on pyrolysis produce some 2-methylpyrrole[79].

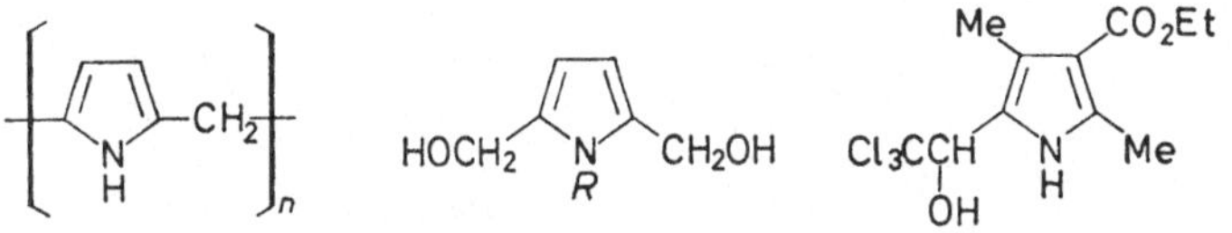

(15)   (16)   (17)

The reaction is analogous to the bakelite polycondensation. It is difficult to halt this reaction at an early stage, but pyrrole is said to react with formalin and a trace of hydrochloric acid, and with other aliphatic aldehydes in the presence of zinc chloride, to give dipyrrylmethanes[80]. If the reaction of pyrrole or 1-methylpyrrole with formalin is carried out in the presence of potassium carbonate, the bis-substituted products (16) result[81]; with acids, (16) undergoes further condensation. Pyrrole is said to react with anhydrous formaldehyde in basic conditions to give at 40–55° 1-hydroxymethylpyrrole, and at 75–90° 2,5-bishydroxymethylpyrrole[82].

Substituted pyrroles such as ethyl 2,4-dimethyl-3- or 2,5-dimethyl-3-carboxylate are hydroxymethylated at the empty nuclear position by formalin and sodium hydroxide or potassium cyanide[68a]. Similarly, the initial carbinols from pyrroles and aldehydes can be isolated if they are stabilized in any other way; an example is the carbinol (17) which, like others of its class, can be obtained from the appropriate pyrrole and chloral hydrate in the presence of hydrochloric acid[83].

The products usually obtained from a pyrrole and an aldehyde or ketone under acid conditions are the dipyrrylmethanes, formed by further reaction of the initial carbinol with a second molecule of the pyrrole. Some examples are the formation of (18) and (19) from propionaldehyde and acetaldehyde with zinc chloride[84], of (20) from acetone with hydrochloric acid[85], and of (21) using formalin containing formic acid[86]. Similarly, glyoxal can be used

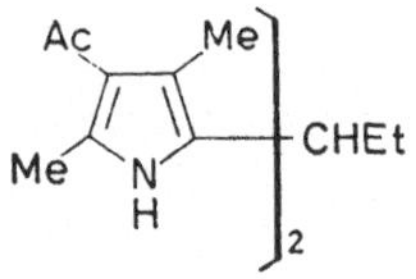

(18) 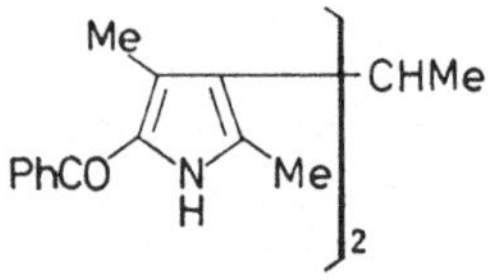   (19) 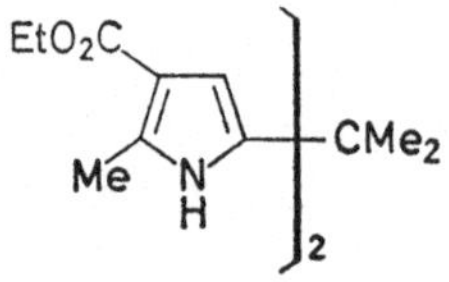   (20)

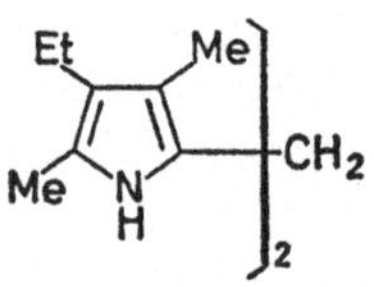

(21)

to synthesize tetrapyrrylethanes[87]. Generally, then, dipyrrylmethanes with *meso*-alkyl- or- aryl substituents result from reaction of a pyrrole, having a free α- or β-position, with an aldehyde or ketone, usually with the assistance of an acid[88]. It should be noted, however, that reaction of a pyrrole with an aldehyde or ketone under severer conditions can give products containing

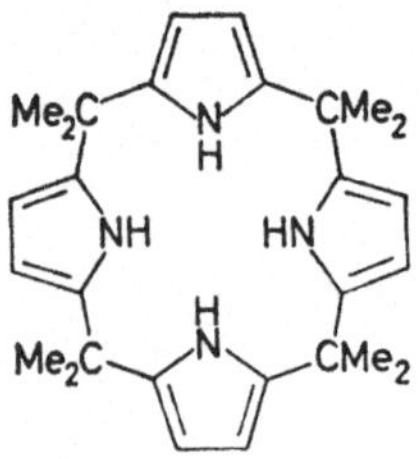

(22) 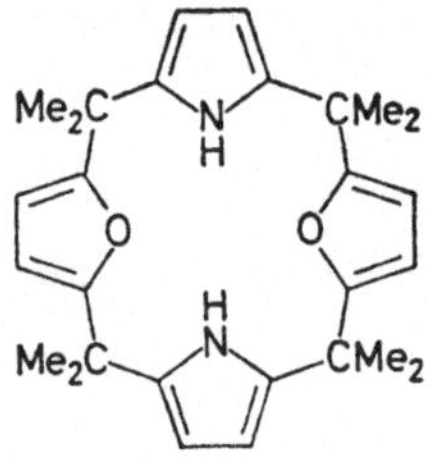   (23)

the porphyrin skeleton[89]. 'Acetone pyrrole', obtained from pyrrole and acetone in the presence of a little acid, possesses the structure[90] (22). Similar reactions have led to 'mixed quaterenes', such as (23), which contain both furan and pyrrole nuclei[91].

72

In certain circumstances the formation of a carbinol from a pyrrole and a carbonyl compound can be followed by a second step different from those already mentioned. For example, 3-formyl-2,5-dimethylpyrrole and 2,5-dimethylpyrrole give, in acid conditions, the first-discovered member (24) of an important class of compounds, the 'pyrromethene' salts[92]. These highly coloured salts are cleary resonance-stabilized, and the reaction, of

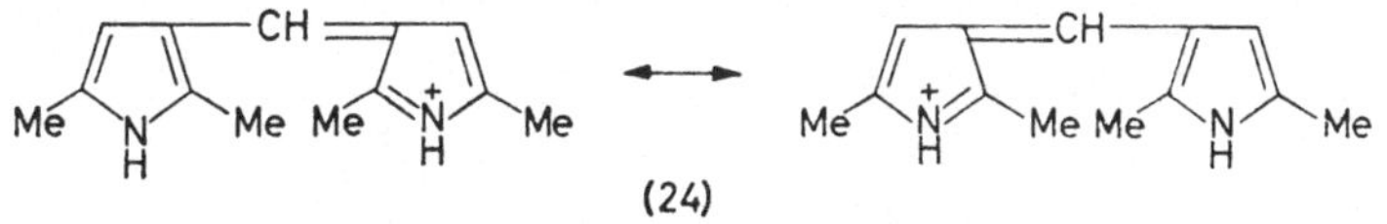

(24)

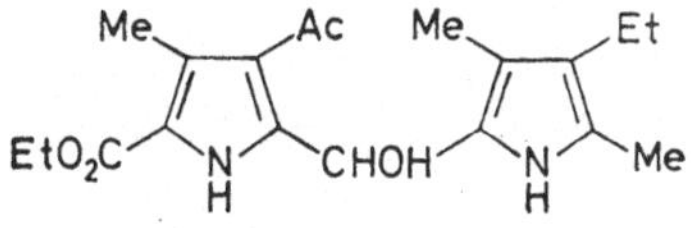

(25)

which many examples are known[93, 251], finds its driving force in this circumstance. The carbinol intermediate in such reactions is rarely encountered, though (25) can be prepared[94] from cryptopyrrole and ethyl 3-acetyl-2-formyl-4-methylpyrrole-5-carboxylate in methanol at $-10°$. The subsequent course of such reactions, leading to pyrromethenes, depends upon the nature of the components, but in some cases certainly involves a tripyrrylmethane intermediate[95]:

$$Py.CHO + Py'H \rightarrow Py.CHOH.Py' \rightarrow Py.CH(Py')_2 \rightarrow PyH + Py'.CH{=}Py'$$
$$(Py \text{ and } Py' \text{ are pyrrole nuclei})$$

The tripyrrylmethane could conceivably provide three pyrromethenes, and this situation has in fact been encountered. Pyrromethenes, usually obtained from a pyrrole aldehyde and a second pyrrole derivative with a free nuclear position, can also be prepared from a pyrrole and ethyl orthoformate by treatment with hydrobromic acid[96].

An interesting reaction is the formation of a pyrromethene from some 2-iodopyrroles and formaldehyde, the iodine atom being replaced[119] (see p. 101).

Fundamentally similar to pyrromethene formation are two of the classical colour reactions associated with pyrroles. The redness produced by pyrroles with an acid-moistened pine shaving is due to reaction with aldehydes in the wood, but it is not specific for pyrroles[97-8]. Ehrlich's reaction[99] depends on the formation of coloured products by the reaction of a pyrrole with p-dimethylaminobenzaldehyde in acid solution. The nature of the reaction is indicated by the isolation of a salt of the cation (26) after treatment of (27) with the Ehrlich reagent, a carboxyl group being eliminated; (26) gives red solutions, the disodium salt of the derived free base produces yellow solutions, and treatment of (26) with excess acid the colourless di-protonated salt[68b].

It is not true, as commonly stated, that all pyrroles with a free α-position give an intense red colour in this test in the cold. In fact, performance in the colour test parallels roughly the ability to couple with diazonium salts[83] (p. 76). Thus, of the four compounds ethyl pyrrole-1-carboxylate, ethyl 4-nitropyrrole-2-carboxylate, diethyl 3-methylpyrrole-2-4- and diethyl 2-methylpyrrole-3,5-dicarboxylate, none of which undergoes coupling, the

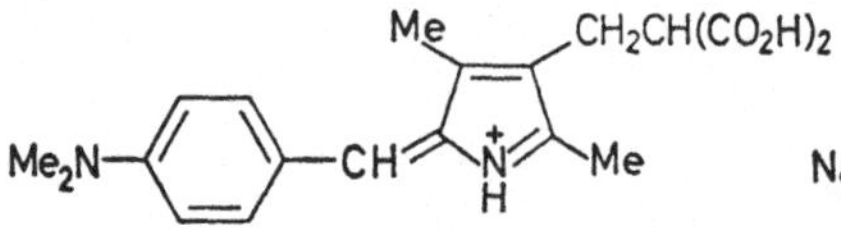

(26) 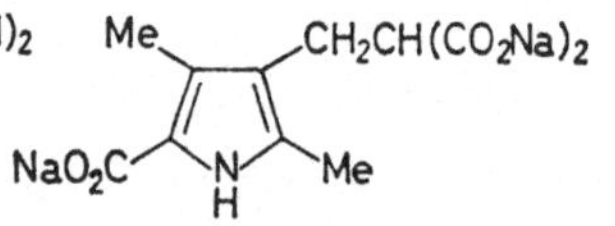   (27)

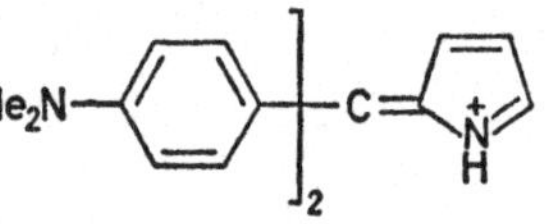

(28)

first three do not give the Ehrlich reaction and the last responds faintly on warming. On the other hand, the Ehrlich response of 2,5-dimethylpyrrole is strong at room temperature, and even 2,3,4,5-tetramethylpyrrole gives the reaction slowly[100], as do some fully substituted pyrrole-sulphonic acids[83b]. Neither 4- nor 5-nitro-2-propionylpyrrole gives the reaction[274]. Clearly, substituents exert the influences to be expected from their electronic characters. The Ehrlich test is not specific to pyrroles, being given by a number of polyhydric phenols. With pyrrole and phosphoryl chloride, Mischler's ketone produces the cation (28), analogous to the Ehrlich type of product and reminiscent of the triphenylmethane dyestuffs[101].

A number of reactions are known[69, 96, 102] which lead to dyestuffs vinylogous with those described above. These are illustrated by the following examples:

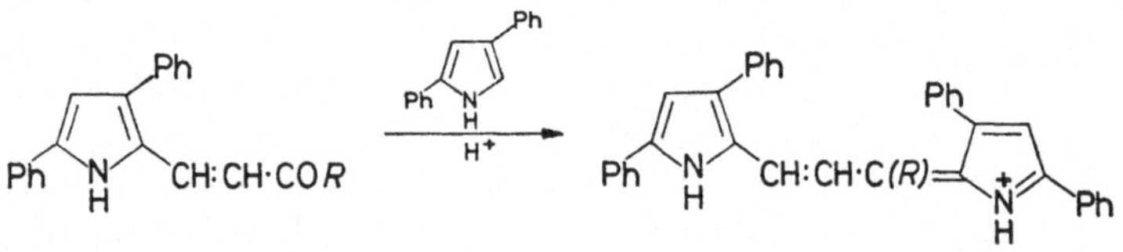

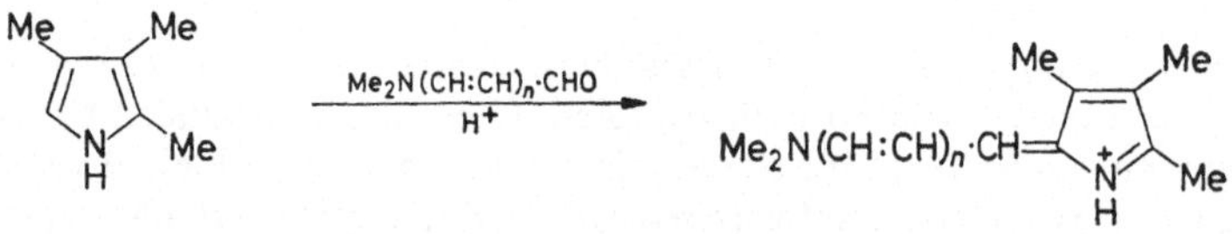

Pyrrole resembles thiophen in its ability to react with isatin and other compounds containing the grouping $\cdot CO \cdot CO$, in the presence of acids, to give deep blue products[103]. Some authors[104] have claimed that in the reaction with isatin two types of 'pyrrole blue' are formed, and it is still not clear whether this is so or not. Pratesi[105] observed that blue products could be prepared from 2,3-dimethylpyrrole and cryptopyrrole, but not from phyllopyrrole, and favoured structures of type (29). However, Steinkopf and

Wilhelm[106], who showed that 1-methylisatin gives (30) with pyrrole in water, took such carbinols to be intermediates in the formation of pyrrole blues to which they ascribed indophenin-like structures (31). The production of such derivatives from 2,3-dimethylpyrrole and cryptopyrrole was explained

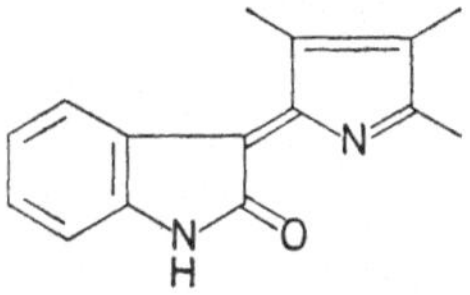

(29) 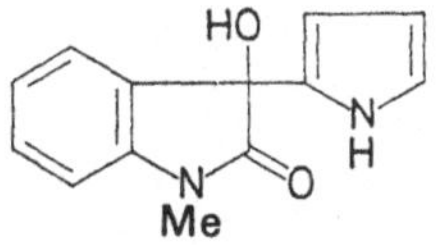   (30) 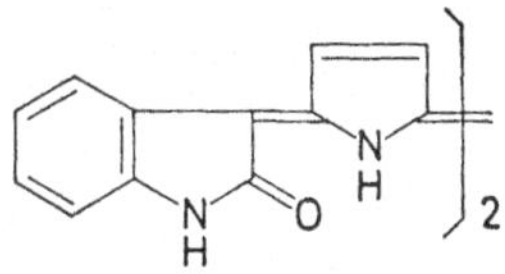   (31)

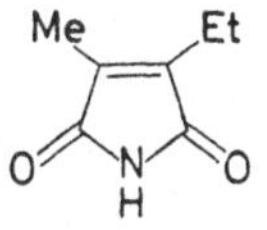

(32) 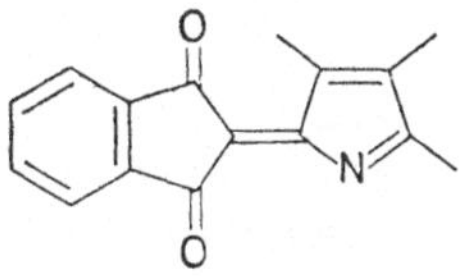   (33) 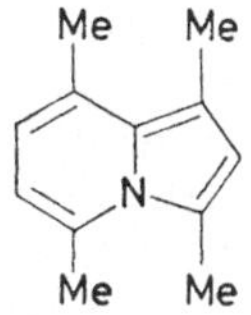   (34)

by methyl migration to the nitrogen atoms of the pyrrole residues. Support for this idea comes from the isolation of (32) after oxidation of the pyrrole blue from cryptopyrrole[107]. Compounds derived from two molecules of isatin and one of pyrrole can also be obtained[106].

Ninhydrin in acetic acid also forms carbinol intermediates with pyrroles. These are converted by stronger acids[108] into blue compounds (33).

With 1,4-dicarbonyl compounds, yet another type of reaction follows the initial condensation. Thus 2,4-dimethylpyrrole, heated with acetonylacetone and zinc acetate in acetic acid, gives[109] 1,3,5,8-tetramethylindolizine (34). However, from pyrrole and acetylacetone the product is 4,7-dimethylindole, and other pyrroles with unsubstituted 2- and 3-positions similarly give derivatives of 4,7-dimethylindole[110].

(e) Deuteration and Protonation

Deuterium exchange between pyrrole and deuterium oxide is slow at pH > 2, and only monosubstitution occurs. At pH 1-2, more than one hydrogen atom is replaced at a measurable velocity, and at pH < 1 rapid exchange of all five hydrogen atoms occurs[111]. Comparison of the monodeuteropyrrole formed at pH > 2 with the compound prepared from deuterium oxide and potassium pyrrole proves it to be 1-d-pyrrole. Clearly, from pyrrole-d_5 symmetrical pyrrole-d_4 can be obtained by reaction with water[112].

With some N-substituted pyrroles in concentrated sulphuric acid the rates of protonation at $C_{(3)}$ are greater than at $C_{(2)}$, though thermodynamically $C_{(2)}$ is favoured (p. 59)[246-7].

The halogen atoms of some iodopyrroles can be replaced by hydrogen through heating with hydriodic acid (p. 101).

(f) Diazo Coupling

The similar reactivities shown by pyrroles and phenols, amply demonstrated in the reactions discussed, are further illustrated by the ability of

D

pyrroles to couple with diazonium cations[113]. In acid or neutral conditions, pyrrole forms 2-monoazo derivatives, but in alkali the 2,5-bis-derivatives result. It is not clear whether the greater reactivity in alkaline solution is to be referred to the anion (35) or to the deactivation in acid solution caused by cation (36) formation[83]. Coupling occurs with 2,5-dimethyl-

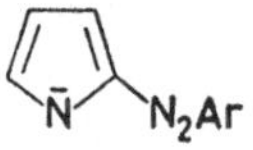

(35) 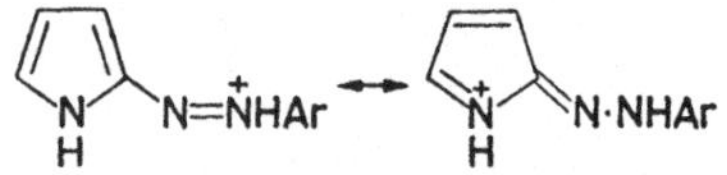   (36) 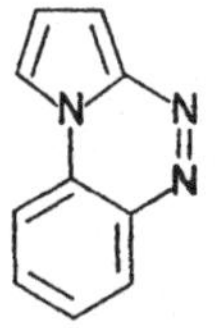   (37)

pyrrole and with 1-alkylpyrroles. Pyrrole-2-carboxylic acid forms with diazonium salts the same mono-aza-derivatives as pyrrole, carbon dioxide being eliminated[114]. 2,3,4,5-Tetramethylpyrrole and phyllopyrrole couple slowly, giving finally an olive-green colour like that obtained with 2,3,4-trimethylpyrrole; presumably a 2-methyl group is eliminated[100]. 2,5-Diphenylpyrrole-3-diazonium chloride has been coupled with pyrroles, β-coupling occurring when both α-positions are occupied[260]. Diazotization of 1-(2-aminophenyl)pyrrole produces (37) by internal coupling[261].

Treibs and Fritz[83a] carried out an extensive survey of the coupling abilities of pyrroles, using three diazonium salts of widely different electrophilic activities. Four pyrroles (38) constituted the most reactive set, which merged into a group of moderately reactive compounds (39), 2,5-dimethylpyrrole, 1-methylpyrrole and pyrrole itself. A less reactive group included ethyl 1,2,4-tri- and ethyl 2,4-dimethylpyrrole-3-carboxylate. The least reactive compounds examined, ethyl pyrrole-1-carboxylate and (40), (41) and (42), failed to couple even with the p-nitrobenzenediazonium cation. These observations are valuable because they demonstrate clearly that substituents

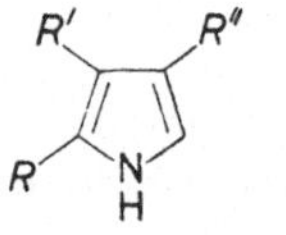

(38; $R = R'' = Me, R' = Et$
$R = H, R' = Et, R'' = Me$
$R = R' = R'' = Me$
$R = Me, R' = R'' = H$)

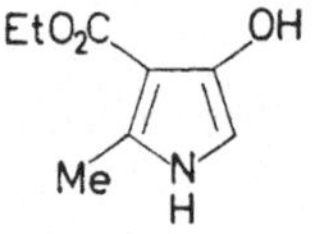

(39)

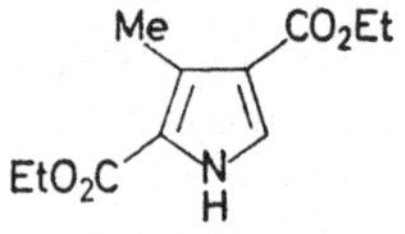

(40)

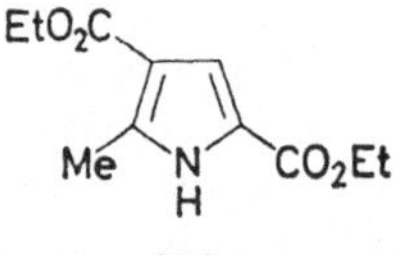

(41)

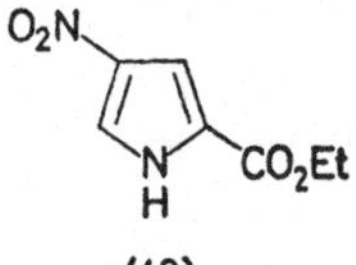

(42)

modify the character of the very reactive pyrrole nucleus in a way which is expected in terms of their electronic properties. The dicyanovinyl substituent is very powerfully deactivating in the coupling test[262].

Treibs and Fritz[83a] also surveyed the problem of the replacement during diazo coupling of groups other than hydrogen. The general requirement for ready replaceability is that a group shall be electron-attracting and able to produce on the nuclear carbon atom to which it is attached a high electron density, being then eliminated as a stable entity. Such is the case with the carboxyl group, the sulphonic acid group[83b] and the halogen atom[83, 119]. Alkoxycarbonyl groups cannot be replaced in coupling reactions, nor in general can acyl groups, although slow coupling appears to occur between 3,4,5-trimethylpyrrole-2-aldehyde and the *p*-nitrobenzenediazonium cation. Within this general framework substituents exert their expected influences. Thus, whilst 3,5-dimethylpyrrole-2-carboxylic acid undergoes eliminative coupling readily, and 4-ethoxycarbonyl-3,5-dimethylpyrrole-2-carboxylic acid almost as readily, 3,5-diethoxycarbonyl-4-methylpyrrole-2-carboxylic acid does not react even with diazotized picramide. In no case is a *β*-carboxyl group eliminated. The effect of substituents upon halogen elimination during diazo coupling makes a very similar pattern.

(g) Halogenation

Halogenation of pyrrole occurs so readily that tetra-substitution is the only easily observed process[62]. Conditions must be mild or decomposition occurs. Tetra-chloro-, -bromo- and -iodo-pyrrole thus result from the action of alcoholic solutions of the halogens, sometimes in the presence of substances such as mercuric oxide which are meant to remove acids, by the action of hypohalite solutions and by that of iodine in potassium iodide solution or in ammonia[115]. If partial substitution is to be observed, other reagents are essential. According to the proportion of reagent used, sulphuryl chloride in ether will convert pyrrole into 2-, 2,5-di-,2,3,5-tri- or 2,3,4,5-tetra-chloro-pyrrole, and finally into pentachloropyrrolenine[116, 282]. The unstable monohalogenopyrroles are said to be formed by treating pyrryl magnesium bromide with bromine or chlorine[263].

Substituted pyrroles can be brominated satisfactorily. Some examples are shown[44a, 264-5]. When methyl groups are present, they themselves can be brominated (p. 96 and below), and to avoid this conditions need to be controlled carefully.

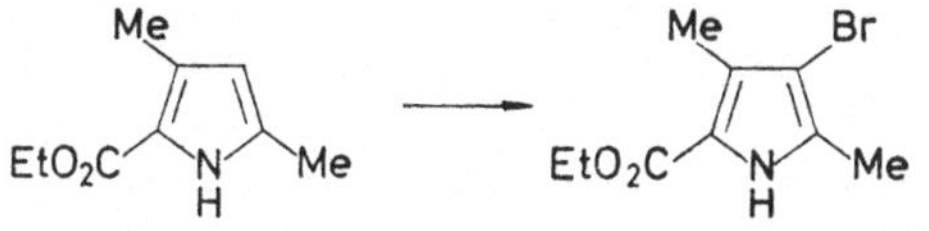

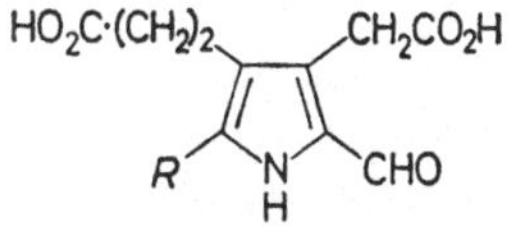

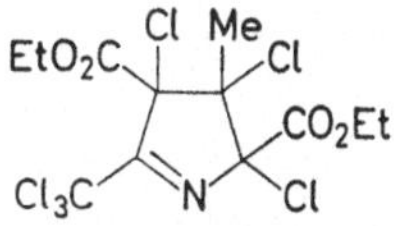

Qualitative experiments indicate that for iodination the range pH 8–9 is best. The difficulty of effecting monosubstitution is illustrated by the case of 3-ethoxycarbonyl-2-methylpyrrole which gives the di-iodo-compound and starting material. From a number of cases studied, only 1,2,5-trimethyl-pyrrole gave a *mono*-iodo-derivative[119]. Clearly, the mono-iodopyrroles are themselves easily iodinated. There is, from the example mentioned, the possible implication that some mono-iodopyrroles are iodinated more rapidly than their parents; this seems improbable unless a change of mechanism, as from substitution into a neutral molecule to substitution into a conjugate base, is happening. Under the circumstances this itself is improbable. Mechanistic information is much needed.

Replacement of certain substituents attached to the pyrrole nucleus is common. The carboxyl group is readily displaced; bromination of pyrrole-2-carboxylic acid gives tetrabromopyrrole[117a], with potassium iodide–iodine in bicarbonate solution (45, $R = \cdot CO_2H$), it gives (45, $R = I$)[266], and other examples are known in both bromination and iodination[264, 267, 272, 350]. Alkoxycarbonyl groups are not affected, though iodination of 1,3-diethoxy-carbonyl-2,4-dimethylpyrrole gives 3-ethoxycarbonyl-5-iodi-2,4-dimethyl-pyrrole[268]. Acetyl groups are sometimes displaced[44a, 120b], as in the conversion of (44) into 2,4-dibromo-3-ethyl-5-methylpyrrole[44a]. Iodination of 2-ethoxycarbonyl-4-formyl-3,5-dimethylpyrrole gives 2-ethoxycarbonyl-4-iodo-3,5-dimethyl-pyrrole[269]. Replacement of functional groups occurs in the reaction between alkaline hypoiodite and pyrrole-2-carboxylic acid or pyrrole-2-aldehyde but the product is complex[269].

Exchange of one halogen for another is mentioned below (p. 101).

Pyrrole homologues suffer both nuclear and side-chain substitution with sulphuryl chloride or bromine[120, 264, 270]. Thus, 2,5-dimethylpyrrole is converted by sulphuryl chloride into 3,4-dichloropyrrole-2,5-dialdehyde, the aldehyde groups arising from hydrolysis of the chlorinated methyl groups[120a], and further reaction of (43) with bromine gives the 2-bromo-methyl compound[264]. The α-methyl group of 2,4-diethoxycarbonyl-3,5-dimethylpyrrole is attacked by sulphuryl chloride, the corresponding aldehyde and carboxylic acid being formed, but a larger proportion of sulphuryl chloride causes nuclear attack, giving[270] (46).

An important reaction is that between bromine and a pyrrole with one free α-position and an α'-methyl group. α-Bromination occurs and subsequent reaction gives 5-bromo-5'-methyl-dipyrrylmethene hydrobromides. This reaction is discussed later (p. 102).

An interesting comparison has been made of the performances of different halogenating agents with methyl pyrrole-2-carboxylate[372]. Electrophilic bromination by bromine in acetic acid occurs mainly at $C_{(4)}$, to a smaller degree at $C_{(5)}$, and also gives some 4,5-disubstitution. The less selective chlorine did not distinguish so noticeably between $C_{(4)}$ and $C_{(5)}$. In both cases, attack at $C_{(3)}$ occurred only when the other two positions had been filled. Radical chlorination by t-butyl hypochlorite in carbon tetrachloride occurred exclusively at $C_{(5)}$, and only subsequently at $C_{(3)}$. These results suggested that chlorination by sulphuryl chloride in ether was partly electrophilic and partly radical. It gave 4- and 5-substitution in about equal amounts, and then 3,5- and 4,5-disubstitution. Initial attack at $C_{(4)}$ was probably an electrophilic substitution and that at $C_{(5)}$ could have been either electrophilic or radical. However, the attack at $C_{(3)}$, after that at $C_{(5)}$, was most probably a radical reaction. The orientations in the electrophilic halogenations resembled those in nitrations (see below). Clearly, two different mechanisms may be at work generally in the chlorination of pyrroles with sulphuryl chloride.

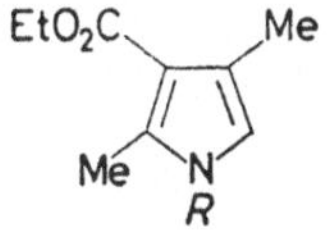

(47; a, R = H;
b, R = Me)

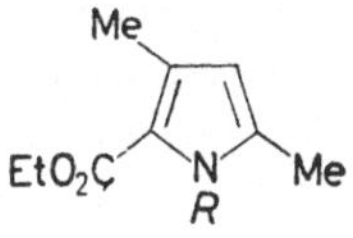

(48; a, R = H;
b, R = Me)

Relative Rates of Iodination

k(l.mole^{-1} sec^{-1})*		Ratios	
47a	15·7	47a:48a	26·2
47b	16·9	47b:47a	1·08
48a	0·60	48b:48a	1·15
48b	0·69	47b:48b	24·5

* In aqueous dioxan (28 wt. per cent dioxan), at pH 6·8–6·9 and 26·5°, with iodine in potassium iodide.

Doak and Corwin[118] carried out an interesting study of the iodination of (47) and (48). Although the free a- and β-positions are not fully equivalent in any of the pairs, they are sufficiently similar to allow acceptance of the conclusions to be drawn from the figures. Thus, a-positions are considerably more reactive than β-positions, and N-methylation produces a modest activation in this electrophilic substitution. The activating effect of methyl groups is smaller than in the benzene series. In these iodinations the reaction rates depend on the concentration of the pyrrole and of the free iodine, but another species, suggested to be hypoiodous acid, is also active. The iodinations are reversible (p. 101).

(h) Nitration and Nitrosation

Since concentrated mineral acids polymerize pyrroles (pp. 59, 83), ordinary nitration methods cannot be used in the series unless deactivating groups are present. It is possible to nitrate pyrrole carboxylic acids and their esters[119, 121] and nitropyrroles[273] with concentrated nitric acid.

Rinkes[122] first showed that pyrrole can be nitrated by acetic anhydride–nitric acid, and he isolated about 21 per cent of 2-nitropyrrole from this reaction. Later it was shown[45] that about 7 per cent of the crude product is 3-nitropyrrole. 1-Methylpyrrole gives about 39 per cent of a mixture of 1-methyl-2- and -3-nitropyrrole in the ratio[45] 2:1. Further nitration of 2-nitropyrrole provides 2,4- and 2,5-dinitropyrrole in the ratio 4:1. 2-Acylpyrroles behave similarly, but methyl pyrrole-2-carboxylate gives about equal amounts of the 4- and 5-nitro-compounds[45, 274]. Nitration of 2-acetyl-1-methylpyrrole favoured 4- over 5-substitution even more than with 2-acetylpyrrole itself. A 2-cyano group is more weakly 4-directing than methoxycarbonyl-, acetyl- or nitro-substituents. 2-Formylpyrrole can be nitrated by means of acetic anhydride and nitric acid at −40°, giving the 4- and 5-nitro-compounds[275]. Thus, in these nitrations the α-directing power of the pyrrole nucleus is not overwhelming. It is weakened in 1-methylpyrrole and sometimes overcome by an electron-attracting group in an α-position.

1-Alkyl- and -aryl-2,5-diphenylpyrroles have been nitrated with acetic acid–nitric acid[273, 276]. Nitration of 1-phenylpyrrole in acetic anhydride gives 2- and 3-nitro-1-phenylpyrrole. However, in sulphuric acid, presumably because of protonation of the pyrrole ring, 1-p-nitrophenylpyrrole is formed[124b].

Elimination of substituents is a very common feature of nitrations in the pyrrole series[119, 121–3]. Thus, pyrrole-2-carboxylic acid gives some 2-nitropyrrole, and 4-nitropyrrole-2-carboxylic acid some 2,4-dinitropyrrole. Acetyl and formyl groups are frequently eliminated. De-iodinative nitration proceeds smoothly when at least one electron-attracting substituent is present[119]. An interesting case is the conversion of ethyl 2,4-dimethyl-pyrrole-3,5-dicarboxylate by nitric acid into ethyl 2,4-dinitropyrrole-3,5-dicarboxylate, which illustrates the great stability of ester groups in this sense. Rinkes[122] suggests that the ease of replaceability increases in the order PhCO < I < MeCO < CO$_2$H.

Besides the reagents already mentioned, ethyl nitrate in ether with sodium or sodium ethoxide has been used in the pyrrole series. It converts pyrrole into 3-nitropyrrole and 2,4-dimethylpyrrole into 2,4-dimethyl-3-nitropyrrole[124a], the yields being poor. The β-orientations in these substitutions are their most interesting aspect. It would be useful to know if the orientation is influenced by heterogeneity under the conditions used.

Similar remarks apply to the nitrosations of pyrroles which have been effected by amyl nitrite in the presence of sodium ethoxide. The products are obtained as their sodium salts, that from pyrrole apparently having the β-orientation but that from 2,4-dimethylpyrrole probably being 2,4-dimethyl-5-nitrosopyrrole[125]. A number of 2,5-disubstituted pyrroles have been smoothly β-nitrosated by means of an alkyl nitrite and sodium ethoxide in ethanol[260, 277–8]. An aryl- or alkoxycarbonyl-substituted pyrrole can be nitrosated by means of sodium nitrite in sulphuric acid[126, 279]. Well established examples of α-nitrosation are the conversions of 2,4-diarylpyrroles into 2,4-diaryl-5-nitrosopyrroles by sodium nitrite in mineral acids[14b].

Nitrosopyrroles are intermediates in the direct conversion of pyrroles into diazopyrroles in buffered nitrous acid solutions[277, 280]. The reaction with a β-nitrosopyrrole, as in the conversion of 2,4,5-triphenylpyrrole into 3-diazo-2,4,5-triphenylpyrrole[277], is faster than with an α-nitrosopyrrole. Thus, in the formation of 2-diazo-3,5-diphenylpyrrole from 2,4-diphenyl-pyrrole, the intermediate stage can easily be isolated[280].

Nitropyrroles have been prepared by reaction of a pyrrole with an alkyl nitrite in ether. Under these circumstances the nitroso compound, first formed, is oxidized to the nitropyrrole[197, 277, 281].

(i) Sulphonation

Pyrrole and its homologues can be safely sulphonated by pyridine-sulphur trioxide, ethylene dichloride being a convenient solvent. Pyrrole,

1-methylpyrrole, 2-chloropyrrole and 2-phenylazopyrrole all give the a- or a'-sulphonic acids[128, 282].

Pyrroles containing electron-attracting substituents have been sulphonated with sulphuric acid or oleum, and also chlorosulphonated[83b].

(j) Thiocyanation

Direct thiocyanation, by electrophilic substitution, is limited to reactive aromatic compounds such as amines and phenols[129]. Accordingly, it is not surprising that thiocyanation of pyrrole is possible; reaction with cupric thiocyanate in methanol at $0°$, or with methanolic thiocyanogen at $-75°$, gives 2-thiocyanatopyrrole, and 2-methyl-5-thiocyanatopyrrole is prepared similarly[130, 283-4]. From pyrrole more vigorous reaction conditions produce 2,5-dithiocyanatopyrrole[285] and will also effect substitution into pyrroles containing electron-attracting substituents[131, 284].

(k) Reactions at the Nitrogen Atom

A number of instances have already been mentioned of reactions in which both N- and C-substitution occur. These are: reactions of pyrroles with acid anhydrides (p. 64), alkylation and alkenylation of metallic derivatives of pyrroles (p. 66) and ethoxycarbonylation of metallic derivatives (p. 66; see also below). Most of the reactions of pyrrole Grignard reagents result in C-substitution, but occasionally N-substitution is also observed (p. 106). The Mannich reaction does not cause N-substitution (p. 70). Deuteration proceeds most easily at the nitrogen atom (p. 75), and reactions with metals (p. 61) cause displacement of hydrogen from nitrogen.

The case of ethoxycarbonylation has also been mentioned (p. 66). Whilst potassium salts of pyrrole and alkyl pyrroles react with ethoxycarbonyl chloride to give 1-ethoxycarbonylpyrroles[257, 291-2], the lithium salts form 2-ethoxycarbonylpyrroles[257]. Potassium pyrrole gives N-acyl derivatives with acyl chlorides and anhydrides[34b, 132-34]), although from acetyl chlorides some 2-acetylpyrrole is also formed. Pyrrole and its homologues, but not ethoxycarbonyl or acetyl derivatives, can be N-benzoylated under Schotten-Baumann conditions[135].

It will be noticed that reaction at the pyrrole nitrogen atom usually involves either the use of potassium pyrrole or of alkaline conditions. It is not surprising, therefore, that in the presence of trimethylbenzylammonium hydroxide, pyrrole undergoes N-cyanoethylation with acrylonitrile[136], and that the sodium salt of 2-nitropyrrole behaves in the same way[293]. Related reactions are the formation of 1-pyridylethylpyrrole from the sodium pyrrole and 2-vinylpyridine[137], and of 1-vinylpyrrole from acetylene and potassium pyrrole[294].

RADICAL SUBSTITUTION

Little is known about the behaviour of pyrroles with free radicals. Phenyl radicals generated from nitrosoacetanilide convert 1-ethoxycarbonyl-pyrrole into 1-ethoxycarbonyl-2-phenylpyrrole, but the methyl ester gives an azo-derivative[127].

It is probable that some of the reactions of pyrroles with sulphuryl chloride (p. 78) involve radicals, but nothing is known about the detailed mechanisms of these reactions.

3. REACTIONS WHICH MODIFY THE PYRROLE NUCLEUS

For general remarks regarding these reactions see p. VII. It is interesting to notice that among them are the small number of instances of nucleophilic attack upon the pyrrole nucleus: the reductions to pyrrolines and the reactions with sodium bisulphite and with hydroxylamine.

(a) Additions

Like other aromatic systems, pyrroles reveal their unsaturated condition in certain addition reactions. The most obvious of these are partial or complete reduction to pyrrolines or pyrrolidines. Sodium and alcohol are without effect on pyrrole[138], but zinc and acid produce Δ^3-pyrrolines[139, 233, 303]. This reaction may involve attack of hydride ion upon the α-protonated pyrrole cation. Many pyrroles have been reduced by a variety of catalytic procedures to the fully saturated pyrrolidines[61, 140], and electrolytic reduction also converts pyrrole into pyrrolidine[141].

Pyrrole forms with sodium bisulphite an addition compound, $C_4H_7NO_6S_2Na_2.2H_2O$, which has been formulated[304] as the dihydrate of the disulphonate (49). The reaction has been represented as a nucleophilic attack upon the polarized pyrrole molecule. Mechanistically it could be very satisfactorily represented as a nucleophilic attack by bisulphite anion upon the highly reactive β-protonated cation of pyrrole in a manner analogous to the formation of pyrrole trimer (p. 86). However, under the conditions used[304], the concentration of cation must be extremely small.

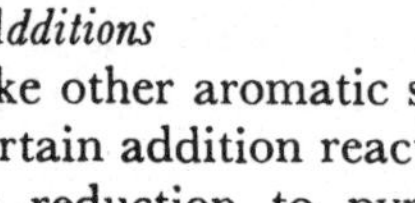

(49) 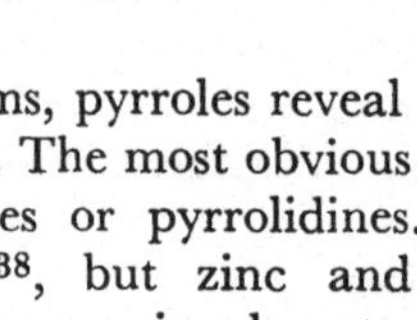 (50) 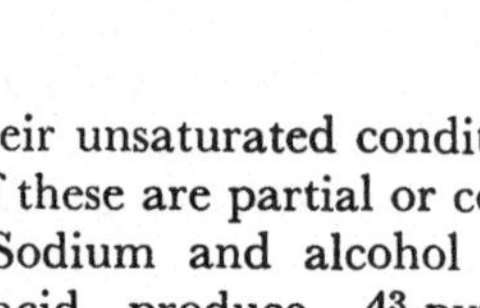 (51)

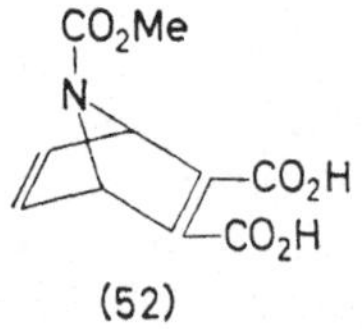

(52) 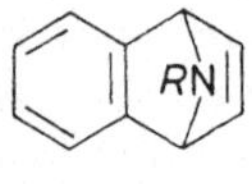 (53, R=Me or PhCH₂)

The formation from pyrroles and dienophiles of substances which are the results of Michael addition has been mentioned (p. 70). In some cases Diels–Alder addition does occur. Thus, with methyl acetylenedicarboxylate, methyl pyrrole-1-carboxylate gives (50), formed by elimination of acetylene from the initial Diels–Alder product[305]. From 1–methylpyrrole, (51) is obtained[142, 306], formed from the initial Diels–Alder product by reaction with a second molecule of methyl acetylenedicarboxylate[306]. 1-Benzylpyrrole behaves similarly. With acetylene dicarboxylic acid, methyl pyrrole-1-carboxylate gives both a Michael addition product (p. 70) and the Diels–Alder product (52)[254]. Products analogous to (52) are obtained from 1-benzyl- and 1-naphthylmethyl-pyrrole[143, 307].

In contrast to pyrrole itself (p. 71), 1-methyl- and 1-benzyl-pyrrole undergo Diels–Alder addition with benzyne giving (53) and substances derived from these[144, 308]. Pyrryl magnesium iodide gives α-naphthylamine formed by aromatization of the Diels–Alder intermediate[308].

The ease with which pyrroles are polymerized by acids depends on the substituents present. This can be seen from their behaviour with ethereal hydrogen chloride. Pyrrole is trimerized[145a, 309–10], 2-phenyl- and some 2-alkyl- and 2,3-dialkyl-pyrroles give hydrochlorides of their dimers (*Table 4.6*), 3-methylpyrrole forms amorphous polymeric material[145b, 180b], and

Table 4.6. Polymers of Pyrroles

Substituents in Monomer	Dimer	Trimer
—	stannichloride[147]	m.p. 99–100°[148] N-acetyl m.p. 174–193° (dec.)[148] hydrochloride[145b] picrate[309]
1-Me[310] 2-Me[37b, 145ab, 297]	hydrochloride, m.p. 190° (dec.) m.p. 62–64 hydrochloride picrate, m.p. 155–6° (dec.)	
2-iPr[37b, 145ab]	b.p. 285–290° hydrochloride picrate, m.p. 146°	
2,3-Me$_2$[22, 145b, 315]	m.p. 84–85° picrate, m.p. 148°	
2,4- and 2,5-Me$_2$ 2-Et-3-Me[22] 3-Et-2-Me[316]	see[12c] picrate, m.p. 137° b.p. 172°/15 mm picrate, m.p. 136°	
2,3,4-Me$_3$[271]	m.p. 135–138° picrate, m.p. 173° (dec.)	
3-Et-2,4-Me$_2$ (cryptopyrrole) 3-Et-2,4,5-Me$_3$ (phyllopyrrole) 2-Ph[110b]	{ see text m.p. 140° hydrochloride, m.p. 202–3° picrate, m.p. 184° } (dec.) methiodide, m.p. 210–11°	

some pyrroles (2,4-dimethyl-, 2,5-dimethyl-, 2,3,4-trimethyl-, and 3-ethyl-2,4-dimethyl-pyrrole) give simple hydrochlorides (*Table 4.1*). Some pyrroles with electron-attracting substituents produce simple salts[14a]. 1-Methylpyrrole appears to give a hydrochloride of the dimer, but the dimer has not been obtained crystalline[310].

A similar division can be made with respect to behaviour with ethereal picric acid solution; picrates of dimers arise from 2-mono- and 2,3-disubstituted pyrroles (*Table 4.6*), and simple picrates are formed in other cases

D*

(*Table 4.1*). 3-Methylpyrrole gives a picrate of uncertain character (*Table 4.1*). The cases of cryptopyrrole and phyllopyrrole (3-ethyl-2,4-dimethyl- and 3-ethyl-2,4,5-trimethyl-pyrrole, respectively) require special mention. They give simple picrates which, when boiled in ethyl acetate, were reported[145c] to be converted into the picrates of dimers. There is some doubt about these conversions[271, 297].

Finally, there is a group of pyrroles from which crystalline salts, either of the monomer or the dimer, appear not to have been obtained. This includes 2,5-dimethyl-[311], 3,4-dimethyl-[311], 2-ethyl-5-methyl-[120c], 3-ethyl-4-methyl-[312] and 3-ethyl-5-methyl-pyrrole[298].

It will be seen that the *easily* dimerized pyrroles have unsubstituted adjacent α- and β-positions. However, the difference between them and pyrroles which form relatively stable simple hydrochlorides is only one of degree, for these simple hydrochlorides can be made to take part in reactions which produce dimers or compounds derived from them (see below) or do themselves give dimeric products when kept (*Table 4.1*).

The dimers are monoacidic bases, derivatives of 5-(2'-pyrrolyl)-1-pyrrolines (54)[145e, 297, 313]. Dimers of pyrroles unsubstituted at $C_{(2)}$ and $C_{(3)}$ can be distilled unchanged, whilst others are depolymerized. The probable mechanism of dimerization will be apparent from the discussion below of the trimerization of pyrrole.

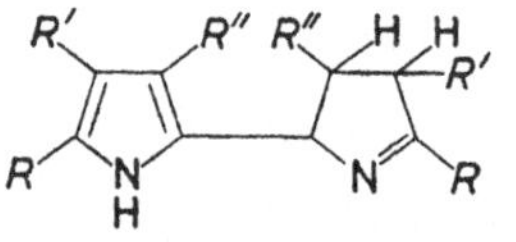

(54) 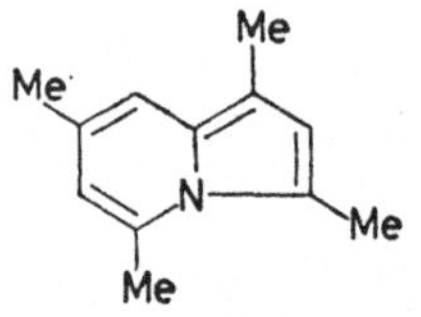   (55) 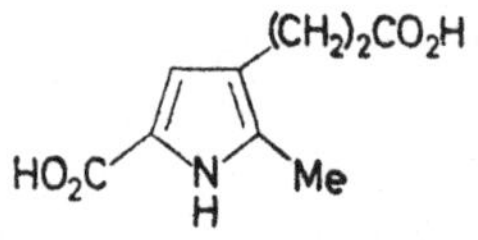   (56)

The dimers of pyrroles having unsubstituted adjacent α- and β-positions are converted by dilute sulphuric acid into indoles. This has been observed with the dimers of 2-methyl-[110b, 145b, 297], 2-isopropyl-, 3- methyl- and 2,3-dimethyl-pyrrole[145b]. The product from 2-methylpyrrole dimer has been proved to be 2,4-dimethylindole[110b, 297], but the others have not been characterized. 2-Phenylpyrrole dimer is unaffected by dilute sulphuric acid[110b].

The conversion of 2-methylpyrrole into 2,4-dimethylindole by treatment with zinc acetate in acetic acid[110b] probably proceeds through the dimer, as does the conversion by these reagents of 2,4-dimethylpyrrole into 1,3,5,7-tetramethyl-indolizine (55)[109, 146]. In neither case is the precise function of the reagent obvious. Originally, Plancher, who considered the reaction to proceed by ring opening of the pyrrole to a 1,4-diketone followed by further condensation (p. 75), ascribed to zinc acetate a hydrolytic function[146]. The pyrrole (56) is converted into an indole when heated with a mixture of formic and hydrobromic acids[44b], the α-carboxyl group being lost. The pattern of all these reactions is clear: 2-alkyl- and 2,3-dialkyl-pyrroles give, through their dimers, indoles, but when the β-position which must be free for indole formation is blocked, then an indolizine results.

The same pattern is seen in the formation of (55) when 2,4-dimethyl-pyrrole and its hydrochloride are heated in benzene. Similarly, cryptopyrrole and its hydrochloride give 2,6-diethyl-1,3,5,7-tetramethylindolizine. From 2,3,4-trimethylpyrrole and its hydrochloride, as well as 1,2,3,5,6,7-hexa-methylindolizine, the pyrrole dimer (54, $R = R' = R'' = Me$) was iso-lated[271].

The formation of 5-(2'-pyrrolyl)-1-pyrrolines (54) by dimerization has not been observed to occur, and indeed could not occur without substituent elimination, with 2,5-disubstituted pyrroles [see the example of the dimeri-zation of (56) above]. The dimerization involves electrophilic attack by a β-protonated pyrrole cation upon the conjugate base, as for example

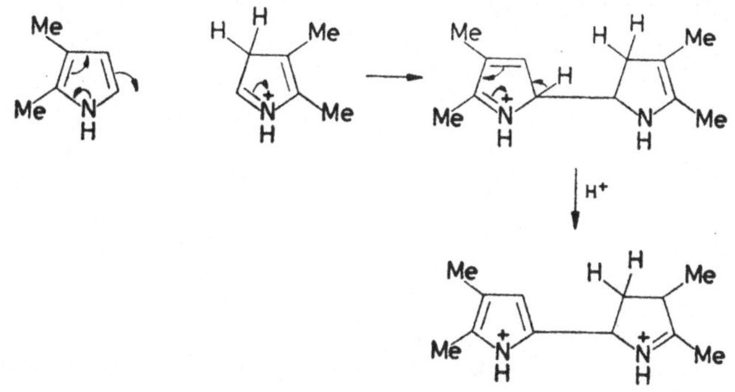

Therefore, special interest attaches to the behaviour of 2,5-dimethylpyrrole in acidic reducing conditions (zinc + acetic acid; tin + hydrochloric acid), for the production of 1,3,4,7-tetramethylisoindoline (57) must involve elec-trophilic attack by the α-protonated cation (58) upon an α-position of 2,5-dimethylpyrrole[314]:

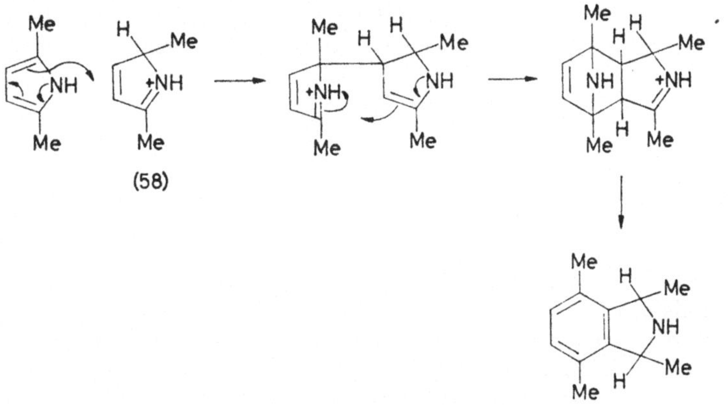

(58)

(57)

The cation (58) is one of the protonated forms of 2,5-dimethylpyrrole (p. 59).

The dimer of pyrrole has been obtained only as its stannichloride, by treating pyrrole in chloroform with stannic chloride[147]. On the other hand, the crystalline trimer (59) is easily obtained by the use of ethereal hydrogen chloride or aqueous hydrochloric acid[37b, 145a, 309-10]. Its formation has been represented as an attack upon pyrrole by β-protonated pyrrole cation, an argument which implies that the β-protonated cation is more powerfully electrophilic than its α-isomer[148, 313]. Further reaction is discouraged by protonation of the central nitrogen atom, but the trimer is reactive and gives more highly polymerized substances of unknown structure in acid solution[313].

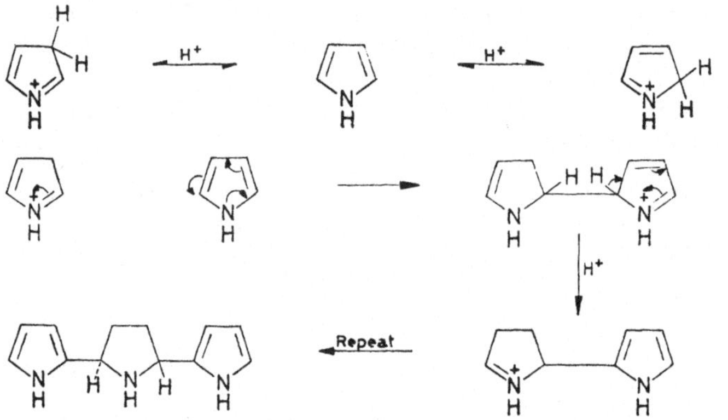

(59)

2-Methylpyrrole and related compounds are thought to give dimers rather than trimers because, for a combination of steric and electronic reasons, the unit $-\overset{+}{N}H=CH-$ is more electrophilic than is $-\overset{+}{N}H=CMe-$[313].

The rarity of free radical substitutions into pyrrole has been remarked upon (p. 81). With triphenylmethyl, addition occurs giving 2,5-di-triphenylmethyl-Δ^3-pyrroline[149a].

Photooxidation of pyrrole[317] and of methanolic 2,3,4,5-tetraphenylpyrrole[149b] gives (60) and (61), respectively. In the second instance some ring opening, reminiscent of that caused by peracids (see below), also occurs, giving α-N-benzoylamino-α'-benzoylstilbene. The instability in air of many

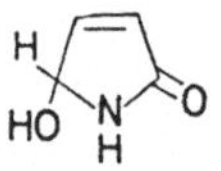

(60) 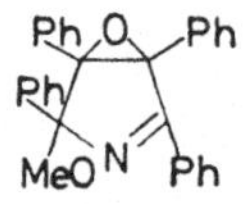   (61) 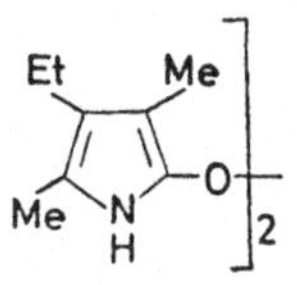   (62) 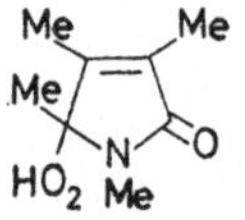   (63)

pyrroles is due to their ease of oxidation. Autoxidations of a number of di- and tri-alkylpyrroles proceed smoothly, giving crystalline products formulated[154] in the manner of (62), and derivatives of aminopyrroles undergo autoxidation (p. 97). Perhydrol in pyridine will oxidize pyrroles with free α-positions, or pyrrole-α-aldehydes, to hydroxypyrroles[154-6], correctly formulated as pyrrolones (p. 104). Sometimes peroxides, again formulated like (62), are also formed. Such structures as (62) are improbable[318]. With hydrogen peroxide, pentamethylpyrrole gives (63) in a reaction represented[318] as an initial electrophilic attack by HO^+.

(b) Disruptions

It is convenient to mention first some reactions which modify pyrrole rings without necessarily disrupting them. Thus, a number of pyrroles have been oxidized by chromic or nitrous acid to maleimides, a reaction which is useful in structural determinations since it eliminates α-substituents[150]. Pyrrole with alkali hypohalites gives dihalogenomaleimides or dihalogeno-maleic acids[115, 151]. Whether or not the ring is disrupted really depends on the ease of hydrolysis of the particular maleimides formed.

Prolonged action of acid upon pyrrole causes the formation, by oxidative condensation, of pyrrole red[152]. The nature of this material and of pyrrole black, the product of other oxidations, is obscure[153].

With some polyphenylpyrroles, opening of the ring by peracids may be initiated by 2,3-epoxide formation[157].

A number of pyrroles react readily with ozone in chloroform at −60°. Pyrrole gives glyoxal (15 per cent) and ammonia, 1-phenylpyrrole produces glyoxal and aniline (38 and 58 per cent, respectively). From 2,5-dimethyl-, 1,2,5-trimethyl- and 1-ethyl-2,5-dimethyl-pyrrole, glyoxal and methyl-glyoxal are obtained, whilst 2,3-dimethylpyrrole gives glyoxal, methyl-glyoxal and biacetyl. The detailed mechanism of ozonization is disputed[158], but these results have been interpreted in terms of the resonance hybrid structure of pyrrole and clearly cannot be accounted for by the classical double-bond representation[159].

A particularly interesting means of opening the pyrrole ring is to treat it with hydroxylamine. The reaction was discovered by Ciamician and Denn-stedt who obtained succindialdoxime by treating pyrrole in alcohol with hydroxylamine hydrochloride and sodium carbonate[160]. Identification of the dicarbonyl compound thus produced as its dioxime provides a means of determining the structure of a substituted pyrrole. The reaction has been regarded as a hydrolysis of the pyrrole nucleus, but this seems unlikely. Findlay[161] showed that hydroxylamine alone does not convert pyrrole into succindialdoxime, and hydroxylamine hydrochloride was also ineffective. However, the latter with alkali sufficient to give a reagent equivalent to $(NH_2OH)_2.HCl$, was very effective, giving 60 per cent of succindialdoxime, and from 2,5-dimethylpyrrole, 90 per cent of acetonylacetone dioxime. The failure with higher proportions of hydrochloric acid is due to polymerization.

The mechanism of the ring opening by hydroxylamine is interesting cause for speculation. As with the addition of bisulphite to pyrrole (p. 82), the reaction could satisfactorily be represented as a nucleophilic attack by hydroxylamine upon a pyrrole cation:

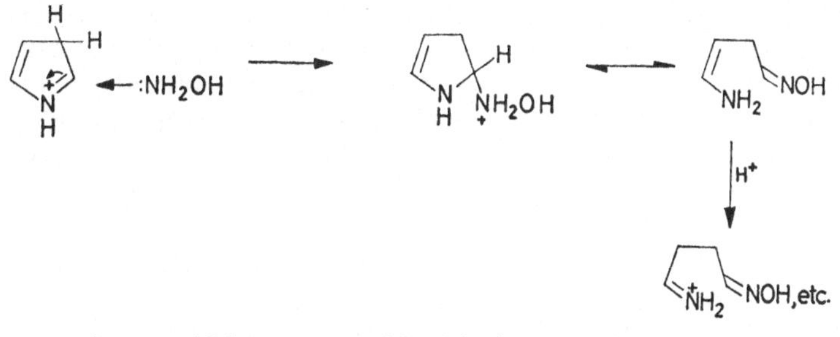

Such a reaction would be more rapid with the more basic (p. 60) dimethyl-pyrrole, but under the recommended conditions the concentrations of pyrrole cations must be extremely small.

A closely related reaction is the conversion of 2,5-dimethylpyrrole into acetonylacetone bis-dinitrophenylhydrazone when it is boiled with 2,4-dinitrophenyl hydrazine in 10 per cent sulphuric acid[110b], and certain reactions with hydrazine may be similar[154, 189, 323].

One or two cases are known of the conversion of pyrroles into diketones by acids alone[72, 162], for example (64)→(65). α- and β-Nitrosopyrroles are susceptible to ring opening by acid hydrolysis. In some cases the reaction

(64)   (65)

with hydroxylamine or semicarbazide may be subsequent to the initial hydrolysis[163, 194a, 280, 319].

Some of the reactions mentioned below as ring expansions certainly proceed through ring opening.

Interesting disruptions of the pyrrole ring occur under the conditions used in mass spectrometry. With 1-methylpyrrole not only does fragmentation to $HC{\equiv}\overset{+}{N}Me$ takes place but some rearrangement to pyridinium may occur· Pyrrole gives a number of ions such as the cyclopropenyl ion ($C_3H_3^+$) and its aza-analogue ($C_2H_2N^+$), and $C_2H_3N^+$ resulting from the loss of acetylene. Acylpyrroles tend to give acylium ions[370].

(c) Expansion

Methylpyrroles have been converted into pyridines by hydrochloric acid under severe conditions, and also by pyrolysis[62] (p. 109). The formation of a 3-chloropyridine derivative from a pyrrole under Reimer–Tiemann conditions has been mentioned (p. 63). This type of reaction was discovered by Ciamician and Dennstedt[164] who treated pyrrole with chloroform in ether and isolated a small yield of 3-chloropyridine. Subsequently, similar reactions were realized with bromoform, carbon tetrachloride, methylene iodide and benzal chloride. Those of several of these reagents with lithium pyrrole in ether and sodium pyrrole under various conditions have been compared. The yields of pyridine derivatives were always low[164]. In submitting 2,5-dimethylpyrrole to the Reimer–Tiemann reaction, Plancher and Ponti[23] isolated a pyrrolenine (7). This and its analogues are not intermediates in the conversion of pyrroles into 3-chloropyridines. The idea that dichlorocarbene is the active reagent in reactions using chloroform[165] is supported by recent work[322]:

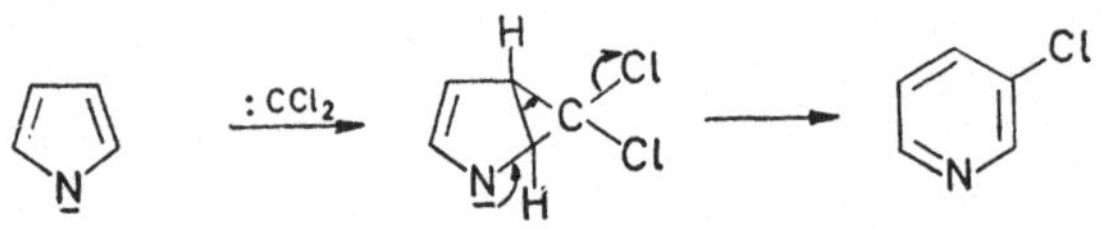

Chlorocarbene, :CHCl, generated from methylene dichloride and methyl lithium converts pyrrole into pyridine in 32 per cent yield[320].

Pyrrole reacts with chloroform at 550° to give 3- (33 per cent) and 2-chloropyridine (2–5 per cent)[321]. A carbene formed by pyrolysis may be involved.

The pyrrolenine (7) undergoes ring expansion to a pyridine derivative (66) when treated with base[322]:

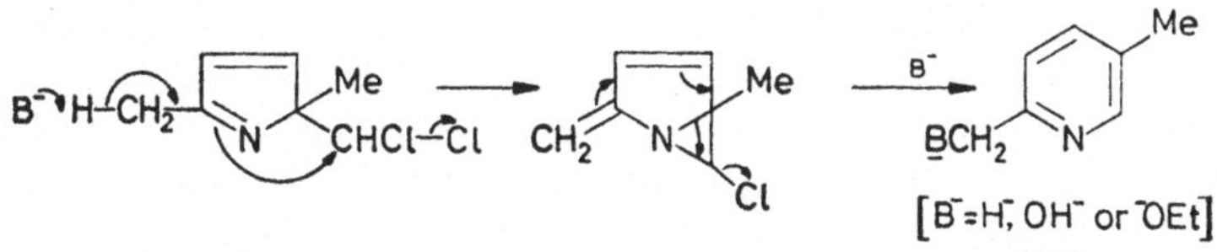

$[B^- = H^-, OH^-$ or $^-OEt]$
(66)

Further products arise from reaction between (66) and the bicyclic intermediate.

Phosphorus pentachloride converts 2,5-di- and 2,4,5-tri-phenyl-3-nitroso-pyrrole into di- and tri-phenyl-hydroxypyrimidine, respectively. The existence of the nitroso compounds in isonitroso forms (p. 107) suggests a similarity between these reactions and Beckmann transformations. However, open-chain intermediates are known to be formed[166].

Under the influence of sunlight, alcoholic ammonia converts 2,4,5-tri-phenylpyrrole into benzamide and 2,4,6-triphenylpyrimidine[324], whilst lead tetra-acetate in chloroform produces 2,3,5,6-tetraphenylpyrazine from 2,3,4,5-tetraphenylpyrrole[325].

(4.) A GENERAL CONSIDERATION OF THE REACTIVITY OF THE PYRROLE NUCLEUS

The structural evidence of aromatic character in pyrrole has been presented (p. 17) and the reactions described above reveal in pyrrole and its derivatives the characteristic aromatic ability to undergo substitutions.

Discussion of the reactivity of pyrroles, except in the most general terms, is made difficult by the almost complete lack of experimental evidence for the mechanisms of reactions in this series:

(1) With substitutions effected in acid media, the possible role of pyrrole cations has not been elucidated. However, in electrophilic substitutions it seems most improbable that any entity other than the neutral molecule should be involved. Not only must it be more reactive to electrophilic attack than are the cations, but also it is difficult to formulate any other mechanism such as, for example, nucleophilic attack upon the cation, followed by electrophilic attack and elimination. The consequences of nucleophilic attack upon the β-protonated pyrrole cation are seen in the trimerization of pyrrole.

(2) With electrophilic substitutions carried out in basic media, the possible role of the pyrrole anion has to be considered. Again, no evidence is available. In reactions carried out with sodium or potassium salts of pyrroles, the observed result might sometimes be ascribable to the pyrrole anion, but often

ion pairs must be involved, and heterogeneity in the reaction medium raises another problem.

(3) Of radical and nucleophilic substitutions, very little is known. In the one or two reactions which look like nucleophilic attacks (addition of bisulphite, attack by hydroxylamine and related compounds), the question of the intervention of pyrrole cation needs investigation.

These difficulties having been recognized, some general points can be made. Reactivity towards electrophilic reagents is high, recalling that of the phenols. This reactivity is modified in the expected way by substituents. Generally, the α-positions are rather more reactive than the β-positions towards electrophilic reagents, but the difference is not so great that it cannot be outweighed by electron-attracting substituents. N-Substitution is unusual except when a preformed metal pyrrole is used or a reaction is carried out in basic media. The few examples known indicate that radicals attack α-positions, and this may be true of nucleophilic reagents, although the evidence is questionable.

It is interesting to see how these facts are met by theory. We have seen (p. 16) that resonance theory describes pyrrole as a hybrid in which important contributing structures have concentrations of electronic charge at either the α- or β-positions. Inspection of these structures gives no help in deciding whether the α- or the β-positions will be the more reactive (from the viewpoint of the Isolated Molecule Theory, p. 34) to electrophilic reagents, nor does it help to consider qualitatively the related Wheland transition states (67) and (68). The semi-quantitative molecular

(67)   (68)

orbital treatment of the Isolated Molecule Theory is also, at first sight, somewhat ambiguous in its implications (p. 35). Brown[167a] has, however, given a more complete discussion of pyrrole using simple molecular orbital theory. He calculated atom localization energies for electrophilic, nucleophilic and radical substitution, π-electron densities, free valencies and bond orders. Calculations were made for different values of the relative electronegativity parameter, $h = (a_N - a_C)/\beta$. The values obtained for q, F and p will be found in *Tables 3.1, 3.2* and *2.3* (pp. 36, 38, 29).

In the free pyrrole molecule, with nitrogen more electronegative than carbon ($h > 0$), the values of q suggest the β-position to be the more reactive in electrophilic substitution, but atom localization energies indicate the opposite. Presumably, then, for an electrophilic reagent which gives a transition state in which bond formation is well developed, that is, one which closely resembles the Wheland transition state, the localization energy is the controlling factor and the α-position will be the more reactive. On the other hand, with a reagent for which the transition state involves little bond-making, the π-electron densities will control the process and β-substitution should occur. The pyrrole anion may be concerned in reactions in alkaline solution, and then the relative electronegativity parameter is taken to be negative, and both the Isolated Molecule and Transition State Theories indicate the α-position to be the more reactive. Similar agreement is noticed in the prediction of α-substitution into the neutral pyrrole molecule by free radicals. The results also suggest that pyrrole would be very unreactive to nucleophilic attack but that the α-position would be more reactive than the β-position.

A different treatment of the pyrrole molecule, using the variable electronegativity self-consistent field molecular orbital method[167b], gives a π-electron distribution consistent with the chemistry of pyrrole. As stated, this is not the case with the Hückel method (above and p. 36), and it was for this reason that Wheland and Pauling introduced auxiliary inductive parameters into that method (pp. 35–36). When it is attempted to reproduce in the Hückel method the π-electron distribution obtained by the variable electronegativity method, it proves necessary to introduce the auxiliary parameter again. If it is true that for all but the most unreactive hetero-aromatic systems, electrophilic substitution is controlled by π-electron distribution rather than by localization energies[167c], these results can be regarded as support for the use of auxiliary inductive parameters (cf. p. 35). The variable electronegativity method applied to pyrrole anion[167d] gives an almost uniform π-electron distribution round the ring.

As stated already, the observed orientations of electrophilic substitution can be accounted for, in the pyrrole molecule, in terms of electrophilic localization energies, without invoking an auxiliary inductive parameter, for positive values of h. However, pyrrole is very reactive, and for this reason orientation would be expected to follow π-electron densities. These fall into line (when the acceptable value $h = +2$ is used) if $h' > 0.19$. The orientation of electrophilic substitution is satisfactorily accounted for[328] when $h = 2$ and $h' = 0.25$.

The theoretical picture is not inconsistent with the available experimental facts but, as indicated, these are unfortunately inadequate to allow us to judge the real status of the theory. In no case have we any knowledge of the detailed mechanisms of pyrrole substitution reactions.

The point of this remark may be seen if one considers the high electron density deduced by theory to be found on the nitrogen atom, and also the relative rareness of N-substitution. These features are connected with ideas frequently met in discussions of pyrrole reactivity.

In the older literature the view that C-substitution is preceded by N-substitution is encountered, and it no doubt sprung from such observations as the rearrangement of N-acyl- or N-alkyl-pyrroles to C-substituted compounds. However, the conditions used in these rearrangements (p. 109) are completely different from those met with in most substitution reactions, and these reactions cannot be admitted as evidence for such a mechanism of C-substitution.

The question of pyrrole cations was mentioned above, and the view expressed that an active role for them in electrophilic substitutions was improbable. Other workers have been impressed by the catalytic influence of acids upon many pyrrole substitutions, and have tried to represent that influence as arising from some form of interaction between the acid and the pyrrole. Treibs[83, 329] attempted to express acid or anionic catalysis in the symbols

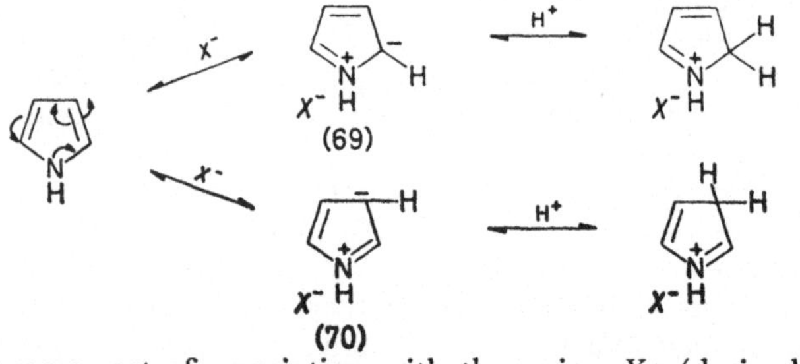

Pyrrole, by some sort of association with the anion X^- (derived from the catalysing acid) was supposed to be activated towards electrophilic attack. The final expressions above represent the particular case of salt formation, the site of protonation depending on the substituents present in the ring. The reactive entities in electrophilic substitution are (69) and (70). In the case of, say, diazo coupling, reactivity in acid solution is attributed to these 'Pyrrolensalz-Ionen'. In neutral media, substitution is referred to the classical polarized form of pyrrole, and in alkaline media the enhanced reactivity is due to the presence of pyrrole anions.

Physically this account of the acid catalysis is obscure. The clue to the true nature of the acid catalysis is contained in the mechanism of pyrrole trimer formation, discussed above: the catalyst functions by activating the entity which is to attack the pyrole nucleus. Thus, the acid catalysis of the reactions of pyrroles with carbonyl compounds is caused by protonation of the carbonyl compound.

(5.) THE PROPERTIES OF FUNCTIONAL GROUPS

Several types of functional groups, when attached to the pyrrole nucleus, show distinctive characteristics which call for comment. These features are usually understandable as consequences of the 'electron-rich' nature of the pyrrole nucleus. Frequently parallels can be drawn with related derivatives of phenols.

(a) Acyl, Carboxyl and Alkoxycarbonyl Groups (Table 4.7)

Whilst formylpyrroles behave in many ways as would be expected, some anomalous properties have attracted attention. Thus, 2-formylpyrrole gives an oxime, arylhydrazones and other typical carbonyl derivatives[117b], and condenses with methyl ketones, hippuric acid and other reactive methylene compounds[165, 168, 169]. It does not give the usual aldehyde reactions with Schiff's, Fehling's or Tollen's reagent[117b], nor does it form a cyanhydrin or take part in the Cannizzaro or Perkin reactions[117b].

The physical properties of the aldehyde are also unusual. It is readily soluble in polar solvents and gives strong evidence of associative dimerization in solution. N-Substituted pyrroles are not associated[170]. However, the dimerization of 2-formylpyrrole can have little relevance to the chemical properties, which are very similar to those of β-aldehydes and N-substituted 2-formylpyrroles[171]; nor is there any support in these facts, or from the

Table 4.7. Some Pyrrole Aldehydes, Ketones and Carboxylic Acids*

Pyrrole	M.p.,°C	Derivatives
2-HCO	50–1	oxime, m.p. 164·5°
2-HCO-1-Me	(b.p. 75–6/12 mm)	semicarbazone, m.p. 207–8°
2-HCO-3-Me	92	phenylhydrazone, m.p. 124°
2-HCO-5-Me	70	oxime, m.p. 153°
2-HCO-3,5-Me₂[337]	90	
2-HCO-4,5-Me₂	126	
2-HCO-1,3,5-Me₃[337]	(b.p. 170–180)	oxime, m.p. 145°
2-HCO-3-Et-4,5-Me₂	85	
2-HCO-4-Et-3,5-Me₂	105–6	
3-HCO-2,5-Me₂	144–144·5	
3-HCO-1,2,5-Me₃	96·5–97	
3-HCO-5-Et-2,4-Me₂	128	oxime, m.p. 118–5°
2,4-(HCO)₂-3,5-Me₂[337]	165–6	
3,4-(HCO)₂-2,5-Me₂[337]	207	oxime, m.p. 268–9°
2,4-(HCO)₂-1,3,5-Me₃[337]	152–53	oxime, m.p. 185–7°
3,4-(HCO)₂-1,2,5-Me₃[337]	160	
2-MeCO	90	oxime, m.p. 145–6°
3-MeCO[339]	115–16	
2-MeCO-5-Me	89	oxime, m.p. 150°
3-MeCO-2,4-Me₂	137	
3-MeCO-2,5-Me₂	89	semicarbazone, m.p. 234° (dec.)
2-EtCO[335]	52	oxime, m.p. 125–6°
3-EtCO[335]	117	semicarbazone, m.p. 180–1°
2-ClCH₂CO[33c]	118–9	
2,5-(MeCO)₂	161–2	
2-PhCO	79	oxime, m.p. 147°
2-MeCO-4-NO₂[45]	195–6	
2-MeCO-5-NO₂[45]	154·5–155	
2-MeCO-1-Me-4-NO₂[45]	141–141·5	
2-MeCO-1-Me-5-NO₂[45]	95·5–96	
2-EtCO-4-NO₂[274]	136–7	semicarbazone, m.p. 229–230°
2-EtCO-5-NO₂[274]	134–5	semicarbazone, m.p. 211–12°
1-CO₂H	95 (dec.)	Et ester, b.p. 180°
2-CO₂H	206 (dec.)	amide, m.p. 174°; Me ester, m.p. 73°; Et ester, m.p. 39°
3-CO₂H	148	Me ester, m.p. 88° Et ester, m.p. 48–49°
2-CO₂H-3-Me†	200 (dec.)	
3-CO₂H-2-Me	168 (dec).	Et ester, m.p. 78–79°
2-CO₂H-3,5-Me₂	136	Et ester, m.p. 125°
3-CO₂H-2,4-Me₂	183 (dec.)	Et ester, m.p. 75–6°
3-CO₂H-2,5-Me₂	213	Et ester, m.p. 117°
2,4-(CO₂H)₂-3,5-Me₂		diethyl ester, m.p. 136°
3-Me-4-p‡	119	opsopyrrolecarboxylic acid**
2,3-Me₂-4-p‡	130–31	haemopyrrolecarboxylic acid**
3,5-Me₂-4-p‡	140	cryptopyrrolecarboxylic acid**
2,3,5-Me₃-4-p‡	86–8	phyllopyrrolecarboxylic acid**

* Most of the data are from [117a]: see also *Tables 4.11* and *4.12*.
† Ref. 340 contains data on almost all of the C-methylpyrrole mono-, di- and tri-carboxylic acids.
‡ p = (CH₂)₂.CO₂H.
** Trivial name.

lack of influence of pH upon the spectrum of 2-formylpyrrole, for the formulation[172] of the aldehyde as a hydroxymethylene structure (71). It is probably best to regard the aldehydes as resonance hybrids in which the

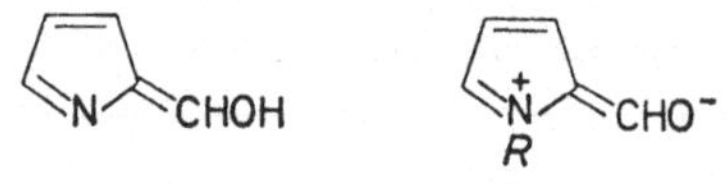

(71)   (72)

form (72) plays an important part. The infra-red spectra and basicities of the aldehydes have already been mentioned (pp. 58 and 61).

With dilute sulphuric acid pyrrole, aldehydes are converted into pyrrole carboxylic acids[330].

Pyrrole aldehydes and ketones are important sources of alkylpyrroles, formed from them by Wolff–Kishner reduction[100, 232-3, 331]. More recently, lithium aluminium hydride has been used for this purpose[32b, 233, 332-5]. Direct addition of the carbonyl compound to the reagent gives the alkyl-pyrrole, whilst inverse addition allows preparation of the carbinol. Carboxylic acids and their esters are also reduced by lithium aluminium hydride to alkylpyrroles. In contrast to pyrrole aldehydes unsubstituted at nitrogen, 2- and 3-formyl-1-methylpyrroles are reduced by lithium aluminium hydride only to the carbinol stage[233]. The difference is explained by the mechanism

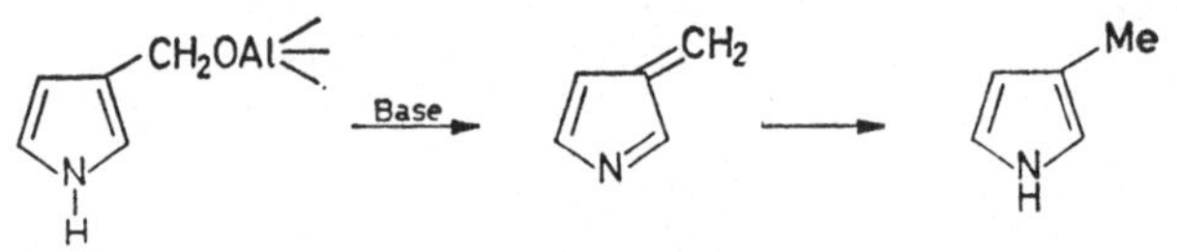

Beckmann rearrangements have been successful with some pyrrole ketoximes[336].

The most striking property of pyrrole-carboxylic acids having the carboxyl group directly attached to the nucleus is their ready decarboxylation. This occurs when the acids are heated under a variety of conditions; the ease varies with the character of other substituents present. Melting is usually accompanied by decarboxylation, and preparative procedures have used decarboxylation by heating at reduced pressure[350], by heating in glycerol, 2-aminoethanol and alkali, and distillation from weakly acid solution[173, 117c]. No quantitative data are available to permit an assessment of the effect of substituents or carboxyl group orientation upon ease of decarboxylation. It seems clear, however, that in the pyrrole series the behaviour observed is similar to that of hydroxybenzoic acids having one or more hydroxyl groups *ortho* and *para* to the carboxyl group[174]. In these cases of decarboxylation

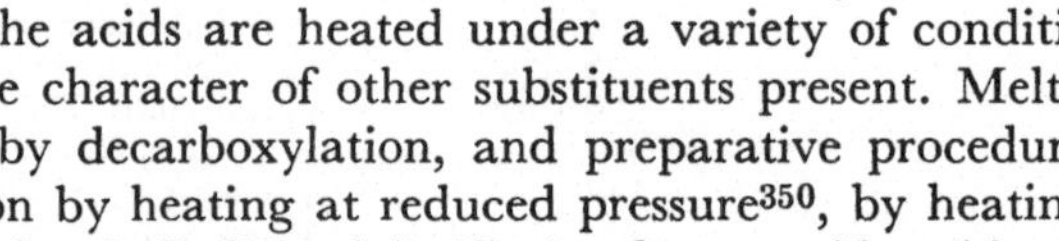

facilitated by electron-releasing groups, it is likely that the mechanism involved is that denoted[175] S_E2, but whether the acid or its anion is involved is not known. Rough qualitative comparisons suggest that pyrrole-2- and -3-carboxylic acids are decarboxylated about as readily as the resorcylic acids[51, 176].

Some pyrrole-2-carboxylic acids, treated with acetic anhydride, give[177] pyrocolls (73), whilst suitable 3-carboxylic acids form[178] quinones (74).

Formation of (73) recalls other N-substitutions brought about by carboxylic acid derivatives (p. 81). The common occurrence of N-substitution under alkaline conditions is exemplified by the production of (75) during the alkaline hydrolysis of the corresponding dipyrrylmethane ester[179].

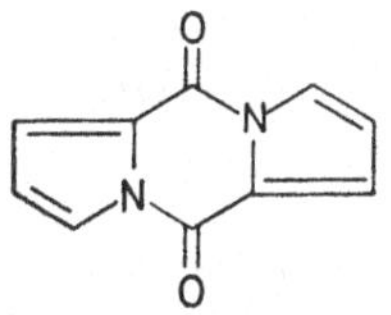

(73) 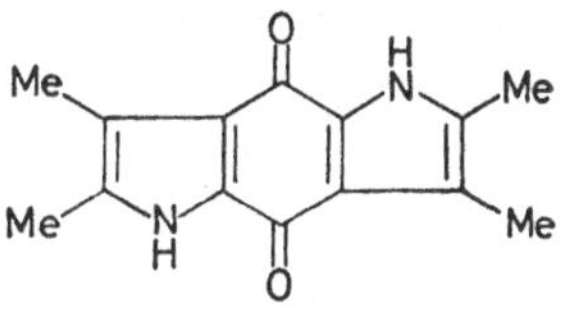　　　　　　　　　　　　　　　　　(74)

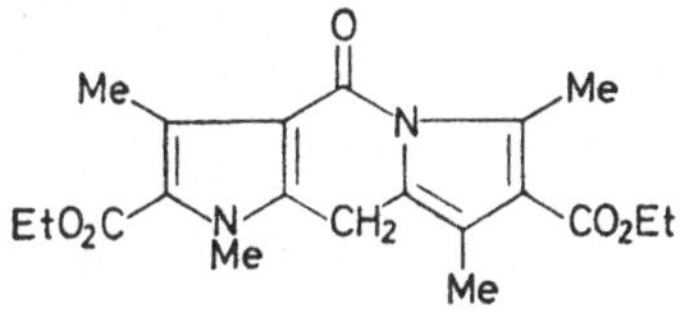

(75)

It is fairly generally established[145d, 178, 180] that diesters of pyrrole-dicarboxylic acids are hydrolysed by alkali to monoesters with a free α-carboxyl group. In such diesters the β-alkoxycarbonyl group is hydrolysed only by alkali in excess of that required for the α-group. Thus, Knorr's pyrrole (76) is hydrolysed by alkali first to 3-ethoxycarbonyl-2,4-dimethyl-pyrrole-5-carboxylic acid. The formation of a number of 3-alkoxycarbonyl-pyrrole-2-carboxylic acids during ring synthesis in alkaline solution which might have been expected to give 2,3-dialkoxycarbonylpyrroles may be examples of the same selectivity. It would be interesting to know the rates of alkaline hydrolysis of 2- and 3-alkoxycarbonylpyrroles.

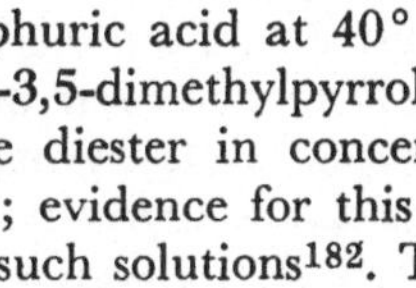

(76) 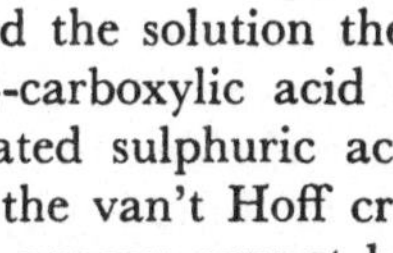　　　　　　　　　　　(77)

Fischer and Walach[181] showed that if Knorr's pyrrole is dissolved in concentrated sulphuric acid at 40° and the solution then poured on ice, 2-ethoxycarbonyl-3,5-dimethylpyrrole-4-carboxylic acid results. Probably, dissolution of the diester in concentrated sulphuric acid generates the acylium ion (77); evidence for this is the van't Hoff cryoscopic factor of 3·5 observed for such solutions[182]. The process cannot be complete at the freezing point, for complete 'acylization' would give[183] a factor of 5. 'Acylization'[184], $A_{Ac}1$, will be facilitated by electron-releasing substituents and will not be subject to steric factors. In the pyrrole series, a balance has to be struck between the facilitation caused by increasing alkyl substitution in the nucleus and the retardation which would accompany salt formation by the nucleus due to its increased basicity. The balance must be complex, as is shown by the results in *Table 4.8*. It is not always a β-alkoxycarbonyl group which is preferentially attacked, nor is steric hindrance necessary for 'acylization' to show itself.

Table 4.8. 'Acylization' of Alkoxycarbonylpyrroles[182]

Pyrrole substituents				Conditions	Product	van't Hoff factor
2	3	4	5	H_2SO_4 for ( ) min at °C		
CO_2H	CO_2Et	Me	CO_2Et	conc. (40) 95–100 / fuming (40) 75	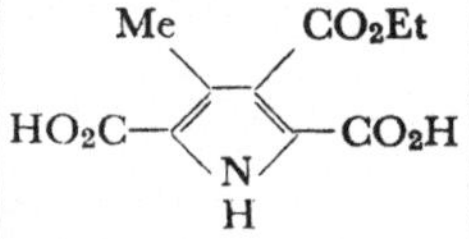 Me CO_2Et / HO_2C— —CO_2H (N, H)	2·5
CO_2H	CO_2Me			fuming (20) 50	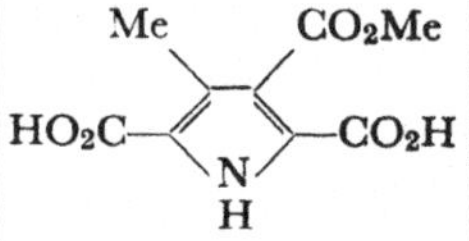 Me CO_2Me / HO_2C— —CO_2H (N, H)	
CO_2Me	CO_2Et	Me	CO_2Et	fuming (20) 50	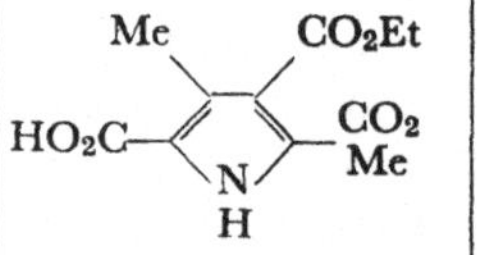 Me CO_2Et / HO_2C— —CO_2Me (N, H)	
H	CO_2Me			conc. (30–45)	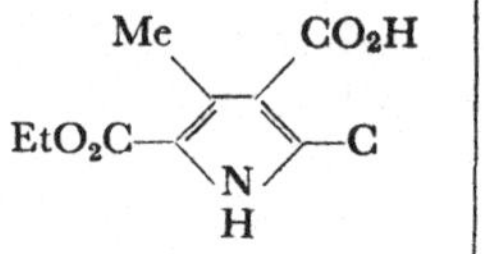 Me CO_2H / EtO_2C— —C (N, H)	3·2
Me	Me	CO_2Et		(not 'acylized' at room temp.) (recovered unchanged)		
Me	CO_2Et	Me	Me			
Me	CO_2Et	Me	CO_2Et	conc. 40	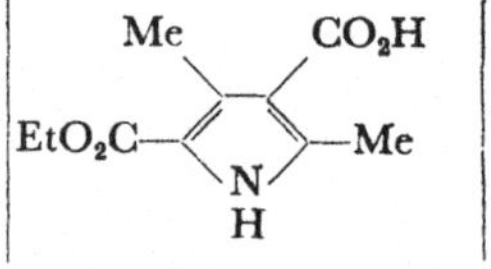 Me CO_2H / EtO_2C— —Me (N, H)	3·5

(b) Alkyl Groups

Some properties of alkylpyrroles are listed in *Table 4.1* (p. 52).

It has already been mentioned (p. 87) that in some oxidations pyrroles are converted into maleimides. In the process, α-alkyl groups are eliminated, and this fact can be used in identifying polyalkylpyrroles.

That halogenation can occur in alkyl side chains as well as in the nucleus[176, 341] has also been mentioned (p. 78). In both chlorination with sulphuryl chloride and bromination, it is α-methyl groups which undergo substitution. Sulphuryl chloride can be used to convert α-methyl groups into di- or tri-chloromethyl groups, and thence into formyl or carboxyl groups[117b], or a combination of bromination and chlorination can be used[350]. The bromination of an α-methylpyrrole with a free α'-position is an important reaction (p. 102).

It seems likely that the chlorination of α-methyl groups by means of sulphuryl chloride is a free radical reaction. Accordingly, it is interesting to note that in free-radical phenylations, methyl groups attached to the pyrrole nucleus show the reactivity order $2 > 3 > 1$[342]. These methyl groups are more reactive than that in toluene, and even more so than those in the picolines.

Lead tetra-acetate converts α-methyl groups into acetoxymethyl groups[343].

The conversion of a 2-alkylpyrrole $R.C_4H_4N$ into the carboxylic acid $R.CO_2H$ by permanganate oxidation has been put to interesting use[344] (p. 107).

(c) Amino, Azo and Diazo Groups (Table 4.9)

Aminopyrroles are usually so unstable in air that a study of their properties, and even their isolation in the free state, is difficult. They are sufficiently

Table 4.9. Amino-, Azo- and Diazo-pyrroles

Substituents	M.p., °C	Derivatives
1-NH_2-2,5-Me_2[347]	52–3	benzoyl, m.p. 184–184·5°
1-NH_2-3-Et-2-4-Me_2[154]	—	picrate, m.p. 208°
2-$NHCO_2$Et[187]	55–6	
2-NH_2-3,5-Me_2[189]	—	acetyl, m.p. 188° (dec.)
2-NH_2-4-Et-3,5-Me_2[186]	—	picrate, m.p. 201°
2-NH_2-3,4-Et_2-5-Me[186]	—	picrate, m.p. 182°
3-NH_2-2,4-Me_2[126]	127	acetyl, m.p. 205°
3-NH_2-2,4,5-Me_3[114c]	186 (dec.)	picrate, m.p. 210–230° (dec.)
2-NH_2-3,5-Ph_2[14b]	155–6	acetyl, m.p. 171–2°
3-NH_2-2,4-Ph_2[200]	178–9	benzoyl, m.p. 218–9°
3-NH_2-2,5-Ph_2[194b, 260]	187–8	
3-NH_2-1,2,5-Ph_3[276]	222	acetyl, m.p. 93°
3-NH_2-2,4,5-Ph_3[194a, 277]	183	acetyl, m.p. 246°[336]
2-NH_2-1-Et-4-CN[192]	34–5	acetyl, m.p. 154°
4-Ac-2-NH_2-3,5-Me_2[114c]	223 (dec.)	picrate, m.p. 175–190° (dec.)
3-NH_2-5-CO_2Et-2,4-Me_2[185]	125·5	picrate, m.p. 191°
3-NH_2-4-CO_2Et-2,4-Me_2[114c]	—	picrate, m.p. 185–195°
2,5-$(NH_2)_2$-3-Et-4-Me[190]	—	picrate, m.p. 195–6°
2-PhN_2[83a]	79	hydrochloride, m.p. 134°
2-α-$C_{10}H_7N_2$[113]	103	
2-β-$C_{10}H_7N_2$[113]	101	
2-PhN_2-1-Me[114d]	(b.p. 140°/21 mm)	picrate, m.p. 150° (dec.)[198]
2-p-$C_7H_7N_2$-1-Me[114f]	99·5–100·5	trinitrobenzene, m.p. 99–100°
2-PhN_2-4-Et-3,5-Me_2[349]	—	hydrochloride decomposes above 120°
2-PhN_2-3,5-Me_2[114d]	118–9	
3-PhN_2-2,5-Me_2[114d]	135	
2-PhN_2-5-Ph[113]	117	
2,5-$(PhN_2)_2$[113]	131	
2,5-$(β$-$C_{10}H_7N_2)_2$[113]	228	
2-N_2-3,5-Ph_2[280]	104 (dec.)	
2-N_2-4-NO_2-3,5-$(Ph)_2$[280]	116–7	
2-N_2-4-Ac-5-Me-3-Ph[280]	102–3	
3-N_2-2,5-Ph_2[194b, 277]	122–3 (dec.)	hydrochloride decomposes[260] above 173°
3-N_2-2,4,5-Ph_2[194b]	158–9 (dec.)	hydrochloride decomposes above 160°
3-N_2-4-NO_2-2,5-Ph_2[277]	144 (dec.)	
3-N_2-5-CO_2Et-2,4-Me_2[277]	79 (dec.)	

basic to form stable, though readily hydrolysed, salts. Neither 2- nor 3-aminopyrrole is known; the properties of the urethane (78) show that these amines would be very unstable. Among substituted derivatives, those containing electron-attracting groups are the most stable. Particularly noteworthy in this respect are the di- and tri-phenyl compounds. The small amount of evidence available suggests that 3-aminopyrroles are more stable than the 2-isomers.

The instability of the 2-amines is indicated by the failure of attempts to prepare them by 'chemical' reduction of azopyrroles[113, 114, 185]. On the other hand, catalytic reduction is sometimes[114c], but not always[186], successful. Attempts to prepare aminopyrroles by the Curtius reaction give interesting indications of the reactivity of the amines and their derivatives. The reaction was used to prepare the highly unstable urethane (78), but attempted

(78)   (79)   (80)

hydrolysis of the latter caused decomposition[187]. Other ethyl urethanes were equally unpromising[188], but the benzyl urethanes (79, $R = $ Ac or CH_2OH) gave the related amines on hydrogenolysis[188]. However, the urethane (79, $R = $ Et) was rapidly autoxidized, and the product gave, on hydrogenolysis, an autoxidation product of the corresponding amine[154]. The azides (80, $R = $ Me or Et) were converted by hot 50 per cent acetic acid into the very unstable amines[186]. In the same reaction, 2,4-dimethyl-5-pyrrolecarboxylic acid azide gave 2-acetamido-3,5-dimethylpyrrole[189], and 2,5-diamino-3-ethyl-4-methylpyrrole was produced in the same way[190]. 2-Nitroso-3,5-diphenylpyrrole has been catalytically reduced to the relatively stable amine[191].

Observations of the reactions of the few known 2-aminopyrroles are sparse. 2-Amino-4-ethyl-3,5-dimethyl- and 2-amino-3,4-diethyl-5-methylpyrrole could not be acetylated, benzoylated or diazotized. It is not clear whether these failures were due to inability to react or to instability[186]. Acetyl and benzoyl derivatives were prepared from 2-amino-1-ethyl- and 2-amino-1-(β-diethoxyethyl)-4-cyanopyrrole[192, 193]. The second of these amines consumed one equivalent of nitrous acid, but it is not clear if a diazonium salt could be detected or not. These reactions of 2-amino-1-(β-diethoxyethyl)-4-cyanopyrrole, its failure to give ammonia with alkali, and the ultra-violet and infra-red spectra of it and its acetyl derivative have been taken to indicate that the compound is a true primary amine and that imine tautomerism is not important[193]. This may be so, but the evidence provided, particularly that of infra-red spectra in Nujol mulls, is quite inconclusive.

The impression that 3-aminopyrroles are more stable than the 2-isomers is derived largely from the behaviour of the di- and tri-phenyl compounds. The purely alkyl-substituted compounds are very unstable. The amines (81, $R = $ Me or CO_2Et), or derivatives, can be isolated from the catalytic reduction of azo compounds in alkaline media[114c]. More frequently, substituted 3-aminopyrroles have been formed by reducing nitroso- or nitro-pyrroles by means of dissolving metals or by catalytic hydrogenation[121a, 126, 185, 194-7, 260, 276, 280, 345-6]. 3-Acetylamino-arylpyrroles have been obtained by Beckmann rearrangement of ketoximes[336].

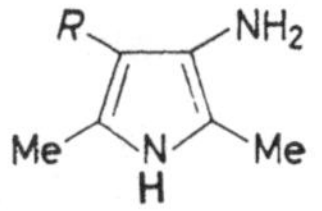

(81) 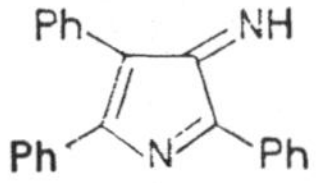   (82) 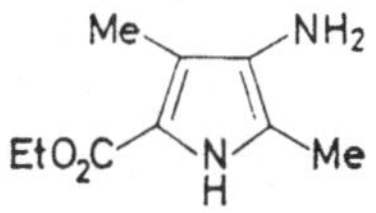   (83)

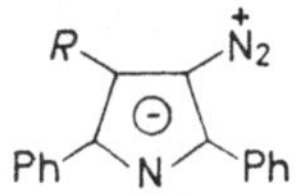

(84) 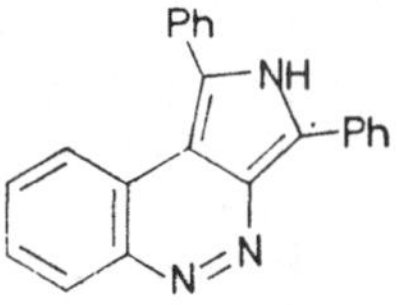   (85)

3-Aminopyrroles can usually be acetylated[126] and benzoylated[194a, 199, 200] (Schotten–Baumann conditions have been used), and have given benzal derivatives[197, 200, 201] and ureas[194a]. In some cases these reactions fail: the amines (81, $R =$ Me or CO$_2$Et) could not be acetylated, benzoylated, methylated or made to react with phenylisocyanate, facts which led to the suggestion that the amines are capable of tautomerism[114c]. Surprisingly, with dimethyl sulphate and alkali, ethyl 3-amino-2,4-dimethylpyrrole-5-carboxylate gave a trimethyl derivative[185]. Several 3-aminopyrroles can be diazotized (see below), but this reaction fails with 3-amino-2,4-dimethyl-pyrrole[126]. 3-Amino-2,4,5-triphenylpyrrole is not so sensitive to oxidation as to be destroyed, even by quite vigorous reagents[195, 196, 202]; perhydrol in acetic acid converts it into the azoxy-compound, and ferricyanide is thought to produce (82).

A few 1-aminopyrroles have been described[154, 186, 203, 347], but little is known of the effect upon the properties of the primary amino group of linking it to nitrogen. Reaction with nitrous acid replaces the amino group by hydrogen[203, 204].

Azo-compounds are the most numerous class of pyrrole nitrogen derivatives. The α-compounds have been stated[114b] to be less stable to air and light than the β-compounds, but this is certainly not generally true[83]. Azo-compounds seem to be more basic than the parent pyrroles, and in some cases to be markedly acidic[114]; neither of these facts is difficult to understand if the structures of the cations (p. 76) and anions are considered. The ability of many α-compounds to couple a second time, as with diazotized p-nitroaniline, is the basis of a colour test for distinguishing them from β-compounds. Azopyrroles give picrates, styphnates, picrolonates and trinitrobenzene complexes, but the usefulness of some of these is limited by their lacking true melting points[114f]. The reduction of azo compounds was mentioned above.

The formation from an aminopyrrole, by diazotization, of a diazonium salt with coupling ability has been observed only with the amine (83)[126, 185]. The crystalline diazonium chloride from (83) is very stable, but in fact the coupling reactions mentioned occur only in sodium bicarbonate solution, and even then to only a small degree[277]. Under more alkaline conditions the diazopyrrole is precipitated, a contingency which prevents coupling under the usual conditions with pyrrole diazonium salts in general. Boiling aqueous solutions of the diazonium salt from (83) decompose slowly, giving ethyl 2,4-dimethylpyrrole-5-carboxylate[126, 185].

Angelico[194b] prepared the first 'diazopyrroles' (84, R = H or Ph) by diazotizing the corresponding amines. Other examples have been reported[277]; as well as by diazotization, they were obtained by direct introduction of the diazo group into a pyrrole (p. 80) and by treating nitrosopyrroles with nitrous acid[277]. 2-Diazopyrroles have only been prepared by the direct introduction of the diazo group[280]. The 3-diazopyrroles are remarkably stable. With mineral acids they give stable diazonium salts[194b, 260, 277, 348]. As mentioned already, coupling with these diazonium salts is difficult to effect, because under the commonly used conditions the markedly acidic diazonium cation is converted into the diazopyrrole[260, 277]. However, coupling can be achieved by treating the diazonium salt with the second component in a neutral solvent such as chloroform[260], by similar use of the diazopyrrole[277] or by fusing the latter with β-naphthol[277].

3-Diazo-2,4,5-triphenylpyrrole resists attack by iodine, hydrosulphite, alcohol and organic acids, is nitrated by nitric acid and converted by prolonged boiling with sulphuric acid, through the diazonium salt, into (85). It can be reduced to the amine[194b, 205-7].

Less is known about the 2-diazopyrroles, but they are less stable and more reactive than the 3-isomers[280].

The analogy between the formation of the diazopyrroles and the conversion of aminophenols into diazo-oxides[208] is obvious. The diazopyrroles are structurally analogous to diazocyclopentadiene[277]. 3-Diazopyrroles show a strong infra-red band at 2,080–2,150 cm^{-1}, and ultra-violet absorption between 320 and 400 mμ[277]. The 2-diazopyrroles absorb at slightly shorter wavelengths[280].

(d) Halogenopyrroles (Table 4.10)

Halogen derivatives of pyrrole, other than the tetrahalogeno compounds, are notable only for their instability. 2-Chloropyrrole is a dense oil which decomposes explosively, and it is doubtful if 2-bromopyrrole has been obtained[132]. The di- and tri-chloropyrroles are somewhat less unstable; 2,3,5-trichloropyrrole has been N-methylated[116b]. Tetra-iodopyrrole is the best known of these compounds. It results from the reaction between tetrachloro- or tetrabromopyrrole and alcoholic potassium iodide[115b]. Whilst tetraiodopyrrole is easily reduced to pyrrole by zinc and alkali, the tetrachloro compound is more resistant and the reduction products always contain chlorine[209]. Tetraiodopyrrole (Iodol) has antiseptic properties[210]. The polyhalogeno compounds are markedly acidic.

Halogen atoms in general have been replaced by hydrogen in pyrroles by zinc and alkali[119] or catalytically[267, 352-3]. The same replacement, brought about by boiling halogen hydracids, is discussed below.

2-Iodopyrroles containing one or more ester groups readily give 2,2'-bipyrroles in the Ullmann reaction[268, 351]. In some cases reaction occurs at room temperature in dimethylformamide[350].

Halogen atoms in pyrroles are unreactive towards alkali and sodium methoxide, even when electron-attracting substituents are present[119, 120b, 211].

Reactions have already been mentioned in which during electrophilic substitution a halogen atom is replaced from the pyrrole nucleus by some

Table 4.10. Halogenopyrroles

Substituents	M.p., °C
2-Cl[132]	dense unstable oil; does not form a picrate
2,5-Cl$_2$[116b]	unstable liquid
3,4-Cl$_2$[358]	74
2,3,5-Cl$_2$[116b]	dense oil
2,3,4,5-Cl$_4$[116a]	110(d)
2,3,4,5-Cl$_4$-1-Me[116b]	118–119
2,3,4,5-Br$_4$[115b]	darkens and decomposes above 120°
2,3,4,5-I$_4$[115c]	decomposes between 140° and 150°

other group; examples are found in those with formaldehyde (p. 73), in diazo coupling (p. 77), in nitration (p. 80), and in the replacement of halogen by hydrogen (p. 75). The last reaction is of considerable interest; its importance was first realized[265] when 2-bromo-4-ethoxycarbonyl-3,5-dimethylpyrrole in acetic acid was found to oxidize iodide to iodine, being converted into 3-ethoxycarbonyl-2,4-dimethylpyrrole very rapidly:

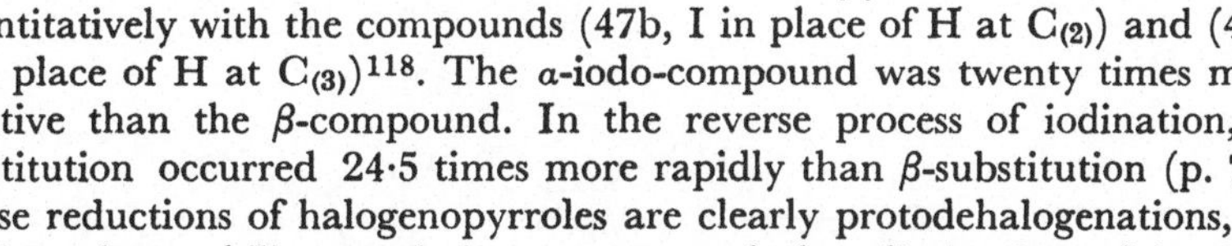

2-Iodopyrroles oxidize hydriodic acid to iodine, giving the iodine-free pyrrole. This reaction is the reverse of the iodination of pyrroles and was studied quantitatively with the compounds (47b, I in place of H at C$_{(2)}$) and (48b, I in place of H at C$_{(3)}$)[118]. The α-iodo-compound was twenty times more reactive than the β-compound. In the reverse process of iodination, α-substitution occurred 24·5 times more rapidly than β-substitution (p. 79). These reductions of halogenopyrroles are clearly protodehalogenations, involving electrophilic attack by protons and the elimination of 'positive halogen'[118, 265]. It is possible quantitatively to titrate some iodopyrroles by treating them with potassium iodide and sulphuric acid and estimating liberated iodine with thiosulphate[119].

Protodehalogenation with the initial formation of 'positive halogen' is presumably involved in the conversion of 3-ethoxycarbonyl-5-iodo-2,4-dimethylpyrrole into the corresponding bromo compound by means of hydrobromic acid, and of the bromo- into the chloro-compound by means of hydrochloric acid[26, 119]. The halogen displaced appears as free halogen; presumably the initially liberated 'positive halogen' (effectively I^+ or Br^+) oxidizes a halogen anion (Br^- or Cl^-), producing an electrophilic substituting reagent which converts the pyrrole formed by protodehalogenation into a bromo- or chloro-pyrrole. It is difficult to discern the character of the reaction mentioned above, in which tetraiodopyrrole is formed from tetrachloro- or tetrabromo-pyrrole and alcoholic potassium iodide.

Fischer and his co-workers observed that the action of bromine or sulphuryl chloride upon trisubstituted pyrroles having one free α-position and an α'-methyl group gave brominated (or chlorinated) pyrromethenes[117d]. The overall reaction is of the form

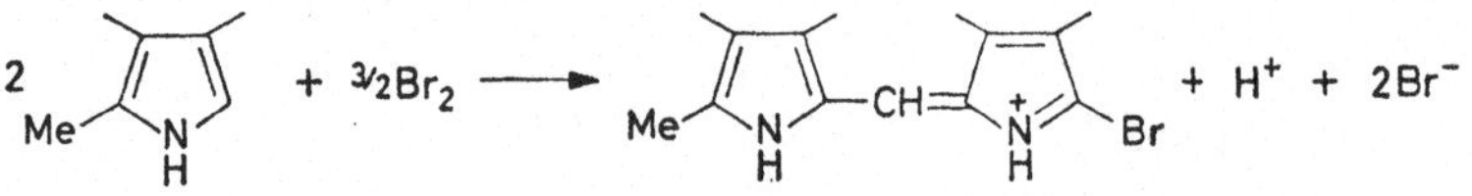

$$2 \quad \underset{Me}{\text{(pyrrole)}} \ + \ \tfrac{3}{2}Br_2 \ \longrightarrow \ \text{(brominated pyrromethene)} \ + \ H^+ \ + \ 2Br^-$$

one molecule of bromine being used for substitution and one atom for oxidation[265]. Fischer believed these reactions to proceed as in the example shown.

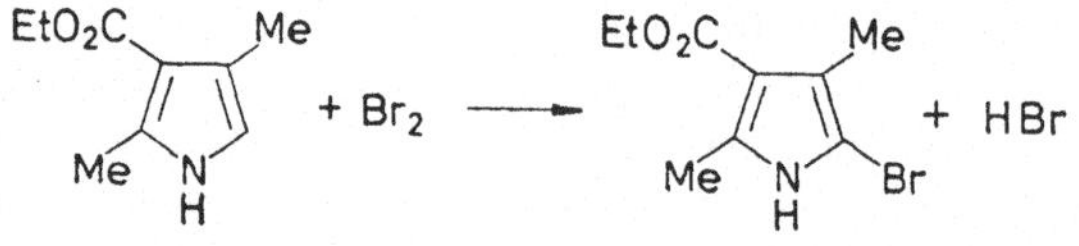

$$(86) \ + \ Br_2 \ \longrightarrow \ (87) \ + \ HBr$$

$$(87) \ + \ Br_2 \ \longrightarrow \ (88) \ + \ HBr$$

$$(86) \ + \ (88) \ \longrightarrow \ (89) \ + \ HBr$$

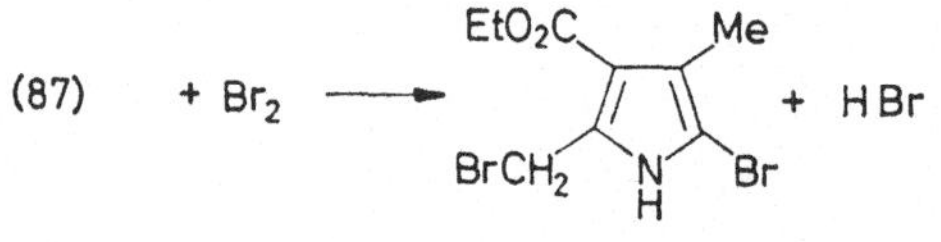

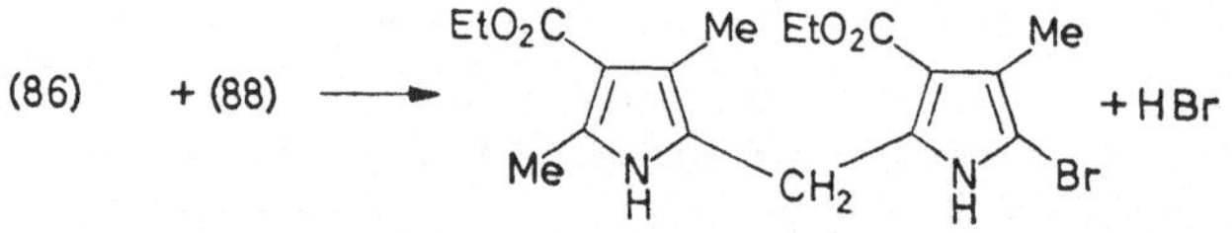

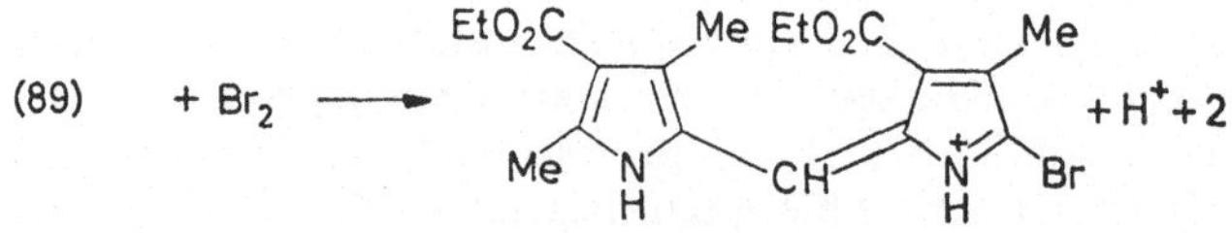

$$(89) \ + \ Br_2 \ \longrightarrow \ (90) \ + \ H^+ + 2Br^-$$

This scheme has been criticized on the grounds that (87) would not react faster with bromine than does (86) [by stopping the reaction after the addition of a molecular equivalent of bromine, (87) can be isolated in good yield] and that, by starting from (87), (90) can be obtained in better yield than by starting with (86)[265]. Accordingly, the following sequence is preferred:

$$(86) + Br_2 \rightarrow (87) + HBr$$
$$(87) + Br_2 \rightarrow (88) + HBr$$
$$(87) + (88) \rightarrow (90)$$

Analogies for the last reaction were realized experimentally, e.g.

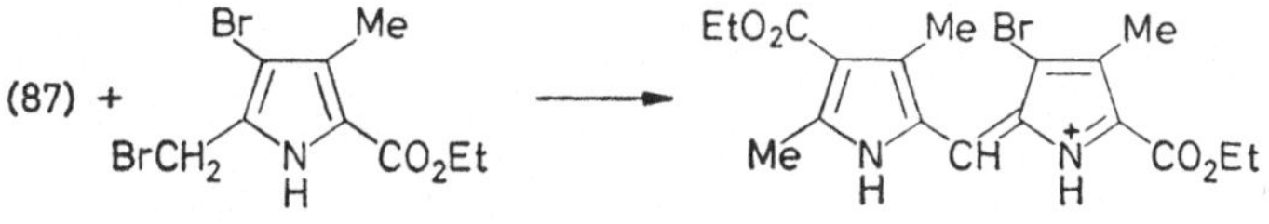

(91)

The reaction is not completely general, for the bromomethyl compound (91, $\cdot C_2Et$ in place of β-bromine) gave no pyrromethene with (87).

Besides the reaction of (87) with bromine to give (88), a much slower acid-catalysed reaction was recognized which also led to pyrromethene formation. Changes in the substituents present in the starting pyrrole might alter the relative importances of these two reactions. The first step in the acid-catalysed process is that of protodehalogenation mentioned above, and the preferred overall sequence is

$$(87) + HBr \rightarrow (86) + (88)$$
$$(87) + (88) \rightarrow (90)$$
$$(86) + (90) \rightarrow (92)$$
$$(92) + HBr \rightarrow (93) + (94)$$

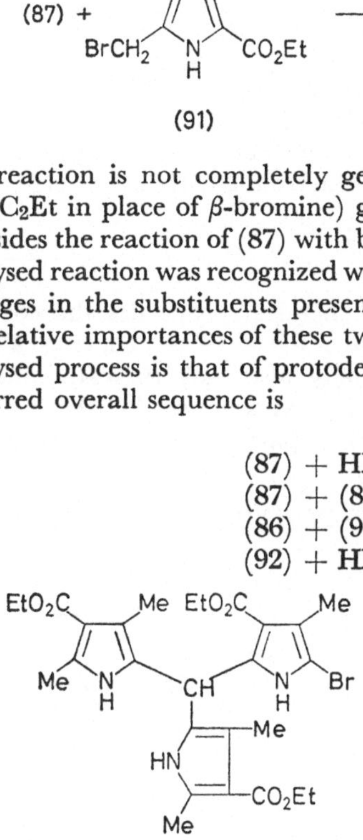

(92)

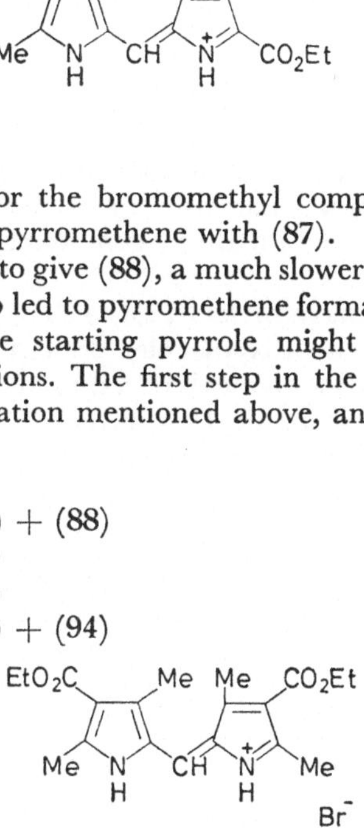

(93)

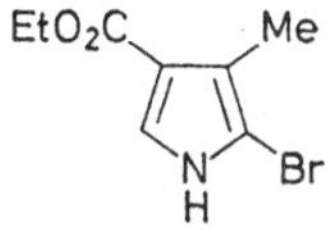

(94)

Experimental analogies for each of these steps are available[265], and the work makes it clear that in any particular example the structure of the pyrromethene obtained requires careful scrutiny.

An interesting reaction occurs with compounds of the type (95, $R = Me$ or Et), but not with their esters[212]. Boiled with alcoholic solutions of halogen hydracids, they undergo decarboxylation and replacement of the bromine atom by an alkoxyl group. Aqueous halogen hydracids give 2-hydroxy-3,4-disubstituted pyrroles. The linking of the decarboxylation and halogen replacement suggests that initial protonation is involved:

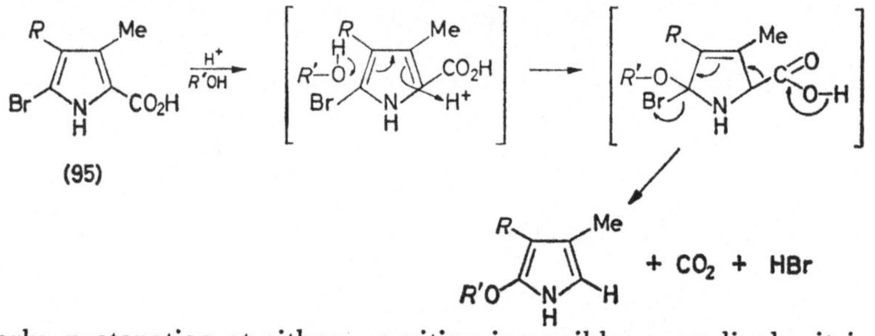

(95)

Clearly, protonation at either α-position is possible; accordingly, it is not surprising that 5-bromo-4-ethyl-3-methylpyrrole-2-carboxylic acid gives a mixture of hydroxy compounds[359].

(e) *Hydroxypyrroles (Table 4.11)*

The hydroxypyrroles present in their chemical properties the confused picture so typical of tautomeric substances[154-6, 212-3]. Lacking alkoxycarbonyl or acyl substituents, they usually also lack typically phenolic properties. They do not, of course, behave as ketones, for the carbonyl function in the α- or β-oxopyrroline form is part of an amidic structure.

Table 4.11. *Hydroxypyrroles*

Substituents	M.p., °C	Derivatives
2-HO(monohydrate)[213j]	83	
2-HO-1,5-Me₂[213h]	62–3	
2-HO-3,4-Me₂[213k]	125	N-acetyl, m.p. 75–6°
		Et ether, b.p. 80–81°/14 mm[356]
3-Et-2-HO-4-Me[359]	98–100	Me ether*
4-Et-2-HO-3-Me[359]	43–5	
2-HO-1,4-Ph₂[213b]	169–170	
4-CO₂Et-2-HO-5-Me[213f]	134	
3-Et-2-HO-4,5-Me₂[213g]	95	
4-Et-2-HO-3,5-Me₂[213g]	84	
4-CO₂H-2-HO-3,5-Me₂[213f]	196 (dec.)	
4-CO₂Et-2-HO-3,5-Me₂[213f]	127	
3-Ac-2-HO-1,4-Ph₂[213b]	111–12	
4-CO₂Et-3-HO-5-Me[213c]	215 (dec.)	
5-CO₂H-3-HO[354]	—	Me ether, m.p. 179–180°
5-CO₂Me-3-HO[354]	—	Me ether, m.p. 85–6°
4-CO₂H-3-HO[354]	—	Me ether, m.p. 203–4°
4-CO₂Et-3-HO[354]	—	Me ether, m.p. 107–9°

* This compound[212] may be a mixture; see above.

Alkoxycarbonyl or acyl substituents usually render the hydroxypyrroles alkali-soluble, capable of coupling with diazonium salts, of being nitrosated, O-acylated and, in a few cases, O-methylated by diazomethane. Hydroxypyrroles carrying alkoxycarbonyl or acyl groups adjacent to the hydroxyl group give colour reactions with ferric ions. Some hydroxypyrroles give benzylidene derivatives. Both acetylation and methylation can in some cases proceed either at the oxygen atom or a nuclear carbon atom.

Chemical evidence is never unequivocal in indicating the position of a tautomeric equilibrium. For a number of both α- and β-hydroxypyrroles, physical data are available. Thus, comparison of the ultra-violet spectra of the compounds (96) with those of their necessarily pyrrolic O-acetyl esters (97) proves clearly the predominance of the oxopyrroline forms in ethanol[214]. In contrast, the chemical evidence is confused: the compounds are soluble in alkali, give no ferric reaction, give benzylidene derivatives, do not react with diazomethane and undergo both O- and C-acetylation. The formulation of '2-hydroxy-3,4-dimethylpyrrole' as 3,4-dimethyl-2-oxo-Δ^3-pyrroline (98) is supported by proton resonance data[355]. In contrast, the formyl group appears to stabilize the hydroxy form (99)[355]. O-Ethylation of (98) is conveniently effected by triethyloxonium borofluoride[356].

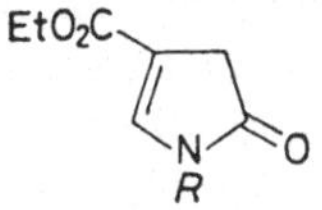

(96) 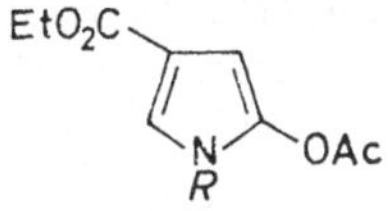   (97) 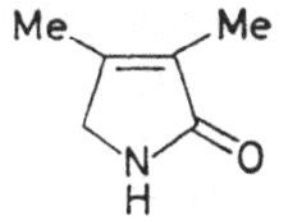   (98)

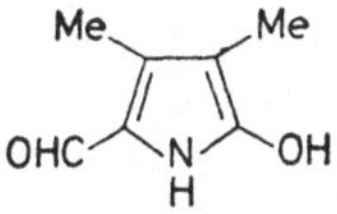

(99) 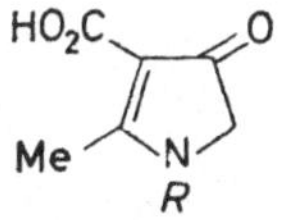   (100) 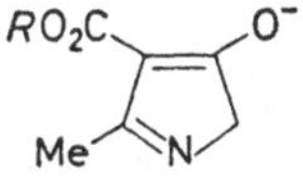   (101)

A more complex picture emerged from the study of a variety of β-hydroxypyrroles[215]. Acids such as (100, R = H or Ph) and their esters show ultra-violet extinction curves very different from those of their O-acetyl, -methyl and -benzyl derivatives. They are therefore taken to possess predominantly oxopyrroline structures in ethanol. However, compounds analogous to (100, R = Ph), but lacking the 2-methyl group, give extinction curves very similar to those of their O-methyl ethers and are therefore predominantly enolic in ethanol. The removal of the 2-methyl group is thought to stabilize the enolic form by reducing the steric hindrance to conjugation between the pyrrole and phenyl nuclei. Infra-red, ultra-violet and proton resonance spectra show esters of (100, R = H) to exist as the 4-oxo-Δ^2-pyrroline form in neutral, acidic and alkaline solution. In the cation, protonation occurs on the 4-oxo-group and the anion has the structure (101)[357].

(f) Metallic Derivatives

Fifty years ago Oddo[18a] showed that when pyrrole was added to methyl magnesium iodide in ether, methane was liberated. Pyrrole was regenerated by treating the solution with water, whilst use of carbon dioxide gave pyrrole-2-carboxylic acid, and of ethyl chloroformate, ethyl pyrrole-2-carboxylate. Addition of pyridine to the ethereal solution gave a complex $C_4H_4MgI.2C_5H_5N$. In the reaction of pyrrole with the Grignard reagent, methane equivalent to one active hydrogen atom is liberated[216].

The pyrrole Grignard reagents are valuable synthetic intermediates. With acyl halides, derivatives of oxalic acid and phosgene they give ketones. Reagents prepared from alkylpyrroles with a free α-position react at that position[47b, 217]. The same has been claimed to be true of the reaction with alkyl chloroformate to give esters[18a, 33b, 218], but pyrryl magnesium bromide gives the 1-, 1,2-di- and 2-substituted products with methyl chloroformate[364]. From 3,4-diethyl-2-methylpyrrole as well as ethyl 3,4-diethyl-2-methyl-pyrrole-5-carboxylate some of the N-substituted isomer was formed[18b]. Carbon dioxide gives the 2-carboxylic acids with the Grignard reagents from pyrrole and 2-allylpyrrole[18a, 217a, 219]. The early report that the passage of carbon dioxide over the pyrrole Grignard reagent heated to 250–270° gave pyrrole-3-carboxylic acid was probably wrong[50].

Early work on the reaction between pyrryl magnesium bromide and alkyl halides[220] has recently been re-examined[253]. Both 2- and 3-alkylpyrroles are formed. Alkylation with tertiary halides was more rapid than with primary halides. When equivalent quantities of alkyl halide and pyrryl magnesium bromide are used, polyalkylation becomes important. 2-5-Dimethylpyrryl magnesium iodide reacts at the β-position with isopropyl iodide and two isomeric di-isopropyl-dimethylpyrrolenines are also formed[220b]. The reactions of pyrryl magnesium bromide with a series of methylating reagents have also been examined[368]. The order of reactivity in the methylating reagents was $MeI < Me_3PO_4 < Me_2SO_4 \cong MeO_3S.C_7H_7$. The results were generally similar to those discussed, but minor amounts of N-methylation and significant amounts of dimethylation were detected.

The alkylation of Grignard reagents from trialkylpyrroles[220c] has also been re-examined[221]. In this reaction, 2,3,5-trimethylpyrryl magnesium iodide gives with ethyl iodide a mixture of the pyrrolenines (102, $R = R'' = $ Me, $R' = $ Et, $R''' = $ H) and (102, $R = R'' = $ Me, $R' = R''' = $ Et). Similarly,

(102)   (103)   (104)

3-ethyl-2,5-dimethylpyrrole gives with methyl iodide (102, $R = R' = $ Me, $R'' = $ Et, $R''' = $ H) and (102, $R = R' = R''' = $ Me, $R'' = $ Et). Thus, in the alkylation of 2,3,5-trialkylpyrryl Grignard reagents, the main result is substitution at the 2- but not the 5-position. Similar methylation of 2,5-dimethyl- or 2,3,4,5-tetramethylpyrrole gave (102, $R = R' = R'' = R''' = $ Me), whilst 2,3,4-trialkylpyrroles give 2,3,4,5-tetra-alkylpyrroles and 2,2,3,4-tetra-alkylpyrrolenines. With methyl iodide, the Grignard reagent from 2,5-dibenzyl-3,4-dimethylpyrrole also reacts exclusively at the α-position[360].

From pyrryl magnesium bromide and allyl bromide, Hess[219a] obtained 2-allyl- and 2,5-diallyl-pyrrole. He supposed the di-substituted product to arise by exchange of the Grignard function between pyrrole and 2-allylpyrrole. Later workers isolated only 2-allylpyrrole[61].

Formic esters convert pyrryl magnesium halides into 2-formylpyrrole. The Grignard reagent from 2,5-dimethylpyrrole likewise gives some 3-formyl-2,5-dimethylpyrrole, but the chief product is the 1-formyl compound.

Esters of other fatty acids have been used to prepare 2-acylpyrroles. Ethyl carbonate reacts with the pyrrole Grignard reagent mainly at the nitrogen atom, but a little ethyl pyrrole-2-carboxylate also results[218c, 222].

With ketones, pyrryl magnesium bromide usually gives *meso*-disubstituted di-2-pyrrylmethanes, though the initially formed carbinol has sometimes been isolated. Aldehydes give pyrromethenes[223].

Pyrryl magnesium iodide, treated in ether with deuterium oxide, gave, after the product had been shaken with water to remove deuterium attached to nitrogen (p. 75), 2- and 3-deuteropyrrole.

It is usually stated that pyrryl magnesium halides react with electrophilic reagents at an α-position (when one is vacant), whilst potassium and sodium pyrrole react at the nitrogen atom. These generalizations are only very roughly true; well established exceptions to both are known (p. 66). The examples of both 1- and 3-substitution encountered with pyrryl magnesium halides are significant for arguments about the structure of the latter. Prepossession with the predominance of α-substitution has led some authors to support structure (103) against (104), whilst supporters of (104) have inclined to view α-substitution as resulting from rearrangement after N-substitution[224]. There is in fact no evidence for structure (103); the imine hydrogen atom would be the one which a Grignard reagent would be expected to replace. Further, it is of course unnecessary to explain α-substitution into (104) by rearrangement after N-substitution. In fact, nuclear magnetic resonance studies show that, in ether, pyrryl magnesium chloride is predominantly in the form (104) or is an ionic, though not necessarily highly dissociated, compound[361]. Further, alkylation in ether of pyrryl magnesium bromide with (−)-2-bromobutane gives optically active 2- and 3-s-butyl-pyrroles with complete inversion of the configuration of the asymmetric carbon atom[344]. In this instance the alkylation resembles S_N2 processes.

The formulation of the pyrrole Grignard reagents as ionic, but not highly dissociated, structures related to (104) enables the predominant C-substitution reactions to be formulated in the same way as the similar processes leading to C-substitution in the alkenylation of other metallic derivatives of pyrrole (p. 67). The reagents should be regarded as sources of more or less dissociated ambient pyrrole anions (p. 68).

(g) Nitro- and Nitrosopyrroles (Table 4.12)

Reference has already been made to the reduction of nitroso- and nitro-pyrroles (p. 99), to their basic (p. 61) and acidic (p. 62) properties and to the action of hydroxylamine upon the nitroso-compounds (p. 88).

Nitrosopyrroles with alkyl substituents have been obtained only as their sodium salts[125]. Liberated from these by carbon dioxide they are very unstable. On the other hand, aryl-substituted nitrosopyrroles[14b, 125b] are relatively stable. Because of their ability to form salts, because of the transient green colours which aqueous solutions of these salts impart to ether when they are acidified, and because derivatives such as (105) have been prepared from the sodium salts[125b], the nitrosopyrroles have frequently been represented as isonitroso compounds (as 105, R =H). This may be true but the evidence does not permit a decision. Marked acidity in nitrosopyrroles would be expected because of the nature of the anion, and such an ambient anion might well give O- rather than N-derivatives, though the structures of (105) are not proved. Finally, although the di- and tri-phenyl-3-nitrosopyrroles are yellow or reddish brown, other compounds, such as 2-acetyl-3,5-dimethyl-

*Table 4.12. Nitro- and Nitrosopyrroles**

Substituents	M.p., °C	Derivatives, m.pt. °C
2-NO$_2$[45, 122]	55–6	1-Me, 29–31[362]
3-NO$_2$[45, 122]	63–4	1-Me, 64–65·5[362]
3,5-Me$_2$-2-NO$_2$[124a]	111	
2,4-Me$_2$-3-NO$_2$[22]	138	
3-NO$_2$-2,5-Ph$_2$[198]	174	
3-NO$_2$-1,2,5-Ph$_3$[273]	180–1	
3-NO$_2$-2,4,5-Ph$_3$[197]	192–4	Me ether, 195
2-MeCO-4-NO$_2$[45]	195–6	1-Me, 141–141·5
2-MeCO-5-NO$_2$[45]	154·5–155	1-Me, 95·5–96
4-CO$_2$H-2-NO$_2$[339]	226	⎫ 204
5-CO$_2$H-2-NO$_2$[122]	161	⎬Me ester 181–2
5-CO$_2$H-3-NO$_2$[122]	217 (dec.)	⎭ 197
		Et ester, 174
		1-Me ester, 199–199·5[45]
4-CO$_2$H-3,5-Me$_2$-2-NO$_2$[123]	231 (dec.)	Et ester, 149·5
5-CO$_2$H-2,4-Me$_2$-3-NO$_2$[123]	240 (dec.)	Et ester, 204
2-CN-4-NO$_2$[45]	150·5–151	1-Me, 157·5–158
2-CN-5-NO$_2$[45]	172·5–173	1-Me, 86–86·5
2-HCO-4-NO$_2$[275]	142	oxime, 228
2-HCO-5-NO$_2$[275]	185	oxime, 186–8
2-NO$_2$-5-EtCO[274]	134–5	
3-NO$_2$-5-EtCO[274]	136–7	
2,4-(NO$_2$)$_2$[122]	152	1-Me, 95·5–96·5[362]
2,5-(NO$_2$)$_2$[122]	173	
3,4-(NO$_2$)$_2$	—	1-Me, 169–170[363]
		1-Et, 107–8[363]
3,4-(NO$_2$)$_2$-1,2,5-Ph$_3$[173, 117c]	238–40	
2-MeCO-3,5-(NO$_2$)$_2$[122]	151	
2-MeCO-4,5-(NO$_2$)$_2$[122]	114	
2,4-(CO$_2$H)$_2$-3,5-(NO$_2$)$_2$[123]	—	di-Et ester, 136 (dec.)[123]
2,5-Me$_2$-3-NO[277]	127 (dec.)	
2-Me-3-NO-5-Ph†	160 (dec.)	
2-NO-3,5-Ph$_2$[14b]	139–140	hydrochloride, 190 (dec.)
3-NO-2,5-Ph$_2$[125b]	204 (dec.)	
3-NO-2,4,5-Ph$_3$[125b]	197–9 (dec.)	Et ether, 125
		benzoyl, 189
2-MeCO-3,5-Me$_2$-4-NO[180d]	148–9 (dec.)	

* For salts of nitropyrroles, see [293].
† The orientation is not certain[277].

4-nitroso-[227] and 2-nitroso-3,5-diphenyl-pyrrole[14b], are green. The nitroso-pyrroles are clearly capable of tautomerism, but the predominant form is in no case known. N-Substituted C-nitrosopyrroles have not been described.

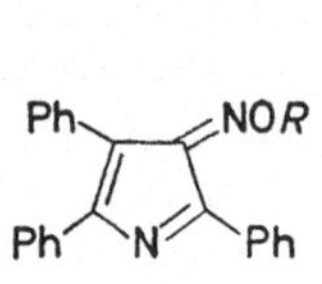

(105, R=Et, PhCO or PhNHCO) 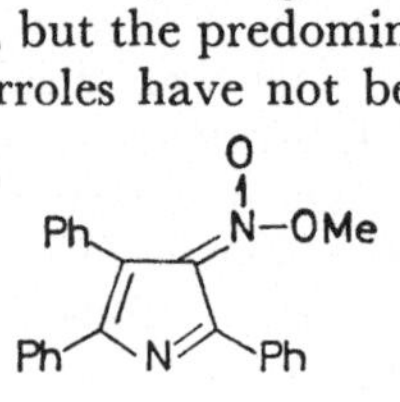   (106)

Like the nitroso compounds, nitropyrroles are readily soluble in alkali, giving yellow or reddish solutions[20c, 45, 121–124a, 197]. 3-Nitro-2,4,5-triphenylpyrrole can be converted into what has been regarded as the methyl nitronate (106)[197]. On the other hand, sodium salts of a number of nitro- and dinitro-pyrroles give with dimethyl sulphate the 1-methyl deriva-tives[45, 362] and also react in this way with acrylonitrile[293].

The acidity of nitropyrroles is readily understood (p. 62), and again this is a group of tautomeric compounds about whose fine structures relatively little is known (see *Table 4.2* for some relevant ultra-violet spectra).

(h) Group Migration

1-Alkyl- and 1-arylpyrroles rearrange to the C-substituted isomers at high temperatures[57, 145a, 228, 229, 231, 365–7]. 1-Arylpyrroles give both the 2- and 3-arylpyrroles in proportions which depend on the time of heating, long times favouring a preponderance of the 2-isomer[229]. This is in agreement with the more fully studied case of alkylpyrroles[231, 233, 365, 366]. The isomeri-zation of 1-methyl- and 1-butyl-pyrrole has been examined, roughly in the range 500–700°. The n-alkyl compounds give the 2-isomers in irreversible homogeneous first-order reactions, and the 2-isomers then isomerize re-versibly into the 3-isomers. These thermal reactions also produce from the alkylpyrroles small amounts of their lower homologues, of pyrrole and of pyridine.

1-Acetylpyrrole gives 2-acetylpyrrole[230] when heated at 250–280°, and 1-benzoylpyrrole rearranges similarly[228a].

Addendum chiefly on work reported during 1965

[*To 1b*] A S.C.F.–L.C.A.O.–M.O. treatment of pyrrole, giving $\pi \rightarrow \pi^*$ transitions in satisfactory agreement with experiment, has been reported[373].

Carbonyl[374–6] and N–H[375–6] stretching frequencies in pyrrole-carboxylic acids and their esters have been examined. The N–H stretching bands of ethyl pyrrole-2-carboxylates show clear evidence of rotational isomerism in these esters[375–6].

The nature of the association existing in 1-methylpyrrole has been examined by n.m.r. spectroscopy, and the results support the idea of association in 'closed cyclic dimers'[377]. The ring current in pyrrole is sug-gested to be a measure of its aromaticity[378].

[*To 1c*] The pK_a values for pyrrole-2- and -3-carboxylic acid (4·39 and 5·00, respectively, in water at 20°), are thought to indicate stabilization of the anion of the former by intramolecular hydrogen bonding to the .NH. group. By comparison with the benzoic acids, these data give apparent σ values for 2- and 3-pyrrolyl of −0·15 and −0·75, respectively[374].

[*To 2a*] In the Vilsmeier reaction, pyrroles can react with amides to give ketones[379].

[*To 2b*] Further examples have been described of the reaction between pyrroles and pyrrolidone in the presence of phosphoryl chloride[380–1]. The use of Δ^3-pyrrolin-2-ones in place of pyrrolidone gives bipyrroles, whilst Δ^4-pyrrolin-2-ones lead to dipyrrylpyrrolines[382].

Dipyrrylmethanes are formed from 2-pyrrylmethyl-pyridinium salts and anions of pyrrole-2-carboxylic acids[383].

[*To 2b and 2e*] In aqueous sodium hydroxide solution, diethyl 2-pyrryl-phosphonate gives pyrrole and 2-ethylpyrrole. The formulation of this reaction as involving the anion is supported by the fact that diethyl 2-(1-methylpyrryl) phosphonate does not react in this way[384].

[*To 2d*] With phthalaldehydic acid, pyrrole gives 3-(2-pyrryl)phthalide. 2,5-Dimethylpyrrole reacts similarly[385] at $C_{(3)}$. With acetonylacetone and zinc acetate in acetic acid, 3,4-diethylpyrrole gives 1,2-diethyl-5,8-dimethyl-indolizine[386].

[*To 2g*] Bromination of 2-formylpyrrole and methyl pyrrole-2-carboxylate occurs at the 4-, 5-, and 4,5-positions[387].

[*To Radical Substitution*] With t-butyl peroxide, 1-methylpyrrole gives (1-pyrryl)methyl radicals; as well as dimerizing these react with 1-methylpyrrole, giving 3-(1-pyrrolyl)methyl-pyrrole[388].

[*To 3a–c*] Pyrroles undergo Diels–Alder additions with benzyne; derivatives of 2-naphthylamine result from rearrangement of the initial products[389].

3,4-Dialkylpyrroles are not trimerized by acids. With zinc acetate and acetic acid they give 1,2,6,7-tetra-alkylindolizines[386].

Perhydrol in pyridine has again been used to convert 3,4-di- and 3,4,5-tri-alkylpyridines into '2-hydroxypyrroles'. The structures of the 'peroxides' formed in some reactions are still uncertain[390].

Further studies of the mass spectroscopy of pyrroles have been made; loss of 1-substituents and ring opening can occur[391].

With dichlorocarbene, some bipyrroles give addition products (107) which on hydrolysis form the corresponding aldehydes[392]. It seems to be agreed that 3-chloropyridines formed in the reaction between some pyrroles and chloroform in the presence of a base, do not arise from 2-dichloromethylpyrrole-nines also formed; both types of product come from a common intermediate[393a]. However, (108) can be rearranged to 3-chloro-2,6-lutidine by means of butyl lithium[393b]. The isolation of (108) and its rearrangement to 2-ethoxyethyl-5-methylpyridine has been described[393c].

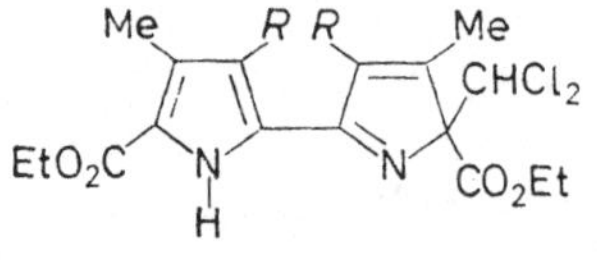

(107) 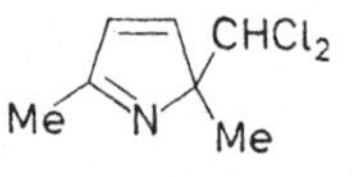   (108) 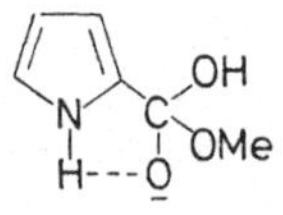   (109)

[*To 5a*] In view of the behaviour of esters of pyrrole-dicarboxylic acids, already reported, it is interesting to notice that methyl pyrrole-2-carboxylate undergoes alkaline hydrolysis considerably faster than the 3-isomer [$10^4 k_2$ (l. mole^{-1} sec^{-1}) (25°), 231, 4·1 and 0·59, for ethyl pyrrole-1-, and methyl pyrrole-2- and -3-carboxylate, respectively, in 56 per cent aqueous acetone]. The effect is attributed to intramolecular hydrogen bonding in the transition state (109)[374] (cf. to 1c above).

Pyrrole carbonyl derivatives have been used in Wittig reactions[394].

[*To 5b*] (2-Pyrryl)methyl-pyridinium salts react with alkali to give 1,2-di-(2-pyrryl)ethylenes[395].

[*To 5d*] Attempts to replace nuclear bromine atoms from ethyl bromo-pyrrole-2-carboxylate by nucleophilic substitutions generally failed, but with cuprous cyanide low yields of cyano esters resulted[387].

[*To* 5*e*] N.M.R. spectroscopy supports the formulation of '2-hydroxy-pyrroles' as pyrrolenones[390].

[*To* 5*f*] Pyrrole magnesium bromide and 2-(1-methylpyrrolyl) lithium give tertiary phosphine oxides with phosphoryl chloride[396].

With acetyl chloride, pyrrole magnesium bromide gives 2- and 3-acetyl-pyrrole in a ratio of about 5:1. With ethyl acetate much less of the 3-isomer is formed[397].

Alkylation of pyrrole magnesium bromide and chloride with γ-chloro-butyronitrile gives 2- and 3-cyanopropylpyrrole (2>3). The isomer distribution is similar to those observed in reactions with butyl chloride, heptyl chloride and δ-methoxybutyl chloride. It is concluded that the constitution of pyrrole magnesium chloride is the same in tetrahydrofuran as in ether[398].

REFERENCES

1 Orgel, L. E., Cottrell, T. L., Dick, W. and Sutton, L. E. *Trans. Faraday Soc.* **47** (1951) 113; Kofod, H., Sutton, L. E. and Jackson, J. *J. chem. Soc.* (1952) 1467

2 (a) Price, W. C. and Walsh, A. D. *Proc. R. Soc.* **A179** (1941) 201; (b) Milazzo, G. *Gazz. chim. ital.* **74** (1944) 152; (c) Walsh, A. D. *A. Rep. Chem. Soc.* **44** (1947) 38; (d) Pickett, L. W., Corning, M. E., Wieder, G. M., Semenov, D. A. and Buckley, J. M. *J. Am. chem. Soc.* **75** (1953) 1618; (e) Milazzo, G. and Miescher, E. *Gazz. chim. ital.* **83** (1953) 782; Milazzo, G. *ibid.* 787; (f) Mason, S. F. *J. chem. Soc.* (1963) 3999

3 Braude, E. A. *A. Rep. Chem. Soc.* **42** (1945) 105

4 (a) Cookson, G. H. *J. chem. Soc.* (1953) 2789; (b) Eisner, U. and Gore, P. H. *ibid.* (1958) 922

5 Pauling, L. *J. Am. chem. Soc.* **58** (1936) 94

6 Halverson, F. *Rev. mod. Phys.* **19** (1947) 87; cf. Zumwalt, L. R. and Badger, R. M. *J. chem. Phys.* **7** (1939) 629

7 Josien, M.-L. and Fuson, N. *J. chem. Phys.* **22** (1954) 1169, 1264; Josien, M.-L., Pineau, P., Paty, M. and Fuson, N. *ibid.* **24** (1956) 1261; Fuson, N., Pineau, P. and Josien, M.-L. in *Hydrogen Bonding*, ed. Hadži, G. D., London (Pergamon) 1959

8 Tuomikoski, P. *J. chem. Phys.* **22** (1954) 2096

9 Bonino, G. B. and Manzoni-Ansidei, R. *Atti Accad. naz. Lincei* **25** (1937) 489; Manzoni-Ansidei, R. *Ricerca scient.* **10** (1939) 328

10 (a) Eisner, U. and Erskine, R. L. *J. chem. Soc.* (1958) 971; (b) Jones, R. A. *Aust. J. Chem.* **16** (1963) 93

11 (a) Sorocco, M. and Nicolaus, R. *Atti Accad. naz. Lincei* **20** (1956) 795; **21** (1956) 103; (b) Roberts, J. D. *Nuclear Magnetic Resonance*, New York (McGraw-Hill) 1959; Abraham, R. J. and Bernstein, H. J. *Can. J. Chem.* **37** (1959) 1056; Schaefer, T. and Schneider, W. G. *J. chem. Phys.* **32** (1960) 1224

12 (a) Naqvi, N. and Fernando, Q. *J. org. Chem.* **25** (1960) 551; (b) Abraham, R. J., Bullock, E. and Mitra, S. S. *Can. J. Chem.* **37** (1959) 1859; (c) Bullock, E. *ibid.* **36** (1958) 1686

13 Ciamician, G. and Zanetti, C. M. *Ber. dtsch. chem. Ges.* **26** (1893) 1711; Ciamician, G. *ibid.* **37** (1904) 4254; Fischer, H. and Treibs, A. *Justus Liebigs Annln Chem.* **450** (1926) 132; Treibs, A. and Dieter, P. *ibid.* **513** (1934) 65; Pratesi, P. *Gazz. chim. ital.* **65** (1935) 658; Stedman, R. J. and MacDonald, S. F. *Can. J. Chem.* **33** (1955) 468

14 (a) Treibs, A. and Kolm, H. G. *Justus Liebigs Annln Chem.* **606** (1957) 166; (b) Rogers, M. A. T. *J. chem. Soc.* (1943) 590

15 Anderson, T. *Justus Liebigs Annln Chem.* **105** (1858) 302; Ciamician, G. and Dennstedt, M. *Ber. dtsch. chem. Ges.* **19** (1886) 173; Lubavin, N. *ibid.* **2** (1869) 99; Bell, C. A. *ibid.* **11** (1878) 1810; Reynolds, J. E. *J. chem. Soc.* **95** (1909) 505

16 Franklin, E. C. *J. phys. Chem., Ithaka* **24** (1920) 81

17 (a) Alexander, E. R., Herrick, A. B. and Roder, T. M. *J. Am. chem. Soc.* **72** (1950) 2760; (b) Shirley, D. A., Gross, B. H. and Roussel, P. A. *J. org. Chem.* **20** (1955) 225

18 (a) Oddo, B. *Gazz. chim. ital.* **39** (1909) 649; (b) Kharasch, M. S. and Reinmuth, O. *Grignard Reactions of Non-metallic Substances*, London (Constable) 1954

19 Herz, W. *J. org. Chem.* **22** (1957) 1260

20 (a) McEwen, W. K. *J. Am. chem. Soc.* **58** (1936) 1124; (b) Fischer, H. and Ernst, P. *Justus Liebigs Annln Chem.* **447** (1926) 139; (c) Novikov, S. S., Belikov, V. M., Egorov, Yu. P., Safonova, E. N. and Semenov, L. V. *Bull. Acad. Sci. USSR* (1959) 1386

21 Bamberger, E. and Djierdjian, G. *Ber. dtsch. chem. Ges.* **33** (1900) 536

[22] Fischer, H., Beller, H. and Stern, A. *Ber. dtsch. chem. Ges.* **61** (1928) 1074
[23] Plancher, G. and Ponti, U. *Atti Accad. naz. Lincei* **18** II (1909) 469
[24] Fieser, L. F. and M. *Organic Chemistry*, 3rd edn. p. 680, New York (Reinhold) 1956
[25] Fischer, H. and Zerweck, W. *Ber. dtsch. chem. Ges.* **55** (1922) 1942
[26] Corwin, A. H. and Kleinspehn, G. G. *J. Am. Chem. Soc.* **75** (1953) 2089
[27] Reichstein, T. *Helv. chim. Acta* **13** (1930) 349
[28] Fischer, H. and Schubert, M. *Ber. dtsch. chem. Ges.* **56** (1923) 1202; **57** (1924) 610
[29] MacDonald, S. F. *J. chem. Soc.* (1952) 4176
[30] Fischer, H. and Zerweck, W. *Ber. dtsch. chem. Ges.* **56** (1923) 519
[31] Fischer, H. and Ernst, P. *Justus Liebigs Annln Chem.* **447** (1926) 139
[32] (a) Smith, G. F. *J. chem. Soc.* (1954) 3842; (b) Silverstein, R. M., Ryskiewicz, E. E., Willard, C. and Koehler, R. C. *J. org. Chem.* **20** (1955) 668
[33] (a) Fischer, H., Schneller, K. and Zerweck, W. *Ber. dtsch. chem. Ges.* **55** (1922) 2390; (b) Fischer, H., Weiss, B. and Schubert, M. *ibid.* **56** (1923) 1194; (c) Blicke, F. F., Faust, J. A., Gearien, J. E. and Warzynski, R. J. *J. Am. chem. Soc.* **65** (1943) 2465
[34] (a) Schiff, R. *Ber. dtsch. chem. Ges.* **10** (1877) 500; (b) Ciamician, G. and Dennstedt, M. *ibid.* **16** (1883) 2348; **17** (1884) 432
[35] Ciamician, G. and Dennstedt, M. *Ber. dtsch. chem. Ges.* **17** (1884) 2944; Ciamician, G. and Silber, P. *ibid.* **19** (1886) 1956
[36] (a) Dennstedt, M. and Zimmermann, J. *Ber. dtsch. chem. Ges.* **20** (1887) 1760; (b) Cooper, W. D. *J. org. Chem.* **23** (1958) 1382
[37] (a) Ciamician, G. and Silber, P. *Ber. dtsch. chem. Ges.* **17** (1884) 1150; (b) Dennstedt, M. and Zimmermann, J. *ibid.* **19** (1886) 2189; **20** (1887) 850
[38] Fischer, H. and Viaud, P. *Ber. dtsch. chem. Ges.* **64** (1931) 193; Fischer, H., Halbig, P. and Walach, B. *Justus Liebigs Annln Chem.* **452** (1927) 268
[39] Ciamician, G. and Silber, P. *Ber. dtsch. chem. Ges.* **20** (1887) 1368
[40] Widman, O. *Justus Liebigs Annln Chem.* **400** (1913) 86; cf. Almström, G. K. *ibid.* **409** (1915) 291
[41] Oddo, B. *Ber. dtsch. chem. Ges.* **47** (1914) 2427
[42] Cornforth, J. W. and Firth, M. E. *J. chem. Soc.* (1958) 1091
[43] Fischer, H. and Schubert, F. *Hoppe-Seyler's Z. physiol. Chem.* **155** (1926) 99
[44] (a) Fischer, H. and Bäumler, R. *Justus Liebigs Annln Chem.* **468** (1929) 58; Fischer, H. and Kutscher, W. *ibid.* **481** (1930) 193; Fischer, H., Goldschmidt, M. and Nüssler, W. **486** (1931) 1; (b) Fischer, H. and Hussong, M. *ibid.* **492** (1932) 128
[45] Anderson, H. J. *Can. J. Chem.* **35** (1957) 21; **37** (1959) 2053
[46] Berlin, A. A. *J. gen. Chem. U.S.S.R.* **14** (1944) 438
[47] (a) Nenitzescu, C. D. and Solomonica, E. *Ber. dtsch. chem. Ges.* **64** (1931) 1924; Diels, O. and König, H. *ibid.* **71** (1938) 1179; (b) Blicke, F. F., Warzynski, R. J., Faust, J. A. and Gearien, J. E. *J. Am. chem. Soc.* **66** (1944) 1675
[48] Treibs, A. and Michl, K.-H. *Justus Liebigs Annln Chem.* **577** (1952) 129
[49] Staudinger, H. and Suter, E. *Ber. dtsch. chem. Ges.* **53** (1920) 1092
[50] Neisser, K. *Ber. dtsch. chem. Ges.* **67** (1934) 2080; Rinkes, I. J. *Recl Trav. chim. Pays-Bas Belg.* **56** (1937) 1224
[51] Corwin, A. H. *Heterocyclic Compounds*, Vol. I, ed. Elderfield, R. C., New York (Wiley) 1950
[52] Fischer, H. and Orth, H. *Justus Liebigs Annln Chem.* **489** (1931) 62; Fischer, H. and Hussong, M. *ibid.* 62
[53] Treibs, A. and Derra, R. *Justus Liebigs Annln Chem.* **589** (1954) 174
[54] Ciamician, G. and Silber, P. *Ber. dtsch. chem. Ges.* **17** (1884) 1437
[55] Dennstedt, M. *Ber. dtsch. chem. Ges.* **23** (1890) 2562; **24** (1891) 2559; **25** (1892) 3636
[56] Ciamician, G. and Anderlini, F. *Ber. dtsch. chem. Ges.* **22** (1889) 656; Plancher, G. and Zambonini, T. *Atti Accad. naz. Lincei* **22** II (1913) 703
[57] de Jong, M. E. A. *Recl Trav. chim. Pays-Bas Belg.* **48** (1929) 1029
[58] Ciamician, G. and Zanetti, C. M. *Ber. dtsch. chem. Ges.* **22** (1889) 659; Zanetti, C. U. *Atti Accad. naz. Lincei* **23** (1890) II, 206
[59] Zanetti, C. U. *Ber. dtsch. chem. Ges.* **22** (1889) 2515
[60] Lubavin, N. *Ber. dtsch. chem. Ges.* **2** (1869) 99; Bell, C. A. *ibid.* **11** (1878) 1810; Ciamician, G. and Dennstedt, M. *ibid.*, **15** (1882) 2579; **17** (1884) 2944; **20** (1887) 1368
[61] Cantor, P. A. and Vanderwerf, C. A. *J. Am. chem. Soc.* **80** (1958) 970
[62] Ciamician, G. *Ber. dtsch. chem. Ges.* **37** (1904) 4200
[63] Khotinsky, E. and Patzewitch, R. *Ber. dtsch. chem. Ges.* **42** (1909) 3104; Illari, G. *Gazz. chim. ital.* **67** (1937) 434; Pieroni, A. and Veremeenco, P. *ibid.* **56** (1926) 455; Meyer, V. and Fischer, P. *J. prakt. Chem.* **82** (1910) 2, 523; Swain, C. G., Kaiser, L. E. and Knee, T. E. C. *J. Am. chem. Soc.* **77** (1955) 4681
[64] Fischer, H. and Bartholomäus, E. *Ber. dtsch. chem. Ges.* **45** (1912) 466; *Hoppe-Seyler's Z. physiol. Chem.* **77** (1912) 185; Fischer, H. and Eismayer, K. *ibid.* **47** (1914) 1820
[65] Cornforth, J. W. and R. H., and Robinson, Sir Robert *J. chem. Soc.* (1942) 682

REFERENCES

66 Knorr, L., and Hess, K. *Ber. dtsch. chem. Ges.* **45** (1912) 2631; Colacicchi, U. and Bertoni, C. *Atti Accad. naz. Lincei* **21 II** (1912/5) 450, 518; Fischer, H. and Hahn, A. *Hoppe-Seyler's Z. physiol. Chem.* **84** (1913) 254

67 Harley-Mason, J. *J. chem. Soc.* (1952) 2433

68 (a) Fischer, H. and Nenitzescu, C. *Justus Liebigs Annln Chem.* **443** (1925) 113; (b) idem, *Hoppe-Seyler's Z. physiol. Chem.* **145** (1925) 295; (c) Fischer, H. and Klarer, J. *Justus Liebigs Annln Chem.* **447** (1926) 48; (d) Fischer, H. and Hussong, M. *ibid.* **492** (1932) 137

69 Johnson, A. W. *J. chem. Soc.* (1947) 1626

70 Möhlau, R. *Ber. dtsch. chem. Ges.* **44** (1911) 3605; Pratesi, P. *Gazz. chim. ital.* **66** (1936) 215; Bullock, E. *Can. J. Chem.* **36** (1958) 1744; Fischer, H., Treibs, A. and Zaucker, E. *Chem. Ber.* **92** (1959) 2026

71 Treibs, A. and Michl, K.-H. *Justus Liebigs Annln Chem.* **589** (1954) 163

72 Diels, O. and Alder, K. *Justus Liebigs Annln Chem.* **470** (1929) 62; **486** (1931) 211; **498** (1932) 1; Norton, J. A. *Chem. Rev.* **31** (1942) 319

73 cf. Badger, G. M. *et al.*, *J. chem. Soc.* (1958) 1179, 4777

74 Herz, W., Dittmer, K. and Cristol, S. J. *J. Am. chem. Soc.* **69** (1947) 1698

75 Herz, W. and Rogers, J. L. *J. Am. chem. Soc.* **73** (1951) 4921

76 Eisner, U., Linstead, R. P., Parkes, E. A. and Stephen, E. *J. chem. Soc.* (1956) 1655; Herz, W. and Settine, R. L. *J. org. Chem.* **24** (1959) 201

77 Eisner, U. *J. chem. Soc.* (1957) 854

78 Burk, W. J. and Hammer, G. N. *J. Am. chem. Soc.* **76** (1954) 1294

79 Pictet, A. and Rilliet, A. *Ber. dtsch. chem. Ges.* **40** (1907) 1166

80 Colacicchi, U. *Gazz. chim. ital.* **42** (1912) I, 10

81 Tschelincev, V. V. and Maksorov, B. V. *Zh. russk. fiz.-khim. Obshch.* **48** (1916) 748

82 *U.S. Patent* 2,492,414 (1949)

83 (a) Treibs, A. and Fritz, G. *Justus Liebigs Annln Chem.* **611** (1958) 162; (b) Treibs, A. and Bader, H. *Ber. dtsch. chem. Ges.* **91** (1958) 2615

84 Colacicchi, U. and Bertoni, C. *Atti Accad. naz. Lincei* **21** (1912) II (5) 600

85 Tronow, B. and Popow, P. *Zh. russk. fiz.-khim. Obshch.* **58** (1927) 745

86 Corwin, A. H. and Quattlebaum, W. M., Jr. *J. Am. chem. Soc.* **58** (1936) 1081

87 Fischer, H. and Beller, H. *Justus Liebigs Annln Chem.* **444** (1925) 238

88 Feist, F. *Ber. dtsch. chem. Ges.* **35** (1902) 1647; Tschelincev, W. *Zh. russk. fiz.-khim. Obshch.* **47** (1916) 1211; Fischer, H. and Schubert, F. *Hoppe-Seyler's Z. physiol. Chem.* **155** (1926) 72; Fischer, H. and Bartholomäus, E. *ibid.* **83** (1913) 50; **87** (1913) 262; **89** (1913) 163; Treibs, A. and Reitsam, F. *Chem. Ber.* **90** (1957) 777

89 Rothemund, P. *J. Am. chem. Soc.* **61** (1939) 2912; Calvin, M., Ball, R. H. and Aronoff, S. *ibid.* **65** (1943) 2259; Aronoff, S. and Calvin, M. *J. org. Chem.* **8** (1943) 205

90 Baeyer, A. v. *Ber. dtsch. chem. Ges.* **19** (1886) 2184; Dennstedt, M. and Zimmermann, J. *ibid.* **20** (1887) 2449; Dennstedt, M. *ibid.* **23** (1890) 1370; Tschelincev, V. V. and Tronow, B. W. *Zh. russk. fiz.-khim. Obshch.* **48** (1916) 105, 127, 1197

91 Brown, W. H. and French, W. N. *Can. J. Chem.* **36** (1958) 371

92 Piloty, O., Krannich, W. and Will, H. *Ber. dtsch. chem. Ges.* **47** (1914) 2531

93 Fischer, H. and Orth, H. *Die Chemie des Pyrrols*, Vol. II, Part 1, Leipzig (Akad. Verlagsges.) 1937

94 Fischer, H. and Fries, G. *Hoppe-Seyler's Z. physiol. Chem.* **231** (1935) 231

95 Corwin, A. H. and Andrews, J. S. *J. Am. chem. Soc.* **58** (1936) 1086; **59** (1937) 1973

96 Cook, A. H. and Majer, J. R. *J. chem. Soc.* (1944) 482

97 Erdmann, E. *Ber. dtsch. chem. Ges.* **35** (1902) 1855

98 Reichstein, T. *Helv. chim. Acta* **15** (1932) 1110

99 Ehrlich, P. *Medsche Woche* (1901) 151

100 Treibs, A. and Derra-Scherer, H. *Justus Liebigs Annln Chem.* **589** (1954) 196

101 Kausche, G. A., Hahn, F. and Schleith, L. *Z. Naturf.* **56** (1950) 87

102 Brooker, L. G. S. and Sprague, R. H. *J. Am. chem. Soc.* **67** (1945) 1869; Strell, M. and Kreis, F. *Ber. dtsch. chem. Ges.* **87** (1954) 1011; Strell, M., Kalojanoff, A. and Brem-Rupp, L. *ibid.* 1019; Strell, M. and Kalojanoff, A. *ibid.* 1025

103 Meyer, V. *Ber. dtsch. chem. Ges.* **16** (1883) 2968; Meyer, V. and Stadler, O. *ibid.* **17** (1884) 1034; Giamician, G. and Silber, P. *ibid.* 142

104 Liebermann, C. and Häse, G. *Ber. dtsch. chem. Ges.* **38** (1905) 2847; Liebermann, C. and Krauss, R. *ibid.* **40** (1907) 2492; Fromm, F. *J. Am. chem. Soc.* **66** (1944) 1227; **67** (1945) 2050

105 Pratesi, P. *Justus Liebigs Annln Chem.* **504** (1933) 258

106 Steinkopf, W. and Wilhelm, H. *Justus Liebigs Annln Chem.* **546** (1941) 211

107 Pratesi, P. and Castorina, G. *Gazz. chim. ital.* **83** (1953) 913

108 Treibs, A., Herrmann, E. and Meissner, E. *Justus Liebigs Annln Chem.* **612** (1958) 229

109 Saxton, J. E. *J. chem. Soc.* (1951) 3239

110 (a) Plancher, G. *Ber. dtsch. chem. Ges.* **35** (1902) 2606; (b) Allen, C. F. H., Young, D. M. and Gilbert, M. R. *J. org. Chem.* **2** (1938) 235

111 Koizumi, M. and Titani, T. *Bull. chem. Soc. Japan* **12** (1937) 107; **13** (1938) 85, 298

[112] Miller, F. A. *J. Am. chem. Soc.* **64** (1942) 1543; Bak, B., Christensen, D., Hansen, L. and Rastrup-Andersen, J. *J. chem. Phys.* **24** (1956) 720

[113] Fischer, O. and Hepp, E. *Ber. dtsch. chem. Ges.* **19** (1886) 2251

[114] (a) Khotinsky, E. and Soloweitschik, M. *Ber. dtsch. chem. Ges.* **42** (1909) 2508; (b) Fischer, H. and Bartholomäus, E. *ibid.* **45** (1912) 1919; (c) Fischer, H. and Rothweiler, F. *ibid.* **56** (1923) 512; (d) Plancher, G. and Soncini, E. *Gazz. chim. ital.* **32** (1902) II, 447; (e) Ciusa, R. *ibid.* **51** (1921) I, 49; (f) Reichstein, T. *Helv. chim. Acta* **10** (1927) 387

[115] (a) Ciamician, G. and Silber, P. *Ber. dtsch. chem. Ges.* **18** (1885) 1763; (b) Kalle & Co., German Patent 38,423; *Friedländers Fortschritte der Theerfarbenfabrikation*, I, 223, Berlin, 1888; (c) Datta, R. L. and Prasad, N. *J. Am. chem. Soc.* **39** (1917) 441

[116] (a) Mazzara, G. *Gazz. chim. ital.* **32** (1902) I, 510; **32** (1902) II, 28; (b) Mazzara, G. and Borgo, A. *ibid.* **34** (1904) I, 253, 414; **35** (1905) I, 477, II, 19; (c) Terent'ev, A. P. and Yanovskaya, L. A. *J. gen. Chem. U.S.S.R.* **21** (1951) 307

[117] Fischer, H. and Orth, H. *Die Chemie des Pyrrols*, Vol. I, (a) p. 88; (b) p. 253; (c) p. 42; (d) Vol. II, Part 1, p. 62, Leipzig (Akad. Verlagsges.) 1934, 1937

[118] Doak, K. W. and Corwin, A. H. *J. Am. chem. Soc.* **71** (1949) 159

[119] Treibs, A. and Kolm, H. G. *Justus Liebigs Annln Chem.* **614** (1958) 176

[120] (a) Colacicchi, U. *Atti Accad. naz. Lincei* **19** (1910) II, 645; (b) Fischer, H. and Scheyer, H. *Justus Liebigs Annln Chem.* **434** (1923) 237; (c) Fischer, H., Sturm, E. and Friedrich, H. *ibid.* **461** (1928) 244

[121] (a) Ciamician, G. and Silber, P. *Ber. dtsch. chem. Ges.* **18** (1885) 1456; **19** (1886) 1079; (b) Anderlini, F. *ibid* **22** (1889) 2503

[122] Rinkes, I. J. *Recl Trav. chim. Pays-Bas Belg.* **53** (1934) 1167; **56** (1937) 1142; **60** (1941) 650

[123] Fischer, H. and Zerweck, W. *Ber. dtsch. chem. Ges.* **55** (1922) 1949

[124] (a) Angeli, A. and Alessandri, L. *Atti Accad. naz. Lincei* **20** (1911) I, 311; Hale, W. J. and Hoyt, W. V. *J. Am. chem. Soc.* **37** (1915) 2538; Oddo, B. *Gazz. chim. ital.* **69** (1939) 10; (b) Dhont, J. and Wibaut, J. P. *Recl Trav. chim. Pays-Bas Belg.* **62** (1943) 177

[125] (a) Spica, M. and Angelico, F. *Gazz. chim. ital.* **29** (1899) II, 49; (b) Angelico, F. and Calvello, E. *ibid.* **31** (1901) II, 4; **34** (1904) I, 38; (c) Angeli, A., Angelico, F. and Calvello, E. *ibid.* **33** (1903) II, 270; (d) Ajello, T. and Cusmano, S. *ibid.* **70** (1940) 512

[126] Fischer, H. and Zeile, K. *Justus Liebigs Annln Chem.* **483** (1930) 257

[127] Rinkes, I. J. *Recl Trav. chim. Pays-Bas Belg.* **62** (1943) 116

[128] Terent'ev, A. P. and Shadkhina, M. A. *Dokl. Akad. Nauk SSSR* **55** (1947) 227; Terent'ev, A. P. and Yanovskaya, L. A. *Zh. obshch. Khim.* **19** (1949) 531, 1365; Terent'ev, A. P. and Yashunskiĭ, V. G. *ibid.* **20** (1950) 510

[129] Wood, J. L. in *Organic Reactions*, Vol. III, New York (Wiley) 1946

[130] Matteson, D. S. and Snyder, H. R. *J. org. Chem.* **22** (1957) 1500

[131] Pratesi, P. *Atti Accad. naz. Lincei* **16** (1932/6) 443

[132] Hess, K. and Wissing, F. *Ber. dtsch. chem. Ges.* **47** (1914) 1416

[133] Ciamician, G. and Silber, P. *Ber. dtsch. chem. Ges.* **18** (1885) 881

[134] Pictet, A. *Ber. dtsch. chem. Ges.* **37** (1904) 2792

[135] Treibs, A. and Michl, K.-H. *Justus Liebigs Annln Chem.* **577** (1952) 115

[136] Blume, R. C. and Lindwall, H. G. *J. org. Chem.* **10** (1945) 255

[137] Reich, H. E. and Levine, R. *J. Am. chem. Soc.* **77** (1955) 4913

[138] Schlink, J. *Ber. dtsch. chem. Ges.* **32** (1899) 947

[139] Ciamician, G. and Dennstedt, M. *Ber. dtsch. chem. Ges.* **16** (1883) 1536; Knorr, L. and Rabe, P. *ibid.* **34** (1901) 3491—but see Sonn, A. *ibid.* **72** (1939) 2150; King, F. E., Marshall, J. R. and Smith, P. *J. chem. Soc.* (1951) 239

[140] Padoa, M. *Gazz. chim. ital.* **36** (1906) II, 317; Wilstätter, R. and Hatt, D. *Ber. dtsch. chem. Ges.* **45** (1912) 1471; Hess, K. *ibid.* **46** (1913) 3113; de Jong, M. and Wibaut, J. P. *Recl Trav. chim. Pays-Bas Belg.* **49** (1930) 237; Signaigo, F. K. and Adkins, H. *J. Am. chem. Soc.* **58** (1936) 709; Rainey, J. L. and Adkins, H. *ibid.* **61** (1939) 1104; Adkins, H. and Coonradt, H. L. *ibid.* **63** (1941) 1563

[141] Sakurai, B. *Bull. chem. Soc. Japan* **11** (1936) 374

[142] Diels, O. and Alder, K. *Justus Liebigs Annln Chem.* **490** (1931) 267

[143] Mandell, L. and Blanchard, W. A. *J. Am. chem. Soc.* **79** (1957) 6198

[144] Wittig, G. and Behnisch, W. *Chem. Ber.* **91** (1958) 2358

[145] (a) Dennstedt, M. and Zimmermann, J. *Ber. dtsch. chem. Ges.* **21** (1888) 1478; (b) Dennstedt, M. *ibid.* **21** (1888) 3429; **22** (1889) 1920; **24** (1891) 2559; (c) Fischer, H. *ibid.* **48** (1915) 401; (d) Piloty, O. and Wilke, K. *ibid.* **45** (1912) 2586; (e) Allen, C. F. H., Gilbert, M. R. and Young, D. M. *J. org. Chem.* **2** (1938) 227

[146] Plancher, G. *Ber. dtsch. chem. Ges.* **35** (1902) 2606

[147] Schmitz-Dumont, O. *Ber. dtsch. chem. Ges.* **62** (1929) 226

[148] Potts, H. A. and Smith, G. F. *J. chem. Soc.* (1957) 4018

[149] (a) Conant, J. B. and Chow, B. F. *J. Am. chem. Soc.* **55** (1933) 3475; (b) Wasserman, H. H. and Liberles, A. *ibid.* **82** (1960) 2086

REFERENCES

[150] Plancher, G. and Cattadori, F. *Atti Accad. naz. Lincei* **13** (1904) I, 489; Plancher, G. and Ravenna, C. *ibid.* **14** (1905) I, 214; Piloty, O. and Quittmann, E. *Ber. dtsch. chem. Ges.* **42** (1909) 4693; Karrer, P. and Smirnoff, A. P. *Helv. chim. Acta* **5** (1922) 832; Ajello, T. and Giambrone, S. *Chem. Abstr.* **49** (1955) 6226

[151] Ciamician, G. and Silber, P. *Ber. dtsch. chem. Ges.* **17** (1884) 1743

[152] Oddo, B. in *Traité de Chimie Organique*, ed. Grignard, V., Vol. 19, p. 13, Paris, 1942

[153] Pratesi, P. *Gazz. chim. ital.* **67** (1937) 188, 199; Nicolaus, R. A., Mangoni, L. and Caglioti, L. *ibid.* **85** (1955) 1397

[154] Metzger, W. and Fischer, H. *Justus Liebigs Annln Chem.* **527** (1937) 1; cf.[186, 190]

[155] Fischer, H., Yoshioka, T. and Hartmann, P. *Hoppe-Seyler's Z. physiol. Chem.* **212** (1932) 146

[156] Fischer, H. and Hartmann, P. *Hoppe-Seyler's Z. physiol. Chem.* **226** (1934) 116

[157] Kuhn, R. and Kainer, H. *Justus Liebigs Annln Chem.* **578** (1952) 227; Sprio, V. and Madonia, P. *Gazz. chim. ital.* **86** (1956) 101

[158] Bailey, P. S. *Chem. Rev.* **58** (1958) 925

[159] Wibaut, J. P. and Guljé, A. R. *Proc.K.ned. Akad. Wet.* **54B** (1951) No. 4, 330; Wibaut, J. P. *Chimia* **11** (1957) 321

[160] Ciamician, G. and Dennstedt, M. *Ber. dtsch. chem. Ges.* **17** (1884) 533; Ciamician, G. and Zanetti, C. U. *ibid.* **22** (1889) 1968; **23** (1890) 1787; Zanetti, C. U. *Gazz. chim. ital.* **21** (1891) II, 25; **22** (1892) II, 269; Oddo, B. and Mameli, R. *ibid.* **44** (1914) II, 162; Ajello, T. *ibid.* **67** (1937) 728; **69** (1939) 207

[161] Findlay, S. P. *J. org. Chem.* **21** (1956) 644

[162] Duden, P. and Heynsius, D. *Ber. dtsch. chem. Ges.* **34** (1901) 3054

[163] Ajello, T. *Gazz. chim. ital.* **67** (1937) 728; Ajello, T. and Cusmano, S. *ibid.* **69** (1939) 207; **70** (1940) 127, 499

[164] Ciamician, G. and Dennstedt, M. *Ber. dtsch. chem. Ges.* **14** (1881) 1153; **15** (1882) 1172; Dennstedt, M. and Zimmerman, J. *ibid.* **18** (1885) 3316; Ciamician, G. and Silber, P. *ibid.* **20** (1887) 191; Bocchi, O. *Gazz. chim. ital.* **30** (1900) I, 89; Alexander, E. R., Herrick, A. B. and Roder, T. M. *J. Am. chem. Soc.* **72** (1950) 2760

[165] Parham, W. E. and Reiff, H. E. *J. Am. chem Soc.* **77** (1955) 1177; Parham, W. E. and Wright, C. D. *J. org. Chem.* **22** (1957) 1473

[166] Ajello, T. *Gazz. chim. ital.* **70** (1940) 504; **72** (1942) 325

[167] (a) Brown, R. D. *Aust. J. Chem.* **8** (1955) 100; (b) Brown, R. D. and Heffernan, M. L. *ibid.* **12** (1959) 319; (c) Brown, R. D. in *Current Trends in Heterocyclic Chemistry*, ed. Albert, A., Badger, G. M. and Shoppee, C. W., p. 13, London (Butterworths) 1958; (d) Brown, R. D. and Heffernan, M. L. *Aust. J. Chem.* **12** (1959) 330

[168] Dittmer, K. and Herz, W. *J. Am. chem. Soc.* **70** (1948) 503; Herz, W. *ibid.* **71** (1949) 3982

[169] Harvey, D. G. *J. chem. Soc.* (1950) 1638

[170] Emmert, B., Diehl, K. and Gollwitzer, F. *Ber. dtsch. chem. Ges.* **62** (1929) 1733; Emmert, B. and Diehl, K. *ibid.* **64** (1931) 130; Pratesi, P. and Berti, V. *Boll. scient. Fac. Chim. Univ. Bologna* (1940) 188; Sardina, M. T. and Bonino, C. *ibid.* **12** (1954) 155; Marinangeli, A. and Bonino, C., Jr. *Annali Chim.* **44** (1954) 949; Chiorboli, P. *Gazz. chim. ital.* **81** (1951) 906; Pratesi, P. and Berti, V. *10th Int. Congr. appl. Chem.* (Rome, 1938) **3** (1939) 313

[171] Fischer, H. and Müller, J. *Hoppe-Seyler's Z. physiol. Chem.* **132** (1923) 181; Fischer, H. and Smeykal, K. *Ber. dtsch. chem. Ges.* **56** (1923) 2368; Herz, W. and Brasch, J. *J. org. Chem.* **23** (1958) 1513

[172] Angeli, A. and Alessandri, L. *Atti Accad. naz. Lincei* **23** (1914) II (5) 93

[173] Fischer, H., Berg, H. and Schormüller, A. *Justus Liebigs Annln Chem.* **480** (1930) 109; Treibs, A. and Schmidt, R. *ibid.* **577** (1952) 105; Ju-Hua Chu, E. and Chu, T. C. *J. org. Chem.* **19** (1954) 266

[174] Brown, B. R., Hammick, D. Ll. and Scholefield, A. J. B. *J. chem. Soc.* (1950) 778

[175] Brown, B. R. *Qu. Rev. chem. Soc.* **5** (1951) 131

[176] Corwin, A. H. and Straughn, J. L. *J. Am. chem. Soc.* **70** (1948) 1416

[177] Ciamician, G. and Silber, P. *Ber. dtsch. chem. Ges.* **17** (1884) 103; Hale, W. J. and Hoyt, W. V. *J. Am. chem. Soc.* **38** (1916) 1065

[178] Piloty, O. and Wilke, K. *Ber. dtsch. chem. Ges.* **46** (1913) 1597

[179] Corwin, A. H. and Ellingson, R. C. *J. Am. chem. Soc.* **66** (1944) 1146; Corwin, A. H. and Buc, S. R. *ibid.* 1151

[180] (a) Knorr, L. *Justus Liebigs Annln Chem.* **236** (1886) 290; (b) Piloty, O. and Hirsch, P. *ibid.* **395** (1913) 63; (c) Magnani, G. *Ber. dtsch. chem. Ges.* **22** (1889) 35; (d) Küster, W. *Hoppe-Seyler's Z. physiol. Chem.* **121** (1922) 135

[181] Fischer, H. and Walach, B. *Ber. dtsch. chem. Ges.* **58** (1925) 2818; Corwin, A. H. and Quattlebaum, W. M., Jr. *J. Am. chem. Soc.* **58** (1936) 1081

[182] Corwin, A. H. and Straughn, J. L. *J. Am. chem. Soc.* **70** (1948) 2968

[183] Newman, M. S., Kuivila, H. G. and Garrett, A. B. *J. Am. chem. Soc.* **67** (1945) 704

[184] Ingold, C. K. *Structure and Mechanism in Organic Chemistry*, Ch. 14, London (Bell) 1953

[185] Fischer, H. and Stern, A. *Justus Liebigs Annln Chem.* **446** (1926) 229

[186] Fischer, H., Guggemos, H. and Schäfer, A. *Justus Liebigs Annln Chem.* **540** (1939) 30

E*

[187] Piccinini, A. and Salmoni, L. *Gazz. chim. ital.* **32** (1902) I, 246
[188] Fischer, H. and Waibel, A. *Justus Liebigs Annln Chem.* **512** (1934) 195
[189] Fischer, H., Süs, O. and Weilguny, F. G. *Justus Liebigs Annln Chem.* **481** (1930) 159
[190] Endermann, F. and Fischer, H. *Justus Liebigs Annln Chem.* **538** (1939) 172
[191] Davies, W. H. and Rogers, M. A. T. *J. chem. Soc.* (1944) 126
[192] Grob, C. A. and Ankli, P. *Helv. chim. Acta* **33** (1950) 273
[193] Grob, C. A. and Utzinger, H. *Helv. chim. Acta* **37** (1954) 1256
[194] Angelico, F. *Atti Accad. naz. Lincei* **14** (1950) (*a*) I, 699; (*b*) II, 167
[195] Ajello, T. *Gazz. chim. ital.* **69** (1939) 453
[196] Ajello, T. *Gazz. chim, ital.* **68** (1938) 602
[197] Ajello, T. *Gazz. chim. ital.* **69** (1939) 315
[198] *Beilstein's Handbuch der Anorganischen Chemie*, Vol. 22, Suppl. 2, p. 495
[199] Rogers, M. A. T. *J. chem. Soc.* (1943) 598
[200] Gabriel, S. *Ber. dtsch. chem. Ges.* **41** (1908) 1127
[201] Rüdenburg, K. *Ber. dtsch. chem. Ges.* **46** (1913) 3555
[202] Ajello, T. and Sigillò, G. *Gazz. chim. ital.* **68** (1938) 681
[203] Bülow, C. *Ber. dtsch. chem. Ges.* **35** (1902) 4311; Blaise, E. E. *C.r. hebd. Séanc. Acad. Sci., Paris* **172** (1921) 221
[204] Bülow, C. and Klemann, E. *Ber. dtsch. chem. Ges.* **40** (1907) 4749
[205] Angelico, F. *Gazz. chim. ital.* **39** (1909) II, 134
[206] Angelico, F. and Labisi, C. *Gazz. chim. ital.* **40** (1910) I, 411
[207] Angelico, F. and Monforte, F. *Gazz. chim. ital.* **53** (1923) 795
[208] Saunders, K. H. *The Aromatic Diazo-Compounds*, p. 28, London (Arnold) 1949
[209] Ciamician, G. and Silber, P. *Ber. dtsch. chem. Ges.* **17** (1884) 553; **19** (1886) 3027
[210] Ciamician, G. *Gazz. chim. ital.* **16** (1886) 543
[211] Fischer, H. and Bartholomäus, E. *Hoppe-Seyler's Z. physiol. Chem.* **87** (1913) 255
[212] Siedel, W. *Justus Liebigs Annln Chem.* **554** (1943) 144
[213] (*a*) Widman, O. and Almström, G. K. *Justus Liebigs Annln Chem.* **400** (1913) 86; (*b*) Almström, G. K. *ibid.* **411** (1916) 350; **416** (1918) 279; (*c*) Benary, E. and Silbermann, B. *Ber. dtsch. chem. Ges.* **46** (1913) 1363; (*d*) Fischer, H. and Herrmann, M. *Hoppe-Seyler's Z. physiol. Chem.* **122** (1922) 1; (*e*) Fischer, H. and Loy, E. *ibid.* **128** (1923) 59; (*f*) Fischer, H. and Müller, J. *ibid.* **132** (1924) 72; (*g*) Fischer, H. and Hartmann, P. *ibid.* **226** (1934) 116; (*h*) Lukeš, R. *Colln. Czech. chem. Commun.* **1** (1929) 119; (*i*) Lukeš, R. and Šperling, V. *ibid.* **8** (1936) 464; (*j*) Langenbeck, H. and Boser, H. *Chem. Ber.* **84** (1951) 526; (*k*) Plieninger, H. and Decker, M. *Justus Liebigs Annln Chem.* **598** (1956) 198; (*l*) Treibs, A. and Ohorodnik, A. *ibid.* **611** (1958) 139, 149
[214] Grob, C. A. and Ankli, P. *Helv. chim. Acta* **32** (1949) 2010, 2023
[215] Davoll, J. *J. chem. Soc.* (1953) 3802
[216] Gilman, H. and Heck, L. L. *J. Am. chem. Soc.* **52** (1930) 4949
[217] (*a*) Oddo, B. *Ber. dtsch. chem. Ges.* **43** (1910) 1012; (*b*) Karrer, P. *ibid.* **50** (1917) 1499; (*c*) Tschelincev, V. V. and Skvorcov, D. K. *Zh. russk. fiz.-khim. Obshch.* **47** (1915) 170; (*d*) Oddo, B. *Gazz. chim. ital.* **50** (1920) II, 258; (*e*) Oddo, B. and Acuto, G. *ibid.* **65** (1935) 1029; (*f*) Ingraffia, F. *ibid.* **63** (1933) 584; Fischer, H., Baumgartner, M. and Plötz, E. *Justus Liebigs Annln Chem.* **493** (1932) 1; (*h*) Fischer, H. and Orth, H. *ibid.* **502** (1933) 237; (*i*) Buu-Hoi, Ng. Ph. and Hoán, Nguyen *Recl Trav. chim. Pays-Bas Belg.* **68** (1949) 5
[218] (*a*) Oddo, B. and Moschini, A. *Gazz. chim. ital.* **42** (1912) II, 244; (*b*) *Org. Synth.* (Coll.) (1943) 198; (*c*) Tschelincev, V. V. and Karmanov, S. G. *Zh. russk. fiz.-khim. Obshch.* **47** (1915) 161
[219] (*a*) Hess, K. *Ber. dtsch. chem. Ges.* **46** (1913) 3125; (*b*) Gilman, H. and Parker, H. H. *J. Am. chem. Soc.* **46** (1924) 2823; Gilman, H. and Pickers, R. M. *ibid.* **47** (1925) 245; McCay, C. M. and Schmidt, C. L. A. *ibid.* **48** (1926) 1933
[220] (*a*) Oddo, B. and Mameli, R. *Gazz. chim. ital.* **43** (1913) II, 504; **44** (1914) II, 162; (*b*) Plancher, G. and Tanzi, B. *Atti Accad. naz. Lincei* **23** (1914) II, 412; (*c*) Hess, K., Wissing, F. and Suchier, A. *Ber. dtsch. chem. Ges.* **48** (1915) 1865; (*d*) Andrews, L. H. and McElvain, S. M. *J. Am. chem. Soc.* **51** (1929) 887; (*e*) Fischer, H., Baumann, E. and Riedl, H. J. *Justus Liebigs Annln Chem.* **475** (1929) 205
[221] Booth, H., Johnson, A. W., Markham, E. and Price, R. *J. chem. Soc.* (1959) 1587
[222] Alessandri, L. *Atti Accad. naz. Lincei* **23** (1914) II, 65; **24** (1915) II, 194; Tschelincev, V. V. and Terentjev, A. *Ber. dtsch. chem. Ges.* **47** (1914) 2647, 2652; Putochin, N. *ibid.* **59** (1926) 1987
[223] Tschelincev, V. V., Tronov, B. and Terentjev, A. *Zh. russk. fiz.-khim. Obshch.* **47** (1915) 1211; Mingoia, Q. *Gazz. chim. ital.* **58** (1928) 673; **62** (1932) 844; Oddo, B. and Perotti, L. *ibid.* **60** (1930) 13; Oddo, B. and Cambieri, F. *ibid.* **70** (1940) 559
[224] Kharasch, M. S. and Reinmuth, O. *Grignard Reactions of Nonmetallic Substances*, New York (Prentice-Hall) 1954
[225] Kornblum, N., Smiley, R. A., Blackwood, R. K. and Iffland, D. C. *J. Am. chem. Soc.* **77** (1955) 6269

REFERENCES

226 Brady, O. L. and Jakobovits, J. *J. chem. Soc.* (1950) 767
227 Küster, W. and Maag, W. *Hoppe-Seyler's Z. physiol. Chem.* **121** (1922) 157
228 (a) Pictet, A. *Ber. dtsch. chem. Ges.* **37** (1904) 2792; (b) Crépieux, P. with **28** (1895) 1904
229 Wibaut, J. P. and Dingemanse, E. *Recl Trav. chim. Pays-Bas Belg.* **42** (1923) 1033;
  Wibaut, J. P. *ibid.* **45** (1926) 657; Oosterhuis, A. G. and Wibaut, J. P. *ibid.* **55** (1936)
  348; Wibaut, J. P. and Gitsels, H. P. L. *ibid.* **57** (1938) 755; **59** (1940) 1093; Wibaut,
  J. P. and Dhont, J. *ibid.* **62** (1943) 272; Späth, E. and Kainrath, P. *Ber. dtsch. chem. Ges.*
  71 (1938) 1276
230 Ciamician, G. and Magnaghi, P. *Recl Trav. chim. Pays-Bas Belg.* **18** (1885) 1828
231 Jacobson, I. A., Jr., Heady, H. H. and Dinneen, G. U. *J. phys. Chem., Ithaka* **62** (1958)
  1563
232 Johnson, A. W., Markham, E., Price, R. and Shaw, K. B. *J. chem. Soc.* (1958) 4254
233 Hinman, R. L. and Theodoropulos, S. *J. org. Chem.* **28** (1963) 3052
234 (a) *Am. Petroleum Inst. Res. Proj.* **44**, Nos. 109 and 734; (b) *ibid.* No. 692
235 Scrocco, M. and Caglioti, L. *Atti Accad. naz. Lincei* **24** (1958) 316
236 Andrisano, R. and Pappalardo, G. *Gazz. chim. ital.* **85** (1955) 1430
237 Elpern, B. and Nachod, F. C. *J. Am. chem. Soc.* **72** (1950) 3379
238 King, S. M., Bauer, C. R. and Lutz, R. E. *J. Am. chem. Soc.* **73** (1951) 2253
239 Badger, G. M., Harris, R. L. N., Jones, R. A. and Sasse, J. M. *J. chem. Soc.* (1962) 4329
240 Kuhn, L. P. and Kleinspehn, G. G. *J. org. Chem.* **28** (1963) 721
241 Bellamy, L. J., Hallam, H. E. and Williams, R. L. *Trans. Faraday Soc.* **54** (1958) 1120;
  Bellamy, L. J. and Hallam, H. E. *ibid.* **55** (1959) 220
242 Mirone, P. and Lorenzelli, V. *Annali Chim.* **49** (1959) 59
243 Happe, J. A. *J. phys. Chem., Ithaka* **65** (1961) 72
244 Abraham, R. J. and Bernstein, H. J. *Can. J. Chem.* **39** (1961) 216
245 Reddy, G. S. and Goldstein, J. H. *J. Am. chem. Soc.* **83** (1961) 5020
246 Whipple, E. B., Chiang, Y. and Hinman, R. L. *J. Am. chem. Soc.* **85** (1963) 26
247 Chiang, Y. and Whipple, E. B. *J. Am. chem. Soc.* **85** (1963) 2763
248 Davies, W. A. M., Pinder, A. R. and Morris, I. G. *Tetrahedron* **18** (1962) 405
249 Kleinspehn, G. G. and Briod, A. E. *J. org. Chem.* **26** (1961) 1652
250 Castro, A. J., Deck, J. F., Hugo, M. T., Lowe, E. J., Marsh, J. P. and Pfeiffer, R. J. *J. org. Chem.* **28** (1963) 857
251 Bullock, E., Grigg, R., Johnson, A. W. and Wasley, J. W. F. *J. chem. Soc.* (1963) 2326
252 Anthony, W. C. *J. org. Chem.* **25** (1960) 2049
253 Skell, P. S. and Bean, G. P. *J. Am. chem. Soc.* **84** (1962) 4655
254 Acheson, R. M. and Vernon, J. M. *J. chem. Soc.* (1963) 1008
255 Rapoport, H. and Castagnoli, N. *J. Am. chem. Soc.* **84** (1962) 2178
256 Treibs, A. and Ohorodnik, A. *Justus Liebigs Annln Chem.* **611** (1958) 149
257 Treibs, A. and Dietl, A. *Justus Liebigs Annln Chem.* **619** (1958) 80
258 Treibs, A. and Fligge, M. *Justus Liebigs Annln Chem.* **652** (1962) 176
259 Plieninger, H., Lerch, U. and Kurze, J. *Angew. Chem. (int. Edn)* **2** (1963) 483; Plieninger,
  H., Bauer, H., Buhler, W., Kurze, J. and Lerch, U. *Justus Liebigs Annln Chem.* **680**(1964) 69
260 Kreutzberger, A. and Kalter, P. A. *J. org. Chem.* **26** (1961) 3790
261 Gross, H. and Gloede, J. *Angew. Chem. (int. Edn)* **2** (1963) 262
262 Badger, G. M. and Harris, R. L. N., Private communication
263 Hess, K. and Wissing, F. *Ber. dtsch. chem. Ges.* **48** (1915) 1884
264 Fischer, H. and Ernst, P. *Justus Liebigs Annln Chem.* **447** (1926) 139
265 Corwin, A. H. and Viohl, P. *J. Am. chem. Soc.* **66** (1944) 1137
266 Arsenault, G. P. and MacDonald, S. F. *Can. J. Chem.* **39** (1961) 2043
267 Kleinspehn, G. G. and Corwin, A. H. *J. Am. chem. Soc.* **75** (1953) 5295
268 Webb, J. L. A. and Threlkeld, R. R. *J. org. Chem.* **18** (1953) 1406
269 Pratesi, P. *Atti Accad. naz. Lincei* **17** (1933) 173
270 Mathewson, J. H. *J. org. Chem.* **28** (1963) 2153
271 Atkinson, J. H., Grigg, R. and Johnson, A. W. *J. chem. Soc.* (1964) 893
272 Kil'disheva, O. V., Lin'kova, M. G. and Knunyants, I. L. *Chem. Abstr.* **52** (1958) 3775
273 Sprio, V. and Fabra, I. *Annali Chim.* **46** (1956) 263
274 Gardner, T. S., Wenis, E. and Lee, J. *J. org. Chem.* **24** (1959) 570
275 Fournari, P. and Tirouflet, J. *Bull. Soc. chim. Fr.* (1963) 484
276 Giambrone, S. and Fabra, J. *Annali Chim.* **50** (1960) 237
277 Tedder, J. M. and Webster, B. *J. chem. Soc.* (1960) 3270
278 Sprio, V. *Gazz. chim. ital.* **86** (1956) 95
279 *U.S. Patent* 2,805,227 (1957)
280 Tedder, J. M. and Webster, B. *J. chem. Soc.* (1962) 1638
281 Sprio, V. and Fabra, I. *Annali Chim.* **49** (1959) 2053
282 Terent'ev, A. P. and Domborvskii, A. V. *Zh. obshch. Khim.* **21** (1951) 278
283 Gronowitz, S., Hörnfeldt, A.-B., Gestblom, B. and Hoffman, R. A. *Ark. Kemi Miner. Geol.*
  18 (1961) 151; *J. org. Chem.* **26** (1961) 2615

[284] Olsen, R. K. and Snyder, H. R. *J. org. Chem.* **28** (1963) 3050
[285] Söderbäck, E., Gronowitz, S. and Hörnfeldt, A.-B. *Acta chem. scand.* **15** (1961) 227
[286] Gronowitz, S., Hörnfeldt, A.-B., Gestblom, B. and Hoffman, R. A. *Ark. Kemi Miner. Geol.* **18** (1961) 133
[287] Ochiai, E., Tsuda, K. and Yokoyama, J. *Ber. dtsch. chem. Soc.* **68** (1935) 2291
[288] *U.S. Patent* 2,488,336 (1949)
[289] *U.S. Patent* 2,797, 228 (1947)
[290] Hobbs, C. F., McMillin, C. K., Papadopoulos, E. P. and VanderWerf, C. A. *J. Am. chem. Soc.* **84** (1962) 43
[291] Ciamician, G. L. and Dennstedt, M. *Ber. dtsch. chem. Ges.* **15** (1882) 2579
[292] Tschelinzeff, W. and Maxoroff, B. *Ber. dtsch. chem. Ges.* **60** (1927) 194
[293] Safonova, E. N., Belikov, V. M. and Novikov, S. S. *Chem. Abstr.* **54** (1960) 1487
[294] Reppe, W. *Justus Liebigs Annln Chem.* **601** (1956) 128
[295] Fischer, H. and Rose, W. *Justus Liebigs Annln Chem.* **519** (1935) 21
[296] Fischer, H. and Höfelmann, H. *Justus Liebigs Annln Chem.* **533** (1938) 216
[297] Booth, H., Johnson, A. W. and Johnson, F. *J. chem. Soc.* (1962) 98
[298] Fischer, H. and Klarer, J. *Justus Liebigs Annln Chem.* **450** (1926) 181
[299] Allen, C. F. H. and Wilson, C. V. *Org. Synth.* **27** (1947) 33
[300] Fehrlin, H. C. *Ber. dtsch. chem. Ges.* **22** (1889) 553
[301] Dilthey, W., Hurtig, G. and Passing, H. *J. prakt. Chem.* **156** (1940) 27
[302] Treibs, A. and Ott, W. *Justus Liebigs Annln Chem.* **577** (1952) 119
[303] Evans, G. G. *J. Am. chem. Soc.* **73** (1951) 5230
[304] Treibs, A. and Zimmer-Galler, R. *Justus Liebigs Annln Chem.* **664** (1963) 140
[305] Acheson, R. M. and Vernon, J. M. *J. chem. Soc.* (1961) 457
[306] Acheson, R. M. and Vernon, J. M. *J. chem. Soc.* (1962) 1148
[307] Mandell, L., Piper, J. U. and Pesterfield, C. E. *J. org. Chem.* **28** (1963) 574
[308] Wittig, G. and Reichel, B. *Chem. Ber.* **96** (1963) 2851
[309] Dennstedt, M. and Voigtländer, F. *Ber. dtsch. chem. Ges.* **27** (1894) 478
[310] Chelintzev, V. V., Tronov, B. V. and Voskresenskii, B. I. *Chem. Abstr.* **9** (1915) 3055
[311] Fischer, H. and Walach, B. *Justus Liebigs Annln Chem.* **450** (1926) 109
[312] Piloty, O. and Stock, J. *Ber. dtsch. chem. Ges.* **46** (1913) 1008
[313] Smith, G. F. *Adv. heterocycl. Chem.* **2** (1963) 287
[314] Bonnett, R. and White, J. D. *J. chem. Soc.* (1963) 648
[315] Piloty, O. and Thannhauser, S. J. *Justus Liebigs Annln Chem.* **390** (1912) 191
[316] Piloty, O., Wilke, K. and Blömer, A. *Justus Liebigs Annln Chem.* **407** (1915) 37
[317] Mayo, P. de and Reid, S. T. *Chemy Ind.* (1962) 1576
[318] Seebach, D. *Chem. Ber.* **96** (1963) 2723
[319] Fabra, I. *Annali Chim.* **50** (1960) 1640
[320] Gloss, G. L. and Schwartz, G. M. *J. org. Chem.* **26** (1961) 2609
[321] Rice, H. L. and Londergan, T. E. *J. Am. chem. Soc.* **77** (1955) 4678
[322] Jones, R. L., Rees, C. W. and Smithen, C. E. *Proc. chem. Soc.* (1964) 217
[323] Ajello, T., Spiro, V. and Vaccaro, G. *Gazz. chim. ital.* **89** (1959) 2232
[324] Capuano, S. and Giammanco, L. *Gazz. chim. ital.* **85** (1955) 217; **86** (1956) 119
[325] Kuhn, R. and Kainer, H. *Justus Liebigs Annln Chem.* **578** (1952) 226
[326] Carles, P. *C.r. hebd. Séanc. Acad. Sci., Paris* **254** (1962) 677
[327] Dahl, J. P. and Hansen, A. E. *Theor. Chim. Acta* **1** (1963) 199
[328] Brown, R. D. and Coller, B. A. W. *Aust. J. Chem.* **12** (1959) 152
[329] Treibs, A. and Bader, H. *Justus Liebigs Annln Chem.* **627** (1959) 188
[330] Ajello, T. and Giambrone, S. *Chem. Abstr.* **48** (1954) 10002
[331] Rips, R. and Buu-Hoi, Ng. Ph. *J. org. Chem.* **24** (1959) 372
[332] Treibs, A. and Scherer, J. *Justus Liebigs Annln Chem.* **577** (1952) 139
[333] Treibs, A. and Derra-Scherer, H. *Justus Liebigs Annln Chem.* **589** (1954) 188
[334] Hertz, W. and Courtney, C. F. *J. Am. chem. Soc.* **76** (1954) 576
[335] Gardner, T. S., Wenis, E. and Lee, J. *J. org. Chem.* **23** (1958) 823
[336] Sprio, V., Madonia, P. and Caronia, R. *Chem. Abstr.* **53** (1959) 16105
[337] Ghighi, E. and Druisiani, A. *Chem. Abstr.* **51** (1957) 6602; **52** (1958) 11818
[338] Smissman, E. E., Graber, M. B. and Winzler, R. J. *J. Am. pharm. Ass.* **45** (1946) 509; *U.S. Patent* 2,898,344 (1960); cf. Raecke, B. *Angew. Chem.* **70** (1958) 1
[339] Rinkes, I. J. *Recl Trav. chim. Pays-Bas Belg.* **57** (1938) 423
[340] Nicolaus, R. A., Mangoni, L. and Misiti, D. *Annali Chim.* **46** (1956) 847
[341] Corwin, A. H., Bailey, W. A. and Viohl, P. *J. Am. chem. Soc.* **64** (1942) 1267
[342] Bridger, R. F. and Russell, G. A. *J. Am. chem. Soc.* **85** (1963) 3754
[343] Bullock, E., Johnson, A. W., Markham, E. and Shaw, K. B. *J. chem. Soc.* (1958) 1430
[344] Skell, P. S. and Bean, G. P. *J. Am. chem. Soc.* **84** (1962) 4660
[345] *U.S. Patent* 2,785, 181 (1957)
[346] *U.S. Patent* 2,962,503 (1961)

REFERENCES

[347] Overberger, C. G., Palmer, P. C., Macks, B. S. and Byrd, N. R. *J. Am. chem. Soc.* **77** (1955) 4100

[348] Kreutzberger, A. and Kalter, P. A. *J. phys. Chem., Ithaka* **65** (1961) 624

[349] Grabowski, J. and Marchlewski, L. *Ber. dtsch. chem. Ges.* **45** (1912) 453

[350] Grigg, R., Johnson, A. W. and Wasley, J. W. F. *J. chem. Soc.* (1963) 359

[351] Webb, J. L. A. *J. org. Chem.* **18** (1953) 1413

[352] see [341]

[353] Treibs, A., Schmidt, R. and Zinsmeister, R. *Chem. Ber.* **90** (1957) 79

[354] Rapoport, H. and Willson, C. D. *J. Am. chem. Soc.* **84** (1962) 630

[355] Plieninger, H., Bauer, H. and Katritzky, A. R. *Justus Liebigs Annln Chem.* **654** (1962) 165

[356] Plieninger, H. and Bauer, H. *Angew. Chem.* **73** (1961) 433

[357] Atkinson, R. S. and Bullock, E. *Can. J. Chem.* **41** (1963) 625

[358] Fischer, H. and Gangl, K. *Hoppe-Seyler's Z. physiol. Chem.* **267** (1941) 188

[359] Hüni, A. and Frank, F. *Hoppe-Seyler's Z. physiol. Chem.* **282** (1947) 96

[360] Booth, H., Johnson, A. W., Johnson, F. and Langdale-Smith, R. A. *J. chem. Soc.* (1963) 650

[361] Reinecke, M. G., Johnson, H. W. and Sebastian, J. F. *J. Am. chem. Soc.* **85** (1963) 2859

[362] Safonova, E. N., Belikov, V. M. and Novikov, S. S. *Chem. Abstr.* **54** (1960) 1486

[363] Novikov, S. S. and Belikov, V. M. *Chem. Abstr.* **54** (1960) 1487

[364] Hodge, P. and Rickards, R. W. *J. chem. Soc.* (1963) 2543

[365] Jacobson, I. A. and Jensen, H. B. *J. phys. Chem., Ithaka* **66** (1962) 1245

[366] Patterson, J. M. and Drenchko, P. *J. org. Chem.* **27** (1962) 1650

[367] Wibaut, J. P. and Gitsels, H. P. L. *Recl Trav. chim. Pays-Bas Belg.* **59** (1940) 1093

[368] Griffin, C. E. and Obrycki, R. *J. org. Chem.* **29** (1964) 3090

[369] Mirone, P. and Lorenzelli, V. *Annali Chim.* **48** (1958) 72; Khan, M. K. A. and Morgan, K. J. *J. chem. Soc.* (1964) 2579

[370] Budzikiewicz, H., Djerassi, C., Jackson, A. H., Kenner, G. W., Newman, D. J. and Wilson, J. M. *J. chem. Soc.* (1964) 1949

[371] Anderson, H. J. and Hopkins, L. C. *Can. J. Chem.* **42** (1964) 1279

[372] Hodge, P. and Rickards, R. W. *J. chem. Soc.* (1965) 459

[373] Solony, N., Birss, F. W. and Greenshield, J. B. *Can. J. Chem.* **43** (1965) 1569

[374] Khan, M. K. A. and Morgan, K. J. *Tetrahedron* (1965) 2197

[375] Jones, R. A. and Moritz, A. G. *Spectrochim. Acta* **21** (1965) 295; Guy, R. W. and Jones, R. A. *ibid.* 1011

[376] Grigg, R. *J. chem. Soc.* (1965) 5149

[377] Anderson, H. J. *Can. J. Chem.* **43** (1965) 2387

[378] Elvidge, J. A. *Chem. Commun.* (1965) 160

[379] Ermili, A., Castro, A. J. and Westfall, P. A. *J. org. Chem.* **30** (1965) 339

[380] Rapoport, H. and Bordner, J. *J. org. Chem.* **29** (1964) 2727

[381] Atkinson, J. H. and Johnson, A. W. *J. chem. Soc.* (1965) 2614

[382] Bordner, J. and Rapoport, H. *J. org. Chem.* **30** (1965) 3824

[383] Jackson, A. H., Kenner, G. W. and Warburton, D. *J. chem. Soc.* (1965) 1328

[384] Griffin, C. E., Peller, R. P. and Peters, J. A. *J. org. Chem.* **30** (1965) 91

[385] Rees, C. W. and Sabet, C. R. *J. chem. Soc.* (1965) 687

[386] Bonnett, R., Gale, I. A. D. and Stephenson, G. F. *J. chem. Soc.* (1965) 1518

[387] Anderson, H. J. and Shu-Fan Lee, *Can. J. Chem.* **43** (1965) 409

[388] Gritter, R. J. and Chriss, R. J. *J. org. Chem.* **29** (1965) 1163

[389] Wolthuis, E., Vander Jagt, D., Mels, S. and de Boer, A. *J. org. Chem.* **30** (1965) 190

[390] Atkinson, J. H., Atkinson, R. S. and Johnson, A. W. *J. chem. Soc.* (1964) 5999

[391] Duffield, A. M., Beugelmans, R., Budzikiewicz, H. and Lightner, D. A. *J. Am. chem. Soc.* **87** (1965) 805

[392] Dolphin, D., Grigg, R., Johnson, A. W. and Leng, J. *J. chem. Soc.* (1965) 1460

[393] Nicoletti, R. and Forcellese, M. L. (a) *Gazz. chim. ital.* **95** (1965) 83; (b) *Tetrahedron Lett.* (1965) 3033; (c) *ibid.* 153

[394] Jones, R. A. and Lindner, J. A. *Aust. J. Chem.* **18** (1965) 875

[395] Hayes, A., Jackson, A. H., Judge, J. M. and Kenner, G. W. *J. chem. Soc.* (1965) 4385

[396] Griffin, C. E., Peller, R. P., Martin, K. R. and Peters, J. A. *J. org. Chem.* **30** (1965) 97

[397] Castro, A., Lowell, J. and Marsh, J. *J. heterocycl. Chem.* **2** (1965) 473; Bean, G. P. *ibid.*

[398] Castro, A. J., Deck, J. F., Ling, N. C. and Marsh, J. P. *J. org. Chem.* **30** (1965) 344

5

PYRIDINE AND ITS DERIVATIVES

(1) PHYSICAL PROPERTIES

(a) General. Dipole Moments

The history of the discovery of pyridine and of the development of its structure was outlined in Chapter 1, and its general character as an aromatic compound, as shown in its geometry and stability, was discussed in Chapter 2. The molecular dimensions of three important pyridine derivatives, 2-pyridone[1a], 2-pyridthione[1b], and nicotinic acid[1c] are indicated in the skeletons (1), (2), and (3). Data are available for pyridine hydrochloride[2a], 1-hydroxypyridinium chloride[2b], 4-nitropyridine 1-oxide[3a] and trans-4,4'-azo-pyridine 1,1'-dioxide[3b].

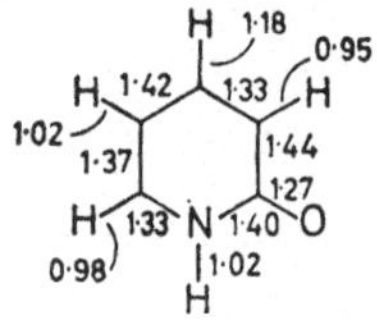

<NCO	121°	<$C_{(3)}C_{(4)}C_{(5)}$	122°
$OC_{(2)}C_{(3)}$	126	$C_{(4)}C_{(5)}C_{(6)}$	116
$NC_{(2)}C_{(3)}$	113	$C_{(5)}C_{(6)}N$	122
$C_{(2)}C_{(3)}C_{(4)}$	122	$C_{(6)}NC_{(2)}$	125

(1)

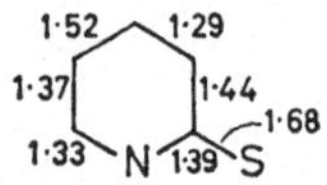

<NCS	119°	<$C_{(3)}C_{(4)}C_{(5)}$	123°
$SC_{(2)}C_{(3)}$	127	$C_{(4)}C_{(5)}C_{(6)}$	111
$NC_{(2)}C_{(3)}$	115	$C_{(5)}C_{(6)}N$	127
$C_{(2)}C_{(3)}C_{(4)}$	123	$C_{(6)}NC_{(2)}$	122

(2)

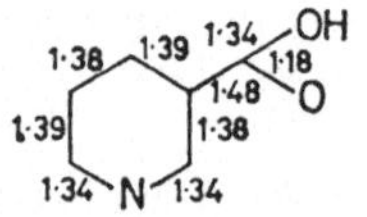

<$NC_{(2)}C_{(3)}$	124°	<$C_{(6)}NC_{(2)}$	117°
$C_{(2)}C_{(3)}C_{(4)}$	118	$C_{(2)}C_{(3)}C$	124
$C_{(3)}C_{(4)}C_{(5)}$	119	$CC_{(3)}C_{(4)}$	118
$C_{(4)}C_{(5)}C_{(6)}$	119	OCO	122
$C_{(5)}C_{(6)}N$	122	(HO)$CC_{(3)}$	114
		$OCC_{(3)}$	124

(Bond lengths in Ångstrom units)

(3)

In *Table 5.1* are listed the melting and boiling points of pyridine, the alkyl and aryl pyridines, and of some of their derivatives.

120

Table 5.1 *Pyridine and its Alkyl, Alkenyl and Aryl Derivatives**

Pyridine	M.p. (or F.p.),°C	B.p.,°C	Derivatives (M.p.,°C)
Parent	−41·7	115·25	picrate (164); styphnate (184·5–185·5); methiodide (118)‡; mercurichloride (177–8); chloroplatinate (264); hydrobromide (206); perchlorate (288)
2-Me†	−66·8	129·4	picrate (169–171); styphnate (180); methiodide (224); mercurichloride (154–5); chloroplatinate (178[195]);
3-Me	−18·25	144·1	picrate (149–150); styphnate (153–4); methiodide (92); mercurichloride (147–9); chloroplatinate (201–2)
4-Me	3·58	145·35	picrate (167–8); methiodide (152) [+ 2I (101); + 4I (63); + 6I (81·5)]; mercurichloride (128–9); chloroplatinate (231)
2-Et	−63·1	148–9	picrate (107·8–108·3)
3-Et	−76·9	165·6–166	picrate (128·1–128·5)
4-Et	−90·5	169·6–170·0/750 mm	picrate (169·4–169·8)
2-n-Pr	(conyrine)	166–8	picrate (74·6–75·1)
3-n-Pr		182–4	picrate (99·8–100·2)
4-n-Pr		184–6	picrate (134)
2-i-Pr		159·8/753 mm	picrate (118·1–118·7)
3-i-Pr		179·3/744 mm	picrate (138·1–138·6)
4-i-Pr		181·5/743 mm	picrate (138·4–139·6)
2-t-Bu	−32·8	169·0/743 mm	picrate (104·6–105·2)
3-t-Bu	−42·9	194·3/742 mm	picrate (153·9–154·4)
4-t-Bu	−39·7	196·3/749 mm	picrate (130·9–131·4)
2,3-Me₂†	−15·3	161·15	picrate (183–4); B.HCl.2HgCl₂ (ca. 120°); B.HCl.5HgCl₂ (191–3)
2,4-Me₂†	−64·0	158·4	picrate (184–5); hydrochloride (195–7, hygroscopic); hydrobromide (189–190)
2,5-Me₂†	−15·6	157·0	picrate (170)
2,6-Me₂†	− 6·16	144·0	picrate (163–4); methiodide (238); hydrochloride (238–9); hydrobromide (210)
3,4-Me₂†	−11·1	179·1	picrate (163); B.HCl.2HgCl₂ (146–8)
3,5-Me₂†	− 6·5	171·9	picrate (235–6)
4-Et-2-Me† (α-collidine)		179–180	picrate (141–2)
5-Et-2-Me† (aldehyde collidine)	−70·3	178·3	picrate (167–8)
6-Et-2-Me†		160–1	picrate (130)
3-Et-4-Me† (β-collidine)		195–6	picrate (148–150)
4-Et-3-Me†		192–3	picrate (144–5)
2,6-Et₂		171/744 mm	picrate (116–7); methiodide (142)
2,6-(n-Pr)₂⁶ᵃ		206–7	picrate (83·5–84)

Pyridine	M.p. (or F.p.),°C	B.p.,°C	Derivatives (M.p.,°C)
2,6-(i-Pr)$_2$	2·5	90/25 mm 194·1–194·5/746 mm	picrate (114); methiodide (118·5); aurichloride (167·0–167·1)
2-t-Bu-6-i-Pr	6·6	94/23 mm	
2,6-(t-Bu)$_2$	2·2	100–101/23 mm	aurichloride (184·2–184·5)
2,3,4-Me$_3$†		187	picrate (165)
2,3,5-Me$_3$†		184	picrate (181)
2,3,6-Me$_3$†		172·8	picrate (144·5–145·5)
2,4,5-Me$_3$†		188	picrate (159–160)
2,4,6-Me$_3$†	−44·46	170·2/749 mm	picrate (156)
2,3,5,6-Me$_4$†	77–8	197–8	picrate (173–4)
2-Vinyl		159–160 69–71/30 mm	picrate (159·5) platinichloride (174 [dec.])
3-Vinyl		162 67–8/18 mm	picrate (143–4)
4-Vinyl		65/15 mm 77–8/14 mm	picrate (198–9[dec.])
2-C⋮CH[6b]			
3-C⋮CH[6d]	38·5	83–4/30 mm	
2-Ph		268–9 140/12 mm	picrate (174–5)
3-Ph		268–270/749 mm	picrate (161–163·5)
4-Ph	77–8	274–5	picrate (195–6)
2-CH$_2$Ph		275–6	picrate (141·5–142)
3-CH$_2$Ph		287–8	picrate (119)
4-CH$_2$Ph		285–6/750 mm	picrate (141–2)
trans-2-Styryl†[6f, g]	92–93·5	194/14 mm	methiodide (229–230)
cis-2-Styryl†[6g]		145–155/10 mm	methiodide (180–3)
trans-3-Styryl†[6h]	72–3		
cis-3-Styryl†[6i]		103–4/0·3 mm	
trans-4-Styryl†[6f]	128		methiodide (218); picrate (213)
cis-4-Styryl†[6f]		105–6/0·4 mm	
2-CH:CH.(2-C$_5$H$_4$N)**[6j]	119		
2-CH:CH.(3-C$_5$H$_4$N)**[6j]	72		
2-CH:CH.(4-C$_5$H$_4$N)**[6j]	72–3		
2-Tolazole†[6l, m]		152/1 mm 136–8/4·5 mm	picrate (151–3)
4-Tolazole†[6e]	95–95·5	109–110/0·5 mm	
2,6-(C⋮CPh)$_2$[6k]	137–8		
2-C⋮C.(2-C$_5$H$_4$N)	69–70		
2-C⋮C.(3-C$_5$H$_4$N)	41–2		
2-C⋮C.(4-C$_5$H$_4$N)	64–5		
2-(2-C$_5$H$_4$N) §	71–2	272·2	picrate (158)
2-(3-C$_5$H$_4$N) §		287–9	picrate (149·5); dipicrate (165–8)
2-(4-C$_5$H$_4$N) §	61·1–61·5	280–2	picrate (215–6)
3-(3-C$_5$H$_4$N) §	68	291–2	dipicrate (232)
3-(4-C$_5$H$_4$N) §	63	295/759 mm	dipicrate (199–201)
4-(4-C$_5$H$_4$N) §	114	305	picrate (257)
2,6-(2-C$_5$H$_4$N)$_2$§	84–5	370	picrate (206)
2,2′,2″,2‴-Quaterpyridyl	219–220		dipicrate (312[dec.])

* For an exhaustive tabulation see[4c]; valuable information on the purification of pyridine and alkylpyridines in[4b, 5].

† The following trivial names are used: methylpyridines—picolines; dimethylpyridines—lutidines; trimethyl-pyridines and ethyldimethylpyridines—collidines; tetramethylpyridines—parvolines; styrylpyridines—stilbazoles; phenethylpyridines—tolazoles.

‡ For quarternary salts of pyridine and its homologues see Kosower and Skorcz[54].

** trans-Isomer. The cis-isomer is a liquid.

§ For polypyridyls see especially[4c].

Pyridine has a higher boiling point than benzene, presumably because of association caused by its greater polarity. Interference with this association by substituents adjacent to the nitrogen atom[8b] causes 2-alkyl- and 2,6-dialkyl-pyridines to boil lower than their isomers. The closeness of the boiling points of a number of the alkylpyridines, as well as the great difficulty of removing traces of water, makes purification difficult. The references to *Table 5.1* describe the preparation of pyridine and a number of alkylpyridines in a high state of purity, and a large range of thermodynamic data for these is available[4a].

The effect of pyridinic nitrogen in rendering the azapyrroles markedly more water-soluble than pyrrole will be dealt with latter on. The effect is startling when pyridine is compared with benzene; pyridine and the picolines are miscible with water in all proportions at all temperatures[7a], whilst the dimethylpyridines and the ethylpyridines are partially miscible with water and give closed-loop solubility curves[7a, b]. Data are given in *Table 5.2*. Whilst 3-methylpyridine is completely miscible with deuterium

Table 5.2. Critical Solution Data for Alkylpyridines in Water[7a, b]

Compound	Lower Consolute Point		Upper Consolute Point	
	Temp.°	Wt. per cent water	Temp.°	Wt. per cent water
Dimethylpyridine, 2,3-	16·5	74	192·6	61·5
2,4-	23·4	74·5	188·7	63
2,5-	13·1	73	206·9	62
2,6-	34·0	70	230·7	59
3,4-	−3·6	75·5	162·5	64
3,5-	−12·5	74·5	192·0	63
Ethylpyridine, 2-	−5·0	66	231·4	58
3-	−35 (est.)		195·6	63
4-	−19·0	72	181·8	63·5

oxide, the 2- and 3-isomers are only partly miscible[7c]. Alkyl groups exert two opposing effects on the solubility of pyridine; they enlarge the hydrophobic part of the molecule but also increase the electron density at the nitrogen atom (more from $C_{(4)}$ than from $C_{(3)}$) and so influence the hydrogen-bonding capability of the molecule. Consequently, a 4-alkyl- is more soluble than a 3-alkyl-pyridine, and 3,5-dimethyl- is more soluble than 3-ethyl-pyridine. For the alkylpyridines in dilute aqueous solution the logarithms of the activity coefficients at a given temperature are additive functions of the groups composing the molecules. The 2-alkylpyridines are more extensively hydrogen-bonded with water than are their isomers, as is shown by the entropies of hydration[8], and the temperature of least miscibility is higher for the 2- than for the 3- and 4-alkylpyridines. Comparison of the thermodynamic properties of aqueous solutions of alkylpyridines with those of other organic solutes shows that the structures of these solutions are not exceptional. Their closed-loop solubility curves result from a fine balance between hydrophobic and hydrophilic properties[7b, 8a].

The dipole moments of some pyridines are given in *Table 5.3*. The results for the methyl- and chloro-pyridines suggest that for these substituents

Table 5.3. Dipole Moments of Pyridines and Pyridine 1-Oxides*

Substituent	Solvent†	D.M. (D)	References
(Pyridine)	Vapour phase	2·15	9
	Benzene, Dioxan	2·22	10, 11
2-Me		1·96	14
3-Me		2·41	12, 15a
4-Me	Benzene	2·61	10, 11, 15a
3-Et		2·47	12
2,6-Me₂		1·87	14
3,5-Me₂		2·55	12
Acyl and carboxyl derivatives			
4-COMe	Benzene	2·41	11
4-CO₂H	Dioxan	2·7	
4-CONH₂	Dioxan	3·88	10
4-CO₂Et	Benzene	2·53	
Amines and substituted amines			
2-NH₂‡	Benzene	2·04	10b
3-NH₂	Benzene	3·12	10b
4-NH₂	Dioxan	4·36	10a
4-NHMe		4·42	
3-Me-4-NHMe	Benzene	4·32	12
3-Et-4-NHMe		4·33	
3,5-Me₂-4-NHMe		3·90	
2-NH₂-3-Me		2·17	
2-NH₂-4-Me		2·27	10b
2-NH₂-5-Me		2·02	
2-NH₂-6-MMe		1·65	
4-NMe₂	Benzene	4·31, 4·24	11, 12
3-Me-4-NMe₂		3·48	
3-Et-4-NMe₂		3·50	12
3-i-Pr-4-NMe₂		3·37	
3,5-Me₂-4-NMe₂		3·28	
2,6-(NH₂)₂		1·46	10b
Aryl substituents			
2-Ph		1·77	
3-Ph	Benzene	2·45	21b
4-Ph		2·50	
Cyanopyridines			
2-CN		5·24	13, 10b
3-CN	Benzene	3·46, 3·48	13, 10b
4-CN		1·65	10, 11, 13
Halogenopyridines			
2-Cl		3·25, 3·22	13, 14
3-Cl		2·02	13
4-Cl		0·78	10a, 11, 13
2-Br		3·21	
3-Br		2·02	13
4-Br		0·89	
2,6-Cl₂		3·65	
3,5-Cl₂	Benzene	0·95	
2,5-Br₂		2·33	
2,6-Br₂		3·54	
3,4-Br₂		1·16	13
3,5-Br₂		1·02	
2,3,6-Br₃		3·11	
2,4,6-Br₃		2·29	
3,4,5-Br₃		0·58	

Table 5.3 (continued)

Substituent	Solvent†	D.M. (D)	References
Hydroxypyridines			
2-OH**	Benzene	1·95	90
2-OH**	Dioxan	2·95	90
1-Me-2-:O	Benzene	4·04	16b
4-OMe	Dioxan	4·07	10a, 11
4-OH	Dioxan	6·0	10a
4-OMe	Benzene	2·96	10a, 11
Mercaptopyridines			
2-SH**	Benzene	2·78§	
2-SH**	Dioxan	5·29	⎱ 16b
1-Me-2-:S	Benzene	5·26	⎰
1-Me-2-:S	Dioxan	5·49	
Nitropyridines			
4-NO$_2$	Benzene	1·58	11

Pyridine 1-Oxides

Pyridine 1-oxide	Dioxan	4·32	16a
Pyridine 1-oxide	Benzene	4·24	16a
4-Me	Benzene	4·74	11
Acyl and carboxyl derivatives			
4-COMe	Benzene	3·19	
4-CO$_2$Et	Benzene	3·80	⎫
Amines and substituted amines			
4-NMe$_2$	Benzene	6·76	
Halogenoderivatives			⎬ 11
4-Cl	Benzene	2·82	
Hydroxy and substituted hydroxy derivatives			
4-MeO	Benzene	5·08	
Nitroderivatives			
4-NO$_2$	Benzene	ca.0	⎭

* At 25° except for 4-hydroxypyridine (50°), 1-methyl-2-pyridone (30°), 2-pyridthione (30°), and 1-methyl-2-pyridthione (30°).

† Dipole moment measurements in carbon tetrachloride and other halogenated solvents reveal weak inter-actions between pyridine and pyridine 1-oxide and these solvents[21c].

‡ Dipole moments have been used to study the self-association of aminopyridines in solution[14].

** Dielectric constant and molecular weight measurements show 2-pyridone, 2-pyridthione and 2-pyridselenone to form hydrogen-bonded dimers in all but the most dilute solutions in benzene and dioxan[16b].

§ Appreciably dimerized solute[16b].

resonance interaction is not very marked[10a, 14]; however, the rather different behaviour of halogen as compared with cyano groups may be due to meso-meric interaction of the substituents and the nuclear carbon atoms to which they are attached[13]. In 4-amino- and 4-methoxy-pyridine, resonance inter-action is very important [see (4)], and in the first case the existence of the imine form is excluded (p. 137). Hydrogen-bonded association of 3- and

4-aminopyridine in benzene or dioxan is also indicated[10b]. On the other hand, the data for 4-ethoxycarbonyl-, 4-cyano- and 4-nitro-pyridine exclude any important contribution from structures such as (5) in pyridines with electron-attracting groups at $C_{(4)}$.

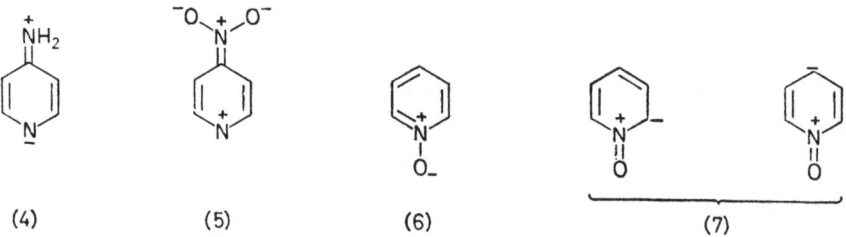

(4)   (5)   (6)   (7)

Linton[16] pointed out that, in aliphatic N-oxides, the bond moment for the group $\overset{+}{N}—\overset{-}{O}$ is 4·38 D, and since in pyridine the negative end of the dipole is towards nitrogen, pyridine 1-oxide should have a dipole moment of about 6·6 D. The observed value is lower (*Table 5.3*), and as well as (6) the forms (7) must be considered. This has important consequences for the chemistry of this compound (p. 274). Similar conclusions follow from an analysis of the dipole moments of some 4-substituted pyridine 1-oxides[11]. The mesomeric moment of a 4-substituted pyridine is

$$\mu_m Py = \mu \left(Z - \left\langle \underline{\quad} \right\rangle N \right) - \left[\mu \left(\left\langle \underline{\quad} \right\rangle N \right) + \mu \left(\text{Alk } Z \right) \right]$$

and that of a 4-substituted pyridine 1-oxide is defined analogously. Because of the importance of (8), the mesomeric moment of 4-substituted pyridines would be greater than those of corresponding benzene derivatives if Z were electron-releasing, and less if Z were electron-attracting. If in the oxide the forms (7) are important, then in all cases their mesomeric moments would be greater than those of the corresponding pyridines. Analysis of the data (*Table 5.3*) in these terms shows that this is so. The difference between the dipole moment of pyridine 1-oxide and of pyridine is increased by electron-donating groups and decreased by electron-attracting groups, i.e. the shifts shown in (9) are greater than those in (10), and those in (11) greater than those in (12) [see above]. The 1-oxide function can generate an excess or a deficit of electrons at $C_{(4)}$. In contrast, complexing of pyridine with boron halides produces[17] only electron deficiency at $C_{(4)}$.

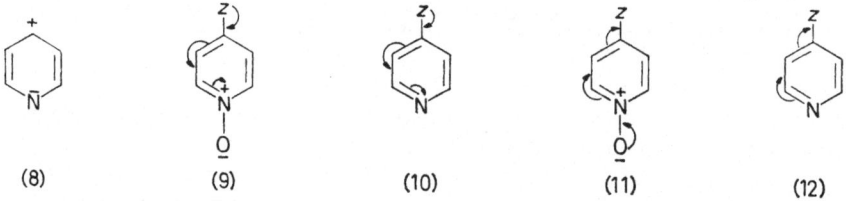

(8)   (9)   (10)   (11)   (12)

In calculations on pyridine using a regular hexagonal model of side 1·4 Å and appropriate σ-bond moments, the σ moment was calculated to be 0·85 D. In the simple molecular orbital treatment the value $h_N = 1$ (see p. 34) thus gave a π moment 1·6 times too great[18]. This could be corrected[19] by taking $h_N = 0·6$ (p. 35). A similar treatment of pyridine 1-oxide has been given[20]. A later, more sophisticated theoretical treatment[21b] of pyridine by the 'self-consistent electronegativity' method attributes most (1·78 D) of the dipole moment of pyridine to the sp^2-hybridized lone-pair orbital, a further small proportion (0·2 D) to the $C_{(4)}$—H bond and only 0·35 D to the π moment.

Similar arguments have been applied to the phenylpyridines, and the dipole moments of these compounds suggest that the two rings are nearly coplanar[21b].

(b) Spectra

Ultra-violet absorption spectroscopy has made contributions of the greatest importance to our knowledge of pyridine chemistry. The ultra-violet absorption spectrum of pyridine, like those of other 6-membered nitrogen heteroaromatics, resembles that of the corresponding carbocyclic compound, i.e. of benzene. The spectrum shows three absorption regions, recalling the α-, p- and β-bands of the homoaromatics[22]. The reduction of symmetry represented by the change from benzene to pyridine causes marked increases in absorption intensity, particularly in the α-band; the transition responsible for this is no longer forbidden and the extinction coefficient is about 30 times that for benzene. The β-band of benzene at 1,790 Å is moved to 1,720 Å in pyridine and is more intense[23], and the p-band is similarly modified.

The ultra-violet extinction curve for pyridine in cyclohexane exhibits a shoulder on the long-wavelength side of the main band system arising from a transition which has no counterpart in the benzene spectrum. Kasha[24] recognized this band as being due to promotion of a nitrogen lone-pair electron to an antibonding π orbital ($n \rightarrow \pi^*$) and pointed out important distinctions between this band and those due to $\pi \rightarrow \pi^*$ transitions (see below). Whilst in the solution spectra of pyridine and, say, the picolines, the longest wavelength $\pi \rightarrow \pi^*$ bands are not separated from the $n \rightarrow \pi^*$ band, vapour-phase studies of the near ultra-violet region reveal two transitions, the lower-frequency one of which is identified as $n \rightarrow \pi^*$ (*Table 5.4*).

Table 5.4. Near Ultra-violet Transitions in Pyridines[26], v_{0-0} cm^{-1}

	$n \rightarrow \pi^*$	$\pi \rightarrow \pi^*$	$n \rightarrow \pi^*$	$n \rightarrow \pi^*$	$\pi \rightarrow \pi^*$
Pyridine	34,769	(38,350)		30,876†	25,500†
α-Picoline	34,753	37,620			
β-Picoline	34,684	37,380	36,475?		
γ-Picoline	35,165	38,320			

† Singlet–triplet transitions.

The most obvious difference between the $n \rightarrow \pi^*$ and $\pi \rightarrow \pi^*$ transitions is seen[24] in the weakness of the former (*Table 5.5*); though not symmetry-forbidden, the $n \rightarrow \pi^*$ transition involves orbitals well separated in space ('spatial forbiddenness'), and only the s component of the lone-pair hybrid orbital contributes to its transition moment[25, 28].

The positions of $\pi \to \pi^*$ bands are not much dependent on solvent character, but $n \to \pi^*$ bands suffer a blue shift as the solvent becomes more polar[24, 25, 27-9]. Promotion of the lone-pair electron to the π orbital is made more difficult by increasing polarity in the solvent and, in protic solvents, by hydrogen bonding. The limit of this process is represented by salt formation, whether by protonation or by quaternization. In the absorption spectrum

Table 5.5. Positions and Intensities of the $n \to \pi^*$ and first $\pi \to \pi^*$ Bands of Pyridine ands Its Derivatives[28a, 31b]

Substituent	Solvent	$n \to \pi^*$ Bands*			$\pi \to \pi^*$ Bands		
		λ_{max} mμ	ϵ_{max}	f †	λ_{max} mμ	ϵ_{max}	f
(Pyridine)	iso-octane			0·0030	252	2,020	0·041
	cyclohexane	270‡	450	—	251	2,000	—
	ethanol	—	—	—	257	2,630	0·049
2-Methyl	⎫	—	—	0·0019	262	2,420	0·047
3-Methyl	⎪	—	—	0·0022	258	2,260	0·045
4-Methyl	⎪	—	—	0·0023	257	1,550	0·032
2-Bromo	⎬ iso-octane	—	—	—	267	2,950	0·047
3-Bromo	⎪	—	—	0·00095	268	2,270	0·042
2-Chloro	⎪	—	—	—	265	2,920	0·048
3-Chloro	⎭	—	—	0·0014	269	2,400	0·044
2-Cyano	⎫	278‡	340	—	265	2,730	—
3-Cyano	⎬ cyclohexane	279‡	430	—	265	2,230	—
4-Cyano	⎭	290‡	500	—	271	2,840	—
2-Fluoro	ethanol	—	—	—	258	3,200	0·05
3-Fluoro	ethanol	—	—	0·0022	263	2,990	0·05

* Values in italics refer to shoulders or inflexions.
† Oscillator strength.
‡ These disappear in aqueous solutions of the neutral molecules or cations.

of the quaternary pyridinium ion, the long-wavelength band represents the lowest singlet π-electron transition, and thus in pyridine the additional absorption which moves with change of solvent must have a different origin, i.e. it must be due to an $n \to \pi^*$ transition[29a, 30]. Salt formation causes a large increase in the intensity of the long-wavelength $\pi \to \pi^*$ band (Table 5.6). For pyridine, the picolines and the 3-halogenopyridines, the intensity of absorption in the region 34–37,500 cm^{-1} decreases when the solvent is changed from iso-octane to ethanol, and that at about 38,500 cm^{-1} increases. In the 2-halogenopyridines, the $n \to \pi^*$ transition cannot be clearly identified because it is shifted to shorter wavelengths and obscured by the $\pi \to \pi^*$ band. According to Sklar's vectorial treatment, the intensities of transitions in 2- and 3-picoline should be equal. This treatment applies fairly well to the bands identified as arising from $\pi \to \pi^*$ transitions but not to the longest-wavelength band, confirming its attribution to an $n \to \pi^*$ transition[28a].

The two types of absorption band differ markedly in the way they are affected by substituents on the pyridine ring. Like those of benzene, the pyridine $\pi \to \pi^*$ bands are usually shifted to the red by both electron-attracting and electron-releasing substituents (Table 5.6) because of the

Table 5.6. A Selection of Ultra-violet Absorption Data for Pyridine and Substituted Pyridines*

Substituents	Organic solvents λ_{max} mμ (log₁₀ ε)	Aqueous solution			Zwitterion λ_{max} mμ (log₁₀ ε)
		Neutral molecule λ_{max} mμ (log₁₀ ε)	Cation λ_{max} mμ (log₁₀ ε)	Anion λ_{max} mμ (log₁₀ ε)	
(Pyridine)†		257(3·46)	256(3·73)		
2-Me*		262(3·55)	263(3·82)		
3-Me*		263(3·50)	262·5(3·74)		
4-Me*		255(3·16)	253(3·65)		
2,3-Me₂[32a]	264(3·45)†a		266(3·85)		
2,3-Cyclopenteno[32a]	272(3·63)†a		274(3·92)		
2,3-Cyclohexeno[32a]	268(3·54)†a		272(3·93)		
2,3-Cyclohepteno[32a]	266(3·53)†a		269(3·96)		
3-Vinyl[32b]	278(3·44), 238(4·00)†b				
Acyl and carboxyl					
2-HCO[f]	320(2·30), 269(3·64)‡c; 233(3·94)				
3-HCO[f]	320(1·48), 269(3·56)‡c; 230(4·03)				
4-HCO[f]	284(3·38), 224(3·97)‡c		262(3·88)		
3-COMe[32b]	267(3·39), 227(3·96)‡a		262(3·70)		
2-CO₂H[47,g]			272(3·66)	265(3·60)	
3-CO₂H[47,g]				262(3·49)	
4-CO₂H[47,g]				266(3·38)	
Amino- and substituted amino[38-41,h]					
2-NH₂	296, 230†b: 288, 231†a	287(3·58), 229(3·97)	300(3·76), 229(3·95)		
2-Imino-1-methyl	362, 255†a				317(3·61), 251(4·07)
3-NH₂	295, 235†b: 292, 232†a	288(3·48), 231(3·91)	315(3·56), 250(3·88)		
4-NH₂	270, 246†b: 260, 233(3·98)†a; 272†b	265(3·38), 241(4·15)	263(4·22)		
4-Imino-1-methyl	242(4·02)†a				268(4·22)
4-NHMe		253(4·20)	271·5(4·24)		
4-NMe₂		261(4·26)	280·5(4·29)		
NMe₂-3-Me		263(3·86)	295(4·21)		
4-NMe₂-3,5-Me₂		272(3·59)	322(4·06)		
2-NHCOMe[43]		273(3·85), 232(4·00)	291(4·07), 229(4·09)		

129

Table 5.6 (continued)

Substituents	Organic solvents	Neutral molecule λ_{max}mμ ($\log_{10}\epsilon$)	Aqueous solution: Cation λ_{max}mμ ($\log_{10}\epsilon$)	Aqueous solution: Anion λ_{max}mμ ($\log_{10}\epsilon$)	Aqueous solution: Zwitterion λ_{max}mμ ($\log_{10}\epsilon$)
Amino- and substituted amino—(*continued*)					
2-NMeCOMe	—	262(3·61), 224(3·72)	293(3·89), 235(3·92)	—	—
2-Acetylimino-1-methyl	—	—	290(3·99), 229(3·99)	—	311(3·87), 265(3·73)
3-NHCOMe	—	271(3·49), 236(3·99)	287(3·72), 247(4·04) 212(4·30)	—	—
3-NMeCOMe	—	260·5(3·56), ***230*(*3·67*)**	280(3·62), 250(3·74), 223(3·94)	—	—
3-Acetylimino-1-methyl	—	—	290(3·72), 249(4·05) 218(4·28)	—	***312*(*3·54*)**, 273(3·95)
4-NHCOMe	—	244(4·25)	266(4·30), 206(4·09)	—	—
4-NMeCOMe	—	253(4·01)	281(4·15), 214(3·92)	—	—
4-Acetylimino-1-methyl	—	—	272(4·34), ***212*(*4·06*)**	—	312(4·26), 269(3·76)
2-$NHSO_2Me$[43,i]	—	—	286(4·06), 221(4·02)	291(3·70), 239(4·14)	310(3·86), 241(4·06)
2-$NMeSO_2Me$	—	265(3·54), 222(3·77)	289(3·89), 227(3·81)	—	—
2-Methanesulphonyl-imino-1-methyl	—	—	283(4·17), 221(4·12)	—	311(3·82), 243(4·04)
3-$NHSO_2Me$	—	320(2·48), 263(3·51)	284(3·62), 234(3·80) 205(4·24)	284(3·62), 292(3·48) 244(4·06)	—
3-$NMeSO_2Me$	—	266(3·43)	240(3·63), 214(3·94)	—	—
3-Methanesulphonyl-imino-1-methyl	—	—	287(3·59), 236(3·78) 208(4·35)	—	325(3·56), 262(4·15)
4-$NHSO_2Me$	—	—	253(4·26)	253(4·24)	281(4·40)
4-$NMeSO_2Me$	—	236(3·90)	263(4·26)	—	—
4-Methanesulphonyl-imino-1-methyl	—	—	257·5(4·27)	—	287(4·44)
Aryl[36,j]					
2-Ph	—	276(3·99), 241(4·09)	294(4·07), 242(3·89)	—	—
3-Ph	—	***266*(*3·86*)**, 244(4·12)	278(3·78), 258(3·98) 231(4·10)	—	—
4-Ph	—	256(4·21)	286(4·24)	—	—

Table 5.6 (continued)

Substituents	Organic solvents λ_{max} mμ (log$_{10}$ ϵ)	Aqueous solution			Zwitterion λ_{max} mμ (log$_{10}$ ϵ)
		Neutral molecule λ_{max} mμ (log$_{10}$ ϵ)	Cation λ_{max} mμ (log$_{10}$ ϵ)	Anion λ_{max} mμ (log$_{10}$ ϵ)	
*Cyano**					
2-CN	—	—	265(3·63)	267(3·91)	—
3-CN	—	—	265(3·42)	265(3·67)	—
4-CN	—	—	276(3·50)	276(3·70)	—
Halogen[35,k,*]					
2-F	258(3·51)‡b	257(3·53)	260(3·77)	—	—
3-F	263(3·49)†b	263(3·52)	263(3·76)	—	—
2-Cl	—	263(3·56)	269(3·86)	—	—
3-Cl	—	266(3·49), 213(3·81), 265(3·57)	270(3·74), 213(3·65), 272(3·88)	—	—
2-Br	—	268(3·40), 217(3·74)	274(3·62), 223(3·60)	—	—
3-Br	—	272(3·60), 228(3·81)	290(3·92), 228(3·73)	—	—
2-I	—	273(3·40), 228(3·81)	291(3·46), 218(3·90)	—	—
3-I	—				
Hydroxyl and substituted hydroxyl[45,l]					
2-OH	—	269(3·51), <205(>3·72)	277(3·84), 209(3·56)	291(3·71), 230(3·95)	293(3·77), 224(3·86)
2-OMe	—		279(3·84), 210(3·55)	—	—
(1-Methyl-2-pyridone)	—		279(3·79), 210(3·54)	—	297(3·76), 226(3·78)
3-OH	—	315(3·48), 278(3·36), 246(3·71)	283(3·77), 222(3·57)	298(3·69), 236(4·04)	—
3-OMe	—	276(3·60), 216(3·92)	284(3·79), 224(3·63)	—	—
(3-Hydroxy-1-methylpyridinium chloride)	—				320(3·76), 249(3·91)
4-OH	—	*235(3·30)*, 222(3·97)	288(3·77), 224(3·58)	*260(3·34)*, 239(4·15)	253(4·17)
4-OMe	—		234(3·99), 235(3·98)	—	—
(1-Methyl-4-pyridone)	—		239(4·07)	—	260(4·28)
Mercapto and substituted mercapto[46,m]					
2-SH	—	293(3·62), 247(3·94)	302(3·94), 238(3·79)	310(3·67), 264(4·10)	345(3·87), 273(4·03)
2-SMe	—		317(3·90), 250(3·86)	—	—
(1-Methyl-2-pyridthione)	—		301(3·96), 240(3·76)	—	341(3·91), 274(4·01)

Table 5.6 (continued)

Substituents	Organic solvents	Aqueous solution			Zwitterion λ_{max}mμ (log$_{10}$ε)
		Neutral molecule λ_{max}mμ (log$_{10}$ε)	Cation λ_{max}mμ (log$_{10}$ε)	Anion λ_{max}mμ (log$_{10}$ε)	
Mercapto and substituted mercapto—continued					
3-SH	—	—	310(3·50), 255(3·89), 221(4·06)	313(3·41), 269(4·13), 219(3·99)	362(3·37), 290(4·09), 232(4·15)
3-SMe	—	294(3·40), 253(3·94)	328(3·45), 268(3·94), 228(4·01)	—	—
(3-Mercapto-1-methylpyridinium chloride)	—	—	314(3·45), 258(3·86), 224(4·07)	—	362(3·34), 294(4·07), 236(4·12)
4-SH	—	264(4·10)	282(4·23), 223(3·90)	287(4·18), 222(4·04)	327(4·34), 275(3·12), 231(4·02)
4-SMe (1-Methyl-4-pyridthione)	—	—	299(4·28), 229(3·94); 286(4·29), 224(3·90)	—	333(4·37), 275(3·05), 231(3·89)
Nitro[37, 805]					
2-NO$_2$	269(3·60)[e]; 242(3·78)[e]	—	—	—	—
3-NO$_2$	264(3·64), 238(3·88)[r]	—	—	—	—
4-NO$_2$	—	284(3·26), 229·5(3·89)	280(3·52), 222(3·85)	—	—
3-Me-4-NO$_2$	—	294(3·30), 222(3·74)	287(3·46), 209(3·82)	—	—
3,5-Me$_2$-4-NO$_2$	—	272(3·31), 201·5(4·02)	276(3·66), 209(3·83)	—	—
Sulphonic acids[32d]					
2-SO$_3$H	—	—	—	254(3·50), 260(3·58), 267(3·45)	262(3·84)
3-SO$_3$H	—	—	—	254(3·40), 260(3·44), 267(3·29)	261·5(3·70)
4-SO$_3$H	—	—	—	—	—
1-Oxides[474, 50-1, n] (Pyridine 1-oxide)	265(4·11), 213(4·22)[+,b]	254(4·07)	259(3·46), 217(3·69)	—	—
3-Me	—	254(4·07), 209(4·30)	263(3·54), 220(3·62)	—	222(3·77), 263(3·70)
4-Me	266(4·17), 212(4·23)[+]	256(4·15), 206(4·27)	254(3·43), 226(3·92)	—	—

Table 5.6 (continued)

Substituents	Organic Solvents	Neutral molecule λ_{max} mμ (log$_{10}$ε)	Aqueous solution Cation λ_{max} mμ (log$_{10}$ε)	Anion λ_{max} mμ (log$_{10}$ε)	Zwitterion λ_{max} mμ (log$_{10}$ε)
1-Oxides—continued					
2-NH$_2$	319(3·70), 248(3·60), 226(4·32)^e	310(3·59), **239(3·84)**, 221(4·37)	301(3·72), 231(3·91)	—	—
3-NH$_2$	—	314(3·43), 252(**4·06**), 234(4·31)		—	—
4-HN$_2$		276(4·28)		—	—
2-OH	303(3·65), 223(3·86)	313(3·79), 237(**3·79**), 210(4·43)	270(4·19)		
3-OH			276(3·75), 209(3·63)	—	—
4-OH	304(3·54), 263·5(4·06)†^{b} 224·5(4·23)	—	—		
4-NO$_2$	268(4·15)†^{b}	313(4·10), 226(3·90)	280(**3·58**), 244(3·91)	—	—
2-Ph	—	276(3·99), 241(4·09)	294(4·07), 242(3·89)	—	—
3-Ph	—	266(3·86), 244(4·12)	278(3·78), 258(3·98), 231(4·10)	—	—
4-Ph	—	256(4·21)	286(4·24)	—	—

Quaternary cations^p	In water	Charge-transfer bands λ_{max}, Å, (ε$_{max}$ apparent)
1-Me	265(3·49), 259(3·64), **253(3·58)**	3796(1210), 2945(1550)^q 3796(1250), 2952(1570) 3778(1200), 2938(1480) 3730(1190), 2890(1500) 3738(1200)^r 3637(870) 3700(1310) 3590(1230)
1,2-Me$_2$	265(3·79)	
1,3-Me$_2$	266(3·67)	
1,4-Me$_2$	262(3·51), **260(3·52)**, 255(3·64), 250(3·60), **247(3·61)**	

* Figures in italics refer to shoulders or inflexions. For more complete spectra of the methyl-, halogeno- and cyano-pyridines and their cations see[33].

† For extensive data on pyridine and its homologues see[38a, 41, 66, 81]. The data for aqueous solutions[81] fit the expression $\lambda_{max} = 257 + 5a + 6b - 2c$ for the neutral molecules, and $\lambda_{max} = 256 + 7a + 6 \cdot 5b + 3c$ for the cations, where a, b, c are the numbers of alkyl groups at the 2-, 3-, and 4-position.

‡ In: (a) cyclohexane; (b) ethanol; (c) dioxan; (d) hexane or cyclohexane; (e) iso-octane.

(f) The weak bands of 2- and 3-formylpyridine are attributed to $n \rightarrow \pi^*$ transitions. In water the aldehydes from acetals and complex equilibria are set up (see p. 311)[33e].

(g) Data for esters and N-methylbetaines are also given[47].

(h) Ref. 41 also gives data for 3- and 3,5-substituted 4-aminopyridines. For dications of the aminopyridines see[48].

(i) 2- and 4-Methanesulphonamidopyridine exist mainly as the Z forms, and the spectral data are recorded accordingly.

(j) Including data for amino- and nitro-phenylpyridines and for dipyridyls.

(k) The very weak basicity of 2-fluoropyridine creates difficulties.

(l) The tautomerism of these compounds is discussed on p.139.

(m) In 2- and 4-mercaptopyridines the Z forms predominate.

(n) The tautomerism of these compounds is discussed on p. 139. The absorption system near 3,300Å in the vapour spectrum of pyridine 1-oxide is attributed to $n \rightarrow \pi^*$ transitions[49].

See also[81, 399, 489, 544].

(p) Further examples[54]; regarding the 1-methylpyridinium ion, see[933].

(q) This and the three following lines of data refer to solutions in 100 per cent chloroform, and chloroform solutions containing 0·27, 0·62 and 0·97 vol. per cent ethanol[54].

(r) This and the remaining data are for the first charge transfer bands, measured with 0·91 × 10⁻⁴ M solutions in chloroform containing 0·90 per cent of ethanol at 26 ± 1°.

134

extension in conjugation which they introduce, and the consequent stabilization of the excited state as compared with the ground state. The effect is
very marked with powerfully conjugating substituents such as –SR,
–OR and –NR$_2$ (see below). It is not true to say that all substituents
produce red shifts in the $\pi \to \pi^*$ bands; thus, whilst the a bands in 2- and
3-picoline are to the red of the pyridine wavelength of maximum absorption,
in 4-picoline there is a 30 Å shift to the blue. Like the π orbitals in non-
alternant hydrocarbons, those in heteroaromatics are not paired in the way
characteristic of bonding and antibonding π orbitals in alternants. The
situation observed with the picolines has been discussed by means of the
simple perturbation methods successful in treating methyl-substituted non-
alternants[31a]. Methyl groups exert inductive and hyperconjugative effects
and, if a and b refer to the highest occupied and lowest unoccupied molecular
orbitals, the change in transition energy consequent on the introduction of a
methyl group into pyridine can be written

$$\Delta\epsilon_{ab} = (\Delta\epsilon_{ab})_{\text{inductive}} + (\Delta\epsilon_{ab})_{\text{hyperconjugative}}$$

If c_{rj} is the atomic orbital coefficient at the rth carbon atom (carrying the
methyl group) in the jth molecular orbital, δa_r the change in the Coulomb
integral of the carbon atom due to the methyl group, and β_{rs} the resonance
integral of the bond between the rth carbon atom of pyridine and the sth
carbon atom of the methyl group, then

$$(\Delta\epsilon_{ab})_{\text{inductive}} = (c_{br}^2 - c_{ar}^2)\delta a_r + \left(\sum_{j \neq b} \frac{c_{br}^2 c_{jr}^2}{\epsilon_b - \epsilon_j} - \sum_{j \neq a} \frac{c_{ar}^2 c_{jr}^2}{\epsilon_a - \epsilon_j}\right)$$

and

$$(\Delta\epsilon_{ab})_{\text{hyperconjugative}} = \beta_{rs}^2\left(\sum_{k \neq b} \frac{c_{br}^2 c_{ks}^2}{\epsilon_b - \epsilon_k} - \sum_{k \neq a} \frac{c_{ar}^2 c_{ks}^2}{\epsilon_a - \epsilon_k}\right)$$

With reasonable values for the parameters, this simple treatment gives fairly
good agreement with experimental figures for the picolines and a number of
lutidines and collidines, predicting a bathochromic shift in all cases except
that of 4-picoline.

As regards the effect of substituents on $n \to \pi^*$ transitions (*Table 5.5*),
the important point is that such transitions remove charge from the nitrogen
atom to the ring in the excited state. Thus, in general, electron-attracting
substituents might be expected to stabilize the excited state, causing a red
shift, and electron-releasing substituents should destabilize the excited state,
causing a blue shift. 4-Substituents produce the biggest effects. The lowest
unoccupied pair of benzene orbitals has been used as a model in discussing
this situation. In benzene these orbitals are degenerate, but aza substitution
lowers the energy of one orbital more than that of the other, removing the
degeneracy. A lone-pair electron, on promotion to what has now become the
lowest unoccupied orbital, is distributed approximately as in (13), $C_{(4)}$
receiving most of the charge. Consequently, a substituent placed at $C_{(4)}$
exercises the largest effect on the position of the $n \to \pi^*$ band.

Molecular orbital theory in various forms has been used to predict the spectroscopic intervals and transition probabilities in pyridine and pyridinium[33] and to interpret $n \to \pi^*$ transitions[34a].

Numerous systematic studies of the ultra-violet absorption characteristics of pyridine derivatives have been made, as e.g. of the halogenopyridines[35] (*Table 5.6*) and the arylpyridines. In the latter series[36], comparison with biphenyl shows that the hetero-atom lowers the transition energy to the first excited state, and change in the hetero-atom is effective in this respect in the order

$$\searrow \atop \diagup N < \quad \searrow \atop \diagup \overset{+}{N} - \bar{O} < \quad \searrow \atop \diagup \overset{+}{N} - H < \quad \searrow \atop \diagup \overset{+}{N} - OH$$

Steric inhibition of resonance is revealed in quaternary salts or N-oxides of 2-arylpyridines[36] (see p. 365) and in 3-alkyl-4-nitropyridines[37a] (pp. 142, 159).

Ultra-violet absorption spectroscopy has been a valuable aid to the study of tautomerism in pyridine derivatives. Tautomerism is possible in several cases [and has frequently been postulated on chemical grounds (180 *et seq.*)] between structures of the general form (14; *NM* = neutral molecule) and (15; *Z* = zwitterion). Precisely analogous possibilities exist for 4-substituted pyridines, and the corresponding forms for 3-substituted pyridines are (16) and (17). In addition, in protic solvents the equilibria are complicated by the presence of the cation (18; Ct, common to both *NM* and *Z*) and the anion (19; An, common to both *NM* and *Z*), and similar equilibria exist for the 4- and 3-substituted compounds. In examining such tautomeric equilibria in solution, the general assumption is made that the ultra-violet absorption

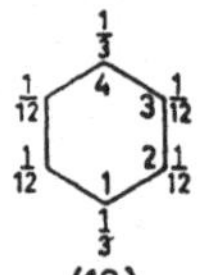

(13) 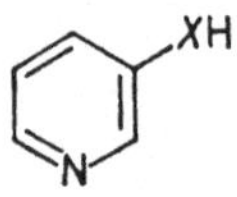   (16) 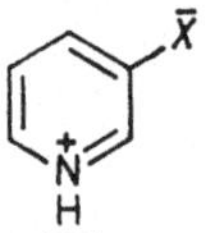   (17)

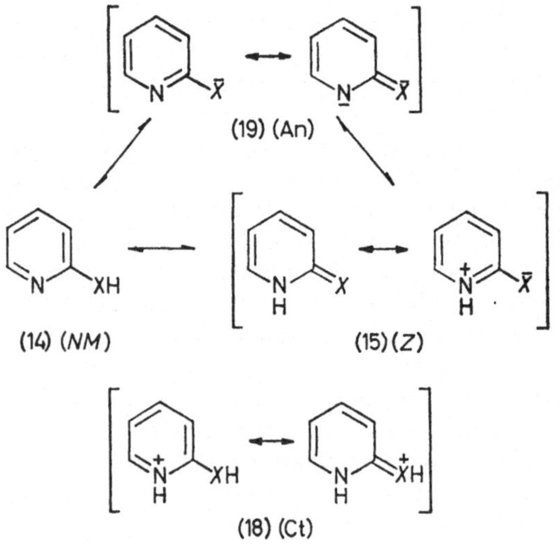

(19) (An)

(14) (NM)   (15) (Z)

(18) (Ct)

shown by (14) and (15) or (16) and (17) will not differ significantly from that of the same structures carrying an alkyl group in place of the mobile proton. Thus, ultra-violet data[38a] for 2-methylpyridine and 1-methyl-2-pyridone-methide (15; $X = CH_2$, Me for H) show clearly the absence of any appreciable degree of tautomerism in 2-methylpyridine (14; $X = CH_2$) in ethereal solution. Ultra-violet, and also infra-red, spectroscopy and ionization constants (see below and *Table 5,12*, p. 154) show the same situation in the ethyl pyridylacetates[34b]. (See also p. 325 for further data on the tautomerism of alkylpyridines).

Comparison of 2- and 4-aminopyridine, 2-dimethylamino- and 4-dimethylamino-pyridine, and 1-methyl-2- and 1-methyl-4-pyridone-imine in ether solution shows the primary amines to exist as such, and not as the imines[38b], and the same is certainly true in alcoholic and aqueous solution[39a] (*Table 5.6*). For the aminopyridines the long wavelength of maximum absorption varies with the molecular species in the order $Z > Ct > NM$, and for a particular species these wavelengths depend on the point of attachment of the substituent to the nucleus in the order $3 > 2 > 4$. The benzyl anion forms a suitable model for the discussion of these facts. In it, according to non-bonding molecular orbital theory (p. 43), the charge distributions in the non-bonding molecular orbital (ψ_N) and in the two lowest unoccupied

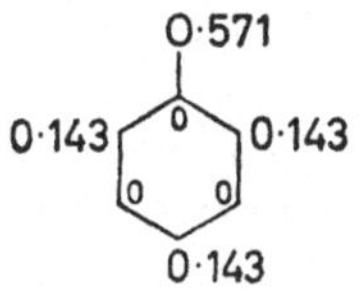

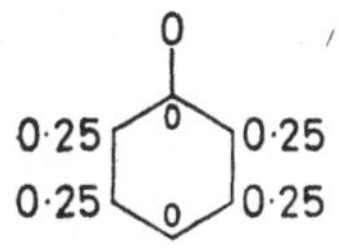

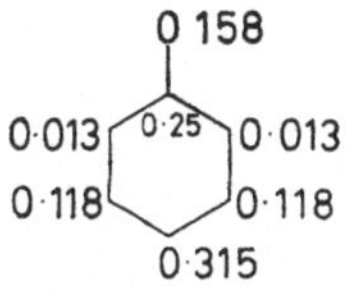

π orbitals (ψ_I and ψ_{II}) are those shown in (20), (21) and (22). For a 2-substituted pyridine the energies of the transitions $\psi_N \rightarrow \psi_I$ and $\psi_N \rightarrow \psi_{II}$ are then

$$E_I = \beta + 0.571\Delta a_x - 0.017\Delta a_n$$

$$E_{II} = 1.26\beta + 0.413\Delta a_x + 0.130\Delta a_n$$

where Δa_x and Δa_n are the increments in the Coulomb integral for the exocyclic and nuclear nitrogen atoms, respectively. For 3- and 4-substituted pyridines the corresponding pairs of equations are

$$E_I = \beta + 0.571\Delta a_x - 0.25\Delta a_n$$

$$E_{II} = 1.26\beta + 0.413\Delta a_x - 0.118\Delta a_n$$

and

$$E_I = \beta + 0.571\Delta a_n + 0.143\Delta a_x$$

$$E_{II} = 1.26\beta + 0.413\Delta a_x - 0.172\Delta a_n$$

Regardless of the character of the exocyclic substituent, the equations show that the absorption maximum due to $\psi_N \rightarrow \psi_I$ must be expected to lie at wavelengths dependent on the position of the substituent in the experimentally observed order $3 > 2 > 4$. In the 4-substituted series, likely values for the parameters to be used with some of the species make E_{II} less than E_I; in fact, only with 4-aminopyridine NM does the transition $\psi_N \rightarrow \psi_I$ show up, as a low-intensity shoulder on the long-wavelength side of the main band due to $\psi_N \rightarrow \psi_{II}$. In species other than NM the weak band is overlaid. The benzyl anion model also indicates that, of the two absorption bands observed in spectra of its heterocyclic analogues, that due to $\psi_N \rightarrow \psi_I$ should be more intense than that due to $\psi_N \rightarrow \psi_I$. Considerations of both energy and intensity therefore suggest that in the aminopyridines the shorter-wavelength maximum is due to $\psi_N \rightarrow \psi_{II}$, and the long-wavelength maximum to $\psi_N \rightarrow \psi_I$. The equations indicate that for a particular ionic species the wavelength of the high-intensity band should vary with the position of the substituent in the order $4 > 3 > 2$, which is observed (*Table 5.6*). Reasonable values for the parameters reproduce the experimental sequence of λ_{max} for the different charged species[39].

The benzyl anion model shows that for both $\psi_N \rightarrow \psi_I$ and $\psi_N \rightarrow \psi_{II}$ charge moves from the substituent into the nucleus. Accordingly, the Z form should be less polar, and the NM form more polar in the excited state. For the Z species, bands due to both transitions show red shifts as the solvent becomes less polar. For the NM form the situation is complicated by the stabilization of the ground state by hydrogen bonding in water. The bathochromic shift caused by the change from water to alcohol is ascribed to the destabilization of the ground state by the diminution of hydrogen bonding[39].

Small changes in the intensity of absorption in aqueous solution at about 240 mμ (*Table 5.6*) by 4-aminopyridines, as an alkyl group at $C_{(3)}$ is increased in size, have been attributed to increasing steric hindrance to solvation of these highly polar molecules. With 4-methylaminopyridines, steric inhibition of resonance also makes itself felt as a result of such changes in 3-substituents, and with 4-dimethylaminopyridines this effect is of major importance[40].

In solutions sufficiently acidic to convert the aminopyridines to their dications, the spectra of the aminopyridines revert to forms closely resembling that of the pyridinium ion[41]. That monoprotonation of the aminopyridines gives cations with ultra-violet absorption spectra quite different from that of pyridine shows that the monocations are constituted as shown in (18), the nuclear nitrogen atom being the dominant basic centre[42] (p. 153).

By the device of using alkylated derivatives as 'fixed' models for the possible tautomers, 2-, 3- and 4-acylaminopyridines have been shown to exist in aqueous solution predominantly in the acyl*amino*-form (23). Similarly, 3-methanesulphonamidopyridine exists predominantly as (24), but the 2- and 4-isomers prefer the imino forms[43a,b] (as 25). In ethanol, 2-nitraminopyridine is mainly in the Z form, but the proportion decreases in dioxan[43c].

(23)   (24)   (25)

The methods discussed above show that 2- and 4-hydroxypyridine exist predominantly as the pyridones (Z forms) in both organic solvents and water[44]. Metzler and Snell[45], by examining the ultra-violet absorption of 3-hydroxypyridine in a range of dioxan–water mixtures, showed that decreasing the polarity of the solvent moved the equilibrium between the NM (16; $X = O$) and Z (17; $X = O$) forms in favour of the former. Mason[44] also showed that in alcohol the spectrum of 3-hydroxypyridine resembled that of 3-methoxypyridine, but that in aqueous solution new bands appeared, corresponding in position to those observed for the zwitterionic N-methyl compound (17; $X = O$; Me for H). The continuous change in spectrum with solvent composition in alcohol–water and dioxan–water mixtures and the presence of a single set of isosbestic points showed that the changes are due to displacement of a single equilibrium process, that between NM and Z forms. The N-methyl compound absorbs at longer wavelengths than 3-methoxypyridine, and solutions of 3-hydroxypyridine made progressively more aqueous develop an absorption band beyond the long-wavelength limit of absorption in ethanol. The intensity of this new band, due to the Z form (17; $X = O$), compared with its intensity in the N-methyl derivative, provides a measure of the tautomeric equilibrium constant $K_t = $ [NH form]/[OH form]. This constant has the value 1·27 at the isoelectric point in aqueous solution. It decreases with rising temperature[44].

Like the aminopyridines, the hydroxypyridines set up complex equilibria in protic solvents, between NM, Ct, An and Z forms (see above). The longest wavelength of maximum absorption lies in the order $Z > $ An $> $ Ct $> NM$. Change from aqueous to non-polar solvents moves the long-wavelength maxima of the Z forms, as represented by their N-methyl derivative, markedly to the red, whilst the methoxypyridines show only minor changes. These facts suggest that in the Z forms the ground state is more polar than the excited state. The treatment of these observations by non-bonding molecular orbital theory, in terms of a benzyl anion model as outlined above for the aminopyridines, is surprisingly successful.

The mercaptopyridines present a picture very similar to that given by the hydroxypyridines; in aqueous solution the Z form greatly predominates over the NM form[46] (*Table 5.6*). The longest wavelength of maximum absorption falls in the order $Z > $ An $> $ Ct $> NM$, and for a particular ionic species in the order $3 > 2 > 4$, just as with the hydroxypyridines. The mercaptopyridines have also been discussed[39] by reference to the benzyl anion model mentioned.

Ultra-violet absorption spectroscopy applied to the pyridine monocarboxylic acids, their methyl esters and N-methylbetaines shows that in aqueous solutions near the isoelectric points the acids are present predominantly in the zwitterionic forms. On the other hand, in ethanol the neutral non-dipolar form is favoured[47].

It remains to discuss the contribution of ultra-violet absorption spectroscopy to our knowledge of two important groups of pyridine derivatives: the 1-oxides and the quaternary salts. The main feature of the pyridine 1-oxide spectrum in aqueous solution (*Table 5.6*) is a band at about the same position as the long-wavelength maximum of pyridine and pyridinium, but

F

of much greater intensity. There is also a band at about 205 mμ. Conversion of the 1-oxide to its cation, 1-hydroxypyridinium (26), causes a drastic reduction in the absorption intensity. 3-Substituents exert little influence on these spectra, but 4-substituents increase absorption intensity and cause red shifts[48]. Examination of the spectra of pyridine 1-oxide and 4-nitro-pyridine 1-oxide in polar and non-polar solvents points to the occurrence of hydrogen bonding in the polar case[49]. Comparison of the spectra of 2-amino- and 2-methylamino-pyridine 1-oxide with those of 2-dimethylaminopyridine 1-oxide, 1-methoxypyrid-2-one-imine (27; R = H), and 1-methoxypyrid-2-one-methylimine (27; R = Me) shows the first two compounds to exist as 1-oxides and not as 1-hydroxypyrid-2-one-imines in aqueous solution[50]. In the 4-aminopyridine 1-oxide series ultra-violet absorption spectroscopy is indecisive on the question of tautomerism[51]. The same is true of its application to 2-hydroxypyridine 1-oxide, though the indications are that in this compound, in alcohol, the 1-hydroxypyrid-2-one form is preferred[51, 52].

Heteroaromatic salts are frequently coloured, both in the solid state and in solution, and the more polarizable the anion the deeper is the colour. Hantzsch[53] showed the presence in the electronic absorption spectra of pyridinium salts, more particularly the iodides, of bands at longer wavelengths than those due to the heterocyclic cation itself—bands which do not obey Beer's law. Hantzsch attributed these to molecules formed by addition of the iodide ion to the 2- or 4-position of the methylpyridinium cation. Kosower[54] confirmed these observations but concluded that the new bands are due to charge-transfer complexes. These are regarded as solvated ion pairs, represented in the ground state by (28), which by a charge-transfer transition give an excited state (29). This conclusion is supported by several

(26)   (27)   (28)   (29)

lines of evidence. Thus, the effect of solvent changes on the charge-transfer bands is very marked. The ground state is more stabilized the more polar the solvent, and it is found (*Table 5.6*) that change of solvent from less to more polar produces a hypsochromic shift in the bands. Some pyridinium iodides in non-polar solvents show two absorption bands which maintain a roughly constant separation in transition energies as the solvent is changed. The closeness between the difference in transition energies for these bands of 1-methylpyridinium iodide in chloroform solution and the energy of photoionization of an iodide ion is regarded as strong evidence for their origin in a charge-transfer complex. As regards the effect of substitution in the pyridine ring, increasing the number of methyl (electron-releasing) groups raises the transition energy, as would be expected. The charge-transfer electronic transition in the ion pair (28) could give an excited state in which the transferred electron was delocalized in the lowest unoccupied π orbital of the heterocyclic nucleus, or in which it occupied the p orbital of a particular carbon atom of the nucleus. The energies of the charge-transfer bands are directly related to the electron affinity of the heterocyclic cation, but not to the nucleophilic atom localization energies (p. 42) of the carbon atoms in the cation. The delocalized model of the excited state for the charge-transfer transition is thus indicated[39b].

The infra-red characteristics of the pyridine nucleus show close parallels with those of benzene[56]. The infra-red and Raman spectra of pyridine and deuterated pyridines have been examined and comprehensive band assignments made[57]. The C—H stretching vibrations occur at 3,100–3,000 cm^{-1} like those in benzene. Overtone and combination frequencies for C—H out-of-plane bonding occur at 2,000–1,650 cm^{-1} and depend on the type of substitution present[58]. There is a general similarity between the positions of C—H deformation bands at about 1,250–1,000 cm^{-1} for pyridines and for benzene derivatives with the same number of hydrogen atoms in similar orientations (*Table 5.7*). C—C in-plane and out-of-plane bending modes produce absorptions at 700–600 and below 550 cm^{-1}, respectively. Complete assignments have been made of bands in the vibrational spectra of the picolines and the halogenobenzenes[55a].

Table 5.7. Infra-red Absorption (cm^{-1}) *of Pyridine Nuclei**

| | Pyridine | Pyridines | | |
		2-substituted	3-substituted	4-substituted
In-plane	1218, 1148	1279± 14, 1147 ± 3	ca. 1190?, 1124 ± 5	ca. 1220,? ?, 1067 ± 3
C—H deformation	1085, 1068	1093 ± 4, 1048 ± 5	1103 ± 5, 1038 ± 7	
		?, 1150 ± 4†, 1106 ± 10, 1044 ± 5	? 1156 ± 2† ? ?	1169 ± 5†, 1101 ± 7, 1033 ± 5
Out-of-plane C—H deformation‡	749, 700	780–740	820–770, 730–690	850–790
	759, 669†	790–750†	820–760, 680–660†	855–820†
Ring-breathing modes	1030, 993	994 ± 4	1025 ± 2	993 ± 2
	1015†	?	1015 ± 2†	?

* For assignments see[56, 58].
† For pyridine 1-oxides.
‡ 2,5-Dialkylpyridines give bands at 828–813 and 735–724 cm^{-1}, 2,3-dialkylpyridines[58] at 813–769 and752 725 cm^{-1}; a useful correlation exists between substitution pattern and C=C and C=N stretching vibrations[55b].

The close similarities in the infra-red spectra of pyridine coordination complexes involving a wide variety of metals and non-metals are believed to indicate a similar electronic density distribution in all these complexes. The evidence is thought to support the idea of back bonding from the metal atom[59] (see p. 159). Pyridinium salts give very different infra-red spectra from those of other pyridine complexes, for the extra hydrogen atom increases the number of vibrational modes. A number of pyridinium salts show three peaks above 3,100 cm^{-1} due to N–H vibrations. In salts where hydrogen bonding is to be expected, these are of low intensity, but in salts such as the tetraphenyl borate, where hydrogen bonding is absent, they are strong.

This criterion, and the presence of an additional intense band well below 3,000 cm^{-1}, point to hydrogen bonding in pyridinium nitrate, bifluoride, methanesulphonate and the halides. Fairly complete assignments have been made for the pyridinium ion vibrational spectrum, that of the N-deuterated form and of 1-methylpyridinium[60a]. Data for a number of monosubstituted pyridines and their cations are also available[60b].

Infra-red studies like those relating to dipole moments reveal weak interactions between pyridines and solvents such as chloroform[60a, 66b, 69].

The stretching frequency of the carbonyl group in aldehydes, ketones and esters of the pyridine series is higher than in corresponding benzene derivatives, and indicates electron-releasing ability in the order phenyl > 4-pyridyl 1-oxide ∼ 3-pyridyl > 4-pyridyl > 3-pyridyl 1-oxide > 3- or 4-C$_5$H$_4$N:BCl$_3$. The situation in 2-substituted pyridines is complicated by steric effects[61, 62]. The cyano group is affected like the carbonyl group[62, 63].

The asymmetric and symmetric N—O stretching bands (1,536 and 1,350 cm^{-1}, respectively) of 4-nitropyridine are close to those for nitrobenzene. Alkyl groups at C$_{(3)}$, or C$_{(3)}$ and C$_{(5)}$, are almost without effect on the asymmetric stretching frequency, whilst they increase the symmetric. The latter reaches 1,369 cm^{-1} in 3,5-dimethyl-4-nitropyridine, a figure approaching that for nitromethane. In the ground state of 4-nitropyridine, conjugation between the substituent and the ring is probably unimportant (p. 159). With symmetric stretching, the ground state (30) moves towards the excited state (31). Twisting of the nitro group by substituents diminishes the possibility of increased conjugation, and the symmetric stretching band moves to higher frequencies. In the asymmetric stretching mode, the hydridization charge caused by stretching one N—O bond is compensated by opposite effects in the other bond, and there is no reason why twisting should influence the asymmetric stretching frequency[37a].

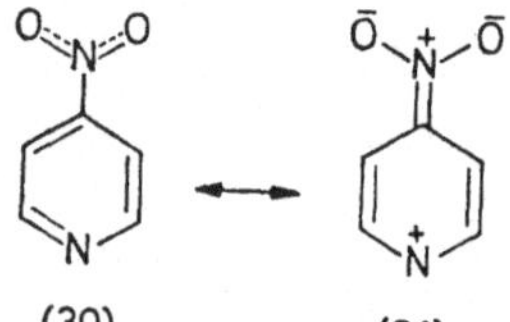

(30)   (31)

The influence of the pyridine nitrogen atom upon the characteristic frequencies of substituents attached to the ring has been represented by a

Hammett relationship (p. 47). σ-Constants for the —N= or ⟩N—Ō

groups in the 4-pyridyl 1-oxide, 3-pyridyl, 4-pyridyl, 2-pyridyl, and 3-pyridyl 1-oxide systems of 0·25, 0·62, 0·93, 1·02 and 1·18 indicate decreasing electron-releasing ability in the sequence of groups given. Conversely, the effect of substituents upon the absorption due to the =N—Ō group is directly related to the σ-constants of the substituents[62].

Infra-red spectroscopy confirms and supplements the evidence from ultra-violet and ionization data concerning the tautomerism of some pyridine derivatives (p. 136). There is no doubt that under the conditions used (*Table 5.8*) the aminopyridines exist predominantly in the *NM*-form (see 14). The imines absorb rather weakly near 3,300 cm^{-1}, and the amines from 5 to 10 times more strongly at markedly higher frequencies[64]. The asymmetric and symmetric N—H stretching frequencies in the primary

Table 5.8. *Infra-red Absorption of Amino- and Iminopyridines*[40, 64, 65a, 66a]

		N—H Stretching, cm^{-1}			Other bands 6μ region in CHCl$_3$
	solvent	ν_a*	ν_s*	$10^5 k$†	
Aniline	CCl$_4$	3481	3395	6·54	1470, 1500, 1605, 1623
2-Aminopyridine	CCl$_4$	3509	3410	6·62	1487, 1570, 1619
	CHCl$_3$	3512	3408	6·62	
3-Aminopyridine	CCl$_4$	3481	3396	6·54	1485, 1586, 1623
	CHCl$_3$	3484	3400	6.56	
4-Aminopyridine	CCl$_4$	3505	3413	6·62	1508, 1572, 1601, 1626
	CHCl$_3$	3512	3415	6·64	
2-Imino-1-ethylpyridine	⎫		3325		
4-Imino-1-ethylpyridine	⎪ CCl$_4$		3278		
2-Methylaminopyridine	⎬		3450		
2-Imino-1-methylpyridine	⎭ CHCl$_3$		3325		1475, 1534, 1572, 1652
4-Imino-1-methylpyridine					1542, 1655

* Asymmetric and symmetric stretching frequencies.
† Stretching force constant, dyn cm^{-1}.

amines, and the N—H stretching frequency in the secondary amines, increase and the bands become more intense as the ring becomes more electron-attracting. Electron-attracting power increases in the order[65a] phenyl < 3-pyridyl < 2-pyridyl ~ 4-pyridyl < 2-pyridyl 1-oxide. The ring nitrogen atom increased the H—N—H bond angle and the *s*-character of the N—H bonds in the order 2 > 4 > 3, and the N—H stretching force constants are high. The special influence of the ring nitrogen atom in 2-aminopyridine is attributed to intramolecular hydrogen bonding and to the powerful inductive effect[64c]. Infra-red data for various substituted aminopyridines support the conclusions from other evidence (pp. 137, 153) regarding their tautomerism[43a, b, 65b]. All three nitraminopyridines in the solid state exist as the Z forms[43c, d].

Infra-red spectra of solid aminopyridines and their solutions clearly indicate dimeric association in these compounds[66a]. Extensive data on the vibration spectra of aminopyridine hydrochlorides are available [66b].

2- and 4-Hydroxypyridine exist in solution predominantly as the pyridones [Z forms, see (15)], whereas in weakly polar solvents 3-hydroxypyridine assumes the enol form (16); accordingly, the former absorb in the amide and N—H stretching vibration region, and the latter shows a sharp O—H stretching band[66c-8] (*Table 5.9*). Similarly, infra-red data support the formulation of 2- and 4-mercaptopyridine, in the solid state and in weakly polar solvents, as thiones. In weakly polar solvents the 3-isomer assumes the thiol structure (16)[46a, 66c, 68].

The vibration spectra of the cations of the methoxypyridines have been measured[75c]. Infra-red spectra probably point to structures of the type (32)

Table 5.9. *Infra-red Absorption of Hydroxypyridines*[68]*

| | N–H *Stretching*, cm^{-1} | | | C=O *Stretching*, cm^{-1} | | O—H *Stretching*, cm^{-1} | |
| | *solution* | | *solid* | *solution* | *solid* | *solution* | *solid* |
	CCl$_4$	CHCl$_3$	KBr disc	CHCl$_3$	KBr disc	CCl$_4$	KBr disc
2-Hydroxypyridine	3,404	3,398	3,198m 3,165s	1,654	1,650s	—	—
3-Hydroxypyridine	—	—	—	—	—	3,595	2,925–2,500mb
4-Hydroxypyridine	—	3,442	3,200m 3,104s	1,638	1,638s	—	—

* *s* = strong, *m* = medium, *b* = broad.

for the salts of pyridones[72b], and not such as (34)[75a]. In the cation of 4-mercaptopyridine one proton is on nitrogen and one on sulphur[75b].

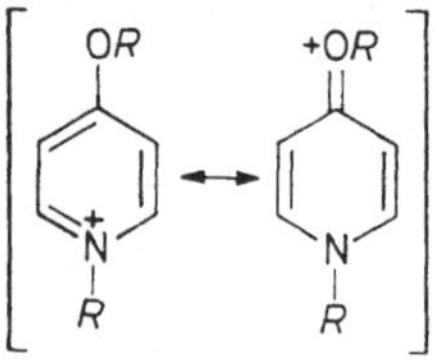

(32)
(R= H or Me)

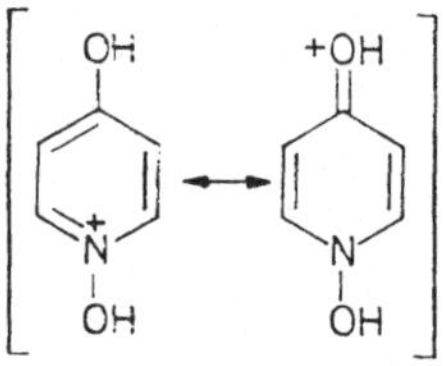

(33)

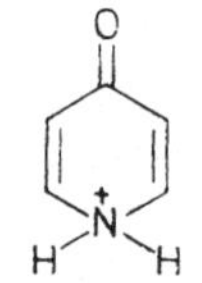

(34)

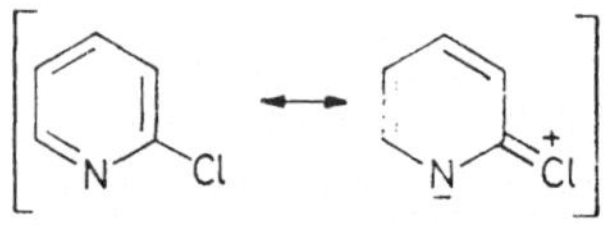

(35)

The $\overset{+}{N}$—$\overset{-}{O}$ stretching mode in pyridine 1-oxides causes absorption[56, 69] at 1,200–1,300 cm^{-1}. Electron-attracting substituents, by increasing the bond order of the group (:$\overset{+}{N}$—$\overset{-}{O}$ → ·$\overset{+}{N}$=O), increase the stretching frequency, inside the given range. This is true also for 3-alkyl groups, a fact not understood[56]. The $\overset{+}{N}$—$\overset{-}{O}$ stretching frequency is lowered by methanol because of hydrogen bonding[62]. The evidence of infra-red studies on the electron-releasing properties of the pyridyl 1-oxide group, and the applicability of the Hammett relationship to substituent effects have already been mentioned. Infra-red data generally support the views derived from ultra-violet and ionization studies about the tautomerism of amino- and hydroxypyridine 1-oxides[51].

Proton magnetic resonance spectra of pyridines have not been extensively studied, though a complete analysis of those of pyridine and some of its homologues has been given, and the pronounced effect of solvents upon the relative chemical shifts of protons in 4-picoline has been ascribed to preferential bonding with the β-protons[70]. The spectrum of deuterated 4-methylpyridine has been examined[71a]. Proton magnetic resonance spectra of 4-substituted pyridines and pyridine 1-oxides indicate the sequence of electron densities at the nuclear positions, $2 < 4 < 3$, and at the oxide nuclear positions, $2 < 3 \sim 4$. The latter sequence does not agree with some theoretical deductions[71a] (p. 275).

The proton magnetic resonance spectrum of the pyridinium ion shows low-field displacement for all the protons, most marked for the γ-proton, almost equally strong for the β-proton and much smaller for the α-proton. These results have been used to estimate the amount by which the charge densities around the pyridine nucleus are altered by cation formation. About 60 per cent of the unit charge deficiency in the pyridinium ion is centred on the nitrogen atom[72a].

An aromatic compound can be defined as one which will sustain an induced ring current. The magnitude of the ring current is a function of the delocalization of π electrons round the ring, and thus is a measure of aromaticity. Since ring currents cause shielding of protons attached to aromatic rings, proton chemical shifts can be used to estimate aromatic character. In these terms, 2-pyridone can be described as having 35 per cent of the aromaticity of benzene[72c]. This aspect of proton resonance spectroscopy promises interesting results.

Proton resonance spectroscopy has been used to study the tautomerism of 4-hydroxypyridine and its 1-oxide, with results in agreement with those obtained by other physical methods discussed here[73]. Resonance spectroscopy also indicated that the cations of 4-hydroxypyridine, 1-methyl-4-pyridone and 4-methoxypyridine are of the type (32), that of 4-hydroxypyridine 1-oxide as (33), and that 2-pyridone cation is protonated on oxygen[73-4].

Nuclear quadrupole resonance studies of chloro- and bromopyridines have been used to examine the charge distribution in the heterocyclic ring. The charge density at a nuclear carbon atom can be related to Hammett substituent constants, and the quadrupole resonance frequencies of halogen atoms at $C_{(3)}$ and $C_{(5)}$ in pyridines can be related satisfactorily to the Hammett constant for the heterocyclic nitrogen atom. However, the relationship breaks down for 2-halogenopyridines, and the quadrupole resonance frequencies indicate[76a], in these cases, an increase in the carbon–halogen bond order (35).

The ^{14}N-nuclear quadrupole resonance frequencies of pyridine and 4-picoline have been discussed in terms of the atomic orbital electron densities of the nitrogen atom[76b].

(c) Ionization Constants (Table 5.10)

Pyridine is a much weaker base than the aliphatic amines ($pK_a \sim 10$), a fact which has been attributed to the greater s character of the nitrogen lone-pair electrons in pyridine[77]. Electron-releasing substituents increase the basic strength, electron-attracting ones decrease it.

The increase in basic strength caused by a 3-methyl group is practically unchanged as the alkyl group is varied through methyl and ethyl to t-butyl.

Table 5.10. Ionization Constants of Pyridine and Pyridine 1-Oxide and their Derivatives

Substituent	Solvent	Temperature, °C	pK$_a$	References
(Pyridine)ᵃ	Water	20	5·23, 5·27*	91, 40
		25†	5·17*, 5·18*, 5·22*	78, 105, 80b
	50%EtOH-H₂O	25	4·38	79a
2-Me	Water	25†	5·96*, 5·97*	80b, 78
	50%EtOH-H₂O	25	5·05	79a
3-Me	Water	20	5·79*	40
		25†	5·52*, 5·63*, 5·68*	105, 80b, 78
4-Me		25†	5·98*, 6·02*, 6·08*	80b, 78, 105
2-Et		25†	5·89*, 5·97*	80b, 78
	50%EtOH-H₂O	25	4·93	79c
3-Et	Water	20	5·80*	40
4-Et		25	5·56*, 5·70*	80b, 78
2-n-Pr			5·87*, 6·02*	80b, 78
2-i-Pr			5·97*	78
			5·83*	78
3-i-Pr	50%EtOH-H₂O	20	4·82	79a
	Water		5·88*	40
4-n-Pr		25	5·72*	78
4-i-Pr		25	6·05*	84g
2-t-Bu		25†	6·02*	78
			5·76*	78
3-t-Bu	50%EtOH-H₂O		4·68	79a
2-Amyl	Water		5·82*	78
2-Hexyl			6·00*	
4-t-Bu			5·95*	81a
2,3-Me₂			5·99*	
2,4-Me₂		25	6·57*	80b,
2,5-Me₂			6·63*, 6·79	80b, 81b
2,6-Me₂			6·40*	80b, 81b
			6·62, 6·72*, 6·75*	80b, 81b 78
3,4-Me₂	50%EtOH-H₂O		5·77	79a
3,5-Me₂	Water		6·46*, 6·48*	80b. 84h
		20	6·23*	40
			6·14*, 6·15*	84h, 80b
2-t-Bu-6-Et	50%EtOH-H₂O		5·36	
2,6-i-Pt₂		25	5·34	79a
2-t-Bu-6-i-Pr			5·13	
2,6-t-Bu₂			3·58	
2,4,6-Me₃	Water		7·45, 7·59	81b, 82e
2,3,5,6-Me₄	Water	20	7·88*	40
Me₅	10%EtOH-H₂O	15	8·75	83d‡
2-Vinyl	Water		4·98*	81a
2-PhCH₂		25	5·13*	81a
4-PhCH₂			5·59*	84h
4-CH:CH.Ph		20	5·92	84d
4-C⦂C.Ph		20	4·62	84d

*Acyl and carboxyl*ᵇ *derivatives*

Substituent	Solvent	Temperature, °C	pK₁	pK₂	References
2-,3- and 4-CHO**	Water				
2-CH:NOH***			3·63	10·14	
2-CH:NOPr			3·58	—	
1-Me-2-CH:NOH			—	8·00	
3-CH:NOH		20	4·10	10·36	88
3-CH:NOPr			4·12	—	
1-Me-3-CH:NOH			—	9·22	
4-CH:NOH			4·77	9·99	
4-CH:NOPr			4·81	—	
1-Me-4-CH:NOH			—	8·57	

Substituent	Temperature, °C	pK$_a$*	References
3-COMe	25	3·18*	82c
4-COMe		3·51*	
3-COPh		3·18*	84h
4-COPh		3·35*	

Substituent	Temp., °C	pK₁	pK₂	pK$_A$§	pK$_B$§	pK$_C$§	pK$_D$§	References
2-CO₂H		1·01*	5·32*	1·04	(2·21)	5·29	4·12	
2-CO₂Me					2·21*			
3-CO₂H‖		2·07*	4·81*	2·11	(3·13)	4·77	3·75	
3-CO₂Me‖					3·13*			47b
1-Me-3-CO₂⁻	22			2·04*				
4-CO₂H‖		1·84*	4·86*	(1·86)	(3·26)	4·84	3·44	
4-CO₂Me‖					3·26*			
3-CO₂H-6-OH			3·82					
3-CO₂Me-6-OH			9·92					41c
1-Methylpyrid-2-one-3-carboxylic acid	20	−1·7	3·84					

Substituent	pK$_a$	References
2-CONH₂	2·10	83a
3-CONH₂	3·35*	36a
4-CONH₂	3·61	83a

146

Table 5.10 (continued)

Substituent	Solvent	Temperature, °C	pK_a						References
Acyl and carboxyl[b] derivatives (continued)									
2-CH$_2$.CO$_2$Et			4·02						
1-Me-2-:CH.CO$_2$Et			10·34						
3-CH$_2$CO$_2$Et		2	4·67						34b
1-Me-3-CHCO$_2$Et			>11·00						
4-CH$_2$CO$_2$Et			4·86						
1-Me-4-:CH.CO$_2$Et			10·18						
Amines[a] and substituted amines[c]									
	Water		pK$_{a1}$ (monocation)		pK$_{a2}$ (dication)				
2-NH$_2$		20	6·86						91
		?			−7·6*				82f
2-:NH-1-Me		21	12·20						89
3-NH$_2$		21	5·98						91
		21			−1·5				41c
		25	6·04*						84h
4-NH$_2$[d]		20	9·17, 9·29*						91, 40
		?			−6·3*				82f
		25	9·12*						84h
4-:NH-1-Me		21	12·50						89
2-NH$_2$-3-Me			7·24						
2-NH$_2$-4-Me			7·48						
2-NH$_2$-5-Me		21	7·22						83b
2-NH$_2$-6-Me			7·41						
2-NH$_2$-4,6-Me$_2$			7·84						
4-NH$_2$-3-Br		20	7·04*						
4-NH$_2$-3-Me		20	9·43*						
4-NH$_2$-3-Et		20	9·51*						
4-NH$_2$-3-i-Pr		20	9·54*						
4-NH$_2$-3,5-Me$_2$		20	9·54						
4-NH$_2$-2,3,5,6-Me$_4$		20	10·58*						
4-NHMe		20	9·66*						
3-Br-4-NHMe		20	7·47*						
3-Me-4-NHMe		20	9·83*						
3-Et-4-NHMe		20	9·90*						40
4-NHMe-3-i-Pr		20	9·96*						
3,5-Me$_2$-4-NHMe		20	9·43*						
2,3,5,6-Me$_4$-NHMe		20	10·06*						
4-NMe$_2$		20	9·71*						
3-Br-4-NMe$_2$		20	6·52*						
3-Me-4-NMe$_2$		20	8·68*						
3-Et-4-NMe$_2$		20	8·66*						
4-NMe$_2$-3-i-Pr		20	8·27*						
3,5-Me$_2$-4-NMe$_2$		20	8·15*						
2-NHAc		20	4·09						
2-:NAc-1-Me		20	7·12						
2-NMeAc		20	2·01						
2-NHCOPh		20	3·33						
2-NMeCOPh		20	1·44						
3-NHAc		20	4·46						
3-$\overline{\text{N}}$Ac-1-Me		20	>11						
3-NMeAc		20	3·52						
3-NHCOPh		20	3·80						
3-$\overline{\text{N}}$COPh-1-Me		20	>11						43a
3-NMeCOPh		20	3·66						
4-NHAc		20	5·87						
4-:NAc-1-Me		20	11·03						
4-NMeAc		20	4·62						
4-NHCOPh		20	5·32						
4-:NCOPh		20	9·89						
4-NMeCOPh		20	4·68						
			pK$_1$	pK$_2$ pK$_A$§	pK$_B$§	pK$_C$§	pK$_D$§		
2-HNSO$_2$Me		20	1·01	8·02 1·1	2·1	8·0	7·0		
2-$\overline{\text{N}}$SO$_2$Me-1-Me		20	−0·33						
2-NMeSO$_2$Me		20			1·73				
3-NHSO$_2$Me		20	3·43	7·02 4·5	2·5	6·0	7·0		43b
3-$\overline{\text{N}}$SO$_2$Me-1-Me		20	4·55						
3-NMeSO$_2$Me		20			3·94				
4-NHSO$_2$Me		20	3·64	9·07 3·6	5·1	9·1	7·6		
4-$\overline{\text{N}}$SO$_2$Me-1-Me		20	3·42						
4-NMeSO$_2$Me		20			5·14				
			pK$_{a1}$ (monocation)		pK$_{a2}$ (dication)				
2,5-(NH$_2$)$_2$		20	6·55		2·13*				41c
3,4-(NH$_2$)$_2$		20	9·08		—				84c
Arylpyridines	Water		pK$_a$						
2-Ph			4·48						
3-Ph		20	4·80						97
4-Ph			5·55						
2-2'-C$_5$H$_4$N		25	4·33						80c

Table 5.10 (continued)

Substituent	Solvent	Temperature, °C	pK_a						References

Arylpyridines (continued)

Substituent	Solvent	Temperature, °C	pK_a	References
2-2'-C$_5$H$_4$N			4·44	
2-3'-C$_5$H$_4$N			4·42	
2-4'-C$_5$H$_4$N		20	4·77	36a
3-3'-C$_5$H$_4$N			4·60	
3-4'-C$_5$H$_4$N			4·85	
4-4'-C$_5$H$_4$N			4·82	

Arylazopyridines — Water

Substituent	Temperature, °C	pK$_{a1}$ (monocation)	pK$_{a2}$ (dication)	References
2-N$_2$Ph		2·0	−3·1	82g
4-N$_2$Ph	25	3·5	−3·1	82g
2-N$_2$C$_6$H$_4$NMe$_2$-p'		4·5	2·0	82d
4-N$_2$C$_6$H$_4$NMe$_2$-p'		5·8	3·4	82d

Cyanopyridines — Water

Substituent	Temperature, °C	pK$_a$	References
2-CN	20	−0·26	31b
3-CN	20	1·35*, 1·36	84h, 31b
3-CN	25	1·45	100
4-CN	20	1·86*, 1·90	84h, 31b
2,6-(CN)$_2$	20	−2·0	84h, 31b

Halogenopyridines — Water (25)

Substituent	pK_a	References
2-F	−0·44*	35c
3-F	2·97*	35c
2-Cl	0·49*, 0·72*	81a, 35c
3-Cl	2·81*, 2·84*	84h, 35c
4-Cl	3·83*	84h
2-Br	0·71*, 0·90*	81a, 35c
3-Br	2·84*, 2·85*	35c, 84h
4-Br	3·75*	84h
2-I	1·82*	35c
3-I	3·25*	35c
3,5-Cl$_2$	0·67	84h
3,5-Br$_2$	0·82	84h
2-Cl-5-NO$_2$	−2·97	923
3-Br-5-OMe	2·60*	84h
2-I-5-NO$_2$	−1·70	923

Hydroxypyridines — Water

Substituent	Temp. °C	pK$_1$	pK$_2$	pK$_A$§	pK$_B$§	pK$_C$§	pK$_D$§	References
2-OH		0·75	11·62	0·75	3·71	11·62	8·66	90, 85
1-Me-2-:O		0·32						90
2-OMe	20	3·28						90
3-OH		4·86	8·72	5·12	5·22	8·46	8·36	90, 85
1-Me-3-Ō		4·96						90
3-OMe		4·88						90
	25	4·78*						84h
4-OH	20	3·27	11·09	3·27	6·56	11·09	7·80	90, 84b
1-Me-4-:O	20	3·33						90
4-OMe	20	6.62						90
	25	6·58*						84h

Substituent	Temp. °C	pK$_{a1}$	pK$_{a2}$	References
2,4-(OH)$_2$	20	6·50	13	90
2,5-(OH)$_2$	20	8·51		41c
2,4,6-(OH)$_3$	20	4·6	9·0	90

Mercaptopyridines — Water (20)

Substituent	pK	pK	References
2-SH	−1·07*	9·97	
1-Me-2-:S	−1·22*		
2-SMe	3·62		
3-SH	2·28	7·01	
1-Me-3-S̄	2·27		46b
3-SMe	4·45		
4-SH	1·43*	8·83	
1-M3-4-:S	1·30		
4-SMe	5·97		

Nitropyridines — Water

Substituent	Temperature, °C	pK$_a$	References
3-NO$_2$	25	0·81	93
4-NO$_2$		1·61*	
3-Me-4-NO$_2$		2·22*	
3-Et-4-NO$_2$	20	2·20*	37a
4-NO$_2$-3-i-Pr		2·21*	
3,5-Me$_2$-4-NO$_2$		2·52*	
2,3,5,6-Me$_4$-4-NO$_2$		4·08*	

Sulphonic acids' — Water (25)

Substituent	pK_a	References
2-SO$_3$H	1·75	
3-SO$_3$H	3·22	
4-SO$_3$H	3·44	32d
2,6-Me$_2$-3-SO$_3$H	4·89	
2,6-Me$_2$-4-SO$_3$H	5·09	
2,6-(t-Bu)$_2$-3-SO$_3$H	4·12	

Table 5.10 (continued)

Substituent	Solvent	Tempera-ture, °C	pK$_a$		References
Pyridine 1-Oxides					
(Pyridine 1-oxide)			0·79*		
3-Me	} Water	} 23·25	1·08*		} 100
4-Me			1·29*		
Aminopyridine oxides	Water				
2-NH$_2$			2·67		
2-NH$_2$-O-Me		20	12·4		} 51
2-NHMe			2·61		
2-NMe$_2$			2·27		
3-NH$_2$		23·25	1·47		100
4-NH$_2$			3·69		
4-NH$_2$-O-Me			>11		
4-NHMe			3·85		} 51
4-NMe$_2$			3·88		
2-NHAc			−0·42		
2-NMeAc			−1·02		
2-NHCOPh		20	−0·44		
2-NMeCOPh			−1·39		} 84e
3-NHAc			0·99		
4-NHAc			1·59		
4-NMeAc			1·36		
4-NMeCOPh			1·70		
4-NH$_2$-3-CO$_2$H^g		?	3·98		81c
Arylpyridine oxides[h]	Water				
2-Ph			0·77		
3-Ph		20	0·74		} 97
4-Ph			0·83		
Pyridine carboxylic	Water		pK$_{a1}$	pK$_{a2}$	
acid oxides					
3-CO$_2$H			0·09	2·7, (2·74)	100, 81c
3-CO$_2$Bu		} 23·25	0·03		100
4-CO$_2$H			−0·48	2·9	100
4-Cl-3-CO$_2$H^g		?		1·80	
4-OH-3-CO$_2$H^g				3·08	} 81c
4-OMe-3-CO$_2$H^g		} 23·25		2·88	
4-NO$_2$-3-CO$_2$H^g				0·50	
Hydroxypyridine oxides	Water				
2-OH			−0·8	5·99	
2-:O-O-Me		} 20	−1·3		} 51
2-OMe			1·23		
3-OH		?		6·4	52a
4-OH			2·45	5·9	51, 52a
4-:O-O-Me		} 20	2·57		51
4-OMe			2·05		51
Mercaptopyridine oxides	Water				
2-SH			−1·95	4·67	
2-S.CH$_2$Ph		} 20	−0·23		} 84e
4-SH			1·53	3·82	
4-S.CH$_2$Ph			2·09		
Nitropyridine oxides	Water				
3-NO$_2$		?	−1·2		518
4-NO$_2$		23·25	−1·7		100

* 'Thermodynamic' values.
† For values similar to those quoted see[82b, 84g].
‡ The reference quotes values for a number of alkylpyridines measured under the conditions mentioned for pentamethylpyridine.
** See p. 311.
*** See also[84f].
|| See also[84g]
§ See p. 152.
(a) For values relating to solutions in nitrobenzene see[84f].
(b) Values (50 per cent EtOH/25°) for a number of 5-substituted picolinic and 6-substituted nicotinic acids are given[83a].
(c) For values for a number of 2-arylsulphonamidopyridines see[85b].
(d) For the dissociation constant of 4-aminopyridinium in water from 0–50° and related thermodynamic quantities see[94].
(e) The first proton goes to the ring nitrogen atom.
(f) Values for the sulphonic acids in 42 per cent ethanol[32d].
(g) For the equilibrium 4-X-3-CO$_2^-$-C$_5$H$_3$NOH$\rightleftharpoons$4-X-3-CO$_2^-$-C$_5$H$_3$NO + H$^+$.
(h) For values of nitro- and aminophenyl derivatives see[97]

149

The greater effect of a 4- than of a 3-methyl group has been attributed to hyperconjugation; varying the 4-alkyl group makes little difference because of the opposite trends of hyperconjugative and inductive effects. The slightly lower effectiveness of a 2- than of a 4-methyl group is attributed to the lower hyperconjugating ability of the former, rather than to steric hindrance to solvation of the 2-picolinium ion, for the second methyl group in 2,6-lutidine causes no enhancement of the effect[78]. Increasing bulk in the 2-alkyl group slightly decreases the augmentation of basic strength; this effect could be due to increasing steric strain involving the proton, or to steric hindrance to solvation of the cations. Additivity is observed for the effect upon pK_a of alkyl groups in some 2- and 2,6-substituted compounds (see *Table 5.10*), even up to 2-t-butyl-6-isopropylpyridine. Additivity breaks down with 2,6-di-t-butylpyridine, which is a weaker base than pyridine. It is argued that, up to this point, neither steric strain involving the proton nor steric hindrance to solvation of the cation can be important, but that in 2,6-di-t-butylpyridine an important steric effect appears. Since steric hindrance to solvation would be expected to make itself felt gradually, the situation encountered in 2,6-di-t-butylpyridine is held to suggest steric strain involving the t-butyl groups and the bound proton[79].

This interpretation is put in doubt by the results of Mortimer and Laidler[80a] who measured the heats of neutralization of pyridine, the picolines and 2,6-lutidine. These were markedly dependent on concentration, and the values were extrapolated to infinite dilution. *Table 5.11* records these values,

Table 5.11. Thermodynamic Quantities at 25·00° for the Ionization of Pyridine and its Homologues[80a]

Compound	ΔG	ΔH kcal/mole	$-T\Delta S$
Pyridine	7·12	5·70 ± 0·30	1·42
2-Picoline	8·13	6·95 ± 0·40	1·18
3-Picoline	7·68	6·70 ± 0·22	0·98
4-Picoline	8·15	7·03 ± 0·13	1·12
2,6-Lutidine	9·17	6·15 ± 0·11	3·02

the values of ΔG (obtained from pK_a values[80b] and the derived values of $T\Delta S$. For the dissociation of the pyridinium ions,

$$(BH^+)_{hydrated} \rightleftharpoons B_{hydrated} + H^+{}_{hydrated}$$

the ΔS values show that the number of free molecules decreases. As we have seen, the ΔG (or the ΔpK_a) values are additive as between mono- and disubstituted compounds. ΔS is roughly constant for pyridine and the picolines, and the increases in basic strength are clearly due to internal electronic factors. However, for 2,6-lutidine ΔH is not additive, and ΔS is markedly low. These features are attributed to steric hindrance to solvation in the 2,6-lutidinium ion. The effect does not appear in ΔG, and so not in pK_a, for the effects of ΔH and $T\Delta S$ in the expression $2\cdot30\,RT\,\mathrm{p}K_a = \Delta G = \Delta H - T\Delta S$ are compensatory. If this view is correct, it is striking that even in 50 per cent ethanol–water this compensation should occur in the complex 2,6-disubstituted pyridines mentioned above, and break down only with 2,6-di-t-butyl-pyridine. It is interesting to enquire if even the latter compound would show the deviation in water. Other workers[82a] interpret the thermodynamic quantities for the neutralization of pyridines to support Brown's views; the only difference between their results and those of Mortimer and Laidler[80a] seems to be that the former did not obtain heats of neutralization at infinite dilution.

Of interest in this connection are the heats of reaction of pyridines with methanesulphonic acid in nitrobenzene[82b]. 3-Alkyl groups increase the heats of reaction in the order t-Bu > i-Pr > Et > Me $\gg$ H, as would be expected from inductive effects. The opposing hyperconjugative and inductive effects of 4-alkyl groups make the heats of reaction of 4-alkylpyridines practically independent of the type of alkyl group present. A 2-methyl group increases the heat of reaction, and the second methyl group in 2,6-lutidine causes an identical further increase. On the other hand, increasing the size of a single 2-alkyl substituent decreases the heat of reaction. Brown and Holmes[82b] conclude that steric hindrance to solvation cannot be important in the methyl compounds but may be so in the other cases. Further, since a linear relationship exists between the heats of reaction in this system and $\mathrm{p}K_a$ values for the bases, they conclude that solvent effects cannot be important in determining the relative strength of these bases. However, from what we have seen above, it is clear that these heats of reaction in nitrobenzene are not linear with ΔH for the ionization in water, and in comparing the former with $\mathrm{p}K_a$, the compensating effect of ΔH and $T\Delta S$ for the water system is obscuring the divergence from linearity which would occur at 2,6-lutidine. It seems reasonable to conclude that, whilst addition of a proton to 2- or 2,6-substituted pyridines may involve no important steric effect, steric hindrance to solvation of the cation shows itself differently in the two systems.

The special position of the proton is brought out by measurements of the heats of reaction of pyridine and its homologues with various 'Lewis acids' in nitrobenzene, and of thermodynamic data for the dissociation of the addition compounds, formed from the pyridines and these acids, in the gas phase[79b, 84a]. 3- or 4-Alkyl substituents exert the expected stabilizing influence on the addition compounds, but a 2-alkyl group has a very different effect. Despite its greater basic strength, 2-picoline gives products with borine, boron trifluoride and trimethylboron which are less stable than those from pyridine. These stabilities decrease still further as the size of the 2-substituent increases. 2,6-Lutidine does not react with trimethylboron, and 2,6-di-t-butylpyridine not even with boron trifluoride. Clearly, in contrast to the proton, these Lewis acids have considerable steric requirements, and the bigger they are, and the more the pyridine nitrogen atom is protected by adjacent substitution, the more strain there is involved in forming an addition product (see p. 190).

Results obtained with triethylaluminium generally resemble those obtained with boron trimethyl, but the former acid is less severe in its steric demands[85a].

Ionization constants provide the most general means of studying the tautomerism illustrated by structures (14)–(15) and (16)–(17). The equilibria involved are shown again, in symbols which do not attempt to represent the resonance possibilities present in each molecule [see structures (14)–(19)].

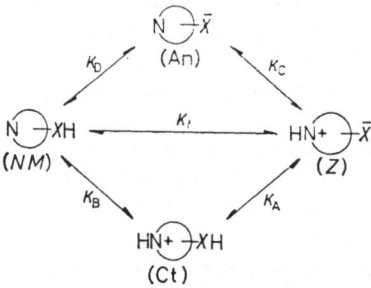

The true ionization constants for the Z (K_A and K_C) and NM (K_B and K_D) tautomers cannot be determined directly. They are, however, related to the two experimentally determined ionization constants by the equations

$$K_1 = K_A + K_B$$

$$1/K_2 = 1/K_C + 1/K_D$$

It can reasonably be assumed that the basic ionization constants of the NM and Z tautomers can approximately be represented by the constants for the X-methyl (K_{MeX}) and N-methyl derivatives (K_{MeN}) which are, of course, directly determinable. Then K_t, the constant governing the tautomeric equilibrium, is given by

$$K_t = \frac{[Z]}{[NM]} = \frac{K_A}{K_B} = \frac{K_D}{K_C} = \frac{K_{MeN}}{K_{MeX}}$$

Further, since

$$K_t = K_1/K_B - 1 = 1/(K_1/K_A - 1)$$

if the basic ionization constant for only one or other of the methyl derivatives is known, we have the approximate relationships

$$K_t = \text{antilog}(pK_{MeX} - pK_1) - 1$$

and

$$K_t = 1/[\text{antilog}(pK_{MeN} - pK_1) - 1]$$

Where it has been possible to compare these different expressions for K_t with values obtained by an independent method, such as that based on spectrometry, the one using both K_{MeN} and K_{MeX} has proved superior to the more approximate forms using only one of these constants[84b]. This method of studying tautomerism was first used with amino acids[86] and applied to heterocyclic compounds by Tucker and Irvin[87]. *Table 5.12* records values for K_t obtained from ionization constants, and *Table 5.10* includes values of the

152

constants K_A, K_B, K_C and K_D derived from the above equations for some of the tautomeric pyridine derivatives.

The conclusions to be reached from these data are, chiefly, that whilst hydroxy-and mercaptopyridines and pyridine carboxylic acids exist predominantly in the Z forms (i.e. as pyridones, pyridthiones and zwitterions), aminopyridines prefer the NM forms (that is, they are mainly amines and not imines). The spectroscopic evidence supporting these conclusions has already been discussed, and this tautomerism is further discussed below.

The above arguments depend on the ability of the tautomers NM and Z to provide a common cation (Ct) and a common anion (An). Spectroscopic evidence for the structures of the monocation derived from amino- and hydroxypridines has been discussed above. The dominant basic centre is the ring nitrogen atom, and the base-strengthening effect (ΔpK_a) of the 2- and 4-amino-groups ($\Delta pK_a = 1\cdot7$ and $4\cdot0$, respectively), which is so much more marked than that produced by a 3-amino-group ($\Delta pK_a = 0\cdot8$), is readily understood in terms of the cation structure (18). The amino group has the electronic character ($-I+M$), and the mesomeric effect is especially important in the 2- and 4-isomers. The additional importance of resonance stabilization[91] of the cations of these compounds is illustrated in (36) for 4-aminopyridine. The situation recalls that found in the amidines.

(36)

The individual ionization constants for pairs of tautomers and the constant for the equilibrium between them can be deduced from the experimentally determined ionization constants in another way. It will be seen (p. 156) that the effect of substituents on the basic strength of the pyridine nitrogen atom can be represented by the Hammett equation. Accordingly, the ionization constant for the NM form (K_B) can be estimated from the Hammett substituent constant of the group $-X$H. Combined with the experimentally determined K_1 this leads to estimates of the remaining quantities. In this way Jaffé[92] deduced that nicotinic and isonicotinic acids exist predominantly as zwitterions in aqueous solution. Similarly, Bryson[93] obtained the values $pK_B \sim 5\cdot0$ and $K_t = 0\cdot78$ for 3-hydroxypyridine (cf. *Tables 5.10* and *5.12*). Hammett correlations have been used in evaluating the tautomeric equilibria in a number of 2-arylsulphonamidopyridines[85b].

✗ The data (*Table 5.10*) for the aminopyridines and the N-methyl derivatives of their imino tautomers illustrate the general point that of two tautomers in equilibrium the weaker base predominates. A study of the temperature coefficient of the ionization constants for 4-dimethylamino-, 4-methylamino- and 4-aminopyridines suggests that the changes in basic strength caused by 3-substituents are due in the tertiary bases to steric inhibition of resonance, in the primary to steric hindrance to solvation, and in the secondary [cf. 155] to both of these effects[40]. 4-Aminopyridine is useful

Table 5.12. Tautomeric Equilibrium Constants (K_t) for Pyridine Derivatives *

Position of substituent	.CH:NOH[88]	.CO$_2$H[47]	.CH$_2$CO$_2$Et[34b]	.NH$_2$[89]	.NHCOMe[43a,b]	.NHCOPh[43a,b]	NHSO$_2$Me[43a,b†]	OH[39b,45,90]	.SH[46]
2	$4{\cdot}2 \times 10^{-5}$	15	5×10^{-7}	5×10^{-6}	10^{-3}	—	ca. 10	912	49×10^{3}
3	$7{\cdot}6 \times 10^{-6}$	10	$<2 \times 10^{-7}$	—	$<3{\cdot}2 \times 10^{-7}$	$<6{\cdot}3 \times 10^{-8}$	$0{\cdot}1$	$0{\cdot}83$	150
4	$1{\cdot}58 \times 10^{-4}$	25	5×10^{-6}	5×10^{-4}	$6{\cdot}3 \times 10^{-6}$	$3{\cdot}2 \times 10^{-5}$	40	1950	35×10^{3}
Pyridine 1-oxides[51]									
2	—	—	—	$<10^{-8}$	—	—	—	?	—
4	—	—	—	$<10^{-7}$	—	—	—	ca. 1	—

* $K_t = (Z)/(NM)$ [see structures (14) and (15)]. These values refer to aqueous solutions at about 20°.
† Values are available for a range of 2-arylsulphonamidopyridines in aqueous alcohol and aqueous dioxan.[86b]

as a titration standard, and the thermodynamic quantities for the dissociation of the 4-aminopyridinium ion in water at 0–50° have been measured[94].

Because of the importance of the Z forms [as (15), X = 0], which confer on the hydroxypyridines an amide-like character, these compounds are only weakly basic or acidic. This effect is very much less marked in 3-hydroxypyridine, for in this case the Z form (17, X = O) does not include amongst its canonical forms one equivalent to the amide-like structure, lacking charge separation, to be found with 2- and 4-hydroxypyridine. The basic strengths of 2- and 4-hydroxypyridine are close to those of 1-methyl-2- and 1-methyl-4-pyridone, respectively, and all are much weaker than the methoxypyridines. Clearly, the enolic (NM) forms are not very important in the tautomeric equilibrium. Examination of pK_D, which represents the dissociation as acids of the enolic (NM) forms of the hydroxypyridines, shows that the ring nitrogen atom exerts an acid-strengthening influence very similar to that of the nitro group in the nitrophenols[90]. Similarly, it can be seen that the ring nitrogen atom strengthens the carboxylic acid group attached to the pyridine ring[47].

If we consider 2- and 4-substituted pyridines in the NM form, it is clear that the substituent can influence the basic strength by stabilizing the conjugate acid through mesomeric electron release, or by destabilizing it by inductive electron attraction. The inductive effect is especially important with 2-substituents, but the two tendencies taken together explain the observed sequence of basic strengths in the 4-substituted series, as the substituent is changed, $\overline{O} > NH_2 > \overline{S} > OR > SR$, and in the 2-substituted series, $\overline{O} > \overline{S} > NR_2 > OR \sim SR$. In the Z forms as their conjugate acids, mesomerism is important in the order $NR > O > S$, and as the free bases, $NR \ll O \sim S$. Accordingly, the observed order of basic strengths in the Z forms is $NR \gg O > S$, and it is clear why the position of tautomeric equilibrium is so similar in hydroxy- and mercaptopyridines, and so different in aminopyridines[46a].

By the use of ionization constants it has been concluded that 2- and 4-aminopyridine 1-oxide exist predominantly as such in aqueous solution, whilst 4-hydroxypyridine 1-oxide is in equilibrium with a comparable amount of 1-hydroxypyrid-4-one. Application of the method to the pyridine 1-oxides is more complicated than in the pyridine series, and it fails with 2-hydroxypyridine 1-oxide[51].

It has been mentioned that the Hammett relationship (p. 47) has been applied to pyridine and its derivatives. The Hammett equation related a rate or equilibrium constant(K_0) for a reaction occurring at a side chain attached to the benzene nucleus to the corresponding constant (K) for a m- or p-substituted derivative by the expression $\log K/K_0 = \sigma\rho$. The reaction constant, ρ, is characteristic of the particular reaction and a measure of the susceptibility of the reaction to the influence of substituents. The substituent constant, σ, is characteristic of the substituent and a measure of the influence of the substituent upon the activation energy of the reaction of the unsubstituted compound, and of the effect of the substituent on the electron density at the reaction site[95]. In the pyridine series the equation has been used in two ways: first, the heterocyclic nitrogen atom has been treated as a substituent in a benzene ring, and σ values derived for it from its effect upon some side-chain reaction, using the reaction constant evaluated for that reaction in the benzene series; secondly, the influence of substituents in the pyridine ring upon a reaction occurring at the nitrogen atom has been expressed by deriving ρ for that reaction, using substituent constants derived from benzene derivatives.

The first kind of application is not readily made through the measurement of ionization constants, for the nitrogen atom is the basic centre. In fact, substituent constants for the heterocyclic nitrogen atom have been obtained from kinetic data[96]. However, from the seond ionization constants for 3- and 4-aminopyridine (*Table 5.10*) it is possible to obtain σ values for the $=\overset{+}{N}H-$ group regarded as a substituent in aniline[92]. Similarly, from the ionization constants of *m*- and *p*-pyridylanilines, constants for the protonated pyridyl ring, and likewise for the pyridyl 1-oxide ring, can be deduced[97]. The results are summarized in *Table 5.13*.

Table 5.13. Substituent Constants for Heterocyclic Groups and Hetero-atoms*

Substituent	Orientation	σ	Substituent	Orientation	σ
$=N-$[96]	m	0·62	$2\text{-}C_5H_4\overset{+}{N}H$[97]	p	0·75
$=N-$[96]	p	0·93	$4\text{-}C_5H_4\overset{+}{N}H$[97]	p	0·65
$=\overset{+}{N}H-$[92]	m	2·1	$2\text{-}C_5H_4\overset{+}{N}-\overset{-}{O}$[97]	m	0·23
$=\overset{+}{N}H-$[92]	p	4·0	$2\text{-}C_5H_4\overset{+}{N}-\overset{-}{O}$[97]	p	0·27
$=\overset{+}{N}-\overset{-}{O}$[92]†	m	1·47	$4\text{-}C_5H_4\overset{+}{N}-\overset{-}{O}$[97]	p	0·33
$=\overset{+}{N}-\overset{-}{O}$[92]†	p	1·34	$[C_6H_5$[99]	m	−0·01]
$=\overset{+}{N}-OH$[92]	m	2·3	$[C_6H_5$[99]	p	+0·06]
$=\overset{+}{N}-OH$[92]	p	3·9	$[NO_2$[95]	m	0·710]
			$[NO_2$[95]	p	0·778

* All these values, except those for $=N-$, come from ionization constants; see also pp. 247, 271, 322.
† Various values are quoted, depending on the substituent concerned, see [92], [98] and pp. 247, 322.

The general implications of these figures will be obvious, and reference will be made to them again in particular connections later on. Though the values for $=\overset{+}{N}H-$ may not have precise significance, it should be noticed that they and those for $=\overset{+}{N}-OH$ are the largest substituent constants ever met (see p. 247 for further data on this point). The resemblance between $=N-$ and $-NO_2$ is striking and gives additional strength to the frequently quoted relationship between pyridine and nitrobenzene. This has already been mentioned (p. 33) and receives ample illustration in the following pages.

Jaffé and Doak[100] first showed that the effect of substituents upon the basic strength of the pyridine nitrogen atom and the $=\overset{+}{N}-\overset{-}{O}$ group of pyridine 1-oxide could be represented by Hammett relationships. The relationships were improved when points for the hydroxyl, amino and 3,4-$(CH)_4$ [in

156

isoquinoline] group were omitted. This is understandable (leaving aside the hydroxyl group where tautomerism is a complication), for in other cases[95] it has been found that readily polarizable substituents interacting with strongly electron-attracting side chains sometimes need smaller substituent constants than usual. The $=\overset{+}{\text{NH}}$— and $=\overset{+}{\underset{|}{\text{NOH}}}$ groups are amongst the most strongly electron-attracting known (see above). The usual quantities associated with a Hammett plot are given for the present cases in *Table 5.14*, and such a plot [40], constructed by using substituents for which resonance is unimportant, is illustrated in *Figure 5.1*. The reaction constant for the dissociation of

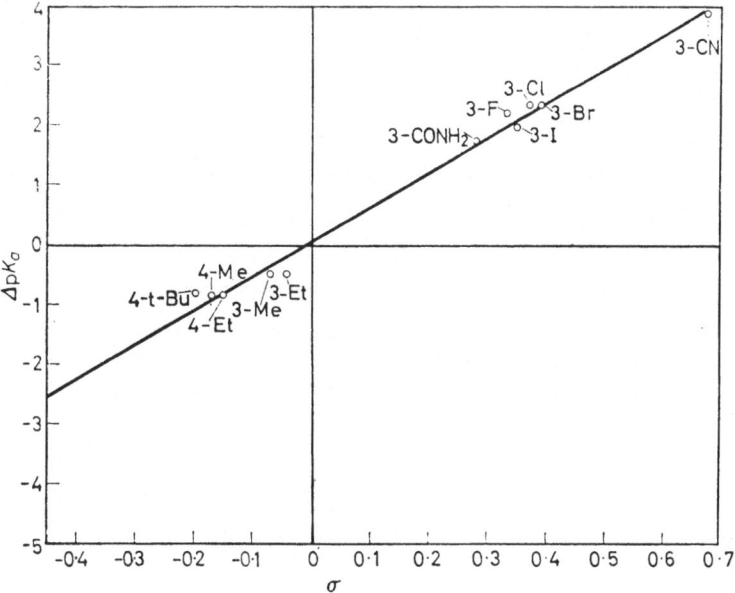

Figure 5.1. Hammett plot of ΔpK_a against σ for monosubstituted pyridines

pyridinium ions is of the same magnitude as those for heterolytic reactions involving nuclear substituents and, as would be expected, much bigger than that for 1-hydroxypyridinium ions. The latter is very close to the reaction constant for phenol ionization[95] ($\rho = 2\cdot11$), which is not surprising since the groups $=\overset{|}{\underset{}{\text{C}}}-\overset{-}{\text{O}}$ and $=\overset{+}{\text{N}}-\overset{-}{\text{O}}$ are isoelectronic.

From *Figure 5.1* the values $\sigma_{4NH_2} = -0\cdot686$ and $\sigma_{4NMe_2} = -0\cdot761$ can be derived[40]. The differing σ values which it has been found necessary to use for a particular electron-releasing group[95], depending on its environment, appear to depend on the degree of importance of p-interaction as shown in (37). The very small σ values for the amino and dimethylamino group at $C_{(4)}$ in pyridine stress the great importance of the delocalization represented in (36), and in the case of the amino group the small substituent constant is in accord with the high stretching force constant of the N—H bonds (p. 143).

Table 5.14. Reaction Constants, ρ, for Pyridinium and 1-Hydroxypyridinium Ion Dissociation in Water

Reaction	Temperature, °C	ρ	s*	r†	n‡
Dissociation of					
$R \cdot C_5H_4\overset{+}{N}H^{100}$	~25	5·714	0·296	0·989	12
$R \cdot C_5H_4\overset{+}{N}H^{100}$	~25	5·685	0·147	0·995	10**
$R \cdot C_5H_4\overset{+}{N}H^{40}$		5·735	0·181	0·995	11
$3R \cdot C_5H_4\overset{+}{N}H^{93}$		6·18		0·998	6§
$3R \cdot C_5H_4\overset{+}{N}H^{93}$		7·00		0·994	6§
$3R \cdot C_5H_4\overset{+}{N}H^{93}$	25	6·36		0·970	12§
$R \cdot C_5H_4\overset{+}{N}OH^{100}$		2·571	0·381	0·967	11
$R \cdot C_5H_4\overset{+}{N}OH^{100}$		2·088	0·121	0·994	8‖

* Standard deviation from regression line.
† Correlation coefficient.
‡ Number of substituents used.
** 4-Aminopyridine and isoquinoline were omitted.
§ The first six substituents were of the type (—I—M), and the second six, (—I+M); the twelve substituents were a combination of these two groups.
‖ 4-Amino- and 4-hydroxypyridines 1-oxides and isoquinoline 2-oxide were omitted.

As mentioned already, 3-substituents in **4-dimethylaminopyridine** cause steric inhibition of resonance. The substituent constant for the dimethyl-amino group in **3,5-dimethyl-4-dimethylaminopyridine**[40] is −0·355.

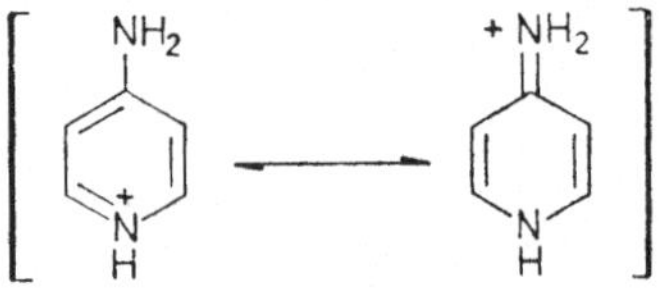

(36)

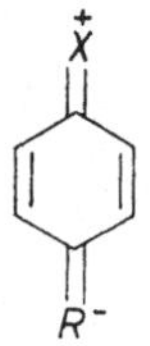

(37)

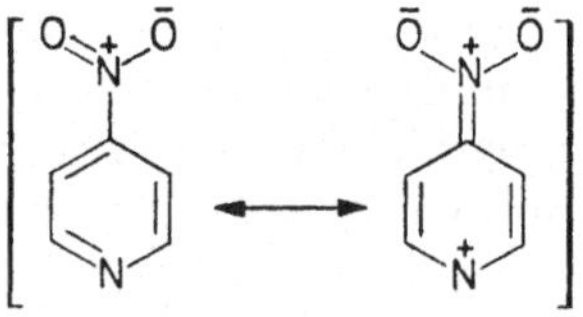

(38)

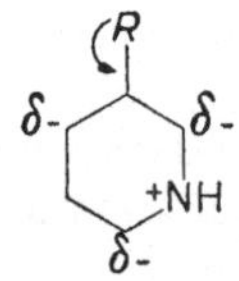

(39)

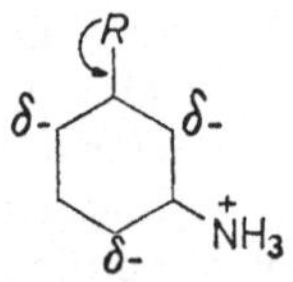

(40)

158

Substituent constants can be divided into a contribution dependent on resonance (σ_R) and one dependent on inductive interaction (σ_I), so that $\sigma = \sigma_I + \sigma_R$. In a particular compound the fractional steric inhibition of resonance is defined as $(\sigma_R^0 - \sigma_R)/\sigma_R^0$, where σ_R and σ_R^0 represent the resonance contributions to the substituent constants of the sterically hindered and the sterically unhindered group, respectively[101]. Applying the value $\sigma_I = +0{\cdot}10$ deduced for the amino group[101] to the dimethylamino group, the fractional steric inhibition of resonance of this group in 3,5-dimethyl-4-dimethylaminopyridine is 0·47. Even in this case, inhibition of the delocalization [(36) and (37)] is far from complete[40]. The Hammett relationship gives $\sigma_{4NO_2} = +0{\cdot}617$, and this value is hardly changed in 3,5-dimethyl-4-nitropyridine; it is, in fact, close to σ_I $(+0{\cdot}63)$ for the nitro group[37a]. These facts indicate the negligible importance of the conjugation represented in (38), and show that the nitro group influences the ring nitrogen atom solely by its inductive effect. (Concerning–M substituents in general, see[84h]).

Bryson[93] analysed the effects of 3-substituents upon the basic strength of pyridine in a different way. Using six substituents of the type $(-I-M)$, six of the type $(-I+M)$, and then both of these groups together, he found a better Hammett 'fit' in the first two cases than in the third (*Table 5.14*). He argued that 3-substituents exercise distinctive effects depending on their $(-M)$ or $(+M)$ character, groups of the second type exerting a greater mesomeric influence as a result of the closeness of the mesomeric moment of the substituent to the $=\overset{+}{N}H-$ pole of the acid (39). For 3-substituents in pyridine the relationship $\sigma = 1{\cdot}02\,\sigma_I + 0{\cdot}40\sigma_R$ is very satisfactory. Comparison with the corresponding expression for *m*-substituted aniline $(\sigma = 1{\cdot}05\,\sigma_I + 0{\cdot}33\,\sigma_R)$ shows the greater importance of the mesomeric component in the pyridines, a situation caused by the mesomeric moment of the substituent being nearer to the functional group in the pyridines than it is in the anilines [cf. (39) and (40)].

(d) Complexes with Metal Ions

Pyridine forms coordination complexes with many metals; only a few of these can be mentioned here. Thus, pyridine is a component of square complexes such as those containing gold[102] and copper[103], $Et_2AuBr.C_5H_5N$ and $CuCl_2.2C_5H_5N$. Aluminium salts containing pyridine in tetrahedral structures are known[104], and the same is true for beryllium, magnesium and boron compounds. Some properties of the boron complexes have already been discussed (p. 151). The cobalt compound, $CoCl_4.2C_5H_5N$, is an example of an octahedral complex. The influence of 3- and 4-substituents in the pyridine ring upon the stabilities of silver complexes has been discussed in terms of π bonding between the unshared $4p$ electrons of the metal and the nitrogen π orbital[105]. This sort of bonding [see (41)] introduces double-bond character between the metal and the nitrogen atoms. It is discouraged by electron-releasing groups such as methyl or methoxyl at $C_{(4)}$, and encouraged by electron-accepting substituents such as .CN.

Most studied of pyridine derivatives as a complexing reagent is 2,2'-bipyridyl[106] which acts as a bidentate chelate molecule forming, for example, with ferrous ions the well-known red octahedral complex $Fe(bipy)_3^{2+}$. In this respect 2,2'-bipyridyl is merely a particular instance of a sizeable group of compounds containing the element of structure :N—C—C—N:. By exerting a steric influence, 6-monosubstitution in 2,2'-bipyridyl weakens its complexing ability, whilst 6,6'-disubstitution completely inhibits it[106]. A second kind of steric interference is seen in 3,3'-dimethyl-2,2'-bipyridyl which gives only a weakly coloured ferrous complex, because of methyl group overlap[107]. The magnetic moments of octahedral bipyridyl complexes of chromium, manganese, iron, cobalt and nickel have been measured and used to discuss the character of the metal–nitrogen bonds. In some of these compounds resonance involving double bonds [cf. (41)] is important[108].

2,6-Di-(2-pyridyl)pyridine and tetrapyridyl also form complexes with metal ions. Whilst the former functions as a tridentate chelate molecule, it is doubtful if the tetrapyridyl is quadridentate[106].

In the sulphonation of pyridine, mercury ions have a catalytic effect. It has been suggested[109a] that in this reaction and in the mercuration of pyridine, which occurs rather easily, complexing permits the $5d$ orbital of mercury to assist in stabilizing the transition state.

The complex of pyridine with chromic anhydride, $CrO_3 \cdot 2C_5H_5N$, which in the crystalline form may be di- or polymeric[110a], is a useful oxidizing agent[110b].

Pyridine oxides form metal complexes[111a], and 2-aminopyridine 1-oxide gives chelates with metals[111b].

Chelate complexes formed by pyridine derivatives are mentioned in the appropriate part of Section 5.

(e) Complexes with Non-metallic Substances

Pyridine forms a variety of complexes with halogen and inter-halogen compounds. All the possible compounds $C_5H_5N.X_2$ have been described, though later workers have been unable to obtain the solid iodine compound[112]. Fluorine forms an unstable complex at low temperatures[113]. The dichloride decomposes rapidly in the presence of moist air. The bromide and inter-halogen compounds[112, 114] are more stable, and the compounds $C_5H_5N.ICl$ and $C_5H_5N.IBr$, prepared in carbon tetrachloride solution, can be recrystallized from alcohol[112]. Besides the 1:1 complexes, others have been described. Examples[115-6] are the compounds $C_5H_5N.Br_4$ and $C_5H_5N.I_4$. Although Williams[112] could not isolate the 1:1 complex of pyridine and iodine, there is no doubt that complexing occurs. Addition of pyridine to iodine solutions causes marked colour changes accompanied by changes in the infra-red spectrum[117]. Some of the pyridine–halogen complexes are listed in *Table 5.15*.

As regards the structures of these compounds, the $C_5H_5N.ICl$ complex in the solid state contains the chlorine, iodine, nitrogen and pyridine $C_{(4)}$ atoms in a straight line, and the interatomic distances suggest that the chlorine atom, the 'outer halogen', does not have a strong negative charge[118a]. In the same state the $C_5H_5N.I_4$ complex contains the centrosymmetric cation, $(C_5H_5N)_2I^+$, and I_3^- anions, to each of the outer atoms of which is linked an iodine molecule[118b]. The same type of cation must be present in the easily prepared[119] compounds $(C_5H_5N)_2BrClO_4$ and $(C_5H_5N)_2BrNO_3$, and in the corresponding iodine derivatives[120].

The vapour pressure of $C_5H_5N.Br_2$ in carbon tetrachloride shows it to be almost completely undissociated[112]. Conductivity measurements on

$C_5H_5N.ICl$ in pyridine were taken to indicate[112] only very slight ionization of a salt-like structure (42). Similar measurements show small concentrations of ions, taken to be $C_5H_5N.I^+$ and I^-, in pyridine solutions of iodine[121]. The situation is complicated by a slow reaction which occurs, causing nuclear substitution and producing I_3^- ions[122].

Table 5.15. Some Pyridine–Halogen Complexes

Compound	*Colour*	*m.p.,°C*
$C_5H_5N.Cl_2$	white[112]	47 (variable)
$C_5H_5N.Br_2$	red[112, 115]	62–3,94–5
$C_5H_5N.BrCl$	white[112]	107–108
$C_5H_5N.ICl$	pale yellow[112, 123b]	134
$C_5H_5N.IBr$	golden yellow[112, 123b]	116–117
$C_5H_5N.Br_4$	red[115]	58·5
$C_5H_5N.I_4$	green[116]	85
2-Me.C_5H_4N.ICl	yellow[123b]	77·5
2-Me.C_5H_4N.IBr		67–68
2,6-Me_2.C_5H_3N.ICl		112–113
2,6-Me_2.C_5H_3N.IBr		106–108

More recently, spectroscopic studies of solutions of $C_5H_5N.ICl$ and 2,2′-$C_5H_4N.C_5H_4N.2ICl$ in acetonitrile have been interpreted in terms of the ionizations

$$2,2'\text{-}C_5H_4N.C_5H_4N.2ICl \rightleftharpoons 2,2'\text{-}C_5H_4N.C_5H_4N.I^+ + ICl_2^-$$

$$2C_5H_5N.ICl \rightleftharpoons (C_5H_5N)_2I^+ + ICl_2^-$$

On the other hand, dissociations in non-polar solvents are of the type

$$\text{Base.IX} \rightleftharpoons \text{Base} + \text{IX}$$

The order of stability of the complexes as the halogen is changed is ICl > IBr > I_2, and as the base is changed, 2-picoline > pyridine > 2,6-lutidine[123]. Charge-transfer bands have been observed in the spectra of dilute solutions of iodine and pyridine in heptane[124], and also in the spectra of the pyridine-iodine monochloride complex[125].

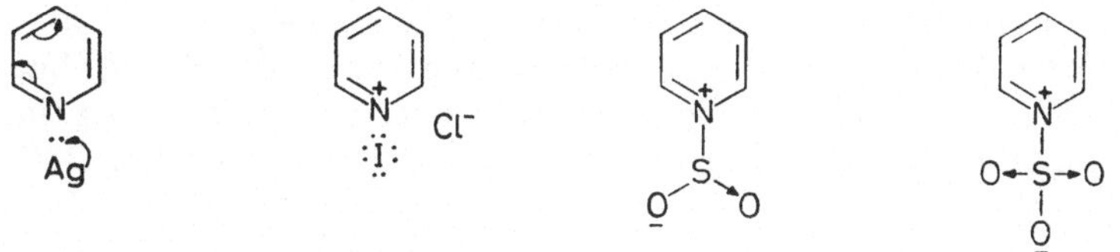

(41) (42) (43) (44)

The pyridine-halogen complexes form salts with acids, the halogen being retained. Pyridine bromide hydrobromide, $C_5H_5\overset{+}{N}H.Br_3^-$, is a useful brominating agent[126]. Crystalline perbromo-hydrobromides of this type have been obtained by adding bromine to pyridine hydrobromide in acetic acid[127a]. Similar in character[127b] are the salts $C_5H_5\overset{+}{N}H.ClI_2^-$ and 2-Me.$C_5H_4\overset{+}{N}H.ClBr_2^-$. Some of these compounds are mentioned again in connection with nuclear halogenation (p. 165). Aminopyridines also form halogen complexes (p. 168).

With dinitrogen pentoxide pyridine gives 1-nitropyridinium nitrate[128a], and with nitronium fluoborate the corresponding fluoborate is formed[128b]. Pyridine forms a 1:1 complex with phosphorus pentachloride[128c], with sulphur dioxide[129] (43) and with sulphur trioxide[130a] (44). The last is useful as a sulphonating agent (p. 80), and is formed from its components in carbon tetrachloride, from pyridine and fuming sulphuric acid, and from pyridine and chlorosulphonic acid. Its reactions are discussed later (p. 177).

It is possible that complexes related to those just mentioned are involved in reactions between pyridine and its derivatives and thionyl chloride (pp. 186–228).

Acylpyridinium salts and acyloxypyridinium salts are mentioned later (pp. 195, 234, 341–2).

(2) BEHAVIOUR IN SUBSTITUTION REACTIONS

Pyridine presents a great contrast to pyrrole as regards its ability to undergo substitution. Electrophilic substitutions are less numerous, varied and important than with the pyrroles. On the other hand, nucleophilic substitutions are of great consequence, and radical substitutions are more widely and significantly represented.

ELECTROPHILIC SUBSTITUTION

(a) Acylation and Carboxylation

Electrophilic C-acylation of pyridine is practically unknown. The exception is the surprising report that pyridine reacts with *N*-methylformanilide and phosphoryl chloride to give a small yield of pyridine-2-aldehyde[130b]. The reported orientation makes the report suspect. The Friedel–Crafts method fails, presumably because the acylium cation is not active enough to attack the pyridine nucleus deactivated by coordination with aluminium chloride (cf. the case of chlorination below). Even the much more reactive 2-hydroxypyridine cannot be C-acylated, though 2-pyridyl benzoate gives a trace of 5-benzoyl-2-hydroxypyridine in the Fries rearrangement[131].

Friedel–Crafts acylation also fails with pyridine 1-oxide[132], but with nicotinic acid 1-oxide, acetic anhydride effects an interesting internal acylation, as well as the commoner nuclear hydroxylation[133a] (p. 234)

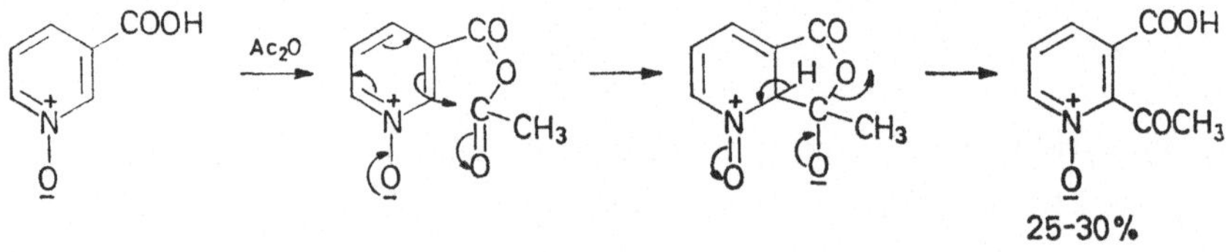

25-30%

The pyridine ring does not prevent Friedel–Crafts acylation of aryl substituents.

Direct carboxylation occurs in the pyridine nucleus when an activating group is present. 2-Hydroxypyridine, as its sodium salt, or with potassium carbonate, reacts with carbon dioxide to give 6-hydroxynicotinic acid, whilst 3-hydroxypyridine gives 3- and 5-hydroxypicolinic acid in proportions which depend on the conditions[134].

(b) Alkylation and Substituted Alkylation

Unambiguously electrophilic C-alkylations of unactivated pyridine nuclei are unknown. A carbonium ion is presumably involved in the conversion of 2-hydroxypyridine into 2-hydroxy-5-triphenylmethylpyridine when it is heated with triphenylmethyl chloride, or with triphenylcarbinol and a little sulphuric acid. 2-Hydroxy-3-methylpyridine reacts similarly, but 2-hydroxy-6-methylpyridine is thought to give 2-hydroxy-6-methyl-3-triphenylmethyl-pyridine. Oddly, 1-methyl-2-pyridone also gives 2-hydroxy-5-triphenyl-methylpyridine, the methyl group being lost[131].

3-Hydroxypyridine, in the Mannich reaction, gives 2-dialkylamino-methyl-3-hydroxypyridines, and 5-hydroxy-2-methylpyridine is substituted at $C_{(6)}$[135a−c]. In contrast, 2-hydroxypyridine, with formaldehyde and alkali, gives 1-hydroxymethyl-2-pyridone[135d]. 2,6-Diaminopyridine reacts at $C_{(3)}$ with Mischler's hydrol[136].

(c) Carbonyl Reactions

Pyridine is too unreactive to suffer electrophilic attack by a carbonyl group. However, a form of electrophilic substitution occurs when picolinic and isonicotinic acids are decarboxylated in the presence of carbonyl compounds. These acids undergo decarboxylation in the zwitterionic forms (p. 319), and when decarboxylation occurs in the presence of carbonyl compounds, carbinols are formed[137−9]. It is not certain whether the reaction proceeds by electrophilic attack by the carbonyl compound on the product of decarboxylation (process A) or whether the carbonyl compound is involved in the decarboxylation (process B). The reaction has preparative value[140a]. See also the discussion of the reactions of aminopyridines with aldehydes (p. 359).

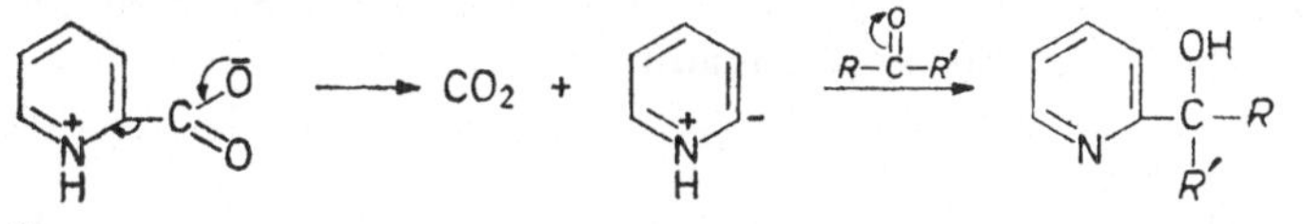

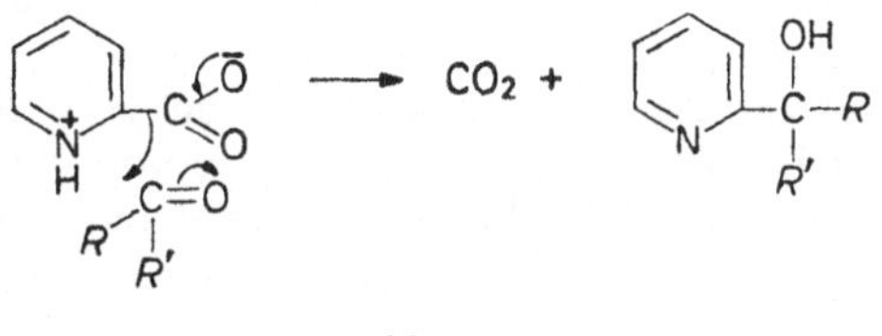

(d) Deuteration and Tritiation

Electrophilic deuteration of pyridine has not been described. Pyridine underwent no exchange with tritiated water after 40 h at 207° in a solution of $H_0 - 12$. However, under similar conditions the 3- and 5-hydrogen atoms of 2,6-lutidine, 2,4,6-collidine and 1,2,4,6-tetramethylpyridinium sulphate were replaced by tritium. The first-order rate constants varied linearly with H_0, showing that the exchange involved the pyridinium ions and not the free bases. This conclusion also followed from comparison with the reaction rate for the quaternary salt. The change from =CH— to $:\overset{+}{N}H$, in going from benzene to pyridine derivatives, reduced the rate about 10^{18} times, and the results suggested clearly that pyridine itself is considerably deactivated by comparison with benzene[140b].

Acid-catalysed deuteration of 2,6-dimethyl- and 2,4,6-trimethyl-pyridine 1-oxides in deuterosulphuric acid occurs at the β-positions and involves the conjugate acids of the oxides. Furthermore, the group $:\overset{+}{N}OH$, present in the cations of the oxides, is somewhat more deactivating than the $:\overset{+}{N}H$ group, present in the cations of pyridines[141a]. The situation should be compared with that occurring in the nitration of pyridine 1-oxide (p. 174).

The base-catalysed α-deuteration of pyridinium salts and pyridones (pp. 379, 397) probably proceeds by way of an ylid.

(e) Diazo Coupling

Aromatic diazonium salts will couple with pyridines only when electron-releasing groups are present in the pyridine nucleus. 2-Aminopyridine and benzene diazonium chloride in bicarbonate solution give 2-benzenediazo-aminopyridine[141b], and nuclear substitution has not been reported. However, numerous examples of coupling at $C_{(3)}$ in 2,6-diaminopyridine are known[142, 143]. Excess of diazonium salt causes 3,5-disubstitution. At low acidity arylazoamino compounds are formed, which rearrange in acid to 3-arylazo-2,6-diaminopyridine[142]. 2,4-Diaminopyridine also couples, probably at $C_{(5)}$[144], and 3,5-diaminopyridine, probably at $C_{(2)}$[145].

2-Hydroxypyridine in alkaline solution couples at $C_{(5)}$[146], as most probably does 2,6-dihydroxypyridine[147].

3-Hydroxypyridine reacts predominantly at $C_{(6)}$, but in slightly alkaline conditions, with p-nitrobenzene-diazonium chloride, a small proportion of coupling also occurs at $C_{(2)}$[134c, 146, 148]. A reported coupling of 3-hydroxy-pyridine at $C_{(4)}$[149] seems questionable. Citrazinic (2,6-dihydroxypyridine-4-carboxylic) acid also couples readily[150].

(f) Halogenation

This is the most studied of electrophilic substitutions in the pyridine series.

Simple substitutive fluorination has not been observed[151]; pyridine gives perfluoropiperidine, $C_5F_{11}N$.

Pyridine hydrochloride reacts with chlorine at 615–175°, giving 3,5-dichloro-, 3,4,5-trichloro- and some pentachloropyridine[152a, b]. Sell and Dootson[152b] seem to have been the first to postulate the intermediate formation of a pyridylpyridinium salt, of which several examples are mentioned later, to account for the presence among the halogenation products of an amino-trichloropyridine. Pyridine hydrochloride slowly absorbs chlorine at room temperature, giving a semi-solid complex; heated rapidly at 160–180°, the latter gives small yields of 3-chloro- and 3,5-dichloro-pyridine[127b]. 2-Methylpyridine, illuminated and heated with chlorine in water, gives 2-trichloromethylpyridine and its 5-chloro-, 3,5-dichloro and 3,4,5-trichloro-derivatives[153].

Phosphorus pentachloride has also been used to chlorinate pyridine, converting it at 210–220° into a mixture of 3,5-dichloropyridine and higher chlorinated compounds[152c]. Carried out at 280–285°, the reaction is a useful source of pentachloropyridine and the tetrachloropyridines[154]. Barium pyridine-3,5-disulphonate is similarly converted into 3,5-dichloro- and 2,3,5-trichloro-pyridine[155], and nuclear chlorination sometimes accompanies the replacement of hydroxyl groups by chlorine atoms through the use of phosphorus chlorides[156a−c].

In the vapour phase at 200° chlorination of pyridine is slow and gives 3,5-dichloro- and 3,4,5-trichloropyridine. By contrast, at 270° the main product is 2-chloropyridine, with some 2,6-dichloropyridine and a small quantity of 2-pyridyl-pyridinium chloride (for hydrolysis of the reaction mixture gives some 2-aminopyridine)[152a]. The change in orientation with rising temperature is usually thought to indicate a transition from electrophilic to radical substitution, but it is possible that the relative thermodynamic stabilities of the different isomers affect the result[156d]. For chlorination, the temperature ranges for each type are nearer together than for bromination (see below). At 320–340° chlorination is more violent than bromination; 2-chloro- and 2,6-dichloropyridine result, but charring occurs[152a].

The pyridine–bromine complex or its hydrochloride, heated at 200°, gives 3-bromo- and 3,5-dibromo-pyridine[157]. Later workers[127b] obtained more complex results, the 3-bromopyridine probably being contaminated with 3,4-dibromopyridine or other reactive halogen derivatives which subsequently decomposed to polymeric salt-like material. More practicable is the preparation of 3-bromo- and 3,5-dibromo-pyridine by adding bromine to pyridine hydrochloride under reflux (a reaction in which iron and mercuric chloride have been used as catalysts)[158a−c]. 3-Bromopyridine in good yield, with some 3,5-dibromopyridine, is obtained by heating pyridine hydrochloride perbromide at 160–170°; pyridine hydrobromide perbromide can be used, but not pyridine sulphate which does not give a stable perbromide[127].

Electrophilic bromination following nucleophilic chlorination seems to be a ready route to 3-bromopyridine (p. 228).

With bromine in oleum containing 65 per cent of sulphur trioxide, at 130°, pyridine gives 3-bromopyridine. The sulphur trioxide oxidizes the hydrogen bromide formed, and the reaction follows the equation

$$2C_5H_5N + 2SO_3 + Br_2 = 2C_5H_4BrN + SO_2 + H_2SO_4$$

Small amounts of dibromopyridines are also formed. When the reaction is carried out in 90 per cent sulphuric acid, comparable amounts of 3-bromo- and 3,5-dibromo-pyridine are formed. In both circumstances the 3-position is first attacked, but the rate is much greater in oleum. The differences arise because in one case the pyridinium ion is undergoing substitution and in the other the more reactive pyridine–sulphur trioxide complex[158d].

Bromine converts pyridine-3-sulphonic acid, in boiling aqueous solution, into 4-amino-3,5-dibromopyridine, presumably through 3,4,5-tribromo-pyridine and a derived 3,5-dibromo-4-pyridyl-pyridinium salt (see p. 266)[159, 160].

Extensive studies have been made of the vapour-phase bromination of pyridine. At 300°, over pumice or charcoal, bromine converts pyridine mainly into 3-bromo- and 3,5-dibromo-pyridine, whilst at 500° 2-bromo- and 2,6-dibromo-pyridine result[127b, 161]. At 400° a mixture results from which 3,5- and 2,6-dibromopyridine were isolated. More complete examination of the reaction at 300° showed that besides the main products there were also formed 2-bromo-, 2,3-, 2,5-, 2,6- and 3,4-dibromo-, 2,3,5- and 3,4,5-tribromo-, but no 2,4-dibromo-pyridine. The main results of the vapour-phase brominations of pyridine and the bromopyridines[161–3] are summarized in *Figure 5.2*.

It is clear that 2- and 2,6-substitution are the characteristic reactions around 500°. 2,6-Dibromopyridine in the presence of pumice and ferrous bromide at 480° gives 2,3,6-tri- and 2,3,5,6-tetra-bromopyridine. Without the specific catalyst, 2,4,6-tribromopyridine is formed, 4-substitution being slow compared with 2- and 2,6-substitution. The conversion of 2,6-dibromo-pyridine into 2,4,6-tribromopyridine over pumice occurs to the extent of 3 per cent at 455°, rising to 36 per cent at 530°. At the higher temperatures, 2,3,4,6-tetra- and pentabromopyridine begin to appear. 4-Bromination of 2,6-dibromopyridine is enhanced by the presence of iodine, and 2,4,6-tribromopyridine is formed in higher yield when a mixture of 2-bromo- and 2,6-dibromopyridine, rather than the latter alone, is brominated. Daylight and ultra-violet light are without effect[163]. As with vapour-phase chlorination, it is usually thought that brominations occurring at about 300°, giving 3- and 3,5-substitution, are electrophilic reactions, whereas those at about 500°, causing 2-, 2,6- and sometimes 4-substitution, are radical reactions but, as mentioned above, thermodynamic factors may be important. However, apart from the fact of the change in orientation with temperature, nothing is known of the detailed mechanism.

3-Chloropyridine and the halogenopicolines cannot be prepared by the direct substitutions discussed. For this reason, and also for its theoretical interest, recent work on halogenation in the presence of aluminium chloride is important. Pyridine and chlorine with about 2·5 molecular equivalents of aluminium chloride gives 3-chloropyridine (33 per cent). 4-Picoline gives 40 per cent of 3-chloro- and 15 per cent of 3,5-dichloro-4-picoline. In the same way, 4-picoline gives 3-bromo-4-picoline, and 2-picoline gives 5- and 3-bromo-2-picoline. Since the total yields are about 50 per cent and since the salt $C_5H_5\overset{+}{N}H\overset{-}{A}lCl_4$ could not be chlorinated, the following mechanism is proposed:

$$2C_5H_5N + 2AlCl_3 \rightarrow 2C_5H_5N:AlCl_3$$

$$2C_5H_5N:AlCl_3 + Cl_2.AlCl_3 \rightarrow Cl.C_5H_4N:AlCl_3 + C_5H_5\overset{+}{N}H\overset{-}{A}lCl_4 + AlCl_3$$

Complexes of the type $C_5H_5N:AlCl_3$ are Lewis salts, and in them deactivation of the nucleus is thought not to be so great as in proton salts. At the same time, the large excess of aluminium chloride is required to convert the chlorine into a particularly active electrophilic reagent, and also to provide a medium of high dielectric constant[164]. The general principle of

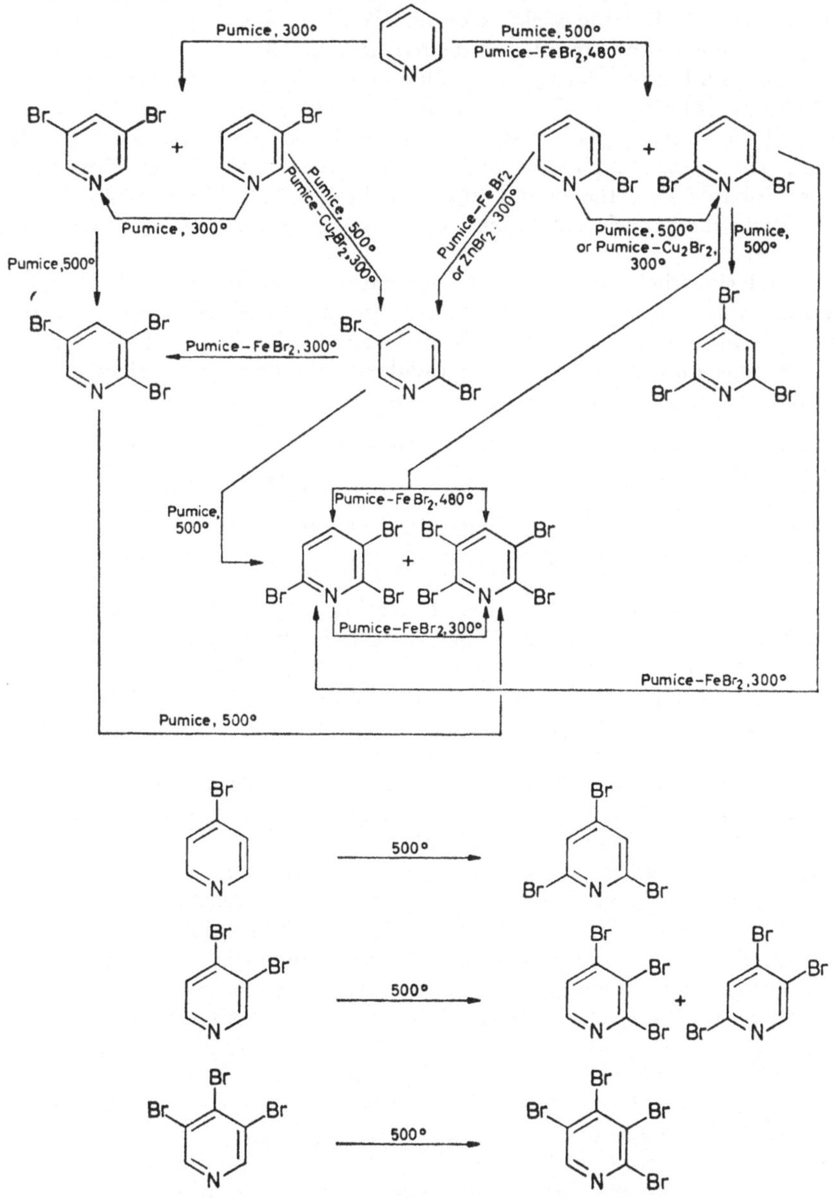

Figure 5.2. The Vapour-Phase Bromination of Pyridine

167

using a catalyst which will activate a substituting agent more than it de-activates the substrate might find wider application (cf. the use of N-bromo-succinimide below).

Pyridine and iodine react in the vapour phase, giving small amounts of 3,5-di- and pentaiodopyridine. The use of iodine monochloride does not improve the yield, and considerable amounts of 2-chloropyridine result[165]. Molten pyridine hydrochloride reacts with iodine to give some penta-iodopyridine[165], but the yield is better[127b] when the very stable pyridine hydrochloride–iodine complex is heated at 280–290°. At 280° the hydro-chloride is said also to form, with iodine or iodine monochloride, 2-pyridyl-pyridinium chloride, via 2-halogenation[165]. Iodine in 50 per cent oleum at 300–320° converts pyridine into 3-iodo- (18 per cent) and some 3,5-di-iodo-pyridine, and 2-picoline at 200–210° gives 5-iodo-2-methylpyridine. It is probable that the oleum functions here rather like the aluminium chloride in the experiments discussed above, generating the reactive iodine cation[165, 166].

Considering the general difficulty of halogenating pyridine, as revealed by the above discussion, it is surprising that pyridine has been converted into 3,5-dibromopyridine (though in unstated yield) by the action of N-bromosuccinimide in carbon tetrachloride[167]. In this experiment alu-minium chloride (about one-half of a molecular equivalent) was used as catalyst. It seems likely that it generates bromine cations [$\diagup$N:Br + AlCl$_3$→ ($\diagup$N:AlCl$_3$)$^-$ + Br$^+$] which can attack the pyridine, the latter being free from the deactivating effect of protonation or coordination [Br$^+$ + C$_5$H$_5$N → Br . C$_5$H$_4$N + H$^+$. ($\diagup$N : AlCl$_3$)$^-$ + H$^+$→ $\diagdown$NH + AlCl$_3$].

Also surprising is the conversion of nicotinic acid chloride hydrochloride into 5-bromonicotinic acid in 87 per cent yield, by heating with bromine at 150–170°. Direct chlorination was much less successful[168]. This may be a direct electrophilic substitution, but the nicotinic acid chloride hydrochloride was prepared from nicotinic acid and thionyl chloride (see p. 322), and it is just possible that the reaction is related to the substitutions into pyridine–thionyl chloride complexes discussed below (p. 228). It might even be that nicotinic acid chloride hydrochloride is not a simple salt but possesses a structure like the pyridine–thionyl chloride complex.

Amino and hydroxyl groups greatly facilitate nuclear halogenation. 2-Aminopyridine in alcohol or dilute sulphuric acid gives 2-amino-5-chloro- and 2-amino-3,5-dichloropyridine[169]. When the 5-position is blocked, as in 2-amino-6-methyl-5-nitropyridine, chlorination occurs[170] at C$_{(3)}$. The same rules control orientation in bromination, and the powerful *para*- and *ortho*-directing ability of a 2-amino group is seen in the bromination in acetic acid of 2-amino-4- and -6-bromopyridine[171]. 2-Aminopyridine with hydri-odic acid and hydrogen peroxide, or with aqueous potassium tri-iodide[172], gives 2-amino-5-iodopyridine. Acid solutions of 2-aminopyridine and iodine monochloride give salts of the 2-aminopyridine-iodine monochloride com-plex; these are converted by alkali to 2-amino-5-iodopyridine[173]. High yields of the latter are obtained by treating 2-aminopyridine in water with iodine, and then alternately basifying and acidifying the solution[174].

In the vapour phase at 500° the bromination of 2-aminopyridine gives 2-amino-3-, -5- and -6-bromopyridine, 2-amino-3,5-, -3,6- and -5,6-dibromo-pyridine and 2-amino-3,5,6-tribromopyridine. The 3- and 5-positions are substituted at about the same rate, and the rate for the 6-position is not very different[175]. At 510° over pumice, 2-aminopyridine and iodine give 2-amino-5-iodopyridine[176].

3-Aminopyridine, with hydrochloric acid and hydrogen peroxide, gives 3-amino-2-chloro- and some 3-amino-2,6-dichloropyridine[177]. 3-Amino-pyridine and iodine monochloride in hydrochloric acid form a complex which, when boiled with water, gives 3-amino-2,6-di-iodopyridine[178].

A 4-amino-group directs chlorination by hydrochloric acid and hydrogen peroxide, and bromination in acetic acid to $C_{(3)}$. Thus, 4-amino-2-halogeno- and 4-amino-2,5-dihalogeno-pyridines are substituted at $C_{(3)}$[161c, 179] and 4-amino-3-nitropyridine at $C_{(5)}$[180]. Iodine monochloride converts 4-amino-2,6-dimethylpyridine into the 3,5-di-iodo compound[181].

2-Hydroxypyridine is readily chlorinated to 5-chloro- and 3,5-dichloro-2-hydroxypyridine[182a]. Bromination[155] and iodination[182a] give the di-sub-stituted compounds, though in mildly alkaline solution 2-hydroxy-5-iodo-pyridine can be prepared[182b]. Substituted 2-hydroxypyridines and 2-alkoxy-pyridines also tend to undergo 3,5-dihalogenation, except of course when one of these positions is filled[183]; the same is true of 1-methyl-2-pyridones[171, 184]. It is not clear whether the conversion of 6-methyl-2-pyridone into 3,5-dibromo-6-methyl-2-pyridone by N-bromosuccinimide in carbon tetra-chloride[171e] is a radical or electrophilic substitution. There is, however, evidence that under conditions favouring radical substitution, bromination of nuclear methyl groups in pyridones occurs[171g].

3-Hydroxypyridine with bromine in pyridine gives 2-bromo-3-hydroxy- or 2,6-dibromo-3-hydroxypyridine, depending on the proportion of bromine used, but in water the product is 2,4,6-tribromo-3-hydroxypyridine[185]. In mildly alkaline solution iodination of 3-hydroxypyridine gives 3-hydroxy-2-iodopyridine[134c, 182b].

As would be expected, 4-hydroxypyridine is readily 3,5-dibrominated[186] and 3,5-di-iodinated[159b], but again, in mildly alkaline solution monosubsti-tution to give 4-hydroxy-3-iodopyridine can be effected[182b].

2,4-Dihydroxy-, 2-ethoxy-4-hydroxy- and 4-ethoxy-2-hydroxy-pyridine with bromine in pyridine are substituted at $C_{(3)}$ first, whereas 2,4-diethoxy-pyridine gives 5-bromo-2,4-diethoxypyridine. 2,4-Dihydroxypyridine exists predominantly as 4-hydroxy-2-pyridone, and bromination at $C_{(3)}$ is thought[187a] to require the presence of the structure $O:C—C=C—OH$ or $O:C—C=C.OR$.

Bromination of 3-chloro-2,4-dihydroxypyridine proceeds, as expected, at $C_{(5)}$, and 5-bromo-2,4-dihydroxypyridine is chlorinated at $C_{(3)}$ by hydro-chloric acid–hydrogen peroxide. However, 3-bromo-2,4-dihydroxypyridine with the same reagents gives 5-bromo-3-chloro-2,4-dihydroxypyridine[179]. This is an example of an interesting rearrangement which is discussed later (p. 371).

Pyridine 1-oxide does not react at 110° with bromine in the presence of iron[132]. Bromine in 90 per cent sulphuric acid with silver sulphate at 200° gives poor yields of 2- and 4-bromopyridine 1-oxide[187b], but reaction with

bromine in fuming sulphuric acid gives mainly 3-bromo-, with some 2,5-
and 3,4-dibromo-pyridine 1-oxide[187c]. The difference is probably caused by
complexing of the oxide with sulphur trioxide, which causes deactivation
and consequent reaction at $C_{(3)}$. The first bromine atom then controls
further substitution. Experiments with 2- and 3-bromopyridine 1-oxide
support this explanation, and the situation should be compared with that
occurring in nitration (p. 174).

The formation of 3,5-dibromopyridine 1-oxide from bromine and pyridine
1-oxide in acetic anhydride (pp. 235, 281) probably involves initial nucleophilic addition.

(g) Hydroxylation

Elbs oxidation of 2- and 3-hydroxypyridine gives in each case mainly
2,5-dihydroxypyridine. In the first case some 2,3-dihydroxypyridine, and
in the second, some 2,3- and 3,4-dihydroxypyridine are also formed. The
kinetics and mechanism of the oxidation of 2-hydroxypyridine have been
examined, and the reaction could involve attack by the anion of the hydroxy
compound upon the peroxy bond of the persulphate ion, with the displacement of sulphate ion[188a]. 2-Aminopyridine gives hydrogen 2-amino-3-
pyridyl sulphate[188b] but 4-aminopyridine behaves differently (p. 359).

(h) Metalation, Halogen–Metal Interchange, etc.

Metalation, the replacement of a nuclear hydrogen atom by a metal atom,
as for example by lithium through the use of an alkyl lithium, is unknown
in the pyridine series. There is some difficulty in classifying such reactions,
but so far (p. 61) they have been recorded, formally, as electrophilic substitutions. There is justification for this in current ideas about these reactions[189a], if attention is centred on the metal atom. However, the special
requirements of the reaction distinguish it from reactions which can be
regarded as without qualification electrophilic, particularly as regards
orientation.

The halogen–metal interconversion, using organolithium compounds[189b],
also presents a problem of classification. One representation of its progress[189b],
shown below, makes it in one sense an electrophilic substitution, and with a
clear recognition of the ambiguity, we shall for classification purposes regard
it as such.

Halogen–metal exchanges are important in pyridine chemistry. The first
reported example[190] was the formation of 3-pyridyl lithium from 3-bromo-
pyridine and n-butyl lithium. In the pyridine series interference from nuclear
metalation is unimportant but addition of the alkyl lithium to the azomethine linkage is a difficulty[191]. It is overcome by carrying out the reaction
at about − 40°, and 3-pyridyl lithium is then formed[190, 192] to the extent
of about 70 per cent. 2-Bromopyridine gives a good yield of 2-pyridyl
lithium[193]. The use of s- or t-butyl lithium does not help to reduce the amount
of addition to the azomethine linkage[191, 193]. 3-Pyridyl lithium is formed

from 3-bromopyridine and various alkyl lithiums at $-20°$, with yields of 30–35 per cent. Under the same conditions[191] 3-iodopyridine gave 40–46 per cent of 3-pyridyl lithium, and 2-bromopyridine over 80 per cent of 2-pyridyl lithium. In both 2,6- and 3,5-dibromopyridine one bromine atom undergoes exchange[191]. 4-Bromopyridine and n-butyl lithium at $-50°$ give more than 90 per cent of 4-pyridyl lithium[194, 195], but 4-chloropyridine does not react[195]. 2-Bromo-3,4,6-triphenylpyridine also exchanges efficiently with butyl lithium[196].

The mercuration of pyridine was originally effected with mercuric acetate at 175–180°, and was said to give 3-chloromercuripyridine (70 per cent) and other substances[197]. Later workers reported these conditions to give 5–10 per cent of monosubstitution and much polymercurated material. The reaction is influenced by water, and by heating an equimolar mixture of pyridine and water with $\frac{1}{8}$ of a molar equivalent of mercuric acetate, at 155° for 2·5 h, about 50 per cent of 3-monosubstituted product was obtained. 2-Methylpyridine reacted more easily (at $C_{(5)}$?)[198]. No 2- or 4-substitution occurs in the reaction with pyridine[199]. The mercuric salts form complexes with pyridine which are thought to play a role in the reaction[198], and complexing with mercury may promote this relatively easy electrophilic substitution (see above, p. 160). 4-Methylpyridine gives a poor yield of 3-acetomercuri-4-methylpyridine[200]. 2-Aminopyridine yields 88 per cent of 2-amino-5-chloromercuripyridine[198b, 201] and 2-hydroxypyridine is also easily mercurated[201, 202].

Conflicting reports have been made regarding the mercuration of pyridine 1-oxide. The substitution with mercuric acetate in acetic acid has been stated to give 76 per cent of 4-chloromercuripyridine 1-oxide[203], but more recently has been convincingly demonstrated to proceed at the 2- and 2,6-positions[204a]. Mercuric halides give mainly 2-substitution, with some of the 3- and 4-isomers, but with mercuric sulphate and sulphuric acid, substitution becomes more difficult, and as the amount of acid is increased, 3-substitution becomes increasingly important[204b]. There is probably a change from substitution into the free base to substitution into the cation of the base (see p. 271).

(i) Nitration and Nitrosation

Pyridine is nitrated only with the greatest difficulty because of the powerful deactivation caused by protonation. Potassium nitrate and fuming sulphuric acid at 330° were said to give 15 per cent of 3-nitropyridine, and with potassium nitrate and sulphuric acid at 290–300°, 13 per cent of the same compound resulted[205]. Under the latter conditions, addition of 0·1 per cent of iron raised the yield[206] to 22 per cent. Den Hertog and Overhoff[207] could not verify this claim and found the nitration of pyridine in 100 per cent sulphuric acid with sodium and potassium nitrate to give, at 300°, 0·5 per cent of 2- and 4·5 of 3-nitropyridine. At 370° the proportions were 2 and 4 per cent, and at 450° only 2·5 per cent of 2-nitropyridine resulted. Potassium nitrate and fuming sulphuric acid at 160° produces 3·6 per cent of 2-methyl-5-nitropyridine from 2-methylpyridine (oxidation also occurs), and at 100° converts 2,6-lutidine and s-collidine into their 3-nitro-derivatives in high yields[208]. Consonant with the view that these reactions involve the cations

G

of the alkylpyridines is the equally ready nitration of 1,2,4,6-tetramethyl-pyridinium ion[208d].

Pyridine is converted into 3-nitropyridine (10 per cent) by dinitrogen tetroxide in the vapour phase[209] at 115–120°.

An amino group should facilitate nitration of the pyridine nucleus. This it does, but the effect is complicated by the fact that protonation of the amino-pyridines at the ring nitrogen atom (p. 153) makes them behave like poly-nitro-anilines. Thus, with mixed acids at 0°, 2-aminopyridine gives 2-nitraminopyridine. This rearranges in sulphuric acid, rapidly at 30–40°, giving 2-amino-3- and 2-amino-5-nitropyridine[210, 215]. Although nuclear nitration of 2-aminopyridine has been carried out[215] in sulphuric acid at 45°, it is not clear that in this and other cases direct substitution rather than nitramine rearrangement has been observed. The only piece of evidence which suggests that nuclear substitution of these primary amines might proceed independently of nitramine formation comes from experiments with 2-amino-3-ethyl-6-methylpyridine; this is stated[216] to give both nitra-mine and nitro-amine when nitrated, yet the nitramine is not rearranged by sulphuric acid. The mechanism of rearrangement of 2-nitraminopyridine has not been investigated, but the high proportion of 2-amino-5-nitropyridine formed sets the case in contrast to that of phenylnitramine, where a dominant intramolecular reaction produces 95 per cent of o-nitroaniline[217]. Other aminopyridines give nitramines, and the behaviour of some of these with sulphuric acid is summarized below.

Nitraminopyridine	Rearrangement products
2-Nitramino	2-amino-3-nitro (20%) + 2-amino-5-nitro (63%)[210-15]
3-Methyl-2-nitramino	2-amino-3-methyl-5-nitro (90%)[214]
4-Methyl-2-nitramino	2-amino-4-methyl-3-nitro (22%) +2-amino-4-methyl-5-nitro (47%)[214]
5-Methyl-2-nitramino	2-amino-5-methyl-3-nitro (33%)[214]
6-Methyl-2-nitramino	2-amino-6-methyl-3-nitro (24%) +2-amino-6-methyl-5-nitro(46%)[214]
3-Ethyl-6-methyl-2-nitramino	(see text)
5-Carboxy-2-nitramino	2-amino-5-carboxy-3-nitro[218]
5-Bromo-2-nitramino	2-amino-5-bromo-3-nitro[171a]
5-Iodo-2-nitramino	destroyed by conc. sulphuric acid[172]
2-Nitramino-3-nitro ⎱ * 2-Nitramino-5-nitro ⎰	2-amino-3,5-dinitro[210, 219b, c]
3-Nitramino	with sulphuric acid gives 3-hydroxypyridine[219a]
4-Nitramino	4-amino-3-nitro[180a, 220]
3-Nitro-4-nitramino	4-amino-3,5-dinitro[180a, 220]
2-(N-Nitromethylamino)	2-methylamino-3-nitro + 2-methylamino-5-nitro (chief product) +· trace of 2-(N-nitrosomethyl-amino)[219b]
3-Nitro-2-(N-nitromethylamino) ⎱ 5-Nitro-2-(N-nitromethylamino) ⎰	2-methylamino-3,5-dinitro[219b]
1-Methyl-2-nitrimino	2-methylamino-3-nitro + 2-methylamino-5-nitro[219b]
3-(N-Nitromethylamino)	3-methylamino-2-nitro[223]

* Cf. [121] and p. 239.

Nuclear nitro groups make the rearrangement more difficult. The be-haviour of 3-nitraminopyridine is exceptional, but substituted 3-nitraminopyridines behave similarly[222]. In view of this feature of the aminopyridines, it is odd that 2-acetamidopyridine resists nitration[224]. For the case of 3-nitramino-2-picolinic acid, see p. 361.

The directing and activating power of the amino group shows clearly in 2-dimethylaminopyridine, where nitramine formation is impossible. Nitration in mixed acids gives about 90 per cent of 2-dimethylamino-5- and 10 per cent of 2-dimethylamino-3-nitropyridine. 3,5-Dinitration is readily effected[225].

It is not surprising that, of the halogenopyridines, only the 3-substituted compounds can be successfully nitrated. 3-Bromo- and 3-iodo-pyridine give 5-nitro derivatives[226].

The hydroxyl group should facilitate nuclear nitration, and hydroxypyridines are in fact readily nitrated. However, 2- and 4-hydroxypyridines exist predominantly as pyridones (p. 154), and the group .NH.CO. is usually electron-attracting. This is shown in the acid-strengthening influence of this group upon a carboxyl or hydroxyl group additionally present in the nucleus[41]. The ready nitration, and halogenation (p. 169), of 2- and 4-hydroxypyridine must therefore be attributed to the electromeric effect, illustrated in (45). The facts are[227] that 2-hydroxypyridine with mixed acids below 40° gives 2-hydroxy-3-nitro-, -3,5-dinitro- and -5-nitro-pyridine, in that order of predominance, the 5-nitro compound being obtained only in very small amounts*. At 80–90°, 2-hydroxy-5-nitropyridine gives 2-hydroxy-3,5-dinitropyridine[228-9]. 2-Hydroxy-4,6-dimethylpyridine, with mixed acids, gives 2-hydroxy-4,6-dimethyl-3-nitropyridine[231]. 2-Alkoxy-pyridines and 1-methyl-2-pyridone produce 5-nitro derivatives[232]. The overall o,p-directing influence, with respect to the oxygen atom, in 2-hydroxy- and 2-alkoxy-pyridines is not surprising, but the apparent preference for the 3-position in the first case and the 5-position in the second would, if substantiated by quantitative studies, be interesting. In the case of the hydroxypyridine, it might indicate chelation with the attacking group, in the transition stage (46)[233]. In that case, orientation of nitration in these compounds may be reagent-dependent, for with some reagents alkoxyl groups in benzene derivatives can take part in chelation, and so lead to dominant o-substitution[234].

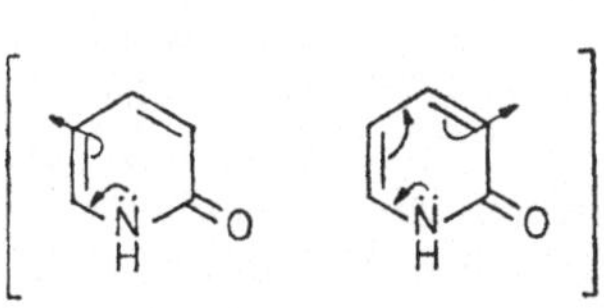

(45) 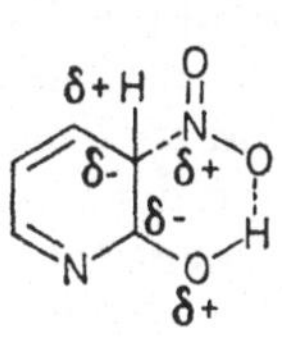   (46)

Nitration of 3-hydroxypyridine in sulphuric acid, with careful avoidance of overheating, gives 3-hydroxy-2-nitropyridine[235], and a number of alkyl-3-hydroxypyridines have been converted into mono-nitro derivatives[236]. Further nitration of 3-hydroxy-2-nitropyridine in acetic anhydride gives 3-hydroxy-2,6-dinitropyridine[237]. The nature of the products isolated from

* The substances, m.p. 286° and 289°, taken to be 2-hydroxy-3,5-dinitropyridine[227], were mixtures of 2-hydroxy-3-nitro- and -3,5-dinitro-pyridine[230]. 2-Hydroxy-3,5-dinitropyridine has[229] m.p. 176°.

an early nitration of 3-acetoxypyridine is not clear[238]. 3-Methoxy-[158b, 239] and 3-ethoxypyridine[158b, 240] give either 2-mono-nitro derivatives or 2,6-dinitro derivatives, depending on experimental conditions. The preference for attack at positions adjacent to the nitrogen atom, rather than at $C_{(4)}$, recalls similar effects of deactivating groups in benzene derivatives. The direction to $C_{(2)}$, rather than $C_{(6)}$, may again be due to chelation in the transition stage.

4-Hydroxypyridine is readily converted to 3-nitro- or 3,5-dinitro-4-hydroxypyridine[241]. 2,4-Dihydroxypyridines are easily nitrated, giving 3-nitro derivatives[242], 2,6-dihydroxypyridine yields 2,6-dihydroxy-3-nitro-pyridine[147], and 3,5-diethoxypyridine gives 3,5-diethoxy-2-nitro- and 3,5-diethoxy-2,6-dinitro-pyridine[158b, 243].

The observation that pyridine 1-oxide is readily nitrated at $C_{(4)}$ in sulphuric acid[162, 244–8] was one of the most important modern developments in pyridine chemistry*. It gave access to a large range of otherwise inaccessible compounds. The conditions of nitration have been fairly widely examined[246], and yields as high as 95 per cent of 4-nitropyridine 1-oxide claimed[162]. 2-Nitropyridine, presumably formed by deoxygenation of 2-nitropyridine 1-oxide, has been detected as a minor product, and some conditions produce 4-nitropyridine by deoxygenation[245, 247]. A 4-alkyl group prevents nitration, but 2-methyl-[244, 249, 250], 3-methyl-[251], other 3-alkyl-[252], 2,6-dimethyl-[247, 253], 3,5-dimethyl-[252] and 2,3,5,6-tetramethyl-pyridine 1-oxide[252] give 4-nitro derivatives. Examination of the literature suggests that the rather variable yields reported, as for example from 3-methylpyridine 1-oxide[251], may be caused by the use of over-severe conditions. 3-t-Butylpyridine 1-oxide is unusual in giving, perhaps for steric reasons, 3-t-butyl-2(or 6)-nitropyridine (deoxygenation also occurring)[252].

2-[250] and 3-Bromopyridine 1-oxide[248, 252, 254] give 4-nitro derivatives, in the second case with some deoxygenation[252]. 3,5-Dibromopyridine 1-oxide produces very high yields of 3,5-dibromo-4-nitropyridine 1-oxide[255].

The case of 2-hydroxypyridine 1-oxide is interesting: nitration in acetic acid gives 2-hydroxy-5-nitropyridine 1-oxide, showing that the hydroxyl group controls the orientation[256–7]. Further reaction leads to 2-hydroxy-3,5-dinitropyridine 1-oxide[257–8a]. In contrast, 2-methoxy- and 2-ethoxy-pyridine 1-oxide are nitrated at $C_{(4)}$[250, 257]. 3-Alkoxypyridine 1-oxides give high yields of 4-nitro derivatives[250, 257], and whilst it is not surprising that 3,5-dialkoxypyridine 1-oxides are nitrated[255, 258b] at $C_{(2)}$, it is unexpected that the only product isolated from 3-bromo-5-methoxypyridine 1-oxide should be 3-bromo-5-methoxy-6-nitropyridine 1-oxide[258b]. 4-Hydroxypyridine 1-oxide gives what is probably 4-hydroxy-3,5-dinitropyridine 1-oxide[259].

The second-order rate constant for the nitration of pyridine 1-oxide in 87·9 per cent sulphuric acid and the Arrhenius parameters for the reaction have been determined[260]. The failure of the 1-methoxypyridinium ion to undergo nitration under these conditions, and the fact that the 1-hydroxy-pyridinium ion should be even less reactive than the pyridinium ion, prove that the nitration of pyridine 1-oxide occurs through the free base. The observed orientation of nitration agrees with theory, but the theory fails in predicting that the oxide should be more reactive than benzene (p. 275).

---

* For nitration of pyridine 1-oxide at $C_{(3)}$ following nucleophilic addition, see p. 235.

As would be expected, a phenyl ring is nitrated in preference to a pyridyl ring or a 1-oxidopyridyl ring, as shown by the results tabulated below. The proportion of *m*-substitution found with 2- and 4-phenylpyridine led to the suggestion[233] that nitration of both the free bases and their conjugate acids

Nitration of Arglyridines

R in R.Ph*	Isomers, per cent			Conditions	References
	o	*m*	*p*		
2-C$_5$H$_4$N	5	34·9	42·3	⎫	⎫
3-C$_5$H$_4$N			64·3	⎬ nitrate added to H$_2$SO$_4$	⎬ 262
4-C$_5$H$_4$N	12·7	28·5	38·0	⎭	⎭
2-C$_5$H$_4$NO	{ 0	58	5	⎫	⎫
	{ 3	54	7		
3-C$_5$H$_4$NO	—	—	38	⎬ mixed acids	⎬ 261
4-C$_5$H$_4$NO	12	51	13	⎭	⎭
2-C$_5$H$_4$N.CH$_2$·	—	10·4	66·7	⎫	
3-C$_5$H$_4$N.CH$_2$·	—	—	63·1	⎬ nitrate added to H$_2$SO$_4$	263
4-C$_5$H$_4$N.CH$_2$·	—	4·8	70·2	⎭	
2-C$_5$H$_4$NO.CH$_2$·	—	2	55	⎫ mixed acids	⎫ 261
4-C$_5$H$_4$NO.CH$_2$·	—	—	9	⎭	⎭
2-C$_5$H$_4$N.(CH$_2$)$_2$·	16·0	3·4	64·5	HNO$_3$ alone or with H$_2$SO$_4$	264

* 4-(*p*-Dimethylaminophenyl)pyridine gives a trinitroderivative, in which one nitro group is thought to be in the pyridine ring[265]. Acetamidophenyl- and methoxyphenylpyridines are nitrated in the benzene ring[266].

was occurring. This now seems unlikely[261]. The proportions of *m*-isomers formed from 2-phenylpyridine and its oxide do not depend very much on the acidity of the nitrating medium. Further, competitive nitration apparently indicates that its oxide is nitrated 10–100 times as fast as 2-phenylpyridine. Thus, if 2-phenylpyridinium were undergoing nitration, we should be observing in the change from that to the oxide a change to a more reactive species, but with increased *m*-substitution, an unknown situation. If 2-phenylpyridine is nitrated as the free base, it is unlikely that the oxide is being nitrated as its cation. The interpretation of these results should be accepted only with caution in view of the far from complete isolation of nitration products. The higher proportion of *p*-nitration observed with 3-phenylpyridine is thought to suggest that the forms (47) and (48) are important, whilst with the oxide, (49) is evidently more important than (50).

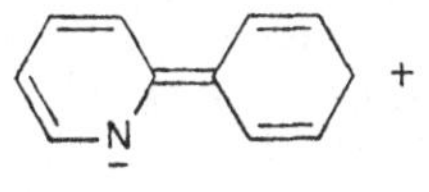

(47) 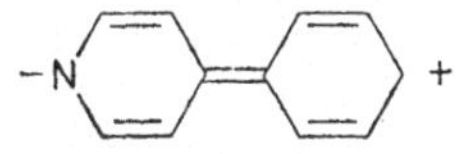   (48)

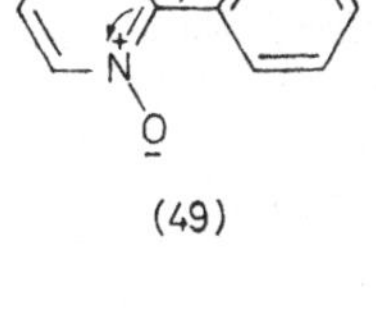

(49) 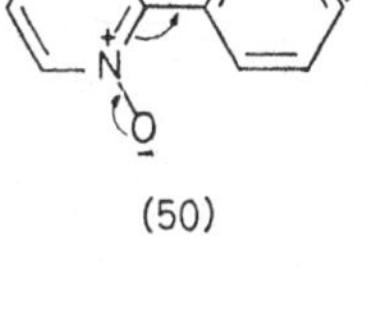   (50)

In the benzylpyridines and their oxides the influence of the heterocyclic ring is much diminished. Little difference is found between the two series. It is not known whether free bases, conjugate acids or both are involved. The heterocyclic ring is even less influential in 2,2′-phenylethylpyridine[264], and the change in $o:p$ ratio with nitrating medium is very small.

The effect upon the $o:p$ ratio of changes in experimental conditions in the nitration of 2-styryl- and 2-phenylacetylenylpyridine have been examined[264] and, for the former, discussed in terms of base-conjugate acid equilibria[233,264]. The significance of the results is not clear at present.

Nitrosation is of little importance in the pyridine series. Two activating groups are necessary for it to occur. Thus, 3-hydroxypyridine does not react with nitrous acid[134c], but 2,6- diamino-[142b, 267a], 2-amino-6-hydroxy-[267a] and 2,6-dihydroxy-pyridine[147] are all nitrosated at $C_{(3)}$. With 2-amino-5-hydroxypyridines and 2,5-dihydroxypyridines, nitrosation also occurs, combined under some conditions with diazotization in the case of the primary amine[148a], as shown in the diagram. 2-Methylaminopyridine gives the N-nitroso derivative which does not rearrange in sulphuric acid[267b].

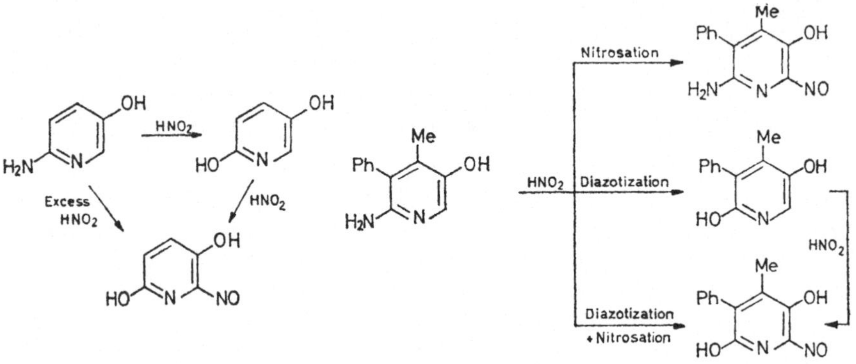

(j) Sulphonation

Pyridine has been sulphonated under a variety of conditions using sulphuric acid or oleum, usually at about 250°, sometimes in open vessels or else in sealed tubes[268], and with aluminium[238], vanadium[269] or mercuric sulphate[134c, 270-4] as catalyst. Important factors in obtaining the best yields of pyridine-3-sulphonic acid are the use of mercuric sulphate, sulphuric acid containing about a molar equivalent of sulphur trioxide, and the correct time of heating at 220–230°; in 24 h, in an open vessel, such conditions gave 71 per cent of pyridine-3-sulphonic acid[275]. Reaction in a sealed tube for 15 h at 260–265°, using 20 per cent oleum and mercuric sulphate, gives 75 per cent yields[274, 276]. At higher temperatures side products are formed[277]. At 270°, in the presence of mercuric sulphate, sulphuric acid produced about 1 per cent of pyridine-4-sulphonic acid, and at the same temperature with fuming sulphuric acid, some pyridine-2-sulphonic acid was detected. At 330°, 35–40 per cent of pyridine was recovered, 25 per cent of it appeared as pyridine-3- and -4-sulphonic acids, 15–20 per cent as 4-hydroxypyridine. Heating pyridine-3-sulphonic acid at 330°

with mercuric sulphate and 100 per cent sulphuric acid gave 50 per cent of pyridine and some 4-hydroxypyridine and pyridine-4-sulphonic acid[276].

The picolines, when sulphonated under conditions like those used for pyridine, appear to give lower yields of monosulphonic acids[269, 272, 275, 278, 279]. α-, β- and γ-Picoline give, respectively, 2-methylpyridine-5-, 3-methylpyridine-5- and 4-methylpyridine-3-sulphonic acid[275]. 4-Ethylpyridine gives 4-ethylpyridine-3-sulphonic acid[280]. Both pyridine and β-picoline have been sulphonated by heating their sulphur trioxide complexes with sulphuric acid and mercury[281].

With sulphur trioxide in sulphur dioxide at $-10°$, pyridine, 2,6-lutidine and 2,6-di-isopropylpyridine give addition complexes (p. 162). In contrast, 2,6-di-t-butylpyridine yields 2,6-di-t-butylpyridine-3-sulphonic acid[79a, 282], and 2-t-butyl-6-isopropylpyridine is also β-sulphonated[282]. This remarkable result is attributed to the prevention by steric hindrance of complexing with 2,6-di-t-butylpyridine, which thus avoids deactivation to the substituting reagent[79a]. Higher temperatures do not alter the course of the sulphonation of 2,6-di-t-butylpyridine, but when the reaction is carried out with sulphur trioxide at 240–250°, the 3-sulphonic acid is converted[283a] into the compound (51).

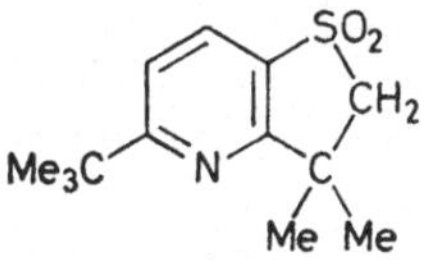

(51)

2,6-Di-t-butyl-4-ethoxypyridine is also readily sulphonated by means of sulphur trioxide[283b].

Disulphonation of pyridine has not been achieved, but piperidine, when heated with sulphuric acid, is said to give pyridine-3-sulphonic acid and pyridine-3,5-disulphonic acid[284].

2-[215, 285], 3-[286] and 4-Aminopyridine[287] have been sulphonated, giving 2-aminopyridine-5-, 3-aminopyridine-2- and 4-aminopyridine-3-sulphonic acid, respectively. Similar orientations are observed in the sulphonation of 3-[286b, 288] and 4-hydroxypyridine[287]. 2-Hydroxypyridine has not been sulphonated, but the quaternary salts from dimethyl sulphate and 1-methyl-2-pyridone, 2-ethoxypyridine or 2-chloropyridine are said[289] to give 1-methyl-2-pyridone-5-sulphonic acid when heated to 200°.

The behaviour of pyridine 1-oxide on sulphonation is in striking contrast to that on nitration (p. 174). With 20 per cent oleum and mercuric sulphate at 220–240°, after 22 h 45 per cent of the oxide is unchanged, and 40–45 per cent of pyridine-3-sulphonic acid 1-oxide, and 2–2·5 per cent of pyridine-4-sulphonic acid 1-oxide are formed. At higher temperatures, more of the 4-sulphonic acid is formed, but above 270° deoxygenation complicates the situation[132, 290a]. The observed orientation is not due to a rearrangement of the 4- to the 3-sulphonic acid[290b] but probably arises from sulphonation of the oxide cation. 2,6-Lutidine 1-oxide is also sulphonated at the 3-position[290b].

(k) Thiocyanation

Direct thiocyanation, by means of a thiocyanate and bromine, cannot be carried out with pyridine, the picolines and 2-aminopyridine[293b], but there are conflicting reports about 3-aminopyridine[293b, 931]. Two activating groups make the reaction possible, and 2,6-diaminopyridine gives 2,6-diamino-3-thiocyanato- and 2,6-diamino-3,5-dithiocyanatopyridine[293c].

(l) Reactions at the Nitrogen Atom

(i) *The attachment of alkyl and substituted alkyl groups*—The formation of quaternary salts by reaction of pyridine and its homologues with ethyl iodide was first observed by Anderson[291]. Although this is the commonest form of quaternization reaction, it is only one of a number of the general type

$$\ce{>N ^\cap C-X -> >\overset{+}{N}:C- + X^-}$$

All quaternizations, with the exception of one or two mentioned below are of this S_N2 type. The classification accords with the retention of activity (doubtless with inversion) in the quaternization of pyridine with an optically active halide[292] and with the kinetic evidence discussed below. The reaction is a particular case of the Menschutkin reaction, the formation of quaternary salts from tertiary bases, and is sometimes given that name.

Pyridine reacts vigorously with methyl iodide[293a]. Early workers rapidly multiplied examples of the reaction[293a, 294], and it was shown that secondary as well as primary halides take part in it, but that tertiary halides are converted into olefins[294b]. Although previously formed pyridinium salts can give olefins when pyrolysed they are not intermediates in these other elimination reactions, in which the pyridine merely acts as a basic catalyst[295]. Quantitative aspects of the reaction are considered below, but we might notice here that primary are generally more reactive than secondary halides[295], that electron-releasing groups at $C_{(3)}$ and $C_{(4)}$ facilitate the reaction whilst hindering it when at $C_{(2)}$, and that the order of reactivity of the derivatives of a particular alkyl group is iodide > bromide > chloride[296]. The effect of substituents at $C_{(2)}$, just mentioned, makes s-collidine especially useful in elimination reactions[297].

Higher alkyl halides (dodecyl, tetradecyl, cetyl, octadecyl, etc. and unsaturated analogues) have been used to quaternize pyridine, its homologues and derivatives[296b, 298] because of the surface-active and germicidal properties of the products[299].

Methylene dihalides, $X \cdot (CH_2)_n \cdot X$, react satisfactorily with pyridine and the picolines, examples with $n = 2, 3, 4, 5$ and 10 having been reported. Methylene dihalides give only the bis-quaternary salts, but in other cases both bis- and mono-quaternary salts have been isolated[300].

Benzyl and diphenylmethyl halides readily quaternize pyridine[301]. Trityl halides also form derivatives with pyridine which appear to be quaternary salts[301–3a]. These examples may be S_N1 quaternizations. The formation of 1-tritylpyridinium perchlorate from pyridine and trityl perchlorate[303a], and the reaction of pyridine with a solution of trityl chloride in nitromethane involve pre-formed carbonium ions and are not strictly S_N1, but S_N2C^+, since the formation of the carbonium ion is not rate-determining[303b].

Halohydrins form quaternary salts with pyridine[300g, 304-5], and ethylene oxides (see below) are not intermediates in the reactions[305].

Phenacyl halides quaternize pyridine[306], whilst 2,6-dibromocyclohexanone[307] and 1,3-dibromacetone[308] give bis-quaternary salts. 1-Phenacyl-pyridinium salts can also be prepared by treating acetophenones with iodine in pyridine[309], and the method has been applied to acetone and the acetyl-pyridines[310]. The steps in the reaction are the familiar base-catalysed enolization of the ketone, followed by rapid iodination and subsequent reaction of the iodoketone with pyridine[311]. The use of bromine in place of iodine works only in a few cases[310]. In view of the similar properties of a methyl group adjacent to carbonyl and one activated by a heterocyclic nucleus (p. 324), it is not surprising that this useful reaction can be applied to 2-methylbenzo-thiazole, 2-methylquinazoline[312], quinaldine, lepidine, 5-methylacridine and analogously, 2,4-dinitrotoluene[310], giving the products (52). The picolines do not take part in the reaction[310], but 1,2- and 1,4-dimethylpyridinium iodide do[313].

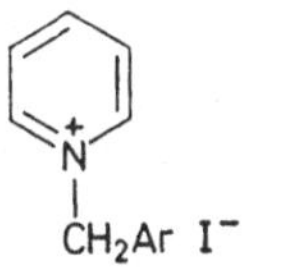

Ar = 2-benzothiazolyl, 2-quinazolyl, 2-quinolyl, 4-quinolyl, 5-acridyl, 2,4-dinitrophenyl

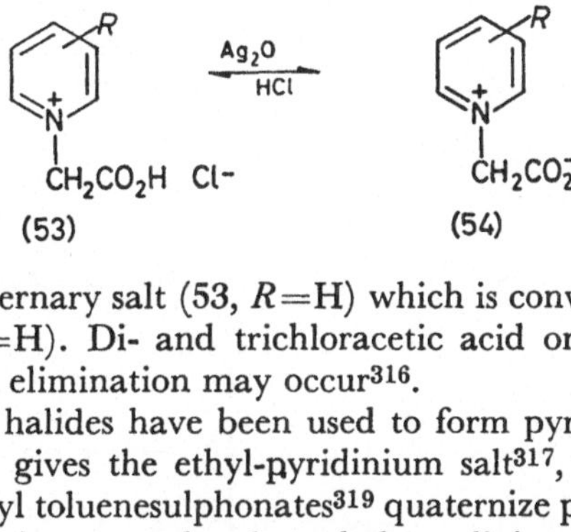

(52)

(53) Ag_2O / HCl   (54)

Chloracetic acid gives a quaternary salt (53, $R=H$) which is converted[314] by bases into betaine (54, $R=H$). Di- and trichloracetic acid only form salts[315]. With β-halogeno-acids, elimination may occur[316].

Alkyl derivatives other than halides have been used to form pyridinium salts. Potassium ethyl sulphate gives the ethyl-pyridinium salt[317], and di-methyl sulphate[184a, 318] and alkyl toluenesulphonates[319] quaternize pyridine. Primary and secondary alkyl nitrates and polymethylene dinitrates have also been used, but with t-butyl and cyclohexyl nitrate only elimination occurs[320]. 2,4,6-Trinitro-anisole and -phenetole convert pyridine into the alkylpyridinium picrates[321].

In general, polar solvents speed up the quaternization reaction[322-4a]. To some extent the effect of solvents follows their dipole moments[322], but the connection is not general, and with solvents of high polarity (tetramethylene sulphone, D = 42; propylene carbonate, D = 65·1) examples are known of quaternizations which are faster in the less polar medium[324b]. The reaction of pyridine with alkyl halides has been used to elucidate the ways in which the parameters of the Arrhenius equation vary with changes in solvent[296d, 323, 324b, 325]. For preparative purposes alcohol is often used as the solvent, but for slow reactions nitrobenzene is useful.

179

As regards the influence of substituents in the pyridine ring upon the quaternization reaction, for which some quantitative data are discussed below, it can be said that they produce the effects to be expected from their electronic characters, modified for 2(6)-substituents by a steric factor. The reaction has been observed with pyridines bearing acyl[298c], alkyl[294c, 298c, 326], alkoxycarbonyl[54e, f, 298c, 327-9], aryl[330], bromo-[328, 331a], carboxamido-[298c, 329], cyano-[54e, f, 328], ethoxycarbonylmethyl[331b], nitro-[208b, 219b] and pyridyl[332] substituents and combinations of these. Certainly, as regards the formation of 1-methylpyridinium compounds, failures are rare but have been reported for 2-chloro-5-nitropyridine[219b], pyridine-2-sulphonic acid and its amide[333] and some 2-substituted derivatives of 3-hydroxypyridine[334]. 2- and 6-Bromo- and -fluoronicotinamides failed to react with ethyl iodide, and they and their chloro analogues also did not with methyl iodide[335]. Whilst 2-fluoropyridine and 2- and 6-fluoro-3-methylpyridine readily form methiodides, 2-bromo- and 2-chloropyridine with methyl iodide give 2-iodopyridine methiodide[328]. This halogen exchange was first observed by Fischer[336], and other examples are known[337, 338]. It does not happen with 3-bromopyridine[328] and is clearly a case of nucleophilic replacement of the activated (p. 230) 2- or 4-bromo- or -chlorosubstituents in the pyridinium salts. It can be avoided by using dimethyl sulphate as the quaternizing reagent[331a, 338].

Substituents on the pyridine ring which are capable of tautomerism require separate consideration. Each component of the tautomeric system (cf. p. 152)

$$N\text{———}X\text{—}H \rightleftharpoons [N\text{———}X]^- + H^+ \rightleftharpoons H\text{—}N\text{———}X$$
$$(NM \text{ form}) \qquad (An) \qquad (Z \text{ form})$$

can conceivably react in two ways:

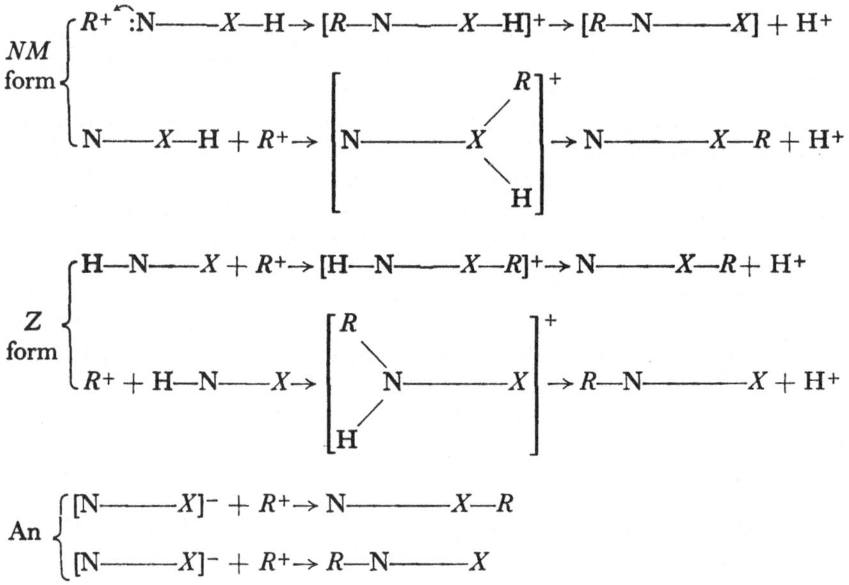

$$NM \text{ form} \begin{cases} R^+ :N\text{———}X\text{—}H \rightarrow [R\text{—}N\text{———}X\text{—}H]^+ \rightarrow [R\text{—}N\text{———}X] + H^+ \\ \\ N\text{———}X\text{—}H + R^+ \rightarrow \begin{bmatrix} N\text{———}X \overset{R}{\underset{H}{\diagup\!\!\!\diagdown}} \end{bmatrix}^+ \rightarrow N\text{———}X\text{—}R + H^+ \end{cases}$$

$$Z \text{ form} \begin{cases} H\text{—}N\text{———}X + R^+ \rightarrow [H\text{—}N\text{———}X\text{—}R]^+ \rightarrow N\text{———}X\text{—}R + H^+ \\ \\ R^+ + H\text{—}N\text{———}X \rightarrow \begin{bmatrix} \overset{R}{\underset{H}{\diagdown\!\!\!\diagup}} N\text{———}X \end{bmatrix}^+ \rightarrow R\text{—}N\text{———}X + H^+ \end{cases}$$

$$An \begin{cases} [N\text{———}X]^- + R^+ \rightarrow N\text{———}X\text{—}R \\ [N\text{———}X]^- + R^+ \rightarrow R\text{—}N\text{———}X \end{cases}$$

where R^+ represents an electrophilic reagent in a purely formal way. That is, we have the possibility of S_E2, S_E2' and S_E2cB reactions. Depending on the intermediate involved, the conditions of the reaction and the character of the substituent, there might result a salt (quaternary or otherwise) or its conjugate base. In the second event, substitutive alkylation has occurred. In all these reactions the ring nitrogen atom is in competition with the substituent. Expect when such reactions are carried out under conditions which make the mesomeric anion of the substituted pyridine the obvious target for attack, we know in general little or nothing of the route by which a particular product arises, that is, whether we are observing the results of S_E2 or S_E2' reactions. The substituents to be considered under this heading are alkyl, amino-, carboxyl, hydroxyl, thiol and sulphonic acid groups. Alkylpyridines exist overwhelmingly as the *NM* forms (p. 324), and quaternization occurs at the nitrogen atom (pp. 180, 186). On the other hand, the anions formed from alkylpyridines are alkylated almost exclusively at the exocyclic carbon atom. For this reason, and since we are concerned here primarily with reactions at the nitrogen atom, these cases will be discussed later (p. 381). For the other types of substituents mentioned, competition between cyclic nitrogen atom and exocyclic substituent leads to much more mixed results.

The preference of 2-, 3- and 4-aminopyridine for the *NM* forms (p. 154) and the greater basicity of the nuclear nitrogen atom (p. 153) make it not surprising that, with these compounds, methyl iodide[339–42b] and the benzyl halides[339, 343], quaternary pyridinium salt formation predominates, although 2-aminopyridine gives small amounts of the 2-alkylaminopyridine also[339, 343]. Again as would be expected, 1-methylpyrid-2-one-imine reacts with methyl-iodide at the exocyclic nitrogen atom, as is proved by the following reactions[339]:

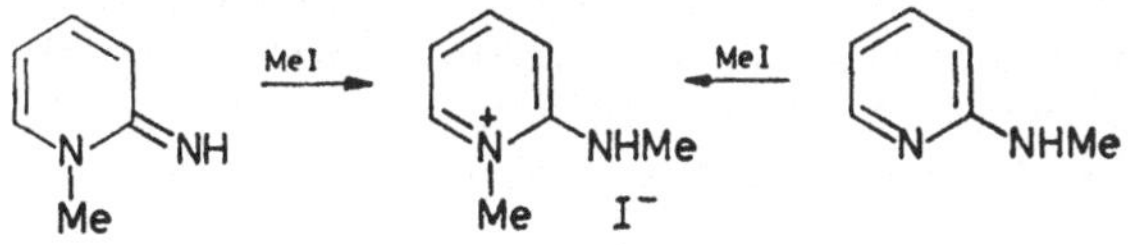

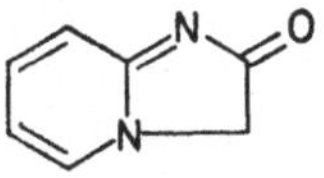

(55)

1-Methylpyrid-2-one-methylimine gives 2-dimethylamino-1-methylpyridinium iodide[344–5] but, surprisingly, 2-dimethylaminopyridine produces 2-pyridyl-trimethylammonium iodide[344, 346]. This reaction was carried out at 100° in a sealed tube for 3 h, and its course has not been satisfactorily explained. In view of it, the formulation of the product from 2-di-isopropyl-aminopyridine and methyl iodide as a pyridinium salt[345] is thrown in doubt.

In general, alkyl halides do not react with 2-aminopyridine so nearly exclusively at the ring nitrogen atom as does methyl iodide; a number of long-chain alkyl halides give 75–85 per cent of reaction at the ring nitrogen atom and 25–15 per cent at the exocyclic nitrogen atom (the products isolated, after basification, being 1-alkylpyrid-2-one-imines and 2-alkylamino-pyridines)[343]. It should be noticed that there is the possibility of ambiguity in results obtained by working up such experiments with alkali, for the Dimroth rearrangement, whereby the nuclear nitrogen atom and the exocyclic nitrogen atom change places (p. 269), could occur. It seems that this rearrangement is responsible for the apparent reaction at the exocyclic nitrogen atoms when 2-amino-3- and -5-nitropyridine are treated with methyl iodide[221, 347-8]. 2-Methylamino-5-nitropyridine reacts mainly at the nuclear nitrogen atom.

Sodium chloroacetate also attacks 2- and 4-aminopyridine at the nuclear nitrogen atom[350-4a] giving the betaines (54, $R = NH_2$). In the case of 2-aminopyridine, subsequent ring closure can give imidazo-[1,2-a] pyridin-2(3H)-ones (55), and this sort of reaction leading to bicyclic compounds through preliminary quaternization of the ring nitrogen atom is general for a variety of α-halogenocarbonyl compounds[354b].

Acylaminopyridines are quaternized at the nuclear nitrogen atom[43a, 342a, 345, 355, 359, 360].

The anion formed from 2-aminopyridine by the action of sodamide or lithium amide in ether or hydrocarbon solvents reacts with alkyl halides to give 2-alkylamino- or 2-dialkylamino-pyridines[225b, 339, 357, 361-4]. Reaction at the ring nitrogen atom is of only minor significance[339]. 2-Alkylamino-, -aralkylamino-[363, 365] and 2-formamido-pyridine[366] behave similarly. Polymethylene dihalides have been used as alkylating reagents[366-7], as have dimethyl sulphate[225a, 267b] and alkyl toluene-p-sulphonates[368].

Although the methylation of 2-nitramino-[369] and 5-nitro-2-nitramino-pyridine[219b] in alkali presumably involves the anion, reaction at the ring nitrogen predominates. Also irregular, probably for steric reasons, are the reactions of benzhydrol[365] and triphenyl carbinol[370] which occur at the exocyclic nitrogen atom of 2-aminopyridine.

The alkylation of 3-aminopyridine has not been reported. 3-Toluene-p-sulphonamidopyridine reacts with dimethyl sulphate, in the presence of potassium carbonate, at the exocyclic nitrogen atom[371], whilst 3-nitramino-pyridine, like its 2-isomer, reacts at the ring nitrogen atom, a betaine being formed[219a].

4-Aminopyridine gives 4-amino-1-methylpyridinium iodide with methyl iodide, and the reaction in the presence of sodamide in ether or benzene follows the same course, though the insolubility of the sodium derivative may influence it[341].

The pyridine monocarboxylic acids react with methyl iodide at 100–150° to give quaternary salts which are easily converted into the betaines (56). The betaines result directly when the reaction is carried out in aqueous alkali[47, 372-7]. Picolinic (homarine) and nicotinic acid methyl betaine (trigonelline) occur naturally. Pyridine-2,6-dicarboxylic acid and 2-methyl-pyridine-6-carboxylic acid do not form betaines[372], although 4-hydroxy-3,5-di-iodopyridine-2,6-dicarboxylic acid is quaternized without difficulty[159b]. The potassium or silver salts of pyridine mono- or dicarboxylic acids are converted to esters by alkylating reagents[317, 372], or sometimes both esterification and quaternization occur[378]. With polymethylene dibromides, nicotinic acid gives either the bis-quaternary salts (57) or the betaines (58); ethylene dibromide yields only the betaine, but longer chains favour the bis-quaternary salts[379].

2-Hydroxypyridine reacts at the nitrogen atom with alkyl or substituted alkyl halides; a salt of a 1-alkylpyrid-2-one results, but the weakly basic 1-alkylpyrid-2-one is usually the product isolated[352, 380-3]. The same result

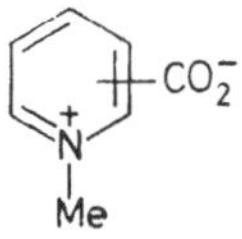

is achieved by carrying out the reaction in alkali with alkyl halides[380, 384-7] or with dimethyl sulphate[385], and it also occurs with a number of substituted 2-hydroxypyridines[388-90], although in some cases O-alkylation also occurs[391]. In contrast, the silver salt of 2-hydroxypyridine is O-alkylated by ethyl iodide, although with methyl iodide both 1-methylpyrid-2-one and 2-methoxypyridine result[380]. N-Alkylation occurs with the silver salts of 2-hydroxypyridine-3- and -6-carboxylic acid[392], but this result may be a consequence of the severer conditions used. Again, the silver salt of 2-hydroxy-5-nitropyridine gives both O- and N-methylation, but the latter predominates[388]. Thus, alkylation in alkali can fairly confidently be used to secure N-alkylation, but use of silver salts does not ensure O-alkylation. The highly variable degrees of ionization to be expected of the silver salts of these ambident ions make the ambiguity not surprising.

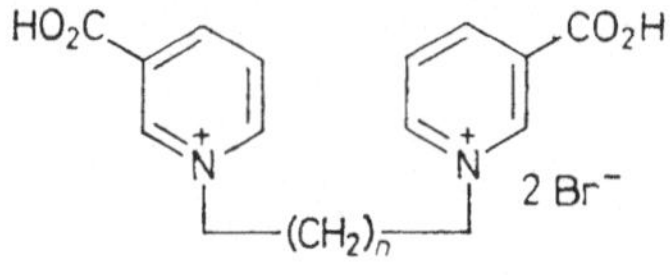

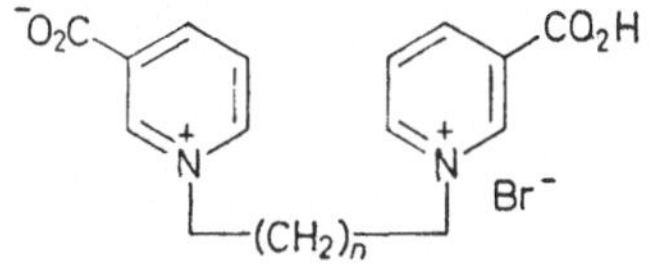

The behaviour of 2-alkoxypyridines resembles that of 2-hydroxypyridine; when heated with an alkyl halide, they give 1-alkyl-2-pyridones, with the loss of the original O-alkyl group[349, 537, 742, 835, 899].

3-Hydroxypyridine is also quaternized by methyl iodide[90, 393-6], chloroacetic acid[381, 397] and β-chloropropionic acid[352]. The quaternary salts are readily converted into betaines, and alkylation in alkali gives these rather than 3-alkoxypyridines[134c, 396, 398-9]. The potassium salt of 3-hydroxypyridine with ethyl bromide in ethanol is said to give 3-ethoxypyridine, but the yield is not stated[393]. So far as it has been examined, 4-hydroxypyridine presents a similar picture to the 2- and 3-isomers[341, 352-3, 400-1]. 1-Alkyl-4-pyridones give 1-alkyl-4-methylpyridinium salts with dimethyl sulphate[977].

Diazomethane converts 2-hydroxypyridine into 2-methoxypyridine[384, 402], whilst 4-hydroxypyridine suffers both O- and N-alkylation[384, 403]. O-Methylation was observed with 3-hydroxypyridines[384, 404], but the result depends on experimental conditions; with ethereal diazomethane, 3-hydroxypyridine gives predominantly N-methylation, whilst reaction in homogeneous solution (t-butanol) at low temperatures gives high yields of 3-methoxypyridine, which fall with rising temperature[399]. The factors affecting the action of diazomethane have been discussed by Gompper[405].

In contrast to the hydroxpyridines, pyridine-2-thiol reacts with alkyl halides[406], methyl sulphate[407], alkyl halides and alkali[407-8] or with diazomethane[407] to give thioethers or their salts. Pyridine-3-thiol in alkali[46] also reacts at the sulphur atom, as do pyridine-4-thiols with alkyl halides[406, 409].

Only when the sulphur atom is already substituted, as in 3-pyridyl thioben-zoate[46], or pyridyl thioethers, does reaction occur at the nitrogen atom[408–10b]. In contrast to 1-alkylpyrid-2-ones, 1-alkylpyrid-2-thiones react readily with alkyl halides[407–11a], and sequences such as those of the diagram are characteristic:

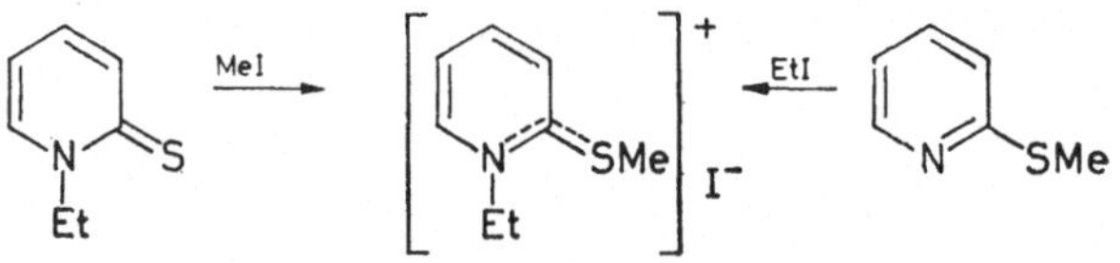

Selenium derivatives behave similarly[410a].

Little work has been done on the pyridine-sulphonic acids. Alkyl halides do not react with pyridine-2-sulphonic acid or its amide, and with its silver salt give the N-alkyl betaines which are incapable of forming salts[333]. With methyl iodide, potassium pyridine-3-sulphonate may have given the methyl ester methiodide, although working up produced the N-methyl-betaine[378].

The use of diazomethane has been mentioned above. Diazoacetophenone reacts with various salts of pyridine to give phenacylpyridinium salts. In these reactions the diazoketone is probably first converted into a phenacyl halide or ester, which then quaternizes the pyridine. However, the formation of methylpyridinium fluoborate from diazomethane and pyridinium fluoborate probably involves nucleophilic attack by the base upon protonated diazo-methane[411b].

(ii) *The attachment of aryl groups*—Pyridine reacts with diphenyliodonium fluoborate to give 1-phenylpyridinium fluoborate[412], and the sodium salt of 2-hydroxypyridine reacts with iodobenzene, in the presence of copper, giving 1-phenylpyrid-2-one[413], but in general N-arylation of pyridines can only be effected by aromatic reagents activated to nucleophilic attack. Some of the products are of great practical importance (pp. 265).

2,4-Dinitrochlorobenzene reacts readily with pyridine, giving 1-(2,4-dinitrophenyl)pyridinium chloride[414]. Since Zincke's original observation of this reaction, several activated halogenobenzenes have been used, and activated nitro groups have also been replaced in quaternary salt formation. Examples are given in *Table 5.16*.

The halogenopyridines are susceptible to nucleophilic substitution (p. 210), and quaternary salts from pyridine and 2-chloro-, -bromo- and -iodo-pyridine have been described[424]. 2-Halogenopyridines and -quinolines react with the sodium salt of 2-hydroxypyridine[133a, 425–6] or 4-hydroxypyridine[427] to give 1-heteroarylpyridones.

The susceptibility of 2- and 4-halogenopyridines to nucleophilic attack has important practical consequences, for these substances are auto-quater-nizing. Thus 4-fluoropyridine decomposes almost immediately, and 4-chloro- and 4-bromo-pyridine do so quickly, giving, by hydrolysis of the products, 1-(4-pyridyl)pyrid-4-one[424, 428]. Several other examples are known[161c, 424].

Groups other than halogen atoms can, of course, be replaced in the formation of pyridylpyridinium salts. Thus 4-acetoxy- and 4-toluene-*p*-sulphonyloxy-pyridine give 1-(4-pyridyl)pyrid-4-one when heated[429–30].

Table 5.16. The Reaction of Benzene Derivatives, activated towards Nucleophilic Substitution, with Pyridine and its Derivatives

Benzene derivative	*Pyridine derivative*	*Product*	*References*
p-Nitrochlorobenzene *p*-Nitrobromobenzene	β-picoline	quaternary salt	415
o-Dinitrobenzene	pyridine	a reaction occurred but no product was isolated	416
2,5-Dichloronitrobenzene	β-picoline	quaternary salt	415
2,4-Dinitrochlorobenzene*	pyridine	quaternary salt	414
	many 3- and 4-substituted and 3,5-disubstituted pyridines	quaternary salt	415 417–9
2,6-Dinitrochlorobenzene	pyridine	quaternary salt	420
3-Nitrophthalic anhydride	pyridine	*N*-(2,3-dicarboxyphenyl) pyridinium betaine	416
	β-picoline	*N*-(2,3-dicarboxyphenyl)-3-methylpyridinium betaine	416
2,3-Dinitrobenzoic acid	pyridine	*N*-(2-carboxy-6-nitrophenyl)pyridinium betaine	416
1,5-Dichloro-2,4-dinitrobenzene	pyridine	bis-quaternary salt + *N*-(5-hydroxy-2,4-dinitrophenyl)pyridinium betaine	421–2
5-Chloro-2,4-dinitrophenol	pyridine	the above betaine	422
2,4,6-Trinitrochlorobenzene 5-Bromo-2,4-dinitrotoluene	β-picoline	quaternary salt	415
2-Chloro-3,5-dinitrobenzoic acid	pyridine	quaternary salt	423
3,6-Dichlorophthalic anhydride	pyridine	the bis-pyridinium betaine	416
	β-picoline	the bis-picolinium betaine	416

* This does not react with 3-chloro-, 3-bromo-, 3-nitro- and 3-ethoxycarbonyl-pyridine[417]. 4-Methoxypyridine gives the quaternary salt, and also 1-(2,4-dinitrophenyl)pyrid-4-one; the latter is also formed from 4-hydroxypyridine[418]. No product was obtained from 2,4-dinitrochloro- or-bromobenzene and 5-ethyl-2-methyl-, 2-phenyl- and 2,3,4-ethoxycarbonyl-pyridine[419].

4-Nitropyridines change similarly when kept, though probably more slowly than 4-halogenopyridines[431–2].

Of greater interest are some reactions, of uncertain mechanism, which produce pyridylpyridinium compounds from pyridine or its derivatives lacking substituents susceptible to nucleophilic displacement. The best known of these, discovered by Koenigs and Greiner[433], is the formation of 1-(4-pyridyl)pyridinium chloride hydrochloride from pyridine and thionyl chloride. The reaction has been used frequently[434-9] and extended to 3-picoline, but is not so successful in this case[200, 252, 435]. A possible mechanism is shown;

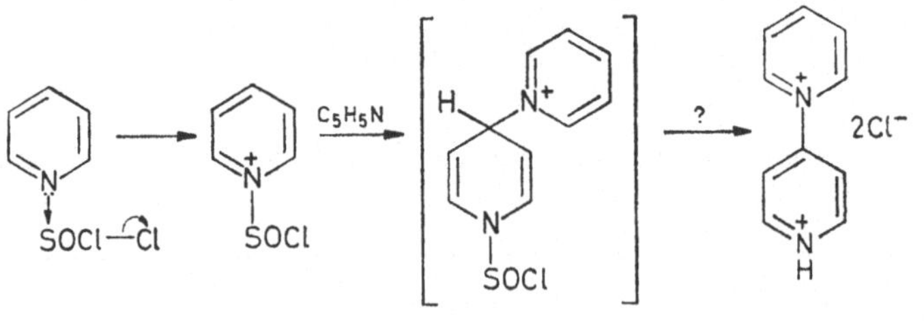

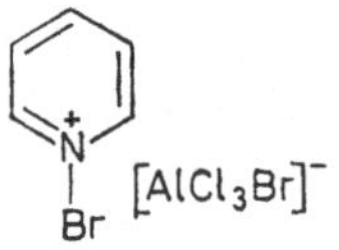

(59)

it represents nucleophilic attack by pyridine upon a pyridine molecule activated by complex formation. Equally possible is the intermediacy of 4-chloropyridine, formed by attack of chloride anion on the initial complex. There is no evidence to distinguish between these possibilities. Closely related is the formation of 1-(4-pyridyl)pyridinium bromide hydrobromide from pyridine and bromine[438, 440-1], which is very satisfactorily carried out in the presence of aluminium chloride[442a]. The catalyst presumably increases the ability of bromine to complex with pyridine (59) but, again, it is not clear if 4-bromopyridine is involved. The reaction of iodine chloride with pyridine hydrochloride at 250° is presumably similar but, oddly, it gives 1-(2-pyridyl)pyridinium iodide[165].

Generally similar to these reactions is that which occurs between pyridine oxides and pyridine in the presence of *p*-toluenesulphonyl chloride. Thus, 2- and 4-chloropyridine 1-oxide give 2-chloro-4- and 4-chloro-2-pyridyl-pyridinium chloride, respectively[442b].

Finally there are a number of examples of quaternary salt formation from pyridine and reactive halogen derivatives of other heterocyclic nuclei[443-5].

(iii) *Quantitative studies*—The Menschutkin reaction in general, and the quaternization of pyridines in particular, have been the frequent subjects of kinetic studies; early workers were interested mostly in establishing orders of reactivity among alkyl halides or amines[294c, 296, 446], but the reactions have been used as sources of information about the effect of solvent changes upon the parameters of the Arrhenius equation[296d, 323, 324b-5], nucleophilic substitution at saturated and aromatic carbon atoms, and the influence of substituents in the pyridine ring upon the nucleophilic power of the nitrogen atom. We are concerned mainly with the last of these topics, but some of the other points will be briefly mentioned.

Table 5.17 contains a selection from the data available, made to illustrate

186

the influence upon the kinetic parameters of the reaction exercised by solvent change, changes in alkyl halide structure, and by substituents in the pyridine ring. As regards the first two factors, the data illustrate the generalizations made above (p. 178). Data for the quaternization of pyridine with p-alkylbenzyl bromides led to the first suggestion regarding hyperconjugative electron release from alkyl groups[447a] and have been analysed[448] in terms of Taft's equation

$$\log(k^R/k^H) = \sigma_R^* \rho^* + nh$$

They are well represented by the expression $\log(k_2^R/k_2^H) = -0 \cdot 15\, \sigma_R^* +0 \cdot 04n$. The low value of h indicates that the polarization of the C—Br bond in the transition state does not produce much unsaturation at the methylene carbon atom and consequently does not stimulate electromeric hyperconjugation from the p-alkyl group. There are other interesting aspects of the reaction of benzyl halides with pyridine. The effect of substituents in the benzyl halide depends on the character of the solvent; as the ionizing power of the solvent is increased, the acceleration of the reaction is accompanied by an increase in activation energy. In any one solvent, differences in velocity depend mainly on the pre-exponential factor of the Arrhenius equation. The increase in activation energy mentioned is attributed to the need to disrupt the coordinate link between water and pyridine in aqueous solvents. Generally, electron accession to the side chain increases the velocity, and electron removal, as by a p-nitro group, decreases it up to a point. Beyond this point, as in 2,4-dinitrobenzyl bromide, an increase again results. Thus in a 'graded polar series' of p-substituents a minimum of reaction velocity occurs which is not associated with a change in reaction kinetics:

2,4-dimethyl > p-alkyl > H > p-nitro ≪ 2,4-dinitro   (in dry acetone)

The various factors operating have been discussed[447, 449] in terms of their influence upon the anionization of the bromine atom (a) and the electrostriction of the pyridine molecule to the methylene group (b)

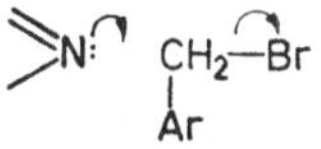

(b)   (a)

The quaternization of pyridine in non-polar solvents has been studied as a means of clarifying the much debated nature of the displacement reaction under such circumstances[450a]. Swain and Eddy[451a] deduced evidence for their theory of specific solvation from the reaction of pyridine and methyl bromide in benzene containing various hydroxylic solutes. Swain and Langsdorf[451b] found the Hammett plot for the reaction of substituted benzyl bromides with pyridine in acetone to be markedly concave, and indeed to fall into two separate lines for $meta$- and $para$-substituents. The curvature and the division illustrate the effects of substituents upon reactions of intermediate character. Ingold and his co-workers[452], from reactions in sulphur dioxide,

Table 5.17. Selected Kinetic Data for the Reaction of Pyridine Bases with Alkyl and Aryl Halides

Substituent	RX	Solvent*	Temp., °C	$k_2 \times 10^6$ l.mole^{-1}sec^{-1}	ΔE kcal.mole^{-1}	$\log_{10}A$	$\Delta H\ddagger$ kcal.mole^{-1}	$-\Delta S\ddagger$ cal/°K	References
[Pyridine]	MeI	1	60	146	14.3	6.50	—	—	296d, 323c
	MeI	2	60	3,490	13.6	6.72	13.3	29.8	453
	MeI	2	25	343	13.9	6.57	—	—	454
	EtI	3	60	6,250	13.4	7.30	—	—	323b
	EtI	4	60.1	2,780	14.7	9.04	—	—	296d, 322b
	EtBr	5	60.1	1,150	18.2	—	—	—	453
	PrBr	1	80.5	41.7	15.7	6.98	15.4	28.8	296d, 322b
	PrBr	1	25	18.3	16.0	—	—	—	454
	i-PrBr	3	80.5	7.25	16.2	—	—	—	453
	n-BuBr	2	80.6	2.82	15.9	6.00	—	—	324b
		6	60	596	15.6	6.93	—	29.5	
		7	25	0.941	17.7	8.46	16.9	30.0	
		8	50	474	16.04	8.44	—	30.0	
		3	60	259	16.40	8.24	—	31.0	454
				160	16.45	6.00	—	—	
	C₃H₅Br	9	20	4,630	12.73	5.38	—	—	447b
	PhCH₂Br			122	12.4	5.63	—	—	
	p-Me.C₆H₄.CH₂Br			202	12.5	5.34	—	—	
	p-Et.C₆H₄.CH₂Br			181	12.2	5.59	—	—	
	p-i-Pr.C₆H₄.CH₂Br			163	12.6	5.52	—	—	
	p-t-Bu.C₆H₄.CH₂Br			165	12.5	5.23	—	—	
	p-NO₂.C₆H₄.CH₂Br			112	12.3	5.60	—	—	
	2,4-di.NO₂.C₆H₃.CH₂Br			228	12.4	6.46	—	—	
2-Methyl-	MeI	2	25	162	14.0	6.70	13.4	31.0	453
	EtI			4.27	16.5	6.80	15.8	30.2	
	i-PrI			0.0509	19.2	5.83	18.5	29.8	
	C₃H₅Br	3	60	753	13.62	5.75	—	—	454
2-t-Butyl-	MeI	2	25	0.080	17.5	6.82	16.8	34.6	453
3-Methyl-	MeI	3	60	712	13.6	6.06	13.0	29.4	454
3-Methyl-	C₃H₅Br	2	25	7,140	12.52	—	—	—	453
3-t-Butyl-	MeI			950	—	—	—	—	

Table 5.17 (continued)

Substituent	RX	Solvent*	Temp., °C	$k_2 \times 10^6$ l.mole⁻¹sec⁻¹	ΔE kcal.mole⁻¹	$\log_{10}A$	$\Delta H^{\ddagger}$ kcal.mole⁻¹	$-\Delta S,^{\ddagger}$ cal/°K	References
4-Methyl-	MeI	2	25	760	13·6	6·86	13·0	29·2	453
	C₃H₅Br	3	60	8,170	12·44	6·07	—	—	454
	n-BuBr	6	50	845	15·95	8·61	—	29·3	324b
		7		470	16·25	8·69	—	29·0	
		8		287	16·25	8·37	—	30·4	
4-t-Butyl-	MeI	2	25	757	13·7	6·91	13·0	29·0	453
2,6-Dimethyl-				14·5	15·1	6·24	14·5	32·2	455
2,4,6-Trimethyl-				37·5	14·8	6·42	14·2	31·4	455
3-Methoxy	C₃H₅Br	3	60	4,130	12·73	5·96	—	—	454
4-Methoxy				9,190	12·38	6·08	—	—	
3-Fluoro†				569	13·76	5·77	—	—	
3-Chloro-				545	13·72	5·74	—	—	
3-Bromo-				587	13·67	5·73	—	—	
3-Ethoxycarbonyl-				898	13·58	5·87	—	—	
4-Ethoxycarbonyl-				1,000	13·51	5·87	—	—	
[Pyridine]	PhCOCH₂Br	9	20	790	11·7	—	—	—	456a
	p-Me.C₆H₄COCH₂Br			740	11·6	—	—	—	
	p-NO₂.C₆H₄COCH₂Br			1,900	—	6·2	—	—	
	2,4-Dinitrochlorobenzene	10	55	11·1	16·7	6·7	—	—	457
3-Methyl				19·9	17·1	6·7	—	—	
4-Methyl				30·9	16·9	—	—	—	

*1 Benzene; 2 nitrobenzene; 3 nitromethane; 4 water (40 g) + acetone (100 ml.); 5 55·7 vol. per cent ethanol in water; 6 tetramethylene sulphone; 7 propylene carbonate; 8 2,4-dimethylsulpholane; 9 acetone; 10 ethanol.
† Recent figures for other halogenopyridines are [460] (substituent, $k_{250} \times 10^4$, ΔE, log PZ): 2-Cl, 2·04, 16·5, 6·41; 3-Cl, 30, 15·1, 6·56; 4-Cl, 73, 14·7, 6·66; 2-Br, 1·1, 16·9, 6·44; 3-Br, 35, 15·2, 6·7; 4-Br, 169, 14·2, 6·63; the solvent was nitrobenzene.

have also been led to consider reaction mechanisms lying between the S_N1 and S_N2 limits.

The most extensive studies of the effects of substituents in the pyridine ring upon the reaction with alkyl halides are due to Brown and Cahn[453] and Clarke and Rothwell[454]. In general, both of the Arrhenius parameters are affected[454], an increase in the activation energy accompanying a decrease in log PZ.

Reversibility becomes an important factor with 2-t-butylpyridine. With respect to a particular alkyl halide, the energies of activation and the frequency factors were essentially identical for the 3- and 4-alkylpyridines. A 3- or 4-methyl group roughly doubled the rate and decreased the activation energy by 0·3–0·4 kcal. Changes in the 3- or 4-alkyl substituent beyond methyl had little additional effect. A 2-methyl group raised the activation energy and decreased the frequency factor, the former seeming to be the more important factor in decreasing the rate. The retardation was attributed to steric strain in the transition state[453, 455]. Characteristically, the strain increased in the series 2-Me < 2-Et < 2-i-Pr, but more markedly in the last step to 2-t-butyl. Under circumstances where 4-triphenylmethylpyridine reacted with ethyl iodide at almost exactly the same rate as did pyridine, the 2-isomer did not react at all[456b].

For 3- and 4-substituted pyridines there is a linear relationship between pK_a (in water) and log k_2 for the quaternization reaction in nitrobenzene[78] or nitromethane[454]. In the second case, 4-alkoxypyridines deviate from the relationship, probably because of the weakly basic ether function and the acidic character of the solvent. The linear relationship enables approximate activation energies for 2-substituted pyridines to be calculated. The difference between the measured activation energy and that calculated ΔE increases[454] with the size of the 2-substituent and of the alkyl halide molecule (*Table 5.18*).

Table 5.18

Alkyl halide	2-Substituent	ΔE, cal
MeI	2-Me	650
	2-Et	880
	2,6-Me$_2$	1,930
C$_3$H$_5$Br	2-Me	1,180
	2-Et	1,480
	2,6-Me$_2$	3,290
Ph.CH$_2$Br	2-Me	1,100

2-Ethoxycarbonyl-, -halogeno-* and -methoxypyridine do not react with allyl bromide or methyl iodide. 2-Methoxypyridine is thus in sharp contrast with 4-methoxypyridine. Consideration of pK_a values suggests that in 2-methoxypyridine, although the inductive effect contributes largely to the '*ortho*-effect', the primary steric factor is more important still[454].

Table 5.17 includes data for quaternizations involving 2,4-dinitrochlorobenzene. These are mainly of interest for comparing the power of the nitro

* But see data of footnotes to *Table 5.17* which relate to reaction in nitrobenzene

group to activate halogen atoms with that of the pyridine nitrogen atom, and will be further discussed in that context (p. 215). It is interesting to notice here the weakness, as nucleophiles, of the pyridine bases relative to primary aromatic amines, despite the lower basic strength of the latter. Data suitable for making the comparisons, for reactions with 2-chloro-5-nitropyridine[457b], are given in *Table 5.19*.

Table 5.19. Reactions with 2-Chloro-5-nitropyridine in 99·8 per cent Ethanol at 25°

Base	-log k_2	ΔE, kcal.mole^{-1}	log$_{10}A$	$\Delta H\ddagger$, cal.mole^{-1}	$-\Delta S\ddagger$, cal.°K	pK_a, *water at 25°*
Aniline	5·80	13·1	3·8	6,100	} 22.8	4·58
m-Toluidine	5.66	12·9	3·8	5,900		4·67
p-Toluidine	5·34	12·7	3·9	5,300	23·2	5·07
Pyridine	6·89	18·1	6·3	8,700	11·0	5·17
3-Picoline	6·61	17·9	6·6	8,100	—	5·68
4-Picoline	6·39	17·5	6·5	7,500	—	6·02

The change in hybridization of the nitrogen atom and the energy liberated by partial bond formation in the transition state cannot facilitate the reactions of the primary amines relative to those of the pyridines. However, solvation of the transition state seems to be stronger for the primary amines, and it is likely that the pyridines encounter a steric hindrance not felt by the primary amines in which the —NH$_2$ group is relatively remote from the aromatic residue. In addition, when the halogen compound contains an *o*-nitro group, the transition state for the reaction with a primary amine may be stabilized by hydrogen bonding (60)[457].

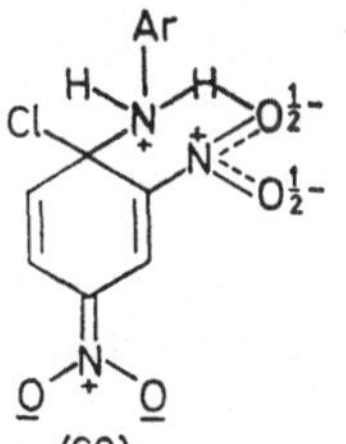

(60)

(iv) *Formation of substituted-alkyl pyridinium salts by addition reactions*—
The first group of reactions to be considered are internal displacements of the type

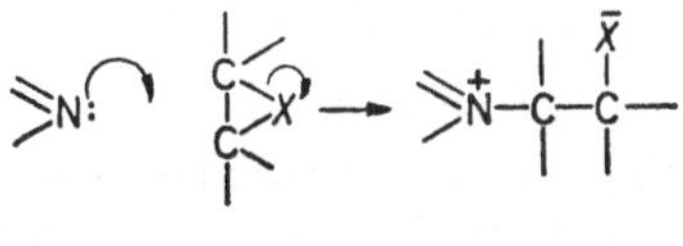

191

The reactions of unsaturated compounds (cyclohexene, stilbene, cinnamic acid, etc.) with pyridine and bromine, or iodine monochloride are of this kind[458]:

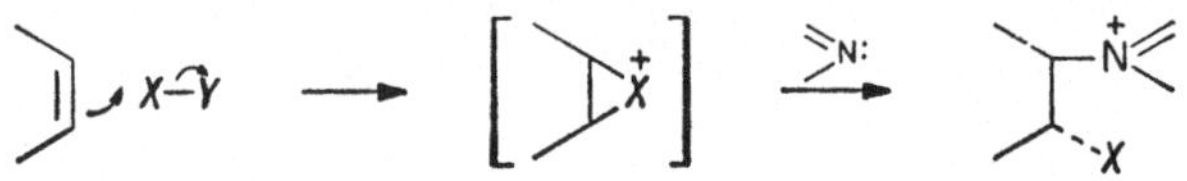

The result is *trans*-addition. Olefin oxides behave similarly[305, 459], as illustrated for the case of styrene oxide

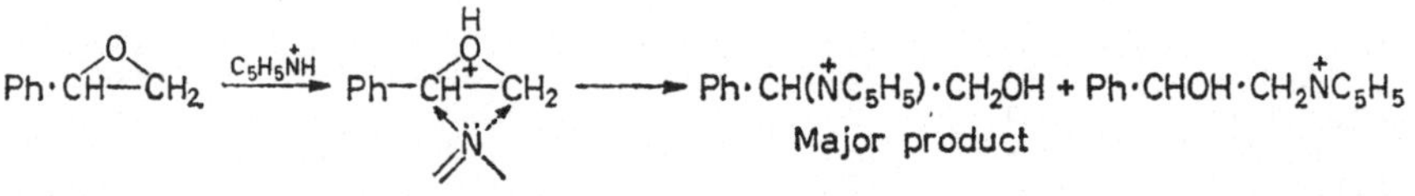

Major product

Halohydrins are not intermediates in this reaction (p. 179). In a related reaction β-propiolactone gives the betaine $C_5H_5\overset{+}{N}.CH_2CH_2.CO_2^-$ with pyridine[460], and β-sultones behave similarly[461], as in the example

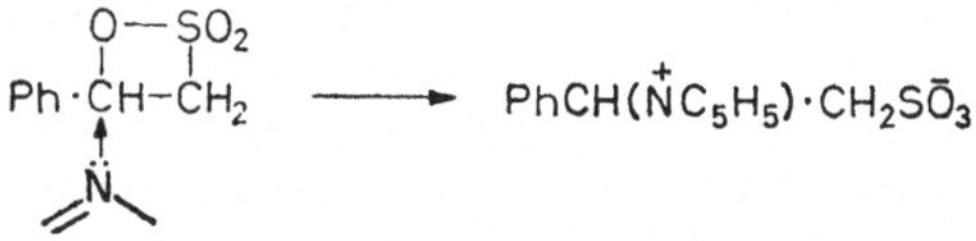

In other reactions, attack on the pyridine nitrogen atom is made by the electrophilic β-carbon atom of an α,β-unsaturated carbonyl compound. An example is the formation of (61) from pyridine and maleic acid; other unsaturated acids behave similarly, but the reaction is not completely general[292, 352, 462]. In some circumstances, cyclization follows the initial reaction, as in the formation of (62) and (63) from 2-hydroxy- and 2-amino-pyridine, respectively, with α-bromacrylic acid[463]. The similarly activated acrylonitrile reacts with 2-hydroxypyridine, giving 1-β-cyanoethylpyrid-2-one[171e, 464].

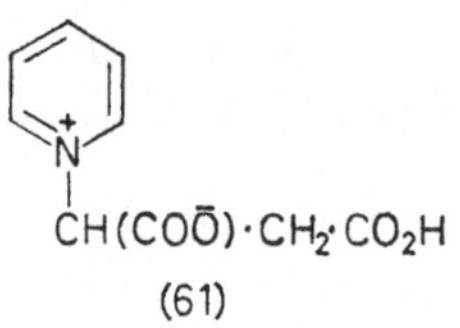

(61) 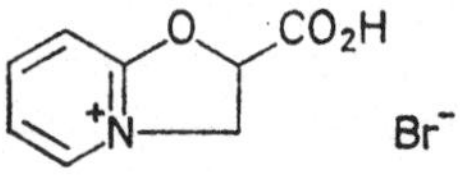   (62) 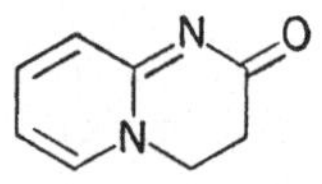   (63)

Another example is the reaction of pyridine with methyl propiolate to give (65), which clearly arises[465] by nucleophilic attack of the acetylenic anion on the first-formed ylid (64, R = H). With 2-aminopyridine cyclization occurs, giving (66) as well as the uncyclized compound (67)[466].

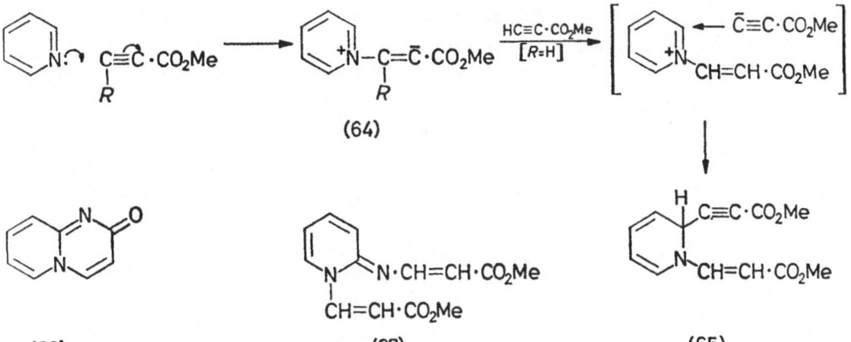

(64)

(66)

(67)

(65)

A similar ylid (64, R = .CO_2Me) is probably the first product in the fascinating reactions of pyridine and its homologues with dimethyl acetylene-dicarboxylate in ether, first investigated by Diels and Alder and their co-workers[467]. The products from pyridine (the 'red labile compound', the 'stable yellow compound' and 'Kashimoto's compound') have recently been re-formulated[468-9]. 'Kashimoto's compound' has the structure (68), and the main course of the reaction is probably represented by the sequence from the initial ylid (64, R = .CO_2Me) shown, i.e. the initial attack on the nitrogen atom is followed by nucleophilic attack upon the ring. The labile red compound is (70, R = R' = H), and the stable yellow product (71, R = R' = H). When β-picoline is used,

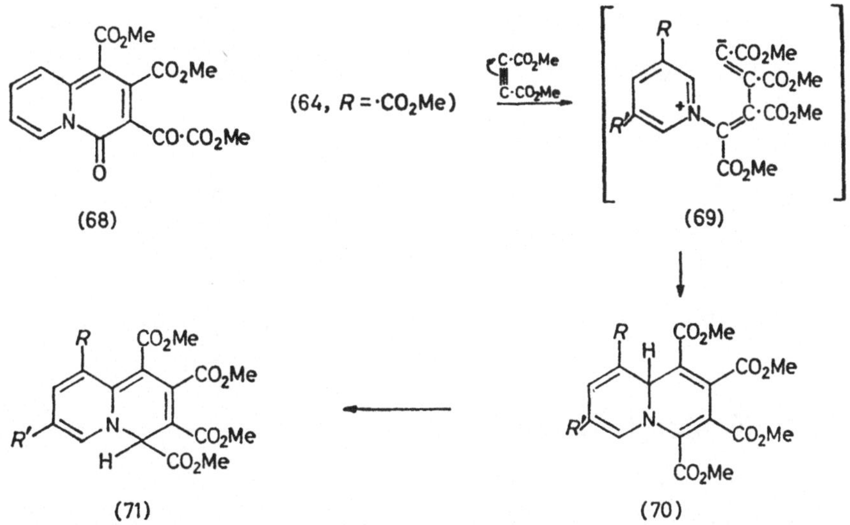

(68)

(64, R = ·CO_2Me)

(69)

(71)

(70)

the labile compound isolated is (70, R = Me, R' = H). It is converted by heat to a stable yellow product (71, R = Me, R' = H). A second stable yellow product isolated from the reaction is (71, R = H, R' = Me), but the corresponding labile intermediate is not found. When β-picoline reacts with the ester in ether at $-$ 10°, a very unstable product can be isolated which may be the true ylid intermediate (69, R = Me, R' = H).

The course of the reaction of pyridine bases with dimethyl acetylenedicarboxylate is dependent on the conditions[470-1], and the ylid intermediate seems to be capable of reacting in a variety of ways[465].

As α,β-unsaturated carbonyl compounds, quinones will react with pyridine and related bases to give quaternary salts. The prototypical reaction is shown. Organic as well as mineral acids will catalyse the reaction[458, 472].

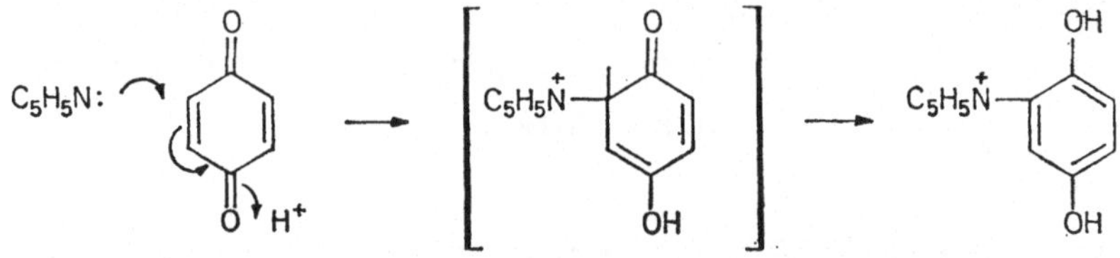

2-Aminopyridine gives, with benzoquinone[473], the cyclic product (72). When reactive methylene compounds are also present, further reaction can occur, as in the formation of (73) from benzoquinone, pyridine and acetylacetone[474], and of (74) from α-naphthoquinone, pyridine and acetoacetic ester[475]. 1,4-Naphthoquinonedibenzenesulphonimide behaves like naphthoquinone in forming a quaternary salt with pyridine and hydrochloric acid[476].

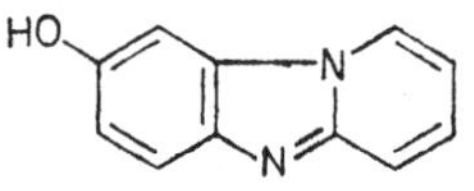

(72)

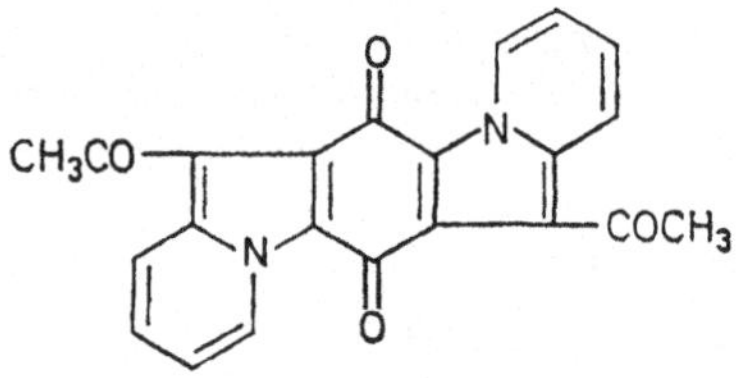

(73)

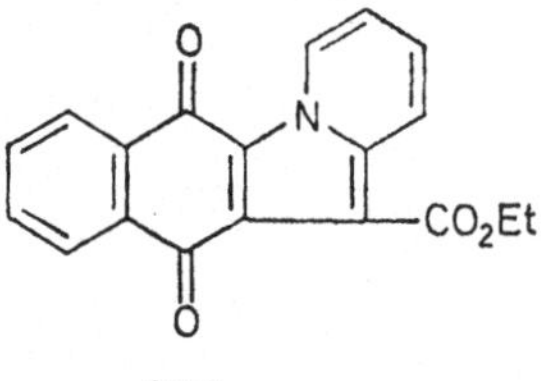

(74)

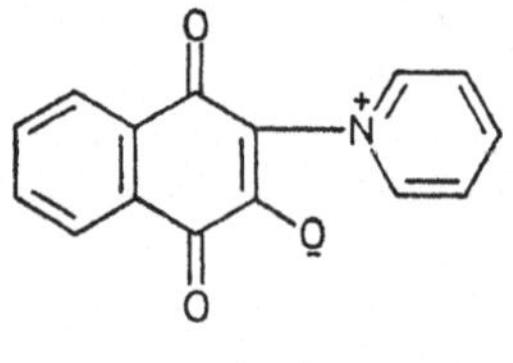

(75)

$$\left[C_5H_5\overset{+}{N}\cdot CH_2\cdot CH_2 \right]_2 \cdot SO_2 \left[2Cl^- \right]$$

(76)

It is appropriate to notice here that halogenoquinones react with pyridine bases to give quaternary salts by halogen displacement. 2,3-Dichloro-1,4-naphthoquinone gives with pyridine, by hydrolysis of the initial product, the betaine (75)[477], and chloranil behaves similarly[478]. 2,3-Dichloro-1,4-naphthoquinone can be used with advantage[475] in place of naphthoquinone in preparing (74).

Groups other than carbonyl can, of course, activate double bonds towards electrophilic attack; consequently, divinyl sulphone reacts with pyridine in the presence of hydrochloric acid[479] to give (76).

(v) *The attachment of acyl and sulphonyl groups*—1-Acyl-,-aroyl- and -sulphonylpyridines are important intermediates in various reactions. The formation of esters from alcohols and phenols, and the conversion of amines into amides by treatment with solutions of carboxylic acid anhydrides or chlorides, or sulphonyl chlorides in pyridine[480], are reactions of preparative value which have been long in use[481]. They have quantitative applications as well, as in the estimation of hydroxyl groups[482]. 1-Acylpyridinium cations are also intermediates in the pyridine-catalysed hydrolysis of acetic anhydride (the acetylation of water):

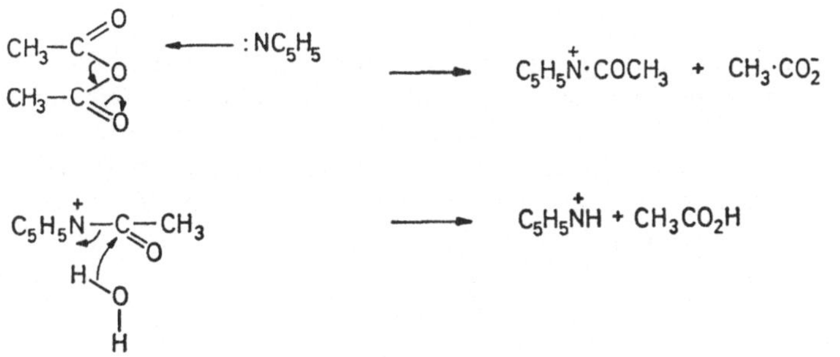

$$CH_3-C{\overset{O}{\diagdown}}_O{\diagup}C-CH_3 \longleftarrow :NC_5H_5 \longrightarrow C_5H_5\overset{+}{N}\cdot COCH_3 + CH_3\cdot CO_2^-$$

$$C_5H_5\overset{+}{N}-\underset{O}{\overset{|}{C}}-CH_3 \quad\underset{H}{\overset{H}{\diagdown}}O \longrightarrow C_5H_5\overset{+}{N}H + CH_3CO_2H$$

2-Methyl- and 2,6-dimethylpyridine have no catalytic effect, and higher acid anhydrides are not so reactive as acetic anhydride, in both cases for steric reasons[483]. Various aspects of acylation processes involving acylpyridinium cations are considered later (pp. 205, 223, 236), whilst at this point are recorded some instances of the actual preparation of acyl- or sulphonylpyridinium salts. These are surprisingly few.

Acetyl-[484–8], benzoyl-[487, 489–90] and various other acylpyridinium chlorides[486–8, 491–3], sometimes of doubtful purity (because of their very ready hydrolysis or of contamination with pyridinium chloride), have been described. Salts of benzoylpyridinium with complex metal anions are also known[494]. Covalent structures, with the chlorine atom at $C_{(2)}$ or $C_{(4)}$ of the pyridine ring, have been suggested for the adducts from pyridine and the nitrobenzoyl chlorides[495], but the ionic character of 1-acetylpyridinium chloride is affirmed by measurements of its electrical conductivity in sulphur dioxide solution[496].

Even less has been reported about sulphonyl compounds, but 1-benzenesulphonylpyridinium and 1-benzenesulphonyl-2-methylpyridinium chloride have been isolated[497].

The presence of tautomerizable substituents in the pyridine nucleus introduces the familiar problem of there being two possible sites for attack by the acylating reagent, and a variety of mechanisms whereby acylation may occur. Nothing is known regarding mechanisms. The hydroxypyridines (or sometimes their sodium salts) give O-acyl derivatives[393, 429, 498–500a] and O-arenesulphonyl derivatives[429, 500b].

The aminopyridines also undergo exocyclic acylation[43a, 220, 345, 359, 501–3]. A second aroyl group can be introduced into 2-monoaroylaminopyridines, best by the Schotten–Baumann method, giving 2-diaroylaminopyridines[359, 504].

Arenesulphonyl chlorides[212] and methanesulphonyl chloride[43b] attack aminopyridines at the exocyclic nitrogen atom, but a 2-arenesulphonamidopyridine with an arenesulphonyl chloride, or 2-aminopyridine with 2 mol. of an arenesulphonyl chloride, can give disulphonyl derivatives with structures (77) or (78) according to conditions[505–6], and rearrangements can occur[506].

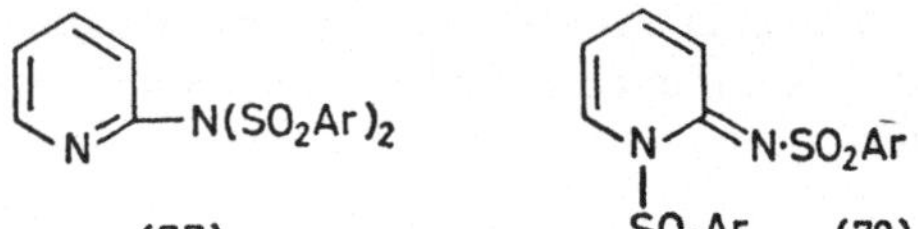

(77)   (78)

The attack of acylating and sulphonylating agents at the exocyclic nitrogen atom of aminopyridines is surprising in view of the greater nucleophilic power of the ring nitrogen atom, as illustrated by its functioning as the basic centre (p. 153). It has been suggested that reaction occurs initially at the ring nitrogen atom, giving an intermediate which is itself a powerful acylating or sulphonylating agent[43a, b].

With benzenesulphonyl chloride, some 1-aryl-4-pyridones give 1-aryl-4-benzenesulphonyloxy-pyridinium salts[977].

For the acylation of pyridthiones see p. 393.

A special kind of acylpyridinium compound may be the initial product in the reaction of ketenes with pyridine. From dimethylketene the isolated product[507] is (79, $R = Me$). In the case of ketene itself, (79, $R = H$) is probably formed, but it undergoes further reaction[508]. A possible sequence for the formation of compounds (79) involves initial attack at nitrogen, followed by nucleophilic substitution, as in the reaction of pyridine with dimethyl acetylenedicarboxylate:

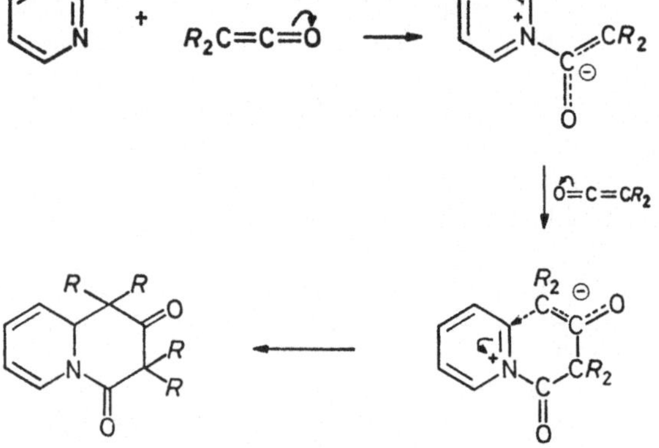

(79)

196

(vi) *N-Oxide formation*—Because of the synthetic value and theoretical interest of the products, the process of N-oxidation is of great importance in the pyridine series. Pyridine 1-oxide was first obtained by Meisenheimer[509] who oxidized pyridine with perbenzoic acid, and later by Bobránski[510] who used perphthalic acid. These reagents have been used to prepare a number of pyridine 1-oxides[52, 162, 248, 250, 254–7, 258b, 511–5], but the most convenient and most frequently used reagent is the solution of 30 per cent hydrogen peroxide in acetic acid, introduced by Ochiai[516]. Pertrifluoracetic acid has been used occasionally[517–8], and hydrogen peroxide in presence of various acids, but hydrogen peroxide alone is ineffective[519].

Qualitatively, the effect of substituents upon the ease of oxide formation is that which would be expected for the electrophilic process[520a] shown. Quantitatively, a limited range of data is available for perbenzoic acid oxidation of pyridine and its homologues and halogen derivatives in 50 per cent aqueous dioxan[520b] at 25°. The second-order rate constants for 3- and 4-substituted pyridines fit a Hammett plot with $\rho = -2.35$ and are also roughly linear with pK_a. Bulky 2-substituents cause deviations but not so markedly as in quaternization (see p. 190).

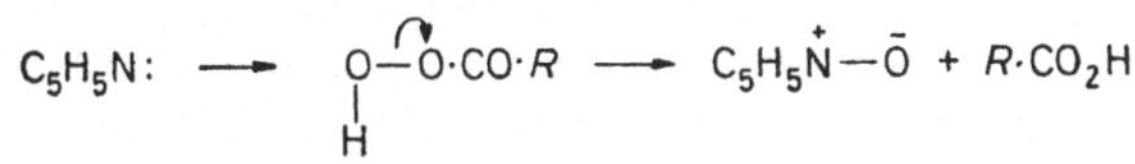

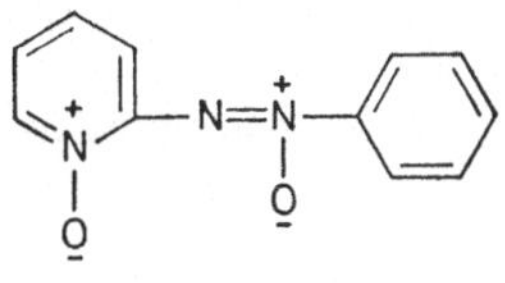

$$C_5H_5N: \longrightarrow \quad \underset{H}{O-\overset{\frown}{O}\cdot CO\cdot R} \longrightarrow C_5H_5\overset{+}{N}-\overset{-}{O} + R\cdot CO_2H$$

(80)

1-Oxides have been prepared from pyridines containing acetamido[100, 521–2], acyl[523], alkoxy[52, 250, 255–7, 258b, 515, 524–5], alkoxycarbonyl[69, 100, 514, 526], alkyl[251a, 252, 511, 519, 524, 527–32], aryl[97, 509, 513], carboxamido[251b], carboxyl[467b, 526, 530, 533], halogen[69, 162, 248, 250, 254–5, 258b, 512, 522, 527], 3-hydroxyl[52a], nitrile[254], and 3-nitro[69, 518, 535] groups. Among more complex compounds which have been successfully *N*-oxidized are the pyridylacrylic acids and their esters[536], styrylpyridines, pyridine-3-acetic acid and the oxime and semicarbazone of pyridine-2-aldehyde[536]. Complete failure to form an oxide is rare, though a few cases have been recorded[245, 515, 517, 536–7], including 2-nitro,- 2,6-dialkoxy-, 2,6-dibromo-, 3-cyano-2-methoxy- and 3-carboxamido-2-methoxy-pyridine. 2,6-Dibromopyridine could only be *N*-oxidized by means of pertrifluoracetic acid[517]. These failures and a number of cases, notably 2,6-diaryl-[509, 513] and dihalogeno-compounds[512], which give poor yields, illustrate the deactivating influence of electron-withdrawing substituents[513] and possibly, from the behaviour of 2,6-dialkylpyridines[252, 511], of a steric factor. In the benzeneazopyridines, the pyridine nitrogen atom has to compete with an exocyclic nitrogen atom for the oxidizing reagent [538–40]. With perbenzoic acid, 2-benzeneazopyridine gives mainly the dioxide (80), with smaller amounts of the two monoxides. 3-Benzeneazopyridine also gives the dioxide, but 4-benzeneazopyridine gives 4-benzeneazopyridine 1-oxide or the dioxide, depending on whether perbenzoic or peracetic acid is used. On the other hand, the N-oxidation of 2- and 3-(p-dimethylamino-benzeneazo)pyridine proceeds preferentially at the dimethylamino group, though from the 4-isomer only the pyridine oxide was recovered[541a]. 2- and 4-(p-Dimethylaminostyryl)pyridine are also oxidized at the dimethylamino-group[541b].

Some other substituents are modified or degraded during N-oxidation. Thus, 2-aminopyridines cannot be converted into 1-oxides[521] (however, 2-dimethylaminopyridine gives the exocyclic oxide[542a]). 3-Aminopyridine, with pertrifluoracetic acid, yields 3-nitropyridine and its 1-oxide[518]. Pyridyl thioethers are converted into sulphoxides[512], and tri-2-pyridyl-phosphine and -arsine give the P- and As-oxides[543]. An aldehyde group is oxidized to carboxyl[536], and acid hydrazides[536] and amides[544], cyanides[251b] (see p. 367) and 2-chloropyridine[251b] can suffer hydrolysis.

(vii) *Miscellaneous additions to the nitrogen atom*—The cases of metallic ions, boron derivatives, halogens, phosphorus pentachloride and other inorganic substances have been mentioned already (pp. 159–60).

The reaction between pyridine and cyanogen bromide[545] gives first 1-cyanopyridinium bromide. From nicotinamide what is probably a 1-cyano-2-hydroxy-1,2-dihydronicotinamide has been isolated[546]. The 1:1 addition product of γ-dipyridyl and cyanogen bromide has likewise been isolated[547]. The use of the cyanogen bromide addition products in the important König reaction is discussed later (p. 268).

Arenesulphonylazides also attack pyridine at the nitrogen atom[41a]:

$$Ar \cdot SO_2 \cdot \bar{N} \overset{+}{-} \overset{\frown}{N} \equiv N \qquad \longrightarrow \qquad Ar \cdot SO_2 \cdot \bar{N} - \overset{+}{N}C_5H_5 \; + \; N_2$$
$$\uparrow$$
$$C_5H_5\overset{+}{N}$$

Pyridines react with potassium hydroxylamine O-sulphonate to give salts of 1-aminopyridinium cations[542b].

The photolysis of diazomethane in pyridine gives 2-methylpyridine. Because of the electrophilic character of carbene, this reaction is thought to involve attack at the nitrogen atom, followed by rearrangement[548].

(m) Intramolecular Electrophilic Substitutions

Cyclizations which, when applied to benzene derivatives, produce quinolines, have been applied with moderate success to pyridine derivatives, giving naphthyridines. The chief results are summarized in *Table 5.20*. Except in one or two favourable cases, the yields of naphthyridine derivatives are only fair, and usually they are low.

The behaviour of 2-aminopyridines suggests that in any particular instance failure to form a naphthyridine may be caused by an insufficiently reactive pyridine ring, or by preferential cyclization at the ring nitrogen. A 6-amino, -acetamido, -ethoxy or -methyl substituent causes naphthyridine formation, presumably by activating $C_{(3)}$, and also perhaps by blocking reactions at the ring nitrogen atom[558]. Little is known about the behaviour of

4-aminopyridines, but clearly, cyclization at $C_{(3)}$ is not easy. 4-Aminopyridine 1-oxide and some of its homologues give very small yields of 1,6-naphthyridine 6-oxides in the Skraup reaction[579].

Table 5.20. Formation of Naphthyridines from Aminopyridines

Pyridine derivative	Malonic ester	Cyclization reaction*						
		Skraup	Combes	Döbner	Döbner–Miller	Emme	Conrad–Limpach	Knorr
2-Amino	N[549-51]	⊖[552]		⊖[553]	⊖[552, 554]	⊖	N[555-6]	N[550, 557]
2-Amino-5-bromo	N[558]					N[556]		
2-Amino-6-bromo						⊖[556]		
2-Amino-4-chloro						⊖[556]		
2-Amino-5-chloro	N[558]					N[556]		
2-Amino-4-ethoxy						N[556]		
2-Amino-6-ethoxy	⊕					⊕[556]		
2-Amino-6-hydroxy						†[556]		
2-Amino-4-methyl	N[558]					N[556]		
2-Amino-5-methyl	N[558]					N[556]		
2-Amino-6-methyl	⊕[551, 558]		‡			⊕[556]		
2-Amino-5-nitro						†[556]		
2,6-Diamino	⊕[558]		⊕[559-63]			⊕[555]	⊕[561]	⊕[561-3, 585]
2-Acetamido-6-amino	⊕[558]							
3-Amino		⊕[552, 566-8]		⊖[569]	⊕[570]	⊕[555, 571]	⊖[566]	⊖[566]
6-Alkoxy-3-amino						⊕[572]	⊕[572]	
3-Amino-6-chloro				⊖[569]	⊕[573]			
3-Amino-2-hydroxy		⊕[574]						
3-Amino-4-hydroxy		⊕[567]						
3-Amino-6-hydroxy		⊕[567, 575]			⊕[570, 575]			
3-Amino-2,6-dimethyl		⊖[576]	⊖[577]		⊖[577]			
3-Amino-1-methylpyrid-6-one					⊕[575]			
3-Aminopyridine 1-oxide		**[534]	†[534]			⊕[534]	†[534]	†[534]
3,6-Diamino		⊕[570]						
4-Amino		⊖[552]	†[566]		⊖[552]	⊕[566]	†[566]	†[566]
4-Amino-2,6-dimethyl						⊕[578]		
4-Aminopyridine 1-oxide		⊕[579]						
4-Amino-2-methyl pyridine 1-oxide		⊕[579]						
4-Amino-3-methyl pyridine 1-oxide		⊕[579]						
4-Amino-2,6-dimethyl pyridine 1-oxide		⊕[579]						

* ⊕ indicates a successful reaction, ⊖ a failure; N denotes cyclization on ring nitrogen *Emme* refers to the well-known quinoline synthesis from ethoxymethylenemalonic ester.
† In these cases the aminopyridine did not even give the uncyclized intermediate of the cyclization reaction.
‡ The uncyclized intermediate was formed but could not be cyclized.
** Deoxygenation of the oxide occurred, and the 3-aminopyridine produced gave 1,5-naphthyridine.

3-Aminopyridines could give 1,5- or 1,7-naphthyridines, by cyclization at $C_{(2)}$ or $C_{(4)}$. In all the successful reactions recorded in *Table 5.20*, except those with 3-amino-2-hydroxypyridine and 3-aminopyridine 1-oxide, reaction at $C_{(2)}$ was preferred.

Though the general situation is what might have been expected, it should be remarked that the study of these reactions has not been systematic.

The acid-catalysed Fischer indole synthesis is an electrophilic substitution, and the evidence suggests that it occurs less readily with pyridylhydrazones than with benzene derivatives. Some failures with derivatives of 2-pyridylhydrazine have been recorded[580-1], but cyclohexanone and desoxybenzoin 2-pyridylhydrazones were cyclized in about 50 per cent yields with polyphosphoric acid. Zinc chloride has been used successfully with methyl isopropyl ketone 2-pyridylhydrazone[582]. The Brunner oxindole synthesis worked poorly with isobutyric acid 2-pyridylhydrazide[583]. Various 3-pyridylhydrazones have been cyclized[584-8]. Usually cyclization occurred at $C_{(2)}$, except when this was blocked[586], but cyclohexanone 3-pyridylhydrazone gave with zinc chloride a very high yield of cyclized material, containing[588] about twice as much of the product of reaction at $C_{(2)}$ as of that at $C_{(4)}$.

NUCLEOPHILIC SUBSTITUTION

From a number of the reactions to be recorded the substitution products arise from decomposition of initial addition products. From other reactions with nucleophilic reagents, pyridine and its derivatives provide addition rather than substitution products. This is especially so when the group which would be replaced is hydride ion. Because addition compounds may be intermediates in nucleophilic processes generally (p. 218), no attempt is made here to classify separately nucleophilic additions and substitutions.

Hydride reductions are important cases of addition reactions which proceed by initial nucleophilic attack. These are treated under 'Reduction' (p. 261).

(a) Acylation

A reaction which has been classified as a nucleophilic substitution is that which occurs between pyridine and an aromatic ester in the presence of magnesium or aluminium[589]. This, a case of 'heteronuclear bimolecular reduction', is an extension of the hydroxyalkylation reaction discovered by Emmert and Asendorf (p. 224):

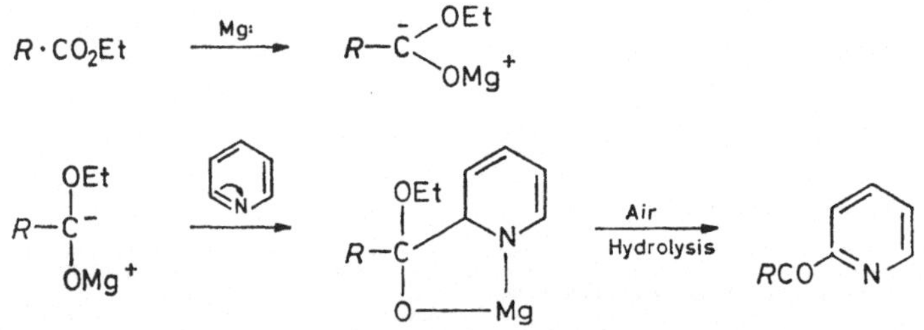

Substitution occurs predominantly at the 2-position, as required by the above mechanism, but also to a small degree at $C_{(4)}$.

(b) Alkylation and Substituted Alkylation

These reactions involve attack on pyridine and its derivatives by organometallic compounds.

Grignard reagents have not proved to be very useful[590] despite early reports to the contrary[591] (see Arylation, p. 220). Alkyl Grignard reagents give very small yields of mixtures of 2- and 4-alkylpyridines[592]. Methyl magnesium iodide attacks only the cyanide group of nicotinonitrile, but n-propyl magnesium bromide also alkylates the 4-position[593]. 3,5-Dicyanopyridine and its methyl derivatives add Grignard reagents fairly readily, giving 1,2- and 1,4-dihydropyridines[594]. Allyl magnesium bromide in ether gives, with pyridine, 9 per cent 4-allylpyridine[595]. 2- and 4-Alkylpyridines have been obtained by reaction of 2- and 4-halogenopyridines with Grignard reagents, but yields are usually poor[595-6].

Reactions of pyridine 1-oxide with Grignard reagents have been little studied. The reported cases[597] give very poor yields of 2-alkylpyridines. Only one example of a Grignard reaction with a pyridone seems to have been recorded; with benzyl magnesium chloride, 1-methyl-2-pyridone gave 2-benzal-1-methyl-1,2-dihydropyridine[598].

The reactions of Grignard reagents with quaternary pyridinium salts, giving unstable 1,2-dihydropyridines, are of practical significance. Early examples used alkyl Grignard reagents[599], but the most important examples have used benzyl Grignard reagents[600]. Characteristic products are the compounds (81–2)

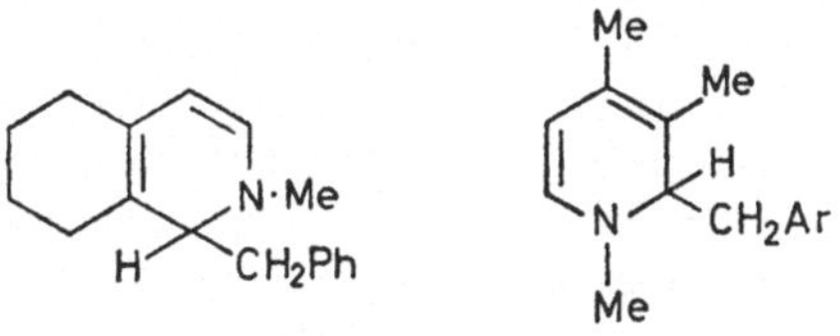

(81)   (82)

Similarly, the reaction of Grignard reagents with quaternary salts from pyridine 1-oxides promises to be of practical value[601]. Thus, 1-ethoxypyridinium bromide reacts with ethyl magnesium bromide to give 78 per cent of 2-ethylpyridine. The picoline 1-oxide quaternary salts react equally efficiently; in the case of 1-ethoxy-3-methylpyridinium bromide, the product is almost exclusively the 2-alkyl-3-methylpyridine. The reaction probably proceeds as shown, but a side-reaction always generates some of the parent base:

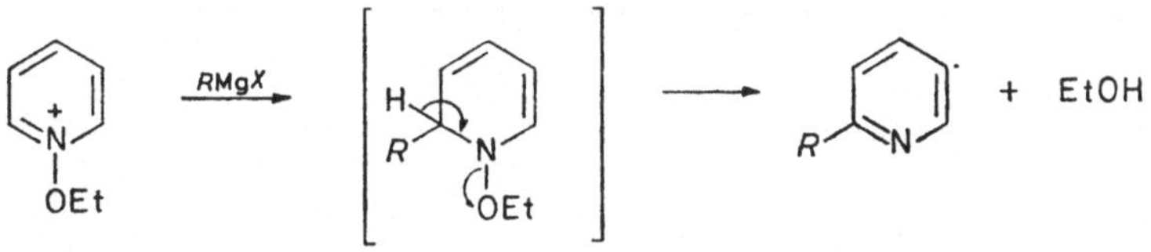

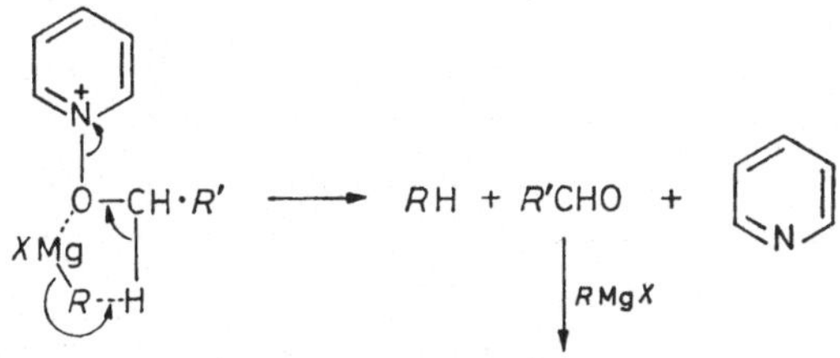

1-Acylpyridinium salts are of great interest in connection with substituted alkylations (see p. 205), but not for their reactions with Grignard reagents, which have been little examined. It is nevertheless noteworthy that 1-benzoylpyridinium chloride gives with s-butyl magnesium bromide a small yield of 4-s-butylpyridine[602]. In contrast, the same compound, with silver phenyl acetylide, produces[603] a high yield of (83)

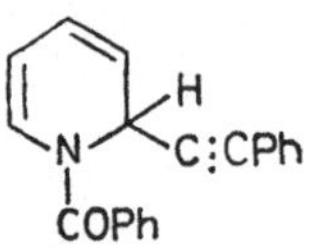

(83) 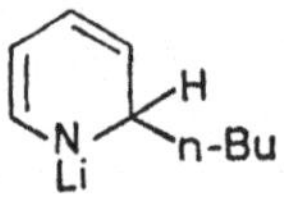   (84) 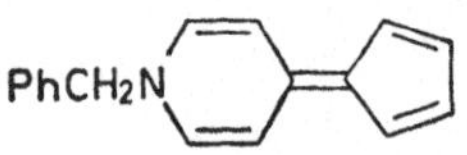   (85)

In contrast to Grignard reagents, alkyl lithium reagents are of great practical importance in pyridine chemistry. Originally, Ziegler and Zeiser[604a] observed that when pyridine and n-butyl lithium in benzene were heated together in a sealed tube at 90–100°, lithium hydride separated. Addition of water to the addition product (84), formed in the cold, gave the unstable dihydropyridine. There can be little doubt that addition products are always formed in these reactions. Subsequent workers have usually carried out the addition reaction in a low-boiling solvent such as ether or ligroin, and have relied on aerial oxidation, occurring during work-up, to convert the dihydropyridine formed by hydrolysis into the desired pyridine. Ziegler and Zeiser[604b] showed that 2-n-butylpyridine, heated with n-butyl lithium in benzene, gave 2,6-di-n-butylpyridine, in contrast to 2-picoline which was 'laterally' metalated, forming 2-pyridylmethyl lithium. Methyl lithium causes mainly lateral metalation of 4-picoline[606]. With organometallic reagents, metalation of an alkyl substituent is always a potential competitor with nuclear substitution, but proper choice of reagent and conditions minimizes this competition, making both reactions of practical value. Some results of nuclear alkylations by means of lithium reagents are collected in *Table 5.21*.

Table 5.21. Nuclear Alkylations by Lithium Alkyls

Substituent in pyridine derivative	Lithium reagent	Conditions	Substituents in product	References
—	s-Bu	ligroin	2-s-Bu	592b
2-n-Bu	n-Bu	benzene, 100°	2,6-di-n-Bu	604b
2-t-Bu	i-Pr	ether	2-t-Bu-6-i-Pr	607
2-t-Bu	t-Bu	ether–ligroin	2,6-di-t-Bu	79a, 607
3-Me	n-Bu		2-n-Bu-3-Me + some 2-n-Bu-5-Me	596b
4-Me	n-Bu	ether,-10°	2-n-Bu-4-Me	608*
3,5-Me₂	Me	ether, then toluene	2,3,5-Me₃	609
4-EtO	t-Bu	ether	2,6-di-t-Bu-4-EtO	283b†

* No reaction occurred at − 80°; reverse addition caused predominant metalation of the 4-methyl group[606].
† Reaction in ligroin at − 75° gave 2-t-butyl-4-ethoxypyridine.

When pyridine is treated with an alkyl halide and lithium, small yields of the 4-alkylpyridine result, and use of fresh magnesium powder in place of lithium gives good yields of the 4-alkylpyridine essentially free from the 2-isomer. These surprising results indicate a mechanism different from that operating in the reactions with pre-formed Grignard reagents and lithium alkyls[605].

Lithium cyclopentadienylide with benzyl pyridinium chloride gives[610] the pyridone methide (85) (see below and p. 328).

Nuclear alkylation by reaction of lithium alkyls with halogenopyridines is of consequence only as a side-reaction which diminishes the efficiency of halogen–metal exchange to produce pyridyl lithiums[191]. This situation differs considerably from that found with some sodium reagents (see below).

Although α-phenylisopropyl potassium reacts with pyridine[604a, 611], sodium and potassium alkyls have not been much applied in pyridine chemistry. Diphenylmethyl sodium and related compounds fail to react with pyridine[622], as does ethyl sodio-isobutyrate[592b].

With sodium acetylide, pyridine 1-oxide gives 2-ethynylpyridine 1-oxide, a rather unusual reaction in leaving the oxide function intact[613].

Like its lithium analogue, sodium cyclopentadienylide gives, with quaternary pyridinium salts, pyridone methides of the type (85)[614-5] The same and related reagents yield analogous products by reaction with 2- and 4-bromo-1-alkylpyridinium salts[615-6], with 4-methoxy- or 4-phenoxy-1-alkylpyridinium salts[617].

2- and 4-Halogenopyridines react readily with the sodium salts of a number of enolate anions. Examples of these and related reactions are collected in *Table 5.22*. Some failures with ethyl sodiomalonates have been reported[627-9]. The use of alkylbarbituric acids in these reactions provides a

Table 5.22. Reactions of Halogenopyridines with Enolate Anions and Related Substances

Substituents in pyridine	Source of substituting anion	Ref.	Substituents in pyridine	Source of substituting anion	Ref.
2-Cl	PhCH$_2$CN PhCH(Et)CN	}618	3-Br 4-Cl 4-Cl	PhCH$_2$CN ArCH$_2$CN PhCH(Et)CN	624 }618
	Ar$_2$CHCN	619	4-halogen	5-alkylbarbi-	625
	Me$_2$CH.CO$_2$Et	592b		turic acids	
2-Br	EtCH(CO$_2$Et)$_2$	620			626
	5-alkylbarbi-	621	4-Cl-3-NO$_2$	CH$_2$(CO$_2$Et)$_2$	
	turic acids		4-Cl-2,6-		
			(CO$_2$Et)$_2$		}627
2-Br (and homologues)	ArCH$_2$SO$_2$R	622		EtCH(CO$_2$Et)$_2$	
2-Cl-5-NO$_2$	CH$_2$(CO$_2$Et)$_2$ EtCH(CO$_2$Et)$_2$	}623			

useful source of 2- and 4-alkylpyridines, obtained from the products by hydrolysis. Groups other than halogen atoms can be replaced in these nucleophilic substitutions; thus 4-methoxy-3-nitropyridine with sodiomalonic ester gives ethyl 3-nitro-4-pyridylmalonate[241c] (but cf.[629]).

The reaction, recorded in *Table 5.22*, in which 3-bromopyridine reacts with phenylacetonitrile, with sodamide in toluene, giving phenyl-3-pyridylacetonitrile (36 per cent), is noteworthy in view of the low reactivity of the

H

3-position towards nucleophilic reagents. It is just possible that the reaction involves 3,4-pyridyne (86). The first clear evidence that 3,4-pyridyne can be formed as an intermediate was the isolation of 4-phenacyl- (13·5 per cent) and 4-amino-pyridine (10 per cent) from the reaction of 3-bromopyridine with sodio-acetophenone in the presence of sodamide and liquid ammonia[630]. The occurrence of 3,4-pyridyne is also clearly demonstrated by the reactions[631]

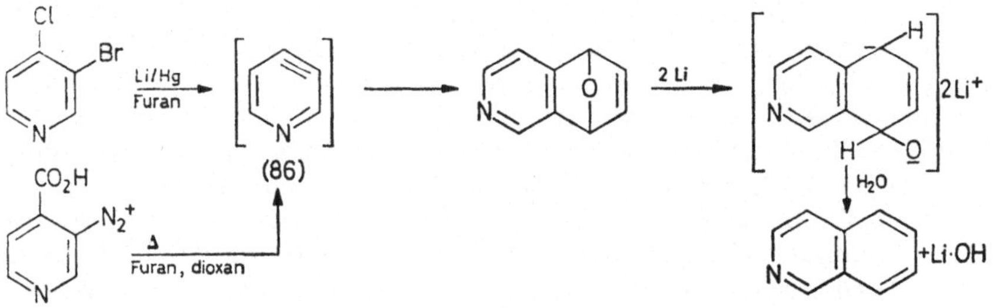

Similarly, the production of 2,3-pyridyne when 3-bromo-2-chloropyridine reacts with lithium amalgam is shown by the isolation of quinoline when this occurs in presence of furan[632].

2-Bromopyridine 1-oxides react with sodium salts of a variety of enolate anions, as in the examples[633]

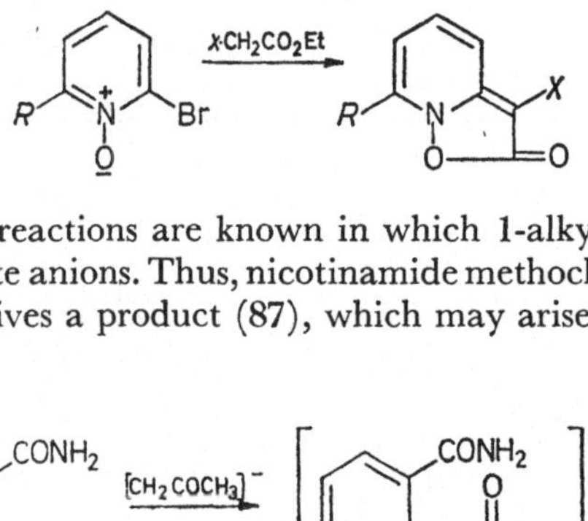

R = H or Me
X = MeCO:, PhCO.:CO$_2$Et

Some reactions are known in which 1-alkylpyridinium salts are attacked by enolate anions. Thus, nicotinamide methochloride with acetone in aqueous potash gives a product (87), which may arise[634] as follows:

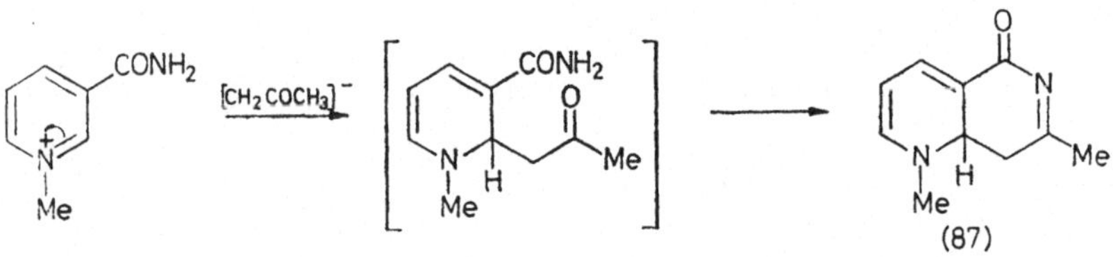

(87)

$[C_5H_5\overset{+}{N}\cdot CH_2\cdot C_6H_3Cl_2]\ [CH_2COCH_3]^-$

(88)

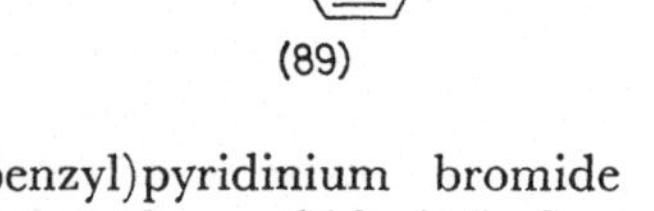

(89)

In similar circumstances 1-(2,6-dichlorobenzyl)pyridinium bromide is said to give a salt (88), which can be oxidized to the methide (89). Several such reactions are known[614]. Closely related are those of diphosphopyridine nucleotide with carbonyl compounds in alkali[635].

The replacement of a 2-halogen atom from a 2-halogeno-1-alkylpyridinium salt by an enolate anion has been observed[636], and closely related is that of a 2- or 4-halogen atom in the familiar synthesis of a cyanine dye[637]:

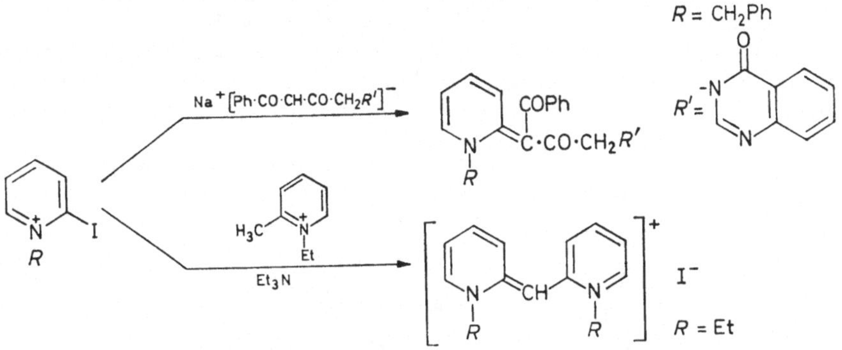

Among the most interesting reactions to be discussed in this Section are those involving an 1-acylpyridinium salt and an enolizable carbonyl compound. The original observation in this connection was made by Claisen and Haase[638]: they found the preparation of acetophenone O-benzoate (see below) by reaction of acetophenone with benzoyl chloride in pyridine to be unsatisfactory, but they isolated a yellow compound, shown by v. Doering and McEwen[639] to have the structure $(90; R = R'' = Ph, R' = H)$:

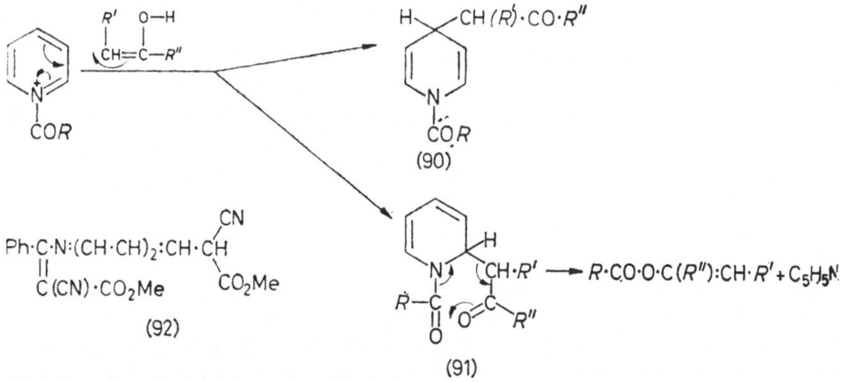

Some other ketones behave similarly, and acenaphthenone gives an analogous product with pyridine and acetic anhydride. Presumably, reaction occurs at $C_{(2)}$ as well as at $C_{(4)}$, but the product (91), which is in equilibrium with (90), rearranges to give the enol ester as shown. When the carbonyl compound is a β-keto-ester, the initial reaction again probably causes C–C bond formation at $C_{(2)}$ and $C_{(4)}$, but the marked acidity of the β-keto-ester makes interconversion between (90) and (91) rapid, and the product is therefore the enol ester. This is the basis of the well-known method for preparing O-acyl derivatives of enols (discussed below). Related reactions occur between 1-benzoylpyridinium chloride and methyl cyanoacetate[640] or nitromethane[641]. In the first case the initial product $(91; R = Ph, R' = CN, R'' = OMe)$ undergoes ring opening, giving (92).

The value of pyridine and an acyl or aroyl chloride as a means of O-acylating β-keto-esters and 1,3-diketones has been known for a long time[642]. The unique efficiency of pyridine in this respect, as compared with other bases[643], probably arises from its ability to effect the acylation by the mechanism outline above. Kinetic studies support this view[644].

The use of pyridine and an acyl chloride or acid anhydride for esterifying alcohols and phenols is a long-established practice[645]. The reaction with phenols is discussed later. As regards alcohols, a kinetic study has been made[646a], but the mechanism is not yet clear.

Related to the above reactions is that in which a pyridine oxide in acetic anhydride is alkylated by ethyl cyanoacetate[646b]:

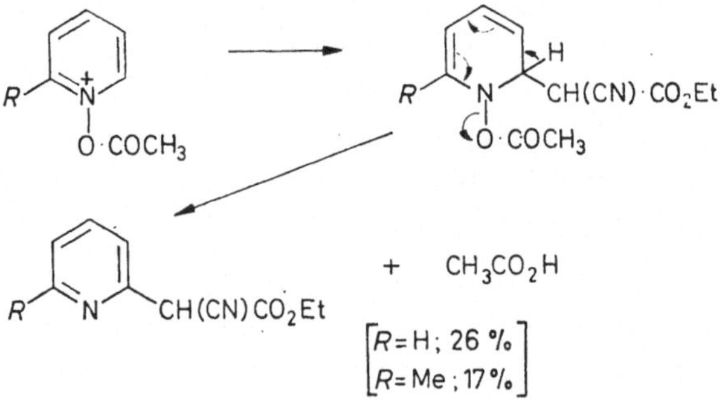

$$\begin{bmatrix} R=H; & 26\% \\ R=Me; & 17\% \end{bmatrix}$$

The alkylating action of alkylpyridinium compounds upon a variety of substances (p. 389) may in some cases depend on initial nucleophilic attack at $C_{(2)}$, but there is no evidence on this point. Some reactions in which C–C bonds are formed by nucleophilic attack on the pyridyl nucleus have already been mentioned (p. 200), for others see 'Carbonyl Reactions' below.

(c) Amination and Substituted Amination

One of the most characteristic and important reactions of pyridine, its homologues and analogues, is that with sodamide, discovered by Tschitschibabin[361]. Heating pyridine with sodamide in toluene, xylene or dimethylaniline, and hydrolysing the product gave 70 per cent of 2-aminopyridine. A little 4-aminopyridine was said to be formed, but later workers did not find this[647]. 2-Picoline gave 2-amino-6-methylpyridine, but the reaction failed with 2,4,6-trimethylpyridine. Numerous examples of the reaction have been described[170, 648–51]. For preparing 2-aminopyridine, and in a number of other cases, dimethylaniline is probably the best solvent[651].

Di-amination can be achieved[239, 361, 650, 652–4], pyridine giving 2,6-diaminopyridine. Oddly, the reaction with 3-hydroxypyridine results in 2,6-diaminopyridine[288]. 2,4,6-Triaminopyridine has also been obtained in the Tschitschibabin reaction[654]. As well as in the di-amination of mono-α-substituted pyridines, substitution at $C_{(4)}$ can also occur when both α-positions are blocked, as with 2,6-dimethylpyridine[648]. A β-substituted pyridine is aminated at the α- rather than the α'-position, as in the cases of nicotinamide[174] and 3,4-dihydroxypyridine[655]. 3-Picoline gives both 2-amino- and 6-amino-3-methylpyridine[656] in the ratio[657a] 9 : 1; the corresponding ratio from 3-ethylpyridine[657b] is 3·5 : 1. 5-Ethyl-2-methylpyridine gives 2-amino-3-ethyl-5-methylpyridine[657c].

Substituted amination has sometimes been effected, as in the conversion of pyridine by sodium anilide into a small yield of 2-anilinopyridine[361], and the related preparation of 2-alkylaminopyridines[659-60]. The reaction with sodium N-methylanilide and lithium dibutylamide failed[612].

Side products of the Tschitschibabin reaction are dipyridyls and dipyridylamines[361, 647] (p. 395).

Pyridine reacts slowly with an excess of potassamide in liquid ammonia, giving 2-aminopyridine, and at 100–130° these reagents give hydrogen and the potassium salts of 2-amino- and 2,6-diamino-pyridine[650]. 2-Picoline, with sodamide or potassamide in a boiling hydrocarbon, gives 2-amino-6-methylpyridine, but in liquid ammonia the picolyl anion is generated[650]. Although sodamide is the most commonly used reagent in the Tschitschibabin reaction, the sodamide–potassamide eutectic has been recommended as possessing advantages of solubility[659].

The conversion of 2-picoline into the picolyl anion mentioned above is a type of reaction which can occur with any pyridine derivative capable of generating an anion in the presence of a strong base. As a consequence, 'lateral' reaction of a suitable alkyl, amino, hydroxyl or similar substituent is always a possible competitor with nuclear alkylation, amination or reaction at the ring nitrogen atom (see pp. 182–3, 379).

As regards the mechanism of the Tschitschibabin reaction, little is definitely known. It has been represented as a direct bimolecular nucleophilic replacement of hydride by amide anion[660]. However, when the situation in related reactions (see Alkylations, p. 202), and evidence from Tschitschibabin reactions in other heterocyclic series are considered, it seems likely that an addition compound is an intermediate[604a, 661]. Barnes[109a] argued that the resulting structure would probably dissociate again, but would be stabilized by conversion into (93), and formulated the conversion of the latter into the reaction products as a bimolecular process:

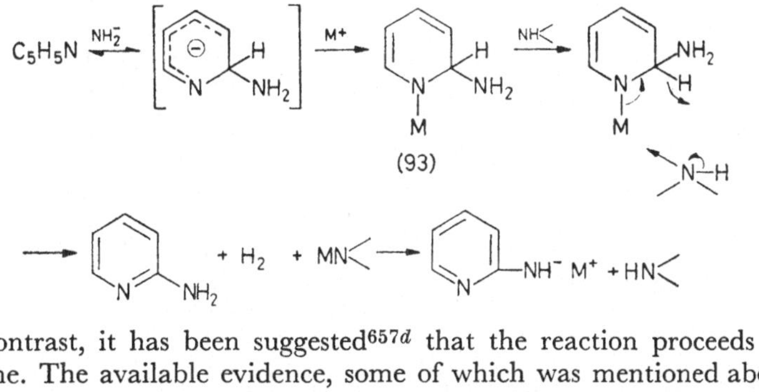

(93)

In contrast, it has been suggested[657d] that the reaction proceeds via a pyridyne. The available evidence, some of which was mentioned above, is sufficient to rule out this as a general mechanism[657e]. In addition, when a mixture of pyridine and pyridine-3D is partially aminated with sodamide, the unreacted material has the same isotopic composition as the starting material, and the same is true when pyridine-2D is used[657a, 658]. Thus, the Tschitschi-babin reaction could not involve formation of a pyridyne with either the loss of proton or hydride anion as the rate-controlling step. The available evidence does not conflict with the addition–elimination mechanism.

A reaction which superficially resembles Tschitschibabin's occurs when pyridine is heated with an alkylamine and powdered sodium in toluene[612c, 662]. Thus, 2-dimethylaminoethylamine gives 79 per cent of 2-2'-dimethylamino-ethylaminopyridine, though yields are not always so high. If this were a variant of the Tschitschibabin reaction, initial formation of a sodium alkyl-amide would occur. However, under the conditions sodium is said not to react with alkylamines. The reactions are believed to be started by a radical process

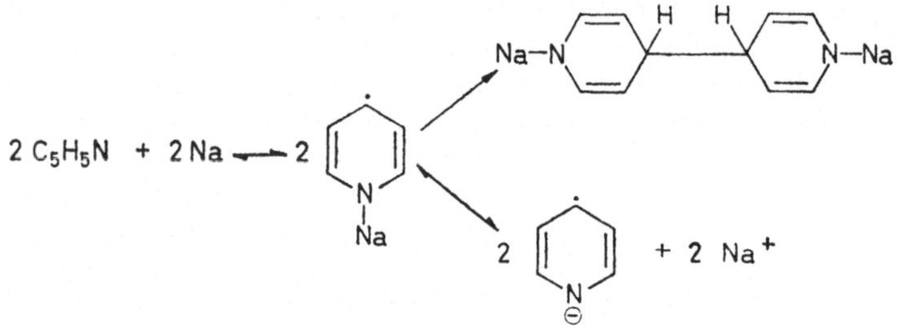

The sodium cations increase the polarization of pyridine, enhancing the ease of nucleophilic addition to it and also assisting the cleavage of hydride ion from the intermediate product. The formation of dipyridyls thus competes with that of Na^+; dipyridyls and tars are formed in these reactions.

When pyridine is heated with an excess of sodium hydrazide in the presence of hydrazine, in a solvent such as benzene or toluene, a red-brown substance is formed and hydrogen is evolved. Hydrolysis gives 2-hydra-zinopyridine[663]. The greater ease of the reaction of $NaNH.NH_2$ and $NaNH.NHMe$ than of $NaNH.NMe_2$ is attributed to the 1,3-dipolar addi-tion which is possible with the first two

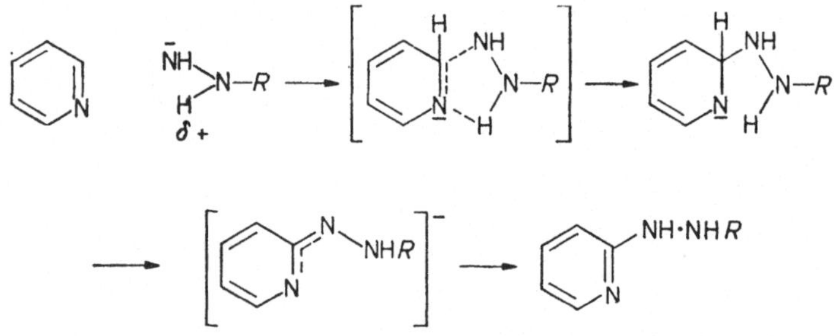

The reaction provides moderate to good yields of various 2-hydrazino-pyridines.

Nucleophilic attack by amines on quaternary pyridinium salts has various consequences. Diphosphopyridine nucleotide is attacked by hydroxylamine in alkaline solution, probably to give a product of type (94)[664]. The possibility that the formation of pyridylpyridinium compounds from pyridine and thionyl chloride and from pyridine, pyridine oxides and *p*-toluenesulphonyl chloride may involve nucleophilic replacement of hydrogen from a quaternary pyridine nucleus has been mentioned (p. 186). An interesting special case of intramolecular nucleophilic amination of a quaternary pyridine nucleus is the reaction[665]

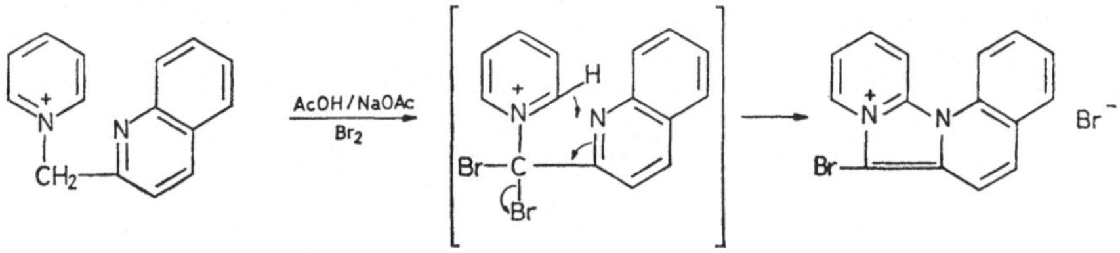

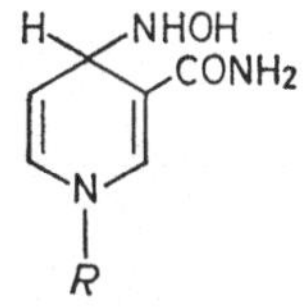

(94)

In contrast, a common consequence of the attack of an amine upon a pyridinium salt is not amination but ring opening

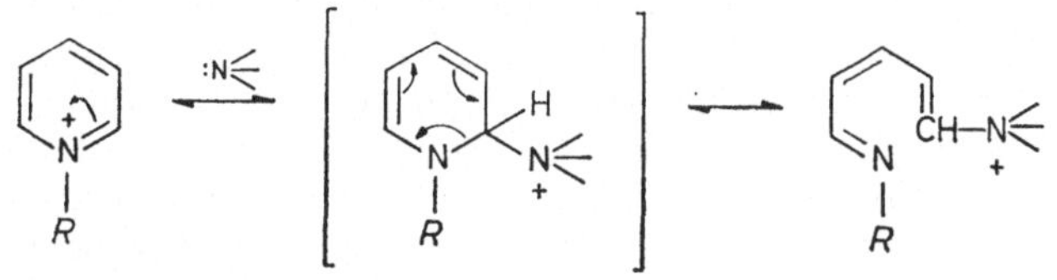

These reactions are discussed later (p. 265).

Surprisingly, 1-methoxypyridinium *p*-toluenesulphate behaves towards amines in quite a different way, methylating them and being converted[666] into pyridine 1-oxide

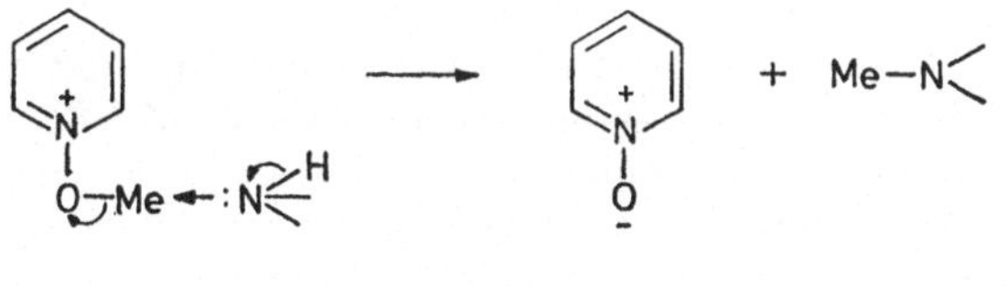

An interesting reaction occurs with another quaternary salt from pyridine 1-oxide; pyridine 1-oxide, *p*-toluenesulphonyl chloride and pyridine give

209

2- and 4-pyridylpyridinium salts. With 2-picoline 1-oxide this type of re-action is in competition with 'lateral' reaction at the methyl group[667]:

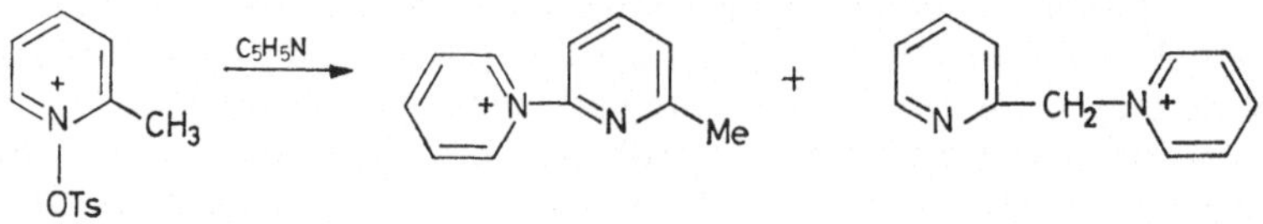

A related intramolecular substitution occurs when a pyridine 1-oxide reacts with 2-bromopyridine or 2-bromoquinoline. Initial quaternary salt formation, assisted by the presence of hydrobromic acid, is followed by substitution[425-6]:

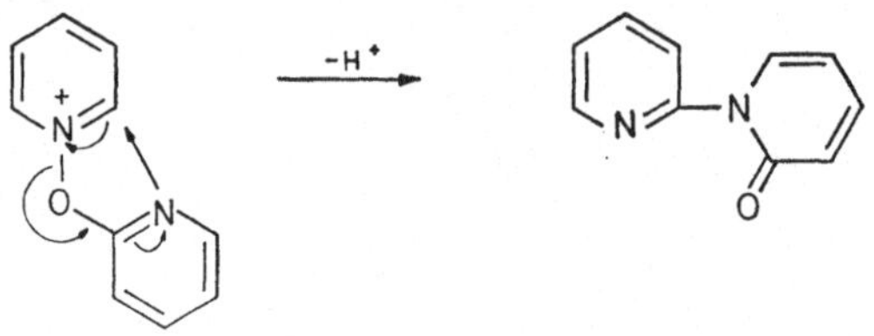

Examples of reactions are known in which an amino- or substituted amino-pyridine is formed by nucleophilic replacement of an already present amino function. Thus, 2-2'-pyridylaminopyridine results when 2-aminopyridine is heated with 2-aminopyridine hydrochloride[668]. Another case is that of amination by replacement of a quaternary function, notably from 4-pyridylpyridinium chloride hydrochloride. Heated with ammonia under pressure, the latter gives a good yield of 4-aminopyridine[433, 436]. 4-Amino-3-methylpyridine has been prepared in the same way[200]. Aminopyridines formed during halogenation of pyridines, or subsequent to auto-quaterni-zation of halogenopyridines, clearly arise in the same way (see pp. 164–5). The useful variation of the original reaction, in which 4-pyridylpyridinium chloride hydrochloride is treated with phenol and an amine, probably proceeds through the intermediate 4-phenoxypyridine (see below). Koenigs and Greiner[433] showed that heating 4-pyridylpyridinium chloride hydro-chloride with aniline gave glutacondialdehyde dianil hydrochloride with small amounts of 4-anilino- and 4-amino-pyridine. If aniline hydrochloride rather than aniline is used, a high yield of 4-anilinopyridine results[439]. Pre-sumably the base converts 4-pyridylpyridinium chloride hydrochloride into the free quaternary chloride and thus swings the balance of electrophilic power from the 4- to the 2'-position of the quaternary nucleus. The difficulty is avoided when the arylamine hydrochloride is used.

2-Pyridylpyridinium chloride also reacts with aniline hydrochloride to give 2-anilinopyridine[667].

In contrast to the cyanopyridines, 4-cyanopyridine methiodide reacts readily with ammonia giving 4-aminopyridine methiodide[669].

Nucleophilic amination by replacement of halogen is among the most im-portant reactions of compounds of the pyridine series. The original observa-tion of this kind of reactivity was made by Marckwald[670a], who drew the analogy between halogenopyridines and halogenonitrobenzenes. Whilst 2-chloropyridine was insufficiently reactive, 6-chloronicotinic acid reacted with concentrated aqueous ammonia at 170° to give 6-aminonicotinic acid. Replacement of a similarly activated 4-chlorine atom was also achieved[670b]. Fischer[336] confirmed the relatively low activity of 2-chloropyridine but, by using the zinc chloride–ammonia complex at 220°, obtained 2-amino-pyridine quantitatively. The influence of the zinc chloride may be an example of acid catalysis of nucleophilic substitution (p. 215). 4-Chloropyridine behaved similarly[671].

Since these early studies, numerous derivatives of 2-chloro- and 2-bromo[161a, 672], and 4-chloro- and 4-bromo-pyridine[673] have been converted into amino- or alkylamino-pyridines under conditions differing little from those mentioned above. Modifications which have been used include the addition of copper sulphate as a catalyst[152a, 672a], reaction in the presence of pyridine[364] and also, in the case of 4-chloropyridine, reaction with primary and secondary aliphatic amines in benzene[674] at 140–180°. The qualitative data from these numerous examples show the effect of other substituents upon 2- and 4-halogen atoms to be as expected. Carboxyl-[670a, 675, 918] and nitro-groups[241a] markedly augment reactivity, and in a compound such as 2-chloro-3,5-dinitropyridine, much milder conditions than usual can be used for amination[676].

2-Bromopyridine has been aminated by reaction with sodamide or potassamide in liquid ammonia (see also p. 214) and, in moderate yield, with sodio-methylaniline and related compounds[612]. In this connection it is interesting that sodium or potassium salts of arenesulphonamides give 2-arenesulphonyl-aminopyridines with 2-halogenopyridines, whilst sulphanilamide itself gives p-(2-pyridylamino)benzenesulphonamide[212, 677].

2- and 4-Halogenopyridines also react satisfactorily with arylamines[336, 543, 678]. Sometimes the reactants have been simply heated together, in other cases barium oxide[678a], potassium carbonate–copper bronze[678b] or copper bronze alone[543, 678c] have been added. Sometimes the aromatic amine has been replaced by its sodium salt[612, 678b]. It is possible that the formation of 3-nitro-4-(2′-hydroxyphenylamino)pyridine from sodium o-aminopheno-late and 4-chloro-3-nitropyridine involves rearrangement of the initially formed 3-nitro-4-(2′-aminophenoxy)pyridine[678h] (p. 218).

Hydrazine and phenylhydrazine are more nucleophilic than ammonia and react more readily with 2- and 4-halogenopyridines[580, 679, 918].

Many examples are known in which a 2,4- or 2,6-dihalogenopyridine is converted by reaction with ammonia either into a diaminopyridine or into an amino-halogenopyridine, depending on the severity of the conditions used, the presence or absence of a catalyst and the character of other substituents present[152a, 161a, 672b, 677a, 680]. The qualitative results of these studies are not satisfactory for comparing the relative reactivities of 2- and 4-halogen atoms, particularly since the presence of further substituents complicates the situation in a number of the examples. Commonly, when both 2- and 4-halogen atoms are present, both possible amino-halogenopyridines are formed; 2,4-dichloropyridine with aqueous ammonia at 170–180° gives 4-amino-2-chloro- and 2-amino-4-chloro-pyridine, the first perhaps predominantly[680b]. This and other examples[171c] suggest that a 4-halogen is slightly more reactive than a 2-halogen atom, an impression which is borne out by quantitative studies (see below).

211

The following examples[680b] illustrate the important point that 3-halogen atoms are less easily replaced than 2- and 4-halogen atoms. Nevertheless, 3-halogenopyridines are useful sources of 3-aminopyridines, since amination

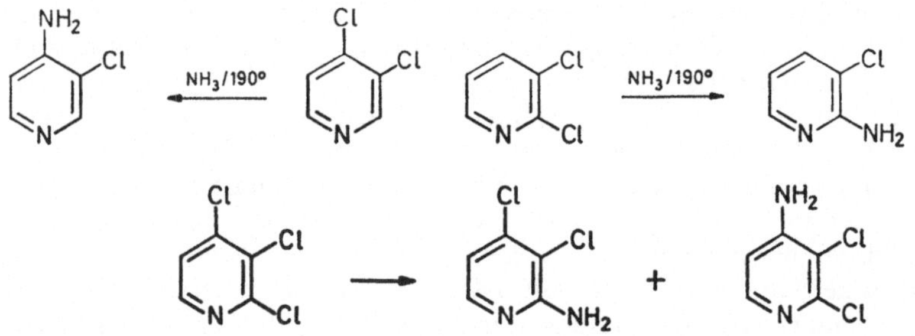

or substituted amination can be effected, provided that a catalyst (copper sulphate) is used[158c, 165b, 166a, 223, 672b, 681]. 3,5-Dihalogenopyridines can be converted into mono- and di-amines. The statement[682] that '. . . 3-bromopyridine reacts more readily with ammonia than the 2-isomer' seems mistaken; a catalyst is unnecessary with 2- but essential with 3-bromopyridine[672b].

Intramolecular cyclization can in some cases accompany substituted amination of 2-halogenopyridines. Examples are the reactions with ethyl β-aminocrotonate[683] and with anthranilic acid[678e, 684], giving (95) and (96), respectively.

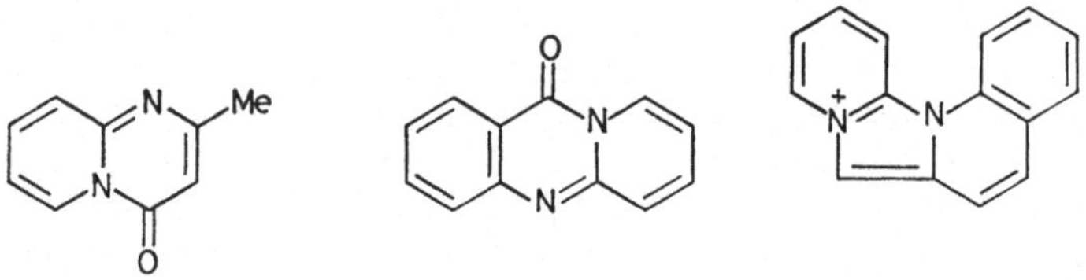

(95)   (96)   (97)

In view of the capacity for auto-quaternization shown by 2- and 4-halogenopyridines, the ready reaction of 4-chloropyridine with pyridine[674] or the formation of (97) from 2-bromopyridine and 2-bromomethylquinoline[665] is not surprising. However, with other tertiary amines the situation is less clear. Thus, with trimethylamine in a sealed tube at 45°, 2-bromopyridine gives in 3 weeks a large amount of solid, containing some trimethyl-(2-pyridyl)ammonium bromide. 2-Chloropyridine reacted more slowly, 3-chloropyridine not at all, and 4-chloropyridine did not give the expected quaternary salt[685]. 4-Chloropyridine did not react with triethylamine[674] at 175°.

Little is known of the relative abilities of halogen atoms and other groups as leaving groups in nucleophilic amination of pyridines. However, when 2-halogeno-4-nitropyridines are heated with aqueous ammonia at 130°, 4-amino-2-halogenopyridines result[686a]. In the reactions shown, a 2-bromo-substituent is in competition with two other replaceable groups. The course of the amination shows a curious solvent dependency[688]. 5-Bromo-2-nitro-pyridine with aqueous or alcoholic ammonia gives a mixture of 2-amino-5-bromo- and 5-amino-2-nitro-pyridine. In water, the bromine atom is

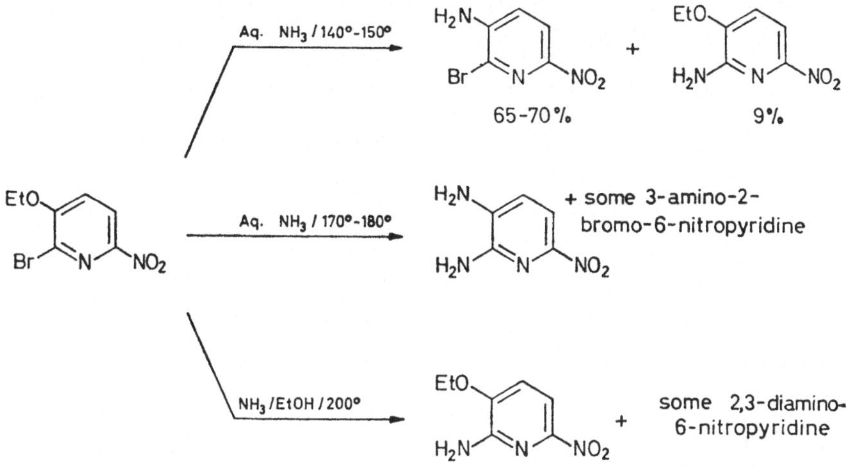

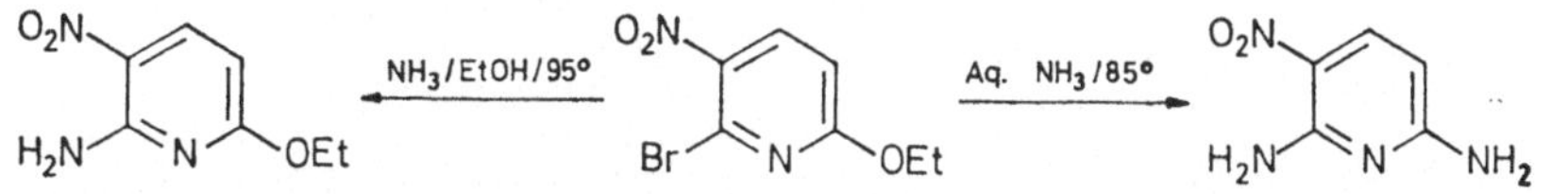

replaced more rapidly than the nitro group, but in alcohol the order is reversed[688]. In 3-chloro-2,6-dinitropyridine the nitro group adjacent to the chlorine atom is replaced in preference to the other groups, both by ammonia and by hydrazine[237].

The halogen atoms in 2-halogenopyridine 1-oxides are more reactive than those in 2-halogenopyridines[686b, 687a], and successful substitutions with secondary amines and 2-chloropyridine 1-oxide have been reported[689]. Under conditions in which 2-halogeno-4-nitropyridines do not react, secondary amines replace the halogen atom in the corresponding 1-oxides[686c]. Similar qualitative results from reactions with secondary amines or hydrazine show the halogen atom in 4-chloropyridine 1-oxide to be more highly activated than that in 4-chloropyridine, and a number of successful aminations of the former have been reported[247, 252, 690]; quantitative studies support this conclusion (see below).

The halogen atom in 3-chloropyridine 1-oxide is very much less activated than those in 2- and 4-halogenopyridine 1-oxides but more so than that in 3-chloropyridine (see below). 3-Fluoropyridine 1-oxide reacts with piperidine or hydrazine more readily than do the chloro- or bromo-compounds[691]. Surprisingly, reaction of amines or hydrazine with 3-halogeno-4-nitropyridine 1-oxides causes replacement of halogen and not of the nitro-group[687b, 692].

As would be expected, 2- and 4-halogen atoms in 1-methylpyridinium salts are readily replaced in reactions with ammonia, with aliphatic and aromatic amines[336, 693] and with benzhydrazide[977].

As mentioned already (p. 204), the formation of 4-phenacylpyridine and 4-aminopyridine from 3-bromopyridine, acetophenone and sodamide in liquid ammonia led Levine and Leake[630] to postulate the occurrence of 3,4-pyridyne as an intermediate in the reaction. The results of other reactions in which a 3,4-pyridyne plays a part are collected in *Table 5.23*. 3-Fluoro-

Table 5.23. Aminations involving Pyridyne Intermediates

Starting pyridine	Reagents	Pyridine formed	Reference
3-Cl, 3-I } 4-Cl, 4-I }	KNH_2/NH_3	3-NH_2 and 4-NH_2 (1:2)	694a
3-Cl, 3-Br	⎧	3- and 4-piperidyl (48:52)	695
3-Br	⎪ Li piperidide/	3- and 4-piperidyl(52:48)	696
3-F	⎨ Et_2O	3- and 4-piperidyl (96:4)	695
4-Cl	⎩	3- and 4-piperidyl (0·4:99·6)	695
4-Br-2-EtO	⎧	3- and 4-NH_2-2-EtO (1-2:99-98)	⎫
4-Br-3-EtO	⎪	3-NH_2-5-EtO	⎪
3-Br-2-EtO	⎪	3- and 4-NH_2-2-EtO (3:97)	⎪
3-Br-4-EtO	⎪	4-EtO, 2-NH_2-4-EtO, and 2-NH_2-5-Br-4-EtO (25:55–60:15–20)	⎪
3-Br-5-EtO	⎨ $KNH_2/NH_3/Et_2O$ — 33°	3-NH_2-5-EtO	⎬ 694
3-Br-6-EtO	⎪	3- and 4-NH_2-6-EtO (35:65)	⎪
2-Br-3-EtO	⎪	2-NH_2-3-EtO and 3-EtO (99:1)	⎪
2-Br-4-EtO	⎪	2-NH_2-4-EtO	⎪
2-Br-5-EtO	⎪	2-NH_2-5-EtO and 3-EtO (90–95:5–10)	⎪
2-Br-6-EtO	⎩	2- and 4-NH_2-6-EtO (80–85:10–15)	⎭

and 4-chloro-pyridine seem to react almost completely by normal substitution with lithium piperidide. The reactions of the bromo-ethoxypyridines suggest that those of the 3- and 4-bromo-derivatives proceed through the 3,4-pyridyne, but with 2-bromo-compounds substitution generally does not go through a 2,3-pyridyne. The exception is 3-bromo-4-ethoxypyridine which evidently gives some of the corresponding 2,3-pyridyne. 2-Amino-5-bromo-4-ethoxypyridine is a by-product; probably some 3-bromo-4-ethoxypyridine is converted into 3,5-dibromo-4-ethoxypyridine [the bromo-ethoxypyridines are prone to rearrangement (p. 371)] which gives 5-bromo-4-ethoxy-2,3-pyridyne and thence 2-amino-5-bromo-4-ethoxypyridine. In a similar way both 3- and 5-bromo-2,4-diethoxypyridine give 6-amino-2,4-diethoxy-pyridine[697a].

Reaction of 2-chloro- or 2-iodo-pyridine with potassamide in liquid ammonia[694a], or of 2-fluoropyridine with lithium piperidide in ether[695] results in direct substitution without pyridyne intermediates. The reaction of 3-fluoropyridine with potassamide in liquid ammonia at —33° takes a surprising course[697b]. Evidently the base abstracts a proton from the 4-

position and the resulting carbanion attacks a second molecule of 3-fluoro-pyridine to give fluoro-dipyridyls.

Pyridynes also arise in the amination of halogenopyridine 1-oxides[694c, 697c]. The situation is here to some extent reversed by comparison to that noted for the pyridines. Thus, with potassamide in liquid ammonia, 3- and 4-chloropyridine 1-oxides give only direct substitution, whilst 2-chloropyridine reacts through the 2,3-pyridyne to give 2- and 3-aminopyridine 1-oxides[697c].

It remains to discuss quantitative aspects of amination by nucleophilic replacement of halogen. Selected data are given in *Table 5.24*. The results are due mainly to Chapman and his co-workers[457, 699-701], and the subsequent discussion follows theirs closely.

There is ample evidence for the occurrence of acid catalysis in the nucleophilic replacement of halogen by amines, an effect first noticed by Banks[698]. The source of the effect is the increased electrophilic character of the halogenopyridines following protonation. As a consequence, the reactions of halogenopyridines with primary amines in which acid is generated, are in general autocatalytic, though the effect is less marked the more basic the primary amine. Aminations with tertiary amines are not autocatalytic. Chapman and his co-workers[457a, 699, 700] encountered the autocatalytic effect with primary amines but were able to find conditions where it was negligible. The reactions obeyed second-order rate laws.

The relatively low reactivity of chloropyridines has centred interest upon derivatives containing activating groups additional to the ring nitrogen atom. 2-Chloropyridine presents regular kinetics in reaction by solvolysis or in methanol with piperidine but not with morpholine[700]. 4-Chloropyridine shows autocatalysis in reaction with either base. These results are a consequence of the balance of base strengths (4-chloropyridine > 2-chloropyridine, and piperidine > morpholine). So far as it goes, the evidence in *Table 5.24* supports the view that a 4-halogen atom is more activated than a 2-halogen atom (cf. 4-chloro-3-nitro- and 2-chloro-3-nitro-pyridine). Also, the 2-bromine atom is more easily replaced than the 2-chlorine atom.

The data permit interesting comparisons of the activating power of a nitro group and a ring nitrogen atom. The reactions of pyridines with 2,4-dinitrochlorobenzene and 4-chloro-3-nitropyridine show that replacement of a *para*-nitro group by a nitrogen atom leaves the activation energy practically unchanged but increases A. The same change at the *ortho*-position increases the activation energy (despite the disappearance of the steric effect of the nitro group) and leaves A unchanged. From the *para*-position, the mainly tautomeric influences of the nitro group and the nitrogen atom seem to be much the same, but from the *ortho*-position the nitro group exercises a powerful inductive effect, whilst the nitrogen atom exerts a weak electromeric one. Movement of a nitro group in a 2-chloropyridine from the *para*- to the *ortho*-position, relative to the chlorine atom, only slightly increases the activation energy despite steric hindrance (4-picoline is anomalous in this respect), whereas the similar shift of a nitrogen atom increases the activation energy and lowers A. A nitro group and a ring nitrogen atom in a *para*-position to the chlorine atom are again similar, whilst an *ortho*-nitro group is more effective than a nitrogen atom in increasing the activation of the chlorine atom. Concentration of interest on pyridine as the nucleophile gives a too

Table 5.24. Nucleophilic Amination of Halogenopyridines

Pyridine	Nucleophile	Conditions	k	E, cal.mole^{-1}	$\log_{10} A$	Ref.
2Cl		EtOH, 20°	4.8×10^{-10}	19,000		700
	Piperidine	Solvolysis	0·0246*	17,100		
2Br			0·200†	16,400		} 703
			k_2, (l.mole^{-1} sec^{-1}) $\times 10^6$			
2Cl-3NO₂	Pyridine		1·03	18,700	6·3	457a
	3-Picoline		1·81	18,500	6·6	
	4-Picoline		3·15	17,400	6·1	} 457b
	m-Toluidine‡		22·4	14,400	5·0	
2Cl-5NO₂	Pyridine		1·97	18,100	6·3	457a
	3-Picoline		4·00	17,900	6·6	
	4-Picoline		6·11	17,500	6·5	} 457b
	m-Toluidine‡	EtOH, 55°	15·8	12,900	3·8	
4Cl-3NO₂	Pyridine		32·1	16,900	6·8	457a
	3-Picoline		39·8	15,600	6·0	} 457b
	4-Picoline		66·4	15,100	5·9	
[2,4-Dinitro-chloroben-zene]	Pyridine		11·1	16,700	6·2	457a
	3-Picoline		19·9	17,100	6·7	} 457b
	4-Picoline		30·9	16,900	6·7	
	Aniline		353	11,200	4·0	457a
			k_2, (l.mole^{-1} sec^{-1}) $\times 10^4$			
2Cl-3NO₂			41·9	12,000	6·2	
2Cl-5NO₂	Piperidine	EtOH, 30°	67·1	11,500	6·1	} 699
2Cl-4Me-5NO₂			12·6	12,400	6·0	
			k_2, (l.mole^{-1} sec^{-1} $\times 10^5$			
2Cl-5NO₂		40°	0·490	14,900	5·06	
2Cl-3CN-5NO₂		10°	385	9,800	5·18	
2Cl-3CN-6Me-5NO₂	Aniline	MeOH 10°	93·2	10,500	5·07	} 701
2Cl-3CN-4,6Me₂-5NO₂		40°	7·79	12,400	4·57	
Pyridine 1-oxides						
2Cl			37·0	14,390	5·49	
3Cl§	Piperidine	MeOH, 80°	0·0104	—	—	} 702
4Cl			10·2	15,180	5·45	

* Pseudo-unimolecular rate constant (h^{-1}) at * 91·6°; † 90·0°. ‡ For data on a number of primary aromatic amines see [457] (cf. *Table 5.19*). § Reaction too slow to permit calculation of accurate Arrhenius parameters.

simple picture of the influence of the *para*-nitrogen atom[457b], and some of the reactions suggest the nitrogen atom to be more effective than the nitro group in lowering the activation energy. However, since the change is accompanied by a fall in A, the rate of reaction is not markedly affected, and structural interpretations are precluded.

It is possible to estimate the rate coefficients for reaction of aniline at 100° with 2-chloro-3-nitro- and 2-chloro-3-cyano-pyridine as about 10^{-3} and 10^{-7} l.mole^{-1} sec^{-1}, respectively. Thus, the nitro group is much more powerfully activating than the cyano group[701]. The introduction of methyl groups into 2-chloro-3-cyano-5-nitropyridine causes a drop in the rate of reaction with aniline, attributed to a combination of the weak polar character of the methyl group with its more important secondary steric effect of forcing the adjacent nitro group out of coplanarity with the pyridine ring[699, 701].

Conversion of the chloropyridines into their 1-oxides increases the reactivity in reaction with piperidine at all three nuclear positions. In these oxides the order of reactivity is 2-chloro > 4-chloro ≫ 3-chloro, the difference between the 2- and 4-position being small. The sequence has been discussed in terms of competing inductive and 'built-in solvation' effects, the latter operating in the transition state for 2-chloropyridine 1-oxide, perhaps by hydrogen bonding[702] (98).

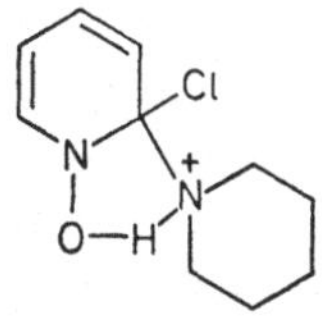

(98) 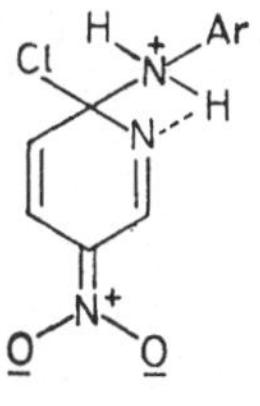   (99)

Like the reaction of 2,4-dinitrochlorobenzene with primary aromatic amines, those involving various 2-chloropyridines follow the Hammett relationship[457a, 701] (p. 47). The reactions of 2-chloro-3-cyano-5-nitro-pyridine and its homologues give a mean value[701] of the reaction constant, $\rho = -3\cdot33$ (methanol at 10°), not very different from the values for those of anilines with various other chloro compounds. The effects of substituents in the primary aromatic amines are those to be expected from their influence upon the nucleophilic power of the amine, as it is measured, for example, by basic strength (*Table 5.19*). Nevertheless, aniline reacts with 2,4-dinitrochlorobenzene faster than the stronger base pyridine, because with the latter there is a strong increase in the energy of activation which is only partly offset by an increase in A. This fact, and its implications for the transition state of the reaction, have already been discussed (p. 191); the nucleophilic superiority of primary aromatic amines depends on a steric factor combined with hydrogen bonding in the transition state. The results with 2-chloro-3-nitro- and 2-chloro-5-nitro-pyridine are qualitatively similar. The position of the nitro group in the former makes it structurally similar to 2,4-dinitrochlorobenzene, and hydrogen bonding is also possible in the transition state from 2-chloro-5-nitropyridine (99). Overall, then, the superiority of the primary amine is attributed[457a] to the relative absence of steric hindrance, coupled with hydrogen bonding to give a 'net *ortho* effect', and the greater solvation of the transition state derived from the primary amine than of that derived from pyridine.

217

Wherever data are available, they show reactions of the present kind to proceed according to strict second-order kinetics. Chapman and his co-workers have generally assumed that the reactions are S_N2 in character, involving a transition state of the Wheland type[700, 701, 704] (p. 41). However, there is strong evidence that in nucleophilic aromatic substitutions the Wheland structure can in fact represent a relatively stable intermediate complex[705]. In certain cases, such as the reaction of 2,4-dinitrofluorobenzene with N-methylaniline, general base catalysis obtains; in such cases, the halogen atom must be lost in a rate-controlling step following the formation of the intermediate complex. The chloro- and bromo-compounds do not show the effect, and with these the formation of the intermediate complex is presumably rate-determining. The reaction of 2-chloro-3-cyano-5-nitro-pyridine with aniline is not catalysed by acetate ions. Thus, whilst the possibility exists of the Wheland structures representing intermediate complexes, and not transition states, in reactions of the halogenopyridines with amines, the evidence at present available is purely analogical. 2-Chloronitropyridines in methanol give yellow colours when mixed with anilines, a phenomenon attributed to charge-transfer complex formation. There is no evidence that these complexes have any significance for the substitution processes[701].

Amination has been effected by replacement of hydroxyl groups and of etherified or esterified hydroxyl groups. Thus, 4-hydroxy-[276] and 4-hydroxy-3-methyl-pyridine[200] give the amino compounds when heated with aqueous ammonia, in the second case with copper sulphate present. 4-Alkoxy-3-nitropyridines react similarly but more readily[180, 692]. Surprisingly, 3-ethoxy-2-nitropyridine, with aqueous ammonia, gives[686a] 3-amino-2-nitro-(65–70 per cent) and 2-amino-3-ethoxy-pyridine (9 per cent). Of more practical importance are aminations of 4-aryloxypyridines. The aryloxy-pyridine is not necessarily isolated, as in the preparation of 4-aminopyridines by heating 4-pyridylpyridinium chloride hydrochloride with phenol in presence of ammonia or an amine[252, 434, 438]. Better yields result if 4-phenoxy-pyridine is aminated in presence of pyridine hydrochloride, or by being heated with the amine hydrochloride[438, 706]. Presumably this is another case of acid catalysis of nucleophilic substitution (p. 215). Semi-quantitative data[439] show that the reactions of 2- and 4-phenoxypyridine with aniline hydrochloride are much faster than those with aniline, 4-phenoxypyridine being somewhat more reactive than 2-phenoxypyridine. 3-Phenoxypyridine reacts hardly at all.

In the reaction of aryloxypyridines with amines, the aryl group is in competition with the pyridine ring for the nucleophile. In fact, 3- and 4-(4'-nitrophenoxy)pyridine in reaction with piperidine give 3- and 4-hydroxy-pyridine and N-(4'-nitrophenyl)piperidine. Dinitrophenyl ethers behave similarly[707]. It would be interesting to know quantitatively the relative success of the pyridyl and aryl groups in these competitive reactions.

4-(Aminophenoxy)pyridine salts can be caused to rearrange to 4-(hydroxy-phenylamino)pyridines. When 4-(2-aminophenoxy)pyridine dihydrochloride is heated with aniline hydrochloride, only 4-(2-hydroxyphenylamino)pyri-dine results, but 4-(3-aminophenoxy)-and 4-(4-aminophenoxy)-pyridine salts give both the corresponding 4-(hydroxyphenylamino)pyridines and

4-anilinopyridine. The second two reactions occur intermolecularly, the first intramolecularly[678h] through the intermediate (100).

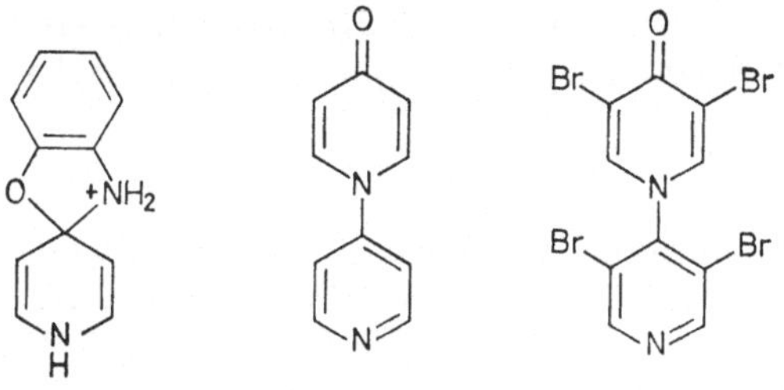

(100)   (101)   (102)

As expected, 4-aryloxy groups in pyridine oxides can be replaced by amines[708]. 4-Alkoxy[709a] and 4-aryloxy-groups[706] in quaternary pyridinium salts are readily replaced. In contrast to 4-phenoxypyridine, 4-phenoxy-pyridine hydrochloride does not react with arylamine hydrochlorides, presumably because of the lowering of the nucleophilic power of the amine by salt formation[706].

As well as alkoxy- and aryloxy-groups, the corresponding sulphur functions have been replaced in the amination of pyridine 1-oxides[708] and quaternary pyridinium salts[706, 709a]. Examples with pyridines are rare[710]. An interesting intramolecular amination occurs in the transformation of 2-pyridyl-2'-(3'-aminopyridyl)thioether into 3-(2'-pyridylamino)pyridine-2-thiol[709b]. 1-Alkylpyrid-2-thiones, warmed with alcoholic ammonia and mercuric oxide, give 1-alkylpyrid-2-one-imines[711].

4-Acetoxy- and 4-p-toluenesulphonyloxy-pyridine both give 1-(4-pyridyl)-pyrid-4-one (101) when heated, reactions illustrating the replacement of esterified 4-hydroxyl groups[429-30].

Several examples are known of amination by replacement of the nitro-group, but the reactions are of little practical value. Piperidine and benzyl-amine react as expected with 2-nitropyridine[712], and the former amine also with 4-nitropyridine[713a]. However, with aqueous ammonia, 4-nitropyridine is said to give 4-hydroxypyridine[713a] whilst 2-nitropyridine does not react[713b]. Hydrazine merely reduces 2-nitro- to 2-amino-pyridine[712]. The nitro group appears to be a relatively poor leaving group by comparison with other substituents, as was seen from examples quoted in connection with amination by replacement of halogen (p. 213) and of the ethoxyl group (p. 218). However, data for direct quantitative comparisons are lacking.

Amination by replacement of the nitro group in 4-nitropyridine 1-oxide has proved surprisingly unsuccessful[690a, 714]. With ammonia, the main product seems to be 4,4'-azopyridine 1,1'-dioxide. Thus, although 4-nitro-pyridine 1-oxides are important sources of a range of 4-aminopyridine 1-oxides and 4-aminopyridines, this is so only because the 4-nitropyridine 1-oxides can be efficiently converted into 4-halogenopyridine 1-oxides (p. 233) and thence into the amines.

In contrast to the nitro compounds, a number of pyridine-2- and -4-sulphonic acids have been found to react readily and efficiently with both ammonia and amines[159b, 276, 282, 290a, 409, 710, 715]. Pyridine-3-sulphonic acid does not react with aqueous ammonia at 180°, a fact which is the basis of a method for analysing mixtures of pyridine–sulphonic acids[276, 282, 290a]. The formation of (102) when 3,5-dibromopyridine-4-sulphonic acid is heated in water[189b] recalls the auto-quaternization of 4-halogenopyridines (p. 184).

(d) Arylation

Phenyl magnesium bromide gave with pyridine a white complex which, when autoclaved at 150–160°, yielded[716] 2-phenylpyridine (44 per cent). The reactions of pyridine with Grignard reagents (see Alkylation, p. 200) have been represented[109a] as

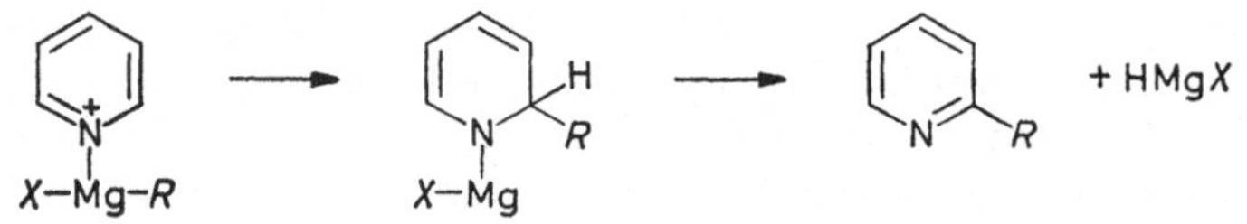

The difficulty of the elimination step is thought to account for the poor yield of 2-substituted pyridine.

3-Benzoylpyridine and phenyl magnesium bromide give 3-benzoyl-4-phenyl-1,4-dihydropyridine[709c].

Little is known of the reactions of aryl Grignard reagents with pyridine 1-oxide. Phenyl magnesium bromide produces 2-phenylpyridine[717], but evidently in poor yield[597]. Benzylpyridinium chloride did not react with phenyl magnesium bromide[718a], but benzoylpyridinium chloride gave 4-phenylpyridine (16 per cent) and a compound (4 per cent) thought[602] to be (103).

The direct 4-alkylation of pyridine by alkyl halides in the presence of magnesium was mentioned above (p. 202). Chlorobenzene takes part in the reaction, giving 4-phenylpyridine[605].

Of far greater importance in the pyridine series are reactions with lithium aryls. As in alkylation the initial product is the addition compound, which can be used as a reducing agent[718b]. Originally Ziegler and Zeiser[604a] eliminated lithium hydride from this by heating it at 100°, thus producing by the use of phenyl lithium, 2-phenylpyridine in 60 per cent yield. Since the original observation, three methods of obtaining the arylpyridine from the initial adduct have been employed. These are: elimination of lithium hydride by use of a high boiling solvent such as toluene or xylene, oxidation by the deliberate passage of air through the reaction mixture or adventitiously during working up, and oxidation by nitrobenzene. *Table 5.25* summarizes most of the known examples which illustrate these techniques.

Two features of the reaction of pyridine derivatives with lithium aryls need separate mention. First, when alkyl substituents are present in the pyridine nucleus, metalation of the alkyl group is in competition with nuclear arylation (cf. p. 327). Ziegler and Zeiser[604b] found that phenyl lithium in ether converted 2-picoline into picolyl lithium. Sometimes, attempts to metalate 2-alkylpyridines have been unsatisfactory because of competing arylation[729–30]. 4-Picoline is both arylated and metalated by phenyl lithium[731].

Table 5.25

Substituents in pyridine derivative	Lithium reagent	Method*	Substituents in product (per cent)	Refs.
—	PhLi	A	2-phenyl (60)	604
—		B	(40–49)	719
—		C	(60)	720
—		D	(40–57)	513†
—		B	2-*p*-tolyl (47)	721
—	*p*-Me.C₅H₄Li	B	(61)	722
—		D	(62)	513
—	*o*-MeO.C₆H₄Li	C	2-*o*-anisyl (?)	720
—		D	(18)	513
—	*p*-MeO.C₆H₄Li	E	2-*p*-anisyl (50)	723
—		D	(35)	513
—	1-C₁₀H₇Li	C	2-(1-naphthyl) (34)	724
—	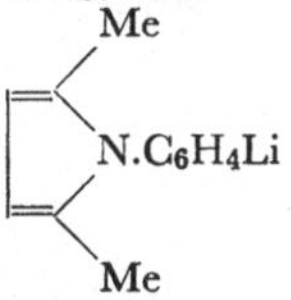 (Me / N.C₆H₄Li / Me)	D	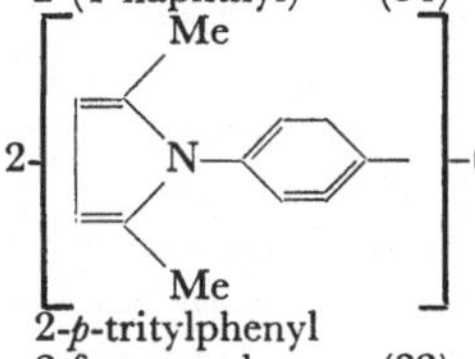 2-[Me/N—⟨⟩—/Me]-(24)	725
—	*p*-Ph₃C.C₆H₄Li	B	2-*p*-tritylphenyl	726a
—	Li ferrocenyl	G	2-ferrocenyl (32)	726b
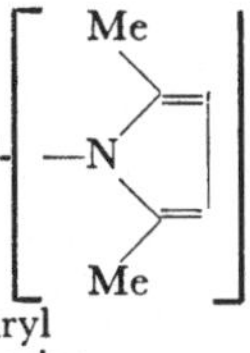 2-[Me/—N/Me]	*p*-Me₂N.C₆H₄Li	D	2-*p*-dimethylaminophenyl-6-(2,5-dimethylpyrryl) (17)	725
2-Aryl	ArLi	D	2,6-diaryl (poor yields)	513
3-Amino	PhLi	B	3-Amino-2-phenyl only	}732
3-Methoxy		B	3-methoxy-2-phenyl only	
3-Ph		F	2,5-diphenyl	728a
3-Me		F	5-methyl-2-phenyl (30)	727‡
			Ratio of 2,3, to 3,6-disubstituted pyridines§	
3-Me	PhLi	B	94·6 : 5·4	733a, 734
3-Et			84·0 : 16·0	
3-iPr			70·0 : 30·0	}733a
3-tBu			4·5 : 95·5	
(Nicotine)			49·6 : 50·4	733a, 734

* A Heating of initial adduct; B reaction started in ether and completed in boiling toluene; C oxidation of dihydro compound with nitrobenzene; D oxidation of dihydro compound with dry air; E as B but using xylene; F reaction in boiling ether; G as B but using cyclohexane.
† This reference describes experiments in which methods B and D are compared.
‡ Repetition of this work[784] gave no arylpyridines.
§ The arylpyridines were formed in total yields of 20–40 per cent. In a comparison of reactions worked up by method B with the same processes worked up by oxidation with oxygen[785b], overall yields were better from the oxidation method, but isomer ratios were substantially unchanged.

The metalation of the 4-methyl group is faster than the nuclear arylation and can be made to predominate[606]. 3-Picoline is arylated rather than metalated[727].

Secondly, there is the problem of orientation. In a reaction between pyridine and phenyl lithium which produced 69 per cent of 2-phenylpyridine, no trace of 4-phenylpyridine was formed[733a]. A 3-substituent usually directs arylation preferentially to the 2-position (*Table 5.25*); only the steric factor introduced by a very bulky 3-substituent (t-butyl) can swing the ratio in favour of 6-substitution. This situation is not easily understood, for a 3-alkyl group would be expected to deactivate the 2-position by virtue of its $+I$ effect as well as by steric influence. With a substituent such as a 3-amino- or a 3-methoxy-group, 2-substitution may be assisted by bonding between the substituent and the reagent (104). The situation generally resembles that noted in electrophilic substitution, for example, nitration (p. 173). By competitive reactions it has been shown that the order of reactivity to phenyl lithium is 3-picoline > pyridine > 3-ethylpyridine, and that a methyl or ethyl group at $C_{(3)}$ activates the 2-position but deactivates the 6-position (relative to pyridine). It has been suggested that even an alkyl group at $C_{(3)}$ can form some sort of loose complex with the reagent analogous to (104), and because of favourable orientation this lowers[733b] the activation energy of substitution at $C_{(2)}$.

The isomer proportions formed in the phenylation of 3-alkylpyridines (*Table 5.25*) do not depend appreciably upon the reaction conditions, and experiments with pyridine-2D and 3-picoline-2D showed that the equilibrium between the adduct and starting material was effectively irreversible or lay far over on the product side. Further, the stage of hydride ion elimination from the adduct was of no importance in determining the isomer ratio. Thus, no satisfactory explanation of the details of orientation has yet been arrived at[728b].

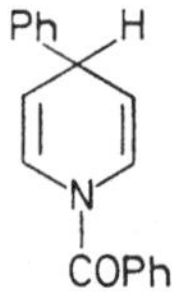

(103) 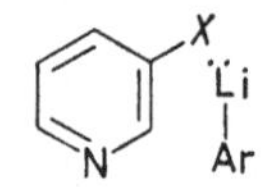   (104) 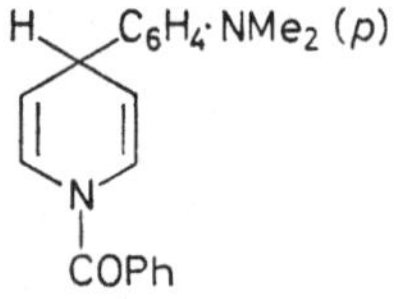   (105)

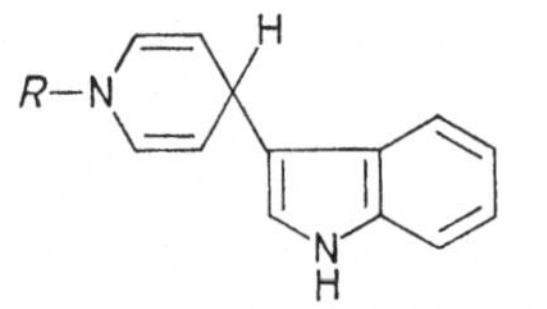

(106; R = PhCO, MeCO or $C_7H_7SO_2$)

The arylation of pyridine 1-oxides or quaternary pyridinium salts by lithium aryls does not appear to have been observed. Phenyl lithium is said to react vigorously with pyridine 1-oxide, but only tars were formed[597].

Reagents containing metals other than magnesium or lithium have hardly been used at all for arylation. An interesting study with phenyl calcium iodide gave a surprising result: with pyridine in tetrahydrofuran at $-20°$, rising to room temperature in 5 h, 2-phenyl- (42 per cent) and 2,6-diphenyl-pyridine (10 per cent) resulted, but at $-60°$, rising to room temperature in 40 h, besides 2-phenylpyridine (4 per cent) there was produced 2,5-diphenyl-pyridine (6 per cent). It appears that the anomalous 2,5-diphenylpyridine was formed under circumstances where hydride elimination was slow, and its formation has been attributed to further attack upon the initial product, before hydride elimination had occurred[735]. In support of this view, 2-phenylpyridine with phenyl calcium iodide gave only 2,6-diphenylpyridine.

Highly nucleophilic aromatic compounds are capable of arylating acyl-pyridinium salts. The first example of this striking reaction was described by Koenigs and Ruppelt[265] who observed the formation of 4-(p-dimethyl-aminophenyl)pyridine from pyridine, benzoyl chloride and dimethyl-aniline in the presence of copper. Benzaldehyde is also formed[265, 736], and the copper is not necessary[736]. The dihydropyridine (105) is probably an intermediate. Other examples of the reaction are known[265, 493], but attempts to isolate the intermediates have failed[736], though that from dimethyl-m-toluidine may have been obtained[265]. In contrast, the dihydropyridines (106) were isolated when indole was the nucleophile. Skatole reacted similarly, at the 2-position of the indole nucleus, giving the fully aromatic 3-methyl-2-(4'-pyridyl)indole. These reactions failed with 2- and 4-picoline[640]. Similar reactions occur between acylpyridinium salts and pyrroles (p. 71).

The acylating power of acylpyridinium salts has been mentioned already (p. 195). It is probable that in the formation of phenol esters, arylation of the 2-position in the pyridinium salt is an essential step:

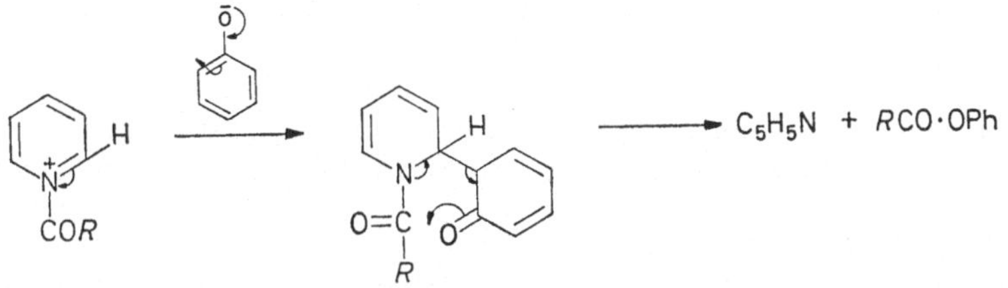

The remarks made about the acylation of 1,3-dicarbonyl compounds (p. 205) are relevant here. No intermediate has been isolated from reactions of this kind, but it is interesting that cinnamoyl–pyridinium chloride gives with phenol a bright yellow solution. The colour fades slowly, and phenyl cinnamate (80 per cent) results[493]. The acylation of aromatic amines may proceed similarly[737].

Pyridine 1-oxide is not arylated by dimethylaniline[666].

The reaction between magnesium or lithium reagents and halogeno-pyridines generally results in halogen–metal exchange, not in C—C bond formation (see Metalation, p. 170). 2-Chloropyridine does not react with phenyl lithium, but 2-bromopyridine, whilst reacting mainly by halogen–metal exchange, gave a low yield of 2-phenylpyridine[513].

Intramolecular arylations involving nucleophilic replacement of 2-halo-gen atoms are known. When a 2-pyridine (107) reacts with phosphoryl chloride, two results are possible: the quaternary salt (108) may be formed or, in appropriate circumstances, the cyclized compound (109). Early reports[738]

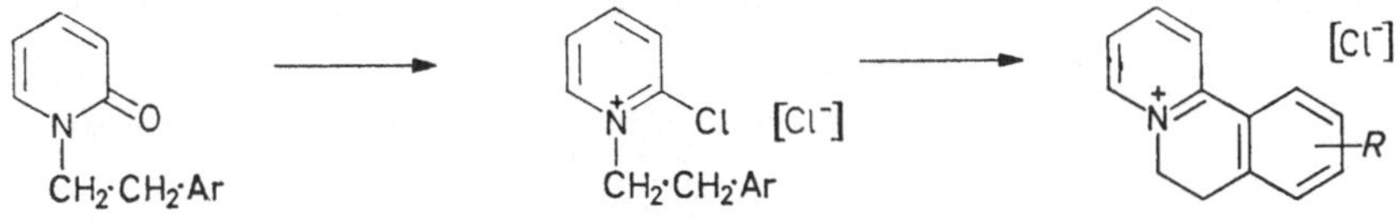

(107) (108) (109)

that this reaction with derivatives of (107) in which Ar=Ph produced the tricyclic products were mistaken[313a, 739-42], and even reactions in which Ar in (107) is activated by the presence of alkoxyl groups stop[740-3] at (108). Appropriate activation of both the benzene and pyridine rings appears to be essential, as in the examples[739]

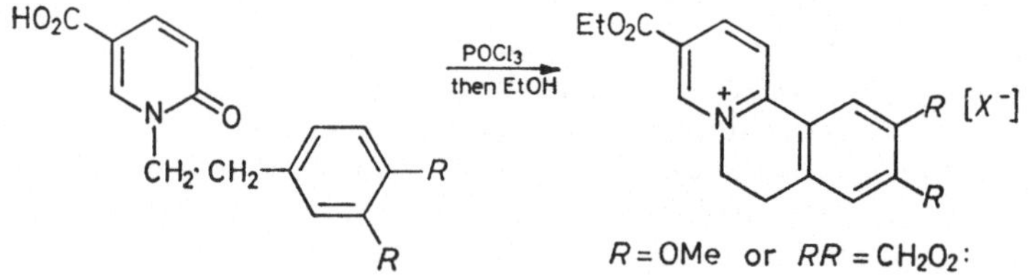

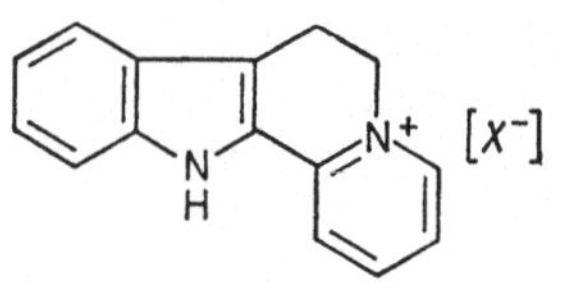

$R = OMe$ or $RR = CH_2O_2$:

(110)

In other cases cyclization occurs more readily, and when 2-halogenopyridines react with 3-(2'-bromoethyl)indole, the cyclized salt, e.g. (110), results[744].

(e) Modified Carbonyl Reactions

Carbonyl compounds are too weakly electrophilic to effect substitution of the pyridine ring. However, in the Emmert–Asendorf reaction (an extension of which has already been discussed, p. 200), ketones and, to a smaller extent, aldehydes have been used to hydroxyalkylate pyridine. The reaction has been classified as a nucleophilic substitution[745], and for that reason is discussed here.

Emmert and Asendorf[746] showed that when ketones were heated with pyridine and amalgamated magnesium or aluminium, 2- and 4-hydroxy-alkylpyridines resulted. The yields from the reaction are moderate to good, and it has been used preparatively (the literature is summarized[745]).

Bachman et al.[745] argued that the nature of the products indicated the reaction to be a nucleophilic substitution, and that the essential property in the reducing agent is the ability to transfer two electrons at a time to the carbonyl compound. This, combined with a minimum reduction potential and coordinating power, makes magnesium and aluminium particularly

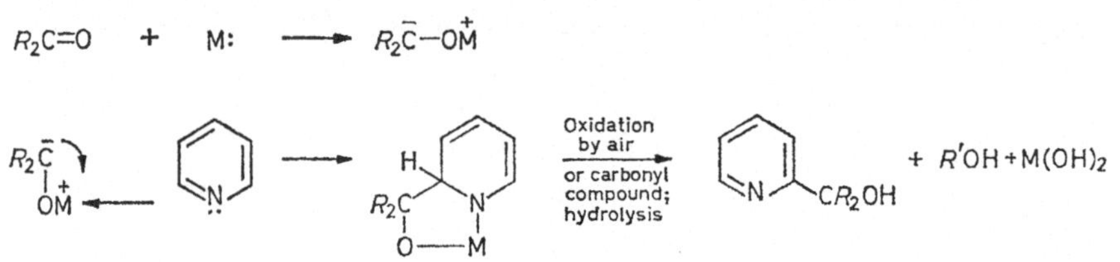

224

effective. Dialkyl, diaryl, alkylaryl and cycloalkyl ketones can be used, and substituents in aromatic ketones have the influence expected from the postulated mechanism. 3- and 4-Picoline react satisfactorily, but with 2-alkylpyridines yields are poor, and with 2,6-dialkylpyridines the reaction fails. 2-Substitution presumably interferes with metal–nitrogen coordination. With aluminium, the ratio of 2- to 4-substitution is about 4:1, whilst magnesium gives only the 2-isomer. The proposed mechanism as illustrated does not make clear how 4-substitution is effected. The reaction strongly recalls the pinacol reduction and has been formulated[109a] as a radical reaction.

(f) Cyanide Formation

Formation of cyanides by replacement of hydrogen does not occur with pyridine, with pyridine 1-oxide[597] or 4-cyanopyridine 1-oxide[747], but pyridinium quaternary salts are sufficiently electrophilic to react. Thus, 3,5-diethoxycarbonyl-1,2,4,6-tetramethylpyridinium methosulphate reacts with potassium cyanide giving 4(2?)-cyano-3,5-diethoxycarbonyl-1,2,4,6-tetramethyl-1,4(2?)-dihydropyridine. Various reactions of this kind have been studied in connection with the properties of diphosphopyridine nucleotide[748]. The cyano-dihydropyridines were isolated in several instances, and the addition process is reversible. The evidence seems to favour attack at $C_{(4)}$ but is not conclusive. Successful reactions have always been with quaternary salts containing electron-attracting substituents, usually at the 3-position. 1-Methyl- and 1,3-dimethyl-pyridinium salts do not react with potassium cyanide either in water or alcohol[748e].

1-Benzoylpyridinium chloride does not form a Reissert compound (p. 226) with potassium cyanide[749].

Of practical importance are the reactions of the quaternary salts of pyridine 1-oxides with aqueous potassium cyanide. The 1-methoxypyridinium methosulphates have usually been used, and with aqueous potassium cyanide, 2- and 4-cyanopyridine result (Table 5.26), the 2-position being somewhat favoured. A 3-substituent usually, but not always, causes 2-substitution to predominate over 6-substitution. The quaternary function is eliminated as an alcohol; its presence probably determines the preference for 2- over 4-substitution, and its ready elimination makes the reaction a practicable source of cyanopyridines, in contrast to the case of other quaternary pyridinium salts (see above) with which reaction stops at the addition stage. These reactions are nucleophilic replacements not of hydride but of alkoxyl:

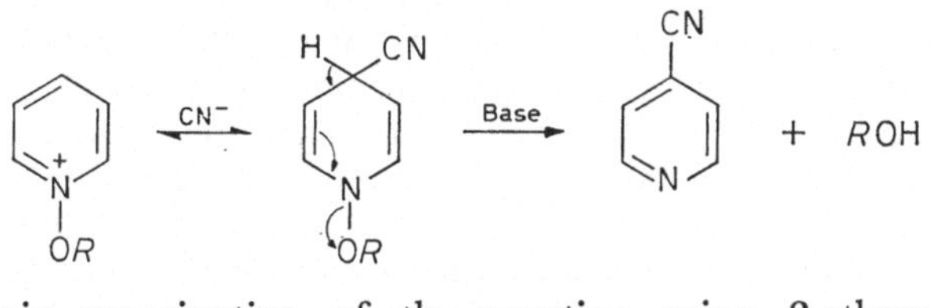

Spectroscopic examination of the reaction using 3-ethoxycarbonyl-1-methoxypyridinium methosulphate suggests that the very rapid initial addition is followed by the slow second step[751c]. Higher temperatures and increase in solvent polarity increase the proportion of 4-substitution which

Table 5.26. Formation of Cyanopyridines from 1-Methoxypyridinium Methosulphates

Substituent in the quaternary salt	Substituents in the pyridine derivative produced (*per cent yield*)	*Ref.*
—	2-CN (49) + 4-CN (32)	747
—	2-CN (50) + 4-CN (25)	750
2-Me	2-CN-6-Me (48) + 4-CN-2-Me (10)	747
2-Me	2-CN-6-Me (45) + 4-CN-2-ME (18)	750
3-Me	2-CN-3-Me (36) + 2-CN-5-Me (6) + 4-CN-3-Me (6)	747, 751*a*
3-Me	2-CN-3-Me (30) + 4-CN-3-Me (15)	750
4-Me	2-CN-4-Me (40)	730, 747
4-Me	2-CN-4-Me (28)	750
4-MeOCH$_2$	2-CN-4-MeOCH$_2$ (51)	752
2,4-Me$_2$	2-CN-4,6-Me$_2$ (73)	}747
2,6-Me$_2$	4-CN-2,6-Me$_2$ (40)	
2,6-Me$_2$	4-CN-2,6-Me$_2$ (13) + 2-CN-6-CH$_2$CN (33)	750
3,5-Br$_2$	3,5-Br$_2$-2-CN (70)	751*a*
2-Cl	2-Cl-6-CN	}751*b*
4-Cl	4-Cl-2-CN + corresponding amide	
2-CN	2,6-(CN)$_2$	751*b*
3-CN	2,3-(CN)$_2$ (27·8) + 2,5-(CN)$_2$ (17·6)	751*a*
4-CN	2,4-(CN)$_2$ (54)	747,751*b*
3-EtO$_2$C	2-CN-5-EtO$_2$C (19) + 4-CN-3-EtO$_2$C (31·6)	751*a, c*
2-MeO	2-CN-6-MeO + 2,6-(CN)$_2$	751*b*
3-MeO	2-CN-3-MeO (67·8)	751*a*
4-MeO	2,4-(CN)$_2$	
2-MeO$_2$C	2-CN-6-MeO$_2$C	
4-MeO$_2$C	2-CN-4-MeO$_2$C	}751*b*
4-NO$_2$	2-CN-4-NO$_2$ + corresponding amide	
4-NMe$_2$	}reaction failed	
2-NH.CO$_2$Et		

occurs. The reaction takes place above pH 8, but pH 11 is optimum. Change of the quaternizing group from methoxy to ethoxy and butoxy changed the 4- to 2-substitution ratio from 0·23 to 2·84 and 2·17. A 2- or 4-methoxyl group can also be replaced by the cyano group under the conditions of this reaction (*Table 5.26*).

Little success has attended attempts to form Reissert compounds from pyridine 1-oxides[753], but 3-nitropyridine 1-oxide does react with benzoyl chloride and silver cyanide to give a small amount of a cyano-3-nitropyridine, a reaction recalling that of 1-methoxypyridinium salts described above[754].

A few 3-cyanopyridines have been prepared, in 40–50 per cent yield, by reaction of a 3-pyridyl diazonium salt with potassium cuprocyanide or cuprous cyanide[218, 755-7]. Though 2- and 4-aminopyridines have been converted into 2- and 4-halogenopyridines via the diazonium reaction, the ready reaction of the 2- and 4-pyridyldiazonium ions with nucleophiles present in the solutions in which they are formed makes their use in cyanation difficult if not impossible. The difference in ease of diazotization between a 2- and a 3-aminopyridine permits the conversion of 2,5-diaminopyridine into 2-amino-5-cyanopyridine by diazotization in dilute acid[218]. The easier diazotization (p. 359) of 4-aminopyridine 1-oxides and the greater stability of the resulting diazonium salts[531] (p. 361) make these compounds useful sources of 4-cyanopyridine 1-oxides[531, 758].

Cyanopyridines can be prepared by heating halogenopyridines with cuprous cyanide. When the reactants are warmed carefully together, a vigorous reaction sets in, and the product is removed immediately by vacuum distillation. The reaction can be erratic, but careful adherence to the conditions worked out for a particular compound usually gives moderate to good yields. 2-Bromopyridine[759] and some of its homologues[649e, 730, 760] and derivatives[230, 760] have been used satisfactorily, as have 3-bromopyridine[254, 761], 2-amino-5-bromo- and 2-amino-5-iodo-pyridine[174]. The reaction might be improved if it could be effected throughout in a solvent. However, an attempt to do this, using potassium cuprocyanide in water at 175°, converted 2-bromopyridine into 2-picolinamide in poor yield[762]. The same reagent, in aqueous alcohol, was used with 4-bromo-5-ethyl-2-methylpyridine, but the product was converted into ethyl 5-ethyl-2-methylisonicotinate without itself being characterized[763]. The reactions of 3-hydroxy-2-iodo- and 4-hydroxy-3-iodo-pyridine with cuprous cyanide have been effected in xylene, toluene or isoamyl alcohol[764].

The conclusion[765] that in this type of cyanation the 3-bromo substituent is more readily replaced than the 2- or 4-bromo substituent, is unjustified and seems to be based on the erratic quality of the reaction mentioned above. Under the circumstances, the more reactive 2- and 4-bromopyridines could be converted into other products (e.g. by autoquaternization), and the yields of cyanopyridines recorded can have little relation to intrinsic reactivities.

Cyanation by replacement of a hydroxyl group has not been reported, but examples have been noted (p. 226) of the replacement of 2- and 4-methoxy groups in 1-methoxypyridinium salts.

Sodium or potassium salts of several pyridine-3-sulphonic acids have given poor to moderate yields of nitriles by fusion with sodium or potassium cyanide[174, 268a, 278, 280, 766-7]. A few pyridine-2- and -4-sulphonic acids have been similarly used[768, 770], as has pyridine-4-sulphonic acid 1-oxide[771].

(g) Halogenation

Nucleophilic replacement of hydrogen is presumably being observed when reduction of a 3-nitropyridine to a 3-aminopyridine by stannous chloride and concentrated hydrochloric acid is accompanied by nuclear chlorination[180, 241a, 772-4]. Thus, 4-amino-3-nitropyridine gives 3,4-diamino-2-chloropyridine[180a] and 4-ethoxy-3-nitropyridine yields 5-amino-2-chloro-4-ethoxypyridine[241a]. Clearly, the pyridinium cations must be involved.

When picolinic acid is heated with thionyl chloride, 4-chloro- or 4,6-dichloro-, and even some 4,5,6-trichloro-picolinic acid result, according to the severity of the conditions[675a, 775]. Isonicotinic acid gives 2-chloro-isonicotinic acid[776]. Quaternary complexes are probably involved in these reactions:

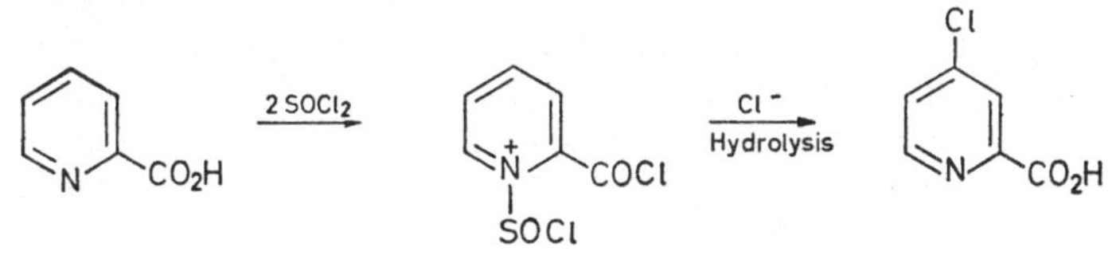

227

The reaction using picolinic acid is catalysed by sulphur dioxide; the effect may be produced either because ionization of the thionyl chloride is assisted or because of complexing of the sulphur dioxide with the pyridine nitrogen atom[777].

The formation of 5-chloro- and 5,6-dichloro-nicotinic acid from nicotinic acid and thionyl chloride[675a, 778] is less easily understood. The orientation is surprising and has been taken to indicate electrophilic substitution[109]. The reaction recalls the formation of 5-bromonicotinic acid in the bromination of nicotinic acid hydrochloride[168] (p. 168). Were it not for the difficulty of discerning the effective electrophile, and also of explaining why nicotinic acid should behave differently from picolinic and isonicotinic acids, a mechanism involving initial nucleophilic addition followed by electrophilic substitution might be considered.

Such a sequence accounts for the convenient reaction in which pyridine is converted into 3,5-dibromopyridine (20–28 per cent) and some 3-bromopyridine when it reacts with bromine and sulphur monochloride or thionyl chloride[779]:

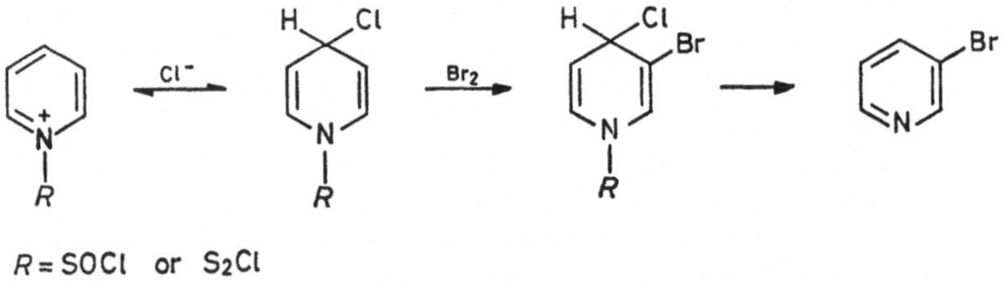

$$R = SOCl \text{ or } S_2Cl$$

The possibility that nuclear chlorination is a step in the formation of 4-pyridylpyridinium chloride from pyridine and thionyl chloride has been mentioned (p. 186).

Bobránski[510] showed that pyridine 1-oxide hydrochloride reacted at 120° with sulphuryl chloride, giving 2- and 4-chloropyridine (the former predominating) and a trace of pentachloropyridine. The reaction has been repeated[780] and applied to substituted pyridine 1-oxides[781-2, 911]. 3-Bromopyridine 1-oxide gave 3-bromo-4-chloro-, 3-bromo-2-chloro- and 3-bromo-6-chloro-pyridine, in that order of yields, but overall α-substitution outweighed γ-substitution[781]. The reaction with 3,5-diethoxypyridine 1-oxide seems to be the only one reported in which chlorination was not accompanied by loss of the oxide function[782].

Phosphorus chlorides have been used in place of sulphuryl chloride[251, 537, 581, 768b, 780, 783-9]. The evidence is insufficient to decide the relative reactivities of the 2- and 4-positions, though from pyridine 1-oxide only 4-chloropyridine was isolated[780]. Whilst 3-methylpyridine 1-oxide gave mainly 4-chloro-3-methylpyridine[768b], nicotinic acid 1-oxide yielded chiefly 2-chloronicotinic acid[251]. 4-Methoxy- and -nitro-groups can be replaced in the reaction[537], and chlorination of a 2-methyl group has been observed[785]. These reactions of sulphuryl and phosphorus chlorides are usually regarded as involving formation of a quaternary complex, followed by nucleophilic substitution by halogen anion[780].

On this basis, other halides might be expected to effect nucleophilic halogenation; in fact, benzoyl chloride converts 4-aminopyridine 1-oxide into 4-benzamido-2-chloropyridine[758]. p-Toluenesulphonyl chloride behaves differently (p. 235).

Replacement of amino groups by halogen atoms does not seem to have been observed, except in cases where the amino group was quaternary; 4-pyridylpyridinium chloride hydrochlorides react with phosphorus pentahalides to give 4-halogenopyridines[438].

The difficulty of diazotizing 2- and 4-aminopyridines, and the great reactivity of the diazonium ions once they are formed (p. 226), does not prevent the use of these amines as sources of halogenopyridines. Indeed, halogenation is one of the reactions which can easily be applied to diazotized 2- and 4-aminopyridines because of the high concentrations of halogen anions which can be introduced into the reaction solution from the very beginning.

2- and 4-Aminopyridine, and many of their derivatives, have been converted into chloro compounds by diazotization in concentrated hydrochloric acid[169a, 170, 171a, 172, 181, 211, 216, 230, 531, 649b, 670, 775b, 790-9]. In a few cases the Sandmeyer method was used[775a, 799]. Yields have varied from very good to very poor, and the hydroxypyridines have occasionally been isolated as by-products. Apart from minor variations, the method has not changed much from that originally used by Marckwald[669b].

2- and 4-Bromopyridines have been produced similarly by diazotization in hydrobromic acid[142a, 171c, 790, 797, 799-801] or by the Sandmeyer process[798, 802]. In the former case, there is evidence[799] that the use of low temperatures (− 15°) greatly increases the yield. An important variation was introduced by Craig[759] who obtained 2-bromopyridine by diazotizing 2-aminopyridine in saturated aqueous hydrobromic acid containing bromine. Over 80 per cent of 2-bromopyridine was obtained[759, 803], but the yield fell off when less than two molecular equivalents of bromine were used. The yield also decreased if the temperature rose above 10°, and some 2,5-dibromopyridine was formed[364]. Craig's method has been used with a number of substituted 2-aminopyridines[171d, 256, 649e, 760] and is usually superior to the earlier method.

2- and 4-Iodopyridines have been prepared by diazotizing the amines in hydriodic acid[797, 800, 802], or by diazotization in sulphuric acid followed by treatment with potassium iodide[171d, 174, 181, 799, 800, 804]. Yields are usually poor.

For preparing 2- and 4-fluoropyridines diazotization of the amines has been effected in aqueous[142a, 424, 790, 805] or anhydrous hydrofluoric acid[113, 806-7]. Alternatively, the Schiemann reaction has been used[805, 807-9] or the diazotization has been carried out in fluorosilicic acid[805, 807]. It is impossible to state which method is generally to be preferred. Yields are usually poor; the best of 2-fluoropyridine (42 per cent) was obtained by using fluorosilicic acid[807]. 4-Fluoropyridine was obtained in poor yield by diazotizing 4-aminopyridine in hydrofluoric acid[424].

Two other modifications of the preparation of 2- and 4-halogenopyridines from amines have been used. 2-Aminopyridines, when treated with sodium ethoxide and amyl nitrite, give diazotates which have been converted by reaction with acids into 2-chloro-, -bromo- and -iodo-pyridines[171a, 790, 810]. 4-Nitraminopyridine, with sodium nitrite in concentrated hydrochloric acid, gave 4-chloropyridine[180a] (heated under pressure with hydrochloric acid, the nitramine gave 4-chloro- and 4-amino-3,5-dichloro-pyridine[680b]), and in hydrobromic acid it gave 4-bromopyridine[673b].

3-Aminopyridines are not only more easily diazotized than the 2- and 4-isomers, but the lower reactivity of the diazonium ions makes the application of the more usual methods for converting these into halogen compounds more satisfactory. Thus, whilst 3-chloropyridines have been prepared by simply diazotizing the amines in concentrated hydrochloric acid[772, 796, 811], both chloro- and bromo-compounds have usually been obtained by the Gattermann[226, 230, 680b, 755, 757, 781, 796] or Sandmeyer procedures[211, 241c, 675a, c, 755, 778, 812]. 3-Iodopyridines can be prepared by diazotizing the amine in hydriodic acid or by diazotizing in some other acid and adding potassium iodide[755, 772, 778, 804, 812-5]. 3-Fluoropyridines have been prepared by diazotization in hydrofluoric acid[755] or fluorosilicic acid[808], but chiefly by some form of the Schiemann reaction[794, 807-9, 816-7]. The best yield of 3-fluoropyridine (50 per cent) was obtained by the use of hydrofluoroboric acid[808], whilst 3-chloro- (65 per cent) and 3-bromo-pyridine (56 per cent) are best prepared by the Gattermann reaction[755]. By diazotizing 3-aminopyridine in dilute hydrochloric acid and treating the solution with potassium iodide, 3-iodopyridine is obtained in 50 per cent yield[755, 804, 813]. The difference in ease of diazotization between α- and β-amino groups makes it possible to convert 2,5-diaminopyridines into 2-amino-5-halogenopyridines[772], though 2,5-diaminopyridine has been converted into 2,5-dichloropyridine by means of the Sandmeyer reaction[675a].

4-Aminopyridine 1-oxide has been diazotized and, by use of the Sandmeyer reaction, converted into 4-chloro- and 4-bromo-pyridine 1-oxide[758]. 4-Iodopyridine 1-oxides have also been prepared through the diazonium reaction[818].

Nucleophilic halogen exchange has been observed frequently with 2- and 4-halogenopyridines. The commonest reaction of this type is that of a 2- or 4-chloropyridine with hydriodic acid[181, 401, 675a, 775, 815, 819-20]. Reductive removal of the halogen substituent sometimes accompanies the use of hydriodic acid[775, 819]. 2-Bromine atoms have similarly been replaced by iodine atoms[821] and by chlorine atoms[230, 795]. In all these reactions halogen–halogen exchange will clearly be facilitated by protonation of the nitrogen atom. When this activation is absent, replacement is more difficult, and 2-bromopyridine does not react with sodium iodide in acetone[821]. 6-Chloronicotinic acid reacts very slowly with the same reagents, but with sodium iodide in methyl ethyl ketone it gives 6-iodonicotinic acid quantitatively[822].

2-Chloro- and 2-bromo-pyridine, and the hydrochloride of the former and the oxide of the latter, do not react with potassium fluoride in dimethylformamide. However, 3- and 5-nitro-2-chloropyridine undergo smooth halogen exchange with these reagents[823]. With potassium fluoride in dimethyl sulphone, 2-chloropyridine yields 15 per cent of 2-fluoropyridine after 430 h at 200°. 2,6-Dichloropyridine gives better yields of 2,4-difluoropyridine in a shorter time[824]. Penta- and tetra-chloropyridines react at 400°–480° with potassium fluoride to give high yields of pentafluoro-, 3-chloro-tetrafluoro- and 3,5-dichloro-trifluoro-pyridine[154].

230

Some cases of halogen–halogen exchange accompanying quaternization of halogenopyridines have been mentioned already (p. 180).

The conversion of 2- and 4-hydroxypyridines into chloropyridines by reaction with phosphorus halides is one of the most familiar and important nucleophilic replacements in pyridine chemistry. Examples are so common and failures so rare as to make complete enumeration not only difficult but unnecessary. The reaction was first described by Haitinger and Lieben[401] who heated 4-hydroxypyridine with phosphorus trichloride at 150°. Early workers used the reaction in all the forms which have been most frequently applied since: heating the hydroxy compound with phosphorus trichloride[401, 825], with phosphorus pentachloride[826] and with mixtures of this compound and phosphoryl chloride[380, 827]. The ease of the reaction varies greatly, but data do not exist to make real comparisons possible. Usually the reactants are heated together under reflux, in a sealed tube or sometimes in a solvent such as chlorobenzene[828] or xylene[170]. Reference has already been made to the fact that nuclear chlorination may accompany the replacement of hydroxyl groups by phosphorus halides (p. 165). Some advantages have been claimed for phenylphosphonic dichloride as a chlorinating reagent, but in comparison with phosphoryl chloride it is far from invariably superior[829].

3-Hydroxypyridine has been converted into 3-chloropyridine by reaction with phosphorus pentachloride in a sealed tube[830], and 3-hydroxy-2,6-dinitropyridine also reacts with phosphorus pentachloride[237], but there can be no doubt that, in general, 2- and 4-hydroxypyridines are more reactive than 3-hydroxypyridines (see below). 2-[380] and 4-Hydroxypyridine[401, 428, 441, 831] give good yields of 2- and 4-chloropyridine.

Hydroxypyridinecarboxylic acids produce chloropyridinecarboxylic acid chlorides in these reactions. Work-up usually hydrolyses these to chloropyridinecarboxylic acids[815, 832], but their presence is shown by the formation of chloropyridinecarboxylic acid amides when ammonia is used[174, 833]. Ester groups survive the reaction[159b, 834], but carboxamides are converted into nitriles[820, 835].

A few examples from the many in the literature in which alkyl-substituted 2- or 4-hydroxypyridines have been converted into alkyl chloropyridines may be quoted[281, 827, 836], but the reactions call for no comment. The same is true of aryl-substituted compounds[835, 837]. Examples in which the pyridine derivatives contained amino[156c, 838] or azo groups[146] are rare, but these groups survive the reaction. Chloro-halogenopyridines have been prepared from halogeno-2- or -4-hydroxypyridines[159b, 174, 181, 795–6, 815], and it is noteworthy that in these reactions bromine or iodine atoms at β-positions are not replaced by chlorine.

Numerous 2- or 4-hydroxynitropyridines have been used in the reaction[170, 212, 230, 241a, 457a, 679g, 786, 794, 839]. The nitro group should, of course, facilitate the substitution, but the preparative conditions used make it difficult to detect this effect. The case of 3-hydroxy-2,6-dinitropyridine mentioned above may illustrate the point.

A number of 2,4-[179, 390, 574, 796, 820, 835, 840] and 2,6-dihydroxypyridines[156b, 841] have been converted into dichloropyridines by reaction with phosphorus halides. Frequently, but not invariably, these reactions have been carried out under pressure: 3-ethyl-4-methyl-2,6-dihydroxypyridine did not react with phosphorus pentachloride and phosphoryl chloride at ordinary pressures[156b], and other failures are known[841b]. 2,4-Dihydroxy-3-nitropyridine gave, with phosphoryl chloride at 100°, a mixture of 2,4-dichloro-3-nitro- and 4-chloro-2-hydroxy-3-nitropyridine[574]. Phosphoryl chloride converted 5-β-ethoxyethyl-2,4-dihydroxypyridine into 5-β-chloroethyl-2,4-dichloropyridine[840a].

The most important modification of the reaction between phosphorus halides and 2- or 4-hydroxypyridines is that in which 1-alkyl- or 1-aryl-2- or -4-pyridones are used. The reaction is especially valuable in its application to 1-alkyl-2-pyridones because of the ready availability of these compounds. Introduced by Fischer[336, 842], it has often been applied[174, 184a, 580, 678a]. Phosphorus chlorides have usually been used, but phosgene and thionyl chloride are also effective[843]. Generally, heating 1-alkyl-2-pyridones with phosphorus chlorides has led to the isolation of 2-chloropyridines, but in some cases the intermediate 1-alkyl-2-chloropyridinium salt has been isolated (p. 223). 1-Methyl-4-pyridone also produced 4-chloro-1-methylpyridinium chloride which gave 4-chloropyridine and methyl chloride when heated[693a]. Other examples of the conversion of 1-substituted 4-pyridones into 4-chloropyridinium quaternary salts are known[977]. Examples of that of 1-aryl-2-pyridones into 2-chloropyridines are rarer but exist[425b].

So far, chlorinations have been mentioned, but by the use of phosphorus bromides, 2- and 4-hydroxypyridines[159b, 171d, 230, 673b, 801, 844], 2,4- or 2,6-dihydroxypyridines[161c, 179, 841e] and 1-alkyl-2-pyridones[184a, 331a, 336, 842, 845] have been converted into bromopyridines. With phosphoryl bromide, 3,4-dihydroxypyridine gives 4-bromo-3-hydroxypyridine[185].

As regards the mechanism of the nucleophilic replacement of hydroxyl by halogen, the illustrated scheme, showing the use of phosphoryl chloride, is reasonable:

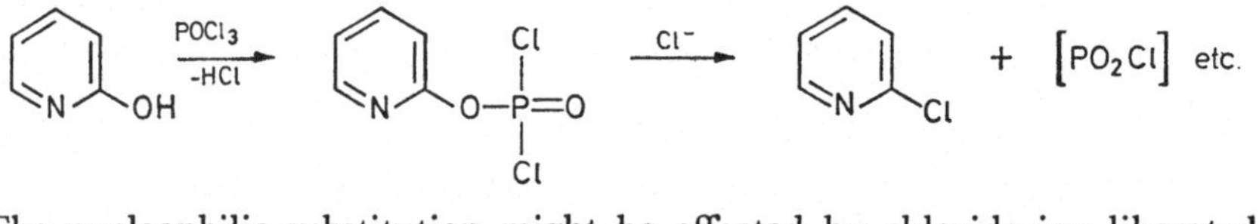

The nucleophilic substitution might be effected by chloride ion liberated in the first step (S_N2) or, internally, by a chlorine atom still attached to phosphorus (S_N1). A similar description could apply to the case of 1-alkyl-2-pyridones, except that the intermediates would be quaternary salts. Clearly, the S_N1 mechanism is not open to 4-hydroxypyridines.

Phosphorus halides can deoxygenate pyridine 1-oxides (p. 387). Consequently, when a 4-hydroxypyridine 1-oxide is treated with phosphoryl chloride, deoxygenation and substitution can give the 4-chloropyridine[846]. The experimental conditions are important: with phosphoryl chloride at 60°, 3-chloro-4-hydroxypyridine 1-oxide gives 3,4-dichloropyridine 1-oxide, but at 90–100° 3,4-dichloropyridine results[911].

Examples are known of the conversion of 2-nitropyridines into 2-halogenopyridines by reaction with halogen hydracids[158b, 170, 185, 240, 847]. Thus, 3-ethoxy-2-nitropyridine with hydrobromic in acetic acid gives 2-bromo-3-ethoxy- or 2-bromo-3-hydroxy-pyridine, according to conditions, whilst hydrochloric acid produces 2-chloro-3-hydroxypyridine[158b, 240]. 2-Nitro-pyridine is said not to react with phosphoryl chloride[713b], whilst from the same reaction with 4-nitropyridine, 4-hydroxypyridine[713a] and 1-4'-pyridyl-4-pyridone[848] have been isolated. It was concluded[713b] that 2-nitropyridine was less reactive than 4-nitropyridine, and the latter less so than 4-nitro-pyridine 1-oxide (see below).

Concentrated halogen hydracids convert 4-nitropyridine 1-oxide into 4-halogenopyridine 1-oxides[162, 849], but with dilute acids products arise from an initial hydrolysis to 4-hydroxypyridine 1-oxide[162, 849-50]. With sulphuryl chloride, 4-nitropyridine 1-oxide gives mainly 2,4-dichloropyridine (35–40 per cent)[248, 796].

Whilst phosphorus trichloride, used under mild conditions, is a valuable reagent for deoxygenating 2- and 4-nitropyridine 1-oxides (p. 387), replacement of the 4-nitro group by a chlorine atom sometimes occurs to a small extent[848, 851]. A mixture of phosphorus tri- and penta-chloride has been used to convert 4-nitronicotinamide 1-oxide into 4-chloronicotinonitrile (with some of the amide)[852], whilst 4-bromo-5-ethyl-2-methylpyridine has been obtained from 5-ethyl-2-methyl-4-nitropyridine 1-oxide and phosphorus tribromide[763]. In contrast, phosphoryl chloride converts 4-nitropyridine 1-oxides fairly efficiently into 4-chloropyridine 1-oxides[247, 254, 690a, 707, 713a, 853].

Acetyl halides are probably the best reagents for converting 4-nitropyridine 1-oxides into 4-halogenopyridine 1-oxides[247, 252, 852-5], and the reaction is of great preparative value. 2-Nitropyridine 1-oxides behave similarly[856]. 3-Nitropyridine 1-oxide does not react with acetyl chloride, but with phosphoryl chloride gives 2- and 6-chlor-3-nitropyridine[518].

Just as the reactions of nitropyridines and nitropyridine 1-oxides with halogen hydracids are presumably acid-catalysed (p. 215), so are those of 4-nitropyridine 1-oxides with phosphoryl and acetyl halides usually considered to involve activation by salt formation (111). However, it has also been suggested[852] that structures such as (112) might be involved.

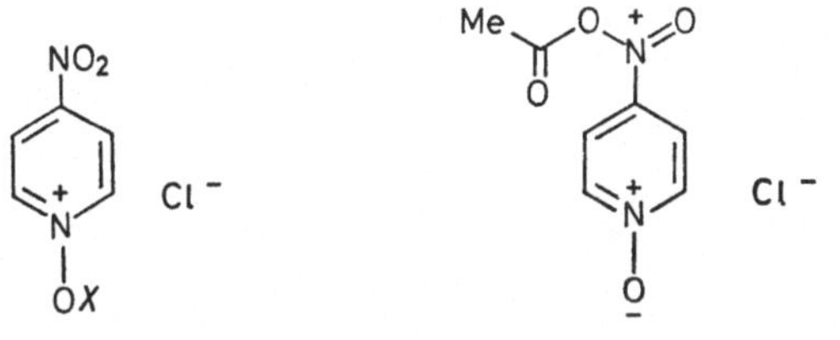

(111; X = POCl₂ or ·COCH₃)   (112)

Examples are known of the conversion of pyridine-sulphonic acids into chloropyridines by reaction with phosphorus pentachloride[409, 857], but they are of no importance. 1-Methyl-2-pyridthione is converted by phosgene into 2-chloropyridine[843].

(h) Hydroxylation and Substituted Hydroxylation

Some of the reactions whereby hydroxyl groups are introduced into pyridine rings require for their completion the presence of an oxidizing agent. The necessity arises because the hydride ion is a poor leaving group, and there is no reason why these reactions should not be included with other substitution processes, since the first step, the attack by the nucleophile, is in no way different[858].

When pyridine is dropped on potassium hydroxide at 280°–320°, a poor yield of 2-hydroxypyridine is formed[859]. The reaction was discovered by Koenigs and Koerner[860] who prepared 6-hydroxyquinolinic acid in this way. 3- and 4-Hydroxypyridine give 2,3- and 2,4-dihydroxypyridine[177], respectively.

The conversion of pyridine 1-oxide into 2-acetoxypyridine (or into 2-hydroxypyridine by hydrolysis of this ester) by reaction with hot acetic anhydride is a valuable reaction[861–2]. Benzoic anhydride produces a similar result, but acetic anhydride has naturally been used more often, and the reaction has been applied to several pyridine 1-oxides[133a, 518, 528, 863]. When a halogen, hydroxyl, methoxycarbonyl or nitro group is present at $C_{(3)}$, hydroxylation occurs mainly at $C_{(2)}$ rather than $C_{(6)}$, whilst a methyl group directs substitution about equally to each position. With nicotinic acid 1-oxide, the course of the reaction is more complicated (p. 162).

2- and 4-Alkylpyridine 1-oxides react with acetic anhydride quite differently (p. 340), but even in these cases small amounts of 2-hydroxypyridines are sometimes formed by substitution at the free α-position[789, 864]. In these cases 3-hydroxypyridines are sometimes formed to a small extent[787, 865, 878–9, 893]. These observations are discussed elsewhere (p. 341).

2-Picolinic acid 1-oxide gives 2-hydroxypyridine when heated with acetic anhydride, but decarboxylation is thought to precede hydroxylation[787]. 2-Ethoxy-, -chloro- and -phenoxy-pyridine 1-oxides give 2-hydroxypyridine 1-oxide, the oxide function surviving in these cases[787]. An unusual example is that of 2-carboxy-5-methoxycarbonylpyridine 1-oxide. With benzoyl chloride in dioxan this gives 6-hydroxynicotinic acid. However, with acetic anhydride or acetyl chloride, methyl 4-acetoxynicotinate results[866].

The reaction between pyridine 1-oxide and acetic anhydride has been represented[862] as a nucleophilic substitution:

(113)

Electron-attracting substituents should assist this reaction. However, with acetic anhydride in acetonitrile, 2-picolinic acid 1-oxide gives mainly pyridine 1-oxide and carbon dioxide, whilst the same reaction carried out under ultra-violet irradiation proceeds similarly but with the formation of a little more 2-hydroxypyridine. The methyl esters of the pyridine-carboxylic acid 1-oxides behave normally, but there is no evidence that the methoxy-carbonyl group promotes the reaction, and 2-cyanopyridine 1-oxide does not react with acetic anhydride[863b]. The results of a kinetic study of the rearrangement of pyridine 1-oxide in acetic anhydride exclude the intramolecular rearrangement of the free cation (113) and also a free radical process. There remain the two possibilities of nucleophilic substitution by reaction between free ions or in ion pairs. The orienting influence of substituents (see above) supports the characterization of the reaction as a nucleophilic substitution[867].

When pyridine 1-oxide reacts with bromine, sodium acetate and acetic anhydride, 3,5-dibromopyridine 1-oxide is formed[868a]. p-Nitrobenzoyl chloride and silver nitrate produce small yields of 3-nitro- and 3,5-dinitro-pyridine 1-oxide[786]. These reactions probably involve nucleophilic addition, followed by electrophilic substitution, and the first step can be represented as occurring at $C_{(4)}$, though not with any compelling reason:

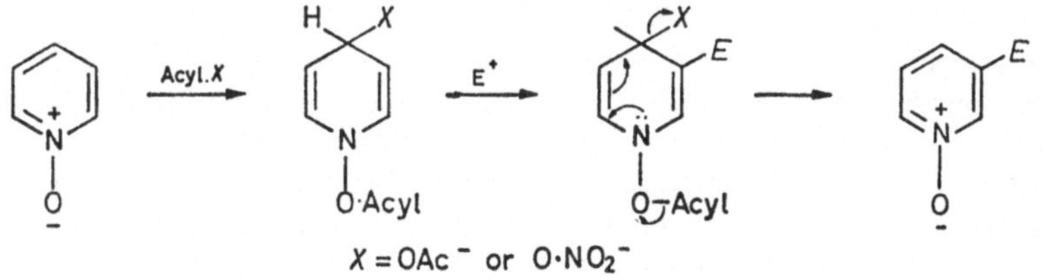

$$X = OAc^- \text{ or } O \cdot NO_2^-$$

The reactions of pyridine 1-oxides with p-toluenesulphonyl chloride have produced complicated results[955]. From pyridine 1-oxide a small amount of 3-pyridyl p-toluenesulphonate resulted, and 3-methylpyridine 1-oxide gave some 3-(5-methylpyridyl) p-toluenesulphonate, but a number of other substances were also formed. The 3-substituted pyridines have been represented as arising from an initial nucleophilic attack by chloride ion, and studies[955e] with labelled p-toluenesulphonyl chloride show that the subsequent step cannot involve a free sulphonate ion but may go through an intimate ion pair:

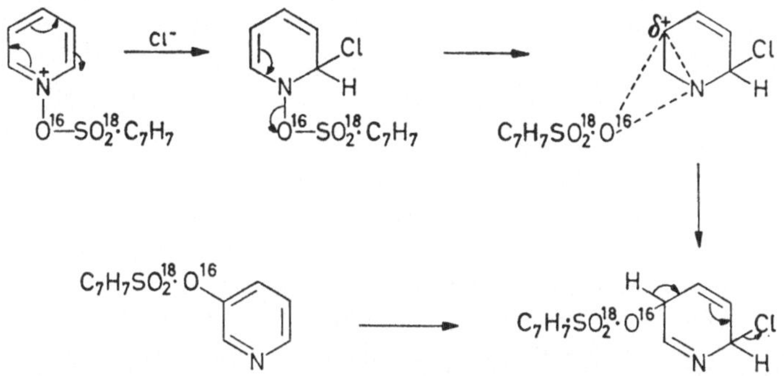

Because of a possible similarity between them and the group of reactions just dealt with, it is convenient to mention here some reactions which have been used to prepare carboxylic acid anhydrides. Minunni[489] observed the formation of benzoic acid anhydride in high yield when benzoylpyridinium chloride was treated with water, and the reaction is generally useful[487, 493]. From the fatty acid series the reaction of heptoyl chloride with pyridine in benzene, followed by the addition of heptoic acid, provides an example of anhydride formation in high yield[803b]. More recently, anhydrides were prepared from acid chlorides and aqueous solutions of the corresponding sodium salts containing catalytic amounts of pyridine[868b]. It seems possible that there is being observed the nucleophilic attack upon an acyl- or aroylpyridinium cation of a carboxylate anion, followed by a splitting off of the anhydride:

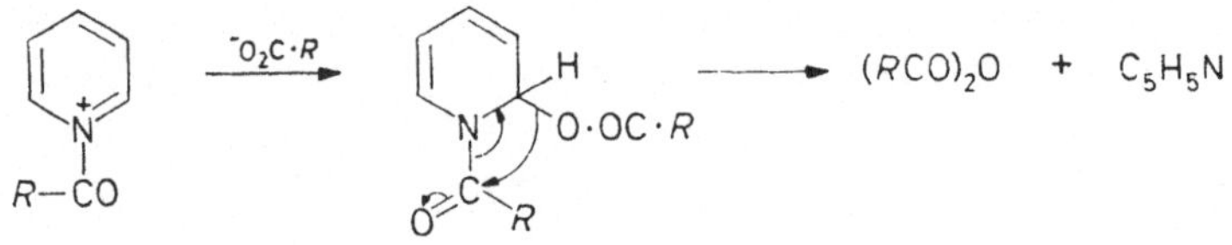

Such a mechanism recalls those postulated for other processes (p. 205). It cannot be the only one possible, for bases other than pyridine have been used. It may account for the special effectiveness of pyridine.

In the discussion of amination (p. 209) it was mentioned that attack by an amine upon some types of quaternary pyridinium salts can cause ring opening. This is true of attack of hydroxyl upon 1-arylpyridinium salts and related compounds (p. 266). There remain, however, the important reactions of 1-alkylpyridinium salts with hydroxyl in the presence of ferricyanide. When 1-methylpyridinium iodide reacts with alkaline potassium ferricyanide, 1-methylpyrid-2-one is formed[869]. The electrical conductivity of a solution containing 1-methylpyridinium and hydroxyl ions shows 1-methylpyridinium hydroxide to be a strong, completely dissociated base, and further, the conductivity does not decrease with time. It follows that the equilibrium between the ions and the pseudo-base (114) lies overwhelmingly to the left[870]. The formation of 1-methylpyridone by the action of ferricyanide is attributed to the continuous removal of the pseudo-base, present in very small concentration[318].

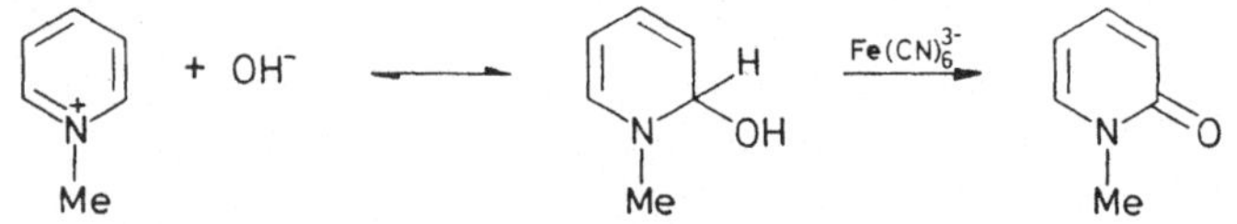

(114)

Decker's reaction gives 1-methylpyrid-2-one in high yield[580, 869a, 871] and has been used to prepare various 1-alkyl- and 1-aralkyl-pyridones[304c, 318, 336, 738b,c, 869a, 872]. The reaction cannot generally be applied to 2- and 4-alkylpyridinium salts having hydrogen atoms on the carbon atom adjacent to the ring, for these give pyridone methides (p. 328). However, although 1,2-dimethylpyridinium methosulphate does not give a pyridone[873], 2-methyl-1-β-phenethylpyridinium bromide gives a poor yield of 6-methyl-1-β-phenethylpyrid-2-one[738c]. 4-Alkyl-groups are sometimes oxidized, as in the formation of 1-methylpyrid-2-one-4-carboxylic acid from 4-ethyl-1-methylpyridinium methosulphate[874]. 4-Formyl- and 4-acetyl-groups, protected as their ethylene glycol derivatives, survive the reaction[874]. 1-Methylisonicotinic acid betaine provides one of the rare failures of the Decker oxidation, but methyl quaternary salts of esters of isonicotinic acid give 1-methylpyrid-2-one-4-carboxylic acid[874-5].

The orientation of substitution observed with 3-substituted pyridinium salts is of interest. A number of results is summarized in *Table 5.27*. It is quite possible that in some of the cases, where only one of the two possible products was obtained, the alternative isomer was formed to some extent

Table 5.27. Oxidation of 3-Substituted Pyridinium Salts by Alkaline Ferricyanide

R	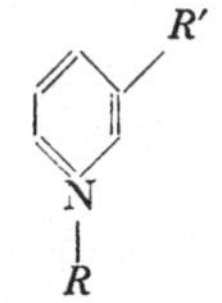 R'	Position of substitution	Ref.
Me	Me	predominantly 2	171f, 876
	Et	2 + 6 (8 : 1)	876a, b, d
	CH₂OAc	2 + some 6	873
	(dioxolane) Ċ.Me	6	876c
	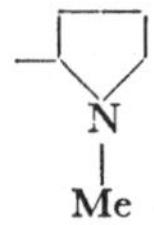 (N-Me pyrrolidine ring) Me	2	880
	aryl and heteroaryl	6	876a
	CO₂H	6	171f, 881
	CONH₂	2 + 6 equally	882
	CN	2 + 6	171f
	Br	2	171f, 318, 876a
Ph(CH₂)₂·	CN	2*	⎫
	Br	2*	⎬ 883
	PhO	6*	⎭
3,4-CH₂O₂C₆H₃·(CH₂)₂	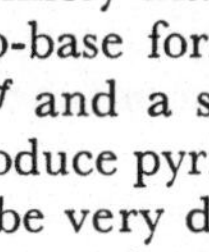 (ring) Me	6	884

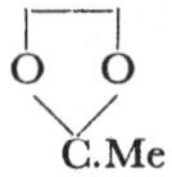

* These orientations were deduced from dipole moments.

but not isolated. It seems likely that at least three factors influence the result: the orientation of pseudo-base formation, the relative rates of oxidation of isomeric pseudo-bases[171f] and a steric factor[876a, 877]. A peculiarity of the reaction is its failure to produce pyrid-4-ones. Since the electrophilic reactivity of the 4-position cannot be very different from those of the 2-positions, the absence of pyrid-4-one formation must be a consequence of some mechanistic aspect of the reaction not revealed in its usual representation. It is clear that the results give no direct indication of the relative electrophilic reactivities of the positions available for attack.

In the above examples of the Decker reaction, hydrogen is removed from

the pyridine ring. When 2,5-dimethoxycarbonyl-1-methylpyridinium methosulphate or pyridine-2,5-dicarboxylic acid 1-methylbetaine are used, 5-
carboxyl-1-methylpyrid-2-one is produced, with some of its methyl ester in
the first case[866].

1-Methylpyridinium sulphate can be oxidized electrolytically to 1-methyl-
pyrid-2-one[184a, 885].

Pyridine 1-oxide and nicotinic acid 1-oxide are unaffected by alkaline
ferricyanide[886].

Reactions in which hydroxypyridines are formed by replacement of acyl
or carboxyl groups are unknown or rare. Examples of replacement of
carboxyl during the Decker reaction have been mentioned, as has the
conversion of 2-carboxy-5-methoxycarbonylpyridine 1-oxide into 6-
hydroxynicotinic acid by benzoyl chloride in dioxan. The latter reaction has
been represented[866] as

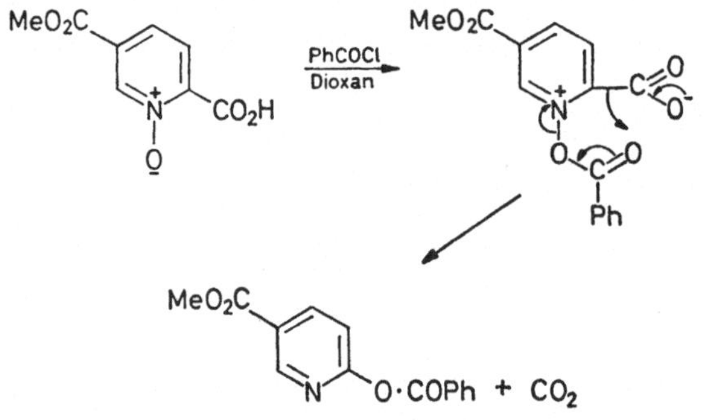

With some 2-alkylpyridinium salts, attack by hydroxyl anion causes methide formation, and subsequent transformations can be complicated (p. 328).
However, in one interesting reaction, 2-1'-pyridylmethylpyridinium salts
are used as sources of 2-pyridones[313a]:

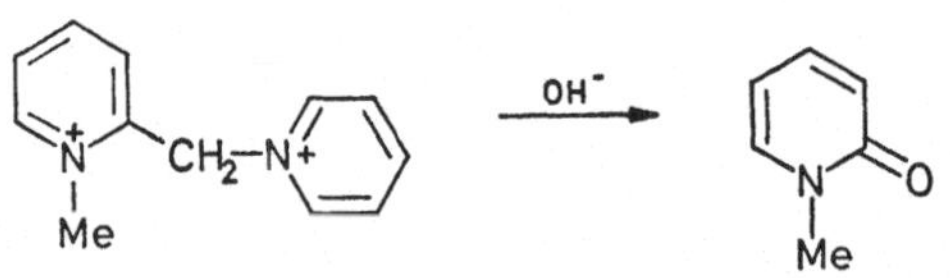

The reaction [—N=C—CH$_2$ ——→ —N=C—OH] is analogous to
[O=C—CH$_3$→ O=C—OH], and resembles the splitting of 1-phenacyl-
pyridinium salts (p. 390). Other examples are known[743, 877]. Although
methide formation must occur in these reactions the methide is not considered
to play a part in pyridone formation[743].

2- or 4-Amino, -methylamino and -dimethylamino groups are readily
replaced by hydroxyl, by treating with dilute sodium hydroxide solution 3-nitropyridines in which they are present[180a, 218, 219b, 887].
Similarly, 1-methyl-2-dimethylaminopyridinium iodide is easily hydrolysed
to 1-methylpyrid-2-one[344]. 1-Alkylpyridinium salts containing 2- or 4-
amino[339, 341, 343, 352, 888], -methylamino[219b] or -nitroamino groups[347, 369]

give 1-alkylpyridone-imines on gentle treatment with alkali. More severe alkaline hydrolysis converts the latter into 1-alkylpyridones.

Quaternary amino substituents on the pyridine ring can also be replaced. The important example is that of 4-pyridylpyridinium chloride hydrochloride which is converted into 4-hydroxypyridine when heated with water[433, 435, 437, 442a]. 4-Hydroxy-3-methylpyridine is prepared in the same way[200]. 4-Alkoxy-[52a, 430] and 4-aryloxy-pyridines[430, 433, 438] are obtained by treating 4-pyridylpyridinium chloride hydrochloride with alkoxides, and 4-phenoxypyridine is probably an intermediate when the salt is converted into 4-aminopyridines by being heated with phenol and ammonia or an amine[252, 434].

Certain other reactions differ from those discussed in being brought about by acid rather than alkali. Thus, although 2-aminopyridine is unaffected by acid[889], 2,6-diaminopyridine is converted by hot 10 per cent hydrochloric acid into 2-amino-6-hydroxypyridine[889–90]. The second step, to 2,6-dihydroxypyridine, is more difficult[889, 891]. 2,6-Diamino-3-nitrosopyridine can be converted into the dihydroxy compound with acid or alkali, but with 2,6-diamino-3-nitropyridine alkali is essential[267a]. The not very convincing explanation usually given for the difference in behaviour between 2-amino- and 2,6-diamino-pyridine is that in the former the amino group is part of a highly mesomeric system and $C_{(2)}$ has much diminished electrophilic character; in the diamine, however, whilst both substituents participate in the mesomerism, neither is involved to the same extent, and hydrolysis is possible.

When 2-nitraminopyridines with one or two free β-positions are heated with sulphuric acid, rearrangement occurs giving a 2-aminonitropyridine (p. 172). A minor product of such reactions is the 2-hydroxypyridine formed by replacement of the nitramino group[218–9a]. The use of acetic acid acetic anhydride converts 2-nitraminopyridine into 2-hydroxypyridine in 60 per cent yield[219a]. 3,5-Disubstituted 2-nitramino- or 2-methylnitramino-pyridines give 3,5-disubstituted 2-hydroxypyridines with sulphuric acid[794, 892]. pyridine does not undergo rearrangement when heated with sulphuric acid 3-Nitramino- or acetic acid–acetic anhydride[219, 222], and by heating 2-nitramino-5- nitropyridine with sulphuric acid some workers[221] obtained 2-hydroxy-5-nitropyridine rather than 2-amino-3,5-dinitropyridine (p. 172). The mechanism of the conversion of nitramines into pyridones by acid is unknown. 2-Nitramino-5-nitropyridine also gives 2-hydroxy-5-nitropyridine and nitrous oxide quantitatively with warm dilute alkali[221].

Replacement of the cyano group rather than its hydrolysis to carboxyl occurs when 4-cyanopyridine methiodide is treated with alkali, giving 1-methylpyrid-4-one[669a]. Similarly, 2-cyanopyridine methiodide gives 1-methylpyrid-2-one and is the intermediate in the conversion of pyridine-2-aldehyde oxime methiodide into 1-methylpyrid-2-one in the range[894] pH 7–13.

The diazotization of aminopyridines has been used extensively as a source of pyridinols (*Table 5.28*). The precise mechanism of the substitution is unknown. Since loss of nitrogen from the protonated pyridine diazonium ion would produce a doubly charged carbonium ion, the carbonium ion mechanism[912] seems unlikely in this series, and S_N2 substitution may be

Table 5.28. Formation of Pyridinols from Aminopyridines by Diazotization

A. From 2-Aminopyridines

Substituents (per cent yield)	References	Substituents (per cent yield)	References
—	501, 790*	3,5-(CO$_2$H)$_2$-4,6-Me$_2$	897
3-Me	656	5-Br-6-Me (82)	} 171e
4-Me (78)	171e, 649b	3,5-Br$_2$-6-Me (97)	} 171e
6-Me (93)	171e, 792	3-Me-5-NO$_2$ ($\sim$100)	794, 898
3-Et-6-Me	216	4-Me-3-NO$_2$ ($\sim$100)	} 817, 898
4-CH:CH.Ph (36)	} 171e	4-Me-5-NO$_2$ ($\sim$100)	} 817, 898
6-CH:CH.Ph (48)	} 171e	5-Me-3-NO$_2$ ($\sim$100)	898
3-CO$_2$H†	887	6-Me-3-NO$_2$ ($\sim$100)*	} 170, 679,
5-CO$_2$H	670b	6-Me-5-NO$_2$ ($\sim$100)*	} 898
4-Cl	775a	3-CO$_2$H-4-Cl	899
5-Cl (83)	169a	3,5-Cl$_2$	169a, 900
6-Cl (70)	895	4,6-Cl$_2$	775b
5-Br	171a	3,5-Br$_2$*	793
6-Br	672b	3,5-I$_2$	174
4-I	775a	5-Cl-3-NO$_2$	585, 901
5-I	172	5-Br-3-NO$_2$	230, 585
3-NO$_2$‡	211, 805	3-Br-6-Me-5-NO$_2$*	170
5-NO$_2$	212, 791	3,5-(NO$_2$)$_2$	901
3-CONH$_2$-6-Me	} 896		
3-CO$_2$H-6-Ph	} 896		
3-CO$_2$H-4,6-Me$_2$	} 896		
3-CO$_2$H-5,6-Me$_2$	} 896		
3-CO$_2$H-4-Me-6-Ph	} 896		

B. From 3-aminopyridines

Substituents (per cent yield)	References	Substituents (per cent yield)	References
—	902	2-CO$_2$H	135c
6-Me	170	4-CO$_2$H	904
6-Et	629	4-CO$_2$Me	905
6-Pr	623a, 629	5-CO$_2$H	778
6-Bu	} 629	5-Br	906
6s-Bu	} 629	4,5-(CO$_2$H)$_2$-2-Me	907
6-Amyl	} 629	5,6-(CO$_2$H)$_2$-2-Me (85)	903
6-Ph(CH$_2$)$_2$	} 629	4,5,6-(CO$_2$H)$_3$-2-Me (86)	908
2,6-Me$_2$	} 208a, 757	2-Cl-6-Me	} 170
2,4,6-Me$_3$	} 208a, 757	6-Cl-2-Me (22)	} 170
2-Me-5,6-(CH$_2$OH)$_2$ (90)	} 903	4,5-(CO$_2$H)$_2$ (50)	675b
2-Me-4,5,6-(CH$_2$OH)$_3$	} 903	2,5-Br$_2$	230

C. From 4-aminopyridines

Substituents (per cent yield)	References	Substituents (per cent yield)	References
—	502	2-Cl	802
3-Me	200	2-MeO	797b
2,6-Me$_2$**	670b	2,6-Br$_2$**	186c
2-CO$_2$H	675a, 798	2,3,5,6-Cl$_4$	910, 911
3-CO$_2$H	909	4-NH$_2$-2-HO-5,6-(CH$_2$)$_3$ (74) §	838
2-CN	798		

---

* In some cases the hydroxypyridine was a byproduct in the preparation of a halogenopyridine.
† From reaction of the nitramino compound with nitrous acid.
‡ Diazotization was carried out[805] in 60 per cent hydrofluoric acid.
** The product was not isolated.
§ By diazotization in dilute sulphuric acid; if hydrochloric acid was used, 4-amino-2-hydroxy-5,6-trimethylene-3-nitropyridine was formed.

occurring. Most of the reactions in *Table 5.28* were effected in dilute sulphuric acid, though in some cases, particularly when the aminopyridine was very weakly basic (e.g. 2-amino-3,5-dinitro- and 4-amino-2,3,5,6-tetrachloro-pyridine), nitrosylsulphuric acid was used. Failures are rare, though 2-amino-3-nitropyridine-5-carboxylic acid is said to be unaffected by nitrous acid[218]. 3-Amino-5-aminomethylpyridines have been converted into 3-hydroxy-5-hydroxymethylpyridines[811, 913], and it is possible to diazotize the two amino groups selectively[811]:

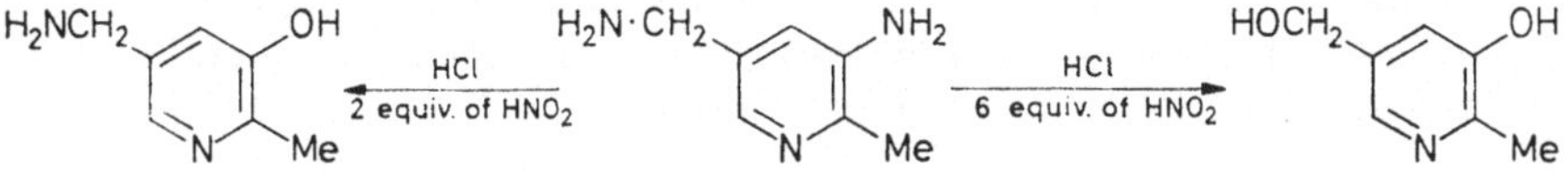

(31%)　　　　　　　　　　　　　　　　　　　　　　　　　　　　(46%)

(115)　　　　　　　　　　　　　　　　　(116)

An interesting case is provided by the compounds (115): (115; $R = $ Cl) gives (116; $R = $ Cl, $R' = $ OH; 44 per cent) when diazotized in hydrochloric acid. However, (115; $R = $ H) gives (116; $R = $ H, $R' = $ Cl; 27 per cent), and the same product results, though in lower yield, when the experiment is done in sulphuric acid[811].

Pyridyl ethers have been prepared from pyridine diazonium salts and alcohols[670b, 815, 914], and with phenol, sodium 2-pyridyldiazotate gives 2-phenoxypyridine as well as coupling products[790].

Hydroxylation by replacement of halogen is of no great importance because synthetically it is more often convenient to pass from a hydroxypyridine to a halopyridine rather than the reverse. Substituted hydroxylation by replacement of halogen is, however, of practical importance.

Halogen atoms in pyridines have been replaced by hydroxyl in both alkaline and acidic media. 2-Chloropyridine gives 2-hydroxypyridine when heated with potassium hydroxide[915] at 170°, and a number of 2-chloropyridines have been similarly converted under conditions which are milder if activating groups are present[679a, 757, 911, 915–6]. With 2- and 4-chloropyridines, the use of alcoholic alkali can give mixtures of ethers and hydroxy compounds[672b, 796]. Alkaline hydrolysis of pentachloropyridine gives 2,3,5,6-tetrachloro-4-hydroxypyridine[911]. Acid hydrolysis has been applied to several 2-halogenopyridines[183e, 230, 675a, e, 757, 915, 917]. Hydrolyses begun in water[626, 918] become acid hydrolyses as the reactions proceed. In all these processes electron-attracting groups facilitate reaction. Nitro groups are noteworthy in this respect; 4-chloro-3-nitropyridine is very easily hydrolysed[626]. By the use of sodium acetate in acetic acid[181, 919a] or silver benzoate[919b], 4-chloropyridines have been converted into 4-acyloxypyridines.

In alkali, acid and carboxylic acid anion reactions, 2,4-dichloropyridines have been converted into chloro-hydroxypyridines[679a, 775b, 840d], usually of uncertain orientation. 2-Iodopyridine methiodide reacts with carboxylic acids in presence of triethylamine, giving 2-acyloxy-1-methylpyridinium ions[920] (p. 379).

In several examples referred to above, halogen atoms at $C_{(3)}$ remain unaffected during the hydrolysis of 2- or 4-substituents. 3-Bromopyridine is converted into 3-hydroxypyridine by caustic soda and a copper catalyst[921] at 200°, and boiling 50 per cent caustic potash converts 3-chloro- into 3-hydroxy-isonicotinic acid[675a].

Alkaline hydrolysis proceeds less readily than acid hydrolysis[335, 918]. The greater ease of replacement of fluorine than of chlorine or bromine [335, 918] is illustrated by the following qualitative results[335].

Effect of 6N-Hydrochloric Acid acting for 24 h on 2-Halogenopyridines

Substituents in Compounds	
hydrolysed	not hydrolysed
2-F; 2-F-3-Me; 2-F-5-Me;	2-Cl; 2-Br
2-F-3-CO₂H; 2-F-5-CO₂H;	2-Br-3-Me; 2-Br-5-Me
2-Br-3-CO₂H; 2-Br-5-CO₂H	

This order of reactivity is commonly met with in nucleophilic aromatic substitutions[922].

The greater effectiveness of acid hydrolysis is due to the protonation of the pyridine nitrogen atom (cf. p. 215). The effect has been studied quantitatively with 2-chloro- and 2-iodo-5-nitropyridine. The results establish the occurrence of acid catalysis, and an overall mechanism has been proposed involving hydration of the intermediate reaction complex[923].

A number of reactions in which halogenopyridines have been converted into alkoxy- or aryloxy-pyridines are collected in *Table 5.29*. Most commonly, alkoxypyridines are prepared from the halogenopyridine and a solution of the appropriate sodium alkoxide or aryloxide in the corresponding alcohol or phenol. Particularly reactive compounds such as 4-chloro-3-nitropyridine react with alcohol in the cold. Occasionally the formation of alkoxypyridines has been observed when the presence of other nucleophiles led to the expectation of other products being formed; thus, 2-chloro-5-nitropyridine with aqueous methanol containing sodium sulphite gave 2-methoxy-5-nitropyridine rather than a sulphonic acid[213], and sodium cyanide in methanol causes methoxylation rather than cyanation[811]. 2-Bromopyridine gives good yields of 2-aryloxypyridines[430] when heated with a phenol and potassium carbonate at 200–210°. In some cases where this method failed, the use of a sodium aryloxide and copper powder succeeded, and this variation was also useful with some sodium aralkoxides[938].

Although 3-halogenopyridines can be converted into 3-pyridyl ethers (*Table 5.29B* and *D*), they are less reactive than 2- and 4-halogenopyridines.

242

Table 5.29. Formation of Alkoxy- and Aryloxy-pyridines from Halogenopyridines

Substituents	Reagent*	References
A. From 2-Halogenopyridines		
(i) *2-fluoropyridines*		
5-CO$_2$Me	} MeONa	} 918
6-CO$_2$Me		
5-NO$_2$		805
(ii) *2-chloropyridines*		
4-CO$_2$H	PrONa; BuONa	924
5-CONH$_2$	BuONa	925
3-CONHPh	Et$_2$N(CH$_2$)$_2$ONa	926
4-NH$_2$	EtONa	796
3-Cl	BuONa	895
5-I	MeONa	927
3-NO$_2$	} RONa; ArONa	629, 844, 928
5-NO$_2$		213, 844,
		928–32
5-SO$_2$NH$_2$	MeONa	131
5-CO$_2$Et-6-Me	EtONa	·183*e*
3-CN-5-Me	MeONa	933
6-Me-3-NO$_2$†	MeONa	679*g*
6-Me-5-NO$_2$	MeONa; ArONa	679*g*, 934
4,6-Me$_2$-5-NO$_2$‡	} MeONa	} 183*f*
4,6-Me$_2$-3,5-(NO$_2$)$_2$‡		
5-CN-4,6-(MeO)$_2$	EtOK	916
3-CN-5-NO$_2$	EtONa	935
3-Cl-5-NO$_2$	ArOK	936
5-Cl-3-NO$_2$	RONa; ArONa	937
3-CN-6-Me-5-NO$_2$‡		
3-CN-4,6-Me$_2$-5-NO$_2$‡	} MeONa	} 183*f*
3-Br-4,6-Me$_2$-5-NO$_2$‡		
(iii) *2-bromopyridines*		
—	PhCH$_2$ONa	430
—	RONa : ArOH/K$_2$CO$_3$	430, 938
—	RONa/Cu; ArONa/Cu	938
6-NH$_2$	EtONa	672*b*
3,4,6-Ph$_3$	BuONa	939
5-EtO	EtONa	158*b*
(iv) *2-iodopyridines*		
5-NO$_2$	MeONa	388
B. From 3-Halogenopyridines		
(i) *3-bromopyridines*		
—	MeONa	158*b*
—	EtONa	940
(ii) *3-iodopyridines*		
—	PhOK/Cu	430
C. From 4-Halogenopyridines		
(i) *4-chloropyridines*		
—	MeONa	401
—	PrONa; PhONa	941
—	RONa; ArONa	942
2,6-Me$_2$	RONa; ArONa	919*b*
2-CO$_2$Me	MeONa	777
3,5-I$_2$**	EtONa	943
3-NO$_2$	MeOH	241*c*
3-NO$_2$	EtOH or EtONa	241*a*, 626
3,5-I$_2$-2,6-Me$_2$	ArONa	944
3-Cl-6-HO	} NaOH/aq.EtOH§	} 796
3-Cl-6-EtO		

I*

Table 5.29 (continued)

Substituents	Reagent*	References
D. From Dihalogenopyridines		
(i) 2,4-dichloropyridines		
—	MeONa	248
—	EtONa	853
6-Me	MeONa ‖	} 945
6-Me	EtONa	
3-CN	MeONa	899
5-CN	MeONa	835
5-Br-6-Me	MeONa ‖	945
3-CN-6-Me-5-NO$_2$	MeOH/NaCN***	811
(ii) 2,6-dichloropyridines		
4-CO$_2$H	RONa	924
(iii) 2,6-dibromopyridines		
—	EtONa ¶	672b
(iv) 3,5-dibromopyridines		
—	EtOK; EtONa ¶	158b, 681c, 940
E. From Trihalogenopyridines		
2,4,6-tribromopyridines		
—	MeONa ††	186c
—	PhONa ‡	946

* Almost invariably a solution of the alkoxide or aryloxide in the corresponding hydroxy compound is used.
† Decomposition resulted.
‡ See text.
** 4-Ethoxy-3,5-di-iodopyridine was a minor product; the main reaction involved 2-carboxythiophenol.
§ Mixtures of hydroxy compounds and ethers resulted.
‖ Only monosubstitution occurred.
*** Monosubstitution occurred, giving 2-chloro-3-cyano-4-methoxy-6-methyl-3-nitropyridine.
¶ Mono- or di-substitution occurred according to conditions.
†† 4-Bromo-2,6-dimethoxy-, 6-bromo-2,4-dimethoxy- or 2,4,6-trimethoxy-pyridine was formed, depending on conditions.

Thus, 4-chloropyridine reacts readily[401] with sodium methoxide at 100°, but 3-chloropyridine is unaffected by these conditions[947]. Further in a number of the tabulated cases of conversions of 2- or 4-halogenopyridines to ethers, halogen atoms at $C_{(3)}$ or $C_{(5)}$ are not replaced. 2,4-, 2,6- or 3,5-Dihalogenopyridines can undergo mono- or di-substitution according to the conditions used. The qualitative evidence favours slightly the view that substitution at $C_{(4)}$ is easier than at $C_{(2)}$. This is seen in the reaction of 2,4,6-tribromopyridine with sodium methoxide, from which 2,6-dibromo-4-methoxy-, 6-bromo-2,4-dimethoxy- or 2,4,6-trimethoxy-pyridine can be obtained, according to the severity of the conditions. This picture is, however, complicated by the reaction of 2,4,6-tribromopyridine with sodium phenoxide, which seems to occur more readily at the α-positions than at the γ-position, except in the presence of water[946].

Electron-attracting groups facilitate the formation of ethers from halogenopyridines. The accumulated effect of two such groups can have another consequence in stabilizing the intermediate addition complex by way of which such aromatic nucleophilic substitutions are widely believed to proceed. This effect is strikingly illustrated by the compounds (117; $R = H$ or Me) which with cold sodium methoxide solution give purple colours. Dilution of these coloured solutions permits the isolation of the 2-methoxypyridines. Similar results are observed with (117; $R = $ Me; NO$_2$ or Br for CN), whilst

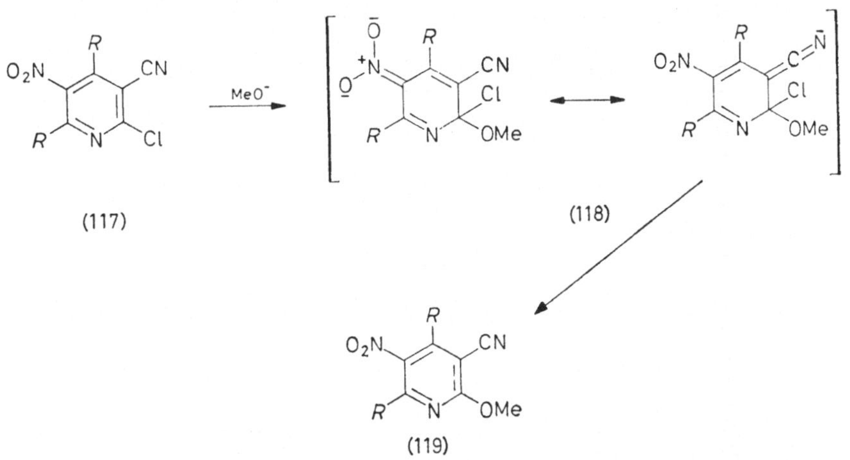

(117)

(118)

(119)

(117; $R = $ Me; H for CN) gives the ether but no colour. The colours produced when electron-attracting groups are present at $C_{(3)}$ and $C_{(5)}$ are attributed to stabilized intermediates of the type (118). Consistently with this explanation, the products (119) react with alkoxides to give coloured solutions and replacement of one alkoxyl group by another[183f].

2- and 4-Halogenopyridine 1-oxides are hydrolysed to hydroxypyridine 1-oxides by alkali[512, 689, 911] or acid[689]. With acetic anhydride, 2-chloropyridine 1-oxide gives 2-hydroxypyridine 1-oxide[787], but 2-chloro-4- or -6-methyl-pyridine 1-oxide reacts mainly at the methyl group[787] (cf. p. 340). Ether formation with sodium alkoxides[247, 689, 690b] and aryloxides[689, 944] has also been described.

Quaternary derivatives of 2- and 4-halogenopyridines are readily hydro-lysed. Thus, alkali converts 2-chloro-1-phenacylpyridinium bromide into 1-phenacylpyrid-2-one[948], and other examples are known[328, 331a, 949]. Acid hydrolysis also occurs[328]; that of 1-ethyl-2-iodo-3- and -5-methylpyridinium iodides is accompanied by the formation of iodinated products[328]. 4-Halo-geno-1-(4'-pyridyl)pyridinium chlorides give 1-(4'-pyridyl)pyrid-4-one[917].

Quantitative data for the reactions of chloropyridines, their 1-oxides and quaternary salts are collected in *Table 5.30*. The order of reactivity between the different series is quaternary salt > 1-oxide > pyridine, for the different positions in the chloropyridines and their 1-oxides, 4 > 2 > 3, and in the quaternary salts, 2 > 4 > 3. In all the series the activation energies would produce the sequence 4 > 2 > 3, but among the quaternary salts this is modified by the entropy factor; in these, the residual charges in the transition state for reaction at $C_{(2)}$ are close together and less favourable for solvation than the well separated charges in that for reaction at $C_{(4)}$, and consequently a more positive value of $S^{\ddagger}$ for reaction at $C_{(2)}$ compensates for a higher activation energy[950]. Differences in the oxide series result from the abnor-mally low entropy for reaction at $C_{(4)}$; it is likely that the negative charge in

Table 5.30. Nucleophilic Alkoxylation and Aryloxylation of Halogenopyridines, Halogenopyridine 1-Oxides and Quaternary Salts

*Pyridine**	*Conditions*	k_2 (l.mole^{-1} sec^{-1}) × 10^6	*E* kcal. mole^{-1}	$\Delta H^{\ddagger}$ kcal. mole^{-1}	$\log_{10}A$	$-\Delta S^{\ddagger}$, e.u.	*References*
2-Cl	MeO^-,MeOH,50°	0·033	28·9		12·1	5·33	950
	EtO^-,EtOH,20°	0·002	26·8	26·2		9·21	700
[*o*-chloronitrobenzene]	EtO^-,EtOH,20°	0·180	23·7	23·0		10·8	700
3-Cl	MeO^-,MeOH,50°	1·09 × 10^{-5}	32·8		11·2	9·21	950
4-Cl	MeO^-,MeOH,50°	0·891	25·2		11·0	10·4	950
	EtO^-,EtOH,20°	0·087		20·2		22·3	700
[*p*-chloronitrobenzene]	EtO^-,EtOH,20°	0·910		19·4		19·5	700
*Pyridine 1-oxides**							
2-Cl	MeO^-,MeOH,50°	640	20·3		10·55	12·4	950
3-Cl	MeO^-,MeOH,50°	1·16	24·6		10·7	11·7	950
4-Cl	MeO^-,MeOH,50°	1,000	19·0		9·8	15·6	950
1-Methylpyridinium salts†							
2-Cl	*p*-$NO_2.C_6H_4O^-$,MeOH,50°	1,390 × 10^4	18·5		13·7	−2·00	950–1
	MeO^-,MeOH,50°	1,530 × 10^8	13·9		14·6	−6·11	950–1
3-Cl	*p*-$NO_2.C_6H_4O^-$,MeOH,50°	0·284	30·2		13·9	−2·91	950–1
	MeO^-,MeOH,50°	3,140	25·6		14·8	−7·03	950–1
4-Cl	*p*-$NO_2.C_6H_4O^-$,MeOH,50°	46·0 × 10^4	17·6		11·5	7·84	950–1
	MeO^-,MeOH,50°	5,080 × 10^6	13·0		12·4	3·73	950–1
[C_6H_5Cl]	MeO^-,MeOH,50°	1·20 × 10^{-10}	39·9		11·1	9·90	950

* See also[952].
† Data for reaction with MeO^- in MeOH are computed from those for reaction with *p*-$NO_2.C_6H_4O^-$ in MeOH.

the transition state is more concentrated on the oxygen atom, producing relatively large solvation and low entropy—an effect more powerful in the unhindered 4-chloropyridine 1-oxide than in the 2-isomer. When allowance is made for the increased entropy of methanolysis of 3-chloro-1-methylpyridinium cation and for the slightly low entropy of reaction of 3-chloropyridine 1-oxide, the remaining activities, due to activation energies, are about equal for the salt and the oxide. It is suggested[950] that there is enhanced electrostatic stabilization in the transition state for reaction at $C_{(3)}$ in the oxide.

From the data for methanolysis of monosubstituted chlorobenzenes at 50°, the Hammett reaction constant $\rho = 8{\cdot}47$ can be obtained[950]. Thence, for the heteroatoms or groups $\overset{+}{N}$, $\overset{+}{N}{-}\overset{-}{O}$ and $\overset{+}{N}{-}Me$ there follow σ^* values at position 4 of 1·165, 1·526 and 2·317, and at position 3 of 0·586, 1·178 and 1·584. These values should be compared with those quoted earlier (p. 156).

Hydroxyl groups might be introduced into the pyridine nucleus by replacement of hydroxyl already present (that is, by oxygen exchange) or by replacement of an ether group. In the latter case, in acid solution two mechanisms are generally possible:

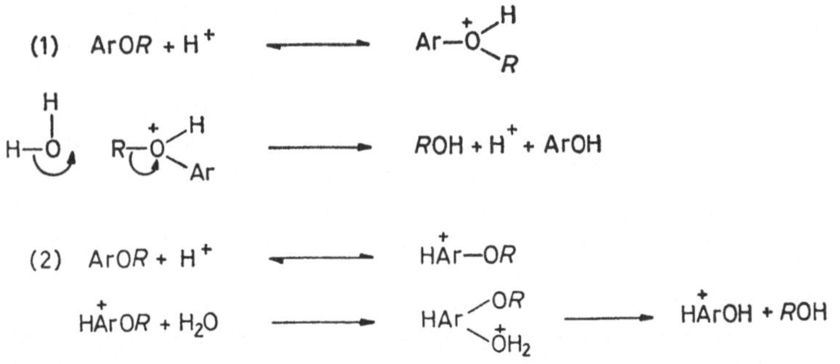

Both are acid-catalysed, but only the second is a nucleophilic replacement at the heterocyclic nucleus. Two related, uncatalysed mechanisms are possible in alkaline cleavage. In the pyridine series there is no evidence to show which of these two general mechanisms prevails in either acid or alkaline conditions. However, with 2-methoxypyrimidine, nucleophilic attack upon the heterocyclic nucleus has been shown[956] to be a major pathway of acid-catalysed ether fission. Pending clarification of reaction mechanisms in the pyridine series, reactions in which pyridyl ethers are converted into hydroxypyridines will be classed together at this point.

A number of 2-[413, 464, 629, 795–6, 895, 935, 957], 3-[158b, 185, 279, 623b, 795, 958], and 4-alkoxy- and -aryloxy-pyridines[186c, 241c, 959], and various dialkoxy-pyridines[158b, 796, 960] have been converted into hydroxypyridines, usually by being heated with halogen hydracids. Some 2-alkoxypyridines were conveniently dealkylated simply by the heating of their hydrochlorides[895]. As well as effecting dealkylation, halogen hydracids can cause nucleophilic replacement of other groups; 2-nitro-[158b] and 2-bromo-groups[795] are examples. Sometimes ether groups surprisingly survive, as in the conversion of 3,5-diethoxy-2,6-dinitropyridine into 2,6-dibromo-3,5-diethoxypyridine by hydrobromic acid in acetic acid[158b] at 100°.

Alkaline cleavages are rare. 3-Ethoxy-6-nitropyridine is very sensitive to alkali[158b], but the result of the reaction is obscured by decomposition. What is happening in the conversion of 5-methoxy-1,2-dimethyl-4-pyridone into 5-hydroxy-1,2-dimethylpyrid-4-one by sodium and amyl alcohol[961] is not clear.

A number of 2-alkoxypyridine 1-oxides have been converted into 2-hydroxypyridine 1-oxides by treatment with hydrochloric acid[52, 256, 524–5]. The reactions go easily; 2-benzyloxypyridine 1-oxide dissolves in 20 per cent hydrochloric acid with the rapid separation of benzyl chloride[52a]. This may well be a case of nucleophilic attack upon the alkyl group. 4-Benzyloxy-pyridine 1-oxide is also easily converted into 4-hydroxypyridine 1-oxide[758].

As would be expected, quaternary derivatives of 2- or -4-alkoxy- or -aryloxy-pyridine can easily be converted into 1-alkyl-2- or -4-pyridones. Several examples are on record[392, 942, 949, 962], in all of which reaction was effected by alkali. 4-Methoxy-1-methylpyridinium iodide is also converted into 1-methyl-4-pyridone by heat[959a].

The formation of hydroxypyridines from nitropyridines is not a well-explored aspect of pyridine chemistry, though 2-hydroxypyridines have been obtained from 2-nitropyridines both by the action of acid[795] and of alkali[237, 713b]. Whilst 4-nitropyridine exposed to the atmosphere is slowly converted by autoquaternization and hydrolysis into 4-hydroxypyridine and 1-(4-pyridyl)-4-pyridone[431], the action of alkali upon 4-nitropyridine appears to be slow; even hot 50 per cent potassium hydroxide solution gives 4-hydroxypyridine slowly[713a]. 2-Nitropyridine is more reactive in this respect[713b]. 2- and 4-Nitropyridine react with sodium ethoxide to give 2- and 4-ethoxy-pyridine[713], but only the latter has been made to do so successfully with sodium phenoxide[713].

In contrast to the nitropyridines, the 4-nitropyridine 1-oxides are useful sources of 4-alkoxy- and -aryloxy-pyridine 1-oxides, into which they are converted by reaction with alkoxides and aryloxides[162, 247, 249, 251c, 537, 714a, 758, 852, 855, 963], or with alcohols and phenols[963a]. With alkali, 4-nitropyridine 1-oxide gives 4,4'-azopyridine[162], but if hydrogen peroxide is present a high yield of 4-hydroxypyridine 1-oxide can be obtained[850]. With halogen hydracids, 4-nitropyridine 1-oxide gives 3-halogeno- or 3,5-dihalogeno-4-hydroxypyridines[162, 849–50].

The nitro group in quaternary salts of 4-nitropyridine is easily replaced. Recrystallization of the methiodide from undried acetone gives 1-methyl-4-pyridone[964]. Reaction of 4-nitropyridine with benzyl chloride yields 1-benzyl-4-pyridone, and with benzyl bromide, 1-benzyl-3,5-dibromo-4-pyridone (nuclear bromination is thought to result from the oxidation of hydrobromic acid by nitrous acid); the experimental description suggests that in these reactions nucleophilic replacement of nitro by halide may occur initially[965]. The consequences of the autoquaternization of 4-nitropyridine have already been mentioned. The formation of 4-hydroxypyridine from 4-nitropyridine and acetic anhydride[713a] presumably involves the acetyl-pyridinium salt. 4-Nitropyridine 1-oxides give with acetic anhydride mainly 4-hydroxy-[966] or 4-acetoxy-3-nitropyridine 1-oxides[251c], but the presence of dimethylaniline suppresses nitration, and 4-hydroxy-[247, 966] or 4-acetoxy-pyridine 1-oxides[251c] are obtained in good yield.

The oxidation of pyridine thiols can give several sorts of compound, but reagents such as nitric and chromic acids, and potassium permanganate, convert some pyridine-2- and -4-thiols into hydroxypyridines[159b, 213, 769]. The formation of 2-methoxypyridines from methanolic ammonia and methyl 2-pyridyl sulphones has also been observed[833]. A few cases are known of the conversion of pyridine-2- and -4-sulphonic acids into hydroxy-[159b, 276, 710, 768a, 967] or alkoxy-pyridines[715a]. Of practical importance, because of the ready availability of pyridine-3-sulphonic acids, is their conversion into 3-hydroxypyridines by fusion with alkali[135a, 238, 282, 967-9]. 3-Alkoxypyridines are formed from sodium pyridine-3-sulphonates by heating with an anhydrous alcohol and copper hydroxide[970]. 2-Alkylthio-pyridine quaternary salts are converted by alkali into 1-alkyl-2-pyridones[411a, 971] but resist acid hydrolysis.

(i) Sulphur Bond Formation

Pyridine, when boiled with 40 per cent sodium bisulphite solution, gives a compound[972] $C_5H_5N.3NaHSO_3.2H_2O$. The structure of this product (p. 264) cannot be regarded as completely established, but carbon–sulphur bonds are not thought to be present.

Diphosphopyridine nucleotide and analogous quaternary pyridinium salts form complexes with bisulphite and with thiols[973-5]. The orientation of these additions is usually uncertain and may vary with the reaction solvent used[974]. The dithionite addition products formed by quaternary pyridinium salts have attracted interest because of their connection with the dithionite reduction of diphosphopyridine nucleotide (p. 259). Thus, the formation of a 1,4-dihydropyridine (122) by reduction of (120) has been represented[976] as proceeding through a 1,2-addition product (121). Regardless of the correctness or otherwise of structure (121), the relevance of the intermediate for diphosphopyridine nucleotide reduction by dithionite has been questioned, and the yellow intermediate formed in this reaction has been formulated[978] as a charge-transfer complex between the pyridinium nucleus and $S_2O_4^{2-}$ (see p. 261).

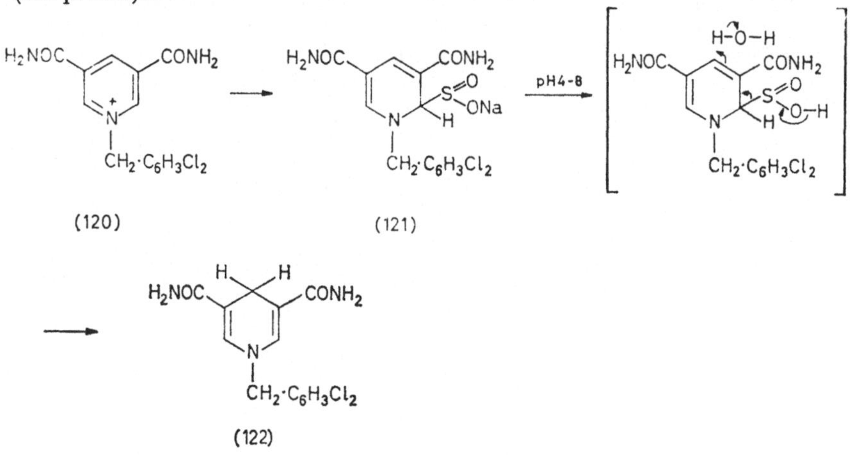

(120)   (121)

(122)

An interesting reaction occurs between 1-ethoxypyridinium ethylsulphate and sodium n-propylmercaptide in propyl mercaptan and ethanol, producing pyridine (70 per cent) and a mixture of 3- and 4-propylmercapto-pyridine in which the former predominates[979a]:

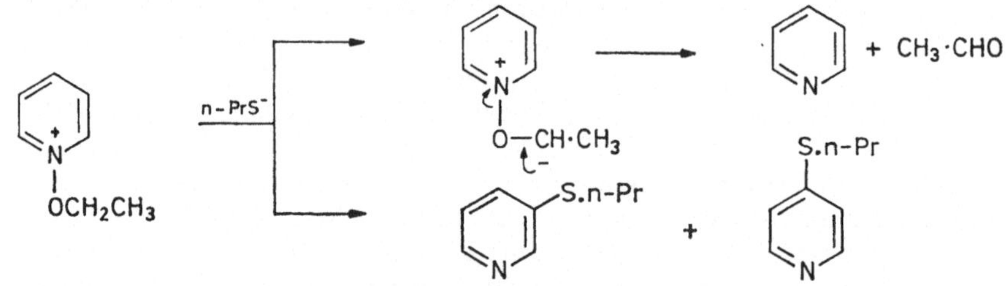

Butyl mercaptan with a number of quaternary derivatives of pyridine 1-oxide gives 2-, 3-, and 4-butylmercaptopyridine[980]. The formation of 2- and 4-substituted compounds is presumably a consequence of nucleophilic substitution, and the 3-substituted one has been represented as arising

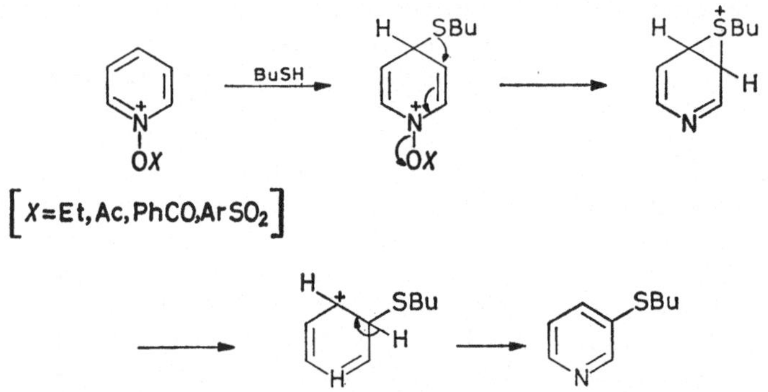

$$[X = Et, Ac, PhCO, ArSO_2]$$

With pyridine 1-oxide, sodium phenylmercaptide gives only pyridine, but with 1-alkoxy-2- and -4-picolinium salts, reaction occurs at the methyl groups producing phenyl-2- and -4-pyridylmethylsulphide, respectively. However, 1-alkoxy-4-picolinium salts react with alkylmercaptides both laterally and at the nucleus, giving by the latter mode both 2- and 3-alkylmercapto-4-methylpyridine. The 2-isomer presumably arises by direct nucleophilic substitution, and the 3-isomer may be formed[979b]

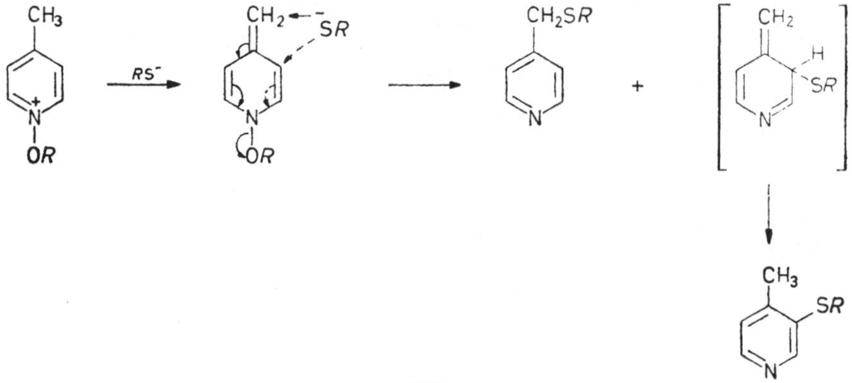

1-(4-Pyridyl)pyridinium chloride hydrochloride is a useful source of pyridine-4-thiol[438], 4-alkyl- and -aryl-mercaptopyridines[46a, 438] which are produced by reaction with hydrogen sulphide or mercaptans, and of pyridine-4-sulphonic acid, formed by reaction with sodium sulphite[290b, 981].

Mercaptans[793, 982] and thiocyanate ion[797b, 802, 983] react satisfactorily with diazotized 3- and 4-aminopyridines. Diazotized 3-aminopyridine does so likewise with potassium ethyl xanthate[979a]. In these reactions, diazotized 4-aminopyridine 1-oxide has not proved very useful[758, 984].

Nucleophilic replacement of halogen atoms is the most commonly used method of introducing sulphur-containing groups into pyridine derivatives. Numerous pyridine-2- and -4-thiols have been obtained in this way, usually by reaction with sodium or potassium hydrosulphide in an alcohol or glycol[159b, 277, 406, 793, 831, 852, 985-7]. β-Halogen atoms are usually unaffected, but pyridine-3-thiol is formed from 3-bromopyridine, potassium hydrosulphide and copper in propylene glycol[334] at 175-190°. Sodium sulphide gives dipyridyl sulphides[988]. 2- and 4-Alkylmercapto-[793, 979, 987] and -arylmercapto-pyridines[407, 929, 979a, 989] are similarly formed by the use of metal mercaptides. A useful variation for preparing pyridine-2- and -4-thiols consists in heating the halogenopyridine with thiourea in alcohol; the resulting thiouronium salts are easily converted into thiols[46a, 986, 990].

Pyridine-2- and -4-sulphonic acids are formed by treating the halogenopyridines with sodium sulphite[290a, 852]; 3-chloropyridine fails to react.

Most of these reactions producing sulphur derivatives from halogenopyridines are also readily effected with 2- and 4-halogenopyridine 1-oxides[290a, 689, 771, 852, 987, 991-2]. 4-Bromopyridine 1-oxide does not react with potassium thiocyanate, but the activation produced by a 3-bromo- or 3-nitro-group makes the reaction possible[991].

Quaternary derivatives of 2- and 4-halogenopyridines react readily with potassium hydrosulphide, giving 1-alkylpyrid-2- or -4-thiones[338, 408, 410a].

2- and 4-Pyridones and their 1-alkyl or 1-aryl derivatives when heated with 'phosphorus pentasulphide' give pyridthiones and 1-substituted pyridthiones[46a, 409, 993].

A few examples are known of the replacement by mercaptide groups of the nitro group from 4-nitropyridine 1-oxide[707, 987, 994] and also of its replacement by a sulphonic acid group[855].

RADICAL SUBSTITUTION

In the reactions to be considered a radical is thought to be the reagent which attacks the pyridine nucleus, forming at some stage a σ-complex from which, by whatever means, a hydrogen atom is lost, leaving a substituted pyridine. As will be seen, the mechanisms involved are disputed. A distinction must be made between radical substitutions such as these, and radical coupling processes involving pyridyl radicals (see e.g. p. 394).

(a) Alkylation and Substituted Alkylation

The earliest radical alkylation of pyridine appears to be that carried out by decomposing benzoylazotriphenylmethane in pyridine[1017]: the triphenylmethylpyridine formed seems not to have been orientated, but its physical properties resemble more closely those reported for 3-[131] than for

2-triphenylmethylpyridine[1018]. The mechanism of this, the Wieland reaction, is not certain, but it has been held to involve attack of the triphenylmethyl radical on the aromatic[1019].

Lead tetra-salts of fatty acids alkylate pyridine, giving 2- and 4-alkyl-pyridines[1020]. With red lead in acetic acid, 3-butylpyridine underwent mono-, di- and poly-methylation. The monomethyl compounds represented about 20 per cent of the product, and in these, substitution was distributed[1021] as shown in (123).

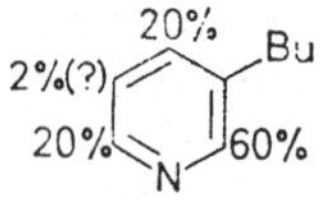

$$
\begin{array}{c}
20\% \quad \text{Bu} \\
2\%(?) \\
20\% \quad 60\% \\
\text{N}
\end{array}
$$

(123)

Diacyl peroxides can also be used to alkylate pyridine. From a study in which products were not isolated, the relative reactivities 4·2 : 3 : 1 for methylation by this method were deduced for chlorobenzene, pyridine and benzene, suggesting a slight activation in pyridine[1022]. The assumptions behind the method have been discussed[1019, 1023].

The results of experiments in which diacyl peroxides were decomposed in pyridine at 100° are given in *Table 5.31*, together with those of electrolysing anhydrous fatty acids in pyridine. The ratios of 2- to 4-alkylation were measured, but it seems likely that the apparent absence of the 3-isomers was a consequence of the analytical method used.

Table 5.31. Alkylation of Pyridine[1024]

Alkyl group	Yield of alkylpyridines*	Ratio of 2- to 4-isomer
Decomposition of diacyl peroxides:		
Me	86	7·6 : 1
Et	87	2·1 : 1
n-Pr	84	2·4 : 1
n-$C_{11}H_{23}$	38	3 : 1
Electrolysis of fatty acids:		
Me	3·5	2·8 : 1
Et	14	1·3–2·8 : 1
n-Pr	4	5·1 : 1

* The authors represent the reactions as radical couplings rather than substitutions, and the yields are based on the consumption of one acyl radical to generate a pyridyl radical, and one acyl radical to provide the alkyl radical for coupling therewith.

Some dialkylation was observed in some of these cases. Clearly, α-substitution predominated, as it did also[1025] in reactions involving peroxides [$MeO_2C.(CH_2)_2.CH_2CO_2$]$_2$.

The methylation of pyridine by passage with methanol over a heated catalyst[1026] may be a radical substitution.

(b) Arylation

More is known about radical arylation than about alkylation. Whilst still complex, it lacks some of the difficulties associated with alkylation. Radical arylation possesses practical importance and theoretical interest.

As regards the former, whilst overall yields are usually not high (*Table 5.32*), the reactions, particularly those using diazonium salts as the sources of radicals, provide easy routes to many compounds not otherwise readily available[1027]. Early preparative experiments created the impression that 2- and 4-substitution occurred, and because of the ease of isolation of the 4-substituted pyridine its importance was generally overestimated. This

Table 5.32. Radical Arylation of Pyridines

Compound	Radical	Source	Total yield*, per cent	Isomers, per cent	References
Pyridine	Ph·	PhN$_2$Cl (solid)	21	2(18), 2 + 4(3)	995
		PhN$_2$+	40	2(10), 3(4), 4(4)	996
		Ph·N(NO)Ac	60	2(14), 3(6), 4(6)	997
		PhN$_3$Me$_2$	51	2 > 3 > 4	998
		Ph$_2$IOH		} 2,3,4	999
		(PhCO$_2$)$_2$	35†		1000
		PhN$_2$CPh$_3$		4 > 3 > 2	1001
	p-Br.C$_6$H$_4$·	} ArN$_2$+	17	} 2 > 4	} 1002
	p-HO$_2$C.C$_6$H$_4$·		83		
	p-Cl.C$_6$H$_4$·		36		
	p-Cl.C$_6$H$_4$·	(ArCO$_2$)$_2$	27	2,4	1000
	p-EtO·C$_6$H$_4$·	ArN$_2$+		2 > 4	1002
	p-HO·C$_6$H$_4$·	Quinone diazide (*hν*)		?	1004
	o-MeO·C$_6$H$_4$·	ArN$_2$+	50	2,3,4	} 1003
	m-MeOC$_6$H$_4$·	ArN$_2$+	30	} 2,4	
	p-MeOC$_6$H$_4$·	ArN$_2$+	54		
	p-MeOC$_6$H$_4$·	(ArCO$_2$)$_2$	27		1000
	o-O$_2$N·C$_6$H$_4$·	} ArN$_2$+	} 35	2,3,4	} 996
	m-O$_2$N·C$_6$H$_4$·			2,3,4	
	p-O$_2$N·C$_6$H$_4$·		75	2(24) > 3(9) > 4(4·5)	
	p-O$_2$N·C$_6$H$_4$·	O$_2$N.C$_6$H$_4$N$_2$ONa	27	2(15) > 3(5) > 4(2)	
	p-O$_2$N·C$_6$H$_4$·	ArN$_3$Me$_2$	50	2,3,4	998
	p-O$_2$N·C$_6$H$_4$·	ArN(NO)Ac		2(14·5) > 3(5) > 4(2·5)	1005
	p-RSO$_2$NHC$_6$H$_4$·	quinone diazide (*hν*)		?	1004
	m-Ph.C$_6$H$_4$·	} ArN$_2$+		} 2,4	} 1006
	p-Ph.C$_6$H$_4$·				
	1-C$_{10}$H$_7$·	(ArCO$_2$)$_2$			} 1000
	2-C$_{10}$H$_7$·	(ArCO$_2$)$_2$		2,3,4	
	2-C$_{10}$H$_7$·	ArN$_3$Me$_2$	41	2,3,4	998
	3-C$_5$H$_4$N·			2(55) > 3(5) > 4(20)	1007
	3-C$_9$H$_6$N·		44	2 > 4	
	5-C$_9$H$_6$N·		26	2	} 1008
	8-C$_9$H$_6$N·		14	2(3,4)	
	8-(6-α-pyridyl)-quinolyl	} ArN$_2$+	21		1009
4-Methyl-pyridine	3-C$_5$H$_4$N·			2(15),3(55)	} 1007
4-Ethyl-pyridine	3-C$_5$H$_4$N·			2(14),3(55)	
Ethyliso-nicotinate	3-C$_9$H$_6$N·			2	1010
Pyridine 1-oxide	Ph·	PhN$_2$NH.Ph			1031

Table 5.32 (continued)

Compound	Radical	Source	Isomers, (per cent)	References
		isomer proportions		
Pyridine	Ph.	Ph in $N_2{}^+(40°)$	2(53·5),3(29·2),4(17·2)	1011
		PhN(NO)Ac (105°)‡	2(46), 3(43), 4(11)	⎫
		Ph.I.O.OC Ph (105°)	2(58), 3(28), 4(14)	⎬ 1012
		$(PhCO_2)_2$ (70°)	2(58), 3(28), 4(14)	1013
		$(PhCO_2)_2$ (105°)	2(54), 3(32), 4(14)	⎫
		PhN_2CPh_3 (105°)**	2(53), 3(31), 4(16)	⎬ 1012
		$Pb(O_2CPh)_4$ (105°)	2(52), 3(32·5), 4(15·5)	⎭
		$PhCO_2H$		
		(electrolysis)	2(56), 3(35), 4(9)	1014
		Ph_3Bi ($h\nu$) §	2(48), 3(31), 4(21)	1015
	o-Me·C_6H_4·	⎫	2(52·7), 3(29·6), 4(17·4)	⎫
	p-Me·C_6H_4·	⎬ $ArN_2{}^+$ (40°)	2(58·1), 3(26·2), 4(15·7)	⎬ 1011
	o-O_2N·C_6H_4·	⎬	2(42·5), 3(51), 4(15)	⎬
	p-O_2N·C_6H_4·	⎭	2(44·0), 3(42·7), 4(12·9)	⎭

* The total yield is often difficult to calculate from the description provided, and frequently refers to crude product.

† The low yield is probably due to the formation of pyridine 1-oxide; with some peroxides no arylation was observed.

‡ In this reaction the true isomer proportions are modified by side reactions.

** The isomer proportions are corrected to allow for the formation of a side product[1012, 1010].

§ The isomer proportions from this reaction are, for several reasons, less reliable in their implications than those from the others[1015].

impression has been corrected by quantitative studies of isomer distributions (*Table 5.32*).

As regards quantitative studies, a number of problems arise. The main question is whether the arylations can be represented by a sequence of the following kind, in which the rate-controlling step produces a σ-complex of the Wheland type, from which a hydrogen atom is removed, probably by a second radical:

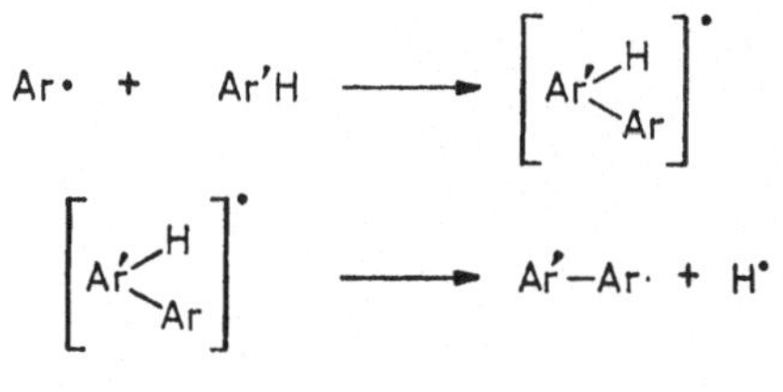

$$Ar \cdot \ + \ Ar'H \ \longrightarrow \ \left[Ar' {\overset{H}{\underset{Ar}{\diagdown}}} \right]^{\cdot}$$

$$\left[Ar' {\overset{H}{\underset{Ar}{\diagdown}}} \right]^{\cdot} \ \longrightarrow \ Ar'-Ar \cdot \ + \ H^{\cdot}$$

The body of evidence from radical arylation of aromatics gives some support to such a scheme[1019, 1023]. The question of the meaningfulness of the isomer ratios, and thence of the partial rate factors for these reactions, is tied up with that of the fate of the intermediate formed in the rate-determining step. The experimentally determined ratios clearly have no direct value as indices of positional reactivity if the σ-complex suffers fates other than aromatization, and the relative reactivities of different aromatics will be obscured by similar causes and also if the aromatic can be removed differentially by a different kind of reaction. The latter possibility certainly exists with pyridine; peroxide can produce pyridine 1-oxide and bipyridyls, as well as arylpyridines[1000]. For these reasons the significance of the quantitative experiments have been disputed[1019, 1023, 1028].

Table 5.33. Relative Reactivities in Arylations[1011, 1019]

Compound	Radical source		Relative reactivity (Benzene, 1·0)	Partial rate factors,		
				$F_{\alpha(or\ o)}$	$F_{\beta(or\ m)}$	$F_{\gamma(or\ p)}$
Pyridine	$(PhCO_2)_2$	(105°)*	1·04	1·7	1·0	0·9
	$(PhCO_2)_2$	(80°)*	1·04†	1·8	0·87	0·87
	$PhN(NO)Ac$		1·0			
	Ph_3Bi		1·2	1·7	1·1	1·5
	PhN_2^+	(40°)	1·14	1·83	1·00	1·18
	$o\text{-}Me\cdot C_6H_4N_2^+$	(40°)	1·72	2·72	1·53	1·80
	$p\text{-}Me\cdot C_6H_4N_2^+$	(40°)	1·44	2·51	1·13	1·36
	$o\text{-}O_2N\cdot C_6H_4N_2^+$	(40°)	0·47	0·60	0·71	0·21
	$p\text{-}O_2N\cdot C_6H_4N_2^+$	(40°)	0·78	1·03	1·00	0·604
$PhNO_2$	$(PhCO_2)_2$	(80°)	4·0	7·5	1·2	6·6
	PhN_2^+	(40°)	2·94			
PhCl	$(PhCO_2)_2$	(80°)	1·4	2·2	1·4	1·6
PhMe	$(PhCO_2)_2$	(105°)	1·23	2·5	0·71	1·0

* The data given for these reactions are composite, cf.[1011, 1029-30].
† For reaction at 70°, 1·5 is given[1011].

With these qualifications in mind, it can be said that the figures (*Table 5.33*) seem to show the α-position to be generally the most reactive, whilst differences between β- and γ-positions are small. The data for attack by *o*-nitrophenyl and *o*-tolyl radicals[1011] are thought to reveal an attraction between the nitro group and the nitrogen lone pair, and a repulsion between the methyl group and the latter in the transition states leading to α-substitution. The nitrogen atom of pyridine appears to cause very slight activation with respect to phenylation, and mild deactivation with respect to attack by the more electrophilic nitrophenyl radicals. In the smallness of the effects which it produces, the nitrogen atom resembles substituents attached to the benzene ring, and this feature of radical arylation distinguishes the reactions markedly from heterolytic substitutions.

An interesting recent development has been the study of radical phenylation of pyridine metal complexes. By comparison with pyridine the complexes show an enhancement of reactivity at $C_{(4)}$ and often at $C_{(2)}$. The effect has been ascribed to back-donation of electrons from the metal to the pyridine ring by *d–p* π-electron conjugation[997].

It is generally accepted that some cyclizations, such as the Pschorr phenanthrene synthesis, can proceed by either a heterolytic or a homolytic process, depending upon the conditions used[1032-3]. Whilst by the former route intramolecular substitution (electrophilic) into a pyridine ring would be unlikely, successful cyclization by the radical process would not be surprising and is most probably observed in the reactions[1034]

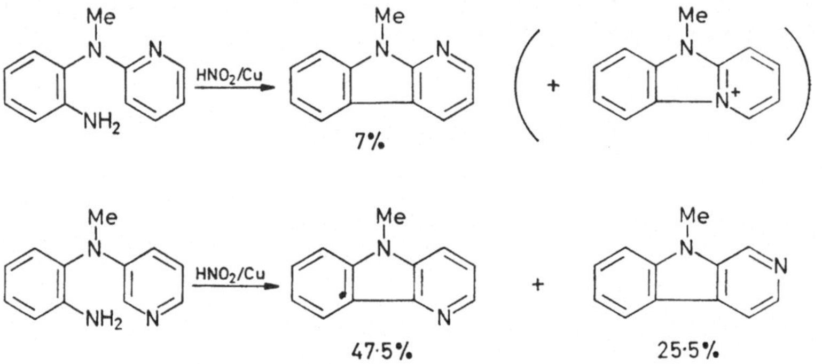

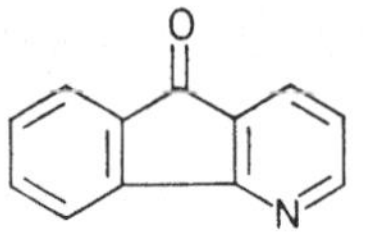

7%

47·5%   25·5%

(124) 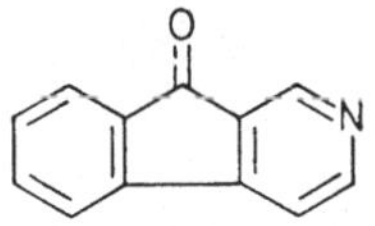   (125) 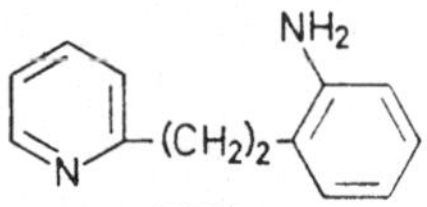   (126)

The formation of small yields of (124) and (125) when 3-(o-amino-benzoyl)pyridine is diazotized in the presence of copper or ultra-violet light also points to a radical substitution process[1035]. When (126) is diazotized in the presence of copper, or decomposed through its N-acetyl-N-nitroso-derivative, a small degree of cyclization occurs[1033]. Other examples are known[1036].

(c) Halogenation and Nitration

The possibility that in electrophilic halogenations and nitrations the use of high temperatures causes the incursion of radical processes has already been mentioned (pp. 165–6, 171).

(d) Miscellaneous Reactions

Pyrolysis of 3-(o-azidophenyl)pyridine gives α- and γ-carboline[1037]. Although the process does not formally involve a free radical (:N.C_6H_4.C_5H_4N), it is reminiscent of a radical substitution. The mechanism of the Graebe–Ullmann reaction, the formation of carbazoles by the pyrolysis of 3-arylbenzotriazoles, has not been elucidated, but some examples of it, notably the formation of carbolines from 3-pyridylbenzotriazoles, suggest the character of a radical substitution.

(3) REACTIONS WHICH MODIFY THE NUCLEUS

Several reactions to be discussed here are not strictly those of the pyridine ring but of products formed from it. The ring openings during some reductions are often reactions of partially reduced pyridine rings. The basic character of the true initial reaction is sometimes discernible as belonging

to one of the three classes of reactions discussed above. Already in the discussion of nucleophilic substitution examples have been noted where the
reaction stops at the addition stage, substitution not being completed.

(a) Additions

Ladenburg[1038] described the reduction of pyridine and some of its
homologues to piperidines by sodium and alcohol, and the method has occasionally been used for this purpose[1039]; it is useful with pyridine-carboxylic
acids and their side-chain vinylogues which give piperidine-carboxylic
acids[1040]. Esters of pyridine-carboxylic and vinylogous side-chain acids give
the fully saturated alcohols[1041].

Pyridine and its homologues have also been reduced to piperidines electro-
chemically[1042]. Dipiperidyls are side products.

Generally, however, sodium-alkanol reductions are not suitable for completely saturating the pyridine nucleus, tetrahydropyridines being present
in the product[280, 841e, 1043, 1116]. Sodium butanol gives better yields of
tetrahydropyridines than does sodium ethanol[280]. The products have in
some cases been identified as $\Delta^{3,4}$-compounds[280, 841e]. For the reduction of
pyridines to piperidines a suitable process is reduction with sodium and
ethanol, followed by removal of the remaining unsaturation by hydrogenation[280, 730, 1044].

Shaw[1045] found that by using sodium and 95 per cent ethanol, rather than
absolute ethanol, little or none of the piperidine was formed and ring-
opened products were obtained which arose from 1,4-dihydropyridines
produced in the reduction. Sodium and alcohol in liquid ammonia have been
used for the same purpose[1046] (see below). The formation of 1,4-dihydro-
pyridines can be represented

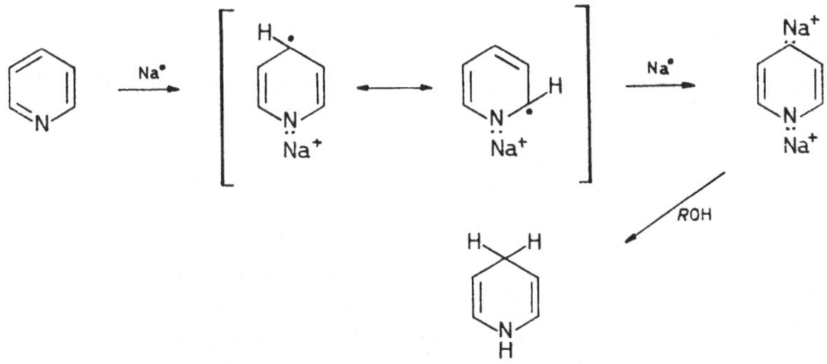

2- and 4-Aminopyridines are reduced by sodium ethanol to give small
amounts of aminopiperidines, but mainly piperidines or open-chain amines
by loss of ammonia[670b, 1047]. Electrolytic reduction is similar[671].

3- and 4-Hydroxypyridines have been reduced by sodium ethanol to
hydroxypiperidines[623a, 1048–9], but yields are usually poor. Alkoxyl groups are
eliminated both by sodium ethanol and electrolytic reduction[670b, 1050].
1-Alkyl- and -aryl-4-pyridones are reduced by sodium ethanol to 4-hydroxy-
piperidines[1051] (though failures have been reported[961]), and sodium in
ammonia has been used for this purpose[1150]. Sodium amalgam has been

employed to convert a 1-alkyl-2-pyridone into the 1-alkyl-2-piperidone[1107].

Electrolytic reduction of pyridine-2- and -4-carboxylic acids can be used to give methylpyridines[1052], and zinc and acetic acid produce either the methylpyridines or the hydroxymethylpyridines[1053]. However, from nicotinic acid the second method produces piperidine, and electrochemical reduction of these acids can be conducted to give piperidine-carboxylic acids, tetrahydropyridine-carboxylic acids and tetrahydro-methylpyridines[1053-4].

Esters of pyridine-dicarboxylic acids are reduced to dihydro-compounds by aluminium and wet ether[1055].

Although substituents attached to the pyridine ring can be reduced by means of lithium aluminium hydride without the ring being affected (e.g. esters of pyridine-carboxylic acids give hydroxymethylpyridines[1056]) but see p. 261), pyridine can nevertheless be attacked by the reagent[1057]. A dihydropyridine may be formed[1058], but small yields of piperidine have also been obtained[1059] (see also p. 367). A crystalline yellow solid prepared from pyridine and lithium aluminium hydride has been shown to contain both 1,2- and 1,4-dihydropyridine groups, and to be lithium tetrakis-(N-dihydropyridyl) aluminate with a structure of type (127). It is a useful specific reducing agent for ketones[1060].

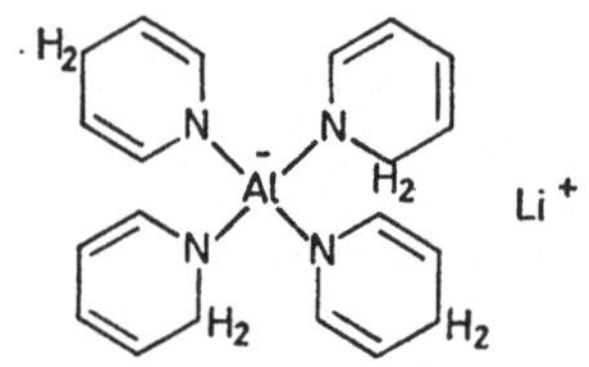

(127)

2-Pyridone gives a small amount of piperidine, as well as some pyridine, but 1-methyl-2-pyridone yields only a trace of methylpyridinium (picrate)[1059].

Reduction of alkylpyridinium salts, electrolytically, with sodium amalgam or by x-ray irradiation in aqueous ethanol, can give 1,1'-dialkyl-1,1',4,4'-tetrahydro-4,4'-dipyridyls[1061-5, 1100, 1108]. Electrolytic reduction of a mixture of 1-methyl- and -benzyl-pyridinium salts gives the three possible 4,4'-dipiperidyls as well as 1-methyl- and 1-benzylpiperidine[1066]. Binuclear compounds produced from quaternary salts of nicotinamide by chromous acetate, magnesium or the zinc–copper couple have been formulated as tetrahydro-2,2'-dipyridyls[1067].

Sodium amalgam reduces 1-phenylpyridinium chloride to give not only the binuclear product but also 1-phenyl-1,4-dihydropyridine[1068-70]. 4-Alkyl-3,5-diethoxycarbonyl-1,2,6-trimethylpyridinium salts give binuclear products, but the 4-aryl analogues result in dihydropyridines[1065, 1071].

Various reducing agents produce from 1-methyl-2,4,6-triphenylpyridinium salts a blue free radical[1072]. The properties of 1,1'-dialkyl-1,1',4,4'-tetrahydro-4,4'-dipyridyls have been interpreted in terms of dissociation into mononuclear free radicals, and the dissociation evidently occurs more readily with the tetrahydro-dipyridyls formed from 4-substituted pyridinium salts[1071, 1073]. When 4-substituents are absent, the properties are now attributed to the conversion of the tetrahydro-dipyridyls into quinonoid structures (128). Two-electron oxidation converts these into the bis-quaternary structures, and the intermediate semi-quinones (129) are known[1068, 1073-5]. Polarographic reduction of the 4,4'-bispyridinium quaternary salts generates the intensely coloured semiquinone (129), and the second step gives dipyridylenes (128) irreversibly. Dithionite also produces the radical, but two-electron reagents such as sodium amalgam and zinc and alkali give the irreversible second stage[1076].

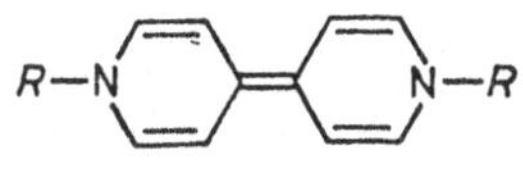

(128) 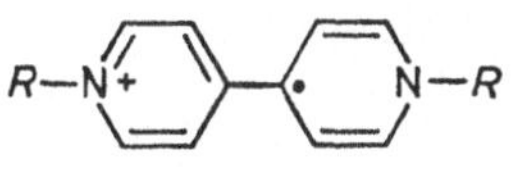   (129) 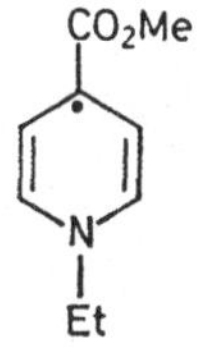   (130)

Reduction of some quaternary salts with chromous acetate gives radicals which quickly dimerize but in some cases are considerably stabilized by resonance. Such a case is the radical from the dichlorobenzyl quaternary salt of isonicotinamide. Dithionite can also be used as the reducing agent[1077]. Similar observations on 1-ethylpyridinium salts were made, using electrochemical methods, and again 4-methoxycarbonyl and 4-carboxamido groups stabilized the radicals[1078]. The radical (130), so observed, has also been prepared from the quaternary perchlorate by reduction with sodium, magnesium, zinc or aluminium in acetonitrile. It is a dark green oil which can be distilled; spectra of it have been recorded[1079] (see p. 394).

The most intensively studied of quaternary salt reductions has been the enzymatic reduction of diphosphopyridine nucleotide, DPN (131) and, because Karrer and Warburg[1080] recognized that this involves the quaternary nicotinamide nucleus, the reduction of simple quaternary derivatives of nicotinamide with dithionite. The nicotinamide nucleus is reduced to the dihydro-stage. In biological oxidation, DPN accepts hydrogen from the substrate being oxidized, giving DPNH which transfers hydrogen to acceptors, regenerating DPN.

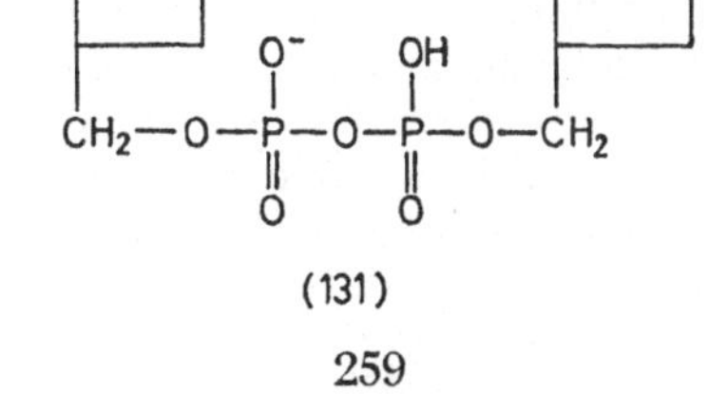

(131)

259

Tracer studies show that, in enzymatic reduction of DPN, the substrate being oxidized transfers hydrogen directly to the coenzyme, and that solvent (water) molecules are not involved[1081]:

$$\text{MeCD}_2\text{OH} + \text{DPN} \xrightarrow[\text{H}_2\text{O}]{\substack{\text{Alcohol}\\ \text{dehydrogenase}}} \text{MeCDO} + \text{DPND} + \text{H}^+$$

Furthermore, the reduction is stereospecific[1082]:

$$\text{DPND} + \text{MeCHO} \underset{\text{Enzyme}}{\overset{}{\longleftrightarrow}} \text{DPN} + \overset{\text{H}}{\underset{\text{D}}{\text{Me}-\overset{|}{\underset{|}{\text{C}^*}}-\text{OH}}}$$

The deuterated coenzyme, DPND, formed from DPN and MeCD$_2$OH in the presence of enzyme, transfers all its deuterium to Me.CHO on enzymatic reoxidation, whereas DPND formed from D$_2$O and dithionite transfers only half of its deuterium in this process; the enzymatic reduction is stereospecific but the dithionite reduction is not[1083]. Quaternization and Decker oxidation of nicotinamide obtained from DPND, formed by both enzymatic and dithionite reduction of DPN, gave 3-carboxamido-1-methyl-2- and -6-pyridone, both of which were still labelled (the enzymatic product half as much as the dithionite product). Thus, the quaternary nicotinamide nucleus of DPN is converted into the 1,4-dihydro form both by enzymatic and dithionite reduction[1084]. The same point was also proved for various quaternary derivatives of nicotinamide and for DPN by observing that the product of dithionite/D$_2$O reduction could enzymatically transfer deuterium to an acceptor, whilst quaternary salts synthesized with deuterium at C$_{(2)}$ or C$_{(6)}$, and then reduced with dithionite/H$_2$O, could not[1085].

The 1,4-dihydro structure of the dithionite reduction product of nicotinamide methochloride has been conclusively demonstrated by these methods[1086] and by nuclear magnetic resonance spectroscopy[1087]. The last technique has also been used to demonstrate the 1,4-dihydro structure of the product from 3,5-di-ethoxycarbonyl-1,2,6-trimethylpyridinium and dithionite[1088].

Before truly reliable criteria for distinguishing between 1,2- and 1,4-dihydropyridines were available, the products of dithionite reduction of many quaternary salts, mainly nicotinamide derivatives, had been examined, notably by Karrer and his co-workers[1089, 1094], and wrongly formulated as 1,2-dihydropyridines. Whilst each example requires careful examination, it seems to be almost generally true that dithionite reduction of pyridinium salts gives 1,4-dihydropyridines[1065, 1077, 1090-2, 1101]. An exception is the salt (132); in this, steric hindrance at $C_{(4)}$ and $C_{(6)}$ is thought to lead to 1,2-reduction[1093].

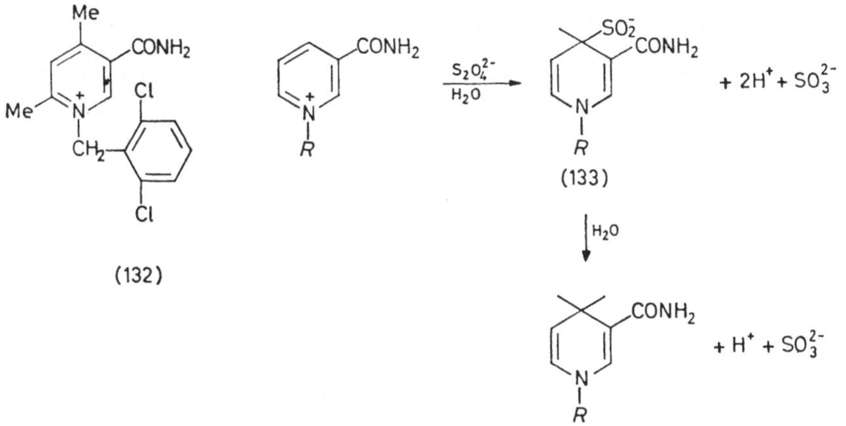

(132)

(133)

The mechanism of dithionite reduction is not fully established. Yellow intermediates formed in these reductions have been represented[1086] as products of nucleophilic attack at $C_{(4)}$, e.g. structure (133). In contrast, a crystalline intermediate isolated in one instance has been formulated as arising from nucleophilic attack at $C_{(2)}$ (p. 249). The site of this attack is dependent upon substituents already present, and it may be that more than one mechanism can operate. The intermediate represented as (133) has also been formulated as a charge transfer complex between the pyridinium nucleus and $S_2O_4^{2-}$, a formulation which is held alone to lead specifically to 1,4-dihydro-products[978] (pp. 273–281).

Reduction of quaternary pyridinium salts by metal hydrides presents interesting features. Sometimes $\Delta^{3,4}$-tetrahydropyridines result from boro-hydride reduction[1054, 1095, 1159]. Borohydride in presence of alkali has been used for this purpose[600, 1096], and some reductions give dihydropyridines in concentrated alkali and tetrahydropyridines in less alkaline media[1097].

1-Phenylpyridinium cation is reduced by borohydride to the 1,2-dihydro-compound, which probably escapes further reduction because of its in-solubility[1070]. Usually borohydride reduction of a quaternary salt gives mainly the 1,2-dihydropyridine, or this together with the 1,4-dihydro-pyridine; with 3-substituted salts, 1,6-reduction predominates[1077, 1091, 1098–1101, 1159]. The cation (132) is unusual in giving the 1,2-dihydropyridine, also formed by dithionite reduction[1093].

Lithium aluminium hydride has occasionally been used in these reductions, giving in one case a dihydropyridine[1094], in another a $\Delta^{3,4}$-tetrahydro-pyridine[1054].

According to Ferles, all borohydride reductions of quaternary salts con-form to a course of reaction initiated by 1,2-attack, followed by the further step of tetrahydropyridine formation[1101-2]. Some detailed evidence about this is available. The production of tetrahydropyridines involves initial nucleophilic attack to give 1,2-dihydropyridines. 1,4- and 1,6-Dihydropyri-dines (in the case of 3-substituted compounds) resist further reductions by

261

borohydride. Electrophilic attack upon the enamine system of a 1,2-di-
hydropyridine will produce an unsaturated ammonium compound which, by
further attack of borohydride, will give the tetrahydropyridine. This is
supported by the isolation of only 1,2-dihydropyridines from reduction of
1,3,5-trisubstituted pyridinium salts, the substituents presenting steric
interference to attack upon the enamine system by the electrophile. Further,
borohydride reduction of pyridinium ions in aprotic solvents gives
dihydropyridines which undergo further reduction only when a proton
source is added:

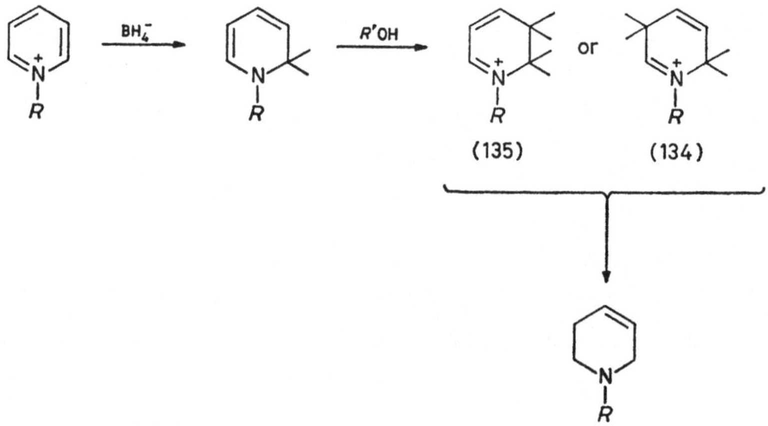

(135)   (134)

That proton attack occurs at the middle of the enamine system giving
(134) rather than terminally giving (135), is proved for the following
example[1103] by the formation of (136) and (137).

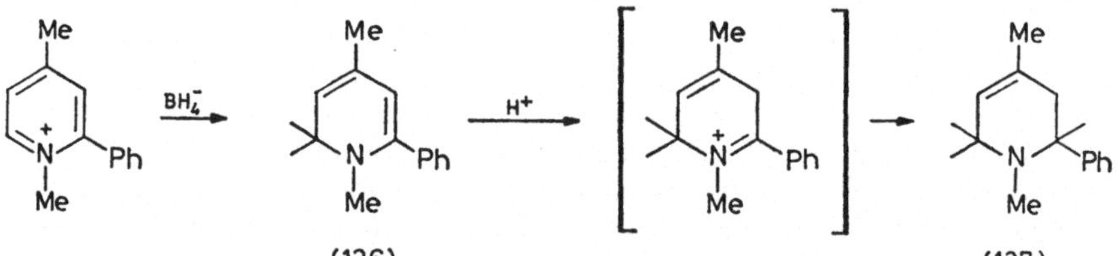

(136)   (137)

Reduction of quaternary pyridinium salts by formic acid containing
potassium formate, which recalls the Leuckart reaction, is basically a nucleophilic attack by hydride ion:

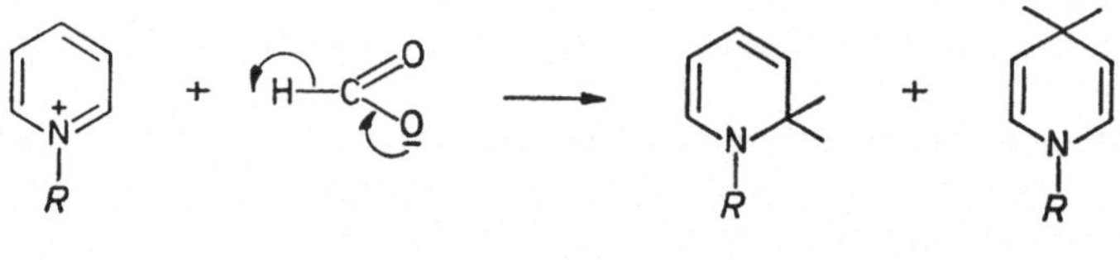

The products are $\Delta^{3,4}$-tetrahydropyridines and piperidines[1104], and sometimes cyclohexenones formed by ring opening of 1,4-dihydropyridines[1039, 1105].
If initial attack occurs at $C_{(2)}$, further reduction of the vinylamine linkage

262

gives a $\Delta^{3,4}$-tetrahydropyridine, but if at $C_{(4)}$, both remaining vinylamine linkages are reduced, giving a piperidine. The ratio of products then indicates the proportion of initial attack at $C_{(2)}$ to that at $C_{(4)}$; it is greater than unity in all the examples examined except that of the 1,2,6-trimethyl-pyridinium ion[1102].

The reduction of pyridine by zinc and acetic anhydride presumably involves the 1-acetylpyridinium cation. 1,1'-Diacetyl-1,1',4,4'-tetrahydro-4,4'-dipyridyl is formed[1160-2], and the reaction has frequently been used, for this product is a useful source of 4-ethylpyridine[1161, 1163]. The reaction has limited usefulness with acid anhydrides other than acetic anhydride[1164] and in preparing 4-alkylpyridines[280, 1164-5], or in some cases 4-acetylpyridines[1167]. With zinc and ethyl chloroformate, pyridine gives 1,1'-diethoxycarbonyl-1,1',4,4'-tetrahydro-4,4'-dipyridyl[1166].

Electrochemical reduction has been mentioned. The polarographic behaviour of various derivatives is reported at the appropriate point in Chapter 6.

The catalytic hydrogenation of pyridines has been effected in many ways[1218]. From pyridine and its homologues piperidines are readily obtained[841c,e, 1109-14]. The ring opening which occurs in some cases is mentioned below (p. 265). Pyridines tend to poison catalysts, but the hydrochlorides are readily hydrogenated[1115]. Hydrogenation in presence of Raney nickel and an alcohol can cause N-alkylation[1116], which is not noticed with ruthenium oxide[1114]. 2-Alkylpyridines are hydrogenated about half as fast as pyridine[1112]. Kinetic studies of the hydrogenation of pyridine with Adams' catalyst have been interpreted to mean that the pyridinium cation undergoes hydrogenation and not pyridine itself. This may be because the nitrogen lone pair causes pyridine to be adsorbed edgeways in a position unfavourable for reaction[1117]. Convenient laboratory conditions for the hydrogenation use platinum oxide in acetic acid or aqueous or alcoholic hydrochloric acid.

Hydrogenation of phenyl- and benzyl-pyridines saturates the heterocyclic rather than the carbocyclic nucleus[6k, 618, 876a, 1110, 1119-22].

In the hydrogenation of the pyridine nucleus, double bonds conjugated with the nucleus are saturated[278, 1123-5].

Ketone groups in side chains (including N-phenacyl groups) may be reduced to carbinol groups before or after hydrogenation of the nucleus[6k, 1126]. Carboxyl, ester and amide groups in side chains survive the hydrogenation of the nucleus[278, 618, 1121-3, 1125, 1127-31]. If a carbonyl or ester group is suitably placed in a side chain, cyclization can accompany hydrogenation, producing derivatives of indolizidine and quinolizidine[1132].

Alkoxyl-substituted side chains are not affected by nuclear hydrogenation[730].

Some functional groups attached to the nucleus may be lost during nuclear hydrogenation. Thus, 2-aminopyridines can give an amino-tetrahydropyridine or piperidine and ammonia[1050, 1118]. 2-Acetylaminopyridine, hydrogenated in acetic anhydride–acetic acid mixture, gives the diacetyl derivative of 2-aminopiperidine[1047b]. Whilst various attempts to hydrogenate 4-aminopyridine failed[1133], 3-aminopyridine readily gave 3-aminopiperidine[1134].

Nuclear-attached carbonyl groups may be reduced during nuclear hydrogenation, as in the conversion of 3-acetylpyridine into 3-ethyl- and 3-hydroxyethyl-piperidine[1111], or may partially survive[1135]. Nuclear carboxyl, ester and amide groups usually survive catalytic hydrogenation, and the method is valuable for preparing piperidine-carboxylic acids and their derivatives[329c, 1053a, 1095, 1111, 1136–41]. Reduction of ethyl pyridine-2-carboxylate with W2 Raney nickel gives 2-hydroxymethylpiperidine[1142].

Nuclear halogen atoms can readily be removed catalytically and, consequently, they do not in general survive nuclear hydrogenation.

Hydrogenation of 2-pyridone or 1-alkyl-2-pyridones gives 1-alkyl-2-piperidones[385, 500a, 876a, 1050, 1143–5] or 1-alkylpiperidines[1146]. 2-Methoxypyridine yields piperidine[1050]. 3-Hydroxypyridines give 3-hydroxypiperidines[969b, 1147], sometimes with some hydrogenolysis of the hydroxyl group[1148], though a failure is known[500a]. Similarly, failure to hydrogenate 4-pyridone and 1-methyl-4-pyridone has been reported[500a, 1149], but under suitable conditions 1-alkyl-4-pyridones give 1-alkyl-4-piperidinols[1150], sometimes with some hydrogenolysis of the hydroxyl group[1152]. Whilst 2- and 4-pyridyl benzoates produce 2- and 4-hydroxypyridine (and then 2-piperidone in the first case) and toluene, and 3-pyridyl benzoate is unchanged[500a], 3-pyridyl diphenylacetate gives piperidine and 3-hydroxypiperidine[1148]. 2-, 3- and 4-Pyridyl p-toluenesulphonates gave pyridine and piperidine, the reduction being fastest and most complete with the 3-isomer[500b].

It was stated above that pyridinium ions undergo hydrogenation more readily than do pyridines. It is not surprising, then, that quaternary pyridinium salts are easily hydrogenated[636, 1115–6, 1153]. The method has been used frequently with both side-chain[1151] and nuclear esters[329c, 1095, 1154]. As well as piperidines, $\Delta^{3,4}$-tetrahydropyridines are sometimes formed[1095, 1154].

Double bonds conjugated with the nucleus are reduced[1120, 1156], whilst si de-chain ketone groups can survive nuclear hydrogenation[1157].

Hydrogenation of 3-hydroxy-1-phenylpyridinium chloride can be effected with or without hydrogenolysis of the hydroxyl group[1158].

Two ring modifications quite different from the reductions discussed above deserve mention. One of these, the formation of an addition product from pyridine and sodium bisulphite, has been mentioned already (p. 249). The product has been formulated as the sodium salt of the trisulphite ester of 2,4,6-trihydroxypiperidine[972], but this structure cannot be regarded as established, and carbon–sulphur bonds may be present.

The other is the photo-dimerization of 1-methyl-2-pyridone[1106, 1168] to give (138). 2-Aminopyridine hydrochloride behaves similarly[1168].

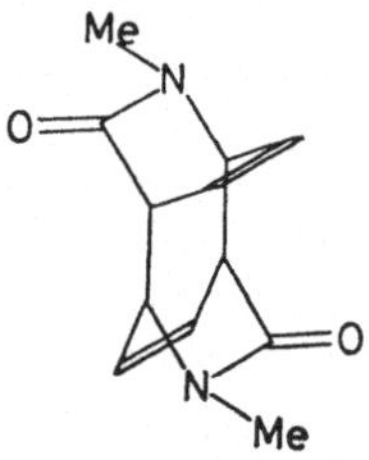

(138)

(b) *Disruptions*

At relatively low temperatures, and also under ultra-violet irradiation, pyridines give bipyridyls. However, pyrolysis at 825°–850° produces quinoline, acetonitrile, benzonitrile, acrylonitrile and benzene. The picolines behave similarly[1169]. Ultrasonic cleavage of pyridine gives acetylene[1170].

Permanganate oxidation slowly converts pyridine into carbon dioxide and ammonia[1171]. Oxidation of phenyl- or benzyl-pyridines with acidic permanganate produces pyridine-carboxylic acids, but with neutral permanganate, benzoic acid[1172]. Photochemical oxidation of pyridine gives the ammonium salt of glutacondialdehyde[1173].

Pyridine and its homologues consume two molecules of ozone for each molecule of heterocycle. The carbon–nitrogen bond is split not by ozonization but by hydrolysis, and the nitrogen is liberated as an amide and then, by hydrolysis, as ammonia[1174]. The rates of ozonization in chloroform of pyridine and a number of its homologues have been measured. The relative rates, $30 : 8.4 : 3.8 : 1$ for toluene, a-picoline, benzene and pyridine, show the activating effect of the alkyl group and, more important, the deactivating effect of the nitrogen atom[1175].

A number of ring openings occurring in the reduction of pyridines are really reactions of intermediate hydropyridines. Only those starting from pyridines will be mentioned here. Catalytic hydrogenation over nickel catalysts can convert pyridine and its homologues into alkylamines or alkanes and ammonia[1113, 1176]. Hydriodic acid also will reduce pyridine to pentane and ammonia[1117]. The formation of ring-opened products from 1,4-dihydropyridines formed by reducing pyridine with sodium and ethanol or sodium and ammonia has already been mentioned (p. 257). Sodium amalgam converts pyridine-carboxylic acids into nitrogen-free products through dihydropyridines[1178].

More interesting and important than these ring openings are those effected by nucleophilic attack upon a quaternary pyridinium compound or upon a substance formed therefrom. Generally such reactions are of the type

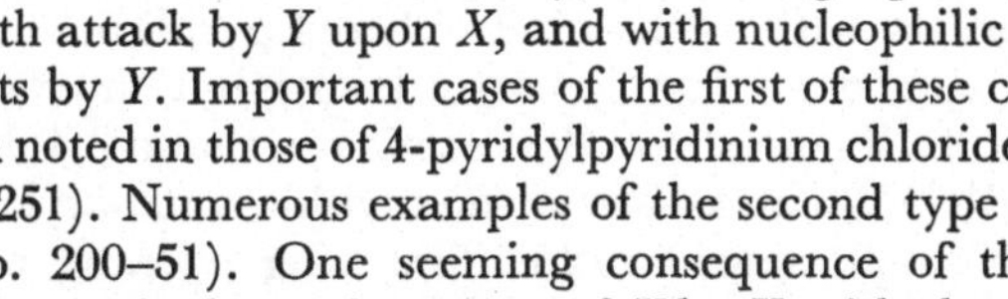

The final consequences of the reaction will depend upon X and Y and upon the experimental conditions. Obviously, such ring openings must be in competition with attack by Y upon X, and with nucleophilic replacement of ring substituents by Y. Important cases of the first of these competing reactions have been noted in those of 4-pyridylpyridinium chloride hydrochloride (pp. 210, 239, 251). Numerous examples of the second type have also been mentioned (pp. 200–51). One seeming consequence of the nucleophilic attack in some cases is the replacement of X by Y, with the production of a new quaternary compound. This happens by recyclization of the ring-opened structure (see below).

The most important cases of nucleophilic ring opening are those discovered by Zincke[414, 1155, 1179] in which $(X = $ 2,4-dinitrophenyl$)$, by König[545] $(X = $ CN$)$ and by Baumgarten[1180] $(X = $ SO$_3)$.

Zincke found 1-(2,4-dinitrophenyl)pyridinium chloride to react with aromatic amines and with alkali to give deep red products

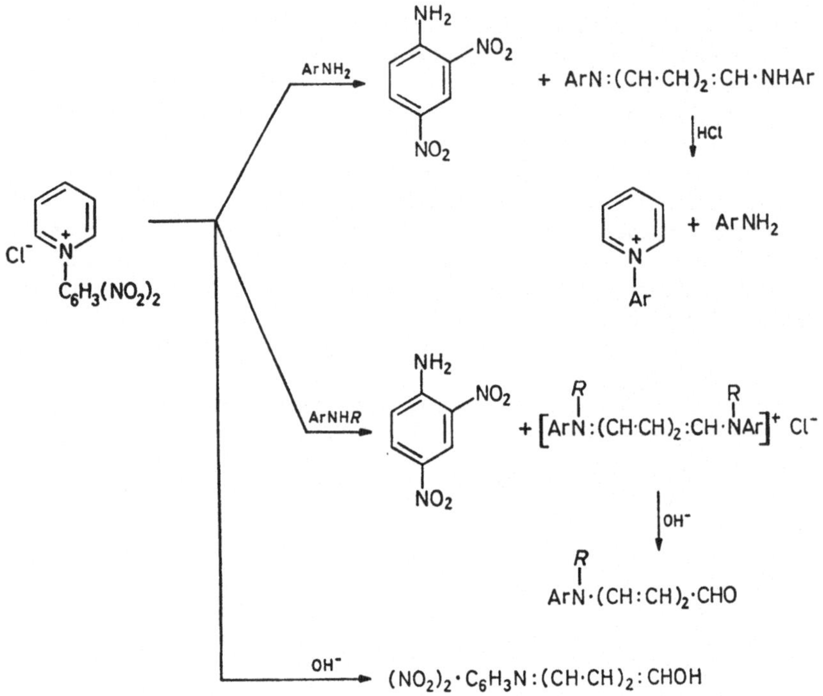

(The tautomeric possibilities in the glutacondialdehyde derivatives are not represented.) The reaction of the dianil with hydrochloric acid illustrates the formation of a new quaternary salt (the replacement of X by Y, mentioned above). Analogous results were reported with aliphatic amines and with phenylhydrazine.

Ring openings of the above kind[414, 1155, 1179, 1181] by amines have been used for synthesizing new quaternary pyridinium salts[1182-3], as a source of the $(CH)_5$ unit for dyestuffs[1184-5] and for azulene synthesis[1186]. Difficulties in preparing dinitrophenyl quaternary salts limit the generality of the reaction. Ring opening occurs with 1-(2,4-dinitrophenyl)pyridinium salts having 3-methyl, 3-methoxy and 3-acetamido substituents, but under some conditions these and other substituents direct the reaction so that a 2,4-dinitrodiphenyl-amine is formed[417]. When 4-alkoxy, -methylthio and -aryloxy substituents are present, their replacement rather than ring opening occurs (p. 219; cf. the reaction with cyanogen bromide, below).

The effect of changing the quaternizing aryl group upon the ring opening by amines has not been extensively investigated. With 4-pyridylpyridinium chloride hydrochloride, a mixture of reactions occurs; aniline produces glutacondialdehyde dianil and 4-aminopyridine by ring opening, and 4-anilinopyridine by attack upon the quaternizing group[433]. The cation (139) and related substances are opened by piperidine[1187] as shown. The extrusion of the quaternizing group as a primary amine makes the ring-opening reaction useful for the preparation of some aromatic amines otherwise difficult to obtain[1188-9]. Salts with aliphatic or alicyclic quaternizing groups would not be expected to undergo ring opening with amines; thus, (140) gives 2-aminocyclohexanone oximes[1190].

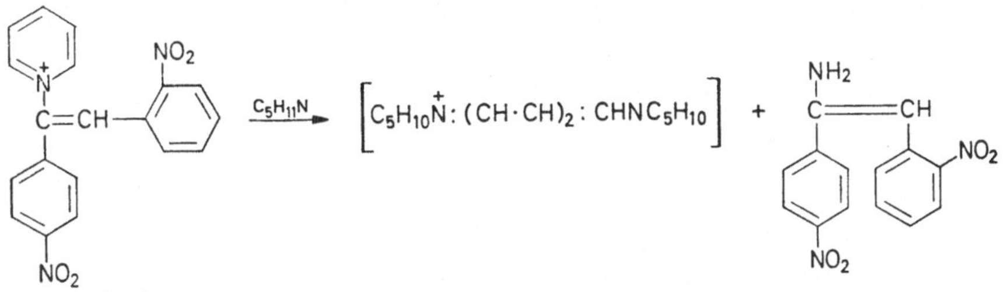

(139)

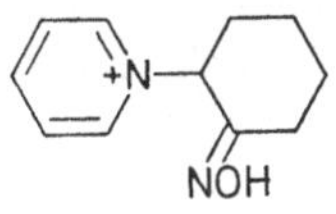

(140)

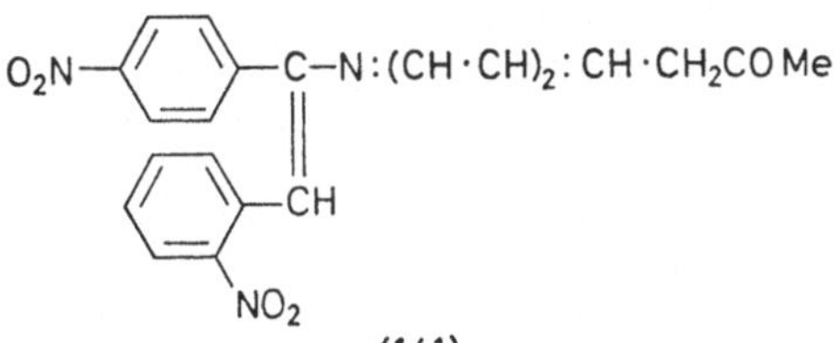

(141)

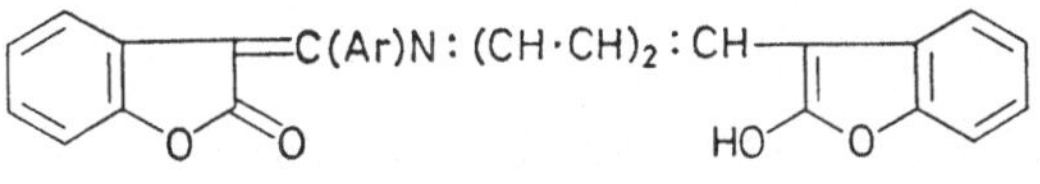

(142)

The Zincke ring opening by alkali has been observed with salts having a variety of aryl quaternizing groups[545, 1183, 1191] and has been used as a source of the five-carbon chain in cyanine synthesis[1192]. It is interesting that whilst 4-pyridylpyridinium chloride hydrochloride gives 4-hydroxy-pyridine with water (p. 239), the isomeric 2- and 3-pyridyl salts undergo ring opening with alkali[1193-4].

Nucleophiles other than amines or hydroxyl can cause ring openings analogous to the Zincke reaction. With acetone and sodium carbonate, (139) gives[1195] (141), and in related reactions 1-(2,4-dinitrophenyl)pyridinium chloride and active methylene compounds yield cyanines, while pyridine, coumaran-2-one and an aroyl chloride give[1196] (142).

Examples have been mentioned in which various nucleophiles produced 4-substituted pyridines from 4-pyridylpyridinium chloride hydrochloride, rather than causing ring opening (pp. 210, 239, 251), and it is interesting to notice that dinitrophenylpyridinium salts can undergo nucleophilic attack

K

without it. Such a process is the basis of Ullmann and Nádai's conversion of dinitrophenols into dinitrochlorobenzenes[1197-8] as shown, and of the extension of this reaction to the preparation of diphenyl ethers[1199].

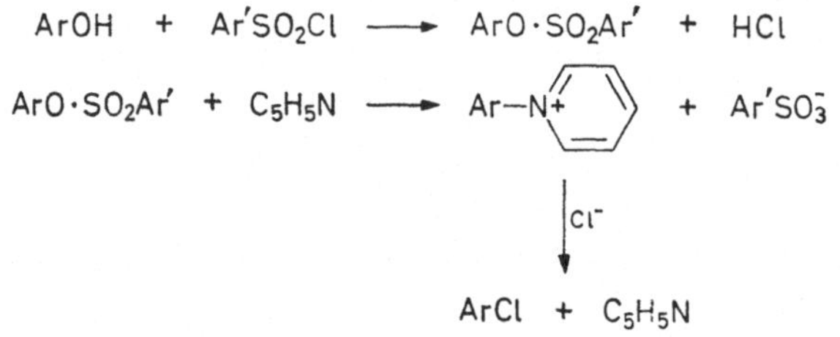

$$ArOH \ + \ Ar'SO_2Cl \ \longrightarrow \ ArO \cdot SO_2Ar' \ + \ HCl$$

$$ArO \cdot SO_2Ar' \ + \ C_5H_5N \ \longrightarrow \ Ar-N^+\!\!\!\langle \ \rangle \ + \ Ar'SO_3^-$$

$$\Big| Cl^-$$

$$\downarrow$$

$$ArCl \ + \ C_5H_5N$$

König found pyridine to give, with cyanogen bromide and arylamines, glutacondialdehyde dianils and cyanamide. The reaction clearly involved the 1-cyanopyridinium ion. Alkylamines can be used[1184]: in an interesting variation[1200], diethylamine hydroperchlorate converts 1-cyanopyridinium bromide into (143). Some substituted pyridines can be used in the reaction;

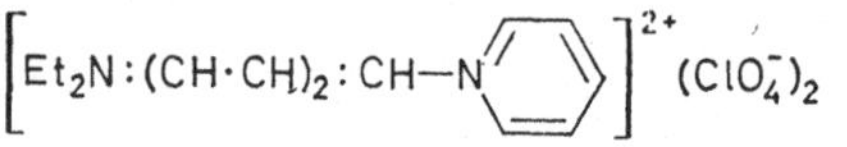

$$\left[Et_2N : (CH \cdot CH)_2 : CH-N\!\!\!\langle \ \rangle \right]^{2+} (ClO_4^-)_2$$

(143)

$$\left[{}^-O_3S \cdot N : (CH \cdot CH)_2 : CHO^- \right] Na_2^+$$

(144)

these include the picolines[545, 1203], 3-hydroxypyridine[1201] and 4,4'-dipyridyl[1202]. The record of successes and failures[1185, 1203] does not permit generalizations, but failure seems to be most frequent with 2-substituted pyridines. Whereas the Zincke reaction causes nuclear substitution when 4-alkoxy-, -methylthio- and -aryloxy-groups are present (see above), pyridines with these substituents are ring-opened by cyanogen bromide and arylamides[706].

Like the Zincke reaction, the König reaction has been used in synthesizing new quaternary salts[545, 1202], for providing the $(CH)_5$ unit for dyestuffs[1184], and for azulene synthesis[1203].

Baumgarten[1180, 1204] showed the pyridine–sulphur trioxide complex to be converted rapidly by alkali into the yellow compound (144). This could be hydrolysed to glutacondialdehyde, and as a source of this compound the reaction has found synthetic use[1205]. The quaternary salt from pyridine and ethyl chlorosulphonate is similarly split by alkali, to glutacondialdehyde and sulphamic acid[1206].

Of the reactions discussed above, only Zincke's has been studied quantitatively[1207]. The rate of reaction of 1-(2,4-dinitrophenyl)pyridinium chloride with aniline in methanol depended on the concentration of the quaternary salt and the square of the aniline concentration. It was suggested that a com-plex of the quaternary cation and aniline was formed rapidly and then re-acted in the rate-determining step with a second molecule of aniline:

$$\text{(pyridinium)} + \text{PhNH}_2 \xrightarrow{\text{Fast}} \text{Complex} \xrightarrow[\text{Slow}]{\text{PhNH}_2} \overset{+}{\text{PhNH}} \colon (\text{CH} \cdot \text{CH})_2 \colon \text{CHNHPh}$$
$$+ \text{PhNH}_2$$

The third-order rate constants for reactions with substituted anilines are given in *Table 5.34*. The substituent effects are those expected for nucleo-philic attack. The electron-attracting power of the quaternizing group should influence the equilibrium of the complex-forming step as well as the rate of the nucleophilic attack, and under the conditions, 1-(p-nitrophenyl)pyri-dinium chloride did not react with aniline. In contrast, 1-(2,4,6-trinitro-phenyl)pyridinium chloride reacted very rapidly with aniline, but the ring-opening process was much slower than the formation of trinitrodiphenyl-amine and pyridine. Qualitatively it has been observed that picrylpyridinium chloride reacts with a molar proportion of aniline to give trinitrodiphenyl-amine, but excess of aniline causes ring opening[1219].

Table 5.34. The Zincke Reaction with 1-(2,4-Dinitrophenyl)pyridinium Chloride

Substituent in ArNH_2	$10^2 k/l^2 .\text{mole}^{-2} \text{ sec}^{-1}$ (MeOH at 30°)
—	1·75
m-Me	4·76
p-MeO	34·3
p-EtO	23·8
p-Cl	0·078
p-NO$_2$	no reaction

As well as the reactions discussed above, there are recorded several less well defined which probably involve similar processes[1208–11]. Thus, pyridine, sodium hydroxide and chloroform (and related compounds) quickly pro-duce an intense red colour[1210]. Dichlorocarbene may be involved. Pyridine homologues give isonitriles of the benzene series under these conditions[1211]. The pyridine–bisulphite compound mentioned above (p. 249) yields glutacon-dialdehyde dianil with caustic soda and aniline[1212].

In the Dimroth rearrangement, which occurs in many heterocyclic systems[1213], an alkylated nuclear nitrogen atom becomes, through ring opening and reclosure, an exocyclic alkylamino group, whilst a nitrogen atom initially present as an amino-substituent is transferred into the nucleus. In the pyridine series, alkali-catalysed examples are known[221, 348, 1214], including the case of the reaction between 2-amino-5-nitropyridine methio-dide and alkali (p. 182), and also some acid-catalysed examples. The latter involve the isomerization of 1-methyl-2-nitriminopyridines by sulphuric acid into 2-methylamino-3- and -5-nitropyridines[219b, 347] (p. 172).

(c) Ring Contractions

Nickel catalysts produce pyrroles from pyridines[1215-6].

Diazo-oxides from 3-amino-2- or -4-hydroxypyridines are converted into pyrroles by light[1217], as in the examples shown.

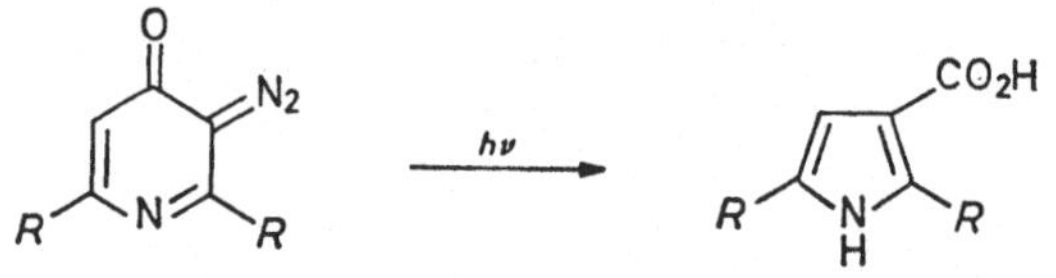

(*R* = H or Me)

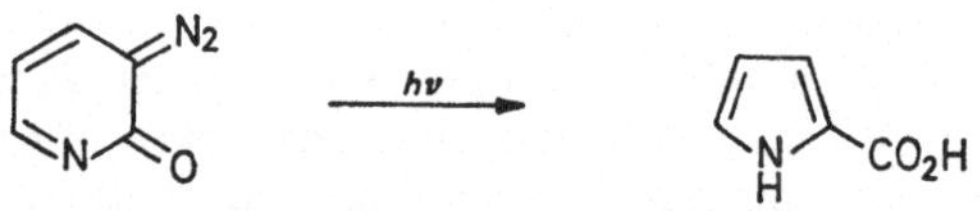

(4) A GENERAL CONSIDERATION OF THE REACTIVITY OF THE PYRIDINE NUCLEUS

Despite the vast amount of information available about substitution reactions involving pyridine and its derivatives, it is difficult to point to much evidence which enables a comparison to be made between experimental and theoretical studies of the reactivity of pyridine.

As regards electrophilic substitution, it is likely that most of the evidence relates to the conjugate acid of pyridine, not to the free base. Observations on the latter are excluded in many situations because the electrophile attaches itself to the lone pair of the nitrogen atom rather than attacking the aromatic system. Examples are the reactions between pyridine and dinitrogen pentoxide (p. 162), nitronium borofluoride (p. 162), sulphur trioxide (p. 162) and halogens at room temperature (p. 160). When the electrophile is generated in a protonating medium, as in tritiation (p. 164), the electrophile is obliged to attack the pyridinium ion. This situation also occurs in nitration and sulphonation. Thus, examples of electrophilic substitution occurring in pyridine itself are difficult to find. The vapour-phase halogenation of pyridine (p. 164) may involve the free base, which suggests that the nitrogen atom directs electrophilic substitution to $C_{(3)}$. Indirect evidence from the sulphonation of 2,6-di-t-butylpyridine, in which the alkyl groups prevent attachment of SO_3 to the nitrogen atom (p. 177), leads to the same conclusion.

Concerning the effect of the nitrogen atom upon the state of activation of the ring, direct evidence from substitution reactions is almost completely lacking. If ozonization be accepted as evidence in this connection, then it shows the ring nitrogen atom to deactivate the ring towards electrophilic attack (p. 265). Other evidence is equally indirect. Thus, by a long extrapolation of results on the tritiation of lutidine and collidine cations (p. 164), it has been estimated[140b] that replacement of :CH· in the benzene ring by :N· produces deactivation by a factor of about 10^5.

Other evidence comes from sources other than substitution reactions. Thus, the rates of pyrolysis of 1-2'-, 1-3'- and 1-4'-pyridylethyl acetate[1220] suggest that the ability to stabilize an adjacent carbonium ion lies in the order Ph $\gg$ 3-C_5H_4N $\gg$ 2-C_5H_4N > 4-C_5H_4N, i.e. all the pyridine positions are deactivated to electrophilic attack but $C_{(3)}$ least so. The σ^+ values for the pyridine positions derived from this work suggest the same. Hammett constants evaluated spectroscopically (p. 142) and from kinetic measurements (pp. 156, 247, 322) are in agreement with this conclusion.

Clearly, as regards pyridine, although direct evidence is lacking, it seems that the nitrogen atom deactivates the ring to electrophilic attack, $C_{(3)}$ being affected least. The σ^+ constants indicate reactivity at $C_{(3)}$ similar to that of a benzene position *meta* to a halogen substituent, and at $C_{(2)}$ and $C_{(4)}$ similar to that of a benzene position *para* to a nitro group[1220].

Common electrophilic substitutions such as nitration in sulphuric acid (p. 171) and sulphonation (p. 176), which lead to 3-substitution, involve the pyridinium ion. Evidence from tritiation studies (p. 164) shows that the $:\overset{+}{N}H\cdot$ group is deactivating by a factor of about 10^{18}. Hammett constants deduced from ionization constants (p. 156) also point to a very powerful deactivating effect from $:\overset{+}{N}H\cdot$.

Although the evidence is not complete, studies of nitration (p. 174) and deuteration (p. 164) make it probable that electrophilic substitutions occurring at $C_{(4)}$ involve the free pyridine 1-oxide molecule, whilst such reactions proceeding at $C_{(3)}$ involve the 1-hydroxypyridinium ion. Nitration in sulphuric acid is an example of the first kind, whilst sulphonation and deuteration are of the second kind. In its effect upon the 3-position the $:\overset{+}{N}(OH)\cdot$ group is somewhat more deactivating than the $:\overset{+}{N}H\cdot$ group. From the nitration experiments, a partial rate factor for $C_{(4)}$ in pyridine 1-oxide of about 10^{-4} is indicated. There is no evidence on which to base comparisons of all three positions in either the free base or the cation of pyridine 1-oxide.

Protonation is the commonest way in which addition of an electrophile to the ring nitrogen atom obscures the behaviour of the free pyridine bases in electrophilic substitutions. However, the phenomenon is general and is further exemplified by halogenation in the presence of aluminium chloride (p. 166). The pyridine–aluminium chloride complex undergoes 3-substitution.

In electrophilic substitutions, then, the basic properties of the pyridine nitrogen atom provide, as it were, first-order difficulties to the study of the reactivity of the neutral molecule. Related ones, and others which might be described as second-order, occur in some nucleophilic substitutions.

All nucleophilic substitutions show that in pyridine and its derivatives as free bases the 2- and 4-positions are more reactive than the 3-position. Organometallic compounds generally effect 2-substitution (pp. 200, 220) and the Tschitschibabin amination and hydroxylation reactions also occur mainly at $C_{(2)}$ (pp. 206, 234). However, these reactions cannot be taken to imply superior electrophilic power in $C_{(2)}$ as compared with $C_{(4)}$, for the issue is confused in each case by mechanistic problems; as well as the precise nature of the reagent, that of the initial step is uncertain, and subsequent stabilization of one transition state rather than another by the formation of an addition compound may be occurring. As evidence about the intrinsic electrophilic properties of individual nuclear positions these reactions are valueless.

271

The basic properties of pyridines make themselves evident in nucleophilic as well as in electrophilic processes; an example is the formation of acyl-pyridinium ions which can react with enolate anions (p. 205). Again the orientation of attack revealed in the isolated product has no bearing on the relative reactivities of $C_{(2)}$ and $C_{(4)}$.

As far as it goes, the evidence from the reactions of halogenopyridines with amines suggests the reactivity order $4 > 2 \gg 3$ (p. 215). Caution is necessary in using data from reactions of this kind, for with primary amines, hydrogen bonding in the transition state formed with a 2-halogenopyridine could be an important factor (p. 217). More immediately relevant still, because examples are available in which no substituents are present other than the halogen atoms being replaced, are the reactions of halogeno-pyridines with sodium alkoxides (p. 245). These indicate clearly the reactivity sequence $4 > 2 \gg 3$.

A number of nucleophilic substitutions starting from pyridine oxides may be in a sense 'cine-substitutions'[858], the oxide group being modified to a quaternary form and then expelled nucleophilically as an anion:

$$R-O-\overset{+}{N}=(CH-CH)_n=C-H \xrightarrow{\quad B^- \quad} R-\overset{\frown}{O}-N-(CH=CH)_n-C\overset{H}{\underset{B}{\diagdown}}$$

$$\longrightarrow \quad RO^- \quad N=(CH-CH)_n=C-B \ + \ H^+$$

Examples are the 2-substitution and loss of the oxide function in the reaction between pyridine, acetic anhydride and ethyl cyanoacetate (p. 206), and the chlorination of pyridine 1-oxides with sulphuryl chloride, though in the latter example the mechanistic necessity for loss of the oxide function is not established. Quaternary salts of pyridine oxides react at $C_{(2)}$ with Grignard reagents and alkoxide ion is eliminated (p. 201), but in the related reaction with cyanide, ion substitution at both $C_{(2)}$ and $C_{(4)}$ can occur (p. 225).

Qualitatively 2- and 4-halogenopyridine 1-oxides are more reactive to-wards amines than are 2- and 4-halogenopyridines, and the 3-isomers are in the same order (p. 213). Quantitative studies give the order $2 > 4 \gg 3$, but involvement of the oxide function in the transition state of the replacement occurring at the 2-position may affect the relative order of the 2- and 4-positions (p. 217). As with the pyridines themselves, the best evidence for the relative positional reactivities in the oxides comes from the reaction of halo-genopyridine 1-oxides and alkoxides (p. 245).

The behaviour of acylpyridinium, acyloxypyridinium and oxide quaternary salts, and the occurrence of acid catalysis in nucleophilic replacements (pp. 215, 233, 242) lead to the expectation of high reactivity towards nucleo-philes in quaternary pyridinium salts. Quantitatively, the reactivity sequence in the reaction of halogen compounds with alkoxides is quaternary salt > oxide > pyridine. The positional sequence is $4 > 2 > 3$ in halogenopyridines and their 1-oxides, and $2 > 4 > 3$ in the 1-methyl-halogenopyridinium salts. However, in all three series activation energies would give the order $4 > 2 > 3$; for the quaternary salts this order is modified by the entropy factor.

Consideration of all these facts reveals the important distinction to lie between the 2- and 4-positions on the one hand and the 3-position on the other. To expect any theoretical treatment based on the familiar reactivity parameters to choose for all cases between $C_{(2)}$ and $C_{(4)}$ as the most reactive site is not reasonable. However, reactions which appear to approach most closely to the kind treated by the usual theoretical approach do indicate that $C_{(4)}$ is probably the most reactive position.

One other point regarding nucleophilic substitution, in particular orientation therein, must be mentioned. It has been suggested that orientation of nucleophilic attack upon pyridinium salts may be connected with the ability of the nucleophile to form a charge-transfer complex with the pyridinium ring; nucleophiles which should easily form charge-transfer complexes will add at $C_{(4)}$, whilst others will add at $C_{(2)}$.[54c, 978] How far this is true is not known, and indeed we know little about the rôle of charge-transfer complexes in substitution processes in general (see p. 281).

In both electrophilic and nucleophilic substitutions, substituents already present in the pyridine ring have the effects which would be expected of them, on electronic grounds, upon both orientation and activation. One feature of orientation commonly observed is the tendency of a substituent at $C_{(3)}$ to direct substitution more to $C_{(2)}$ than to $C_{(6)}$. Interaction between the substituent and the substituting reagent, in which the nitrogen lone pair may play a part, has been held[233] to be responsible for this effect (p. 173). Substituent–reagent interactions account for similar features of benzene substitutions[234].

The position as regards radical substitution is obscured by mechanistic uncertainties. The evidence from alkylation suggests the 2-position to be more reactive than the 4-position (p. 252). In arylation, the 2-position is again the most reactive, and differences between the 3- and 4-positions are small (p. 255). Interaction between substituents on the aryl radical and the nitrogen lone pair may be important (a 'second-order effect' like those mentioned in connection with nucleophilic substitution); whereas in phenylation the nuclear nitrogen causes slight activation, in nitrophenylation there is slight deactivation (p. 255).

Let us now examine the theoretical position. In qualitative terms, mesomerism leads us to expect overall deactivation of the pyridine nucleus with respect to electrophilic attack, activation with respect to nucleophilic attack, and a distinction in each case between $C_{(3)}$ on the one hand and $C_{(2)}$ and $C_{(4)}$ on the other, arising from the $-I$ and $-M$ effects of the nitrogen atom (p. 33). The same situation would be foreseen, in a more exaggerated form, for the pyridinium ion. It should be noticed that activation of pyridine by a $+E$ effect from the nitrogen atom is not possible[1220].

The same general position is indicated by the non-bonding molecular orbital treatment of pyridine (p. 44).

Numerous computations of the various reactivity indices of molecular orbital theory have been made. As regards π-electron densities, a number of results have been summarized already in *Table 3.1*. Generally, the Hückel method gives π-electron densities in the order $3 > 2,4$ when the auxiliary inductive parameter is small or zero*, and to that extent the results correctly

* When $h_i = 0$, although the positional order of reactivities is that expected $(3 > 2,4)$, $q_3 > 1$, and this would imply activation of $C_{(3)}$ relative to benzene.

273

indicate the expected order of reactivity in both electrophilic and nucleo-philic substitution, but do not allow $C_{(2)}$ and $C_{(4)}$ to be set in a definite order. The results would be similar for the pyridinium ion. Some recent[1220] Hückel-method calculations, additional to those of *Table 3.1*, are given in *Table 5.35*. This work (p. 270) indicates that the parameters $h_N = 0.5$, $h_i = 0.085$ are particularly satisfactory.

Table 5.35. *π-Electron Densities in Pyridine*[1220] (*Hückel method*, $h_N = 0.5$)

h_i	q_2	q_3	q_4
0·0	0·923	1·004	0·950
0·05	0·943	0·991	0·950
0·075	0·953	0·985	0·951
0·085	0·957	0·982	0·951
0·095	0·961	0·980	0·951

Although these simple molecular orbital theory results agree with the usual view of pyridine and pyridinium as having their highest electron density at $C_{(3)}$, more advanced calculations make π-electron densities of doubtful significance. The orders $q_3 > q_4 > q_2$ and $q_2 > q_4 > q_3$ have been given[33a, 1221], and one treatment states $q_2 > q_3 > q_4$ for pyridine and $q_3 > q_2 > q_4$ or $q_2 > q_3 > q_4$ for pyridinium[33d]. Nuclear magnetic resonance spectroscopy[72a] shows that protonation of pyridine withdraws electrons chiefly from $C_{(4)}$ and $C_{(3)}$ (p. 145).

In terms of mesomerism, pyridine 1-oxide would be expected to undergo electrophilic substitution at $C_{(2)}$ and $C_{(4)}$ (145). The evidence from dipole

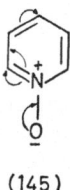

(145)

moments suggests the character $+M$ for the oxide group, so that relative to pyridine the oxide should be activated towards electrophilic substitution. Protonation of the oxygen atom would clearly remove the activating effect and produce a situation like that in pyridinium as regards orientation.

Two Hückel molecular orbital treatments of pyridine 1-oxide have been given[20, 1222]. In both, the assumptions and conclusions were similar, and data from the earlier work[20] will be discussed. As regards π-electron distributions, the results depend markedly on the value selected for β_{NO}, the resonance integral of the oxygen–nitrogen bond (*Table 5.36*).

Comparison of these results with experiment through the dipole moment showed the value $\beta_{NO} \sim 0.8\beta$ to give the best results, and calculations were made with $\beta_{NO} = 0.75\beta$. With this value, the model indicates the order of reactivity to electrophilic attack $C_{(2)} > C_{(4)} > C_{(3)}$. In fact, so far as nitration is concerned, $C_{(4)}$ is the most reactive position, as has been seen. The π-electron densities give no guide at all to the state of activation or reactivity towards nucleophilic attack. As Jaffé has pointed out[98], pyridine 1-oxide evidently violates Brown's non-crossing rule (p. 46).

Table 5.36. π-Electron Densities in Pyridine 1-Oxide[20]
($h_N = 0.6$; $h_i = 0.06$; $h_O = 1.0$)

β_{NO}	q_O	q_N	q_2	q_3	q_4
0.5	1.949	0.188	0.962	0.988	0.963
0.75	1.880	0.145	1.003	0.987	0.993
1.0	1.808	0.122	1.039	0.985	1.023
1.5	1.625	0.054	1.131	0.981	1.097

Before leaving π-electron densities, it should be mentioned that an attempt has been made, using the Hückel method, to evaluate the way in which an electron-releasing substituent modifies the nuclear charge distribution[1223].

Overall, π-electron densities succeed in matching with the experimental facts of substitution only in a limited way. This is, of course, not surprising (see Chapter 3). Localization energies, now to be considered, are more generally successful despite serious shortcomings. Table 5.37 lists their values for electrophilic (A_e), nucleophilic (A_n) and radical substitution (A_r), and also indicates those of the relevant parameters adopted in the calculations.

The general success of the method as applied to pyridine was indicated in Chapter 3 (p. 42) by references to Wheland's original calculations. The A values of Table 5.37 differ from Wheland's localization energies, since in the evaluation of the former overlap was neglected.

The implications of the data of Table 5.37 for orientation and for activation should be considered separately. For electrophilic substitution in both pyridine and pyridinium (the situation in which is represented by giving h_N a high value), $C_{(3)}$ is always indicated to be the most reactive position. The relative order of $C_{(2)}$ and $C_{(4)}$ (for which there is no direct evidence) depends upon the assumptions made, in particular as to whether the auxiliary inductive effect operates at $C_{(2)}$ when that is the position of localization. For pyridine 1-oxide $C_{(2)}$ is always the most reactive position, and with the preferred[20] parameters, the complete sequence is $C_{(2)} > C_{(4)} > C_{(3)}$. In contrast, for the protonated oxide, $C_{(3)}$ is the favoured position, and these results led Barnes[20] correctly to conclude that in electrophilic substitutions of pyridine 1-oxide which proceeded at $C_{(4)}$ the free base was involved, as against the conjugate acid in those which proceeded at $C_{(3)}$. In the former case, $C_{(4)}$ rather than $C_{(2)}$ was attacked because of a steric factor.

For nucleophilic substitution also the results for pyridine, pyridinium and pyridine 1-oxide correctly reproduce the observed order, $C_{(2)}$, $C_{(4)} > C_{(3)}$. For pyridine and its oxide the order of $C_{(2)}$ and $C_{(4)}$ depends on the choice of parameters, but for pyridinium the sequence $C_{(2)} > C_{(4)} > C_{(3)}$ is given.

Table 5.37. Localization Energies

Substance	Parameters h_N	h_i	Reference	$A_e(-\beta)$ 2	3	4	$A_n(-\beta)$ 2	3	4	$A_r(-\beta)$ 2	3	4
Pyridine	0·5	0·0	1220, 1224	2·67	2·54	2·70	2·35	2·54	2·37	2·51	2·54	2·54
	1·0	0·0	1225*	2·73	2·52	2·85	2·16	2·52	2·22	2·45	2·52	2·54
	0·5	0·05	20, 1220	2·62	2·57	2·69	2·40	2·51	2·36	2·51	2·54	2·53
	0·5	0·085	1220	(2·72)[†]	2·59	2·69	—	—	—	—	—	—
				2·59								
				(2·76)[‡]	2·59	2·69	—	—	—	—	—	—
				2·71[†]								
				2·70[‡]								
				2·69[‡]								
				2·68[‡]								
				2·67[‡]								
Pyridinium	1·0	0·1	20	2·85	2·62	2·84	2·21	2·48	2·25	2·53	2·55	2·55
	2·0	0·0	1224	2·71	2·56	3·07	1·86	2·56	2·01	2·28	2·56	2·54
	0·6	0·06		2·51	2·58	2·60	2·42	2·50	2·39	2·47	2·54	2·50
Pyridine 1-oxide	$[h_O = 1·0; \beta_{NO} = 0·5]$	0·5]	20	2·36	2·58	2·44	2·48	2·50	2·46	2·42	2·53	2·47
		= 0·75]		2·22	2·59	2·34	2·54	2·49	2·54	2·38	2·53	2·44
		= 1·0]		1·92	2·60	2·09	2·58	2·52	2·75	2·28	2·56	2·42
		= 1·5]										
1-Hydroxy-pyridinium	0·65	0·065		2·70	2·58	2·67	2·39	2·50	2·36	2·56	2·54	2·51
	$[h_O = 1·5; \beta_{NO} = 0·5]$											

* These figures should be compared with those given for pyridinium.
† Calculated by assuming the auxiliary inductive effect not to operate at the localized centre ($C_{1\alpha}$).
‡ Calculated by assuming the auxiliary inductive effect to be partially effective at the localized centre ($C_{1\alpha}$) and that there h_i becomes δh_i where $\delta = 0·3, 0·35, 0·4, 0·45$ and $0·5$ successively.

The variation in A_r with position is small for pyridine but overall, $C_{(2)}$ is predicted to be slightly the most reactive, as is found by experiment.

Thus, as regards orientation, the localization energies quite successfully reproduce the results of what are probably the most relevant experiments.

When we consider the predictions regarding activation, the situation is not so happy. For benzene, A_e ($= A_n = A_r$) $= 2.54$, and thus for pyridine and pyridinium the localization energies do not indicate the powerful de-activating effect of $:\overset{+}{N}\cdot$ and fail very much more seriously for $:\overset{+}{N}H$ and $:\overset{+}{N}\cdot OH$. For pyridine oxide, the theory actually indicates some activation of $C_{(4)}$ when $\beta_{NO} > 0.6\beta$, as it is in the preferred calculations. It is interesting to note here the similarity between the $:\overset{+}{N}-\overset{-}{O}$ group and halogen atoms in halogenobenzenes; both groups are o,p-directing with deactivation in electro-philic substitutions.

The values of A_n show some degree of activation at $C_{(2)}$ and $C_{(4)}$ and clearly indicate this to be greatest in the pyridinium salts.

If the magnitudes of A_r for the pyridine positions suggest anything, it can only be that, as regards activation, the compound should be but little different from benzene, as it appears to be.

For the special interest of the method which it uses to evaluate the necessary parameters, through the use of Hammett substituent constants (p. 156), the work of Jaffé[98] on pyridine and pyridine 1-oxide should be mentioned. Considering both orientation and activation, it is about as successful overall in dealing with electrophilic and nucleophilic substitution in these compounds as are the studies already discussed.

It remains to consider the performance of Dewar's method (p. 45) for evaluating localization energies. The essential equation, already discussed there, is

$$\Delta E_{naza} = 2\beta(a_{0s} + a_{0t}) \pm \Sigma a_{0r}^2 a_r$$

Dewar and Maitlis[122b] applied this equation to the case of the nitration of nitrogen heteroaromatics in sulphuric acid. It is necessary to select values for β, for a_N and for the auxiliary inductive parameter. If β is given its usual value, -20 kcal mole^{-1}, h is given that of from 1 to 2, and h' is $\frac{1}{3}$, agreement between prediction and experiment is not obtained. Earlier work on the nitra-tion of aromatic hydrocarbons in acetic anhydride suggested that β in the above equation is not the normal carbon–carbon resonance integral. Some modification is needed because the transition state does not possess the Wheland structure but lies nearer to the reactants; β is to be replaced by β_X, a constant appropriate for the particular reagent X, and smaller the more reactive X is. For hydrocarbon nitration in acetic anhydride, an appropriate value was $\beta_X = -6$ kcal mole^{-1}. Since nitration in sulphuric acid is argued to produce a more reactive reagent, the value $\beta_X = -4$ kcal mole^{-1} was adopted for it.

Experimental partial rate factors for isoquinoline and the equation $(p_5/p_8)_T = \exp(\Delta E_8 - \Delta E_5)/RT$ give $(\Delta E_8 - \Delta E_5) = -1.27 \pm 0.06$ kcal mole^{-1} in isoquinoline. At the same time, from the above equation $(\Delta E_8 - \Delta E_5)$ is found to be $0.032a_N$, whence $a_N = -39.7 \pm 2$ kcal mole^{-1}.

This value is appropriate for nitration in sulphuric acid, i.e. for circumstances in which the nitrogen atom is protonated. Comparison with values derived for the nitrogen atom from nucleophilic and radical substitution[1226] show that the heterocyclic component of the ring should make the greatest effect upon electrophilic substitution and the least upon radical substitution, as is the case.

With the values of a_N ($- 40$ kcal mole^{-1}) and β_X ($- 4$ kcal mole^{-1}) decided, values of ΔE_π can be calculated. For nucleophilic substitution, the corresponding parameters $a_N = - 22$ kcal mole^{-1} and $\beta_X = - 10$ kcal mole^{-1} (with again, an auxiliary inductive factor of $\frac{1}{3}$) have been used[1227]. ΔE_π values (kcal mole^{-1}) for the two cases are given in (146) and

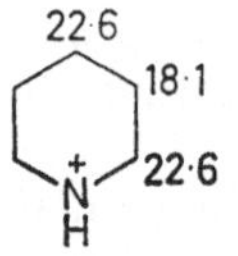

(146) 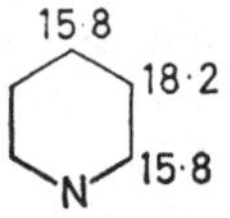   (147)

(147), respectively. The method is thus astonishingly successful in reproducing the facts of orientation. It does in fact maintain this success over a wide range of heterocycles but fails badly over the question of activation in the case of nitration in sulphuric acid.

Finally, we may notice that free valencies (cf. Chapter 3) have been used in discussions of the reactivity of pyridine towards free radicals, but not with very great success[1019].

When the many approximations used in the molecular orbital treatment are remembered, and when the wide divorce between experimental conditions and theoretical models is considered, the success of theory is startling. As regards orientation in substitution reactions, this is particularly so. The failure to deal with the problem of activation is most marked in electrophilic substitution, and is general for cases where heterocycles are reacting as their conjugate acids. A large electrostatic factor may be a major source of difficulty in these cases.

Addendum chiefly on work reported during 1965

[*To* 1*a*] X-Ray crystallographic studies of α-pyridoin[1228a] and picolinic acid hydrochloride[1228b] have been reported.

Dipole moments and ultra-violet spectra have been used in the discussion of the conformations of nitrophenylpyridines[1229a]. Dipole moments of *cis*- and *trans*-azopyridines have been re-measured: the new value for trans-4,4'-azopyridine (0·36 *D*) differs considerably from the old value[1229b] (2·0 *D*).

Dipole moment measurements show 2-pyridone, 2-pyridthione and 2-pyridselenone to be powerfully hydrogen-bonding and dimeric[1230].

A modified Hückel M.O. method has been used to calculate the dipole moment of pyridine[1231].

[*To* 1*b*] A further theoretical treatment of the electronic spectrum of pyridine has been given[1232].

Ultra-violet spectroscopic data for many amino-, diamino- and hydroxy-pyridines, and their nitro- and halogeno-derivatives have been published[1233]. Ultra-violet spectroscopy has been used in elucidating the tautomerism of phenyl 2-, 3- and 4-picolyl sulphones and phenacylpyridines[1234a] (see 1c).

Fresh assignments of peaks in the infra-red spectra of pyridine[1235a], its homologues[1235b] and 1-oxides[1236] have been made. The infra-red spectra of complexes of 1-methyl-2-pyridone with Lewis acids are consistent with interaction at the oxygen atoms[1237].

Proton resonance spectroscopy has been used to determine the sites of protonation of thiopyridines and aminopyridines[1234b].

[To 1c] Ionization constants have been reported for a number of amino-, diamino- and hydroxy-pyridines, and nitro- and halogeno-derivatives of these[1233]. Ionization constant measurements show that, in aqueous solution, phenyl 2-, 3- and 4-picolyl sulphones exist predominantly as such ($K_T = 10^{7.6} - 10^{9.6}$ in their favour). 3- and 4-Phenacylpyridine are chiefly ketonic, but 2-phenacylpyridine contains appreciable amounts of the chelated enol in non-polar media and in the solid state[1234a].

The applicability of the Hammett acidity function to the protonation of pyridines and pyridine 1-oxides in sulphuric acid has been confirmed. Ionization constants for several polyhalogeno-pyridines and a number of negatively substituted pyridine 1-oxides were determined[1238].

The Hammett equation has been applied to pK_a values for the excited states of pyridines and pyridine 1-oxides[1239a].

1-Nitro- and 1-nitroso-pyridinium borofluoride have been further examined[1239b] (see 3b).

Evidence for the existence in mixtures of nitrobenzene and carbon tetrachloride of two types of complex between iodine and collidine ($C_8H_{11}N.I_2$ and $(C_8H_{11}N)_2I^+I_9^-$) has been obtained by nuclear magnetic resonance spectroscopy[1240].

[To 2.A. Electrophilic Substitution] Bromination of pyridine homologues in fuming sulphuric acid gives excellent yields of 3-bromopyridines[1241].

There is now some evidence which suggests that the rearrangement of 2-nitraminopyridine in sulphuric acid may be intramolecular[1242a].

Our knowledge of the true substrate undergoing electrophilic substitution in reactions of a number of pyridine derivatives has been considerably augmented. Acid-catalysed hydrogen exchange occurs with the free base for 3-hydroxypyridine and with the neutral oxo forms in the cases of 4-pyridone and 1-methyl-4-pyridone. Exchange at $C_{(2)}$ in 3,5-dimethyl-pyridine 1-oxide involves the free base above $H_0 - 3.5$ and the conjugate acid at higher acidities[1243].

Bromination of 5-substituted 2-aminopyridines in aqueous sulphuric acid goes through the free bases, and comparison with anilines[1243] shows the pyridinic nitrogen atom to be deactivating by the surprisingly small factor of about 0.5.

The nitration of 2,4,6-collidine in sulphuric acid involves the cation, and the $\overset{+}{:}\text{NH}.$ group deactivates the 3-position 10^{12} times[1243].

2,6-Dichloropyridine is mononitrated at $C_{(3)}$ as the free base, whilst the mononitration of 2,6-dimethoxypyridine goes through the cation. The subsequent nitration of 2,6-dimethoxy-3-nitropyridine involves the free base. In these examples the ring nitrogen atom is deactivating by a factor of 20–200[1248].

279

The nitration of 4-pyridone, 4-methoxypyridine and 1-methyl-4-pyridone occurs on the conjugate acids[1243]. 2,6-Dichloro-, 3,5-dichloro- and 3,5-dimethyl-pyridine 1-oxide[1243], as well as 2,6-dimethylpyridine 1-oxide[1244] are nitrated as the free bases, whilst nitration of 4-methoxy-2,6-dimethylpyridine 1-oxide occurs at $C_{(3)}$ on the conjugate acid[1243].

Two points are particularly worthy of comment. First, the unprotonated pyridinic nitrogen atom exerts only a surprisingly small deactivating influence, and it may be that the pyridine ring elicits from substituents electron-releasing abilities superior to those which they exhibit when attached to benzene rings. In the case of 2,6-lutidine 1-oxide, each methyl group appears to activate the 4-position about 30 times, compared with about 3 times in benzene[1244]. Secondly, whilst in hydrogen ion exchange 4-pyridone reacts through the free base, and in nitration through the conjugate acid as might be expected, the reverse situation is found with the pyridine 1-oxides.

[*To* 2.1] 4-(*p*-Nitrobenzyl)pyridinium cation reacts with diazoalkanes giving the alkyl quaternary salts[1245].

The kinetics of quaternization of pyridine with ethyl methanesulphonate in water[1246a] and with ethyl iodide in nitrobenzene[1246b] have been studied. Hammett analysis of the latter reaction shows that $-M$ substituents at $C_{(4)}$ in pyridine influence the nitrogen atom only by their inductive effects.

The reactions of pyridines with dimethyl acetylenedicarboxylate continue to receive attention[1247]. The structure of the product formed with ketene has also been re-examined[1246c].

Tetracyanoethylene oxide reacts with pyridine, giving 1-(dicyanomethylene)pyridinium[1248], and with 2-pyridone di-(trifluoromethyl)acetylene gives 1-(di-trifluoromethyl)vinyl-2-pyridone[1249]. In the same reaction 1-methyl-2-pyridone yields only tars, and no Diels–Alder reaction occurs.

A relatively stable acylpyridinium salt is 1-dimethylcarbamoylpyridinium chloride[1250a] (see Chapter 6).

With *p*-nitrophenyl phosphate and *p*-nitroanisole under illumination, pyridine gives the 1-arylpyridinium nitrites[1250b].

With phosphoramidate monoanion ($\overset{+}{N}H_3.PO_3^{2-}$), pyridine and 4-picoline give anions of the type $C_5H_5\overset{+}{N}-PO_3^{2-}$ which then undergo hydrolysis[1250c].

[*To* 2.2. *Nucleophilic Substitution*] Many characteristic nucleophilic substitution reactions proceed at $C_{(4)}$ with pentafluoropyridine. Competitive amination with ammonia shows the sequence of reactivity 3,5-dichloro-trifluoropyridine > 3-chloro-tetrafluoropyridine > pentafluoropyridine[1251a, c].

Fluoropyridines have been prepared from pentachloropyridine[1251d]. Nucleophilic substitutions occur at $C_{(4)}$ in the latter[1252], and at $C_{(2)}$ in 4-bromo-tetrafluoropyridine[1251b].

A full description of work on the Tschitschibabin amination already mentioned has been published[1253]. Piperidine replaces the halogen atom from 3-fluoro- and 3-chloro-4-nitropyridine 1-oxide, whilst with cold sodium methoxide the former gives 3,4-dimethoxypyridine 1-oxide[1254].

3-Bromo-4-ethoxypyridine with lithium piperidide produces mainly 3-bromo-4-piperidylpiperidine, with a small amount of 2,4-dipiperidylpyridine, formed via the pyridyne. 2-Bromo-6-ethoxypyridine reacts about equally at each substituent[1255]. Amination of 2- and 4-chloropyridine 1-oxide goes through the aryne[1256].

In the presence of aluminium chloride, 1-acylpyridinium salts give 4-(p-dialkylaminophenyl)pyridines with dialkylanilines[1242b].

Pyridine 1-oxide and phenyl magnesium bromide in tetrahydrofuran gives the dihydro-product, which when heated gives 2-phenylpyridine[1257].

Evidence has been presented that, in the reaction of quaternary pyridinium salts with cyanide ion, kinetically controlled attack occurs at $C_{(2)}$, and that $C_{(4)}$-substituted products arise by subsequent equilibration. The initial attack is controlled by the electron density at $C_{(2)}$ and it is unnecessary to invoke π-complexes in discussing orientation[1258a].

In acetic anhydride, pyridine 1-oxide and bromine give 3,5-dibromo-pyridine 1-oxide, by electrophilic substitution into 1,4-diacetoxy-1,4-dihydropyridine[1259].

The reactions of pyridine 1-oxide[1260a] and picoline 1-oxide[1261] with carboxylic acid anhydrides continue to attract attention, particularly as regards mechanism (see Chapter 6).

[To 2.2. Radical Substitution] The phenylation of pyridine by radicals from the decomposition of 1-phenylazopyridinium tetrafluoborate has been described[1260b], and a full description of the Pschorr cyclization of diazotized 3-(2-aminobenzoyl)pyridine, already mentioned, has been published[1262].

In accordance with theoretical predictions, protonation of pyridine enhances the proportion[1258b] of radical phenylation which occurs at $C_{(2)}$.

[To 3a] It has been confirmed that borohydride reduction of quaternary pyridinium salts can produce piperidines. 1,2,5,6-Tetrahydropyridines result from initial attack at $C_{(2)}$ and piperidines from attack at $C_{(4)}$. Thus, the relative amounts of piperidine and tetrahydropyridines produced are an indication of the relative rates of attack of hydride at $C_{(4)}$ and $C_{(2)}$. As the quaternizing alkyl group is increased in size, the relative amount of piperidine formed increases[1258c].

Borohydride reduction of 3-aminocarbonyl-1-benzylpyridinium cation gives the 1,2- and 1,6-dihydro-compounds and some 3,4-tetrahydro-product[1263a]. Catalytic hydrogenation of 1-alkyl-3-alkoxycarbonylpyridinium salts produces 1,2,3,4-tetrahydropyridines[1263b]. Electrochemical reduction has also been studied[1264].

With trimethylsilane, pyridine gives 1-trimethylsilyl-1,4-dihydropyridine which, with methanol, forms 1,4-dihydropyridine[1263c].

[To 3b] The influence of substituents upon pyridine ring opening by cyanogen bromide and amines or phenols has been further examined[1265]. 1-Nitro- and 1-nitroso-pyridinium ions react with alkali or amines to give derivatives of glutacondialdehyde[1239b].

3-Amino-1-arylpyridinium salts, when diazotized, suffer ring opening, presumably because of nucleophilic hydroxylation at $C_{(6)}$, giving[1266] triazoles (148)

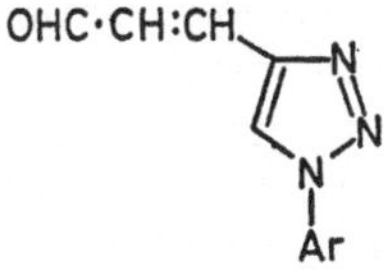

(148)

The known decomposition of 1-alkoxypyridinium cations by alkali, yielding an aldehyde and pyridine, is a slow and irreversible reaction which is accompanied by a rapid and reversible ring opening to glutacondialdehyde mono-O-alkyloximes[1267].

APPENDIX

NUCLEAR MAGNETIC RESONANCE SPECTROSCOPY OF PYRIDINES

Pyridine and many monosubstituted pyridines give highly complex spectra, the interpretation of which is of considerable interest from the spectroscopic point of view. Recently the field has been comprehensively covered by Brügel[1268] who has investigated the 40 and 60 Mc/s spectra of 154 pyridines and provided a detailed list of values of chemical shifts and spin coupling constants for these compounds. More detailed analyses of spectra of certain pyridines have also been performed by Rao and Venkateswarlu[1269] on 2-, 3- and 4-picoline, and by Kowalewski and de Kowalewski on 2- and 3-substituted pyridines[1270]. The latter workers employed perturbation approximation methods as well as exact iterative solutions in their analyses. Recently, the method of sub-spectral analysis[1271] has been applied to the $ABB'XX'$ spectrum of pyridine, and it was shown that certain of the spectral parameters could be evaluated without complete analysis of the spectrum. Homo- and hetero-nuclear double irradiation techniques have been applied to pyridines both to aid spectral analysis by removing N-quadrupolar broadening[1272a] and to determine the relative signs of spin coupling constants[1272b]. [19]F resonance studies have been performed on pentafluoropyridine and a number of its substituted derivatives[1273]. For these compounds the chemical shifts of the ring fluorine atoms, in order of increasing magnetic field, are 2-F < 4-F < 3-F, which is the same order as for pyridine and its derivatives. In addition, a clear dependence of ring chemical shift upon the nature and position of substituents was found.

The effects of solvents on the nuclear magnetic resonance spectra of pyridines are very pronounced[1274]. For pyridine and methylpyridines, change from a dilute solution in carbon tetrachloride to the pure liquid or to a dilute solution in benzene shifts the signal due to the proton or the methyl group at $C_{(2)}$ to high field to a smaller extent than the signals due to protons or methyl groups at $C_{(3)}$ and $C_{(4)}$. These differential shifts have been explained by Murrell and Gil[1275a] in terms of specific association with benzene and self-association in the pure liquid.

The relationship between ring proton chemical shifts and π-electron densities of aromatic molecules and ions is important. S.C.F. molecular orbital calculations have been used to evaluate ring currents and associated chemical shifts for pyridine and other heterocycles[1276a]. When corrections are made for polar solvent effects and ring current effects, a roughly linear relationship between chemical shift and electron density is obtained for pyridine and other heterocycles[1276b]. With 4-substituted pyridines, a simple additivity relationship of substituent contribution to the proton shieldings in these compounds can be applied, as for substituted benzenes[1277]. Correlations of proton chemical shifts and π-electron densities have been made for a series of imidazo(1,2-a)pyridines[1278]. Gil and Murrell[1275b] have shown that the difference between the nitrogen chemical shifts in pyridine and the pyridinium ion is due mainly to the large paramagnetic term arising from the n$\to \pi^*$ transition of pyridine. These workers also calculated the contributions to the proton chemical shift in pyridine of the magnetic anisotropy of the nitrogen atom and the polarization of the C—H bonds by the nitrogen lone pair. Further work[1279] on π-electron densities in the pyridinium cation has taken into account the effect of the counterion on the observed chemical shifts.

All of the foregoing has been primarily concerned with studies of the ^{1}H resonance spectra of pyridines. However, studies of ^{13}C resonance[1280] and ^{14}N resonance[1272a] have also been made with pyridine.

REFERENCES

[1] Penfold, B. R. *Acta crystallogr.* **6** (1953) (a) 591 (b) 707; (c) Wright, W. B. and King, G. D. S. *ibid.* 305

[2] (a) Rérat, C. *Acta crystallogr.* **15** (1962) 427; (b) Tsoucaris, G. **14** (1961) 914

[3] Eichhorn, E. L. *Acta crystallogr.* (a) **9** (1956) 787; (b) **12** (1959) 746

[4] Biddiscombe, D. P., Coulson, E. A., Handley, R. and Herington, E. F. G. *J. chem. Soc.* (1954) 1957; Coulson, E. A., Cox, J. D., Herington, E. F. G. and Martin, J. F. (1959) 1934

[5] (a) Coulson, E. A. and Jones, J. I. *J. Soc. chem. Ind., Lond.* **65** (1946) 169; (b) Jones, J. I. **69** (1950) 99; (c) Coulson, E. A., Hales, J. L., Holt, E. C. and Ditcham, J. B. *J. appl. Chem., Lond.* **2** (1951) 71; (d) Brown, H. C. and Murphy, W. A. *J. Am. chem. Soc.* **73** (1951) 3308; (e) Brown, H. C. and Kanner, B. *U.S. Pat.* 2,780,626 (1957); (f) Kyte, C. T., Jeffery, G. H. and Vogel, A. I. *J. chem. Soc.* (1960) 4454

[6] (a) Lukeš, R. and Pergál, M. *Chemické Listy* **52** (1958) 68; (b) Leaver, D., Gibson, W. K. and Vass, J. D. R. *J. chem. Soc.* (1963) 6053; (c) Burstall, F. H. *ibid.* (1938) 1662; (d) Alberts, A. A. and Bachman, G. B. *J. Am. chem. Soc.* **57** (1935) 1284; (e) Smith, J. M., Stewart, H. W., Roth, B. and Northey, E. H. *ibid.* **70** (1948) 3997; (f) Williams, J. L. R., Adel, R. E. *et al., J. org. Chem.* **28** (1963) 387; (g) Williams, J. L. R., Webster, S. K. and van Allen, J. A. *ibid.* **26** (1961) 4893; (h) Clarke, F. H., Felock, G. H., Silverman, G. B. and Watnick, C. M. *ibid.* **27** (1962) 533; (i) Beard, J. A. T. and Katritzky, A. R. *Recl Trav. chim. Pays-Bas Belg.* **78** (1959) 592; (j) Jerchel, D. and Melloh, W. *Justus Liebigs Annln Chem.* **622** (1959) 53; (k) Schewing, G. and Winterhalter, L. *ibid.* **473** (1929) 126; (l) Katsumoto, T. and Houda, A. *Chem. Abstr.* **59** (1963) 15,254; (m) *Germ. Pat.* 594,849 (1934)

[7] Andon, R. J. L. and Cox, J. D. *J. chem. Soc.* (a) (1952) 4601; (b) Cox, J. D. (1954) 3183; (c) (1952) 4606

[8] (a) Andon, R. J. L., Cox, J. D. and Herington, E. F. G. *J. chem. Soc.* (1954) 3188; *Trans. Faraday Soc.* **53** (1957) 410; (b) Sacconi, L., Paoletti, P. and Ciampolini, M. *J. Am. chem. Soc.* **82** (1960) 3828

[9] DeMore, B. B., Wilcox, W. S. and Goldstein, J. H. *J. chem. Phys.* **22** (1954) 876

[10] (a) Leis, D. G. and Burran, B. C., *J. Am. chem. Soc.* **67** (1945) 79; (b) Barassin, J. and Lumbroso, H. *Bull. Soc. chim. Fr.* (1961) 492

[11] Katritzky, A. R., Randall, E. W. and Sutton, L. E. *J. chem. Soc.* (1957) 1769

[12] Essery, J. M., Schofield, K. and Sutton, L. E., unpublished results

[13] Cumper, C. W. N. and Vogel, A. I. *J. chem. Soc.* (1960) 4723

[14] Rogers, M. T. and Campbell, T. W. *J. Am. chem. Soc.* **75** (1953) 1209

[15] (a) Cumper, C. W. N., Vogel, A. I. and Walker, S. *J. chem. Soc.* (1956) 3621; (b) Barassin, J. *Annls Chim.* **8** (1963) 637

[16] (a) Linton, E. P. *J. Am. chem. Soc.* **62** (1940) 1945; (b) Krackov, M. H., Lee, C. M. and Mautner, H. G. **87** (1965) 892

[17] Bax, C. M., Katritzky, A. R. and Sutton, L. E. *J. chem. Soc.* (1958) 1254

[18] Orgel, L., Cottrell, T., Dick, W. and Sutton, L. E. *Trans. Faraday Soc.* **47** (1951) 113

[19] Löwdin, P. *J. chem. Phys.* **19** (1951) 1323

[20] Barnes, R. A. *J. Am. chem. Soc.* **81** (1959) 1935

[21] (a) Brown, R. D. and Heffernan, M. L. *Aust. J. Chem.* **10** (1957) 493; (b) Sobczyk, L. *Trans. Faraday Soc.* **57** (1961) 1041; (c) Sharpe, A. N. and Walker, S. *J. chem. Soc.* (1961) 2974; (1962) 157

[22] Clar, E. *Aromatische Kohlenwasserstoffe*, 2nd ed., Berlin, 1952; Badger, G. M., Pearce, R. S. and Pettit, R. *J. chem. Soc.* (1951) 3199^v

[23] Price, W. C. and Walsh, A. D. *Proc. R. Soc.* **A191** (1947) 22

[24] Kasha, M. *Discuss. Faraday Soc.* **9** (1950) 14

[25] Mason, S. F. *J. chem. Soc.* (1959) 1240

[26] Sponer, H. and Rush, J. H. *J. chem. Phys.* **20** (1952) 1847; Reid, C. **18** (1950) 1673

[27] Fischer, H. and Steiner, P. *C. r. hebd. Séanc. Acad. Sci., Paris* **175** (1922) 882

[28] (a) Stephenson, H. P. *J. chem. Phys.* **22** (1954) 1077; (b) Platt, J. R. **19** (1951) 101

[29] (a) Sidman, J. W. *Chem. Rev.* **58** (1958) 689; (b) Coppens, G., Gillet, C., Nasielski, J. and van der Donckt, E. *Spectrochim. Acta* **18** (1962) 1441

[30] Halverson, F. and Hirt, R. C. *J. chem. Phys.* **19** (1951) 711

[31] (a) Coulson, C. A. *Proc. phys. Soc. Lond.* **65A** (1952) 933; Longuet-Higgins, H. C. and Sowden, R. G. *J. chem. Soc.* (1952) 1404; Chandra, A. K. and Basu, S. (1959) 1623; (b) Mason, S. F. *ibid.* 1247

[32] (a) Godar, E. and Mariella, R. P. *J. Am. chem. Soc.* **79** (1957) 1402; (b) Swain, M. L., Eisner, A., Woodward, C. F. and Brice, B. A. **71** (1949) 1341; (c) Bobbitt, J. M. and Scola, D. A. *J. org. Chem.* **25** (1960) 560; (d) Evans, R. F. and Brown, H. C. **27** (1962) 3127

— [33] (a) McWeeny, R. with Peacock, T. E. *Proc. phys. Soc. Lond.* **70** (1957) 41; (b) *ibid.* 593; (c) Murrell, J. N. *Molec. Phys.* **1** (1958) 384; (d) Brown, R. D. and Heffernan, M. L. *Aust. J. Chem.* **12** (1959) 554; (e) Favini, G., Vandoni, I. and Simonetta, M. *Theor. Chim. Acta* **3** (1965) 45

— [34] Orgel, L. E. *J. chem. Soc.* (1955) 121; Goodman, L. and Harrell, R. W. *J. chem. Phys.* **30** (1959) 1131; (b) Jones, R. A. and Katritzky, A. R. *Aust. J. Chem.* **17** (1964) 455

[35] (a) Spiers, C. W. F. and Wibaut, J. P. *Recl Trav. chim. Pays-Bas Belg.* **56** (1937) 573; (b) Miller, W. K., Knight, S. B. and Roe, A. *J. Am. chem. Soc.* **72** (1950) 1629; (c) Brown, H. C. and McDaniel, D. H. **77** (1955) 3752

[36] (a) Krumholz, P. *J. Am. chem. Soc.* **73** (1951) 3487; (b) Katritzky, A. R. and Simmons, P. *J. chem. Soc.* (1960) 4901; (c) Sobczyk, L. *Bull. Acad. pol. Sci. Sér. Sci. tech.* **9** (1961/4) 237

[37] (a) Essery, J. M. and Schofield, K. *J. chem. Soc.* (1963) 2225; for infra-red data on nitro-pyridines and nitropyridine 1-oxides, cf. Katritzky, A. R. and Simmons, P. *Recl Trav. chim. Pays-Bas Belg.* **79** (1960) 361; (b) Godfrey, M. and Murrell, J. N. *Proc. chem. Soc.* (1961) 171

[38] Anderson, L. C. and Seegers, N. V. *J. Am. chem. Soc.* **71** (1949) (a) 343; (b) 340

[39] Mason, S. F. *J. chem. Soc.* (1960) (a) 219; (b) 2437

[40] Essery, J. M. and Schofield, K. *J. chem. Soc.* (1961) 3939

[41] (a) Ashley, J. N., Buchanan, G. L. and Easson, A. P. T. *J. chem. Soc.* (1947) 60; (b) Hirayama, H. and Kubota, T. *J. pharm. Soc. Japan* **73** (1953) 140; (c) Albert, A. *J. chem. Soc.* (1960) 1020

[42] Steck, E. A. and Ewing, G. W. *J. Am. chem. Soc.* **70** (1948) 3397

[43] Jones, R. A. and Katritzky, A. R. *J. chem. Soc.* (a) (1959) 1317; (b) (1961) 378; (c) Sheinker, Y. N., Peresleni, E. M., Zosimova, N. C. and Pomerantsev, Y. T. *J. gen. Chem. U.S.S.R.* **33** (1959) 303; (d) Taurius, A. *Can. J. Chem.* **36** (1958) 465

[44] Specker, H. and Gawrosch, H. *Ber. dtsch. chem. Ges.* **75** (1942) 1338; Mason, S. F. *J. chem. Soc.* (1957) 5010

[45] Metzler, D. E. and Snell, E. E. *J. Am. chem. Soc.* **77** (1955) 2431; Mason, S. F. *J. chem. Soc.* (1959) 1253

[46] (a) Jones, R. A. and Katritzky, A. R. *J. chem. Soc.* (1958) 3610; (b) Albert, A. and Barlin, G. B. (1959) 2384

[47] (a) Evans, R. F., Herington, E. F. G. and Kynaston, W. *Trans. Faraday Soc.* **49** (1953) 1284; (b) Green, R. W. and Tong, H. K. *J. Am. chem. Soc.* **78** (1956) 4896; (c) Stephenson, H. P. and Sponer, H. **79** (1957) 2050

[48] Jaffé, H. M. *J. Am. chem. Soc.* **77** (1955) 4451

[49] Ito, M. and Hata, N. *Bull. chem. Soc. Japan* **28** (1955) 260; Ito, M. and Mizushima, W. *J. chem. Phys.* **23** (1956) 495; Hirayama, H. and Kubota, T. *Chem. Abstr.* **51** (1957) 8532; but see^{54d}

[50] Katritzky, A. R. *J. chem. Soc.* (1957) 191

[51] Gardner, J. N. and Katritzky, A. R. *J. chem. Soc.* (1957) 4375

[52] (a) Shaw, E. *J. Am. chem. Soc.* **71** (1949) 67; (b) Cunningham, K. G. Newbold, G. T., Spring, F. S. and Stark, J. *J. chem. Soc.* (1949) 2091

[53] Hantzsch, A. *Ber. dtsch. chem. Ges.* **44** (1911) 1776, 1783

[54] (a) Kosower, E. M. *J. Am. chem. Soc.* **77** (1955) 3883; (b) with Klinedinst, P. E. **78** (1956) 3493; (c) **78** (1956) 3497; (d) with Burbach, J. C. *ibid.* 5838; (e) **80** (1958) 3253; (f) with Skorcz, J. A. *et al.* **82** (1960) 2188; (g) *ibid.* 2195; (h) with Hofmann, D. and Wallenfels, K. **84** (1962) 2755

[55] (a) Green, J. H. S., Kynaston, W. and Paisley, H. M. *Spectrochim. Acta* **19** (1963) 549; (b) Abramovitch, R. A., Seng, G. C. and Notation, A. D. *Can. J. Chem.* **38** (1960) 624

REFERENCES

[56] Katritzky, A. R. *Qu. Rev. chem. Soc.* **13** (1959) 353

[57] Corssin, L. and Lord, R. C. *J. chem. Phys.* **21** (1953) 1170; Wilmshurst, J. K. and Bernstein, H. J. *Can. J. Chem.* **35** (1957) 1183; Long, D. A. with Murfin, F. S. and Thomas, E. L. *Trans. Faraday Soc.* **59** (1963) 12; with Thomas, E. L. *ibid.* 783; with Bailey, R. T. *ibid.* 599

[58] Cook, G. L. and Church, F. M. *J. phys. Chem., Ithaka* **61** (1957) 458; Podall, H. E. *Analyt. Chem.* **29** (1957) 1423

[59] Nuttall, R. H., Sharp, D. W. A. and Waddington, T. C., *J. chem. Soc.* (1960) 4965; Evans, R. F. and Kynaston, W. (1962) 1005; Gill, N. S., Nuttall, R. H., Scaife, D. E. and Sharp, D. W. A. *J. inorg. nucl. Chem.* **18** (1961) 79

[60] (a) Cook, D. *Can. J. Chem.* **39** (1961) 2009; Spinner, E. *J. chem. Soc.* (1963) 3870; (b) (1963) 3860; (c) Katritzky, A. R. with Gardner, J. N. (1958) 2198; with Hands, A. R. *ibid.* 2202

[61] Katritzky, A. R., Monro, A. M., Beard, J. A. T., Dearnaley, D. P. and Earl, N. J. *J. chem. Soc.* (1958) 2182

[62] Shindo, H. *Pharm. Bull., Tokyo* **5** (1957) 472; **6** (1958) 117; **7** (1959) 791

[63] Katritzky, A. R. *Recl Trav. chim. Pays-Bas Belg.* **78** (1959) 995; Sensi, P. *Gazz. chim. ital.* **85** (1955) 235

[64] (a) Angyal, C. L. and Werner, R. L. *J. chem. Soc.* (1952) 2911; (b) Goulden, J. D. S. (1952) 2939; (c) Mason, S. F. (1958) 3619; (1959) 1281

[65] Katritzky, A. R. and Jones, R. A. *J. chem. Soc.* (a) (1959) 3674; (b) *ibid.* 2067; (1960) 676, 4497

[66] (a) Ramiah, K. V. and Puranik, P. G. *J. molec. Spectrosc.* **7** (1961) 89; (b) Spinner, E. *J. chem. Soc.* (1962) 3119; (c) (1960) 1237

[67] Sensi, P. and Gallo, G. G. *Annali Chim.* **44** (1954) 232; Gibson, J. A., Kynaston, W. and Lindsey, A. S. *J. chem. Soc.* (1955) 4340

[68] Mason, S. F. *J. chem. Soc.* (1957) 4874; cf. Albert, A. and Spinner, E. (1960) 1221, and Katritzky, A. R. and Jones, R. A. *ibid.* 2947

[69] (a) Wiley, R. H. and Slaymaker, S. C. *J. Am. chem. Soc.* **79** (1957) 2233; (b) Katritzky, A. R. with Gardner, J. N. *J. chem. Soc.* (1958) 2192; (c) with Hands, A. R. (1958) 2195; (d) with Beard, J. A. T. and Coats, N. A. (1959) 3680

[70] Bernstein, H. J., Pople, J. A. and Schneider, W. G. *Can. J. Chem.* **35** (1957) 65, 1487; Scheefer, T. and Schneider, W. G. *J. chem. Phys.* **32** (1960) 1224

[71] (a) Biddiscombe, D. P., Herington, E. F. G., Lawrenson, I. J. and Martin, J. F. *J. chem. Soc.* (1963) 444; (b) Katritzky, A. R. and Lagowski, J. M. (1961) 43

[72] (a) Smith, I. C. and Schneider, W. G. *Can. J. Chem.* **39** (1961) 1158; (b) Cook, D. **41** (1963) 515, 2575; (c) Elvidge, J. A. and Jackman, L. M. *J. chem. Soc.* (1961) 859

[73] Jones, R. A. Y., Katritzky, A. R. and Lagowski, J. M. *Chemy Ind.* (1960) 870

[74] (a) Katritzky, A. R. and Jones, R. A. Y. *Proc. chem. Soc.* (1960) 313; (b) Katritzky, A. R. and Reavill, R. E. *J. chem. Soc.* (1963) 753; (c) van der Haak, P. J. and de Boer, Th. J. *Recl Trav. chim. Pays-Bas Belg.* **83** (1964) 186

[75] Spinner, E. *J. chem. Soc.* (a) (1960) 1226; (b) (1962) 3127; (c) with White, J. C. B. (1962) 3115

[76] (a) Segel, S. L., Barnes, R. G. and Bray, P. J. *J. chem. Phys.* **25** (1956) 1286; Bray, P. J., Moskowitz, S., Hooper, H. O., Barnes, R. G. and Segel, S. L. **28** (1958) 99; Negita, H., Sato, S., Yonezawa, T. and Fukui, K. *Bull. chem. Soc. Japan* **30** (1957) 721; Dewar, M. J. S. and Lucken, E. A. C. *J. chem. Soc.* (1958) 2653; (b) Lucken, E. A. C. *Trans. Faraday Soc.* **57** (1961) 729

[77] Ingold, C. K. *Structure and Mechanism in Organic Chemistry*, p. 174, London (Bell) 1953

[78] Brown, H. C. and Mihm, X. R. *J. Am. chem. Soc.* **77** (1955) 1723

[79] (a) Brown, H. C. with Kanner, B. *J. Am. chem. Soc.* **75** (1953) 3865; (b) with McDaniel, D. H. and Häfliger, O. in *Determination of Organic Structures by Physical Methods*, ed. Braude, E. A. and Nachod, F. C. New York (Academic Press) 1955; (c) *ibid.* p. 597

[80] (a) Mortimer, C. T. and Laidler, K. J. *Trans. Faraday Soc.* **55** (1959) 1731; (b) Andon, R. J. L., Cox, J. D. and Herington, E. F. G. **50** (1954) 918; (c) Baxendale, J. H. and George, P. **46** (1950) 55

[81] (a) Linnell, R. H. *J. org. Chem.* **25** (1960) 290; (b) Gero, A. and Markham, J. J. **16** (1951) 1835; (c) Driscoll, J. S., Pfleiderer, W. and Taylor, E. C. **26** (1961) 5230

[82] (a) Sacconi, L., Paoletti, P. and Ciampolini, M. *J. Am. chem. Soc.* **82** (1960) 3831; (b) Brown, H. C. and Holmes, R. R. **77** (1955) 1727; (c) Hall, N. F. and Sprinkle, M. R. **54** (1932) 3469; (d) Klotz, I. M., Fiess, H. A., Chen Ho, J. Y. and Mellody, M. **76** (1954) 5136; (e) Brown, H. C., Johnson, S. and Podall, H. *ibid.* 5556; (f) Bender, M. L. and Chow, Y. L. **81** (1959) 3929; (g) Cilento, G. and Miller, E. C. and J. A. **78** (1956) 1718; Toffani, A. and M. R. *Atti Accad. naz. Lincei* **23** (1957) 60

[83] (a) Jellinek, H. and Urwin, J. *J. phys. Chem., Ithaka* **58** (1954) 548; (b) Fastier, F. *Aust. J. exp. Biol. med. Sci.* **36** (1958) 491; (c) Otsuji, Y., Koda, Y. and Hirai, J. *Nippon Kagaku Zasshi* **80** (1959) 3929; (d) Ikekawa, N., Sato, Y. and Maeda, T. *Chem. Abstr.* **50** (1956) 994

[84] (a) Brown, H. C. *J. chem. Soc.* (1956) 1248; (b) Mason, S. F. (1958) 674; (c) Albert, A. and Pedersen, C. (1956) 4683; (d) Katritzky, A. R., Short, D. J. and Boulton, A. J. (1960) 1516; (e) Jones, R. A. and Katritzky, A. R. *ibid.* 2937; (f) Hanania, G. I. H. and Irvine, D. H. (1962) 2745; (g) Feakins, D., Last, W. A. and Shaw, R. A. (1964) 2387; (h) Fischer, A., Galloway, W. J. and Vaughan, J. (1964) 3591

[85] (a) Hoeg, D. F. and Liebman, J. *J. org. Chem.* **28** (1963) 1554; (b) Mastrukova, T. A., Sheinker, Y. N. et al., *Tetrahedron* **19** (1963) 357

[86] Ebert, L. Z. *phys. Chem.* **121** (1926) 385; Edsall, J. T. and Blanchard, M. H. *J. Am. Chem. Soc.* **55** (1933) 2337

[87] Tucker, G. F. and Irvin, J. L. *J. Am. chem. Soc.* **73** (1951) 1923

[88] Mason, S. F. *J. chem. Soc.* (1960) 22

[89] Angyal, S. J. and Angyal, C. L. *J. chem. Soc.* (1952) 1461

[90] Albert, A. and Phillips, J. N. *J. chem. Soc.* (1956) 1294

[91] Albert, A., Goldacre, R. and Phillips, J. *J. chem. Soc.* (1948) 2240

[92] Jaffé, H. H., *J. Am. chem. Soc.* **77** (1955) 4445

[93] Bryson, A. *J. Am. chem. Soc.* **82** (1960) 4871

[94] Bates, R. G. and Hetzer, H. B. *J. Res. natn. Bur. Stand.* **64A** (1960) 427

[95] Jaffé, H. H. *Chem. Rev.* **53** (1953) 191

[96] Jaffé, H. H. *J. chem. Phys.* **20** (1952) 1554

[97] Katritzky, A. R. and Simmons, P. *J. chem. Soc.* (1960) 1511

[98] Jaffé, H. H. *J. Am. chem. Soc.* **76** (1954) 3527

[99] McDaniel, D. H. and Brown, H. C. *J. org. Chem.* **23** (1958) 420

[100] Jaffé, H. H. and Doak, G. O. *J. Am. chem. Soc.* **77** (1955) 4441

[101] Taft, R. W. and Evans, H. D. *J. chem. Phys.* **27** (1957) 1427

[102] Gibson, C. S. and Simonsen, J. L. *J. chem. Soc.* (1930) 2531

[103] Cox, E. G., Sharratt, E., Wardlaw, W. and Webster, K. C. *J. chem. Soc.* (1936) 129

[104] Eley, D. D. and Watts, H. *J. chem. Soc.* (1952) 1914

[105] Murmann, R. K. and Basolo, F. *J. Am. chem. Soc.* **77** (1955) 3484

[106] Brandt, W. W., Dwyer, F. P. and Gyarfas, E. C. *Chem. Rev.* **54** (1954) 958; cf. Bergh, A., Offenhartz, P. O'D., George, P. and Haight, G. P. *J. chem. Soc.* (1964) 1533

[107] Cagle, F. W. and Smith, G. F. *J. Am. chem. Soc.* **69** (1947) 1860; Irving, H. and Hampton, A. *J. chem. Soc.* (1955) 430

[108] Burstall, F. H. and Nyholm, R. S. *J. chem. Soc.* (1952) 3570

[109] Barnes, R. A. in *Pyridine and Its Derivatives*, Part 1, ed. Klingsberg, E., New York (Interscience) 1960; (b) Part 2, p. 191, 1961

[110] (a) Sisler, H. H., Bush, J. D. and Accountius, O. E. *J. Am. chem. Soc.* **70** (1948) 3827; (b) Poos, G. I., Arth, G. E., Beyler, R. E. and Sarett, L. H. **75** (1953) 422

[111] (a) Quagliano, J. V., Fujita, J., Franz, G., Phillips, D. J., Walmsley, J. A. and Tyree, S. Y. *J. Am. chem. Soc.* **83** (1961) 3770; Carlin, R. L. *ibid.* 3773; Shupack, S. I. and Orchin, M. **85** (1963) 902; **86** (1964) 586; Heller, A., Marcus, Y. and Eliezer, I. *J. chem. Soc.* (1963) 1579; Kida, S., Quagliano, J. V., Tyree, S. Y. and Walmsley, J. A. *Spectrochim. Acta* **19** (1963) 189; Kakuiti, Y., Kida, S. and Quagliano, J. V. *ibid.* 201; (b) Sigel, H. and Brintzinger, H. *Helv. chim. Acta* **46** (1963) 701; with Erlenmeyer, H. *ibid.* 712

[112] Williams, D. M. *J. chem. Soc.* (1931) 2783

[113] Simons, J. H. and Herman, D. F. *Abstr. N. Y. Meet. Am. chem. Soc.* (1947) 13J

[114] Zappi, E. V. and Fernandez, M. *Chem. Abstr.* **34** (1940) 3741

[115] Trowbridge, P. F. and Diehl, O. C. *J. Am. Chem. Soc.* **19** (1897) 558

[116] Prescott, A. B. and Trowbridge, P. F. *J. Am. chem. Soc.* **17** (1895) 859

[117] Morcillo, J. and Heranz, J. *Chem. Abstr.* **48** (1954) 8655

[118] (a) Hassel, O. with Rømming, C. *Act. chem. scand.* **10** (1956) 696; and Tufte, T. **15** (1961) 967; (b) with Hope, H. *ibid.* 407

[119] Kauffman, G. B. and Stevens, K. L. *Inorg. Synth.* **7** (1963) 169, 172-3, 176

[120] Carlsohn, H. *Ber. dtsch. chem. Ges.* **68** (1935) 2209; *Über eine neue Klasse von Verbindungen des positiv einwertigen Jods*, Leipzig, 1932

[121] Kortüm, G. and Wilski, H. *Z. phys. Chem.* **202** (1953) 35

[122] Audrieth, L. F. and Birr, E. J. *J. Am. chem. Soc.* **55** (1933) 668; Zingaro, R. A., VanderWerf, C. A. and Kleinberg, J. **73** (1951) 88; Kleinberg, J., Colton, E., Saltizahn, J. and VanderWerf, C. A. **75** (1953) 442

[123] (a) Popov, A. I. with Pflaum, R. T. *J. Am. chem. Soc.* **79** (1957) 570; (b) with Rygg, R. H. *ibid.* 4622

[124] Reid, C. and Mulliken, R. S. *J. Am. chem. Soc.* **76** (1954) 3869

[125] Person, W. B., Humphrey, R. E., Deskin, W. A. and Popov, A. I. *J. Am. chem. Soc.* **80** (1958) 2049

[126] Djerassi, C. and Scholz, C. R. *J. Am. chem. Soc.* **70** (1948) 417

[127] (a) Englert, S. M. E. and McElvain, S. M. *J. Am. chem. Soc.* **51** (1929) 863; (b) McElvain, S. M. and Goese, M. A. **65** (1943) 2227

REFERENCES

[128] (a) Foster, R. W., *Ph.D. Thesis*, London University, 1954; (b) Jones, J. and J. *Tetrahedron Lett.* (1964) 2117; (c) Beattie, I. R. and Webster, M. *J. chem. Soc.* (1961) 1730

[129] Feigl, F. and Feigl, E. *Z. Anorg. allg. Chem.* **203** (1932) 57

[130] (a) Baumgarten, P. *Ber. dtsch. chem. Ges.* **59** (1926) 1166; **64** (1931) 1505; with Marggraff, I., *ibid.* 1582; (b) Mingoia, Q. and Ferreira, P. C. *Chem. Abstr.* **49** (1955) 7566

[131] Adams, R., Hine, J. and Campbell, J. *J. Am. chem. Soc.* **71** (1949) 387

[132] Mosher, H. S. and Welch, F. J. *J. Am. chem. Soc.* **77** (1955) 2902

[133] (a) Bain, B. M. and Saxton, J. E. *J. chem. Soc.* (1961) 5216: (b) Villani, F. J. and Papa, D. *J. Am. chem. Soc.* **72** (1950) 2722

[134] (a) Tschitschibabin, A. E. and Kirsanov, A. W. *Ber. dtsch. chem. Ges.* **57** (1924) 1161; (b) Baine, O., Adamson, G. F. *et al.*, *J. org. Chem.* **19** (1954) 510; (c) Bojarska-Dahlig, H. and Urbański, T. *Chem. Abstr.* **48** (1954) 1337

[135] (a) Brown, R. F. and Miller, S. J. *J. org. Chem.* **11** (1946) 388; (b) Stempel, A. and Buzzi, E. C. *J. Am. chem. Soc.* **71** (1949) 2969; (c) Urbański, T., *J. chem. Soc.* (1946) 1104; (1947) 132; (d) *Swiss Pat.* 243,101 (1946)

[136] Kahn, H. J. and Petrov, V. A. *J. chem. Soc.* (1945) 858

[137] Dyson, P. and Hammick, D. Ll. *J. chem. Soc.* (1937) 1724; Ashworth, M. R. F., Daffern, R. P. and Hammick, D. Ll. (1939) 809; Brown, B. R. and Hammick, D. Ll. (1949) 173, 659; with Thewlis, B. H. *Nature, Lond.* **162** (1948) 73

[138] Mislow, K. *J. Am. chem. Soc.* **69** (1947) 2559

[139] Cantwell, N. H. and Brown, E. V. *J. Am. chem. Soc.* **74** (1952) 5967; **75** (1953) 1489, 4466

[140] (a) Sperber, N., Papa, D., Schwenk, E. and Sherlock, M. *J. Am. chem. Soc.* **71** (1949) 887; (b) Katritzky, A. R. and Ridgewell, B. J. *J. chem. Soc.* (1963) 3753

[141] (a) Katritzky, A. R., Ridgewell, B. J. and White, A. M. *Chemy Ind.* (1964) 1576; (b) Tschitschibabin, A. E. and Persitz, R. L. *Zh. russk. fiz.-khim. Obshch.* **57** (1925) 301

[142] (a) Tschitschibabin, A. E. with Zeide, O. A. *Zh. russk. fiz.-khim. Obshch.* **46** (1914) 1216; (b) **50** (1918) 522; (c) *ibid.* 512; Ostromislensky, I. *U.S. Pat.* 1,680,109; 1,680,111 (1928); 1,724,305 (1929); 1,809,352; 1,820,483 (1931); Tisza, E. T. and Joos, B. *U.S. Pat.* 1,856,602 (1932); 2,029,315 (1936); Renshaw, R. R. and Tisza, E. T., 2,135,293 (1938); Tschitschibabin, A. E. and Hoffmann, C. *C. r. hebd. Séanc. Acad. Sci., Paris* **205** (1937) 153; Charrier, G. and Jorio, M. *Gazz. chim. ital.* **68** (1938) 640

[143] Tschitschibabin, A. E. and Ossetrowa, E. D. *J. Am. chem. Soc.* **56** (1934) 1711

[144] Ostromislensky, I. *J. Am. chem. Soc.* **56** (1934) 1713

[145] *Germ. Pat.* 543,288 (1928); 617,187 (1935); Binz, A. and Schickh, O. v. *U.S. Pat.* 2,156,141

[146] Mills, W. H. and Widdows, S. T. *J. chem. Soc.* **94** (1908) 1372

[147] Gattermann, L. and Skita, A. *Ber. dtsch. chem. Ges.* **49** (1916) 494

[148] (a) Moore, J. A. and Marascia, F. J. *J. Am. chem. Soc.* **81** (1959) 6049; (b) Urban, R. and Schnider, O. *Helv. chim. Acta* **47** (1964) 363

[149] Ochiai, E. and Morishita, E. *J. pharm. Soc. Japan* **76** (1956) 531

[150] *Swiss Pat.* 222,727; 222,728 (1942)

[151] Haszeldine, R. N. *J. chem. Soc.* (1950) 1638, 1966; (1951) 102; Simmons, T. C., Hoffmann, F. W. *et al.*, *J. Am. chem. Soc.* **79** (1957) 3429

[152] (a) Wibaut, J. P. and Nicolai, J. R. *Recl Trav. chim. Pays-Bas Belg.* **58** (1939) 709; (b) Sell, W. J. and Dootson, F. W. *J. chem. Soc.* **75** (1899) 979; (c) **73** (1898) 432

[153] McBee, E. T., Hass, H. B. and Hodnett, E. M. *Ind. Engng Chem. analyt. Edn* **39** (1947) 389; cf. Sell, W. J. *J. chem. Soc.* **93** (1908) 1993; Seyfferth, E. *J. prakt. Chem.* **34** (1886/2) 241

[154] Chambers, R. D., Hutchinson, J. and Musgrave, W. K. R. *J. chem. Soc.* (1964) 3573

[155] Koenigs, W. and Geigy, R. *Ber. dtsch. chem. Ges.* **17** (1884) 589

[156] (a) Steinhäuser, E. and Diepolder, E. *J. prakt. Chem.* **93** (1916) 387; (b) Ruzicka, L. and Fornasir, V. *Helv. chim. Acta* **2** (1919) 338; (c) Chase, B. H. and Walker, J. *J. chem. Soc.* (1953) 3548; (d) Engelsma, J. W. and Kroogman, E. C. *Proc. chem. Soc.* (1958) 238

[157] Hofmann, A. W. *Ber. dtsch. chem. Ges.* **12** (1879) 984; Ciamician, G. and Silber, P. **18** (1885) 721

[158] (a) Blau, F. *Mh. Chem.* **10** (1889) 372; (b) Koenigs, E., Gerdes, H. C. and Sirot, A. *Ber. dtsch. chem. Ges.* **61** (1928) 1022; (c) Maier-Bode, H. **69** (1936) 1534; (d) den Hertog, H. J., van der Does, L. and Landheer, C. A. *Recl Trav. chim. Pays-Bas Belg* **81** (1962) 864

[159] (a) Fischer, O. and Riemerschmid, C. *Ber. dtsch. chem. Ges.* **16** (1883) 1183; (b) Dohrn, M. and Diedrich, P. *Justus Liebigs Annln Chem.* **494** (1932) 284

[160] den Hertog, H. J. and Wibaut, J. P. *Recl Trav. chim. Pays-Bas Belg.* **51** (1932) 948

[161] (a) den Hertog, H. J., with Wibaut, H. P. *Recl Trav. chim. Pays-Bas Belg.* **51** (1932) 381, 940; (b) **64** (1945) 55; (c) *ibid.* 85; (d) *Germ. Pat.* 574,655 (1933); (e) *Dutch Pat.* 29,614 (1933)

[162] den Hertog, H. J. and Combé, W. P. *Recl Trav. chim. Pays-Bas Belg.* **70** (1951) 581

[163] den Hertog, H. J., Combé, W. P. and Kolder, C. R. *Recl Trav. chim. Pays-Bas Belg.* **77** (1958) 66

287

[164] Pearson, D. E., Hargrove, W. W., Chow, J. K. T. and Suthers, B. R. *J. org. Chem.* **26** (1961) 789

[165] (a) Rodewald, Z. and Plažek, E. *Roczn. Chem.* **16** (1936) 444; (b) *Ber. dtsch. chem. Ges.* **70** (1937) 1159

[166] (a) Plažek, E. and Rodewald, Z. *Roczn. Chem.* **21** (1947) 150; (b) Symons, M. C. R. *J. chem. Soc.* (1957) 387

[167] Buu-Hoï, Ng. Ph. *Recl Trav. chim. Pays-Bas Belg.* **73** (1954) 197

[168] Bachman, G. B. and Micucci, D. D. *J. Am. chem. Soc.* **70** (1948) 2381

[169] (a) Tschitschibabin, A. E. and Egorov, A. F. *Zh. russk. fiz.-khim. Obshch.* **60** (1928) 683; (b) English, J. P., Clark, J. H. *et al.*, *J. Am. chem. Soc.* **68** (1946) 453

[170] Parker, E. D. and Shive, W. *J. Am. chem. Soc.* **69** (1947) 63

[171] (a) Tschitschibabin, A. E. and Tyazhelova, V. S. *Zh. russk. fiz.-khim. Obshch.* **50** (1920) 483, 492; (b) Wibaut, J. P. and Kraay, G. M. *Recl Trav. chim. Pays-Bas Belg.* **42** (1923) 1084; (c) den Hertog, H. J. **65** (1946) 129; (d) Case, F. H. *J. Am. chem. Soc.* **68** (1946) 2574; (e) Adams, R. and Schrecker, W. **71** (1949) 1186; (f) Bradlow, H. L. and VanderWerf, C. A. *J. org. Chem.* **16** (1951) 73; (g) Cook, D. J., Bowen, R. E., Sorter, P. and Daniels, E. **26** (1961) 4949

[172] Magidson, O. and Menschikoff, G. *Ber. dtsch. chem. Ges.* **58** (1925) 113

[173] Dohrn, M. and Dirksen, R. *U.S. Pat.* 1,723,457 (1929); 1,793,683 (1931)

[174] Caldwell, W. T., Tyson, F. T. and Lauer, L. *J. Am. chem. Soc.* **66** (1944) 1479

[175] den Hertog, H. J. and Bruin, P. *Recl Trav. chim. Pays-Bas Belg.* **65** (1946) 385

[176] Wibaut, J. P. and den Hertog, H. J. *Germ. Pat.* 574,655 (1933)

[177] Schickh, O., Binz, A. and Schultz, A. *Ber. dtsch. chem. Ges.* **69** (1936) 2593

[178] Rodewald, Z. and Plažek, E. *Roczn. Chem.* **16** (1936) 130

[179] den Hertog, H. J. and Schogt, J. C. M. *Recl Trav. chim. Pays-Bas Belg.* **70** (1951) 353

[180] (a) Koenigs, E., Mields, M. and Gurlt, H. *Ber. dtsch. chem. Ges.* **57** (1924) 1179; (b) Bremer, O. *Justus Liebigs Annln Chem.* **518** (1935) 274

[181] Ochiai, E. and Fujimoto, M. *Pharm. Bull., Tokyo* **2** (1954) 131

[182] (a) Dohrn, M. and Dirksen, R. *U.S. Pat.* 1,706,775 (1929); (b) Brockman, F. W. and Tendelov, H. J. C. *Recl Trav. chim. Pay-Bas Belg.* **81** (1962) 107

[183] *Chlorinations:* (a) Graf, R. and Stauch, J. *J. prakt. Chem.* **148** (1937) 13; (b) den Hertog, H. J. and de Bruyn, J. *Recl Trav. chim. Pays-Bas Belg.* **70** (1951) 182; (c) Kolder, C. R. and den Hertog, H. J. **72** (1953) 285
Brominations: (d) Wibaut, J. P. and van Wagtendonk, H. M. *ibid.* **60** (1941) 22; (e) Ramirez, F. and Paul, A. P. *J. org. Chem.* **19** (1954) 183; (f) Mariella, R. P., Callahan, J. J. and Jibril, A. O. **20** (1955) 1721;
Iodination: (g) Sugii, Y. and Shindo, H. *J. pharm. Soc. Japan* **51** (1931) 416

[184] (a) Fischer, O. and Chur, M. *J. prakt. Chem.* **93** (1916/2) 363; (b) Dohrn, M. and Thiele, A. *German Pat.* 500,915 (1927)

[185] den Hertog, H. J., Schepman, F. R., de Bruyn, J. and Thysse, G. J. E. *Recl Trav. chim. Pays-Bas Belg.* **69** (1950) 1281

[186] (a) Lerch, *Mh. Chem.* **5** (1884) 367; (b) Lieben, A. and Haitinger, L. **4** (1883) 339; **6** (1885) 279; (c) bromination of 2,6-dibromo-4-hydroxypyridine: den Hertog, H. J. *Recl Trav. chim. Pays-Bas Belg.* **67** (1948) 381

[187] (a) Kolder, C. R. and den Hertog, H. J. *Recl Trav. chim. Pays-Bas Belg.* **79** (1960) 474 (cf. [161c]); (b) van der Plas, H. C., den Hertog, H. J., van Ammers, M. and Haase, B. *Tetrahedron Lett.* (1961) 32; (c) van Ammers, M., den Hertog, H. J. and Haase, B. *Tetrahedron* **18** (1962) 227

[188] (a) Behrman, E. J., with Pitt, B. M. *J. Am. chem. Soc.* **80** (1958) 3717; with Walker, P. P. **84** (1962) 3454; (b) Boyland, E. and Sims, P. *J. chem. Soc.* (1958) 4198

[189] (a) Gilman, H. *Org. React.* **8** (1954) 258; (b) Jones, R. G. and Gilman, H. **6** (1951) 339

[190] Gilman, H. and Spatz, S. M. *J. Am. chem. Soc.* **62** (1940) 446

[191] Gilman, H. and Spatz, S. M. *J. org. Chem.* **16** (1951) 1485

[192] Murray, A., Foreman, W. W. and Langham, W. *J. Am. chem. Soc.* **70** (1948) 1037

[193] Spatz, S. M. *Iowa St. Coll. J. Sci.* **17** (1942) 129

[194] Murray, A. and Langham, W. H. *J. Am. chem. Soc.* **74** (1952) 6289

[195] Wibaut, J. P. and Heeringa, L. G. *Recl Trav. chim. Pays-Bas Belg.* **74** (1955) 1003

[196] Gilman, H. and Melstrom, D. S. *J. Am. chem. Soc.* **68** (1946) 103

[197] Sachs, G. and Eberhartinger, R. *Ber. dtsch. chem. Ges.* **56** (1923) 2223; McCleland, N. P. and Wilson, R. H. *J. chem. Soc.* (1932) 1263

[198] (a) Shreve, R. N. and Swaney, M. W. *U.S. Pat.* 2,206,309 (1940); 2,216,140 (1941); 2,297,636 (1943); (b) Swaney, M. W., Skeeters, M. J. and Shreve, R. N. *Ind. Engng Chem. analyt. Edn* **32** (1940) 360

[199] Hurd, C. D. and Morissey, C. J. *J. Am. chem. Soc.* **77** (1955) 4658

[200] Clemo, G. R. and Swan, G. A. *J. chem. Soc.* (1948) 198

[201] *Austr. Pat.* 112,128 (1929)

[202] Andersen, C. N. *U.S. Pat.* 2,085,063 (1937)

REFERENCES

[203] Ukai, T., Yamamoto, Y. and Hirano, S. *J. pharm. Soc. Japan* **73** (1953) 823
[204] van Ammers, M. and den Hertog, H. J. *Recl Trav. chim. Pays-Bas Belg.* (a) **77** (1958) 340; (b) **81** (1962) 124
[205] Friedl, F. *Ber. dtsch. chem. Ges.* **45** (1912) 428; *Mh. Chem.* **34** (1913) 759
[206] Kirpal, A. and Reiter, E. *Ber. dtsch. chem. Ges.* **58** (1925) 699
[207] den Hertog, H. J. and Overhoff, J. *Recl Trav. chim. Pays-Bas Belg.* **49** (1930) 552; cf. [93]
[208] (a) Plažek, E. *Ber. dtsch. chem. Ges.* **72** (1939) 577; (b) van Rijn, P. J. *Recl Trav. chim. Pays-Bas Belg.* **45** (1926) 267; (c) Brown, E. V. and Neil, R. H. *J. org. Chem.* **26** (1961) 3546; (d) Katritzky, A. R. and Ridgewell, B. J. *J. chem. Soc.* (1963) 3882
[209] Schorigin, P. and Toptschiew, A. *Ber. dtsch. chem. Ges.* **69** (1936) 1874
[210] Tschitschibabin, A. E. and Rasorenov, B. A. *Zh. russk. fiz.-khim. Obshch.* **47** (1915) 1286
[211] Tschitschibabin, A. E. and Bylinkin, I. G. *Zh. russk. fiz.-khim. Obshch.* **50** (1920) 471
[212] Phillips, M. A. *J. chem. Soc.* (1941) 9
[213] Caldwell, W. T. and Kornfeld, E. C. *J. Am. chem. Soc.* **64** (1942) 1695
[214] Pino, L. N. and Zehrung, W. S. *J. Am. chem. Soc.* **77** (1955) 3154
[215] Korte, F. *Chem. Ber.* **85** (1952) 1012
[216] Tschitschibabin, A. E. and Widonowa, M. S. *Zh. russk. fiz.-khim. Obshch.* **53** (1921) 238
[217] Hughes, E. D. and Jones, G. T. *J. chem. Soc.* (1950) 2678
[218] Räth, C. and Prange, G. *Justus Liebigs Annln Chem.* **467** (1928) 1
[219] Tschitschibabin, A. E., with Kirsanov, A. W. *Ber. dtsch. chem. Ges.* (a) **60** (1927) 2433; (b) **61** (1928) 1223; (c) **58** (1925) 1707
[220] Koenigs, E., Kinne, G. and Weiss, W. *Ber. dtsch. chem. Ges.* **57** (1924) 1172
[221] Räth, C. and Prange, G. *Ber. dtsch. chem. Ges.* **58** (1925) 1208
[222] Czuba, W. *Roczn. Chem.* **34** (1960) 905, 1639, 1647
[223] Plažek, E., Marcinikow, A. and Stammer, C. *Chem. Abstr.* **30** (1936) 1377
[224] Plažek, E. and Sucharda, E. *Ber. dtsch. chem. Ges.* **61** (1928) 1813
[225] Tschitschibabin, A. E. and Knunianz, I. L. *Ber. dtsch. chem. Ges.* (a) **61** (1928) 427; (b) **62** (1929) 3053
[226] Plažek, E., Sorokowska, A., and Tolopka, D. *Chem. Abstr.* **33** (1939) 3379
[227] (a) Tschitschibabin, A. E. and Schapiro, S. A. *Zh. russk. fiz.-khim. Obshch.* **53** (1921) 233; (b) Binz, A. and Maier-Bode, H. *Angew. Chem.* **49** (1936) 486
[228] Takahashi, T. and Yamamoto, Y. *J. pharm. Soc. Japan* **69** (1949) 409
[229] Plažek, E. *Recl Trav. chim. Pays-Bas Belg.* **72** (1953) 569
[230] Berrie, A. H., Newbold, G. T. and Spring, F. S. *J. chem. Soc.* (1952) 2042
[231] Collie, J. N. and Tickle, T. *J.* (*Trans.*) *chem. Soc.* **73** (1898) 229
[232] *Germ. Pat.* 568,549 (1932)
[233] Schofield, K. *Qu. Rev. chem. Soc.* **4** (1950) 382
[234] Norman, R. O. C. and Radda, G. K. *J. chem. Soc.* (1961) 3030
[235] Plažek, E. and Rodewald, Z. *Chem. Abstr.* **31** (1937) 3918
[236] *U.S. Pat.* 1,889,303; 1,957,089 (*Chem. Abstr.* **27** (1933) 1366; **28** (1934) 4073); *French Pat.* 705,113 (1930)
[237] Czuba, W. and Plažek, E. *Recl Trav. chim. Pays-Bas Belg.* **77** (1958) 92
[238] Weidel, H. and Murmann, E. *Mh. Chem.* **16** (1895) 749
[239] Bernstein, J., Stearns, B., Shaw, E. and Lott, W. A. *J. Am. chem. Soc.* **69** (1947) 1151
[240] den Hertog, H. J., Jouwersma, C., van der Wal, A. A. and Willebrands-Schogt, E. C. C. *Recl Trav. chim. Pays-Bas Belg.* **68** (1949) 275
[241] (a) Koenigs, E. and Freter, K. *Ber. dtsch. chem. Ges.* **57** (1924) 1187; (b) Crowe, W. H. *J. chem. Soc.* **127** (1925) 2029; (c) Bremer, O. *Justus Liebigs Annln Chem.* **529** (1937) 290
[242] (a) Lapworth, A. and Collie, J. N. *J.* (*Trans.*) *chem. Soc.* **71** (1897) 838; (b) Kögl, F., van der Want, G. M. and Salemink, C. A. *Recl Trav. chim. Pays-Bas Belg.* **67** (1948) 29; (c) *Br. Pat.* 259,961; 629,439; (d) *Swiss Pat.* 260,573 (1949)
[243] den Hertog, H. J. and van Weeren, J. W. *Recl Trav. chim. Pays-Bas Belg.* **67** (1948) 980
[244] Ochiai, E., Arima, K. and Ishikawa, M. *J. pharm. Soc. Japan* **63** (1943) 79
[245] Ochiai, E., Hayashi, E. and Katada, M. *J. pharm. Soc. Japan* **67** (1947) 79
[246] Ochiai, E. and Hayashi, E. *J. pharm. Soc. Japan* **67** (1947) 157
[247] Ochiai, E. *J. org. Chem.* **18** (1953) 534
[248] den Hertog, H. J. and Overhoff, J. *Recl Trav. chim. Pays-Bas Belg.* **69** (1950) 468
[249] (a) Ishikawa, M. *Chem. Abstr.* **45** (1951) 8529; (b) Suzuki, I. *J. pharm. Soc. Japan* **68** (1948) 126
[250] den Hertog, H. J., Kolder, C. R. and Combé, W. P. *Recl Trav. chim. Pays-Bas Belg.* **70** (1951) 591
[251] (a) Herz, W. and Tsai, L. *J. Am. chem. Soc.* **76** (1954) 4184; (b) Taylor, E. C. and Crovetti, A. J. *J. org. Chem.* **19** (1954) 1633; (c) Itai, T. and Ogura, H. *J. pharm. Soc. Japan* **75** (1955) 292
[252] Essery, J. M. and Schofield, K. *J. chem. Soc.* (1960) 4953
[253] Ochiai, E., Ishikawa, M. and Arima, K. *J. pharm. Soc. Japan* **63** (1943) 83
[254] Jujo, R. *J. pharm. Soc. Japan* **66** (1946) 21

[255] den Hertog, H. J., Henkens, C. H. and Dilz, K. *Recl Trav. chim. Pays-Bas Belg.* **72** (1953) 296

[256] Lott, W. A. and Shaw, E. *J. Am. chem. Soc.* **71** (1949) 70

[257] den Hertog, H. J. and van Ammers, M. *Recl Trav. chim. Pays-Bas Belg.* **74** (1955) 1160

[258] (*a*) van Ammers, M. and den Hertog, H. J. *Recl Trav. chim. Pays-Bas Belg.* **75** (1956) 1259; (*b*) with Schukking, S. **74** (1955) 1171

[259] Hayashi, E. *J. pharm. Soc. Japan* **70** (1950) 142

[260] Moodie, R. B., Schofield, K. and Williamson, M. J. *Chemy. Ind.* (1964) 1577

[261] Hands, A. R. and Katritzky, A. R. *J. chem. Soc.* (1958) 1754

[262] Forsyth, R. and Pyman, F. L. *J. chem. Soc.* (1926) 2912 (including interesting notes by Flürscheim, Ingold and Robinson)

[263] Bryans, F. and Pyman, F. L. *J. chem. Soc.* (1929) 549

[264] Shaw, B. D. and Wagstaff, E. A. *J. chem. Soc.* (1933) 79; cf. Wagstaff, E. A. (1934) 276

[265] Koenigs, E. and Ruppelt, E. *Justus Liebigs Annln Chem.* **509** (1934) 142

[266] Coates, H., Cook, A. H. *et al.*, *J. chem. Soc.* (1943) 406

[267] (*a*) Titov, A. I. *Chem. Abstr.* **33** (1939) 4248; (*b*) Tschitschibabin, A. E. and Knunjanz, I. L. *Ber. dtsch. chem. Ges.* **61** (1928) 2215

[268] (*a*) Fischer, O. *Ber. dtsch. chem. Ges.* **15** (1882) 62; (*b*) with Renouf, E. **17** (1884) 755

[269] Meyer, H. and Ritter, W. *Mh. Chem.* **35** (1914) 765

[270] *French Pat.* 685,202 (*Chem. ZentBl.* **101** (1930) II, 2576)

[271] Wulff, O. *Germ. Pat.* 541,036 (*Chem. Abstr.* **26** (1932) 1945)

[272] Wulff, O. *U.S. Pat.* 1,880,646 (*Chem. Abstr.* **27** (1933) 515; on some of the results, see 109*b*

[273] Machek, G. *Mh. Chem.* **72** (1938) 77

[274] Möller, E. F. and Birkofer, L. *Ber. dtsch. chem. Ges.* **75** (1942) 1108

[275] McElvain, S. M. and Goese, M. A. *J. Am. chem. Soc.* **65** (1943) 2233

[276] den Hertog, H. J., van der Plas, H. C. and Buurman, D. J. *Recl Trav. chim. Pays-Bas Belg.* **77** (1958) 963

[277] van Gastel, A. J. P. and Wibaut, J. P. *Recl Trav. chim. Pays-Bas Belg.* **53** (1934) 1031

[278] Webb, J. L. and Corwin, A. H. *J. Am. chem. Soc.* **66** (1944) 1456

[279] Heyns, K. and Vogelsang, G. *Chem. Ber.* **87** (1954) 13

[280] Wawzonek, S., Nelson, M. F. and Thelen, P. J. *J. Am. chem. Soc.* **74** (1952) 2894

[281] *Br. Pat.* 602,882 (1948)

[282] van der Plas, H. C. and den Hertog, H. J. *Chem. Weekbl.* **53** (1957) 560; *Tetrahedron Lett.* **1** (1960) 13

[283] (*a*) van der Plas, H. C., with Crawford, T. H. *J. org. Chem.* **26** (1961) 2611; (*b*) with den Hertog, H. J. *Recl Trav. chim. Pays-Bas Belg.* **81** (1962) 841

[284] Machek, G. *Mh. Chem.* **73** (1939) 180

[285] Tschitschibabin, A. E. and Tjashelowa, V. S. *Zh. russk. fiz.-khim. Obshch.* **50** (1918) 495; Naegeli, C., Kündig, W. and Brandenburger, H. *Helv. chim. Acta* **21** (1938) 1746; Skrowaczewska, Z. *Chem. Abstr.* **48** (1954) 7568

[286] Plažek, E. (*a*) with Marcinków, A. *Chem. Abstr.* **29** (1935) 2535; (*b*) **31** (1937) 4669

[287] Koenigs, E. and Jungfer, O. *Ber. dtsch. chem. Ges.* **57** (1924) 2080

[288] Plažek, E. *Chem. Abstr.* **31** (1937) 1808

[289] *Germ. Pat.* 597,452 (*Chem. Abstr.* **28** (1934) 5083)

[290] (*a*) van Ammers, M. and den Hertog, H. J. *Recl Trav. chim. Pays-Bas Belg.* **78** (1959) 586; (*b*) Evans, R. F. and Brown, H. C. *J. org. Chem.* **27** (1962) 1329

[291] Anderson, T. *Justus Liebigs Annln Chem.* **94** (1855) 358

[292] Lutz, O. *Ber. dtsch. chem. Ges.* **43** (1910) 2636

[293] Prescott, A. B. *J. Am. chem. Soc.* **18** (1896) 91; (*b*) Maggiolo, A. **73** (1951) 5815; (*c*) Baker, J. A. and Hill, S. A. *J. chem. Soc.* (1962) 3464

[294] (*a*) Oechsner de Coninck, W. *J. chem. Soc.* (*Abstr.*) **46** (1884) 612; (*b*) Lippert, W. *Justus Liebigs Annln Chem.* **276** (1893) 181; (*c*) Menschutkin, N. A. *Zh. russk. fiz.-khim. Obshch.* cf. **34** (1902) 411; (*d*) Klages, A. and Keil, R. *Ber. dtsch. chem. Ges.* **36** (1903) 1632

[295] Noller, C. R. and Dinsmore, R. *J. Am. chem. Soc.* **54** (1932) 1025

[296] (*a*) Clarke, H. T. *J. chem. Soc.* **97** (1910) 416; (*b*) Knight, G. A. and Shaw, B. D. (1938) 682; (*c*) Tronov, B. V., Akivis, A. I. and Orlova, V. N. *Zh. russk. fiz.-khim. Obshch.* **61** (1929) 345; (*d*) Winkler, C. A. and Hinshelwood, C. N. *J. chem. Soc.* (1935) 1147

[297] Butenandt, A., Manioli, L. *et al.*, *Ber. dtsch. chem. Ges.* **72** (1939) 1617

[298] (*a*) Macovski, E. *Bull. Soc. chim. Fr.* **3** (1936) 498; (*b*) Shelton, R. S., van Campen, M. G. *et al.*, *J. Am. chem. Soc.* **68** (1946) 757; (*c*) Harris, G. H., Shelton, R. S., van Campen, M. G. *et al.* **73** (1951) 3959

[299] Lawrence, C. A. *Surface-Active Quaternary Ammonium Germicides*, New York (Academic Press) 1950

[300] (*a*) Davidson, J. *Justus Liebigs Annln Chem.* **121** (1862) 254; (*b*) Flintermann, R. F. and Prescott, A. B. *J. Am. chem. Soc.* **18** (1896) 28; (*c*) Baer, S. H. and Prescott, A. B. *ibid.* 988; (*d*) Schmidt, E. *Arch. Pharm., Berl.* **251** (1913) 183; (*e*) Kröhnke, F. *Ber. dtsch.*

REFERENCES

chem. Ges. **66** (1933) 1386; (*f*) Hartwell, J. L. and Pogorelskin, M. A. *J. Am. chem. Soc.* **72** (1950) 2040; (*g*) Colichman, E. L., Vanderzanden, W. R. and Lui, S. K. **74** (1952) 1953

301 Tschitschibabin, A. E. *Zh. russk. fiz.-khim. Obshch.* **33** (1901) 249; **34** (1901) 133

302 Norris, J. F. and Culver, L. R. *Am. chem. J.* **29** (1903) 129; Kraus, C. A. and Rosen, R. *J. Am. chem. Soc.* **47** (1925) 2739; Meyer, E. v. and Fischer, P. *J. prakt. Chem.* **82** (1910/2) 523; Hantzsch, A. and Meyer, K. H. *Ber. dtsch. chem. Ges.* **43** (1910) 337; Helferich, B. with Moog, L. and Jünger, A. **58** (1925) 872; with Sieber, H. **59** (1926) 600; Rebek, M. **62** (1929) 2508; with Kramaršič, V. *ibid.* 477

303 (*a*) Hantzsch, A. and Burawoy, A. *Ber. dtsch. chem. Ges.* **63** (1930) 1181; (*b*) Gelles, E., Hughes, E. D. and Ingold, C. K. *J. chem. Soc.* (1954) 2918

304 (*a*) Roithner, E. *Mh. Chem.* **15** (1894) 665; (*b*) Litterscheid, F. M. *Arch. Pharm., Berl.* **240** (1902) 77; (*c*) Gautier, J.-A. *C. r. hebd. Séanc. Acad. Sci., Paris* **198** (1934) 1430; **203** (1936) 794; (*d*) Kröhnke, F. *Ber. dtsch. chem. Ges.* **67** (1934) 656; (*e*) King, L. C. and Brownell, W. B. *J. Am. chem. Soc.* **72** (1950) 2507; (*f*) Hansson, J. *Svensk Kem. Tidskr.* **62** (1950) 185

305 King, L. C., Berst, N. W. and Hayes, F. N. *J. Am. chem. Soc.* **71** (1949) 3498

306 (*a*) Bamberger, E. *Ber. dtsch. chem. Ges.* **20** (1887) 3338; (*b*) Schmidt, E. and Hartong van Ark, H. *Arch. Pharm., Berl.* **238** (1900) 321; (*c*) Babcock, S. H., Nakamura, F. I. and Fuson, R. C. *J. Am. chem. Soc.* **54** (1932) 4407; (*d*) Krollpfeiffer, F. and Müller, A. *Ber. dtsch. chem. Ges.* **68** (1935) 1169; (*e*) Kröhnke, F. *ibid.* 1177

307 Galinovsky, F., Schoen, C. and Weiser, R. *Mh. Chem.* **80** (1949) 288

308 Kröhnke, F. and Lüderitz, O. *Chem. Ber.* **33** (1950) 60

309 King, L. C. *J. Am. chem. Soc.* **66** (1944) 894; with McWhirter, M. and Barton, D. M. **67** (1945) 2089; and Rowland, R. L. **70** (1948) 239

310 Kröhnke, F. and Gross, K. F. *Chem. Ber.* **92** (1959) 22

311 Pearson, R. G. *J. Am. chem. Soc.* **69** (1947) 3100

312 Reid, W. and Bender, H. *Chem. Ber.* **89** (1956) 1893

313 (*a*) Berson, J. A. and Cohen, T. *J. Am. chem. Soc.* **78** (1956) 416; (*b*) Kröhnke, F., Leister, H. and Vogt, T. *Chem. Ber.* **90** (1957) 2792

314 Gerichten, E. v. *Ber. dtsch. chem. Ges.* **15** (1882) 1251; Bezzi, S. *Chem. Abstr.* **33** (1939) 6311

315 Reitzenstein, F. *Justus Liebigs Annln Chem.* **326** (1903) 305

316 Pfeiffer, P. and Langenberg, A. *Ber. dtsch. chem. Ges.* **43** (1910) 2926

317 Meyer, H. *Mh. Chem.* **15** (1894) 164

318 Decker, H. and Kaufmann, A. *J. prakt. Chem.* **84** (1911) 425

319 Ferns, J. and Lapworth, A. *J. (Trans.) chem. Soc.* **101** (1912) 273; Marvel, C. S., Scott, E. W. and Amstutz, K. L. *J. Am. chem. Soc.* **51** (1929) 3638; Tipson, R. S. *Advances in Carbohydrate Chemistry*, Vol. 8, p. 118, New York (Academic Press) 1953

320 Lane, E. S. *J. chem. Soc.* (1953) 1172

321 Kohn, M. and Grauer, F. *Mh. Chem.* **34** (1913) 1751; Walther, R. v. *J. prakt. Chem.* **91** (1915) 329

322 (*a*) Hawkins, J. A. *J. (Trans.) chem. Soc.* **121** (1922) 1170; (*b*) Norris, J. F. and Prentiss, S. W. *J. Am. chem. Soc.* **50** (1928) 3042; (*c*) Kerr, R. N. *J. chem. Soc.* (1929) 239; (*d*) Mukhin, G. E. and Zilberfarb, M. I. *Chem. Abstr.* **24** (1930) 2942

323 (*a*) Pickles, N. J. T. and Hinshelwood, C. N. *J. chem. Soc.* (1936) 1353; (*b*) Fairclough, R. A. and Hinshelwood, C. N. (1937) 1573; (*c*) Laidler, K. J. with Hinshelwood, C. N. (1938) 858; (*d*) *ibid.* 1786

324 (*a*) Reinheimer, J. D., Harley, J. D. and Meyers, W. W. *J. org. Chem.* **28** (1963) 1575; (*b*) Coleman, B. D. and Fuoss, R. M. *J. Am. chem. Soc.* **77** (1955) 5472; Kronick, P. L. and Fuoss, R. M. *ibid.* 6114; Hirsch, E. and Fuoss, R. M. *ibid.* 6115; Watanabe, M. and Fuoss, R. M. **78** (1956) 527

325 Hinshelwood, C. N. *The Kinetics of Chemical Change*, p. 250, Oxford University Press, 1940

326 Birchenough, M. J. *J. chem. Soc.* (1951) 1263

327 Mumm, O. *Justus Liebigs Annln Chem.* **443** (1925) 272

328 Bradlow, H. L. and Vanderwerf, C. A. *J. org. Chem.* **16** (1951) 1143

329 (*a*) Pollak, F. *Mh. Chem.* **16** (1895) 45; (*b*) Ternájgó, L. **21** (1900) 456; (*c*) Grob, C. A. and Renk, E. *Helv. chim. Acta* **37** (1954) 1672

330 Ziegler, K. and Fries, F. A. *Ber. dtsch. chem. Ges.* **59** (1926) 244

331 (*a*) Wibaut, J. P., Speekman, B. W. and van Wagtendonk, H. M. *Recl Trav. chim. Pays-Bas Belg.* **58** (1939) 1100; (*b*) Hartmann, M. and Bosshard, W. *Helv. chim. Acta* **24** (1941) 28E

332 Beyerman, H. C. and Bontekoe, J. S. *Recl Trav. chim. Pays-Bas Belg.* **74** (1955) 1395

333 Gautier, J.-A. and Renault, J. *C. r. hebd. Séanc. Acad. Sci., Paris* **237** (1953) 733

334 Wuest, H. M. and Sakal, E. H. *J. Am. chem. Soc.* **73** (1951) 1210

335 Bradlow, H. L. and Vanderwerf, C. A. *J. org. Chem.* **14** (1949) 509

336 Fischer, O. *Ber. dtsch. chem. Ges.* **32** (1899) 1297

337 Michaelis, A. and Hanisch, R. *Ber. dtsch. chem. Ges.* **35** (1902) 3156

[338] Michaelis, A. *Justus Liebigs Annln Chem.* **366** (1909) 324
[339] Tschitschibabin, A. E., Konowalowa, R. A. and A. A. *Ber. dtsch. chem. Ges.* **54** (1921) 814
[340] Turitsyna, N. F. and Vompe, A. F. *Dokl. Akad. Nauk SSSR* **74** (1950) 509
[341] Tschitschibabin, A. E. and Ossetrowa, E. D. *Ber. dtsch. chem. Ges.* **58** (1925) 1708
[342] (a) Koenigs, E., Friedrich, H. and Jurany, H. *Ber. dtsch. chem. Ges.* **58** (1925) 2571; (b) Tomita, K. *J. pharm. Soc. Japan* **71** (1951) 1053
[343] Sharp, T. M. *J. chem. Soc.* (1939) 1855
[344] Tschitschibabin, A. E. and Konowalowa, R. A. *Ber. dtsch. chem. Ges.* **59** (1926) 2055
[345] Magidson, O. and Menschikoff, G. *Ber. dtsch. chem. Ges.* **59** (1926) 1209
[346] Gol'dfarb, Y. L., Setkina, O. N. and Danyushevskii, Y. L. *Zh. obshch. Khim.* **18** (1948) 124
[347] Tschitschibabin, A. E. and Konowalowa, R. A. *Ber. dtsch. chem. Ges.* **58** (1925) 1712
[348] Goerdeler, J. and Roth, W. *Chem. Ber.* **96** (1963) 534
[349] Beak, P. and Benham, J. *Tetrahedron Lett.* (1964) 3083
[350] Reindel, F. *Ber. dtsch. chem. Ges.* **57** (1924) 1381
[351] Reindel, F. and Rauck, H. *Ber. dtsch. chem. Ges.* **58** (1925) 393
[352] Kirpal, A. and Wojnar, B. *Ber. dtsch. chem. Ges.* **71** (1938) 1261
[353] Kirpal, A. and Poisel, F. *Ber. dtsch. chem. Ges.* **70** (1937) 2367
[354] Tschitschibabin, A. E. *Ber. dtsch. chem. Ges.* (a) **57** (1924) 2092; (b) **58** (1925) 1704; (c) **59** (1926) 2048
[355] Schilling, K., Kröhnke, F. and Kickhöffen, B. *Chem. Ber.* **88** (1955) 1093
[356] Kröhnke, F., Kickhöffen, B. and Thoma, C. *Chem. Ber.* **88** (1955) 1117
[357] Adams, R. and Dix, J. S. *J. Am. chem. Soc.* **80** (1958) 4618
[358] Mosby, W. L. *Heterocyclic Systems with Bridgehead Nitrogen Atoms*, Pt. I, p. 460 *et seq.* New York (Interscience) 1961
[359] Tschitschibabin, A. E. and Bylinkin, J. G. *Ber. dtsch. chem. Ges.* **55** (1922) 998
[360] Tomita, K. *J. pharm. Soc. Japan* **71** (1951) 220
[361] Tschitschibabin, A. E. and Seide, O. A. *Zh. russk. fiz.-khim. Obshch.* **46** (1914) 1216
[362] *Brit. Pat.* 265,167 (1927)
[363] Huttrer, C. P., Djerassi, C. *et al.*, *J. Am. chem. Soc.* **68** (1946) 1999
[364] Whitmore, F. C., Mosher, H. S., Goldsmith, D. P. J. and Rytina, A. W. *J. Am. chem. Soc.* **67** (1945) 393
[365] Kaye, I. A., Kogon, I. C. and Parris, C. L. *J. Am. chem. Soc.* **74** (1952) 403
[366] Blicke, F. F. and Tsaó, M. U. *J. Am. chem. Soc.* **68** (1946) 905
[367] Sharp, T. M. *J. chem. Soc.* (1938) 1191
[368] Slotta, K. H. and Franke, W. *Ber. dtsch. chem. Ges.* **63** (1930) 678
[369] Tschitschibabin, A. E. and Menschikow, G. P. *Ber. dtsch. chem. Ges.* **58** (1925) 406
[370] Adams, R. and Campbell, J. B. *J. Am. chem. Soc.* **71** (1949) 3539
[371] Clark-Lewis, J. W. and Thompson, M. J. *J. chem. Soc.* (1957) 442
[372] Meyer, H. *Ber. dtsch. chem. Ges.* **36** (1903) 616; *Mh. Chem.* **24** (1903) 199
[373] Kirpal, A. *Mh. Chem.* **24** (1903) 519
[374] Turnau, R. *Mh. Chem.* **26** (1905) 537
[375] Winterstein, E. and Weinhagen, A. B. *Hoppe-Seyler's Z. physiol. Chem.* **100** (1917) 170
[376] Hoppe-Seyler, T. A. *Hoppe-Seyler's Z. physiol. Chem.* **222** (1933) 105
[377] Sarett, H. P., Perlzweig, W. A. and Levy, E. D. *J. biol. Chem.* **135** (1940) 483
[378] Hantzsch, A. *Ber. dtsch. chem. Ges.* **19** (1886) 31
[379] Gautier, J.-A. and Leroi, E. *C. r. hebd. Séanc. Acad. Sci., Paris* **216** (1943) 669; **218** (1944) 200
[380] Pechmann, H. v. and Baltzer, O. *Ber. dtsch. chem. Ges.* **24** (1891) 3144
[381] Kirpal, A. *Mh. Chem.* **29** (1908) 471
[382] Kirpal, A. *Ber. dtsch. chem. Ges.* **57** (1924) 1954
[383] Adams, R. and Jones, V. V. *J. Am. chem. Soc.* **71** (1949) 3826
[384] Meyer, H. *Mh. Chem.* **26** (1905) 1311
[385] Binz, A. and Räth, C. *Justus Liebigs Annln Chem.* **489** (1931) 107
[386] Gautier, J.-A. and Renault, J. *Bull. Soc. chim. Fr.* (1954) 1463
[387] Alberti, C. *Gazz. chim. ital.* **86** (1956) 1181
[388] Binz, A. and Räth, C. *Justus Liebigs Annln Chem.* **484** (1930) 52
[389] Maier-Bode, H. *Z. angew. Chem.* **44** (1931) 835
[390] Schroeter, C., Seidler, C., Sulzbacher, M. and Kanitz, R. *Ber. dtsch. chem. Ges.* **65** (1932) 432
[391] Takahashi, T. and Kawashima, M. *J. pharm. Soc. Japan* **63** (1943) 546
[352] Späth, E. and Koller, G. *Ber. dtsch. chem. Ges.* **56** (1923) 880
[393] Fischer, O. and Renouf, E. *Ber. dtsch. chem. Ges.* **17** (1884) 1896
[394] Williams, R. R. *J. ind. Engng Chem.* **13** (1921) 1107
[395] Harris, S. A., Webb, T. J. and Folkers, K. *J. Am. chem. Soc.* **62** (1940) 3198
[396] Shapiro, S. L., Weinberg, K. and Freedman, L. *J. Am. chem. Soc.* **81** (1957) 5140
[397] Kirpal, A. *Ber. dtsch. chem. Ges.* **57** (1924) 1954
[398] Fürst, H. and Dietz, H. J. *J. prakt. Chem.* **4** (1956/4) 147

REFERENCES

[399] Prins, D. A. *Recl Trav. chim. Pays-Bas Belg.* **76** (1957) 58
[400] Lieben, A. and Haitinger, L. *Ber. dtsch. chem. Ges.* **17** (1884) 1507
[401] Haitinger, L. and Lieben, A. *Mh. Chem.* **6** (1885) 279
[402] Pechmann, H. v. *Ber. dtsch. chem. Ges.* **28** (1895) 1624
[403] Peratoner, A. and Azzarello, E. *Atti Accad. naz. Lincei* **15** (1906) [v] i, 139
[404] Marion, L. and Cockburn, W. F. *J. Am. chem. Soc.* **71** (1949) 3402
[405] Gompper, R. *Chem. Ber.* **93** (1960) 187,198
[406] Marckwald, W., Klemm, W. and Trabert, H. *Ber. dtsch. chem. Ges.* **33** (1900) 1556
[407] Renault, J. *Annls Chim.* **10** (1955) 135
[408] Fry, D. J. and Kendall, J. D. *J. chem. Soc.* (1951) 1716
[409] King, H. and Ware, L. L. *J. chem. Soc.* (1939) 873
[410] (a) Michaelis, A. and Hölken, A. *Justus Liebigs Annln Chem.* **331** (1904) 245; (b) Bradsher, C. K., Quin, L. D., Le Bleu, R. E. and McDonald, J. W. *J. org. Chem.* **26** (1961) 4944
[411] (a) Fischer, O. *Ber. dtsch. chem. Ges.* **35** (1902) 3674; (b) Daniels, R. and Kormendy, C. G. *J. org. Chem.* **27** (1962) 1860
[412] Makarova, L. G. and Nesmeyanov, A. N. *Chem. Abstr.* **40** (1946) 4686
[413] Tschitschibabin, A. E. and Jeletzky, N. P. *Ber. dtsch. chem. Ges.* **57** (1924) 1158
[414] Zincke, T. *Justus Liebigs Annln Chem.* **330** (1904) 361; **338** (1905) 107
[415] Lukeš, R. *Colln. Czech. chem. Commun.* **12** (1947) 263
[416] Balfe, M. P., Doughty, M. and Kenyon, J. *J. chem. Soc.* (1953) 2470
[417] Vompe, A. F. and Turitsyna, N. F. *Dokl. Akad. Nauk SSSR* **64** (1949) 341
[418] Vompe, A. F., Turitsyna, N. F. and Levkoev, I. I. *Dokl. Akad. Nauk SSSR* **65** (1949) 839
[419] Lukeš, R. and Vaculík, P. *Chemické Listy* **45** (1951) 264
[420] Borsche, W. and Rantscheff, D. *Justus Liebigs Annln Chem.* **379** (1911) 152
[421] Reitzenstein, F. and Rothschild, J. *J. prakt. Chem.* **73** (1906/2) 257
[422] Zincke, T. and Weispfenning, G. *J. prakt. Chem.* **82** (1910/2) 1
[423] Zincke, T. *J. prakt. Chem.* **82** (1910/2) 17
[424] Wibaut, J. P. and Holmes-Kamminga, W. *J. Bull. Soc. chim. Fr.* (1958) 424
[425] (a) Takeda, K., Hamamoto, K. with Tone, H. *J. pharm. Soc. Japan* **72** (1952) 1427; (b) **73** (1953) 1158; (c) Hamamoto, K. and Kubota, T. *ibid.* 1162
[426] Ramirez, F. and von Ostwalden, P. W. *J. Am. chem. Soc.* **81** (1959) 156
[427] de Villiers, P. A. and den Hertog, H. *J. Recl Trav. chim. Pays-Bas Belg.* **76** (1957) 647
[428] Wibaut, J. P. and Broekman, F. W. *Recl Trav. chim. Pays-Bas Belg.* **58** (1939) 885
[429] Arndt, F. and Kalischek, A. *Ber. dtsch. chem. Ges.* **63** (1930) 587; Arndt, F. **65** (1932) 92
[430] Renshaw, R. R. and Conn, R. C. *J. Am. chem. Soc.* **59** (1937) 297
[431] den Hertog, H. J., Broekman, F. W. and Combé, W. P. *Recl Trav. chim. Pays-Bas Belg.* **70** (1951) 105
[432] Taylor, E. C. and Driscoll, J. S. *J. org. Chem.* **26** (1961) 3001
[433] Koenigs, E. and Greiner, H. *Ber. dtsch. chem. Ges.* **64** (1931) 1049
[434] Albert, A. *J. chem. Soc.* (1951) 1376
[435] Bowden, K. and Green, P. N. *J. chem. Soc.* (1954) 1795
[436] Wibaut, J. P., Herzberg, S. and Schlatmann, J. *Recl Trav. chim. Pays-Bas Belg.* **73** (1954) 140
[437] Bak, B. and Christensen, D. *Acta chem. scand.* **8** (1954) 390
[438] Jerchel, D., Fischer, H. and Thomas, K. *Chem. Ber.* **89** (1956) 2921
[439] Jerchel, D. and Jacob, L. *Chem. Ber.* **91** (1958) 1266
[440] Haack, E. *Germ. Pat.* 598,879 (1932); 600,499 (1934)
[441] Rubtsov, M. V. and Klimko, V. T. *J. gen. Chem. U.S.S.R.* **16** (1946) 1860
[442] (a) Baker, W. and Briggs, A. S. *J. Soc. chem. Ind., Lond.* **62** (1943) 189; (b) Hamara, M. and Funakoshi, K. *J. pharm. Soc. Japan* **84** (1964) 23
[443] Saure, S. *Chem. Ber.* **83** (1950) 335
[444] Taylor, E. C., Paudler, W. W. and Cain, C. K. *J. org. Chem.* **20** (1955) 264
[445] Spickett, R. G. W. and Timmis, G. M. *J. chem. Soc.* (1955) 4354
[446] Long, F. S. *J. chem. Soc.* **99** (1911) 2164; Thomas, E. R. **103** (1913) 594
[447] Baker, J. W. and Nathan, W. S. *J. chem. Soc.* (1935) (a) 1844; (b) 519, 1840
[448] Baker, J. W. *Tetrahedron* **5** (1959) 135
[449] Baker, J. W. *J. chem. Soc.* (a) (1932) 2631; (1933) 1128; (b) (1936) 1448
[450] (a) Streitwieser, A. *Chem. Rev.* **56** (1956) 603; (b) Coppens, G., Declerck, F., Gillet, C. and Nasielski, J. *Bull. Soc. chim. Belg.* **72** (1963) 25
[451] (a) Swain, C. G., with Eddy, R. W. *J. Am. chem. Soc.* **70** (1948) 2989; (b) with Langsdorf, W. P. **73** (1951) 2813
[452] Bird, M. L., Hughes, E. D. and Ingold, C. K. *J. chem. Soc.* (1954) 634; Bunton, C. A., Greenstreet, C. H., Hughes, E. D. and Ingold, C. K. *ibid.* 647
[453] Brown, H. C. and Cahn, A. *J. Am. chem. Soc.* **77** (1955) 1715
[454] Clarke, K. and Rothwell, K. *J. chem. Soc.* (1960) 1885
[455] Brown, H. C., Gintis, D. and Podall, H. *J. Am. chem. Soc.* **78** (1956) 5375
[456] (a) Baker, J. W. *J. chem. Soc.* (1938) 445; (b) Benkeser, R. A. and Clark, F. S. *J. org. Chem.* **27** (1962) 3727

457 (a) Bishop, R. R., Cavell, E. A. S. and Chapman, N. B. *J. chem. Soc.* (1952) 437; (b) the latter two, (1953) 3392

458 Barnett, E. de B., Cook, J. W. and Peck, W. C. *J. chem. Soc.* **125** (1924) 1035; Hayes, F. N., Suzuki, H. K. and Peterson, D. E. *J. Am. chem. Soc.* **72** (1950) 4524

459 Hayes, F. N., King, L. C. and Peterson, D. E. *J. Am. chem. Soc.* **78** (1956) 2527

460 Gresham, T. L., Jansen, J. E. *et al.*, *J. Am. chem. Soc.* **73** (1951) 3169

461 Bordwell, F. G., Peterson, M. L. with Rondestvedt, C. S. *J. Am. chem. Soc.* **76** (1954) 3945; *ibid.* 3957

462 Lutz, O., Klein, R. and Jirgenson, A. *Justus Liebigs Annln Chem.* **505** (1933) 307

463 Adams, R. and Pachter, I. J. *J. Am. chem. Soc.* **74** (1952) 4906, 5491; Lappin, G. R. *J. org. Chem.* **23** (1958) 1358

464 Adams, R. and Jones, V. V. *J. Am. chem. Soc.* **69** (1947) 1803

465 Crabtree, A., Johnson, A. W. and Tebby, J. C. *J. chem. Soc.* (1961) 3497

466 Lappin, G. R. *J. org. Chem.* **26** (1961) 2350

467 (a) Diels, O., with Alder, K. *et al.*, *Justus Liebigs Annln Chem.* **498** (1932) 16; (b) **505** (1933) 103; (c) **510** (1934) 87; (d) with Moeller, F. **516** (1935) 45; (e) with Schrum, H. **530** (1937) 68; (f) with Pistor, H. *ibid.* 87; (g) *Ber. dtsch. chem. Ges.* **75** (1942) 1452; (h) with Kock, U. *Justus Liebigs Annln Chem.* **556** (1944) 38

468 Jackman, L. M., Johnson, A. W. and Tebby, J. C. *J. chem. Soc.* (1960) 1579

469 Acheson, R. M. and Taylor, G. A. *J. chem. Soc.* (1960) 1691

470 Diels, O. and Meyer, R. *Justus Liebigs Annln Chem.* **513** (1934) 129

471 Wiley, R. H. and Knabenschuh, L. H. *J. org. Chem.* **18** (1953) 836

472 Ortoleva, G. with di Stefano, G. *Gazz. chim. ital.* **31** (1901/2) 256; **32** (1902/1) 447; Barnett, E. de B., Cook, J. W. and Driscoll, E. P. *J. chem. Soc.* **123** (1923) 503; Koenigs, E. and Greiner, H. *Ber. dtsch. chem. Ges.* **64** (1931) 1045; Buchta, E. **70** (1937) 2339; Diels, O., with Kassebart, R. *Justus Liebigs Annln Chem.* **530** (1937) 51; with Preiss, H, **543** (1940) 94

473 Schmid, L. and Czerny, H. *Mh. Chem.* **83** (1952) 31

474 Islam, A. M. and Raphael, R. A. *Chemy Ind.* (1955) 1635

475 Pratt, E. F., Luckenbaugh, R. W. and Erickson, R. L. *J. org. Chem.* **19** (1954) 176

476 Adams, R. and Pomerantz, S. H. *J. Am. chem. Soc.* **76** (1954) 702

477 Ullmann, F. and Ettisch, M. *Ber. dtsch. chem. Ges.* **54** (1921) 259

478 Schönberg, A. and Ismail, A. F. A. *J. chem. Soc.* (1940) 1374

479 Stahmann, M. A., Golumbic, C., Stein, W. H. and Fruton, J. S. *J. org. Chem.* **11** (1946) 719

480 Houben-Weyl, *Methoden der Organischen Chemie*, 4th ed. (1955), Vol. IX, pp. 388, 612, 666, 671

481 Deninger, A. *Ber. dtsch. chem. Ges.* **28** (1895) 1322; Verley, A. and Bölsing, F. **34** (1901) 3354; Freundler, P. *C. r. hebd. Séanc. Acad. Sci.*, *Paris* **136** (1903) 1553; **137** (1903) 712; Behrend, R. and Roth, P. *Justus Liebigs Annln Chem.* **331** (1904) 359

482 Smith, D. M. and Bryant, W. M. D. *J. Am. chem. Soc.* **57** (1935) 61, 841

483 Bafra, S. L. and Gold, V. *J. chem. Soc.* (1953) 1406; Gold, V. and Jefferson, E. G. *ibid.* 1409; Butler, A. R. and Gold, V. (1961) 4362; (1962) 976

484 Dennstedt, M. and Zimmermann, J. *Ber. dtsch. chem. Ges.* **19** (1886) 75

485 Dehn, W. M. *J. Am. chem. Soc.* **34** (1912) 1399

486 Freudenberg, K. and Peters, D. *Ber. dtsch. chem. Ges.* **52** (1919) 1463

487 Adkins, H. and Thompson, Q. E. *J. Am. chem. Soc.* **71** (1949) 2242

488 Prey, V. *Ber. dtsch. chem. Ges.* **75** (1942) 537

489 Minunni, G. *Gazz. chim. ital.* **22** (1892/2) 213

490 Dehn, W. M. and Ball, A. A. *J. Am. chem. Soc.* **36** (1914) 2091

491 Staudinger, H. and Stockmann, H. *Ber. dtsch. chem. Ges.* **42** (1909) 3485

492 *Germ. Pat.* 109,933 (1898)

493 Baumgarten, H. E. *J. Am. chem. Soc.* **75** (1953) 1239

494 Paul, R. C., Chander, K. and Singh, G. *J. Indian chem. Soc.* **35** (1958) 869

495 Bogoslovskiǐ, B. M. *Chem. Abstr.* **31** (1937) 4319

496 Klages, F. and Zange, E. *Justus Liebigs Annln Chem.* **607** (1957) 35

497 Gebauer-Fülnegg, E. and Riesenfeld, F. *Mh. Chem.* **47** (1926) 185; Schwartz, G. L. and Dehn, W. M. *J. Am. chem. Soc.* **39** (1917) 2444

498 Tschitschibabin, A. E. and Szokow, P. G. *Ber. dtsch. chem. Ges.* **58** (1925) 2650

499 Tschitschibabin, A. E. and Oparina, O. P. *Zh. russk. fiz.-khim. Obshch.* **56** (1915) 153

500 Cavallito, C. J. and Haskell, T. H. *J. Am. chem. Soc.* **66** (1944) (a) 1166; (b) 1927

501 Camps, R. *Arch. Pharm.*, *Berl.* **240** (1902) 345

502 Räth, C. *Justus Liebigs Annln Chem.* **486** (1931) 95

503 Huntress, E. H. and Walker, H. C. *J. org. Chem.* **13** (1948) 735

504 Johnson, A. W., King, T. J. and Turner, J. R. *J. chem. Soc.* (1960) 1509

505 Dorn, H., Hilgetag, G. and Rieche, A. *Angew. Chem.* **73** (1961) 560

506 Dorn, H. and Hilgetag, G. *Chem. Ber.* **97** (1964) 695

REFERENCES

507 Staudinger, H., Klever, H. W. and Kober, P. *Justus Liebigs Annln Chem.* **374** (1910) 1
508 Berson, J. A. and Jones, W. M. *J. Am. chem. Soc.* **78** (1956) 1625
509 Meisenheimer, J. *Ber. dtsch. chem. Ges.* **59** (1926) 1848
510 Bobrański, B., Kochańska, L. and Kowalewska, A. *Ber. dtsch. chem. Ges.* **71** (1938) 2385
511 Ishikawa, M. and Sai, Zai-Ren *J. pharm. Soc. Japan* **63** (1943) 78
512 Shaw, E., Bernstein, J., Losee, K. and Lott, W. A. *J. Am. chem. Soc.* **72** (1950) 4362
513 Gilman, H. and Edward, J. T. *Can. J. Chem.* **31** (1953) 457
514 Colonna, M. and Runti, C. *Annali Chim.* **43** (1953) 87
515 Ames, D. E. and Grey, T. F. *J. chem. Soc.* (1955) 631
516 Ochiai, E. and Sai, Zai-Ren *J. pharm. Soc. Japan* **65** (1945) 73
517 Evans, R. F., van Ammers, M. and den Hertog, H. J. *Recl Trav. chim. Pays-Bas Belg.* **78** (1959) 408
518 Taylor, E. C. and Driscoll, J. S. *J. org. Chem.* **25** (1960) 1716
519 Ochiai, E., Katada, M. and Hayashi, E. *J. pharm. Soc. Japan* **67** (1947) 33
520 (a) Davies, A. G. *Organic Peroxides*, p. 135, London (Butterworths) 1961; (b) Dondoni, A., Modena, G. and Todesco, P. E. *Gazz. chim. ital.* **91** (1961) 613
521 Adams, R. and Miyano, S. *J. Am. chem. Soc.* **76** (1954) 2785
522 Brown, E. V. *J. Am. chem. Soc.* **79** (1957) 481
523 Kanno, S., *J. pharm. Soc. Japan* **73** (1953) 120
524 Newbold, G. T. and Spring, F. S. *J. chem. Soc.* (1948) 1864
525 Adams, R. and Miyano, S. *J. Am. chem. Soc.* **76** (1954) 3168
526 Clemo, G. R. and Koenig, H. *J. chem. Soc.* (1949) S231
527 Ochiai, E., Ishikawa, M. and Sai, Zai-Ren *J. pharm. Soc. Japan* **65** (1945) 72
528 Boekelheide, V. and Linn, W. J. *J. Am. chem. Soc.* **76** (1954) 1286
529 Bullitt, O. H. and Maynard, J. T. *J. Am. chem. Soc.* **76** (1954) 1370
530 Jerchel, D. and Jacobs, W. *Angew. Chem.* **66** (1954) 298
531 Berson, J. A. and Cohen, T. *J. org. Chem.* **20** (1955) 1461
532 Ginsburg, S. and Wilson, I. B. *J. Am. chem. Soc.* **79** (1957) 481
533 Ghigi, E. *Ber. dtsch. chem. Ges.* **75** (1942) 1318
534 Murray, J. G. and Hauser, C. R. *J. org. Chem.* **19** (1954) 2008
535 Ochiai, E. and Kaneko, C. *Pharm. Bull., Tokyo* **5** (1957) 56
536 Katritzky, A. R. and Monro, A. M. *J. chem. Soc.* (1958) 150
537 Taylor, E. C. and Crovetti, A. J. *J. Am. chem. Soc.* **78** (1956) 214
538 Colonna, M. and Risaliti, A. *Gazz. chim. ital.* **85** (1955) 1148
539 Colonna, M., Risaliti, A. and Pentimalli, L. *Gazz. chim. ital.* **86** (1956) 1067
540 Pentimalli, L. and Risaliti, A. *Annali Chim.* **46** (1956) 1037
541 (a) Pentimalli, L. *Gazz. chim. ital.* **89** (1959) 1843; *Tetrahedron* **9** (1960) 194; (b) **14** (1961) 151; *Gazz. chim. ital.* **91** (1961) 991
542 (a) Wieczorek, J. S. and Plažek, E. *Recl Trav. chim. Pays-Bas Belg.* **83** (1964) 249; (b) Gösl, R. and Meuwsen, A. *Chem. Ber.* **92** (1959) 2521
543 Mann, F. G. and Watson, J. *J. org. Chem.* **13** (1948) 502
544 Newbold, G. T. and Spring, F. S. *J. chem. Soc.* (1949) S133
545 König, W. *J. prakt. Chem.* **69** (1904/2) 105; **70** (1904/2) 19
546 Brusse, E. S. *Pharm. Weekbl. Ned.* **85** (1950) 569; Douzou, P. and Le Clerc, A.-M. *Analyt. chim. Acta* **12** (1955) 239
547 König, W. *J. prakt. Chem.* **85** (1912/2) 353; with Ebert, G. and Centner, K. *Ber. dtsch. chem. Ges.* **56** (1923) 756
548 Daniels, R. and Salerni, O. L. *Proc. chem. Soc.* (1960) 286
549 Tschitschibabin, A. E. *Ber. dtsch. chem. Ges.* **57** (1924) 1168
550 Seide, O. *Ber. dtsch. chem. Ges.* **58** (1925) 352
551 Buu-Hoï, Ng. Ph. and Declerq, M. *Recl Trav. chim. Pays-Bas Belg.* **73** (1954) 376
552 Bobrański, B. and Sucharda, E. *Ber. dtsch. chem. Ges.* **60** (1927) 1081; *Roczn. Chem.* **7** (1927) 192
553 Allen, C. F. H., Spangler, F. W. and Webster, E. R. *J. org. Chem.* **16** (1951) 17
554 Schmid, L. and Bangler, B. *Ber. dtsch. chem. Ges.* **58** (1925) 1971; **59** (1926) 1360
555 Adams, J. T., Bradsher, C. K. *et al.*, *J. Am. chem. Soc.* **68** (1946) 1317
556 Lappin, G. R. *J. Am. chem. Soc.* **70** (1948) 3348
557 Crippa, G. B. and Scerola, E. *Gazz. chim. ital.* **67** (1937) 327
558 Lappin, G. R., Petersen, Q. R. and Wheeler, C. E. *J. org. Chem.* **15** (1950) 377
559 Mangini, A. *Chem. Abstr.* **36** (1942) 5476
560 Ochiai, E. and Mujaki, K. *Ber. dtsch. chem. Ges.* **74** (1941) 1115
561 Miyagi, K. *J. pharm. Soc. Japan* **62** (1942) 26
562 Mangini, A. and Colonna, M. *Chem. Abstr.* **37** (1943) 3096; *Gazz. chim. ital.* **73** (1943) 323, 330
563 Petrow, V., Rewald, E. L. and Sturgeon, B. *J. chem. Soc.* (1947) 1407
564 Hauser, C. R. and Weiss, M. J. *J. org. Chem.* **14** (1949) 453
565 Seide, O. *Ber. dtsch. chem. Ges.* **59** (1926) 2465

[566] Hauser, C. R. and Reynolds, G. A. *J. org. Chem.* **15** (1950) 1224
[567] Hart, E. P. *J. chem. Soc.* (1954) 1879; Rapoport, H. and Batcho, A. D. *J. org. Chem.* **28** (1963) 1753
[568] *Germ. Pat.* 507,637 (1926)
[569] Weiss, M. J. and Hauser, C. R. *J. Am. chem. Soc.* **68** (1946) 722
[570] Miyaki, K. *J. pharm. Soc. Japan* **62** (1942) 257
[571] Price, C. C. and Roberts, R. M. *J. Am. chem. Soc.* **68** ((1946) 1204
[572] Goldberg, A. A., Theobald, R. S. and Williamson, W. *J. chem. Soc.* (1954) 2357
[573] Takahashi, T., Yatsuka, T. and Senda, S. *J. pharm. Soc. Japan* **64** (1944) No. 7/8A, 9
[574] Albert, A. and Hampton, A. *J. chem. Soc.* (1952) 4985
[575] Petrow, V. and Sturgeon, B. *J. chem. Soc.* (1949) 1157
[576] Allen, C. F. H. *Chem. Rev.* **47** (1950) 275
[577] Gulland, J. M. and Robinson, R. *J. chem. Soc.* **127** (1925) 1493
[578] Okuda, S. *Pharm. Bull., Tokyo* **5** (1957) 460
[579] Kato, T., Hamaguchi, F. and Oiwa, T. *Pharm. Bull., Tokyo* **4** (1956) 178
[580] Fargher, R. G. and Furness, R. *J. chem. Soc.* **107** (1915) 688
[581] Okuda, S. and Robison, M. M. *J. Am. chem. Soc.* **81** (1959) 740
[582] Ficken, G. E. and Kendall, J. D. *J. chem. Soc.* (1959) 3202
[583] Ficken, G. E. and Kendall, J. D. *J. chem. Soc.* (1961) 747
[584] *Brit. Pat.* 259,982 (1925)
[585] Takahashi, T., Saikachi, H., Goto, H. and Shimamura, S. *J. pharm. Soc. Japan* **64** (1944) 7
[586] Clemo, G. R. and Holt, R. J. W. *J. chem. Soc.* (1953) 1313
[587] Ficken, G. E. and Kendall, J. D. *J. chem. Soc.* (1961) 584
[588] Abramovitch, R. A. and Adams, K. A. H. *Can. J. Chem.* **40** (1962) 864
[589] Bachmann, G. B., with Schisla, R. M. *J. org. Chem.* **22** (1957) 1302; with Karickhoff, M. **24** (1959) 1696
[590] Goetz-Luethy, N. *J. Am. chem. Soc.* **71** (1949) 2254
[591] Kharasch, M. S. and Reinmuth, O. *Grignard Reactions of Nonmetallic Substances*, p. 1257 *et seq.*, New York (Prentice-Hall) 1954
[592] (a) Veer, W. L. C. and Goldschmidt, S. *Recl Trav. chim. Pays-Bas Belg.* **65** (1946) 793; (b) Doering, W. von E. and Pasternak, V. Z. *J. Am. chem. Soc.* **72** (1950) 143; (c) Benkeser, R. A. and Holton, D. S. **73** (1951) 5861
[593] Frank, R. L. and Weatherbee, C. *J. Am. chem. Soc.* **70** (1948) 3482
[594] Lukeš, R. and Kuthan, J. *Angew. Chem. (int. Edn)* **72** (1960) 919
[595] Gilman, H., Eisch, J. and Soddy, T. *J. Am. chem. Soc.* **79** (1957) 1245
[596] (a) Mariella, R. P. and Kvinge, V. *J. Am. chem. Soc.* **70** (1948) 3126; (b) Leonard, N. J. and Ryder, B. L. *J. org. Chem.* **18** (1953) 598; (c) Hank, R. quoted by Thomas, K. and Jerchel, D. *Angew. Chem. (int. Edn)* **70** (1958) 719
[597] Lowman, V. C., quoted in *Heterocyclic Compounds*, ed. Elderfield, R. C., Vol. 4, p. 243, New York (Wiley) 1952
[598] Decker, H. *Ber. dtsch. chem. Ges.* **38** (1905) 2493
[599] Freund, M. and Boole, G. *Ber. dtsch. chem. Ges.* **42** (1909) 1746; Karrer, P. and Widmer, A. *Helv. chim. Acta* **9** (1926) 461
[600] (a) Grewe, R. and Mondon, A. *Chem. Ber.* **81** (1948) 279; (b) Schnider, O. and Grüssner, A. *Helv. chim. Acta* **32** (1949) 821; (c) May, E. L. and Fry, E. M. *J. org. Chem.* **22** (1957) 1366; (d) Eddy, N. B., Murphy, J. G. and May, E. L. *ibid.* 1370; (e) May, E. L. and Ager, J. H. **24** (1959) 1432; (f) **25** (1960) 984; (g) Fullerton, S. E., Ager, J. H. and May, E. L. **27** (1962) 2554
[601] Červinka, O. *Colln. Czech. chem. Commun.* **27** (1962) 567
[602] Lowman, V. C., quoted by McEwen, W. E. and Cobb, R. L. *Chem. Rev.* **55** (1955) 541
[603] Agawa, T. and Miller, S. I. *J. Am. chem. Soc.* **83** (1961) 449
[604] Ziegler, K. and Zeiser, H. (a) *Ber. dtsch. chem. Ges.* **63** (1930) 1847; (b) *Justus Liebigs Annln Chem.* **485** (1931) 174
[605] Bryce-Smith, D., Morris, P. J. and Wakefield, B. J. *Chemy Ind.* (1964) 495
[606] (a) Osuch, C. and Levine, R. *J. org. Chem.* **22** (1957) 939; (b) Wibaut, P. J. and Hey, J. W. *Recl Trav. chim. Pays-Bas Belg.* **72** (1953) 513
[607] *U.S. Pat.* 2,780,626 (1957)
[608] Gilman, H. and Broadbent, H. S. *J. Am. chem. Soc.* **70** (1948) 2809
[609] Bohlmann, F., Englisch, A., Politt, J., Sander, H. and Weise, W. *Chem. Ber.* **88** (1955) 1831
[610] Kursanov, D. N. and Baranetskaya, N. K. *Izv. Akad. Nauk SSSR, Otdel. khim. nauk* (1961) 1703; with Setkina, V. N. *Dokl. Akad. Nauk SSSR* **113** (1957) 116
[611] Ziegler, K. and Wollschitt, H. *Justus Liebigs Annln Chem.* **479** (1930) 125
[612] (a) Bergmann, E. and Rosenthal, W. *J. prakt. Chem.* **135** (1932/2) 267; (b) Hauser, C. R. and Weiss, M. J. *J. org. Chem.* **14** (1949) 310; (c) Vajda, T. and Kovács, K. *Recl Trav. chim. Pays-Bas Belg.* **80** (1961) 47
[613] *U.S. Pat.* 2,874,162 (1959)

REFERENCES

614 Kröhnke, F., Ellegast, K. and Bertram, E. *Justus Liebigs Annln Chem.* **600** (1956) 176
615 Berson, J. A. and Evleth, E. M. *Chemy Ind.* (1961) 1362
616 Berson, J. A., Evleth, E. M. and Hamlet, Z. *J. Am. chem. Soc.* **82** (1960) 3793
617 Boyd, G. V. and Jackman, L. M. *J. chem. Soc.* (1963) 548
618 Panizzon, L. *Helv. chim. Acta* **27** (1944) 1748
619 Sury, E. and Hoffmann, K. *Helv. chim. Acta* **37** (1954) 2133; Heer, J., Sury, E. and Hoffmann, K. **38** (1955) 134
620 Walter, L. A. and McElvain, S. M., *J. Am. chem. Soc.* **57** (1935) 1891
621 Frank, R. L. and Phillips, R. R. *J. Am. chem. Soc.* **71** (1949) 2804
622 Lewis, T. R. and Archer, S. *J. Am. chem. Soc.* **73** (1951) 2109
623 Gruber, W. and Schlögl, K. *Mh. Chem.* (*a*) **80** (1949) 499; (*b*) **81** (1950) 473
624 Mosher, H. S. and Tessieri, J. E. *J. Am. chem. Soc.* **73** (1951) 4925
625 *Germ. Pat.* 638,596 (1936); 644,193 (1937)
626 Koenigs, E. and Fulde, A. *Ber. dtsch. chem. Ges.* **60** (1927) 2106
627 Koenigs, E. and Jaeschke, W. *Ber. dtsch. chem. Ges.* **54** (1921) 1351
628 Kuhn, C. S. and Richter, G. H. *J. Am. chem. Soc.* **57** (1935) 1927
629 Gruber, W. *Can. J. Chem.* **31** (1953) 1181
630 Levine, R. and Leake, W. W. *Science, N.Y.* **121** (1955) 780
631 Kaufmann, T. and Boettcher, F.-P. *Chem. Ber.* **95** (1962) 949
632 Martens, R. J. and den Hertog, H. J. *Tetrahedron Lett.* (1962) 643
633 Adams, R. and Reifschneider, W. *J. Am. chem. Soc.* **79** (1957) 2236
634 Huff, J. W. *J. biol. Chem.* **167** (1947) 451
635 Burton, R. M. and Kaplan, N. O. *J. biol. Chem.* **206** (1954) 283
636 Baker, B. R. and McEvoy, F. J. *J. org. Chem.* **20** (1955) 118
637 (*a*) Hamer, F. M. and Kelly, M. I. *J. chem. Soc.* (1931) 777; (*b*) Brooker, L. G. S. and Keyes, G. H. *J. Am. chem. Soc.* **57** (1935) 2488; (*c*) with White, F. L. *ibid.* 2492; (*d*) Sprague, R. H. and Brooker, L. G. S. **59** (1937) 2697
638 Claisen, L. and Haase, E. *Ber. dtsch. chem. Ges.* **36** (1903) 3674
639 Doering, W. von E. and McEwen, W. E. *J. Am. chem. Soc.* **73** (1951) 2104
640 Dobeneck, H. von and Goltzsche, W. *Chem. Ber.* **95** (1962) 1484
641 Terss, R. H. and McEwen, W. E. *J. Am. chem. Soc.* **76** (1954) 580
642 Claisen, L. *Justus Liebigs Annln Chem.* **291** (1896) 25; **297** (1897) 1; with Haase, E. *Ber. dtsch. chem. Ges.* **33** (1900) 1242; Dieckman, W. and Stein, R. **37** (1904) 3370; McElvain, S. M. and Kundiger, D. *J. Am. chem. Soc.* **64** (1942) 254; McEwen, W. E. and Cobb, R. L. *Chem. Rev.* **55** (1955) 541
643 Wright, P. E. and McEwen, W. E. *J. Am. chem. Soc.* **76** (1954) 4540
644 Gilkerson, W. R., Argersinger, W. J. and McEwen, W. E. *J. Am. chem. Soc.* **76** (1954) 41; cf. Stutz, R. L., Reynolds, C. A. and McEwen, W. E. *J. org. Chem.* **26** (1961) 1684
645 Einhorn, A. *Justus. Liebigs Annln Chem.* **301** (1898) 95; Denninger, A. *Ber. dtsch. chem. Ges.* **38** (1905) 1322
646 (*a*) Stutz, R. L. *Dissert. Abstr.* **22** (1962) 4191; (*b*) Hamana, M. and Yamazaki, M. *Chem. pharm. Bull., Tokyo* **4** (1963) 415
647 Wibaut, J. P. and Dingemanse, E. *Recl Trav. chim. Pays-Bas Belg.* **42** (1923) 240
648 Tschitschibabin, A. E. *Zh. russk. fiz.-khim. Obshch.* **47** (1915) 835
649 (*a*) Tschitschibabin, A. E. and Vidonova, M. S. *Zh. russk. fiz.-khim. Obshch.* **53** (1921) 238; (*b*) Seide, O. *Ber. dtsch. chem. Ges.* **57** (1924) 791; (*c*) Tschitschibabin, A. E. and Kirsanov, A. W. *ibid.* 1163; (*d*) Bergstrom, F. W. and Fernelius, W. C. *Chem. Rev.* **20** (1937) 413; (*e*) Solomon, W. *J. chem. Soc.* (1946) 934; (*f*) Lecocq, J. *Bull. Soc. chim. Fr.* (1950) 188; (*g*) Levine, R. and Fernelius, W. C. *Chem. Rev.* **54** (1954) 537; (*h*) Roe, A. M. *J. chem. Soc.* (1963) 2195
650 Bergstrom, F. W. and Fernelius, W. C. *Chem. Rev.* **12** (1933) 154
651 Leffler, M. T. *Org. React.* **1** (1942) 91
652 *U.S. Pat.* 1,789,022 (1931); 2,062,680 (1936); 2,461,119 (1949)
653 *French Pat.* 815,373 (1937); *Chem. Abstr.* **32** (1938) 1713
654 *Germ. Pat.* 663,891 (1938)
655 Belonosov, L. S. *Zh. prikl. Khim., Leningr.* **22** (1949) 1103
656 Seide, O. *Ber. dtsch. chem. Ges.* **57** (1924) 1802; *U.S. Pat.* 2,456,379 (1948); *Chem. Abstr.* **43** (1949) 7050
657 (*a*) Abramovitch, R. A., Helmer, F. and Saha, J. G. *Chemy Ind.* (1964) 659; (*b*) Ban, Y. and Wakamatsu, T. *ibid.* 710; (*c*) Childs, R. F. and Johnson, A. W. *ibid.* 542; (*d*) Levitt, L. S. and B. W. (1963) 1621; (*e*) Barrett, G. C. and Schofield, K. *ibid.* 1980
658 Abramovitch, R. A., Helmer, F. and Saha, J. G. *Tetrahedron Lett.* (1964) 3445
659 Bergstrom, F. W., Sturz, H. G. and Tracy, H. W. *J. org. Chem.* **11** (1946) 239
660 Deasy, C. L., *J. org. Chem.* **10** (1945) 141
661 Kirsanov, A. V. with Ivastchencko, Y. N. *Bull. Soc. chim. Fr.* **2** (1935/5) 2109; with Poliakova, I. **3** (1936/5) 1600
662 Kovács, K. and Vajda, T. *Acta chim. hung.* **29** (1961) 245

297

[663] Kauffmann, T., Hansen, J., Kosel, C. and Schoeneck, W. *Justus Liebigs Annln Chem.* **656** (1962) 103

[664] Burton, R. M. and Kaplan, N. O. *J. biol. Chem.* **211** (1954) 447

[665] Brown, B. R. and Wild, E. H. *J. chem. Soc.* (1956) 1158

[666] Coats, N. A. and Katritzky, A. R. *J. org. Chem.* **24** (1959) 1836

[667] Hamana, M. and Funakoshi, K. *J. pharm. Soc. Japan* **82** (1962) 518

[668] Tschitschibabin, A. E. and Preobrashensky, W. A. *Ber. dtsch. chem. Ges.* **61** (1928) 199

[669] Poziomek, E. J. *J. org. Chem.* **28** (1963) 590

[670] Marckwald, W. *Ber. dtsch. chem. Ges.* (a) **26** (1893) 2187; (b) **27** (1894) 1317

[671] Emmert, B. and Dorn, W. *Ber. dtsch. chem. Ges.* **48** (1915) 687

[672] (a) *Germ. Pat.* 510,432 (1930); (b) den Hertog, H. J. and Wibaut, J. P. *Recl Trav. chim. Pays-Bas Belg.* **55** (1936) 122; (c) Kaye, I. A. and Kogan, I. C. *J. Am. chem. Soc.* **73** (1951) 5891

[673] (a) Kalthod, G. G. and Linnell, W. H. *Qu. Jl Pharm. Pharmac.* **21** (1948) 63; (b) Wibaut, J. P., Overhoff, J. and Geldof, H. *Recl Trav. chim. Pays-Bas Belg.* **54** (1935) 807

[674] Wibaut, J. P. and Broekman, F. W. *Recl Trav. chim. Pays-Bas Belg.* **80** (1961) 309

[675] (a) Meyer, H. and Graf, R. *Ber. dtsch. chem. Ges.* **61** (1928) 2202; (b) Reed, L. J. and Shive, W. *J. Am. chem. Soc.* **68** (1946) 2740; (c) Bachman, G. B. and Micucci, D. D. **70** (1948) 2381; (d) *Swiss Pat.* 227,124 (*Chem. Abstr.* **43** (1949) 3471); (e) Bäumler, J., Sorkin, E. and Erlenmeyer, H. *Helv. chim. Acta* **34** (1951) 496

[676] Talik, Z. and Plažek, E. *Recl Trav. chim. Pays-Bas Belg.* **79** (1960) 193; Talik, Z. *Roczn. Chem.* **34** (1960) 165, 465

[677] (a) Meyer, H. and Beck, F. R. v. *Mh. Chem.* **36** (1915) 731; (b) *Swiss Pat.* 212,060 (1941)

[678] (a) Steinhäuser, E. and Diepolder, E. *J. prakt. Chem.* **93** (1916/2) 387; (b) Wibaut, J. P., with La Bastide, G. L. C. *Recl Trav. chim. Pays-Bas Belg.* **52** (1933) 493; (c) with Tilman, G. *ibid.* 987; (d) Mangini, A. and Frenguelli, B. *Gazz. chim. ital.* **69** (1939) 86, 97; (e) Petrow, V. A. *J. chem. Soc.* (1945) 927; (f) Kermack, W. O. and Weatherhead, A. P. (1942) 726; (g) *U.S. Pat.* 2,435,392 (1948) (h) Jerchel, D. and Jakob, L. *Chem. Ber.* **92** (1959) 724

[679] (a) Marckwald, W. *Ber. dtsch. chem. Ges.* **31** (1898) 2496; (b) with Rudzik, K. **36** (1903) 1111; (c) Koenigs, E., Weiss, W. and Zscharn, A. **59** (1926) 316; (d) *U.S. Pat.* 1,733,695 (1929); (e) Schroeter, G. and Finck, E. *Ber. dtsch. chem. Ges.* **71** (1938) 671; (f) Tarbell, D. S., Todd, C. W. *et al.*, *J. Am. chem. Soc.* **70** (1948) 1381; (g) Baumgarten, H. E. and Chien-Fan Su, H. **74** (1952) 3828; (h) Kimura, M. and Takano, Y. *Chem. Abstr.* **53** (1959) 18030

[680] (a) *U.S. Pat.* 2,129,294 (1938); (b) den Hertog, H. J. with Schogt, J. C. M., de Bruyn, J. and de Klerk, A. *Recl Trav. chim. Pays-Bas Belg.* **69** (1950) 673; (c) with Jouwersma, C. **72** (1953) 44

[681] (a) Marcinkow, A. and Plažek, E. *Roczn. Chem.* **16** (1936) 136; (b) Späth, E. and Eiter, K. *Ber. dtsch. chem. Ges.* **73** (1940) 719; (c) den Hertog, H. J., Falter, A. W. M. and van der Linde, A. *Recl Trav. chim. Pays-Bas Belg.* **67** (1948) 377; (d) Zwart, C. and Wibaut, J. P. **74** (1955) 1062

[682] *Pyridine and its Derivatives*, ed. Klingsberg, E., Pt. 3, p. 5, New York (Interscience) 1962

[683] Antaki, H. and Petrow, V. *J. chem. Soc.* (1951) 551

[684] Späth, E. and Kuffner, F. *Ber. dtsch. chem. Ges.* **71** (1938) 1657

[685] Reynolds, C. F. *Ph.D. thesis*, University of Exeter, 1963

[686] Talik, Z. *Bull. Acad. pol. Sci. Sér. Sci. tech.* (1961) No. 9 (a) 567; (b) 571; (c) 561

[687] Talik, Z. *Roczn. Chem.* **36** (1962) (a) 1183; (b) 1465

[688] den Hertog, H. J. and Jouwersma, C. *Recl Trav. chim. Pays-Bas Belg.* **72** (1953) 125

[689] Hamana, M. and Yamazaki, M. *J. pharm. Soc. Japan* **81** (1961) 612

[690] (a) Ochiai, E., Itai, T. and Yoshino, K. *Proc. imp. Acad. Japan* **20** (1944) 141; (b) Katritzky, A. R. *J. chem. Soc.* (1956) 2404

[691] Bellas, M. and Suschitzky, H. *J. chem. Soc.* (1963) 4007

[692] Clark-Lewis, J. W. and Singh, R. P. *J. chem. Soc.* (1962) 2379

[693] (a) Fischer, O. and Demeler, K. *Ber. dtsch. chem. Ges.* **32** (1899) 1307; (b) Michaelis, A. and Hillmann, O. *Justus Liebigs Annln Chem.* **354** (1907) 91; (c) *Germ. Pat.* 595,361 (1934) (*Chem. Abstr.* **28** (1934) 4069)

[694] (a) Pieterse, M. J. and den Hertog, H. J. *Recl Trav. chim. Pays-Bas Belg.* **80** (1961) 1376; (b) den Hertog, H. J., Pieterse, M. J. and Buurman, D. J. **82** (1963) 1173; (c) Martens, R. J. and den Hertog, H. J. **83** (1964) 621

[695] Kauffmann, T. and Boettcher, F.-P. *Chem. Ber.* **95** (1962) 1528

[696] Huigsen, R. and Herbig, K. in *Organometallic Chemistry*, p. 78, ed. Zeiss, H. H., New York (Reinhold) 1960

[697] (a) Pieterse, H. J. and den Hertog, H. J. *Recl Trav. chim. Pays-Bas Belg.* **81** (1962) 855; (b) Martens, R. J., den Hertog, H. J. and van Ammers, M. *Tetrahedron Lett.* (1964) 3207; (c) Kato, T., Niitsuma, T. and Kusaka, N. *J. pharm. Soc. Japan* **84** (1964) 23

[698] Banks, C. K. *J. Am. chem. Soc.* **66** (1944) 1127; Maggiolo, A. and Phillips, A. P. *J. org. Chem.* **16** (1951) 376; Morley, J. S. and Simpson, J. C. E. *J. chem. Soc.* (1949) 1014

REFERENCES

699 Chapman, N. B. and Rees, C. W. *J. chem. Soc.* (1954) 1190

700 Chapman, N. B. and Russell-Hill, D. Q. *J. chem. Soc.* (1956) 1563

701 Chapman, N. B., Chaudhury, D. K. and Shorter, J. *J. chem. Soc.* (1962) 1975

702 Coppens, G., Declerck, F., Gillet, C. and Nasielski, J. *Bull. Soc. chim. Belg.* **70** (1961) 480

703 Brower, K. R., Way, J. W., Samuels, W. P. and Amstutz, E. D. *J. org. Chem.* **19** (1954) 1830; Young, T. E. and Amstutz, E. D. *J. Am. chem. Soc.* **73** (1951) 4773

704 Chapman, N. B. *Spec. Publs chem. Soc. No.* 3 (1955) 155

705 Bunnett, J. F. and Randall, J. J. *J. Am. chem. Soc.* **80** (1958) 6020

706 Vompe, A. F., Monitch, N. V., Turitsyna, N. F. and Ivanova, L. V. *Tetrahedron* **2** (1958) 361

707 Takahashi, T. and Shibasaki, J. *Chem. Abstr.* **49** (1955) 12459

708 Itai, T. *Chem. Abstr.* **49** (1955) 327

709 (a) King, L. C. and Ozoz, F. J. *J. org. Chem.* **20** (1955) 448; (b) Rodig, O. R., Collie, R. E. and Schlatzer, R. K. **29** (1964) 2652; (c) Lyle, R. E. and Nelson, D. A. **28** (1963) 169

710 Schmidt, U. and Giesselmann, G. *Chem. Ber.* **93** (1960) 1590

711 Gautier, J. A. and Renault, J. *C. r. hebd. Séanc. Acad. Sci., Paris* **234** (1952) 2081

712 Mangini, A. and Colonna, M. *Boll. scient. Fac. Chim. Univ. Bologna* **20** (1941) 1

713 Katada, M. *J. pharm. Soc. Japan* **67** (1947) (a) 56; (b) 59

714 (a) Ochiai, E. and Katada, M. *J. pharm. Soc. Japan* **63** (1943) 265; (b) Itai, T. **65** (1945/No. 9/10A) 8; (c) Takeda, K. and Tokuyama, M. **75** (1955) 286, 620

715 (a) Mangini, A. and Colonna, M. *Gazz. chim. ital.* **75** (1943) 313; (b) Suzuki, Y. *J. pharm. Soc. Japan* **81** (1961) 1146

716 Bergstrom, F. W. and McAllister, S. H. *J. Am. chem. Soc.* **52** (1930) 2845

717 Colonna, M. *Boll. scient. Fac. Chim. Univ. Bologna* **19** (1940/4) 134; Ochiai, E. and Arima, K. *J. pharm. Soc. Japan* **69** (1949) 51

718 (a) Snyder, H. R., Eliel, E. L. and Carnahan, R. E. *J. Am. chem. Soc.* **73** (1951) 970; (b) Abramovitch, R. A. and Vig, B. *Can. J. Chem.* **41** (1963) 1961

719 *Org. Synth. Coll.* **2** (1943) 517

720 Geissman, T. A., Schlatter, M. J., Webb, I. D. and Roberts, J. D. *J. org. Chem.* **11** (1946) 741

721 Meek, J. S., Merrow, R. T. and Cristol, S. J. *J. Am. chem. Soc.* **74** (1952) 2667

722 Cymerman-Craig, J. and Loder, J. W. *J. chem. Soc.* (1956) 100

723 Papa, D., Sperber, N. and Sherlock, M. *J. Am. chem. Soc.* **73** (1951) 1279

724 Bradsher, C. K. and Beavers, L. E. *J. Am. chem. Soc.* **78** (1956) 2459

725 Gilman, H., Stuckwisch, C. G. and Nobis, J. F. *J. Am. chem. Soc.* **68** (1946) 326

726 (a) Hey, D. H., Stirling, C. J. M. and Williams, G. H. *J. chem. Soc.* (1955) 3963; (b) Schlögl, K. and Fried, M. *Mh. Chem.* **94** (1963) 536

727 Miller, A. D., Osuch, C., Goldberg, N. N. and Levine, R. *J. Am. chem. Soc.* **78** (1956) 674

728 (a) Wiley, R. H., Jarboe, C. H., Callahan, P. X. and Nielson, N. J. *J. org. Chem.* **23** (1958) 780; (b) Abramovitch, R. A. and Giam, Choo-Seng *Can. J. Chem.* **41** (1963) 3127

729 Burger, A. and Ullyot, G. E. *J. org. Chem.* **12** (1947) 342

730 Moynehan, T. M., Schofield, K., Jones, R. A. Y. and Katritzky, A. R. *J. chem. Soc.* (1962) 2637

731 Prijs, B., Lutz, A. H. and Erlenmeyer, H. *Helv. chim. Acta* **31** (1948) 571

732 Abramovitch, R. A. and Notation, A. D. *Can. J. Chem.* **38** (1960) 1445

733 Abramovitch, R. A. and Giam, Choo-Seng *Can. J. Chem.* (a) **40** (1962) 213; (b) **42** (1964) 1627

734 Abramovitch, R. A., Giam, Choo-Seng and Notation, A. D. *Can. J. Chem.* **38** (1960) 761

735 Bryce-Smith, D. and Skinner, A. C. *J. chem. Soc.* (1963) 577

736 McEwen, W. E., Terss, R. H. and Elliott, I. W. *J. Am. chem. Soc.* **74** (1952) 3605

737 Reynolds, C. A., Walker, F. H. and Cochran, E. *Analyt. Chem.* **32** (1960) 983

738 (a) Sugasawa, S. and Akaboshi, S. *Chem. Abstr.* **47** (1953) 6957; (b) with Suzuki, M. *J. pharm. Soc. Japan* **72** (1952) 1273; (c) Govindachari, T. R. and Thyagarajan, B. S. *Proc. Indian Acad. Sci.* **39A** (1954) 232

739 Wiley, R. H., Smith, N. R. and Knabeschuh, L. H. *J. Am. chem. Soc.* **75** (1953) 4482

740 Sugasawa, S., Akaboshi, S. and Ban, Y. *Chem. pharm. Bull., Tokyo* **7** (1959) 263

741 Ban, Y., Yonemitsu, O., Oishi, T., Yokoyama, S. and Nakagawa, M. *Chem. pharm. Bull., Tokyo* **7** (1959) 609

742 Paquette, L. A. and Nelson, N. A. *J. org. Chem.* **27** (1962) 1085

743 Berson, J. A. and Walia, J. S. *J. org. Chem.* **24** (1959) 756

744 Ban, Y. and Seo, M. *Tetrahedron* **16** (1961) 5, 11; Büchi, G., Manning, R. E. and Hochstein, F. A. *J. Am. chem. Soc.* **84** (1962) 3393

745 Bachman, G. B., Hamer, M., Dunning, E. and Schisla, R. M. *J. org. Chem.* **22** (1957) 1296

746 Emmert, B., with Asendorf, E. *Ber. dtsch. chem. Ges.* **72** (1939) 1188; with Pirot, E. **74** (1941) 714

747 Feely, W. E. and Beavers, E. M. *J. Am. chem. Soc.* **81** (1959) 4004

L

[748] (a) Meyerhof, O., Ohlmeyer, P. and Möhle, W. *Biochem. Z.* **297** (1938) 113; (b) Colowick, S. P., Kaplan, N. O. and Ciotti, M. M. *J. biol. Chem.* **191** (1951) 447; (c) San Pietro, A. **217** (1955) 579; (d) Marti, M., Viscontini, M. and Karrer, P. *Helv. chim. Acta* **39** (1956) 1451; (e) Lamborg, M. R., Burton, R. M. and Kaplan, N. O. *J. Am. chem. Soc.* **79** (1957) 6173; (f) Anderson, A. G. and Berkelhammer, G. *J. org. Chem.* **23** (1958) 1109

[749] Reissert, A. *Ber. dtsch. chem. Ges.* **38** (1905) 3415; McEwen, W. S. and Cobb, R. L. *Chem. Rev.* **55** (1955) 511

[750] Okamoto, T. and Tani, H. *Pharm. Bull.*, *Tokyo* **7** (1959) 130, 925

[751] Tani, H. *Yakugaku Zasshi* (a) **81** (1961) 141; (b) **80** (1960) 1418; (c) *Pharm. Bull.*, *Tokyo* **7** (1959) 930

[752] Swan, G. A. and Thomas, P. R. *J. chem. Soc.* (1963) 3440

[753] Ochiai, E. and Nakayama, I. *Chem. Abstr.* **45** (1951) 8529

[754] Kaneko, C. *Pharm. Bull.*, *Tokyo* **8** (1960) 286

[755] Binz, A. and Räth, C. *Justus Liebigs Annln Chem.* **486** (1931) 95

[756] Räth, C. and Schiffmann, F. *Justus Liebigs Annln Chem.* **487** (1931) 127

[757] Batkowski, T. and Plažek, E. *Roczn. Chem.* **36** (1962) 51

[758] Ochiai, E., Teshigawara, T., Oda, K. and Naito, T. *Chem. Abstr.* **45** (1951) 8527

[759] Craig, L. C. *J. Am. chem. Soc.* **56** (1934) 231

[760] Case, F. H. and Kasper, T. J. *J. Am. chem. Soc.* **78** (1956) 5842

[761] McElvain, S. M. and Goese, M. A. *J. Am. chem. Soc.* **63** (1941) 2283

[762] Brode, W. R. and Bremer, C. *J. Am. chem. Soc.* **56** (1934) 993

[763] Lee, T. B. and Swan, G. A. *J. chem. Soc.* (1956) 771

[764] Broekman, F. W., van Veldhuizen, A. and Janssen, H. *Recl Trav. chim. Pays-Bas Belg.* **81** (1962) 792

[765] *Pyridine and its Derivatives*, ed. Klingsberg, E., Pt. 2, pp. 347, 355, New York (Interscience) 1961

[766] McElvain, S. M. and Goese, M. A. *J. Am. chem. Soc.* **65** (1943) 2233

[767] Craig, L. C. *J. Am. chem. Soc.* **55** (1933) 2855

[768] Suzuki, Y. *Pharm. Bull.*, *Tokyo* **5** (1957) (a) 13; (b) 78

[769] Klingsberg, E. and Papa, D. *J. Am. chem. Soc.* **71** (1949) 2373

[770] Ochiai, E. and Suzuki, I. *Pharm. Bull.*, *Tokyo* **2** (1954) 247

[771] Suzuki, Y. *J. pharm. Soc. Japan* **81** (1961) 917

[772] Tschitschibabin, A. E. and Kirsanov, A. W. *Ber. dtsch. chem. Ges.* **60** (1927) 766

[773] Bremer, O. *Justus Liebigs Annln Chem.* **514** (1934) 279

[774] Bremer, O. *Justus Liebigs Annln Chem.* **539** (1939) 276

[775] (a) Graf, R. *Ber. dtsch. chem. Ges.* **64** (1931) 21; (b) *J. prakt. Chem.* **133** (1932/2) 36

[776] Fox, H. H. and Gibas, J. T. *J. org. Chem.* **23** (1958) 64

[777] Mosher, H. S. and Look, M. *J. org. Chem.* **20** (1955) 283

[778] Graf, R. *J. prakt. Chem.* **138** (1933/2) 244

[779] Garcia, E. E., Greco, C. V. and Hunsberger, I. M. *J. Am. chem. Soc.* **82** (1960) 4430

[780] Murakami, M. and Matsumura, E. *J. chem. Soc. Japan (Pure Chem.)* **70** (1949) 393

[781] den Hertog, H. J. and Boelrijk, N. A. I. M. *Recl Trav. chim. Pays-Bas Belg.* **70** (1951) 578

[782] den Hertog, H. J. and Hoogzand, C. *Recl Trav. chim. Pays-Bas Belg.* **76** (1957) 261

[783] Kato, T. and Ohta, M. *J. pharm. Soc. Japan* **71** (1951) 217

[784] Colonna, M., Risaliti, A. and Serra, R. *Gazz. chim. ital.* **85** (1955) 1508

[785] Kato, T. *J. pharm. Soc. Japan* **75** (1955) 1236

[786] Ochiai, E. and Kaneko, C. *Pharm. Bull.*, *Tokyo* **8** (1960) 28

[787] Hamana, M. and Yamazaki, M. *J. pharm. Soc. Japan* **81** (1961) 574

[788] Gadient, F., Jucker, E., Lindenmann, A. and Taeschler, M. *Helv. chim. Acta* **45** (1962) 1860

[789] Morris, I. G. and Pinder, A. R. *J. chem. Soc.* (1963) 1841

[790] Tschitschibabin, A. E. and Rjasanjew, M. D. *Zh. russk. fiz.-khim. Obshch.* **47** (1915) 1571

[791] Tschitschibabin, A. E. *Zh. russk. fiz.-khim. Obshch.* **46** (1914) 1236

[792] Seide, O. A. *Zh. russk. fiz.-khim. Obshch.* **50** (1920) 534

[793] Tschitschibabin, A. E. and Woroshtzow, N. N. *Ber. dtsch. chem. Ges.* **66** (1933) 364

[794] Hawkins, G. F. and Roe, A. *J. org. Chem.* **14** (1949) 328

[795] den Hertog, H. J. and de Bruyn, J. *Recl Trav. chim. Pays-Bas Belg.* **70** (1951) 182

[796] Kolder, C. R. and den Hertog, H. J. *Recl Trav. chim. Pays-Bas Belg.* **72** (1953) 285

[797] Talik, T. and Plažek, E. *Roczn. Chem.* **33** (1959) (a) 387; (b) 1343

[798] Talik, T. and Plažek, E. *Roczn. Chem.* **35** (1961) 463

[799] Profft, E. and Richter, H. *J. prakt. Chem.* **9** (1959/4) 164

[800] Wibaut, J. P. and la Bastide, G. *Chem. Abstr.* **21** (1927) 3619

[801] Willink, H. D. T. and Wibaut, J. P. *Recl Trav. chim. Pays-Bas Belg.* **53** (1934) 417

[802] Talik, T. and Plažek, E. *Roczn. Chem.* **29** (1955) 1019

[803] *Org. Synth.* **26** (1946) (a) 16; (b) 1

[804] Gergely, E. and Iredale, T. *J. chem. Soc.* (1953) 3226

[805] Gruber, W. *Can. J. Chem.* **31** (1953) 1020
[806] Ferm, R. L. and Vander Werf, C. A. *J. Am. chem. Soc.* **72** (1950) 4809
[807] Beaty, R. D. and Musgrave, W. K. R. *J. chem. Soc.* (1952) 875
[808] Roe, A. and Hawkins, G. F. *J. Am. chem. Soc.* **69** (1947) 2443
[809] Finger, G. C., Storr, L. D., Roe, A. and Link, W. J. *J. org. Chem.* **27** (1962) 3965
[810] Tschitschibabin, A. E. *Zh. russk. fiz.-khim. Obshch.* **50** (1918) 502
[811] Perez-Medina, L. A., Mariella, R. P. and McElvain, S. M. *J. Am. chem. Soc.* **69** (1947) 2574
[812] Graf, R. *J. prakt. Chem.* **133** (1932/2) 19
[813] *Brit. Pat.* 251,578 (1924)
[814] *Germ. Pat.* 491,681 (1924)
[815] Graf, R. and Stauch, J. *J. prakt. Chem.* **148** (1937) 13
[816] *U.S. Pat.* 2,516,830 (1950)
[817] Roe, A. and Seligman, R. B. *J. org. Chem.* **20** (1955) 1729
[818] Suzuki, Y. *J. pharm. Soc. Japan* **81** (1961) 1206
[819] Sell, W. J. and Dootson, F. W. *J. chem. Soc.* **77** (1900) 233
[820] Bruce, W. F. and Perez-Medina, L. A. *J. Am. chem. Soc.* **69** (1947) 2571
[821] Baker, W., Curtis, R. F. and Edwards, M. G. *J. chem. Soc.* (1951) 83
[822] Klingsberg, E. *J. Am. chem. Soc.* **72** (1950) 1031
[823] Finger, G. C. and Starr, L. D. *J. Am. chem. Soc.* **81** (1959) 2674
[824] Hamer, J., Link, W. J., Jurjevich, A. and Vigo, T. L. *Recl Trav. chim. Pays-Bas Belg.* **81** (1962) 1058
[825] Leben, J. A. *Ber. dtsch. chem. Ges.* **29** (1896) 1673
[826] Collie, J. N. *J. chem. Soc.* **71** (1897) 299
[827] Conrad, M. and Epstein, W. *Ber. dtsch. chem. Ges.* **20** (1887) 162
[828] *Brit. Pat.* 686,012 (1953)
[829] Robison, M. M. *J. Am. chem. Soc.* **80** (1958) 5481
[830] Tamari, K. and Morino, K. *J. agric. chem. Soc. Japan* **22** (1948) 18
[831] Koenigs, E. and Kinne, G. *Ber. dtsch. chem. Ges.* **54** (1921) 1357
[832] (a) Bardhan, J. C. *J. chem. Soc.* (1929) 2223; (b) Tracy, A. H. and Elderfield, R. C. *J. org. Chem.* **6** (1941) 70
[833] Forrest, H. S. and Walker, J. *J. chem. Soc.* (1948) 1939
[834] (a) Reider, M. J. and Elderfield, R. C. *J. org. Chem.* **7** (1942) 286; (b) Scott, M. L., Norris, L. C., Heuser, G. F. and Bruce, W. F. *J. Am. chem. Soc.* **67** (1945) 157
[835] Taylor, E. C., Crovetti, A. J. and Loux, H. M. *J. Am. chem. Soc.* **77** (1955) 5445
[836] Prelog, V. and Geyer, U. *Helv. chim. Acta* **28** (1945) 1677
[837] Petrenko-Kritschenko, P. and Schöttle, S. *Ber. dtsch. chem. Ges.* **42** (1909) 2020; Plati, J. T. and Wenner, W. *J. org. Chem.* **15** (1950) 1165
[838] Schroeder, H. E. and Rigby, G. W. *J. Am. chem. Soc.* **71** (1949) 2205
[839] (a) *Austr. Pat.* 112,127 (1929); (b) Saikachi, H. *J. pharm. Soc. Japan* **64** (1944) 201; (c) Childress, S. J. and McKee, R. L. *J. Am. chem. Soc.* **73** (1951) 3504; (d) Baumgarten, H. E., Chien-fan Su, H. and Krieger, A. L. **76** (1954) 596; (e) Mariella, R. P., Callahan, J. J. and Jibril, A. O. *J. org. Chem.* **20** (1955) 1721
[840] (a) Tracy, A. H. and Elderfield, R. C. *J. org. Chem.* **6** (1941) 54; (b) Prelog, V., with Szpilfogel, S. *Helv. chim. Acta* **25** (1942) 1306; **28** (1945) 1684; (c) with Hinden, W. **27** (1944) 1854; (d) Wilson, A. N. and Harris, S. A. *J. Am. chem. Soc.* **73** (1951) 2388
[841] (a) Levelt, W. H. and Wibaut, J. P. *Recl Trav. chim. Pays-Bas Belg.* **48** (1929) 466; (b) Stevens, J. R. and Beutel, R. H. *J. Am. chem. Soc.* **65** (1943) 449; (c) Prelog, V. and Metzler, O. *Helv. chim. Acta* **29** (1946) 1163; (d) Abe, K., Kitagawa, Y. and Ishimura, A. *J. pharm. Soc. Japan* **73** (1953) 969; (e) Ayerst, G. G. and Schofield, K. *J. chem. Soc.* (1958) 4097; (f) Bailey, A. S. and Brunskill, J. S. A. (1959) 2554
[842] Fischer, O. *Ber. dtsch. chem. Ges.* **31** (1898) 609
[843] Binz, A. and Räth, C. *Justus Liebigs Annln Chem.* **486** (1931) 71
[844] Binz, A. and Schikh, O. v. *Ber. dtsch. chem. Ges.* **68** (1935) 315
[845] Wibaut, J. P. and Wagtendonk, H. M. *Recl Trav. chim. Pays-Bas Belg.* **60** (1941) 22
[846] Ochiai, E., Suzuki, I. and Futaki, K. *J. pharm. Soc. Japan* **74** (1954) 666
[847] den Hertog, H. J. and Mulder, B. *Recl Trav. chim. Pays-Bas Belg.* **67** (1948) 957
[848] Hamana, M. and Yoshimura, H. *J. pharm. Soc. Japan* **72** (1952) 1051
[849] Ochiai, E., Ito, T. and Okuda, T. *J. pharm. Soc. Japan* **71** (1951) 591
[850] den Hertog, H. J. and Combé, W. P. *Recl Trav. chim. Pays-Bas Belg.* **71** (1952) 745
[851] Hamana, M. *J. pharm. Soc. Japan* **75** (1955) 123
[852] Wieland, T. and Biener, H. *Chem. Ber.* **96** (1963) 266
[853] Itai, T. *Chem. Abstr.* **45** (1951) 8525
[854] Taylor, E. C. and Driscoll, J. S. *J. Am. chem. Soc.* **82** (1960) 3141
[855] Profft, E. and Steinke, W. *J. prakt. Chem.* **13** (1961) 58
[856] Brown, E. V. *J. Am. chem. Soc.* **79** (1957) 3565
[857] Koenigs, W. and Geigy, R. *Ber. dtsch. chem. Ges.* **17** (1884) 1832

858 Bunnett, J. F. and Zahler, R. E. *Chem. Rev.* **49** (1951) 273
859 Tschitschibabin, A. E. *Ber. dtsch. chem. Ges.* **56** (1923) 1879; *Germ. Pat.* 406,208 (1924)
860 Koenigs, W. and Koerner, G. *Ber. dtsch. chem. Ges.* **16** (1883) 2152
861 Katada, M. *J. pharm. Soc. Japan* **67** (1947) 51
862 Ochiai, E. and Okamoto, T. *J. pharm. Soc. Japan* **68** (1948) 88
863 (a) Cava, M. P. and Weinstein, B. *J. org. Chem.* **23** (1958) 1616; (b) Boekelheide, V. and Lehn, W. L. **26** (1961) 428
864 Kobayashi, G. and Furukawa, S. *Pharm. Bull.*, *Tokyo* **1** (1953) 347; Furukawa, S. *J. pharm. Soc. Japan* **59** (1959) 487
865 Berson, J. A. and Cohen, T. *J. Am. chem. Soc.* **77** (1955) 1281; Biemann, K., Büchi, G. and Walker, B. H. **79** (1957) 5558
866 Peterson, M. L. *J. org. Chem.* **25** (1960) 565
867 Markgraf, J. H., Brown, H. B., Mohr, S. C. and Peterson, R. G. *J. Am. chem. Soc.* **85** (1963) 958
868 (a) Hamana, M. and Yamazaki, M. *Pharm. Bull.*, *Tokyo* **9** (1961) 414; (b) Smalley, R. K. and Suschitzky, H. *J. chem. Soc.* (1964) 755
869 Decker, H. *Ber. dtsch. chem. Ges.* **25** (1892) 443; *J. prakt. Chem.* **47** (1893/2) 28
870 Hantzsch, A. and Kalb, M. *Ber. dtsch. chem. Ges.* **32** (1899) 3109; Aston, J. G. and Lasselle, P. A. *J. Am. chem. Soc.* **56** (1934) 426
871 *Org. Synth., Coll.* **2** (1943) 419
872 Ochiai, E., Tsuda, K. and Yokoyama, J. *Ber. dtsch. chem. Ges.* **68** (1935) 2291; Rohmann, C. and Zietan, K. **71** (1938) 296; Sugasawa, S., with Sugimoto, N. **72** (1939) 977; with Shigehara, H. **74** (1941) 459; Gautier, J. A. and Renault, J. *Recl Trav. chim. Pays-Bas Belg.* **69** (1950) 421
873 Bohlmann, F., Ottawa, N. and Keller, R. *Justus Liebigs Annln Chem.* **587** (1954) 162
874 Sugasawa, S. and Kirisawa, M. *Pharm. Bull.*, *Tokyo* **6** (1958) 615
875 Fronk, M. H. and Mosher, H. S. *J. org. Chem.* **24** (1959) 196; Tomisawa, H. *Chem. Abstr.* **54** (1960) 3416
876 Sugasawa, S., with Kirisawa, M. *Pharm. Bull.*, *Tokyo* (a) **3** (1955) 187; (b) **4** (1956) 139; (c) **3** (1955) 190; (d) with Ban, Y. *J. pharm. Soc. Japan* **72** (1952) 1336
877 Podrebarac, E. G. and McEwen, W. E. *J. org. Chem.* **26** (1961) 1386
878 Sato, Y., Iwashige, T. and Mujalera, T. *Chem. Abstr.* **55** (1961) 12401; Rimek, H.-J. *Justus Liebigs Annln Chem.* **670** (1963) 69
879 Furukawa, S. *Pharm. Bull.*, *Tokyo* **3** (1955) 413; Kobayashi, G., Furukawa, S. and Kawada, Y. *J. pharm. Soc. Japan* **74** (1954) 7990; Kato, T. **75** (1955) 1228
880 Karrer, P., with Widmer, R. *Helv. chim. Acta* **8** (1925) 364; with Takahashi, T. **9** (1926) 458
881 Huff, J. W. *J. biol. Chem.* **171** (1947) 639; Holman, W. I. M. and Wiegand, C. *Biochem. J.* **43** (1948) 423
882 Pullman, M. E. and Colowick, S. P. *J. biol. Chem.* **206** (1954) 121
883 Tomisawa, H. *Chem. Abstr.* **54** (1960) 3417
884 Sugasawa, S. and Tatsuno, T. *J. pharm. Soc. Japan* **72** (1952) 248
885 Fischer, O. and Neundlinger, K. *Ber. dtsch. chem. Ges.* **46** (1913) 2544
886 Hamana, M. and Yamazaki, M. *Pharm. Bull.*, *Tokyo* **10** (1962) 51
887 Carboni, S. *Gazz. chim. ital.* **83** (1953) 637
888 Knunjanz, I. L. *Ber. dtsch. chem. Ges.* **68** (1935) 397
889 Seide, O. A. and Titow, A. I. *Ber. dtsch. chem. Ges.* **69** (1936) 1884
890 Tusza, E. T. and Joos, B. *Chem. Abstr.* **26** (1932) 4347
891 Titov, A. I. and Levin, B. B. *J. gen. Chem. U.S.S.R.* **11** (1941) 9
892 Tschitschibabin, A. E. and Kirsanov, A. W. *Ber. dtsch. chem. Ges.* **61** (1928) 1236
893 Oae, S., Kitaoka, Y. and Kitao, T. *Tetrahedron* **20** (1964) 2677; *J. Am. chem. Soc.* **84** (1962) 3362
894 Ellin, R. I. *J. Am. chem. Soc.* **80** (1958) 6588
895 Cava, M. P. and Bhattacharyya, N. K. *J. org. Chem.* **23** (1958) 1287
896 Dornow, A. and Neuse, E. *Ber. dtsch. chem. Ges.* **84** (1951) 296
897 Kirpal, A. and Reimann, K. *Mh. Chem.* **38** (1917) 249
898 Lappin, G. R. and Slezak, F. B. *J. Am. chem. Soc.* **72** (1950) 2806
899 Späth, E. and Koller, G. *Ber. dtsch. chem. Ges.* **56** (1923) 2454
900 Sell, W. J. *J. chem. Soc.* **23** (1908) 2001
901 Berrie, A. H., Newbold, G. T. and Spring, F. S. *J. chem. Soc.* (1951) 2590
902 Pollak, F. *Mh. Chem.* **16** (1895) 45
903 Jones, R. G. *J. Am. chem. Soc.* **74** (1952) 1489
904 Blanchard, K. C., Dearborn, E. H., Lasagna, L. C. and Buhle, E. L. *Chem. Abstr.* **47** (1953) 10535
905 Fox, H. H. *J. org. Chem.* **17** (1952) 547
906 Marcinkow, A. and Plažek, E. *Chem. Abstr.* **31** (1937) 2216
907 Itiba, A. and Emoto, S. *Chem. Abstr.* **35** (1941) 6960

REFERENCES

908 Jones, R. G. *J. Am. chem. Soc.* **73** (1951) 5610
909 Kirpal, A. *Mh. Chem.* **23** (1902) 239
910 Sell, W. T. and Dootson, F. W. *J. chem. Soc.* **73** (1898) 777
911 den Hertog, H. J., Maas, J., Kolder, C. R. and Combé, W. P. *Recl Trav. chim. Pays-Bas Belg.* **74** (1955) 59
912 Zollinger, H. *Azo and Diazo Chemistry*, p. 138, New York (Interscience) 1961
913 Wibaut, J. P., Uhlenbroek, J. H., Kooyman, E. C. and Kettenes, D. K. *Recl Trav. chim. Pays-Bas Belg.* **79** (1960) 481
914 *Brit. Pat.* 588,806 (*Chem. Abstr.* **41** (1947) 6893)
915 *Brit. Pat.* 288,628 (1927); *U.S. Pat.* 1,778,784 (1930)
916 Schroeter, G. and Seidler, C. *J. prakt. Chem.* **105** (1922) 165
917 Wibaut, J. P., Haayman, P. W. and van Dijk, J. *Recl Trav. chim. Pays-Bas Belg.* **59** (1940) 202
918 Finger, G. C., Starr, L. D., Roe, A. and Link, W. J. *J. org. Chem.* **27** (1962) 3965
919 (a) Ochiai, E., Fujimoto, M. and Ichimura, S. *Pharm. Bull., Tokyo* **2** (1954) 137; (b) Kato, T. and Hamaguchi, F. *Chem. Abstr.* **51** (1957) 11342
920 Sutherland, J. K. and Widdowson, D. A. *J. chem. Soc.* (1964) 4650
921 Maier-Bode, H. *Ber. dtsch. chem. Ges.* **69** (1936) 1534
922 Bunnett, J. F. *Qu. Rev. chem. Soc.* **12** (1958) 1
923 Reinheimer, J. D., Gerig, J. T., Garst, R. and Schrier, B. *J. Am. chem. Soc.* **84** (1962) 2770
924 Büchi, J., Labhart, P. and Ragaz, L. *Helv. chim. Acta* **30** (1947) 507
925 Kushner, S., Dalalian, H. *et al.*, *J. org. Chem.* **13** (1948) 834
926 *Brit. Pat.* 355,017 (1930); *Chem. Abstr.* **26** (1932) 5574
927 Magidson, O. and Menschikoff, G. *Ber. dtsch. chem. Ges.* **58** (1925) 113
928 Shibasaki, J. *Chem. Abstr.* **46** (1952) 8090
929 *Germ. Pat.* 550,327 (1930)
930 Friedman, H. L., Braitberg, L. D., Tolstooukov, A. V. and Tisza, E. T. *J. Am. chem. Soc.* **69** (1947) 1204
931 Yamamoto, Y. *Chem. Abstr.* **46** (1952) 8109
932 Shibasaki, J. *Chem. Abstr.* **47** (1953) 6403
933 Mariella, R. P. and Havlik, A. J. *J. Am. chem. Soc.* **74** (1952) 1915
934 Takahashi, T., Shibasaki, J., Hamada, Y. and Okuda, J. *Chem. Abstr.* **47** (1953) 6404
935 Fanta, P. E. and Stein, R. A. *J. Am. chem. Soc.* **77** (1955) 1045
936 Takahashi, T. and Shibasaki, J. *Chem. Abstr.* **47** (1953) 6403
937 Shibasaki, J. *Chem. Abstr.* **47** (1953) 6404
938 Hill, A. J. and McGraw, W. J. *J. org. Chem.* **14** (1949) 783
939 Kohler, E. P. and Allen, C. F. H. *J. Am. chem. Soc.* **46** (1924) 1522
940 Weidel, H. and Blau, F. *Mh. Chem.* **6** (1885) 651
941 Koenigs, E. and Neumann, L. *Ber. dtsch. chem. Ges.* **48** (1915) 956
942 Shinkai, J. H. and Parks, L. M. *J. Am. pharm. Ass.* **39** (1950) 283
943 Archer, S., Hoppe, J. O., Lewis, T. R. and Haskell, M. N. *Chem. Abstr.* **46** (1952) 468
944 Fujimoto, M. *Pharm. Bull., Tokyo* **4** (1956) 1
945 Woodburn, H. M. and Hellman, M. *Recl Trav. chim. Pays-Bas Belg.* **70** (1951) 813
946 den Hertog, H. J. and de Jonge, A. P. *Recl Trav. chim. Pays-Bas Belg.* **67** (1948) 385
947 Pechmann, H. von and Mills, W. H. *Ber. dtsch. chem. Ges.* **37** (1904) 3829
948 Kröhnke, F. and Heffe, W. *Ber. dtsch. chem. Ges.* **70** (1937) 864
949 *Germ. Pat.* 596,821 (1934); *Chem. Abstr.* **28** (1934) 5078
950 Liveris, M. and Miller, J. *J. chem. Soc.* (1963) 3486
951 Liveris, M. and Miller, J. *Aust. J. Chem.* **11** (1958) 297
952 Boxer, R. J. *Dissert. Abstr.* **22** (1961) 66
953 Spinner, E. *J. chem. Soc.* (1953) 3855
954 Spinner, E. *Aust. J. Chem.* **16** (1963) 174
955 (a) Murakami, M. and Matsumura, E. *J. chem. Soc. Japan (Pure Chem.)* **70** (1949) 393; (b) Matsumura, E. **74** (1953) 446; (c) de Villiers, P. A. and den Hertog, H. J. *Recl Trav. chim. Pays-Bas Belg.* **75** (1956) 1303; **76** (1957) 747; (d) with Buurman, D. J. **80** (1961) 325; (e) Oae, S., Kitao, T. and Kitaoka, Y. *Tetrahedron* **19** (1963) 827
956 Daniels, R., Grady, L. T. and Bauer, L. *J. org. Chem.* **27** (1962) 4711
957 *Germ. Pat.* 596,728; 597,974 (1934); (*Chem. Abstr.* **28** (1934) 5083, 7268); Mariella, R. P. and Belcher, E. P. *J. Am. chem. Soc.* **73** (1951) 2616
958 Peratoner, A. and Tamburello, A. *Gazz. chim. ital.* **36** (1906) I, 56
959 (a) Haitinger, L. and Lieben, A. *Ber. dtsch. chem. Ges.* **18** (1885) 929; (b) den Hertog, H. J. and Buurman, D. J. *Recl Trav. chim. Pays-Bas Belg.* **75** (1956) 257
960 den Hertog, H. J., Wibaut, J. P., Schepman, F. R. and van der Wal, A. A. *Recl Trav. chim. Pays-Bas Belg.* **69** (1950) 700
961 Armit, J. W. and Nolan, T. J. *J. chem. Soc.* (1931) 3032
962 Conrad, M. and Eckhardt, F. *Ber. dtsch. chem. Ges.* **22** (1889) 73
963 (a) Katada, M. *J. pharm. Soc. Japan* **67** (1947) 61; (b) Ochiai, E. and Katoh, T. **71** (1951) 156 (*Chem. Abstr.* **45** (1951) 9537, 9542)

964 Essery, J. M. and Schofield, K., unpublished results
965 Kröhnke, F. and Schäfer, H. *Chem. Ber.* **95** (1962) 1104
966 Hayashi, E. *J. pharm. Soc. Japan* **70** (1950) 145 (*Chem. Abstr.* **44** (1950) 5881)
967 van der Plas, H. C. and den Hertog, H. J. *Tetrahedron Lett.* (1960) 13
968 *Brit. Pat.* 335,818; *French Pat.* 685,583; *Swiss Pat.* 147,793 (1929); *Germ. Pat.* 541,681 (1931); *U.S. Pat.* 1,880,645 (1932)
969 (a) Duesel, B. F. and Scudi, J. V. *J. Am. chem. Soc.* **71** (1949) 1866; (b) Marion, L. and Cockburn, W. F. *ibid.* 3402; (c) Jacobs, W. A. and Sato, Y. *J. biol. Chem.* **191** (1951) 71
970 *Brit. Pat.* 591,392 (1947)
971 Renault, J. *C. r. hebd. Séanc. Acad. Sci., Paris* **233** (1951) 182
972 Bucherer, H. J. and Schenkel, J. *Ber. dtsch. chem. Ges.* **41** (1908) 1346
973 Kaplan, N. O. and Ciotti, M. M. *J. biol. Chem.* **221** (1956) 823; van Eys, J. and Kaplan, N. O. **228** (1957) 305
974 Wallenfels, K. and Schüly, H. *Justus Liebigs Annln Chem.* **621** (1959) 86
975 Dittmer, D. C. and Kolyer, J. M. *J. org. Chem.* **28** (1963) 1720
976 Wallenfels, K. and Schüly, H. *Justus Liebigs Annln Chem.* **621** (1959) 178
977 Hünig, S. and Köbrich, G. *Justus Liebigs Annln Chem.* **617** (1958) 181, 203, 216
978 Kosower, E. M. and Bauer, S. W. *J. Am. chem. Soc.* **82** (1960) 2191
979 Bauer, L. and Gardella, L. A. *J. org. Chem.* **28** (1963) (a) 1320; (b) 1323
980 Bauer, L. and Dickerhofe, T. E. *J. org. Chem.* **29** (1964) 2183
981 *U.S. Pat.* 2,330,641 (1944)
982 Plažek, E. and Sucharda, E. *Ber. dtsch. chem. Ges.* **59** (1926) 2282
983 Talik, Z., with Plažek, E. *Roczn. Chem.* **34** (1960) 165; **35** (1961) 475
984 Ochiai, E., with Teshigawara, T. *J. pharm. Soc. Japan* **65B** (1945) 435; with Naito, T. *ibid.* 441
985 (a) Bittner, K. *Ber. dtsch. chem. Ges.* **35** (1902) 2933; (b) Koenigs, E. and Kantrowitz, H. **60** (1927) 2097; (c) *U.S. Pat.* 1,753,658 (1930); (d) Räth, C. *Justus Liebigs Annln Chem.* **487** (1931) 105; (e) Sucharda, E. and Troskiewicz, C. *Roczn. Chem.* **12** (1932) 493; (f) Thirtle, J. R. *J. Am. chem. Soc.* **68** (1946) 342
986 Phillips, M. A. and Shapiro, H. *J. chem. Soc.* (1942) 584
987 Profft, E. and Rolle, W. *Chem. Abstr.* **55** (1961) 1609; *J. prakt. Chem.* **11** (1960) 22
988 Surrey, A. R. and Lindwall, H. G. *J. Am. chem. Soc.* **62** (1940) 173
989 Brooker, L. G. S., Keyes, G. H. *et al.*, *J. Am. chem. Soc.* **73** (1951) 5326; Adams, R. and Ferrelli, A. **81** (1959) 4927
990 Surrey, A. R. and Lindwall, H. G. *J. Am. chem. Soc.* **62** (1940) 1697; Boarland, M. P. V. and McOmie, J. F. W. *J. chem. Soc.* (1951) 1218; Markees, D. G. *J. org. Chem.* **28** (1963) 2530
991 Suzuki, Y. *J. pharm. Soc. Japan* **81** (1961) 1151
992 Itai, T. *J. pharm. Soc. Japan* **69** (1949) 545; Hayashi, E.; Yamanaka, H. and Iijima, C. *Chem. Abstr.* **55** (1961) 546
993 Guthzeit, M. and Epstein, W. *Ber. dtsch. chem. Ges.* **20** (1887) 2111; Gutbier, A. **33** (1900) 3358; Renault, J. *C. r. hebd. Séanc. Acad. Sci., Paris* **232** (1951) 77; *Bull. Soc. chim. Fr.* (1953) 1001; Musante, C. and Fabbrini, L. *Gazz. chim. ital.* **84** (1954) 584; *Germ. Pat.* 936,071 (1955); Katritzky, A. R. and Jones, R. A. *J. chem. Soc.* (1960) 2947; Elkaschef, M. A.-F., Nosseir, M. H. and Abdel-Kader, A. (1963) 4647
994 Takahashi, T., Yamashito, I. and Iwai, J. *Chem. Abstr.* **52** (1958) 20144
995 Möhlau, R. and Berger, R. *Ber. dtsch. chem. Ges.* **26** (1893) 1994
996 Haworth, J. W., Heilbron, I. M. and Hey, D. H. *J. chem. Soc.* (1940) (a) 349; (b) 372
997 Gritter, R. J. and Godfrey, A. W. *J. Am. chem. Soc.* **86** (1964) 4724
998 Elks, J. and Hey, D. H. *J. chem. Soc.* (1943) 441
999 Sandin, R. B. and Brown, R. K. *J. Am. chem. Soc.* **69** (1947) 2253
1000 Hey, D. H. and Walker, E. W. *J. chem. Soc.* (1948) 2213
1001 Adams, W. J., Hey, D. H., Mamalis, P. and Parker, R. E. *J. chem. Soc.* (1949) 3181
1002 Butterworth, E. C., Heilbron, I. M. and Hey, D. H. *J. chem. Soc.* (1940) 355
1003 Haworth, J. W., Heilbron, I. M. and Hey, D. H. *J. chem. Soc.* (1940) 358
1004 Süs, O., Möller, K. and Heiss, H. *Justus Liebigs Annln Chem.* **598** (1956) 123
1005 Lythgoe, B. and Rayner, L. S. *J. chem. Soc.* (1951) 2323
1006 Heilbron, I. M., Hey, D. H. and Lambert, A. *J. chem. Soc.* (1940) 1279
1007 Frank, R. L. and Crawford, J. V. *Bull. Soc. chim. Fr.* (1958) 419
1008 Coates, H., Cook, A. H., Heilbron, I. M., Hey, D. H., Lambert, A. and Lewis, F. B. *J. chem. Soc.* (1943) 401; Abramovitch, R. A. (1954) 3839
1009 Cook, A. H., Heilbron, I. M., Hey, D. H., Lambert, A. and Spinks, A. *J. chem. Soc.* (1943) 404
1010 Hey, D. H. and Williams, J. M. *J. chem. Soc.* (1950) 1678
1011 Abramovitch, R. A. and Saha, J. G. *Tetrahedron Lett.* (1963) 301; *J. chem. Soc.* (1964) 2175

1012 Hey, D. H., Stirling, J. M. and Williams, G. H. *J. chem. Soc.* (1955) 3963
1013 Dannley, R. L., Gregg, E. C., with Phelps, R. E. and Coleman, C. B. *J. Am. chem. Soc.* **76** (1954) 445; *ibid.* 2997
1014 Bunyan, P. J. and Hey, D. H. *J. chem. Soc.* (1960) 3787
1015 Hey, D. H., Shingleton, D. A. and Williams, G. H. *J. chem. Soc.* (1963) 5613
1016 Grashey, R. and Huisgen, R. *Chem. Ber.* **92** (1959) 2641
1017 Wieland, H., Ploetz, T. and Indest, H. *Justus Liebigs Annln Chem.* **532** (1937) 166
1018 Dirstine, P. H. and Bergstrom, F. W. *J. org. Chem.* **11** (1946) 55
1019 Williams, G. H. *Homolytic Aromatic Substitution*, London (Pergamon) 1960
1020 *U.S. Pat.* 2,502,174 (1950)
1021 Hardegger, E. and Nikles, E. *Helv. chim. Acta* **40** (1957) 2421
1022 Levy, M. and Szwarc, M. *J. Am. chem. Soc.* **77** (1955) 1949
1023 Norman, R. O. C. and Radda, G. K. in *Advances in Heterocyclic Chemistry*, Vol. 2, ed. Katritzky, A. R., New York (Academic Press) 1963
1024 Goldschmidt, S. and Minsinger, M. *Chem. Ber.* **87** (1954) 956
1025 Goldschmidt, S. and Beer, L. *Justus Liebigs Annln Chem.* **641** (1961) 40
1026 Cullinane, N. M., Chard, S. J. and Meatyard, R. *J. Soc. chem. Ind., Lond.* **67** (1948) 142
1027 Bachmann, W. E. and Hoffman, R. A. *Org. React.* **2** (1944)
1028 DeTar, D. F. and Long, R. A. J. *J. Am. chem. Soc.* **80** (1958) 4742
1029 Augood, D. R., Hey, D. H. and Williams, G. H. *J. chem. Soc.* (1952) 2094
1030 Hey, D. H. and Williams, G. H. *Discuss. Faraday Soc.* **14** (1953) 216
1031 Pausacker, K. H. *J. chem. Soc.* (1961) 18
1032 DeTar, D. F., with Ballentine, A. R. *J. Am. chem. Soc.* **78** (1956) 3916; with Sagmanli, S. V. **72** (1950) 965
1033 Hey, D. H. and Osbond, J. M. *J. chem. Soc.* (1949) 3164, 3172
1034 Abramovitch, R. A., with Hey, D. H. and Mulley, R. D. *J. chem. Soc.* (1954) 4263; *Can. J. Chem.* **38** (1960) 2273
1035 Abramovitch, R. A. and Tertzakian, G. *Tetrahedron Lett.* (1963) 1511
1036 Plieninger, H. and Schach von Wittenau, M. *Chem. Ber.* **91** (1958) 1905; Herz, W. and Murty, D. R. K. *J. org. Chem.* **26** (1961) 418
1037 Smith, P. A. S. and Boyer, J. H. *J. Am. chem. Soc.* **73** (1951) 2626
1038 Ladenburg, A. *Justus Liebigs Annln Chem.* **247** (1888) 50
1039 Lukeš, R. and Jizba, J. *Chemické Listy* **46** (1952) 622
1040 Rabe, P. and Kindler, K. *Ber. dtsch. chem. Ges.* **52** (1919) 1842; Koenigs, E. and Ottmann, W. **54** (1921) 1343; Clemo, G. R. and Ramage, G. R. *J. chem. Soc.* (1931) 437; with Ormston, J. *ibid.* 3185; with Raper, R. *ibid.* 3190; with Metcalfe, T. P. (1937) 1523
1041 Löffler, K. and Flügel, M. *Ber. dtsch. chem. Ges.* **42** (1909) 3420; Sandborn, L. T. and Marvel, C. S. *J. Am. chem. Soc.* **50** (1928) 563; Renshaw, R. R., Ziff, M., Brodie, B. B. and Kornblum, N. **61** (1939) 638
1042 Ahrens, F. B. *Z. Elektrochem.* **2** (1895) 577; Pincussohn, L. *Z. anorg. allg. Chem.* **14** (1897) 379; Tafel, J. *Z. phys. Chem.* **34** (1900) 220; Emmert, B. *Ber. dtsch. chem. Ges.* **46** (1913) 1716; Hackmann, J. T. and Wibaut, J. P. *Recl Trav. chim. Pays-Bas Belg.* **62** (1943) 229
1043 Koenigs, W. and Bernhart, K. *Ber. dtsch. chem. Ges.* **38** (1905) 3042; Tschitschibabin, A. E. *ibid.* 3834; Koenigs, W. **40** (1907) 3199
1044 Lukeš, R. and Ernest, I. *Colln. Czech. chem. Commun.* **15** (1950) 107
1045 Shaw, B. D. *J. chem. Soc.* **125** (1924) 1930; **127** (1925) 215; (1937) 300
1046 Birch, A. J. *J. chem. Soc.* (1947) 1270
1047 (*a*) Tschitschibabin, A. E. and Gertschuk, M. P. *Ber. dtsch. chem. Ges.* **63** (1930) 1153; (*b*) Kirsanov, A. V. and Ivastchenko, Y. N. *Bull. Soc. chim. Fr.* **3** (1936/5) 2279
1048 Benary, E. and Bitter, G. A. *Ber. dtsch. chem. Ges.* **61** (1928) 1057
1049 Renshaw, R. R. and Conn, R. C. *J. Am. chem. Soc.* **60** (1938) 745
1050 Grave, T. B. *J. Am. chem. Soc.* **46** (1924) 1460
1051 Borsche, W. and Bonacker, I. *Ber. dtsch. chem. Ges.* **54** (1921) 2678; Riegel, E. R. and Reinhard, M. C. *J. Am. chem. Soc.* **48** (1926) 1334; Toomey, R. F. and Riegel, E. R. **74** (1952) 1492; Mills, W. H., Parkin, J. D. and Ward, W. J. V. *J. chem. Soc.* (1927) 2613
1052 Wibaut, J. P. and Boer, H. *Recl Trav. chim. Pays-Bas Belg.* **68** (1949) 72
1053 (*a*) Šorm, F. *Colln. Czech. chem. Commun.* **13** (1948) 57; (*b*) with Šedivý, L. *ibid.* 289
1054 Ferles, M. and Prystaš, M. *Colln. Czech. chem. Commun.* **24** (1959) 3326
1055 Mumm, O. and Beth, W. *Ber. dtsch. chem. Ges.* **54** (1921) 1591
1056 *Brit. Pat.* 631,078 (1949); *U.S. Pat.* 2,520,037 (1950); Jones, R. G. and Kornfeld, E. C. *J. Am. chem. Soc.* **73** (1951) 107; Mićović, V. M. and Mihailović, M. L. *Recl Trav. chim. Pays-Bas Belg.* **71** (1952) 970; Davoll, H. and Kipping, F. B. *J. chem. Soc.* (1953) 1395
1057 Hochstein, F. A. *J. Am. chem. Soc.* **71** (1949) 305
1058 Bohlmann, F. *Chem. Ber.* **85** (1952) 390
1059 de Mayo, P. and Rigby, W. *Nature, Lond.* **166** (1950) 1075
1060 Lansbury, P. T. and Peterson, J. O. *J. Am. chem. Soc.* **85** (1963) 2236

1061 Hofmann, A. W. *Ber. dtsch. chem. Ges.* **14** (1881) 1497
1062 Emmert, B. *Ber. dtsch. chem. Ges.* **42** (1909) 1997; **52** (1919) 1351
1063 Mumm, O., Roder, O. and Ludwig, H. *Ber. dtsch. chem. Ges.* **57** (1924) 865
1064 Emmert, B., Jungck, G. and Häffner, H. *Ber. dtsch. chem. Ges.* **57** (1924) 1792
1065 Mumm, O. and Diederichsen, J. *Justus Liebigs Annln Chem.* **538** (1939) 195
1066 Ochiai, E. and Kawagaye, N. *Chem. Abstr.* **45** (1951) 5153
1067 Wallenfels, K. and Gelbrich, M. *Chem. Ber.* **92** (1959) 1406
1068 Weitz, E., König, T. and Wistinghausen, L. v. *Ber. dtsch. chem. Ges.* **57** (1924) 153
1069 Karrer, P., Schwarzenbach, G. and Utzinger, G. E. *Helv. chim. Acta* **20** (1937) 72
1070 Saunders, M. and Gold, E. H. *J. org. Chem.* **27** (1962) 1439
1071 Mumm, O. and Ludwig, H. *Ber. dtsch. chem. Ges.* **59** (1926) 1605
1072 Ziegler, K. and Fries, F. A. *Ber. dtsch. chem. Ges.* **59** (1926) 242
1073 Weitz, E. *Angew. Chem.* **66** (1954) 658
1074 Weitz, E. and Nelken, A. *Justus Liebigs Annln Chem.* **425** (1921) 187
1075 Müller, E. and Bruhn, K. A. *Chem. Ber.* **86** (1953) 1122
1076 Elofson, R. M. and Edsburg, R. L. *Can. J. Chem.* **35** (1957) 646
1077 Wallenfels, K. and Gelbrich, M. *Justus Liebigs Annln Chem.* **621** (1959) 198
1078 Schwarz, W. M., Kosower, E. M. and Shain, I. *J. Am. chem. Soc.* **83** (1961) 3164
1079 Kosower, E. M. and Poziomek, E. J. *J. Am. chem. Soc.* **85** (1963) 2035; **86** (1964) 5515
1080 Karrer, P. and Warburg, O. *Biochem. Z.* **285** (1936) 297
1081 Fisher, H. F., Conn, E. E., Vennesland, B. and Westheimer, F. H. *J. biol. Chem.* **201** (1953) 687
1082 Loewus, F. A., Westheimer, F. H. and Vennesland, B. *J. Am. chem. Soc.* **75** (1953) 5018
1083 Fisher, H. F., Ofner, P., Conn, E. E., Vennesland, B. and Westheimer, F. H. *Fedn Proc. Fedn Am. Socs exp. Biol.* **11** (1952) 211
1084 Pullman, M. E., San Pietro, A. and Colowick, S. P. *J. biol. Chem.* **206** (1954) 129
1085 Mauzerall, D. and Westheimer, F. H. *J. Am. chem. Soc.* **77** (1955) 2261; Loewus, F. A., Vennesland, B. and Harris, D. L. *ibid.* 3391
1086 Rafter, G. W. and Colowick, S. P. *J. biol. Chem.* **209** (1954) 773
1087 Hutton, R. F. and Westheimer, F. H. *Tetrahedron* **3** (1958) 73; Dubb, H. E., Saunders, M. and Wang, J. H. *J. Am. chem. Soc.* **80** (1958) 1767
1088 Sims, A. F. E. and Smith, P. W. G. *Proc. chem. Soc.* (1958) 282
1089 Karrer, P., with Schwarzenbach, G., Benz, F. and Solmssen, U. *Helv. chim. Acta* **19** (1936) 811; with Ringier, B. H., Büchi, J., Fritzsche, H. and Solmssen, U. **20** (1937) 55; with Kahnt, F. W., Epstein, R., Jaffé, W. and Ishii, T. **21** (1938) 223; with Blumer, F. **30** (1947) 1157; Kuhnis, H., Traber, W. and Karrer, P. **40** (1957) 751
1090 Anderson, A. G. and Berkelhammer, G. *J. Am. chem. Soc.* **80** (1958) 992
1091 Wallenfels, K. and Schüly, H. *Justus Liebigs Annln Chem.* **621** (1959) 106
1092 Wallenfels, K., Gelbrich, M. and Kubowitz, F. *Justus Liebigs Annln Chem.* **621** (1959) 137
1093 Wallenfels, K. and Schüly, H. *Justus Liebigs Annln Chem.* **621** (1959) 215
1094 Kuss, L. and Karrer, P. *Helv. chim. Acta* **40** (1957) 740
1095 Lyle, R., Perlowski, E. F., Troscianiec, H. J. and Lyle, G. G. *J. org. Chem.* **20** (1955) 1761
1096 Saito, S. and May, E. L. *J. org. Chem.* **27** (1962) 948
1097 Panouse, J. J. *C. r. hebd. Séanc. Acad. Sci., Paris* **233** (1951) 260, 1200
1098 Mathews, M. B. and Conn, E. E. *J. Am. chem. Soc.* **75** (1953) 5428
1099 Stein, G. and Stiassny, G. *Nature, Lond.* **176** (1955) 734
1100 Piass, Y. and Stein, G. *J. chem. Soc.* (1958) 2905
1101 Traber, W. and Karrer, P. *Helv. chim. Acta* **41** (1958) 2066
1102 Ferles, M. *Colln. Czech. chem. Commun.* **23** (1958) 479
1103 Lyle, R. E., Nelson, D. A. and Anderson, P. S. *Tetrahedron Lett.* (1962) 553; Anderson, P. S. and Lyle, R. E. (1964) 153
1104 Lukeš, R. *Colln. Czech. chem. Commun.* **12** (1947) 71; with Pliml, J. **15** (1950) 463; and Jizba, J. and Štěp, V. **19** (1954) 949
1105 Lukeš, R. and Jizba, J. *Colln. Czech. chem. Commun.* **19** (1954) 941
1106 Paquette, L. A. and Slomp, G. *J. Am. chem. Soc.* **85** (1963) 765
1107 Sugasawa, S., Sakurai, K. and Okayama, T. *J. pharm. Soc. Japan* **62** (1942) 77
1108 Stein, G. and Swallow, A. J. *J. chem. Soc.* (1958) 306
1109 Zelinsky, N. and Borissow, P. *Ber. dtsch. chem. Ges.* **57** (1924) 150; Borissow, P. **63** (1930) 2278
1110 Overhoff, J. and Wibaut, J. P. *Recl Trav. chim. Pays-Bas Belg.* **50** (1931) 957
1111 Adkins, H., Kuick, L. F., Farlow, M. and Wojcik, B. *J. Am. chem. Soc.* **56** (1934) 2425
1112 Ushakov, M. I., with Bronevskiĭ, A. I. *Chem. Abstr.* **31** (1937) 5799; with Yakovleva, E. V. *ibid.* 5799; with Promyslov, M. S. **42** (1948) 4583
1113 Jones, J. I. and Lindsey, A. S. *J. chem. Soc.* (1952) 3261
1114 Freifelder, M. and Stone, G. R. *J. org. Chem.* **26** (1961) 3805
1115 Hamilton, T. S. and Adams, R. *J. Am. chem. Soc.* **50** (1928) 2260
1116 Ayerst, G. G. and Schofield, K. *J. chem. Soc.* (1960) 3445

REFERENCES

[1117] Skomoroski, R. M. and Schriesheim, A. *J. phys. Chem., Ithaka* **65** (1961) 1340

[1118] Freifelder, M., Mattoon, R. W. and Ng, Y. H. *J. org. Chem.* **29** (1964) 3730

[1119] Adkins, H. and Cramer, H. I. *J. Am. chem. Soc.* **52** (1930) 4349

[1120] Lee, J. and Freudenberg, W. *J. org. Chem.* **9** (1944) 537

[1121] Hartmann, M. and Panizzon, L. *Chem. Abstr.* **44** (1950) 8379

[1122] Scholz, K. and Panizzon, L. *Helv. chim. Acta* **37** (1954) 1605

[1123] Rabe, P., Huntenburg, W., Schultze, A. and Volger, G. *Ber. dtsch. chem. Ges.* **64** (1931) 2487

[1124] Kleiman, M. and Weinhouse, S. *J. org. Chem.* **10** (1945) 562

[1125] King, J. A., Hofmann, V. and McMillan, F. H. *J. org. Chem.* **16** (1951) 1100

[1126] Truitt, P. with Hall, B. and Arnwine, B. *J. Am. chem. Soc.* **74** (1952) 4552; with Bryant, B., Goode, W. E. and Arnwine, B. *ibid.* 2179

[1127] Späth, E. and Galinovsky, F. *Ber. dtsch. chem. Ges.* **71** (1938) 721

[1128] Rubtsov, M. V. *J. gen. Chem. U.S.S.R.* **16** (1946) 461

[1129] Clemo, G. R., Fletcher, N., Fulton, G. R. and Raper, R. *J. chem. Soc.* (1950) 1140

[1130] McMillan, F. M. and King, J. A. *J. Am. chem. Soc.* **73** (1951) 3165

[1131] Rhodes, H. J. and Soine, T. O. *J. Am. pharm. Ass.* **45** (1956)746

[1132] Mosby, W. L. *Heterocyclic Systems with Bridgehead Nitrogen Atoms*, Pt. I, p. 305 *et seq.*; Pt. II, p. 1016 *et seq.*, New York (Interscience) 1961

[1133] Orthner, L. *Justus Liebigs Annln Chem.* **456** (1927) 225

[1134] Nienburg, H. *Ber. dtsch. chem. Ges.* **70** (1937) 635

[1135] Andersson, N. E. and Soine, T. O. *J. Am. pharm. Assoc.* **39** (1950) 463

[1136] McElvain, S. M. and Adams, R. *J. Am. chem. Soc.* **45** (1923) 2738

[1137] Feldkamp, R. F., Faust, J. A. and Cushman, A. J. *J. Am. chem. Soc.* **74** (1952) 3831

[1138] Clarke, R. L., Mooradian, A., Lucas, P. and Slauson, T. J. *J. Am. chem. Soc.* **71** (1949) 2821

[1139] Andersson, N. E. and Soine, T. O. *J. Am. pharm. Assoc.* **39** (1950) 460

[1140] Recklow, W. A. and Tarbell, D. S. *J. Am. chem. Soc.* **74** (1952) 4961

[1141] Doyle, F. P., Mehta, M. D., Sach, G. S., Ward, R. and Sherman, P. S. *J. chem. Soc.* (1964) 578

[1142] Adkins, H. and Pavlic, A. A. *J. Am. chem. Soc.* **69** (1947) 3039

[1143] Ruzicka, L. *Helv. chim. Acta* **4** (1921) 472

[1144] Covert, L. W., Connor, R. and Adkins, H. *J. Am. chem. Soc.* **54** (1932) 1651

[1145] Gautier, J. A. *C. r. hebd. Séanc. Acad. Sci., Paris* **205** (1937) 614; with Renault, J. **225** (1947) 880

[1146] Galinovsky, F. and Stern, E. *Ber. dtsch. chem. Ges.* **77** (1944) 132

[1147] *Germ. Pat.* 571,227 (1933)

[1148] Biel, J. H., Friedman, H. L., Leiser, H. A. and Sprengeler, E. P. *J. Am. chem. Soc.* **74** (1952) 1485

[1149] Ruzicka, L. and Fornasir, V. *Helv. chim. Acta* **3** (1920) 806

[1150] Campbell, K. N., and B. K., and Ackerman, J. F. *J. org. Chem.* **15** (1950) 337

[1151] Clemo, G. R., Morgan, W. McG. and Raper, R. *J. chem. Soc.* (1935) 1743

[1152] Ishii, T. *J. pharm. Soc. Japan* **71** (1951) 1097

[1153] Barltrop, J. A. and Taylor, D. A. H. *J. chem. Soc.* (1951) 108

[1154] Supniewski, J. V. and Serafinowna, M. *Chem. Abstr.* **33** (1939) 7301

[1155] Zincke, T., Heuser, G. and Möller, W. *Justus Liebigs Annln Chem.* **333** (1904) 296

[1156] Phillips, A. P. *J. Am. chem. Soc.* **72** (1950) 1850

[1157] Wibaut, J. P. and Kloppenberg, C. C. *Recl Trav. chim. Pays-Bas Belg.* **65** (1946) 100

[1158] Koelsch, C. F. and Carney, J. J. *J. Am. chem. Soc.* **72** (1950) 2285

[1159] Schenker, K. and Druey, J. *Helv. chim. Acta* **42** (1959) 1971

[1160] Dimroth, O. and Heene, R. *Ber. dtsch. chem. Ges.* **54** (1921) 2934

[1161] Wibaut, J. P. and Arens, J. F. *Recl Trav. chim. Pays-Bas Belg.* **60** (1941) 119

[1162] Frank, R. L., Pelletier, F. and Starks, F. W. *J. Am. chem. Soc.* **70** (1948) 1767

[1163] *Org. Synth.* **27** (1947) 38

[1164] Arens, J. F. and Wibaut, J. P. *Recl Trav. chim. Pays-Bas Belg.* **61** (1942) 59

[1165] Wibaut, J. P. and van der Vennen, D. *Recl Trav. chim. Pays-Bas Belg.* **66** (1947) 236

[1166] van Dorp, D. A. and Arens, J. F. *Recl Trav. chim. Pays-Bas Belg.* **66** (1947) 189

[1167] Kutney, J. P. and Tabata, T. *Can. J. Chem.* **41** (1963) 695

[1168] Taylor, E. C. and Kan, R. O. *J. Am. chem. Soc.* **83** (1961) 4484

[1169] Hurd, C. D. and Simon, J. I. *J. Am. chem. Soc.* **84** (1962) 4519

[1170] Currell, D. L., Wilheim, G. and Nagy S. *J. Am. chem. Soc.* **85** (1963) 127

[1171] Delépine, M. *Bull. Soc. chim. Fr.* **41** (1927) 390

[1172] Tschitschibabin, A. E. *Ber. dtsch. chem. Ges.* **37** (1904) 1373; Tronov, B. V. and Nikonova, L. S. *Zh. russk. fiz.-khim. Obshch.* **61** (1929) 541

[1173] Freytag, H. *Ber. dtsch. chem. Ges.* **69** (1936) 32

[1174] Shive, W., Ballweber, E. G. and Ackermann, W. N. *J. Am. chem. Soc.* **68** (1946) 2144; Wibaut, J. P. *Chimia* **11** (1957) 321

307

[1175] Sixma, F. L. J. *Recl Trav. chim. Pays-Bas Belg* **71** (1952) 1124
[1176] Sabatier, P. and Mailhe, A. *C. r. hebd. Séanc. Acad. Sci., Paris* **144** (1907) 874; Jones, J. I. *J. chem. Soc.* (1950) 1392
[1177] Hofmann, A. W. *Ber. dtsch. chem. Ges.* **16** (1883) 590
[1178] Mumm, O. and Broderson, K. *Ber. dtsch. chem. Ges.* **56** (1923) 2295
[1179] Zincke, T. and Würker, W. *Justus Liebigs Annln Chem.* **341** (1905) 365
[1180] Baumgarten, P. *Ber. dtsch. chem. Ges.* **59** (1926) 1166
[1181] König, W. and Becker, G. A. *J. prakt. Chem.* **85** (1912/2) 353
[1182] Lettré, H., Haede, W. and Ruhbaum, E. *Justus Liebigs Annln Chem.* **579** (1953) 123
[1183] Grigor'eva, N. E., with Yavlinskiĭ, M. D. *Chem. Abstr.* **48** (1954) 11411; with Gintse, I. K. and Rozenberg, M. I. **49** (1955) 4640
[1184] Knunyants, I. L. and Kefeli, T. Y. *Chem. Abstr.* **40** (1946) 6079
[1185] Strell, M., Braunbruck, W. B., Fühler, W. F. and Huber, O. *Justus Liebigs Annln Chem.* **587** (1954) 177
[1186] Hafner, K. *Justus Liebigs Annln Chem.* **606** (1957) 79
[1187] Kröhnke, F. and Vogt, I. *Justus Liebigs Annln Chem.* **589** (1954) 26, 45
[1188] *Germ. Pat.* 592,202 (1934)
[1189] Hünig, S. and Requardt, K. *Angew. Chem.* **68** (1956) 152
[1190] Birch, A. J. *J. chem. Soc.* (1944) 314
[1191] Allan, D. and Loudon, J. D. *J. chem. Soc.* (1949) 821
[1192] Fisher, N. I. and Hamer, F. M. *J. chem. Soc.* (1933) 189
[1193] Baumgarten, P. and Dammann, E. *Ber. dtsch. chem. Ges.* **66** (1933) 1633
[1194] Baumgarten, P. *Ber. dtsch. chem. Ges.* **69** (1936) 1938
[1195] Kröhnke, F., Meyer-Delius, M. and Vogt, I. *Justus Liebigs Annln Chem.* **597** (1955) 87
[1196] Pfeiffer, P. and Enders, E. *Chem. Ber.* **84** (1951) 313
[1197] Ullmann, F. and Nádai, G. *Ber. dtsch. chem. Ges.* **41** (1908) 1870
[1198] Borsche, W. and Feske, E. *Ber. dtsch. chem. Ges.* **60** (1927) 157
[1199] Borrows, E. T., Clayton, J. C., Hems, B. A. and Long, A. G. *J. chem. Soc.* (1949) S190
[1200] Schwarzenbach, G., with Weber, R. *Helv. chim. Acta* **25** (1942) 1628; **26** (1943) 418
[1201] Zincke, T. and Mühlhausen, G. *Ber. dtsch. chem. Ges.* **38** (1905) 3824
[1202] König, W., Ebert, G. and Centner, K. *Ber. dtsch. chem. Ges.* **56** (1923) 751
[1203] Hafner, K. and Asmus, K.-D. *Justus Liebigs Annln Chem.* **671** (1964) 31
[1204] Baumgarten, P. *Ber. dtsch. chem. Ges.* **60** (1927) 1174
[1205] Klages, F. and Träger, H. *Chem. Ber.* **86** (1953) 1327
[1206] Baumgarten, P. *Ber. dtsch. chem. Ges.* **57** (1924) 1622
[1207] Oda, R. and Mita, S. *Bull. chem. Soc. Japan* **36** (1963) 103
[1208] Reitenstein, F. and Breuning, W. *J. prakt. Chem.* **83** (1911/2) 97
[1209] König, W. and Bayer, R. *J. prakt. Chem.* **83** (1911/2) 325
[1210] Treibs, A. *Justus Liebigs Annln Chem.* **497** (1932) 297
[1211] Ploquin, J. *Bull. Soc. chim. Fr.* (1947) 901
[1212] Schenkel, J. *Ber. dtsch. chem. Ges.* **43** (1910) 2598
[1213] Brown, D. J. *Nature, Lond.* **189** (1961) 828; with Harper, J. S. *J. chem. Soc.* (1963) 1276; Perrin, D. D. *ibid.* 1284
[1214] Gol'dfarb, Y. L. and Danyushevskiĭ, Y. L. *Chem. Abstr.* **48** (1954) 679
[1215] Granelli, C. *Chem. Abstr.* **33** (1939) 4245
[1216] Sasse, W. H. F. *J. chem. Soc.* (1959) 3046
[1217] Süs, O., with Glos, M., Möller, K. and Eberhardt, H.-D. *Justus Liebigs Annln Chem.* **583** (1953) 150; with Möller, K. **593** (1955) 91; *Germ. Pat.* 949,467 (1956)
[1218] Freifelder, M. *Adv. Catalysis* **14** (1963)
[1219] Grochowski, J. W. and Okon, K. *Roczn. Chem.* **37** (1963) 1429
[1220] Taylor, R. *J. chem. Soc.* (1962) 4881
[1221] Pritchard, H. O. and Sumner, F. H. *Proc. phys. Soc. Lond.* **A235** (1956) 136
[1222] Tsoucaris, G. *J. Chim. phys.* **58** (1961) 613
[1223] Cartier, J. P. and Sandorfy, C. *Can. J. Chem.* **41** (1963) 2759
[1224] Brown, R. D. and Heffernan, M. L. *Aust. J. Chem.* **9** (1956) 83
[1225] Yvan, P. *C. r. hebd. Séanc. Acad. Sci., Paris* **229** (1949) 622
[1226] Dewar, M. J. S. and Maitlis, P. M. *J. chem. Soc.* (1957) 2521
[1227] Ridd, J. H. in *Physical Methods in Heterocyclic Chemistry*, ed. Katritzky, A. R., Vol. I, New York (Academic Press) 1963
[1228] (a) Ashida, T., Hirokawa, S. and Okaya, Y. *Acta crystallogr.* **18** (1965) 122; (b) Laurent, A. *ibid.* 799
[1229] Bullock, D. J. W., Cumper, C. W. N. and Vogel, A. I. *J. chem. Soc.* (1965) (a) 5311; (b) 5316
[1230] Krackov, M. H., Lee, C. M. and Mautner, H. G. *J. Am. chem. Soc.* **87** (1965) 892
[1231] Kier, L. B. *Tetrahedron Lett.* (1965) 3273
[1232] Pujol, L. and Julg, A. *Tetrahedron* **21** (1965) 717
[1233] Barlin, G. B. *J. chem. Soc.* (1964) 2150

REFERENCES

[1234] (a) Golding, S., Katritzky, A. R. and Kucharska, H. Z. *J. chem. Soc.* (1965) 3090; Katritzky, A. R., with Kucharska, H. Z. and Rowe, J. D. *ibid.* 3093; (b) with Reavill, R. E. *ibid.* 3825

[1235] (a) Cunliffe-Jones, D. B. *Spectrochim. Acta* **21** (1965) 747; (b) Medhi, K. C. and Mukherjee, D. K. *ibid.* 895

[1236] Jones, R. A. and Rao, R. P. *Aust. J. Chem.* **18** (1965) 583

[1237] Cook, D. *Can. J. Chem.* **43** (1965) 741

[1238] Johnson, C. D., Katritzky, A. R., Ridgewell, B. J., Shakir, N. and White, A. M. *Tetrahedron* **31** (1965) 1055

[1239] (a) Jaffé, H. H. and Jones, H. L. *J. org. Chem.* **30** (1965) 964; (b) Olah, G. A. and J. A. and Overchuk, N. A. *ibid.* 3373

[1240] Larsen, D. W. and Allred, A. L. *J. Am. chem. Soc.* **87** (1965) 1219

[1241] van der Does, L. and den Hertog, H. J. *Recl Trav. chim. Pays-Bas Belg.* **84** (1965) 951

[1242] (a) Geller, A. and Samosvat, L. S. *J. gen. Chem. U.S.S.R.* **34** (1964) 614; (b) Kost, A. N., Sheinkman, A. K. and Kazarinova, N. F. *ibid.* 2059

[1243] Katritzky, A. R., private communication

[1244] Gleghorn, J., Moodie, R. B., Schofield, K. and Williamson, M. J. *J. chem. Soc.* (*B*)(1966) 870

[1245] Preussmann, R., Hengy, H. and Druckney, H. *Justus Liebigs Annln Chem.* **684** (1965) 57

[1246] (a) Hudson, R. F. and Withey, R. J. *J. chem. Soc.* (1964) 3513; (b) Fischer, A., Galloway, W. J. and Vaughan, J. *ibid.* 3596; (c) Taylor, G. A. (1965) 3332

[1247] Acheson, R. M., with Plunkett, A. O. *J. chem. Soc.* (1964) 2676; with Feinberg, R. S. and Gagan, J. M. F. *ibid.* 948; with Goodall, D. M. and Robinson, D. A. *ibid.* 2633; Winterfeldt, E. *Chem. Ber.* **98** (1965) 3537

[1248] Linn, W. J., Webster, O. W. and Benson, R. E. *J. Am. chem. Soc.* **87** (1965) 3651

[1249] Paquette, L. A., *J. org. Chem.* **30** (1965) 2107

[1250] (a) Johnson, S. L. and Rumon, K. A. *J. phys. Chem., Ithaka* **68** (1964) 3149; *J. Am. chem. Soc.* **87** (1965) 4782; (b) Letsinger, R. L., Ramsay, O. B. and McCam, J. H. *ibid.* 2945; (c) Jencks, W. P. and Gilchrist, M. *ibid.* 3199 *et seq.*

[1251] Chambers, R. D., Hutchinson, J. and Musgrave, W. K. R. *J. chem. Soc.* (a) (1964) 3736, 5634; (b) (1965) 5040; (c) Banks, R. E., with Burgess, J. E., Cheng, W. M. and Haszeldine, R. N. *ibid.* 575; (d) and Latham, J. V. and Young, I. M. *ibid.* 594

[1252] Roedig, A. and Grohe, K. *Chem. Ber.* **98** (1965) 923

[1253] Abramovitch, R. A., Helmer, F. and Saha, J. G. *Can. J. Chem.* **43** (1965) 725

[1254] Bellas, M. and Suschitzky, H. *J. chem. Soc.* (1965) 2096

[1255] van der Plas, H. C., Hijwegen, T. and den Hertog, H. J. *Recl Trav. chim. Pays-Bas Belg.* **84** (1965) 53

[1256] Kauffmann, T. and Wirthwein, R. *Angew. Chem. (int. Edn)* **3** (1964) 806

[1257] Kato, T. and Yamanaka, H. *J. org. Chem.* **30** (1965) 910

[1258] (a) Lyle, R. E. and Gauthier, G. J. *Tetrahedron Lett.* (1965) 4615; (b) Dou, H. J. M. and Lynch, B. M. *ibid.* 897; (c) Anderson, P. S., Krueger, W. E. and Lyle, R. E. *ibid.* 4011

[1259] Yamazaki, M., Chono, Y., Noda, K. and Hamana, M. *J. pharm. Soc. Japan* **85** (1965) 62

[1260] (a) Oae, S. and Kozuka, S. *Tetrahedron* **21** (1965) 1971; (b) Abramovitch, R. A. and Saha, J. G. *ibid.* 3297

[1261] Ford, P. W. and Swan, J. M. *Aust. J. Chem.* **18** (1965) 867; Traynelis, V. J. and Gallagher, A. I. *J. Am. chem. Soc.* **87** (1965) 5710

[1262] Abramovitch, R. A. and Tertzakian, G. *Can. J. Chem.* **43** (1965) 940

[1263] (a) Büchi, G., Coffen, D. L. *et al.*, *J. Am. chem. Soc.* **87** (1965) 2073; (b) Wenkert, E., Dave, K. G. and Haglid, F. *ibid.* 5461; (c) Cook, N. C. and Lyons, J. E. *ibid.* 3283

[1264] Burnett, J. H. and Underwood, A. L. *J. org. Chem.* **30** (1965) 1154

[1265] Vompe, A. F., Levkoev, I. I. *et al.*, *J. gen. Chem. U.S.S.R.* **34** (1964) **34** 1772; Podkletnov, N. E. *ibid.* 3403

[1266] König, W., Coenen, M., Lorenz, W., Bahr, F. and Bassl, A. *J. prakt. Chem.* **30** (1965) 96

[1267] Eisenthal, R. and Katritzky, A. R. *Tetrahedron* **21** (1965) 2205

[1268] Brügel, W., *Z. Elektrochem.* **2** (1962) 159

[1269] Rao, B. D. N. and Venkateswarlu, P. *Proc. Indian Acad. Sci.* **54** (1961) 305

[1270] Kowalewski, V. J. and de Kowalewski, D. G. *J. chem. Phys.* (a) **37** (1962) 2603; (b) **36** (1962) 266

[1271] Diehl, P., Jones, R. G. and Bernstein, H. J. *Can. J. Chem.* **43** (1965) 81

[1272] (a) Randall, E. W. and Baldeschwieler, J. D. *Proc. chem. Soc.* (1961) 303; (b) Rao, B. D. N. and Baldeschwieler, J. D. *J. chem. Phys.* **37** (1962) 2473

[1273] Lee, J. and Orrell, K. G. *J. chem. Soc.* (1965) 582; Chambers, R. D., Hutchinson, J. and Musgrave, W. K. R. (1964) 5634

[1274] Schaefer, T. and Schneider, W. G. *J. chem. Phys.* **32** (1960) 1218, 1224; Nakagawa, N. and Fujiwara, S. *Bull. chem. Soc. Japan* **34** (1961) 143; Hatton, J. V. and Richards, R. E. *Molec. Phys.* **5** (1962) 153

[1275] Murrell, J. N. and Gil, V. M. S. *Trans. Faraday Soc.* (a) **61** (1965) 402; (b) **60** (1964) 248

[1276] (a) Hall, G. G., Hardisson, A. and Jackman, L. M. *Discuss. Faraday Soc.* No. **34** (1962) 15; (b) Dailey, B. P., Gawer, A. and Neikam, W. C. *ibid.* 18

309

[1277] Wu, T. K. and Dailey, B. P. *J. chem. Phys.* **41** (1964) 3307
[1278] Paudler, W. W. and Blewitt, H. L. *Tetrahedron* **21** (1965) 353
[1279] Kotowycz, G., Schaefer, T. and Bock, E. *Can. J. Chem.* **42** (1964) 2541
[1280] Stothers, J. B. *Qu. Rev. chem. Soc.* **19** (1965) 144

6

PYRIDINE (CONTINUED)

THE PROPERTIES OF FUNCTIONAL GROUPS

(a) *Acyl, Carboxyl and Alkoxycarbonyl Groups*

The simpler properties of a number of pyridine aldehydes are recorded in *Table 6.1*. Spectra and ionization constants have been mentioned before (pp. 129, 142, 146; see also below).

The compounds behave generally like typical aromatic aldehydes, forming oximes, semicarbazones and arylhydrazones. They usually react normally with compounds containing reactive methylene groups and are easily oxidized to pyridine-carboxylic acids.

The bisulphite addition products deserve mention, for the free hydroxy-sulphonic acids from pyridine-2- and -4-aldehydes are exceptionally stable, possibly because their structures are zwitterionic[1-3](1). The derivative of pyridine-3-aldehyde is known only as its sodium salt[4].

Hydration of the aldehydes occurs readily, though pyridine-4-aldehyde is the only one which readily gives a crystalline hydrate[5]. The complex equilibria set up by a number of pyridine aldehydes in aqueous solution have been evaluated by means of ultra-violet spectroscopy[6, 7]. For pyridine-2-aldehyde the equilibria have been represented as

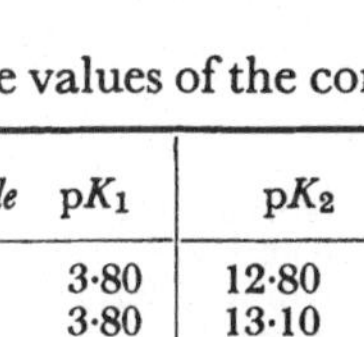

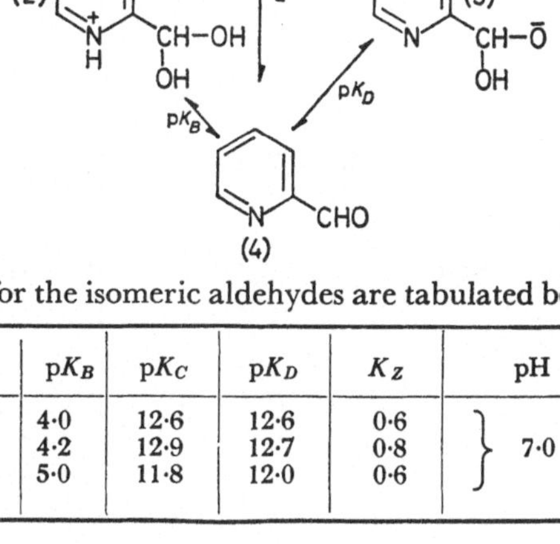

The values of the constants for the isomeric aldehydes are tabulated below.

Aldehyde	pK₁	pK₂	pK_A	pK_B	pK_C	pK_D	K_Z	pH
2	3·80	12·80	4·2	4·0	12·6	12·6	0·6	
3	3·80	13·10	4·0	4·2	12·9	12·7	0·8	7·0
4	4·77	12·20	5·2	5·0	11·8	12·0	0·6	

Table 6.1. Some Pyridine and Pyridine 1-Oxide Aldehydes* and Ketones†

Pyridine	B.p., °C	M.p., °C	Derivatives (m.p., °C)
2-CHO	181/760 mm		'bisulphite derivative' (210)[3]
	70–1/16 mm[1]		semicarbazone (196)
			oxime (114)[8]
			dinitrophenylhydrazone (239–240)[9]
3-CHO	203/760 mm		semicarbazone (214)
	84–6/12 mm[5]		oxime (150–1)[8]
			dinitrophenylhydrazone (259)[10]
4-CHO	195/760 mm		'bisulphite derivative' (243)[3]
	82–3/16 mm[11]		semicarbazone (224–5)[5]
			oxime (syn: 132–3; anti: 165–7)[12]
			hydrate (78)[13a]
2-CHO-3-Me	83–4/12–13 mm[8]	8	semicarbazone (197)
	52–4/0·2 mm		oxime (152–4)[8]
2-CHO-4-Me	41/0·5 mm		
2-CHO-5-Me	70–2/0·7 mm	41·5	semicarbazone (209)
			oxime (158·5)
2-CHO-6-Me	77–8/12 mm[13c]	33	hydrochloride (147)
			oxime (170–1)[8]
3-CHO-2-Me	94/12 mm[14]		semicarbazone (209)
3-CHO-4-Me	62–4/3 mm		oxime (177·5–180·5)[15]
			dinitrophenylhydrazone (255·5–6)[15]
3-CHO-6-Me	92–3/15 mm[5]		semicarbazone (234)[5]
4-CHO-2-Me	53–4/0·5 mm		hydrate (83)
2-CHO-4,6-Me₂	75/3 mm	12·5	phenylhydrazone (178·5)
4-CHO-2,6-Me₂	59–60/0·5 mm		phenylhydrazone (186)
2-CHO-6-CO₂H		165	oxime (220)
2-CHO-3-HO	72–4/12–14 mm	83[13b]	oxime (173–5)[8]
			O-methyl ether (oxime, 198–9)[8]
3-CHO-2-NH₂		99	semicarbazone (216)
			oxime (163·5)
3-CHO-6-NH₂		161	semicarbazone (230)
			oxime (217–8)
2,3-(CHO)₂*			bis-dinitrophenylhydrazone (265)[17]
3,4-(CHO)₂*			bis-dinitrophenylhydrazone (279)[17]
2,4-(CHO)₂		73·5, 72[17]	hydrate (88·5); bis-dinitrophenylhydrazone (331)[17]
2,5-(CHO)₂			bis-dinitrophenylhydrazone (339)[17]
2,6-(CHO)₂	152–4/103 mm[13c]	124	hydrochloride (90)
			oxime (211·5)
			phenylhydrazone (199·5)[13c]
2,4-(CHO)₂-6-Me		115	
2,6-(CHO)₂-4-Me		158·5	
Pyridine 1-Oxide 2-CHO[19]		92‡	hydrate (82)
			oxime (216)
3-CHO[19]		138	hydrate (100)
			oxime (216)
4-CHO[19]		152	hydrate (138)
			oxime (217)

Pyridine	B.p., °C	M.p., °C	Derivatives (m.p., °C)
2-CHO-4-Me[18]		127–9	semicarbazone (212–4)
2-CHO-6-Me[19]		84	hydrate (69)
			oxime (201)
Pyridine			
2-COMe	187–190[20]		oxime (120)[21]
	78/12 mm[21]		phenylhydrazone
			(154·5)[22]
3-COMe**	217–8[20]		oxime (130·5)[23]
	106/12 mm[23]		phenylhydrazone
			(201·5–202[dec.])
4-COMe	211–2[20]		oxime (157·5–158)
	90/8 mm[24]		phenylhydrazone
			(146·5–147·5)[24]
			semicarbazone (218·9)[26]
2-COEt	94–8/17 mm[25]		phenylhydrazone
			(141–2)[22]
	73/5 mm[22]		semicarbazone (163–4)[25]
3-COEt	115–7/18 mm[25]		oxime (117–8)[25]
	96–9/5 mm[27]		phenylhydrazone (145)[28]
4-COEt	133–6/23 mm[25]		picrate (100–3)[25]
2-COPh	315–9/750 mm[29a]		oxime [*syn*-pyridyl,
			165–7][29b]
	170–5/10 mm[30]		[*anti*-pyridyl,
			150·5–152·5]
			p-nitrophenylhydrazone
			(199–199·5)[29b]
3-COPh	138–140/1 mm[30]		oxime (161)[29c]
	180/12 mm[31]		picrate (164–6)[30]
4-COPh	170–2/10 mm[30]	72[29a]	phenylhydrazone
			(180–2)[32]
			semicarbazone (208)[33]
2-COPy-2		53·5–54[22]	dinitrophenylhydrazone
			[250–1(dec.)][22]
			semicarbazone
			(219–220)[22]
2-COPy-3		72[22]	dinitrophenylhydrazone
			[233–233·5(dec.)][22]
			semicarbazone (156·5–
			157·5)[22]
2-COPy-4		122–122·5[34a]	oxime (219)[22]
			semicarbazone (241·5–
			242·5)[34b]
3-COPy-3		117·6–118·8[34b]	phenylhydrazone
			(170–1)[34b]
			semicarbazone (196–7)[22]
3-COPy-4		124[34b]	oxime (176·5–177·5)[22]
			semicarbazone
			[227·5(dec.)][22]
4-COPy-4		136·2–137·5[34a]	phenylhydrazone (250·3–
			250·8)[34a]
2-COCH₂Ph	138–142/2 mm[35]		oxime (157)[35]
			semicarbazone (151–2)[36a]
3-COCH₂Ph[37a]		58–9	oxime (124·5–125·5)
2-COCH₂Py-2[38a]	149–150/1·2 mm	86–7	dipicrate (154–5)
2-COCH₂Py-3[39]	—	—	
3-COCH₂Py-2	158·5–160/1·3 mm[38a]	71·5–73[37a]	dipicrate (187·3–187·8)[38a]
			oxime (135–136·5)[37a]
3-COCH₂Py-3[37b]	169–173/3 mm	79·8–80·6	dipicrate (199·5–200)
4-COCH₂Py-2[38a]	145–7/2·5 mm	114·7–115·2	monopicrate (182–3)
			dipicrate (200·5–201)
4-COCH₂Py-3[37b]	169–170/3 mm	65·8–66·8	
4-COCH₂Py-4[36b]		116–116·8	

313

Table 6.1 (continued)

Pyridine	B.p., °C	M.p., °C	Derivatives (m.p., °C)
2-CH$_2$COMe	102–102·5/10·3 mm[38b] 74–7/1·5 mm[38b]		picrate (141–2)[40]
3-CH$_2$COMe[37a]	119–123/1 mm		oxime (117·5–119) semicarbazone (184·5–185)
4-CH$_2$COMe	75·5/0·1 mm[41]		picrate (167·5–168·1)[36b]
2-CH$_2$COPh	145–153/2 mm[38b]	54[35]	oxime (120)[35] phenylhydrazone (84–5)[38b]
3-CH$_2$COPh[37b]	170–5/3 mm	48·6–49·5	oxime (154·2–155·2) picrate (168·8–169·6)
4-CH$_2$COPh[42a]		115	oxime (157)
2,6-(CH$_2$ COPh)$_2$[42b–3]		92	oxime (188)
2-CO.CH (OH)Py-2		156[44a]	osazone (168)[44c] phenylhydrazone (138)[44c]
2-CO.CO.Ph		72–3[39]	picrate (87–8)[45]
2-CO.CO.Py-2[44a]		155	oxime (215)
2-CO.CO.Py-3		96–8[39]	
3-CO.CO.Py-3		79–80[39]	

* Unless otherwise indicated data are from [13d] and [16a].
† See[16].
‡ M. p. 74–6° (oxime, m.p. 220–1°) given[18].
** The m.p. reported for the oxime and phenylhydrazone vary considerably[16].

It has been suggested that, as the pH is increased through zero to 5·8, the proportion of (2) rapidly decreases and that of (3) and (4) increases. Between pH 5·9 and 10·8, only (3) and (4) are present. A simpler interpretation[46] is that the proton ionizes from nitrogen in (2) at pH 4–5, giving an equilibrium mixture of the unprotonated hydrate and (4), and above pH 12–13 (5) is formed. Polarographic studies have also been used to evaluate the hydration equilibria[47].

Pyridine-2-aldehyde reacts with potassium cyanide at pH 3·5 and −10° to give the cyanhydrin, but at ordinary temperatures 2-pyridoin cyanhydrin is formed. The behaviour of other pyridine aldehydes depends upon the substituents present[48].

The benzoin reaction of pyridine-2-aldehyde[1] has been frequently examined. Merely by heating, or with potassium cyanide-acetic acid, the aldehyde gives 2-pyridoin[49a]. Boron halides convert the aldehyde into pyridoin and pyridil[49b], and it gives pyridoin quantitatively with acetic acid without the need for cyanide ions[50]. Under the last conditions, pyridine-3- and -4-aldehyde do not react.

Pyridine-3-aldehyde with aqueous potassium cyanide gives a yellow solution, but the pyridoin cannot be isolated. With acetic anhydride it yields the pyridoin enediol diacetate[44b]. In the benzoin reaction, pyridine-4-aldehyde gave isonicotinic acid and 1,2-di-(4-pyridyl)ethylene glycol[13a, 44b], but 4-pyridoin has also been obtained[39]. Pyridine-2-aldehyde 1-oxide gives the corresponding pyridoin[39].

Pyridine aldehydes undergo the Cannizzaro reaction normally[1, 13c, 39, 51], and also the Wittig reaction[52].

With diazoalkanes, the aldehydes give good yields of ketones[25] or, sometimes, ethylene oxides[53].

Pyridine-2-aldehyde oximes have attracted attention because of their ability to form chelate complexes with metals[54-5.] Complexing with ferrous ions increases the acid strength of the oxime group[55] (see *Table 5.10*). Generally, the aldehyde oximes give with methyl iodide the quaternary salts[8], but from 6-methylpyridine-2-aldehyde oxime, the oxime N-methyl ether results[56].

In most respects the pyridyl ketones (*Table 6.1*) behave as normal aromatic ketones, and their reactions call for no special comment.

The catalytic reduction of ketones has been mentioned (p. 263). The Wolff–Kishner reduction is frequently used, and kinetic studies have been made with hydrazones of the benzoylpyridines (and pyridine aldehydes). The 2-pyridyl compounds react faster than their isomers, and the ring nitrogen atom probably plays an important role in the transition state[57].

Like the pyridine aldoximes, the ketoximes give metal chelate complexes[54]. The *anti*-pyridyl oxime of 2-benzoylpyridine gives such complexes whilst the *syn*-pyridyl oxime does not. The stereochemical assignments are based on the Beckmann rearrangement of the oximes[29b], and the behaviour of the two oximes with metal ions is that to be expected from the structures of chelates in this series[54]. The Beckmann rearrangement has been used several times in the pyridine series[29b, 58-9]. The oximes prepared from 2-phenacyl- and 2,6-diphenacyl-pyridine give pyridine-2-acetic acid anilide and pyridine-2,6-diacetic acid dianilide, respectively, and must be the *syn*-pyridyl compounds[60-1].

The *p*-toluenesulphonate esters of 2-, 3- and 4-acetylpyridine oximes undergo the Neber reaction[21, 62].

2-Picolyl alkyl and aryl ketones have attracted attention because of their chelating[38, 40, 63a] and bactericidal properties[40]. In the solid state they exist in the enolic forms (6) and in solution as equilibrium mixtures[63a]. Some C-aroyl-2-phenacylpyridines exist in two hydrogen-bond-stabilized enolic forms[63b], (7) and (8). Both C- and O-alkylation and -acylation of these

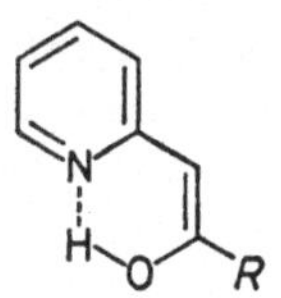

(6) 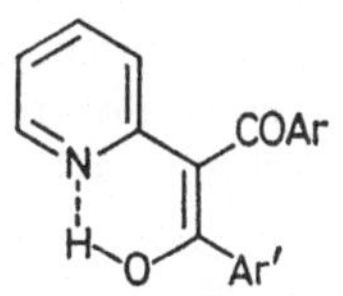   (7) 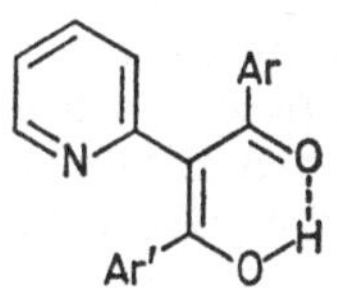   (8)

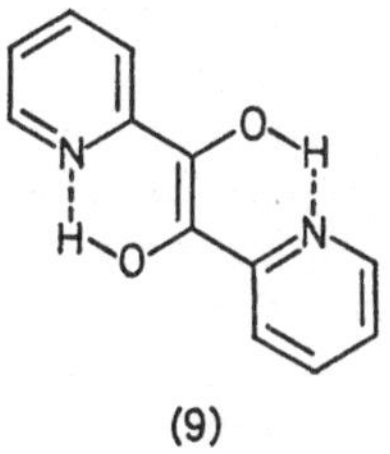

(9)

ketones have been observed[64], and picolyl aryl ketones are oxidized by selenium dioxide to diketones[39].

Of the pyridoins, 2,2'-pyridoin has been most studied. It is oxidized very readily, by Tollens' reagent, by nitric acid and by the action of oxygen on the methanolic solution to 2,2'-pyridil[44a]. It gives the indophenol reaction, and a mono- or a di-acetate can be prepared. Benzoylation yields the dibenzoate[40c]. It gives a phenylhydrazone and the phenylosazone, though less readily than a normal acyloin[44c]. With p-chlorophenylisocyanate a bis-urethane is formed, and diazonium salts split the molecule[65]. These characteristics and other reactions[66] have been adduced strongly to support the suggestion[44c] that 2,2'-pyridoin exists in the solid state and in neutral or weakly acidic solution as the hydrogen-bond-stabilized enediol (9). Although convincing enough for the view that 2,2'-pyridoin can react as the enediol, the evidence is far from complete on the structural side. The spectroscopic evidence[59, 65, 67] is more relevant (the infra-red spectrum of the solid and of the solution in carbon tetrachloride shows hydroxyl but not carbonyl bands), but a thorough study of the tautomerism is lacking. In accordance with the structure (9) there is evidence that enediol stability follows the sequence[49a]

$$\alpha\text{-}C_5H_4N \cdot C(OH):C(OH) \cdot \alpha\text{-}C_5H_4N >$$

$$\alpha\text{-}C_5H_4N \cdot C(OH):C(OH) \cdot Ph > Ph \cdot C(OH):C(OH)Ph$$

2,2'-Pyridoin forms chelate complexes with metal ions[68a].

The pyridils are not especially remarkable. Some examples have been reported to undergo the benzilic acid rearrangement[39, 68b], but others with ethanolic potash give an aldehyde (which subsequently undergoes the Cannizzaro reaction) and a carboxylic acid[39]. Heated with lead oxide, 2,2'-pyridil gives di-(2-pyridyl) ketone[69]. An x-ray crystallographic study of 2,2'-pyridil has been made[70].

Alcoholysis of aryl nicotinyl and alkyl nicotinyl β-diketones causes fission on both sides of the methylene group[71].

Concerning the pyridine-carboxylic acids (*Table 6.2*), there has already been mentioned their tautomerism (p. 154), betaine and ester formation with alkylating reagents (p. 182), and the Hammick reaction (p. 163).

The pyridine-2-carboxylic acids are distinguished by their ability to form chelate complexes with metal ions[72]. The property has been known since Skraup[73] tested all 19 of the pyridine-carboxylic acids and observed that those with 2-carboxyl groups gave characteristic colours with ferrous ions. The colour is due to an absorption band of low intensity at $\sim$4,000 Å. 2,6-Dicarboxylic acids absorb at longer wavelengths[74a].

Metal complexes of picolinic acid 1-oxide have also been described[74b].

The ready decarboxylation of pyridine-carboxylic acids was also early appreciated. Historically, the reaction played an important role in the orientation of quinoline, isoquinoline and the benzoquinolines[75]. Decarboxylation occurs more readily than with benzene-carboxylic acids, and in the sequence 2 $\gg$ 4 > 3 (see p. 367). The decarboxylation temperatures of solid pyridine-dicarboxylic acids depend roughly on their strengths as acids— the stronger the acid the lower the temperature[76]. At 185°–190°, pyridine-2,3,4-tricarboxylic acid gives pyridine-3,4-dicarboxylic acid, which above

Table 6.2. The Pyridine-carboxylic Acids*

Substituent	M.p., °C	Derivatives (m.p., °C)
2-CO$_2$H (picolinic acid)	136[78a]	Me ester (14) (b.p. 232°) Et ester (0–2) (b.p. 240–1°) amide (107) chloride (45–7) anhydride (124) anilide (76–7) hydrazide (101–2) nitrile (26) (b.p. 222·5–223·5°; 118–120°/25 mm)
3-CO$_2$H (nicotinic acid)	235·5–236·5	Me ester (38) (b.p. 204°) Et ester (—) (b.p. 223–4°) amide (128·5–129·5) chloride hydrochloride (155·5– 156·5) anhydride (124) hydrazide (159–161) nitrile (50–1) (b.p. 204–8°)
4-CO$_2$H (isonicotinic acid)	317	Me ester (b.p. 207–9°; 104°/21 mm)[79] Et ester (b.p. 219–220; 110°/15 mm)[80a] amide (156–7) chloride hydrochloride (143–4) anhydride (114) hydrazide (168·5–170·5)
2-CO$_2$H-3-Me	109[78a]	
2-CO$_2$H-4-Me	137[78a]	
2-CO$_2$H-5-Me	153[78a]	amide (179)
2-CO$_2$H-6-Me	129[78a]	amide (116)
3-CO$_2$H-2-Me	226–7	amide (158)
3-CO$_2$H-4-Me	215–6	amide (167–167·5)
3-CO$_2$H-5-Me	215–6	
3-CO$_2$H-6-Me	212–3	amide (194–6)
4-CO$_2$H-2-Me	292	amide (163)
4-CO$_2$H-3-Me		Et ester (b.p. 120°/20 mm)
Amino acids		
3-NH$_2$-2-CO$_2$H	210	amide (184)
4-NH$_2$-2-CO$_2$H	260	Me ester (129)
5-NH$_2$-2-CO$_2$H	218–9	Et ester (131–2)
6-NH$_2$-2-CO$_2$H	315	
2-NH$_2$-3-CO$_2$H	310	amide (199)
4-NH$_2$-3-CO$_2$H	338–341	amide (229·5–230·5)
5-NH$_2$-3-CO$_2$H	292–4	Me ester (137)
6-NH$_2$-3-CO$_2$H	312 (dec.)	amide [243–4 (312)]
2-NH$_2$-4-CO$_2$H	300–350 (dec.)	Me ester (149·5–151)
3-NH$_2$-4-CO$_2$H	319–320	amide (151–2)
Halogeno acids		
2-CO$_2$H-3-Cl	121 (dec.)	amide (140)
2-CO$_2$H-4-Cl	182	amide (158)
2-CO$_2$H-5-Cl	169–170	amide (200–1)
2-CO$_2$H-6-Cl	190	
3-CO$_2$H-2-Cl	194	amide (164–5)
3-CO$_2$H-4-Cl	164	
3-CO$_2$H-5-Cl	171	amide (205–6)
3-CO$_2$H-6-Cl	199	amide (213·5–214·2)
4-CO$_2$H-2-Cl	245	Me ester (30–1)
4-CO$_2$H-3-Cl	235	Me ester (32)

Table 6.2 (continued)

Substituent	M.p., °C	Derivatives (m.p., °C)
Hydroxy acids		
2-CO₂H-3-OH	215	Me ester (73–4)
2-CO₂H-4-OH	257–8 (dec.)	Et ester (124–6)
2-CO₂H-5-OH	269–270 (dec.)	Me ester (191–2)
2-CO₂H-6-OH	280	
3-CO₂H-2-OH	255	Et ester (139)
3-CO₂H-4-OH	250 (dec.)	Et ester (64)
3-CO₂H-5-OH	299 (dec.)	
3-CO₂H-6-OH	302	Me ester (164)
4-CO₂H-2-OH	318–325 (dec.)	Me ester (209–212)
4-CO₂H-3-OH	312 (dec.)	Me ester (78·5–80·5)
Nitro acids		
2-CO₂H-3-NO₂	105	amide (211)
2-CO₂H-4-NO₂	152	
2-CO₂H-5-NO₂	211–2	amide (246–7)
2-CO₂H-6-NO₂	168	
3-CO₂H-2-NO₂	156	
3-CO₂H-4-NO₂	120	
3-CO₂H-5-NO₂	172	
3-CO₂H-6-NO₂	183	
4-CO₂H-2-NO₂	175	Me ester (80–81)
4-CO₂H-3-NO₂	222	
Polycarboxylic acids†‡		
2,3-(CO₂H)₂	190	amide (209)
(quinolinic acid)		
		Me₂ ester (55–6)
		3-Me ester (102–3)[83a]
		2-Me ester (122–3)[83a]
		imide (233)[80b]
		anhydride (134)[78b]
2,4-(CO₂H)₂	248–250	amide (254–5)
(lutidinic acid)		Me₂ ester (58)
2,5-(CO₂H)₂	256–8	amide (319–321)
(isocinchomeronic acid)		Me₂ ester (164)
2,6-(CO₂H)₂	236–7 (252)	Me₂ ester (124–5)
(dipicolinic acid)		
3,4-(CO₂H)₂	266–8	amide (175-6)
(cinchomeronic acid)		4-monoamide (170–1)
		imide (231–232)[81b]
		anhydride (77–8)[84]
3,5-(CO₂H)₂		
(dinicotinic acid)	320–3	amide (303–4)
		monoamide (265–6)
		Me₂ ester (84–5)
2,3,4-(CO₂H)₃	249–250	Me₃ ester (101–2)
2,3,5-(CO₂H)₃	324	
2,3,6-(CO₂H)₃	245 (dec.)	
2,4,5-(CO₂H)₃	249	
(berberonic acid)		
2,4,6-(CO₂H)₃	227	
3,4,5-(CO₂H)₃	261	
2,3,4,5-(CO₂H)₄[82]		
2,3,4,6-(CO₂H)₄	236	Me₄ ester (135–6)
2,3,5,6-(CO₂H)₄	200	Me₄ ester (118–9)
		dianhydride (277–280[dec.])
2,3,4,5,6-(CO₂H)₅[82]		
Side-chain acids		
2-CH₂CO₂H	98	picrate (141)
		hydrochloride (131)[78c]

Substituent	M.p., °C	Derivatives (m.p., °C)
3-CH₂CO₂H	145	picrate (100)
		hydrochloride (154)
4-CH₂CO₂H		hydrochloride (131)
2-CH:CH.CO₂H	203 (dec.)	picrate (223)
		hydrochloride (220)
		Et ester (25)[83b]
3-CH:CH.CO₂H	233	Et ester (b.p. 159°/14 mm, m.p.
		18°)[83b]
4-CH:CH.CO₂H	296 (dec.)	hydrochloride [190 (dec.)]
		Et ester (65–6)[83b]
Pyridine 1-oxides		
2-CO₂H	162 (dec.)	amide (161–2)[83d]
		Et ester (b.p. 152–160°/1mm)[83b]
3-CO₂H[83c]	249 (dec.)	Me ester (97)
		Et ester (99·5)
4-CO₂H	272 (dec.)[86]	Et ester (68–70)[83b]
2-CH:CH.CO₂H §		Et ester (70–71)[83b]
3-CH:CH.CO₂H §		Et ester (100–1)[83b]
4-CH:CH.CO₂H §		Et ester (144–6)[83b]

* Unless otherwise indicated, data are from[1b].
† For the influence of thermal decomposition on m.p. see[81a].
‡ See especially[82].
§ M.p. of these acids have been reported but cover a range of temperatures because of decomposition[83b].

its m.p. produces mainly nicotinic with some isonicotinic acid[77]. The easier removal of an α- than of a β-carboxyl group has important practical consequences, for the oxidation of quinoline at 150°–190° with sulphuric and nitric acid and a catalyst gives quinolinic acid, but at > 210° nicotinic acid results[87]. For the same reason, 'aldehyde collidine' (5-ethyl-2-methylpyridine) is also a valuable source of nicotinic acid[88]. Decarboxylations have been effected under various conditions: heating with lime, soda lime, copper, etc., in a solvent such as nitrobenzene[89a] or with steam under pressure[89b].

A kinetic study of the decarboxylation of picolinic acid, molten and in solution, showed decarboxylation to be easier in basic than in acid media, and the reaction was concluded to be of the S_E1 type[90]:

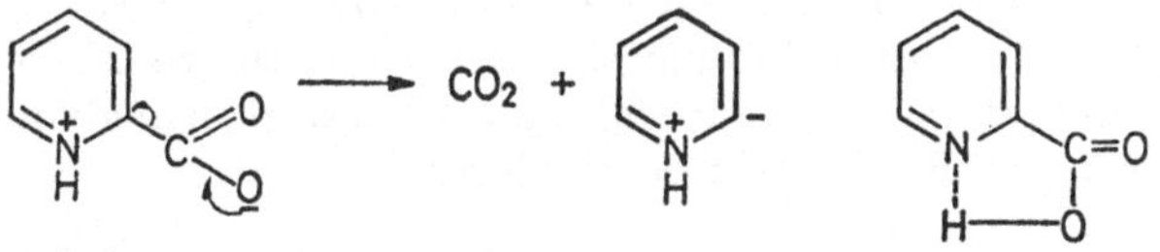

(10) CO_2 +   (11)

The Hammick reaction (p. 163) provides some evidence for this view. Kinetic data for the decarboxylation of picolinic and methylpicolinic acids are given in *Table 6.3*. Methyl groups have the expected effect, but the results and those for reaction in a range of solvents do not permit choice between the zwitterion (10) and the form (11) as the entity undergoing decarboxylation[78a].

Data for decarboxylation in polar solvents (*Table 6.3*)[91] suggest that the mechanism is similar to that found for malonic acid. Increased nucleophilicity of solvent decreases $\Delta H^{\ddagger}$. The changes in $\Delta S^{\ddagger}$ reflect the differing steric effects of the solvents. In the melt, complexing probably occurs between the electrophilic carbonyl group of the zwitterion and the nitrogen atom in a neighbouring molecule or ion. The mechanism of decarboxylation in the melt resembles that in non-polar solvents. All the solvents examined lowered $\Delta H^{\ddagger}$ but the rate of reaction was always faster in the melt than in solution; it is probable that steric hindrance is greater in the attack of the zwitterion on nucleophilic atoms in the solvent molecules than in uniting with picolinic acid molecules or ions (cf. $\Delta S^{\ddagger}$ values).

Table 6.3. Decarboxylation of Picolinic Acids

Substituent	E, kcal mole^{-1} [78a]	log$_{10}A$[78a]	k, sec^{-1} × 10^4 (°C)[78a]	
—	31·1	15·63	2·155	(171·5)
3-Me	32·1	16·66	8·471	(171·2)
4-Me	34·6	17·35	1·858	(169·6)
5-Me	40·0	20·66	1·510	(175·4)
6-Me	35·0	17·41	1·439	(170·0)
4,6-Me$_2$	38·7	19·27	1·547	(171·0)

Picolinic acid in various solvents[91]

Solvent	$\Delta H^{\ddagger}$, kcal mole^{-1}	$\Delta S^{\ddagger}$, e.u. mole^{-1}	$\Delta F^{\ddagger}_{180°}$, kcal mole^{-1}
[Melt]	39·8	+13·2	33·8
p-Cresol	35·8	+3·4	34·3
Phenetole	35·9	+4·5	33·8
Nitrobenzene	34·4	+0·5	34·1
β-Chlorophenetole	33·1	−2·0	34·0
p-Dimethoxybenzene	32·0	−4·6	34·0
Aniline	31·9	−4·4	33·9

The ease of decarboxylation of pyridine-2- and -4-acetic acids is also noteworthy[92–4] and much greater than that of the β-acetic acid[95]. Optically active ethyl-methyl-(2-pyridyl)acetic acid gave on decarboxylation in boiling, neutral aqueous solution racemic 2-s-butylpyridine. Thus, decarboxylation must proceed through a symmetrical intermediate, and since the acid is stable in boiling hydrochloric acid and in strongly alkaline solution, the neutral molecule must be involved:

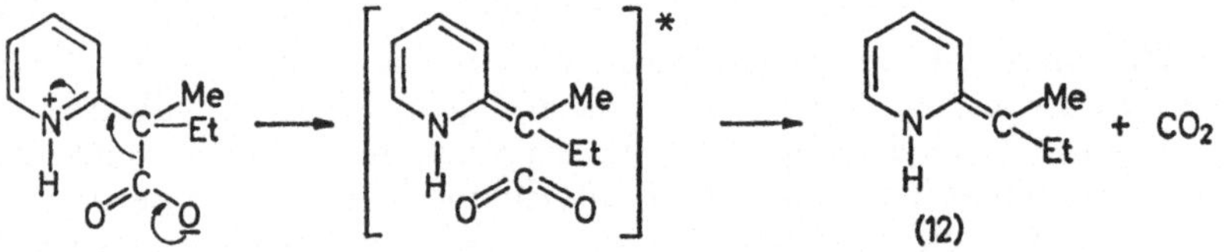

(12)

The 2-isomer reacts more readily than the 3-isomer because it can give a transition state (see above) stabilized by incipient formation of an enamine and carbon dioxide. It is argued that pyridine-4-acetic acid can also react through a transition state resembling an enamine. Other possibilities not excluded by the evidence are[78c]

The rates of reaction of the pyridine-carboxylic acids with diphenyldiazomethane fall in the order 4>2>3, the order of electron deficiency at the nuclear positions. Similarly, the rates of reaction of the pyridine-1-oxide carboxylic acids are in the order 3>4>2, pointing to an order[96] of electron availability 2>4>3.

The reduction of pyridine-carboxylic acids has been mentioned already (pp. 258, 263), and it has been seen that electrolytic reduction or reduction with zinc and acetic acid can give methyl- or hydroxymethylpyridines. Whilst the electrolytic reduction of benzoic acid to benzyl alcohol is well known, benzene-carboxylic acids are generally not so readily reduced as the pyridine-carboxylic acids. Two other additional cases might be quoted. A dichloropyridine-carboxylic acid has been reduced by phosphorus and hydriodic acid to a dichloromethylpyridine[97], and 2,6-dichloropyridine-4-carboxylic acid with zinc and acetic acid gives 2,6-dichloro-4-hydroxymethylpyridine[98]. Isonicotinic acid is reduced to the alcohol by tin and hydrochloric acid[98].

The amides and hydrazides of pyridine-carboxylic acids are of some importance, though their chemistry is not marked by unusual properties. Nicotinamide is, of course, an important compound, and isonicotinic acid hydrazide (isoniazid) is an antitubercular drug. Substituted derivatives are used as antidepressants. In general, however, these compounds show normal chemical behaviour. The amides undergo hydrolysis, dehydration and Hofmann bromination without difficulty. Their reduction has been much studied as a route to pyridine aldehydes. The Sonn–Müller reduction is not very satisfactory in this series, but the McFadyen–Stevens reaction is useful[99]. Nicotinic acid diethylamide gives only poor yields of the aldehyde upon reduction with lithium aluminium hydride, but yields from the methylphenylamide are high. Most satisfactory is the reduction of nicotinic acid dimethylamide with lithium diethoxyaluminium hydride[100].

The general reactions of the esters of pyridine-carboxylic acids are in no way exceptional; their reduction has been mentioned (pp. 258, 263).

Kinetic data for the alkaline hydrolysis of ethyl pyridine-carboxylates and their 1-oxides are summarized in *Table 6.4*. All the esters of pyridine monocarboxylic acids and their oxides are hydrolysed with lower activation energies than that associated with ethyl benzoate. The oxide esters are hydrolysed about 1,000 times faster than ethyl benzoate but rather because of high $\log_{10} A$ values than because of low activation energies. In these reactions the mechanism is probably $B_{Ac}2$; in such processes E is reduced by increased electron-attracting power in the group attached to the ethoxycarbonyl group. The expected order as between 3- and 4-pyridyl is observed, but 2-pyridyl falls out of place, perhaps for 'steric' reasons. Clearly, in the oxides powerful electron-attracting influences are operating which cannot be purely conjugative since they effect the 3- more than the 4-position. The dominant factor is a direct or inductive effect. In the oxide series, the 2-position differs only slightly from the 3-position, possibly because of the

Table 6.4. Alkaline Hydrolysis of Ethyl Pyridine-carboxylates and Pyridyl-acrylates *[83b]

Isomer	$\log_{10} k(k$ in l. mole^{-1} sec$^{-1})25°$ carboxylate	acrylate	$\log_{10} A$ carboxylate	acrylate	E, kcal mole^{-1} carboxylate	acrylate
2-Py	−1·189	−1·855	9·3	8·7	14·4	14·5
3-Py	−1·416	−1·903	8·6	9·4	13·7	15·5
4-Py	−0·692	−1·478	8·0	9·3	12·0	14·6
1-Oxides						
2-Py	0·277	−1·228	9·7	7·7	13·0	12·2
3-Py	0·258	−1·314	10·5	7·9	14·1	12·7
4-Py	−0·026	−1·340	9·7	8·2	13·3	13·0
[Ph]	−3·031		8·7		16·1	

* In 70·55 per cent (w./w.) ethanol–water, in the range 15–35°.

steric effect of the oxygen atom upon the formation of the transition state. The mesomeric electron release may deactivate $C_{(4)}$ but not $C_{(2)}$ because the ethoxycarbonyl group at $C_{(2)}$ is twisted out of the plane of the ring. The situation with the pyridyl-acrylic esters is similar. Again, the 2-pyridyl group is not so effective as would have been expected on electronic grounds, and the transmission of electronic effects from this group to the ester grouping again suffers from steric hindrance. These results have been analysed by a modified Hammett treatment[83b].

Esters of pyridine-carboxylic acids react normally with compounds containing activated methylene groups. The reactions are valuable, for example as routes to acylpyridines. The ethoxide-catalysed condensation of ethyl pyridine carboxylates with ethyl acetate has often been described; early work suggested that esters of nicotinic acid gave lower yields than their isomers, but this is not so[101]. Many esters other than ethyl acetate have been used and a number of substituted pyridine esters[27, 102a]. Condensations with picolines to give desoxypyridoins are of practical value (p. 380).

Reactions of the types just mentioned are base-catalysed, and the rate-controlling step is probably the nucleophilic attack by an enolate anion upon the pyridine-carboxylic ester. The ester with the most electrophilic carbonyl group should be the most reactive. Yields from reactions of methyl pyridine carboxylates with acetone, acetophenone or pinacolone in the presence of sodium methoxide give the sequence $2\text{-Py}.CO_2Me > 4\text{-Py}.CO_2Me > 3\text{-Py}.CO_2Me$. The electrophilic character of the ester carbonyl group is augmented by adjacency to the electron-deficient 2- and 4-positions[103a].

Pyridine-carboxylic acid chlorides, usually prepared by the use of thionyl chloride [a reaction in which nuclear chlorination sometimes occurs (p. 227)], are in most respects normal though rather unstable. Their reactions with alcohols provide some of the best ester preparations in this series. The Rosenmund reduction fails with all three of the pyridoyl chlorides[51, 104], though it

322

works well with some polyhalogenated acid chlorides[51]. A common side reaction is loss of the chlorocarbonyl group; 4,5,6-trichloropicolinoyl chloride gives the aldehyde and 2,3,4-trichloropyridine[51]. Nicotinoyl chloride forms the aldehyde in high yield when reduced with lithium tri-t-butoxyaluminium hydride[103b]. A curious reaction occurs when comenamic acid (4,5-dihydroxy-pyridine-2-carboxylic acid) is treated with phosphorus pentachloride and the product is then reduced with tin and hydrochloridic acid: 4,5-dihydroxy-2-methylpyridine results[105a].

Derivatives of some pyridine-dicarboxylic acids are of interest and value. Quinolinic acid anhydride gives, with ammonia, 2-aminocarbonylnicotinic acid[105b], which in the Hofmann reaction yields 2-aminonicotinic acid. Similarly, cinchomeronic anhydride gives 4-aminocarbonylnicotinic acid and thence 4-aminonicotinic acid[106]. Use of quinolinic imide in the Hofmann reaction produces mainly 3-aminopicolinic acid[80b, 84, 107], and cinchomeronic imide gives 3-amino-isonicotinic acid[81b, 106, 108], whilst diamides of quinolinic and cinchomeronic acid yield the copazolines (13) and (14), respectively[108-9].

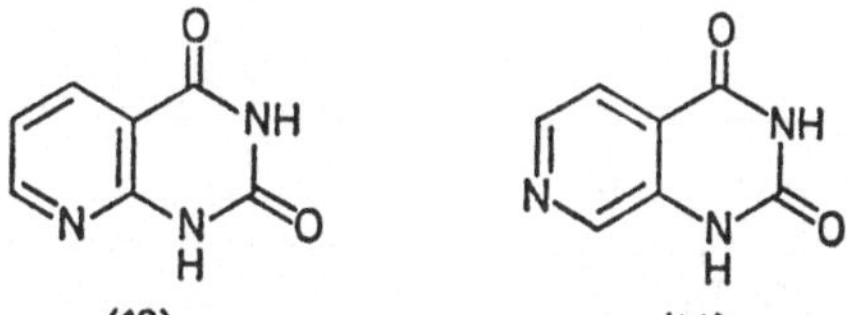

(13)   (14)

Paralleling the reaction with ammonia, reactions of cinchomeronic anhydride with alcohols give 4-alkoxycarbonylnicotinic acids almost exclusively[79, 84]. This reaction, and those above involving ammonia, can be understood as the attacks of nucleophiles upon the more electrophilic of two carbonyl groups.

For this reason, the reaction of quinolinic anhydride with water is surprising[83a]. With methanol, both possible esters are formed but the proportion of 2-methoxycarbonylnicotinic acid increases with time (the 3-methyl ester gives the isomer almost completely when boiled in ethyl acetate). With one molecular equivalent of methanol at room temperature, the anhydride produces the pseudo-ester (15) which forms a crystalline hydrate (16). The 3-methyl ester arises from (15), and the relationships of these compounds have been represented[83a] by the scheme

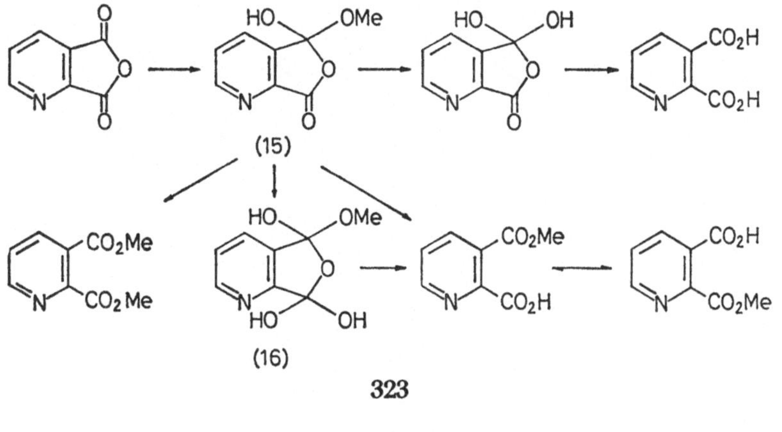

(15)

(16)

The use of an optically active ester showed that an alkoxyl, and not an alkyl, group migrated. The formation of (15) as the primary product is surprising, for the carbonyl group at $C_{(2)}$ might have been expected to be the more electrophilic (see p. 322). However, the formation of (15) would have been expected from Wegscheider's rule[83a], according to which attack should occur at the carbonyl group corresponding to the stronger acid (see *Table 5.10*). An intermediate corresponding to (15) but from cinchomeronic anhydride has not been reported.

Partial saponification of quinolinic acid diesters gives 3-alkoxycarbonyl-picolinic acids, whilst cinchomeronic diesters form 3-alkoxycarbonylisonicotinic acids[79].

(b) C- and N-*Alkyl and Substituted Alkyl Groups*

Simple properties of some members of this group of compounds have been collected in *Table 5.1*. Their dipole moments (p. 123), spectra (p. 135) and ionization constants (p. 145), and the bearing of some of these on tautomerism in the alkylpyridines have been discussed already; the latter is further mentioned below.

At high temperatures (500°) in the presence of a catalyst, 2- and 4-methylpyridines are converted into 3-methylpyridines[110], but these transformations have not been evaluated quantitatively.

Polarographic studies of the picolines have been reported[111].

The chemistry of alkylpyridines in which the substituent is primary or secondary is dominated by the acidity of the hydrogen atom or atoms attached to the carbon atom adjacent to the nucleus. The degree to which ionization

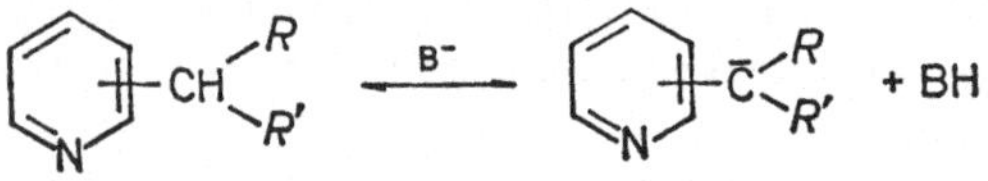

(17)   (18)

occurs must vary immensely in the wide range of reactions which are a consequence of its possibility, from full anion or ion-pair formation to synchronous reaction with an electrophile and removal of the proton. Qualitative evidence for anion or ion-pair formation in the presence of strong bases is

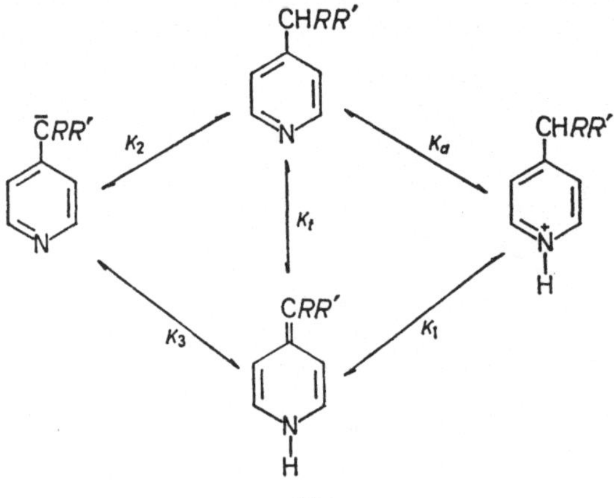

324

plentiful (see p. 327), but quantitative data about the acidity of alkyl groups are almost completely lacking.

The acidity is, of course, increased when R and R' in (17) are electron-attracting, and when the nitrogen atom is protonated or quaternized. Some comparative data are available for phenyl-substituted compounds in the last condition[112]. The system of ionic and tautomeric equilibria shown occurs with, for example, a 4-alkylpyridine. K_a has, of course, been measured in many cases, and K_t is commonly very large in favour of the NM (alkyl-pyridine) form rather than the Z (methide) form (pp. 137, 154). The remaining constants K_1, K_2 and K_3 are unknown. K_1 can be approximated by the equivalent constant for the quaternary system:

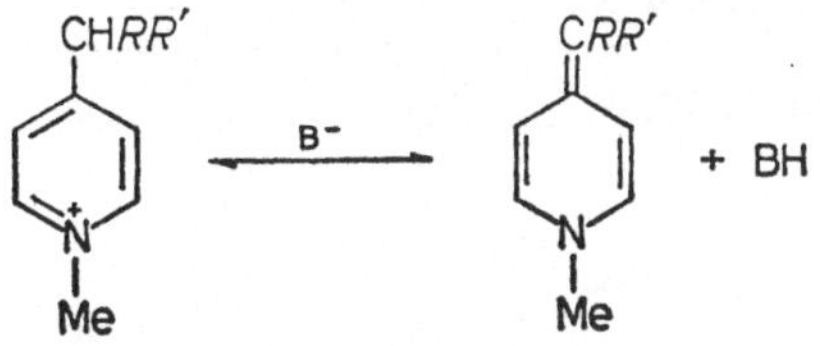

This constant is experimentally accessible. With the methiodide of 4-diphenylmethylpyridine as a standard, some approximate values of K_1 and K_t have been measured, using ultra-violet spectroscopic methods (*Table 6.5*).

Table 6.5. *Ionization and Tautomerization Data for some Substituted Alkyl-pyridines*[112] (21 °C)

Pyridine	pK$_a$ ($\pm$ 0·2) (50 per cent EtOH-H$_2$O)	pK$_1$* (Et$_3$N-MeOH)	pK$_1$†	pK$_t$†
2-CHPh$_2$	3·46	2·0 $\pm$ 0·5	15·2	11·7
3-CHPh$_2$	3·61	>3·5	>16·7	>13
4-CHPh$_2$	4·01	0·0	13·2‡	9·2
2-CH$_2$Ph	4·21	2·5 $\pm$ 0·5	15·7	11·5
4-CH$_2$Ph	5·33	1·8 $\pm$ 0·5	15·0	10·5
2-Me		>3·5	>16·7	>12
4-Me		>3·5	>16·7	>12

* Values for each 1-methyl-methide relative to that of the methide from 1-methyl-4-diphenylmethyl-pyridinium.
† Calculated values.
‡ Experimental value (50 per cent EtOH-H$_2$O) used to correct relative values.

As noticed before (p. 137), the tautomeric equilibrium vastly favours the alkylpyridine (NM) forms, and even with electron-attracting acyl groups present the alkyl hydrogen atom is extremely weakly acidic. The evidence, so far as it goes, gives the order of acidity $CH_{(4)} > CH_{(2)} > CH_{(3)}$.

Although equilibrium measurements of acidity in simple alkylpyridines have not been made, data from reactivity studies (see below) have often been used as indices of alkyl group acidity and generally reveal the order $Me_{(4)} > Me_{(2)} \gg Me_{(3)}$. Competitive metalation and side-chain alkylation (see below) of the picolines is said[113a] to give the order of acidity $Me_{(4)} > Me_{(2)}$. Evidence of this kind is almost always only semi-quantitative, and for this reason studies of proton–deuterium exchange in alkylpyridines are of great interest. Acid-catalysed exchange does not seem to have been observed with

C-alkylpyridines. 1-Methylpyridinium does not exchange with D_2O or D_2SO_4 even at 140°–150°, but some exchange can be effected[114] in the methylene group of $(C_5H_5\overset{+}{N}.CH_2.\overset{+}{N}C_5H_5)$. However, 1-allylpyridinium exchanges four hydrogen atoms of the allyl group for tritium in the presence of pyrrolidine, through the ylid intermediate[113b], and base-catalysed deuterium exchange of the protons in methyl groups in the picolines is well established[115]. *Table 6.6* gives kinetic data for the picolines and their oxides

Table 6.6. Deuterium Exchange of Picolines and their 1-Oxides[116]

Pyridine	Conditions	Temp. (°C)	$k \times 10^6$, sec^{-1}	E, kcal mole^{-1}	log$_{10}$ A
2-Me	} MeOD-Et₃N	160	3·4	18·2	4·3
2-Me 1-Oxide		160	6·9	23·6	2·44
2-Me		90	17·0	24·7	9·86
2-Me 1-Oxide		10	66·0	19·9	11·05
3-Me	} EtOD-EtOK	100	0·37	31·0	11·73
3-Me 1-Oxide		30	26·0	22·5	11·30
4-Me		70	59·0	20·3	8·66
4-Me 1-Oxide		10	38·0	15·8	7·8

taking part in deuterium exchange with MeOD in the presence of triethylamine, and with EtOD in that of EtOK. The triethylamine had substantially no influence on the speed of exchange, and for this reason the medium was convenient for examining the very rapidly reacting 2-picoline 1-oxide. The view that exchange in the ethoxide-catalysed reaction goes by the anion is supported by the development of colour in the solutions.

The results show the sequence of reactivities in the picolines $Me_{(4)} > Me_{(2)} \gg Me_{(3)}$ and in the oxides $Me_{(2)} > Me_{(4)} > Me_{(3)}$. Further, the oxides are more reactive than the parent bases, a situation noticed already in reactions such as the replacement of nuclear halogen atoms in nucleophilic amination (p. 213). Differences in π-electron energy between starting compound (17) and the assumed anionic intermediate (18), and also π-electron densities at the nuclear carbon atoms carrying the methyl groups, were found to be acceptable as measures of reactivity. It should be noticed, however (*Table 6.6*), that in these reactions differences in reactivity arise sometimes as much from changes in log$_{10}$ A as in those in the activation energy; the reaction of 4-picoline 1-oxide has an activation energy considerably lower than that of 2-picoline 1-oxide, and yet 2-picoline 1-oxide reacts faster because of the larger pre-exponential factor associated with it. Kinetic isotope effects were examined by comparing rates of deuterium exchange with those of tritium exchange. The results were considered normal for a reaction in which a carbon–hydrogen bond is broken in the rate-determining step. The absence of a kinetic isotope effect in the cases of 3-picoline and its oxide was taken to be connected with the low stability of the anions from these compounds and with control of the overall reaction rate in these cases by diffusion of the solvent in the neutralization of the anion.

It remains to enquire if the rates of exchange discussed reflect the relative acidities of the substrates. Analogy with results in the benzene series suggests that they do[117a]. The detailed character of the mechanism of reaction also deserves attention to define the nature of the anion or ion-pair intermediate. In this connection, the stereochemical consequences of exchange need to be known. It is interesting to note that exchange in the case of ethylbenzene-*a-d* proceeds with a large degree of retention of configuration[117b]. For the exchange reactions of pyridine compounds nothing is known.

In this general connection the properties of (−)-*a*-s-butylpyridine are interesting[78c]. This compound is optically stable in boiling 1,2-propanediol (b.p. 189°), concentrated hydrochloric acid and ethanolic sodium ethoxide. Racemization is caused by a solution of potassium in triethylcarbinol, slowly at 100°, rapidly at 142°. Comparison of racemization and exchange rates in such a compound would be valuable. As noted above, quaternization must activate the alkyl hydrogen atom, and the methiodide of the s-butylpyridine is racemized by boiling in propanediol.

The qualitative evidence for anion or ion-pair formation from alkylpyridines is extensive. Ziegler and Zeiser[118] made the original observation of metal–hydrogen exchange between 2-picoline and methyl lithium in ether. The solution of lithium 2-picolyl is reddish brown. Phenyl lithium gives the same result[118-9]. 4-Picoline gives a deep red anion in ether (metalation being accompanied by 2-phenylation unless precautions are taken)[120a, 121]. Solutions of lithium 2-picolyl have been prepared by using lithamide[120b], and of lithium 4-picolyl with lithium diethylamide[120a]. With phenyl lithium, 2,4-lutidine gives a mixture of the two possible monolithium derivatives[120b]. 2,6-Lutidine yields an orange-red or brown solution[120c, 122a]; with excess of phenyl lithium, the ethereal solution is dark red[122b], and there is evidence for the formation of the dilithium derivative[123].

Of higher homologues of the picolines, 2-ethyl-, 2-isobutyl, and 2-n-pentyl-pyridine have been metalated with methyl or phenyl lithium[78c,122a,124], but 2-isopropyl- and 2-s-butylpyridine did not react[78c, 124]. (−)-s-Butylpyridine was, in fact, unracemized by phenyl lithium. 3-Methylpyridine merely undergoes phenylation with phenyl lithium (p. 221).

Although lithium salts have so frequently been prepared in solution in anhydrous aprotic solvents, they have not been obtained free, and practically nothing is known of their properties. The lithium derivative of ethyl 6-methylnicotinate is a bright yellow powder said to resemble a 'true' salt[125].

Tschitschibabin[124] originally prepared sodium derivatives of 2- and 4-methyl- or -ethyl-pyridine by grinding them with sodamide. 3-Ethylpyridine did not react. Usually, however, sodium or potassium derivatives of 2- and 4-picoline have been obtained by reaction with sodamide or potassamide in liquid ammonia[126]. The solutions of 2- and 4-picolyl sodium or potassium in liquid ammonia are deep or brownish red. The addition of a picoline to sodamide ammonia instantly produces the characteristic intense colour, and Brown and Murphey[127d] were the first to observe that this occurs with 3- as well as 2- and 4-picoline. Before this important observation the methyl group at $C_{(3)}$ was considered to be unreactive[128]. As well as 2-, 3- and 4-methyl and -ethyl[124, 127] groups, 2-isopropyl[127], 2-benzyl[129] and 3-isopropyl[127] groups have been metalated with sodamide or potassamide. Triphenyl sodium was without effect on 2-s-butylpyridine[78c].

Nothing is known of the properties of the lithium and sodium compounds.

Their magnesium counterparts have been prepared[130] by exchange between 2-picolyl lithium and magnesium bromide or iodide, and exchange also occurs between 2-picoline and ethyl magnesium bromide[131]. 2-Picolyl lithium provides the cadmium derivative by reaction with cadmium chloride. This fails with 4-picoline, but in this case reaction with mercuric chloride evidently gives the picolyl mercuric compound[122c].

As already mentioned, in quaternary pyridinium salts the acidity of hydrogen atoms attached to carbon adjacent to the ring is increased. With 2- and 4-substituted compounds the substances formed by removal of the proton can often be isolated. They are enamines, anhydro-bases or pyridone methides, coloured substances whose formation is reversible. They are strong bases, giving alkaline solutions in water (cf. *Tables 5.10* and *6.5*). Decker[132a] observed the first example when he treated 2-benzylpyridine methiodide with caustic soda and was able to extract the orange methide into benzene. In the simplest case, that of 1,2-dimethylpyridinium, addition of caustic soda and extraction with ether gave yellow solutions of the methide, but it could not be isolated[132b]. The aqueous equilibrium in the case of 1,2-dimethylpyridinium greatly favours the cation[132d]. *Table 6.7* lists some examples of these highly reactive methides (see below). As would be expected, methides from pyridines in which the alkyl group is substituted by phenyl, nitrophenyl and other electron-attracting substituents, are more stable than compounds from which these are absent.

Table 6.7. Some Pyridone Methides *

Quaternary cation	Methide (m.p.,°C)	Derivative with PhNCS
1,2-Me$_2$†		dark red, m.p. 153°[132b]
1,4-Me$_2$	yellow[132b]	brown-yellow, m.p. 218°[132b]
1-Me-2-CH$_2$Ph†		red, m.p. 147–8°[132e]
1-Me-4-CH$_2$Ph	orange oil[132a, e]	yellow, darkens 145°, m.p. 162°[132e]
1-Me-2-CH$_2$C$_6$H$_4$.NO$_2$(p)†	dark blue flecks (160)[132e]	
1-Me-4-CH$_2$C$_6$H$_4$.NO$_2$(p)	(dec. 50)[132e]	
1-Me-2-CH$_2$C$_6$H$_3$(NO$_2$)$_2$(2′,4′)	blue crystals (201)[132g]	
1-Me-2-CHPh$_2$	red-violet crystals (147)[132f]	
1-Me-4-CHPh$_2$	yellow (113)[132f]	
1-Me-2-C$_5$H$_5$‡ (cyclopentadienyl)	orange (56–7 and 74–5)[133c]	
1-CH$_2$Ph-4-C$_5$H$_5$** (cyclopentadienyl)	yellow (> 200 dec.)[134, 135a]	
1-CH$_2$C$_6$H$_3$.Cl$_2$(2′,6′)-4-C$_5$H$_5$ (cyclopentadienyl)	red prisms (199–200)[136a]	
1,2,4,6-Me$_4$-3,5-(CO$_2$Et)$_2$***	yellow and red§[132c]	
1-Me-2-CH$_2$CO$_2$Et	yellow needles (52–4)[135b]	
1-Me-4-CH$_2$CO$_2$Et	(110–1)[135b]	
1-CH$_2$C$_6$H$_3$.Cl$_2$(2′,6′)-4-CH$_2$COMe	yellow crystals (203–4)[136a]	
1-CH$_2$C$_6$H$_3$.Cl$_2$(2′,6′)-4-CH$_2$COPh	yellow (166–7)[42a, 136a]	

* The name seems to have first been used by Mumm[131c].
† For the ultra-violet spectrum of the methide see [133a, b].
‡ Methide(s) not prepared from the quaternary salt (p. 203).
** Related compounds [133c, 136a, c, d].
*** Related compounds[131c, 136b].
§ Both possible isomers appear to be formed.

The methide (19), corresponding to the quaternary cation 2-cyclo-pentadienyl-1-methylpyridinium (though it was not prepared from this cation, see p. 203), is of some interest. The dipole moment (5·2 *D*) indicates a large contribution to its structure from the form (20), and a similar situation occurs with related compounds[135a]. Protonation of (19) gives only two of the possible cations[136c], in the proportions shown:

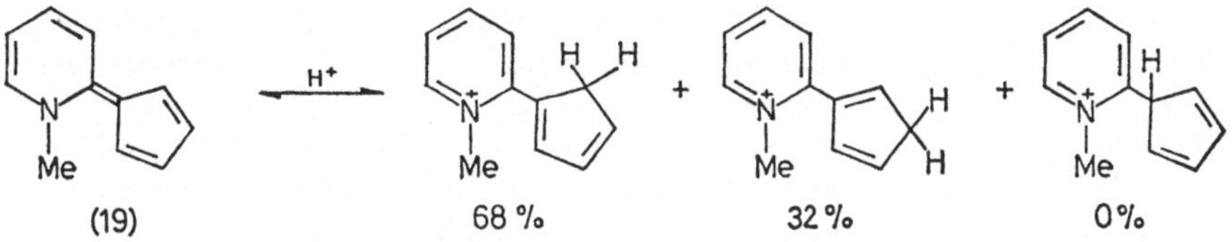

| (19) | 68 % | 32 % | 0 % |

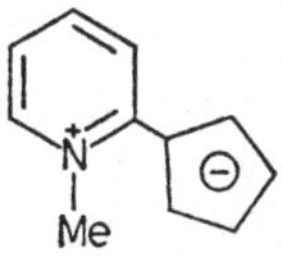

(20)

Indications that reactivity of the type under discussion might be found in primary or secondary alkyl groups attached to the ring nitrogen atom in quaternary pyridinium salts arose from some of the deuterium exchange studies mentioned. In fact, this possibility seems to be realized only when other activating factors additional to the positively charged nitrogen atom are present. Thus, 1-(9-fluorenyl)pyridinium bromide gives an indigo-blue solution when treated with alcoholic caustic soda. The colour is due to the betaine (21) which cannot be isolated intact[137]. The analogous cyclo-pentadiene derivative is a coppery red-brown compound (m.p. 350°) which gives colourless solutions of the quaternary pyridinium cation in acid[138] and whose colour in various solvents has been related to intramolecular charge-transfer transitions[139].

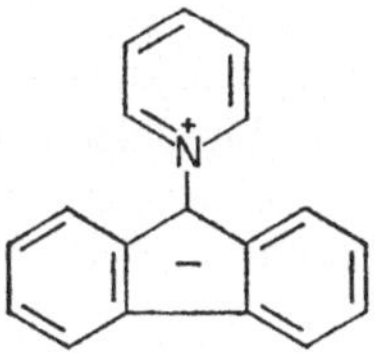

(21)

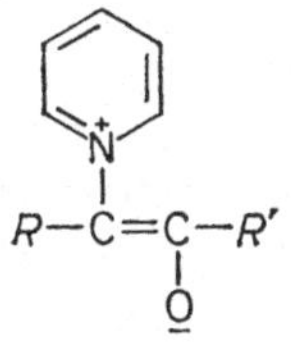

$$R-C=C-R'$$
with O below

(22)

$$R-CH-C-R'$$
with O below

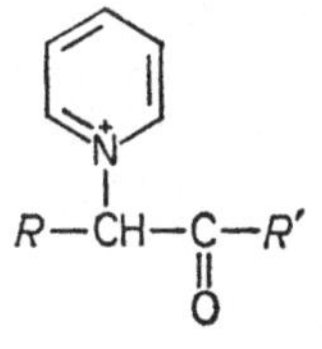

(23)

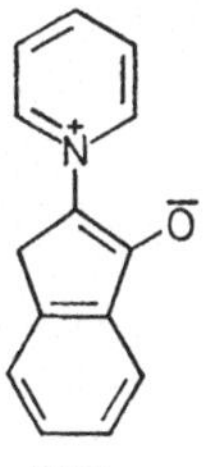

(24)

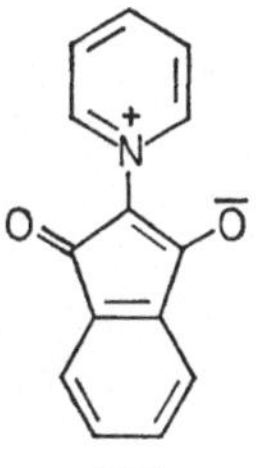

(25)

More extensively studied are the enol betaines (22; the mesomeric possibilities are not represented) from quaternary salts of type (23). One of the first examples[140a] was (22; $R = H$, $R' =$ 2-ethylthio-5-methylphenyl). From 1-phenacylpyridinium bromide, Kröhnke[154c], by using sodium carbonate rather than sodium hydroxide, was able to isolate (22; $R = H$, $R' = Ph$), avoiding the splitting (p. 390) caused by the stronger base. This compound is yellow when hydrated, orange when anhydrous and with acids gives back the colourless 1-phenacylpyridinium salts. Numerous related examples were described. The quaternary salt from pyridine and ethyl bromomalonate gave in the same way the enol betaine (22; $R = CO_2Et$, $R' = OEt$), but the corresponding one from the cation (23; $R = H$, $R' = OEt$) could not be isolated[140b]. Two interesting enol betaines of the present type are (24) and (25), the former being red or orange, depending on its state of hydration, the latter yellow[141]. The ultra-violet spectra of phenacylpyridinium enol betaines have been studied[142]; their reactions, which are strongly reminiscent of those of the anions of 1,3-dicarbonyl compounds, will be discussed below.

The data referred to which relate to acidity in the picolines and their derivatives, and also evidence from reactions depending on their prototropic activity (some of it to be dealt with below) show an order of activity $C-H_{(4)}$, $C-H_{(2)} \gg C-H_{(3)}$. Further, the nitrogen atom in 3-picoline does exert an activating effect upon the methyl group, for despite the inability of 3-picoline to take part in numerous reactions characteristic of 2- and 4-picoline it is nevertheless ionized by the amide anion in conditions under which toluene is not affected. In considering their discovery of this fact, Brown and Murphey[127d] suggested that the weak but positive activation in 3-picoline arose from stabilization of the anion by resonance of the kind represented in (26). Such stabilization would, of course, be much weaker than the resonance stabilization of the 2- and 4-picolyl anions. Brown and Dewar[143] pointed out that, translated into molecular orbital terms, such an explanation implied that the neglect of resonance integrals between non-adjacent atoms (p. 23) is not justified. An examination of this possibility by the L.C.A.O. method, with reasonable values for the usual parameters and a ratio of 10:1 for the C–C resonance integral between adjacent and *meta*-situated atoms, showed that the inclusion of *meta*-interactions had almost no effect upon the situation. An explanation in terms of the inductoelectromeric effect, caused by the unusually electronegative 2- and 6-carbon atoms *ortho* and *para* to the 3-position, gave a reasonable account of the activation. The effect is represented by (27) and (28).

(26)   (27)   (28)

In discussing the chemical reactions of alkylpyridines, a problem of classification arises. A number of processes in which they react through conversion by a strong base into an anion might be regarded as reactions of the corresponding organometallic reagents. As mentioned above, it is never clear

330

to what degree initial anion formation proceeds. The arbitrary choice has been made to classify those reactions of alkylpyridines which can be so regarded as reactions of the corresponding organo-metallics; they are dealt with in Section (*i*) below. The fission of quarternizing groups in some pyridinium salts by bases is discussed below (pp. 389–90).

Acylation—Acylations of alkyl groups other than base-catalysed ones are not numerous. An interesting case is the reaction of 4-picoline with dimethylformamide and phosgene or phosphoryl chloride. The product (29) gives the dialdehyde on alkaline hydrolysis[144].

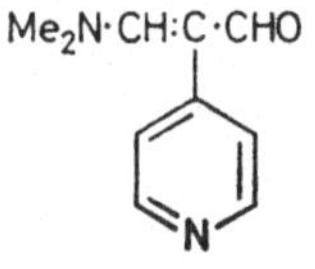

(29) 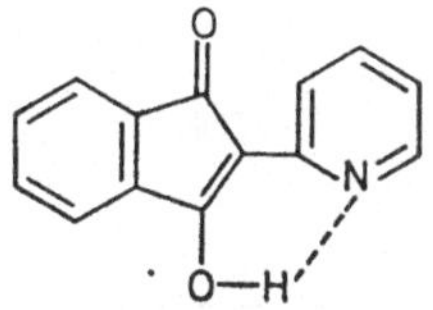   (30) 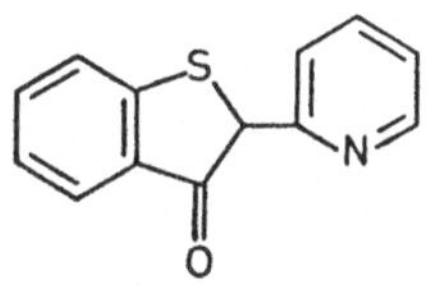   (31)

2-Picoline reacts with phthalic anhydride in the presence of zinc chloride or acetic anhydride to give 'pyrophthalone'[145]. Much work has been done on the structure of this compound[146]. Perhaps the best formulation[147] is (30). Similar reactions occur with naphthalic and diphenic anhydride[148] and in the cases of lutidines and collidines are assumed to occur at the 2-methyl group. Under similar conditions 2-picoline gives (31) with benzenesulphonylbenzisothiazolone[149].

In the Scholtz indolizine synthesis[150] 2-picoline is heated with acetic anhydride, giving diacetylindolizine. The reaction has been represented as going by diacetylation of the methyl groups

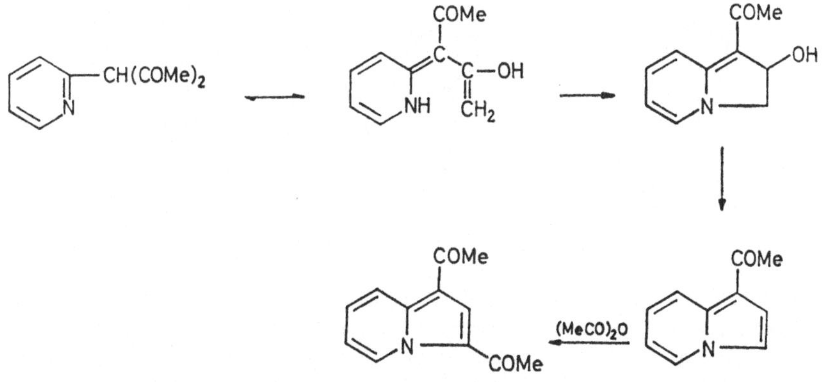

Other examples are known, and the reaction proceeds with propionic anhydride[151].

The use of zinc chloride and related substances, or acetic anhydride, is a common feature of many reactions of alkylpyridines. The first type of reagent probably exercises a generalized acid catalysis, augmenting the reactivity of alkyl groups by coordination with the nitrogen atom. Acetic anhydride

M

could exert a similar influence by generating acetylpyridinium ions. 1-Benzoyl-2-methylpyridinium ion is presumably involved in the formation of 2-phenacylpyridine from 2-picoline, benzoyl chloride and silver phenyl-acetylide[152a] (cf. p. 202).

The acylation of C-alkyl groups in quaternary pyridinium salts has not otherwise been reported, but pyridone methides undergo substituted acylations with phenyl isocyanate and isothiocyanate. Some products of such reactions are recorded in *Table 6.7*. Some interesting examples of methide acylations are shown[153]:

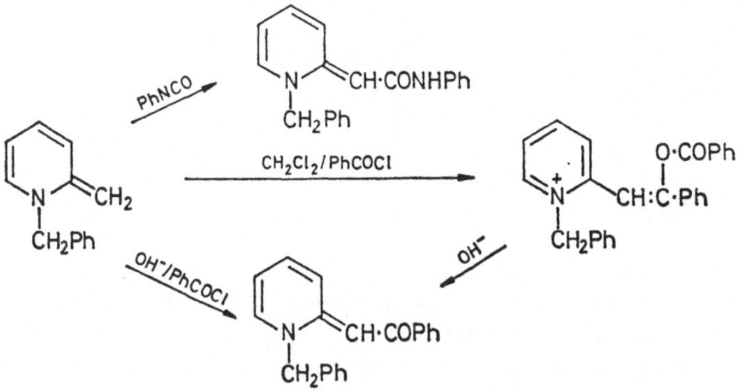

The enol betaines derived from 1-phenacylpyridinium salts undergo analogous reactions. Characteristic examples are the formation of (32) and (33) from phenacylpyridinium enol betaine and phenyl isocyanate or isothiocyanate[154a] and benzoyl chloride[154b] or benzoic anhydride[154c], respectively. In a related case[154d] reaction with carbon disulphide gives (34).

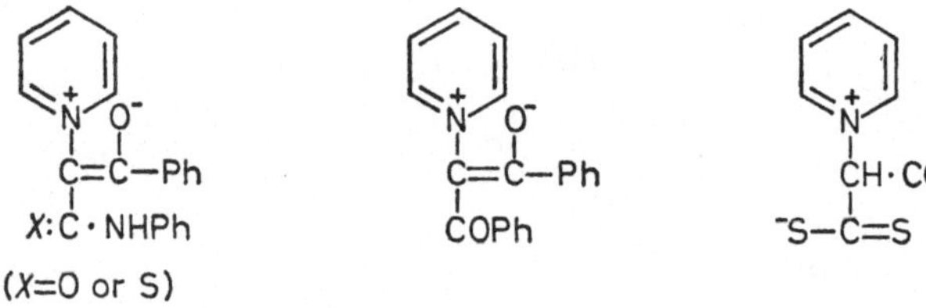

Alkylation and Substituted Alkylation—Reactions in this group (other than hydroxyalkylations and benzylations effected by carbonyl compounds, dealt with below) are not numerous. Mannich aminomethylations have been carried out with 2- and 4-picoline and 2-ethylpyridine, formaldehyde and a number of amines[155], but yields are poor[156a].

γ-2-Pyridyl- and -4-pyridyl-butyronitrile are conveniently obtained from 2- and 4-picoline and acrylonitrile reacting in the presence of acetic acid, copper and copper acetate[157].

The pyridone methide generated from 2-picoline methiodide by sodium methoxide[156b] or piperidine[158] gives the bis-cyanoethyl derivative (35)

332

with acrylonitrile. In the piperidine-catalysed reaction[158] 4-picoline methiodide gives the tris-cyanoethyl compound. The differing behaviour of the 2- and 4-picoline derivatives and of their homologues has been attributed to the discouragement of methide formation in some cases by steric factors:

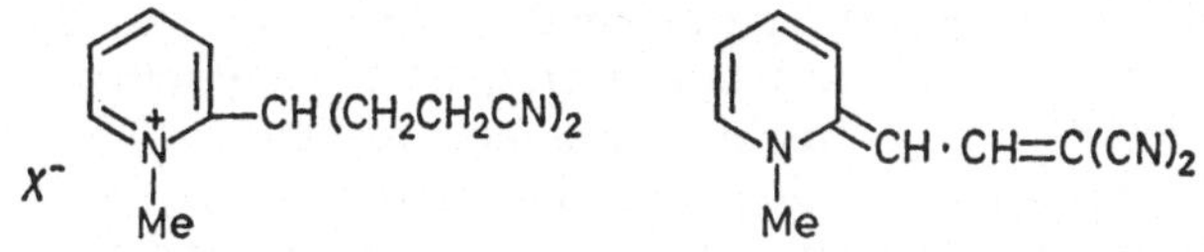
CH (CH2CH2CN)2
X⁻
N
Me
N
CH · CH=C(CN)2
Me

(35) (36)

With ethoxymethylene malononitrile, the methide from 2-picoline gives[159] (36).

Alkylation of N-alkyl groups has not been reported, but interesting examples are known involving phenacylpyridinium enol betaine:

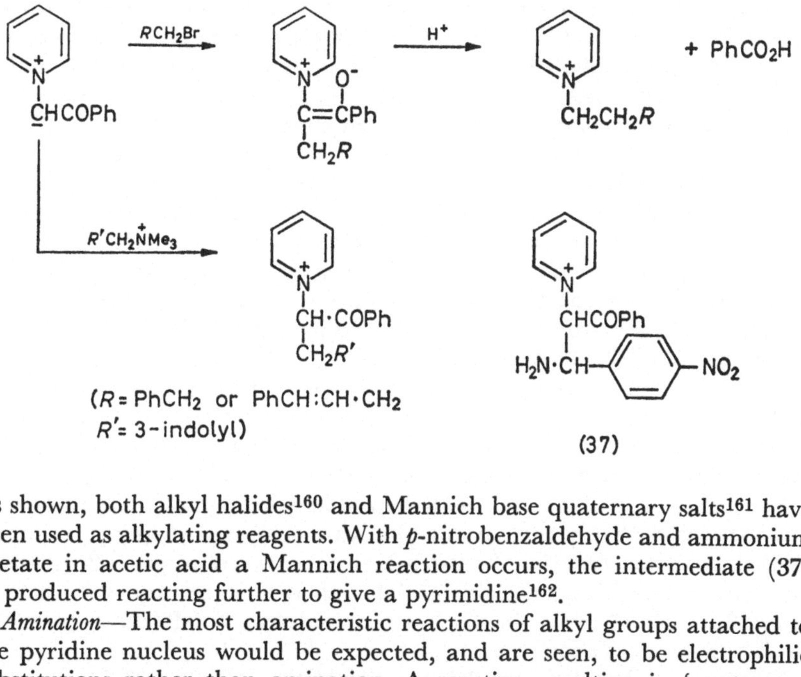
RCH2Br
H⁺
+ PhCO2H
CHCOPh
C=CPh
CH2R
CH2CH2R
R'CH2NMe3
CH·COPh
CH2R'
(R= PhCH2 or PhCH:CH·CH2
R'= 3-indolyl)
CHCOPh
H2N·CH
NO2

(37)

As shown, both alkyl halides[160] and Mannich base quaternary salts[161] have been used as alkylating reagents. With *p*-nitrobenzaldehyde and ammonium acetate in acetic acid a Mannich reaction occurs, the intermediate (37) so produced reacting further to give a pyrimidine[162].

Amination—The most characteristic reactions of alkyl groups attached to the pyridine nucleus would be expected, and are seen, to be electrophilic substitutions rather than amination. A reaction resulting in 'quaternary amination' of the methyl group in 2-picoline 1-oxide has been mentioned (p. 210). Its mechanism has not been clarified.

Arylation—The only important case is that of cyanine dye formation. This has been mentioned under the heading of substituted alkylation by replacement of halogen in a quaternary pyridinium salt. The relevant case here is the complementary one in which a picoline quaternary salt reacts with a halogen-substituted heterocyclic quaternary salt (p. 205). Both 2- and 4-picolinium salts have been used satisfactorily in such reactions[152b] which proceed in the presence of triethylamine and presumably involve the pyridone methides.

Carbonyl reactions—Their aldol reactions with carbonyl compounds are perhaps the most important undergone by primary or secondary alkyl groups attached to the pyridine ring. Ladenburg[75d] first observed the reaction of 2-picoline with paraldehyde, which gave, at 250°–260°, 2-propenylpyridine. At lower temperatures, α-(2-picolyl)ethanol is formed, but in low yield[163a, b-4]. Formaldehyde is more reactive. β-(2-Pyridyl)ethanol has been prepared by heating 2-picoline and formalin or formaldehyde under pressure[163a, c, 165a], and this reaction and that with 4-picoline has been used as a path to vinylpyridines, formed by dehydrating the alcohols[166]. Small amounts of β-(2-pyridyl)propan-1,3-diol are also formed[163d, 167a, c]. The maximum reported yield[168] of β-(2-pyridyl)ethanol is 50 per cent. 4-Picoline is more reactive than the 2-isomer and can give all three possible alcohols[167a, 169].

In similar circumstances, 3-ethyl-4-methylpyridine gives the dihydroxymethyl derivative[167b], 2,4-lutidine yields β-(2-methyl-4-picolyl)ethanol[170], and 2,6-lutidine produces β-(6-methyl-2-picolyl)ethanol and 2,6-di(hydroxyethyl)pyridine[167a, 171]. Whereas 2-ethylpyridine gives more of the alcohol than of the diol[172a], the position is reversed with 4-ethylpyridine[173]. 2- and 4-Benzylpyridine behave similarly[174a]. In a number of these reactions the alcohols are accompanied by their dehydration products.

Although 2,4,6-trimethylpyridine reacts at the 4-methyl group, 2,4,6-trimethyl-3-nitropyridine does so at the 6-methyl group[168].

As indicated above for acetaldehyde, higher aldehydes react less satisfactorily than formaldehyde. With 2-picoline, propionaldehyde gives a very poor yield of the mono-adduct[172b]. The highly electrophilic chloral is exceptional, reacting readily with 2-picoline to give a high yield of α-(2-picolyl)-β,β,β-trichloro-ethanol[165a, 175]. 4-Picoline[176a, 177] and 3-ethyl-4-methylpyridine[176b] behave similarly.

Reactions with ketones generally fail, but with acetone and zinc chloride at 260°, 2-picoline gives some 2-isobutenylpyridine[174b]. More reactive carbonyl compounds of the type $O{:}CR_2$ ($R = CO_2Et, PhCO$) give reasonable yields of the corresponding alcohols[165b].

The reaction of 2-picoline with benzaldehyde in the presence of zinc chloride was probably the earliest observation of methyl group reactivity in the pyridine series[145]. At 220°–225°, the reaction gave 2-stilbazole (2-styrylpyridine) in good yield[178a]. Though the major product is *trans*-2-stilbazole, some of the *cis*-isomer has been isolated[179]. 4-Picoline reacts[178b], but 3-picoline does not[178c].

Numerous stilbazoles have been made by this method[180-1], e.g. from 2- and 4-benzylpyridine[132e]. 2,4-Lutidine produces 2,4-di- and 2-monostyryl derivatives[182a, b, 183a], 2,6-lutidine gives mono- and di-styryl ones[182c, d, 183b], and 2,4,6-collidine yields 2-mono[182c, 184] and tri-styryl derivatives[184]. Hydroxyl groups in the pyridine ring diminish the reactivity of picolines[185]. With the nitrobenzaldehydes, 2-picoline reacts without a catalyst[186a].

Sometimes in these reactions the formation of the intermediate carbinol ('alkine') is observed[182a, 186b, c]. Isatin gives the carbinol (38) and the dehydrated product merely by boiling with 2-picoline[187]. The carbinols can be obtained by heating the picoline and aldehyde with water at a lower

(38)   (39)

temperature (130°–150°) than is used in stilbazole formation[186d–h]. With benzaldehyde and water at 125°, 2,6-lutidine gives the monocarbinol[188]. The yields (per cent) of carbinols from 2-picoline and various aldehydes under standard conditions were: p-tolualdehyde (5), benzaldehyde (11), m-nitrobenzaldehyde (25), o-nitrobenzaldehyde (37), p-nitrobenzaldehyde (48). Alkine formation is reversible, and when heated with water at 140°–200°, alkines give 2-picoline and an aldehyde. Under these conditions stilbazoles are unchanged[186h]. The carbinols are dehydrated by acetic anhydride[186g, h] or zinc chloride[186f]. In these dehydrations the carbinols (39; $R = NO_2$ or NH_2) have been claimed to give both cis- and trans-stilbazoles[186f], but usually only the trans-isomers are obtained[186h] (see below).

Boiling an aromatic aldehyde with a picoline and acetic anhydride provides the second general method for preparing stilbazoles[186h, 189]. As would be expected, electron-attracting substituents in the aldehyde increase the ease of reaction, and electron-releasing groups decrease it; 2,4-dinitrobenzaldehyde reacts especially easily. Numerous applications of the method have been reported[181d, e, 190–1]. Pyridones[192] and hydroxy- or acetylamino-pyridines[193] give poor yields of stilbazoles. 2,3-Dimethylpyridine forms 3-methyl-2-stilbazoles[186g], and 2,6-lutidine can give mono- or di-stilbazoles[122b]. From 2,6-di-n-propylpyridine and p-hydroxybenzaldehyde the distilbazole was obtained[122b]. Ethyl pyridyl-4-acetate reacts readily with benzaldehyde under these conditions[177]. It is interesting that in the presence of alkali 5-acetyl-2-methylpyridine reacted with benzaldehyde at the ketone methyl group, but in acetic anhydride the stilbazole was the main product[194].

The acetic anhydride method has been modified by the inclusion of potassium acetate in the reaction mixture. Under these conditions, 2-picoline and benzaldehyde give a good yield of 2-stilbazole, 3-picoline does not react, 2,6-lutidine forms the di- and some mono-stilbazole[195], and 2,4,6-collidine gives the tri-stilbazole[193].

Some condensations have been effected in the presence of piperidine. By this method 2-hydroxy-4,6-dimethyl-3-nitropyridine gives the distyryl derivative[185], and ethyl pyridyl-2- and -4-acetate react readily[196a]. Pyridyl-3-acetic acid reacts slowly with benzaldehyde when heated in a mixture of pyridine and piperidine[196b] at 115°.

It has been reported that the best 'acidic' conditions for stilbazole preparation require reaction of picoline with benzaldehyde (1·2 molar equivalents) in acetic anhydride (1 molar equivalent) and acetic acid (1 molar equivalent) with the exclusion of air, whilst in the best 'basic' conditions picoline and benzaldehyde (1·4 molar equivalents) were heated with piperidine (0·2 molar equivalents) in methanol. Yields suggested that, in the 'acidic' conditions, methyl group reactivity decreased in the sequence 4-picoline > 4-picoline methiodide > 2-picoline > 2-picoline methiodide. In the 'basic' conditions, quaternization enhanced the reactivity of both 2- and 4-methyl groups[193].

Although the acetic anhydride method has frequently been preferred to the zinc chloride method because it is more convenient, the claim that it gives better yields of cleaner products[186h] is not generally true. A comparison of the three methods, zinc chloride, Phillips' (in which a quaternary salt is used and the stilbazole obtained by pyrolysis of the stilbazole methiodide) and the acetic anhydride one, showed the first two to be superior to the third[52]. All three methods give the *trans*-stilbazoles[52, 197a] (but see above), and under the conditions of the acetic anhydride method *cis*- are converted into *trans*-stilbazoles (p. 348). An exception appears to be the formation of the *cis*-stilbazole from 2-picoline and salicaldehyde with acetic anhydride[197b].

2- and 4-Picoline evidently give the stilbazoles when treated with potash in benzyl alcohol, but further reaction gives the phenethylpyridines[198].

Reactions using picoline 1-oxides have been less successful than those using picolines. 2- and 4-Picoline 1-oxides did not react with aromatic aldehydes under various acidic and basic conditions[193, 199], but poor yields of stilbazole oxides were obtained using piperidine acetate in boiling toluene[199]. Under these conditions, the picolines performed no better than their oxides. 3-Methyl-4-nitropyridine 1-oxide reacted with benzaldehyde in presence of piperidine, circumstances under which 3-methyl-4-nitropyridine did not react[193]. More successful have been reactions of the oxides catalysed by metal alkoxides or hydroxides (p. 383).

The reaction between aromatic aldehydes and picoline quaternary salts has been much studied because suitably substituted products from such reactions are photosensitizing agents related to the cyanines. The methiodide of 2-*p*-dimethylaminostyrylpyridine was first prepared by Mills and Pope[200] who heated 2-picoline methiodide with *p*-dimethylaminobenzaldehyde and piperidine in ethanol. These conditions have been applied with little variation to numerous cases of the reaction involving aromatic aldehydes (or cinnamaldehyde) and 2-picoline[201–3b, 204] or 4-picoline[202b, 203c, 205] quaternary salts. Pyrrole-[206a] and indole-aldehydes[187] have also been used. There is evidence to suggest the 4-methyl group to be more reactive than the 2-methyl group[193]. 3-Picoline methiodide does not react under these conditions[206b] (but see below), and 2,3-dimethylpyridine methiodide reacts only at the 2-position[186g]. 2,4-[205, 207] and 2,6-Lutidine methiodides[207–8a] give distyryl derivatives, and 2,4,6-collidine methiodide forms the tri-styryl derivative[208a].

Schiff bases have been used in place of aldehydes in reactions of the present type[208b].

When the reaction between 2-picoline methiodide and benzaldehyde in the presence of piperidine is carried out at 15°–18°, the intermediate carbinol can be isolated. Heating converts it into the stilbazole methiodide[208c]. Nitro and halogeno benzaldehydes behave similarly[204].

The picoline quaternary salts form in these reactions, carried out under the conditions described, the quaternary salts of *trans*-stilbazoles[52, 197a]. It is interesting to notice that in the same circumstances ethyl pyridyl-2-acetate methiodide with benzaldehyde gives the product of configuration opposite to that obtained by reaction of ethyl pyridyl-2-acetate with benzaldehyde[196a] (see above):

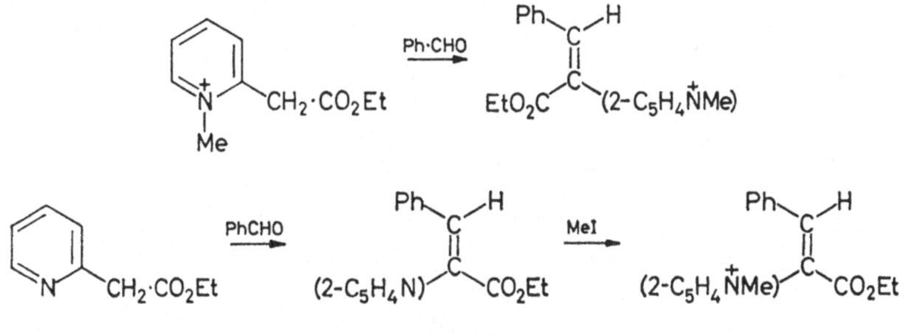

The alkyl groups of picoline quaternary salts do not react with ketones, except in the important intramolecular case in which a 2-picoline and a α-haloketone give an indolizine. This is Tschitschibabin's indolizine synthesis[209]. The reaction is of some generality[151, 210], and though a base is usually used it is not always necessary.

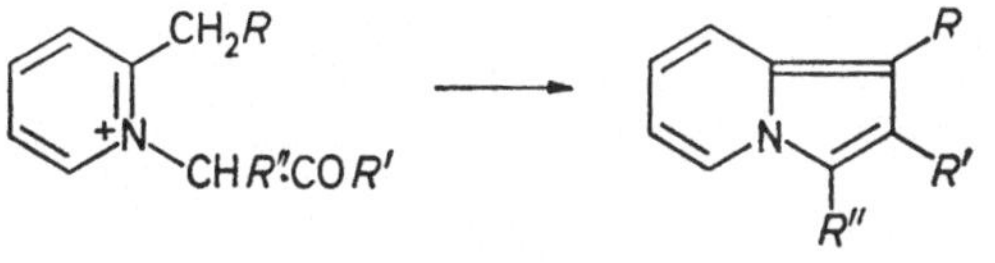

As regards the mechanism of the formation of stilbazole quaternary salts from picoline quaternary salts and aldehydes in the presence of piperidine, it seems probable that pyridone methides are the reacting entities. Mills and Raper[211] showed that the methide from 2-methyl-5,6-benzoquinoline methiodide reacted with p-dimethylaminobenzaldehyde to give the allene (40) in the absence of a catalyst. Protonation of such a structure would then give the stilbazole quaternary salt. However, there is no reason to believe an allene to be an intermediate in the reaction as ordinarily carried out. Such a supposition would imply that higher homologues of the picolines would be incapable of reaction. In fact, 2-ethyl- and 2-β-phenethyl-pyridine methiodides did not react with p-dimethylaminobenzaldehyde in the presence of piperidine, though this failure was attributed to steric factors[203b]. Despite speculation[203b] there are no real grounds for writing anything more than a generalized representation such as shown below. It is relevant to note that stilbazole quaternary salt formation is reversible; heating stilbazole methiodide with alkali gives benzaldehyde and 1-methylpyrid-2-one methide[132e].

There remains the case of the reaction of N-alkyl and N-substituted alkyl groups with carbonyl compounds. 1-Methylpyridinium bromide reacts slowly with benzaldehyde in ethanolic piperidine to give[212a, d, f] the carbinol (41; $R = H$). The reaction is fairly general ($R = Me$, Ar, $ArCH_2$), though the catalyst has usually been 10N sodium hydroxide. Both aliphatic and aromatic aldehydes can be used, and their sequence of reactivities depends on substituents in the usual way. Benzylpyridinium salts have

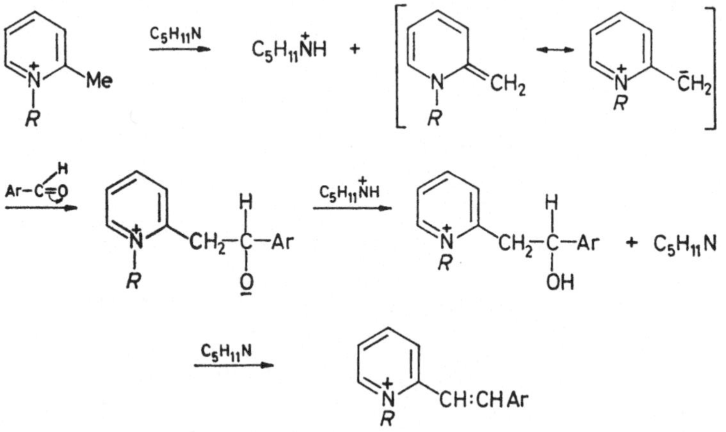

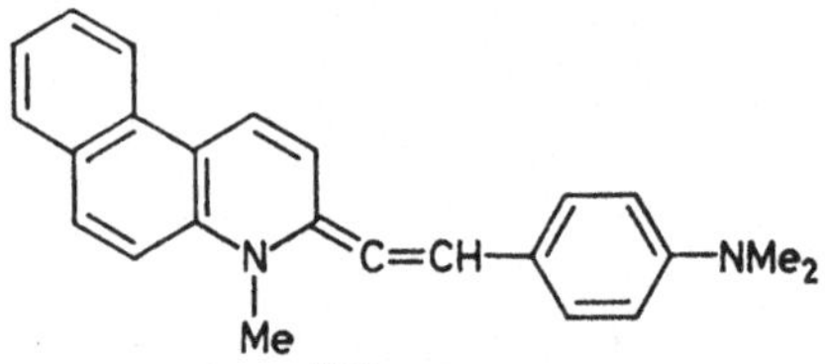

(40)

also been combined with aldehydes under Perkin conditions[212b]. If the pyridinium salt is of the type $C_5H_5\overset{+}{N}CH_2R$, where R is an acyl group, the latter is split off under the conditions of the reaction[212c-e]. Thus phenacyl-pyridinium, reacting through its enol betaine, gives with benzaldehyde (41; $R = H$) and benzoic acid (such reactions are common in this series [p. 390]). With pyridinium ions $C_5H_5\overset{+}{N}CH_2CN$ and $C_5H_5\overset{+}{N}CH_2CS_2Me$ reaction occurs with aldehydes in the presence of piperidine, but the products lose water to give vinylpyridinium salts[212d]. Formaldehyde reacts with phenacylpyridinium in neutral solution and the cation (42) can be isolated as its picrate[212d].

An interesting case is that of the reaction of 3-picoline methiodide with benzaldehyde in the presence of piperidine. Reaction in methanol is very slow (see above), but in s-butanol the alcohol (41; $R = H$; Me at $C_{(3)}$) results. The complete sequence of reactivities of methyl groups in quaternary salts (see above) is thus[193] 4 Me > 2 Me > 1 Me > 3 Me.

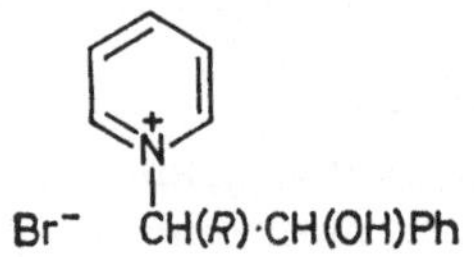

Br⁻ CH(R)·CH(OH)Ph   HOCH₂·CH·CO·Ph
       (41) 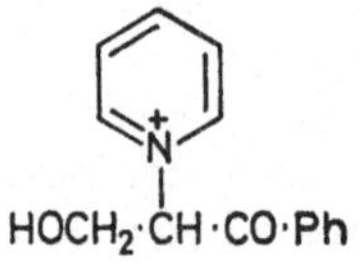   (42)

338

A recent quinolizine synthesis used the reactivity of both C- and N-alkyl groups in picolinium quaternary salts[213a]:

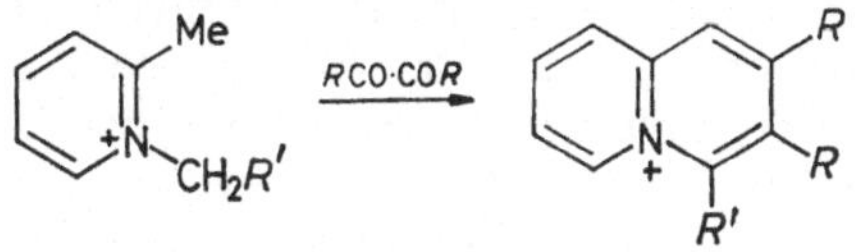

Deuteration and tritiation—These processes were discussed in connection with the acidity of alkyl groups (p. 325).

Diazo coupling—Under conditions where 2-picoline failed to react with diazotized sulphanilic acid or *p*-aminobenzoic acid, diazotized *p*-nitroaniline gave 7 per cent of pyridine-2-aldehyde *p*-nitrophenylhydrazone[213b]. In acetic acid containing sodium acetate, the yield[214] was 58 per cent. Pyridyl-2-acetic acid and its homologues take part efficiently in the Japp–Klingemann reaction[213b]

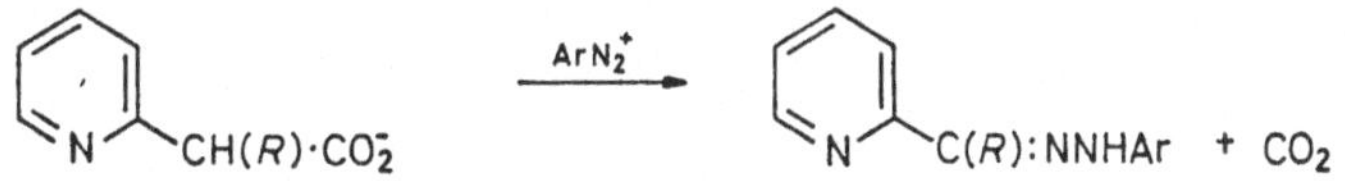

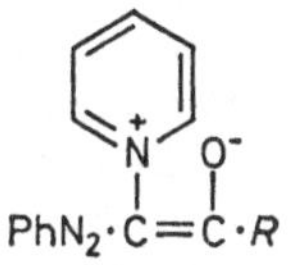

(43; *R*= Me or Ph)

Phenacyl- and acetonyl-pyridinium bromide react with benzenediazonium cation in the presence of sodium acetate, giving the betaines (43) and their salts[215].

Halogenation—Some instances of halogenation occurring in methyl groups attached to the pyridine nucleus have been noted already under the headings both of electrophilic and nucleophilic halogenation. The chlorination of 2-picoline and its homologues in both the nucleus and the methyl group[216] are examples. In modifying Sell's conditions of chlorination, which converted 2-picoline into 3,5,ω,ω,ω-penta- and 3,4,5,ω,ω,ω-hexa-chloropyridine (p. 165), Dyson and Hammick[2] chlorinated 2-picoline in acetic acid heavily buffered with sodium acetate and obtained 2-trichloromethylpyridine. 4-Picoline behaved similarly, but 3-picoline did not react[217a], and direct ω-bromination could not be achieved. Evidence from experiments with quinaldine suggests these chlorinations to be analogous to acid-catalysed halogenation of ketones, and the second and third substitution to be faster than the first[217b].

Although 2-chloromethylpyridine cannot be prepared under the conditions described, chlorination in carbon tetrachloride in the presence of anhydrous sodium carbonate gives this product in good yield (65 per cent), with 5 per cent of 2-dichloromethylpyridine and some of the trichloromethyl compound[218]. This is easily the most convenient method for preparing the monochloro compound.

3-Ethylpyridine with chlorine in hydrochloric acid under illumination gave 3-trichloroacetylpyridine in what may be a radical process[219a].

In some early experiments, 3-picoline and 5-ethyl-2-methylpyridine were converted into 3-bromomethyl- and 5-α-bromoethyl-2-methylpyridine, respectively, by means of bromine in hydrochloric acid[220a]. With bromine in acetic acid, 4-picoline gives a tribromo-derivative which is not 4-tribromomethylpyridine[221]. 2- and 4-Phenacylpyridine[222] and ethyl 2-pyridylacetate[223] are easily monobrominated in the methylene groups.

Some side-chain halogenations occurring in pyridine 1-oxides are essentially nucleophilic in character. Thus 4-picoline 1-oxide gives with phosphoryl chloride predominantly 4-chloromethylpyridine[224a], in a reaction recalling those in which alkylthio-groups are inserted into methyl substituents (see below):

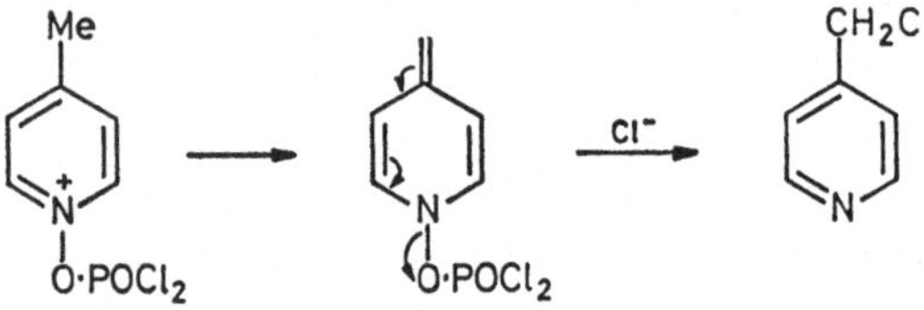

With arenesulphonyl chlorides in benzene, 2-picoline 1-oxide and its homologues give crystalline products (the quaternary salts) which when heated give the corresponding 2-chloromethylpyridines[224b-5]. These cases should be compared with the reactions occurring when the methyl group is absent, or with that of 3-picoline 1-oxide which gives 5-methyl-3-pyridyl toluene-*p*-sulphonate[226] (p. 235).

The King reaction, in which quaternary pyridinium salts are prepared from iodine, pyridine, and a compound containing a reactive methyl group has already been mentioned. It was seen that the picolines are not sufficiently reactive to take part in the reaction, but that 1,2- and 1,4-dimethylpyridinium salts do react (p. 179).

Radical bromination of the side chain is useful with the picolines, and is mentioned below.

Hydroxylation and substituted hydroxylation—Important reactions of this category, notably acetoxylation, occur with picoline oxides. In the parent example, 2-picoline 1-oxide is converted by boiling acetic anhydride into 2-pyridylmethyl acetate[227-8a]. 4-Picoline reacts similarly[227, 229] but not 3-picoline[227a] (p. 234). Higher primary alkyl[227, 230], aralkyl[219b] and hydroxyalkyl groups[219c, 231] at the 2-position react satisfactorily, and the reaction has been applied to 2- and 4-alkylpyridines with nuclear acetoxyl[8], alkoxyl[232], β-alkyl[8, 227b, 230, 233], amino[228b], aryl[234], cyano[228b], ethoxycarbonyl[125, 235-6], halogen[237a] and hydroxymethyl[238a] groups. The lutidines give the monoesters[227a, 237b], in the case of 2,4-lutidine both 4-methyl-2-pyridylmethyl and 2-methyl-4-pyridylmethyl acetate being formed[239]. A related example is the application of the reaction to 2,6-decamethylenepyridine[240]. 4-Benzyl-2,6-dimethylpyridine 1-oxide reacts at the benzyl group[237c].

An important application of the reaction is that in which the pyridylmethyl acetate first formed is converted into the 1-oxide, and this in turn is boiled with acetic anhydride to give a pyridine aldehyde diacetate[8, 227a, 235-6, 238b]. Thus, 2- or 4-picoline 1-oxide forms 2- or 4-pyridine-aldehyde diacetate, and thence the aldehyde[227a]. The reaction fails with α-2-pyridylethyl acetate 1-oxide, and 6-methyl-2-pyridylmethyl acetate 1-oxide gives 2,6-di-(acetoxymethyl)-pyridine[227a].

In all these reactions, acetic anhydride is a convenient and effective reagent, but it is not unique; benzoic[228a] and butyric anhydride[241a] have been used. As mentioned earlier (p. 234), from these reactions of picoline oxides small amounts of pyridones and 3-hydroxypyridines have sometimes been isolated.

The mechanisms of these reactions have excited much interest. Early considerations were centred on two possibilities: nucleophilic attack on the methylene group of an anhydro base and a radical chain reaction. The first might have been intra- or inter-molecular, the latter being necessary for 4-picoline oxides. The intra-molecular case is represented by (A), the inter-molecular case—which also permits representation of the way in which 3-substitution arises[229]—by (B) and the radical chain reaction by (C).

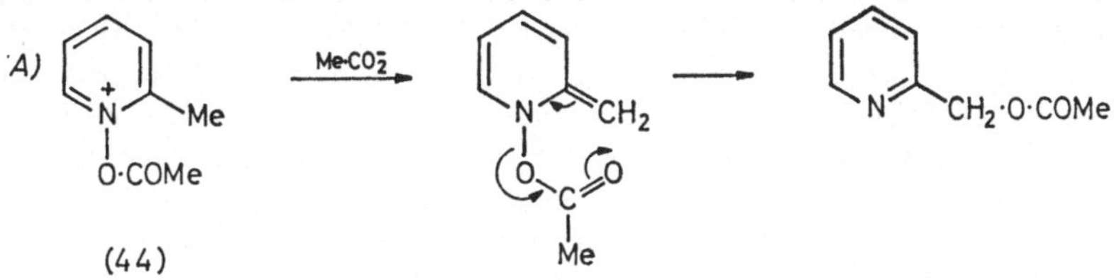

341

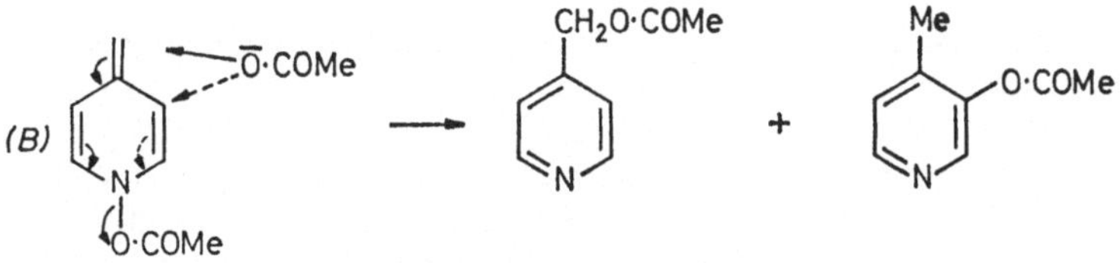

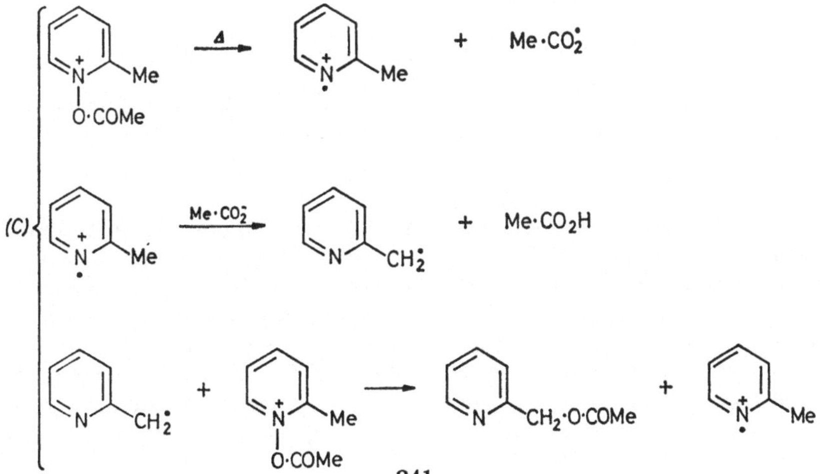

The radical chain mechanism was considered because these reactions have induction periods[227a] and because styrene added to the reaction mixture is polymerized[242]. However, in the reaction of 2-picoline 1-oxide with acetic anhydride, whilst the formation of carbon dioxide and methane demonstrates the presence of free radicals, these are irrelevant to the main reaction, the addition of radical scavengers being without effect on the yield of 2-pyridylmethyl acetate. Further, conversion of 2-picoline 1-oxide by butyric anhydride into 2-pyridylmethyl butyrate is not influenced by the presence of acetate ions. These facts point to the intra-molecular production of radical or ion pairs in a solvent cage as the essential characteristic of the reaction[243a]. Similar experiments in which ^{18}O-labelled acetic anhydride was used and the effect of solvents studied, point to the same conclusion and favour the formation of a radical pair in a solvent cage for the reaction both with and without solvents[241c, d].

Examination of the reactions of 4-picoline 1-oxide with labelled acetic anhydride and labelled butyric anhydride, both of which cause substitution at the methyl group and at $C_{(3)}$, suggested three processes to be at work: first, homolytic cleavage of the N—O bond in the anhydro base followed by radical recombination within a solvent cage; second, similar cleavage followed by recombination after some radical transfer with solvent acid; third, heterolytic cleavage of the N–O bond followed by nucleophilic attack of n-butyrate on the anhydro base. The relative importance of each of these three processes varies with the solvent used[241a, b, d].

In all these considerations it has always been assumed that quaternary derivatives of the kind (44) were the first-formed species and that they give rise to anhydro bases, as in (45) and (46). In seeking to substantiate these assumptions, Traynelis[243b] observed that aryl acetates in which the aryl group carried electron-attracting substituents reacted with 2-picoline 1-oxide to give 2-pyridylmethyl acetate and a phenol. With picryl acetate, 2-picoline 1-oxide gave the salt (47). The picrate anion is too weakly basic to initiate the subsequent changes, but trialkylamine causes the reaction to proceed:

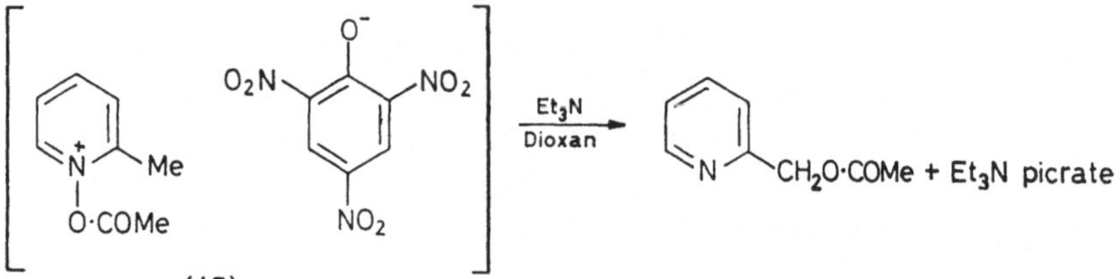

(47)

These results lend general support to the two suppositions mentioned. Further, salts of the type (48) can be prepared[133b, 244]. When these were treated with triethylamine, it proved impossible to detect spectroscopically the formation of the anhydro base. Clearly, during the rapid reaction the concentration of the anhydro base is low, i.e. $k_{-1} \gg k_2$ or $k_2 \gg k_{-1}$. In fact, the second condition holds, for when (48; $R = $ Ph) in which the methylene group was deuterated, was treated with sodium acetate–acetic acid, there had been no loss of deuterium in the starting material after 50 per cent reaction. There is, therefore, rate-controlling conversion of (48) into an-hydro base and rapid rearrangement of the latter to product:

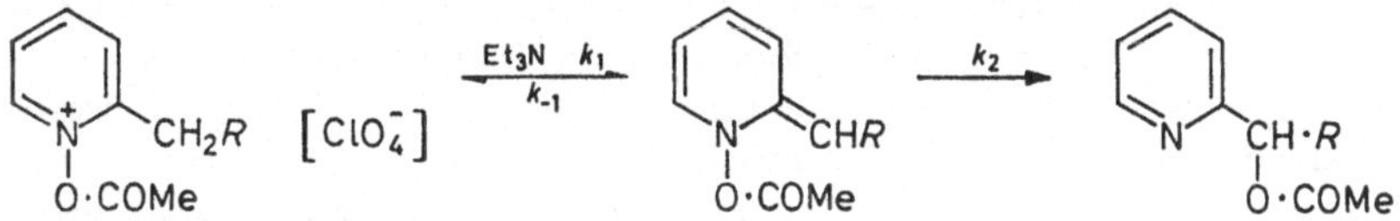

(48; R = H, PhCH$_2\cdot$,
or p-O$_2$N$\cdot$C$_6$H$_4\cdot$CH$_2\cdot$)

In a reaction related to those just discussed, 4-picoline 1-oxide is converted by ketene into poor yields of 4-hydroxymethylpyridine, 3-hydroxy-4-methylpyridine and 4-picoline. 2-Picolines behave in the same way[245], and some acid chlorides produce similar reactions[224b].

Nitration and nitrosation—These processes have rarely been observed[246a]. With boiling nitric acid, 2- and 4-isobutyl-3,5-di-isopropylpyridine are said to be nitrated at the secondary carbon atoms of the isopropyl groups[246b].

Nitrosation is almost equally rare. 2-Picoline failed to react with butyl nitrite under either basic or acidic catalysis. The sodium derivative does react[247a], and the reaction is mentioned under Section (*i*) below. 4-Pyridyl-acetonitrile has been nitrosated at the methylene group, but details have not been given[247b].

Schiff base formation—Anils or Schiff bases of pyridine aldehydes cannot be prepared from picolines and aromatic nitroso compounds. However, 2- and 4-picoline quaternary salts do react with nitroso compounds and the products are of interest because of their photosensitizing properties and because of their possible value as a source of pyridine aldehydes[248a].

2-Picoline methiodide reacted with *p*-dialkylamino-nitrosobenzenes in piperidine[248b] to give the anils (49). Other workers similarly obtained anils from 2- or 4-picoline methiodide and *p*-diethylamino-nitrosobenzene but claimed that the latter gave[249] nitrones of the type (50). In reactions of this kind, two processes are presumably at work

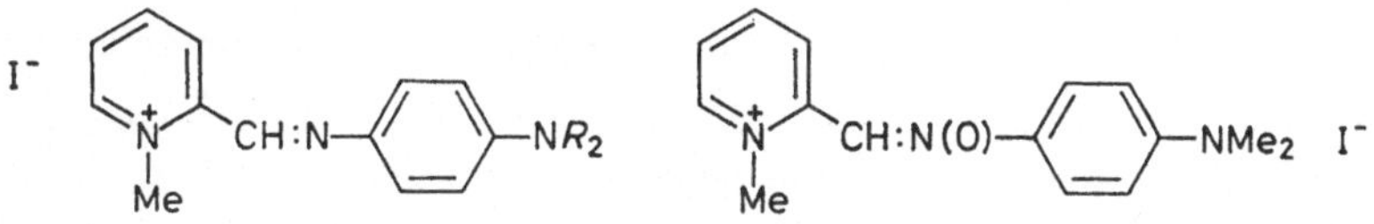

(49) (50)

which have been represented[212d] as

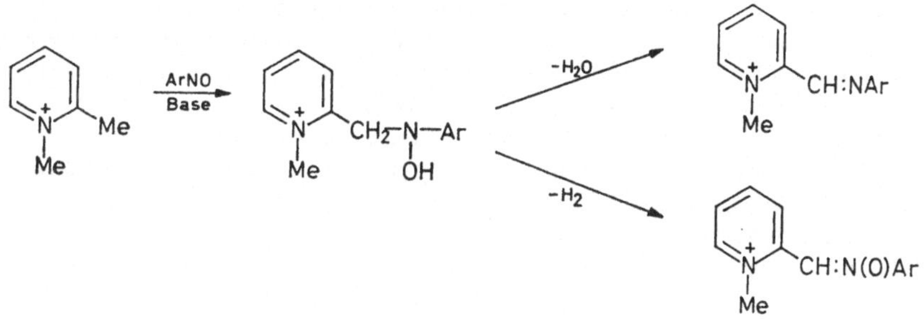

Nitrone formation should be favoured by the use of an excess of the nitroso compound as an oxidizing agent. In re-examining the reaction between *p*-dimethylamino-nitrosobenzene and 2- and 4-picoline methiodides, Kröhnke and his co-workers[250a] found that, although raising the proportion of nitroso compound did increase that of nitrone formed, nevertheless the anil always predominated. Sodium hydroxide was a better catalyst than piperidine; using it and 4 molar equivalents of *p*-dimethylamino-nitroso-benzene with 2-picoline methiodide gave 9 per cent of the product, about a quarter of it nitrone. In contrast, with *p*-dimethylamino-nitrosobenzene in piperidine, the iodide of (51) gave 96 per cent of the nitrone:

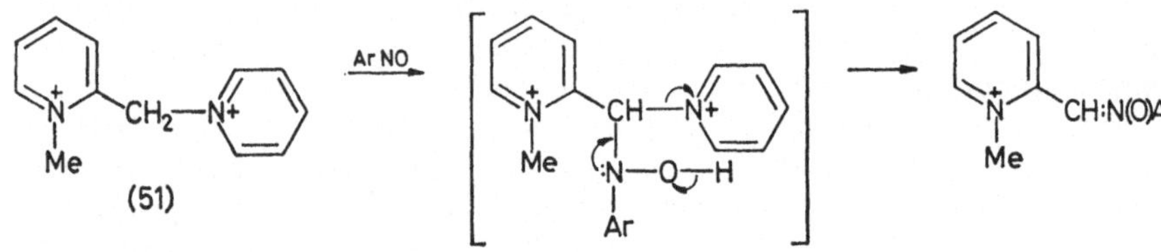

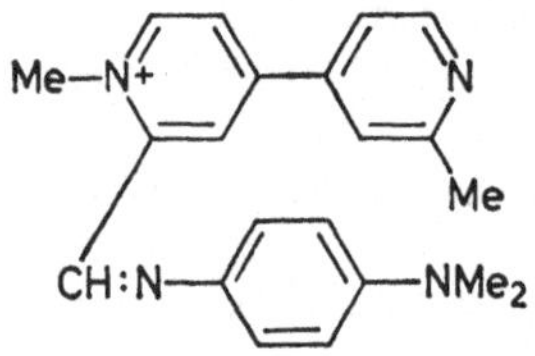

(52)

The pyridinium group leaving with its lone pair, acts as a 'built-in' oxidizing agent[212d, 250a]. Reactions of 2- and 4-picoline methiodides and of (51) and its 4-isomer with other nitroso compounds produce a similar pattern of anil and nitrone formation[251], and anils of the type (52) have been prepared[252].

The reaction of the compound (51) is a special case of the general behaviour of pyridinium salts, $C_5H_5\overset{+}{N}CH_2R$, in which the N-methylene group is suitably activated. Thus, phenacylpyridinium salts (or rather, the enol betaines derived from them) commonly react with *p*-nitrosodimethylaniline to give nitrones of carbonyl compounds. Since the carbonyl compound can be obtained by hydrolysis, this reaction is of general importance as a synthesis of aldehydes and ketones, keto-aldehydes and keto-acids[212d, 250b].

Sulphur bond formation—Interesting examples of substitutions into alkyl side chains in picoline 1-oxide quaternary salts during reactions with thiol anions have been mentioned already (p. 250).

Especially noteworthy is the production in good yield of 1-(4-pyridyl-)-1-benzenesulphonyl-ethane from 4-ethylpyridine and benzenesulphonyl chloride in the presence of triethylamine. 2-and 3-Alkylpyridines do not react in this way[253].

Radical substitutions—The reactions of picoline 1-oxides with acetic anhydride may have radical character (p. 341).

Towards phenyl[254] and t-butoxyl radicals[255] the methyl groups of 3- and 4-picoline are less reactive than that of toluene.

Radical chlorination of 2,6-lutidine with thionyl chloride at 120° gives 2,6-di-(trichloromethyl)pyridine, and 2-methylpyridine-carboxylic acids behave similarly[256].

Radical bromination of the alkyl groups in pyridine homologues by means of N-bromosuccinimide in carbon tetrachloride, in the presence of benzoyl peroxide, is a useful reaction[257]. By this means, 2-picoline is converted into 2-bromomethylpyridine, and to a small extent into the dibromomethyl compound[258-9a]. 2-Picolines with electron-attracting substituents also give the bromomethyl compounds[258, 260a, 261]. 4-Picoline yields 4-tribromomethylpyridine under conditions in which 3-picoline did not react. The order of reactivity[259a] is 4Me > 2Me > 3Me. 2,6-Lutidine gives a number of the possible products[123, 258]. Electron-releasing substituents in 2,4-lutidine make it impossible to direct bromination exclusively into the methyl groups[262].

Higher alkylpyridines have been brominated by this method. 2-Ethylpyridine gave 2-a-bromoethylpyridine[260b], 4-ethylpyridine either the a-bromoethyl[259b] or the a,a'-dibromoethyl compound[259a]. 3-n-Propylpyridine formed the a-bromo-n-propyl derivative[263].

Photoreactions—Some benzylpyridines show photochromotropism connected with the mobility of the protons of the methylene group. Tschitschibabin and his co-workers[132g] noticed that 2-(2,4-dinitrobenzyl)pyridine, a pale yellow compound, turns deep blue in light and back to yellow in the dark. Because the methide from 2-(2,4-dinitrobenzyl)-1-methylpyridinium is dark blue (*Table 6.7*), Tschitschibabin suggested that the colour change was caused by tautomerization in light to the form (53). Crystals of 4-(2,4-dinitrobenzyl)pyridine did not show the phenomenon. A study of crystals of the 2-compound showed[264a] that under radiation they develop dichroic absorption (maxima at 5,450 and 6,200 Å), and it was suggested that the tautomerization postulated by Tschitschibabin occurs during the bending vibration of the $-CH_2-C=N-$ group in the excited molecule. It is now known that solutions of 2-(2,4-dinitrobenzyl)pyridine show the phenomenon, but at room temperature fading is so rapid that it is scarcely visible[264b]. With solutions of 4-(2,4-dinitrobenzyl)pyridine, photochromotropism can be demonstrated at low temperatures. If in this case fading is a unimolecular process, as it is for the 2-isomer, it is unlikely that a form corresponding to (53) is involved, but rather[265] that the tautomer is (54). Extensive studies

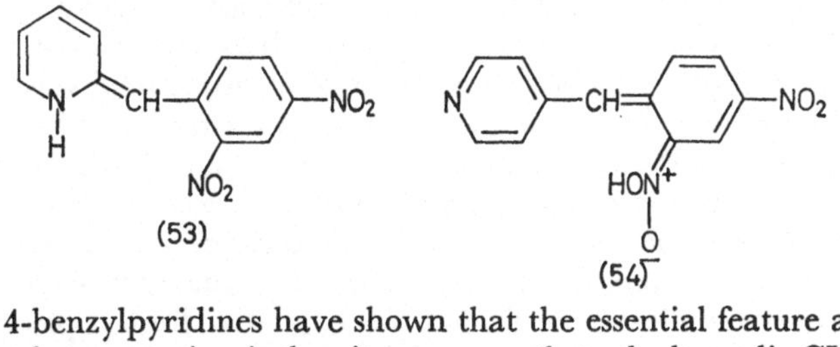

(53)   (54)⁻

of 2- and 4-benzylpyridines have shown that the essential feature associated with photochromotropism is the nitro group *ortho* to the benzylic CH– group, and the weight of evidence seems to point to the aci-nitro form as the tautomer responsible for the colour change[266].

Oxidation and dehydrogenation—The oxidation of alkylpyridines has been extensively studied for two main purposes. Early workers[267] used oxidation in proving the structures of both alkylpyridines and the pyridine-carboxylic acids. Oxidation for the purpose of producing pyridine-carboxylic acids[268] remains a major aim, particularly because of the importance of nicotinic acid. Only recently have methods been devised for oxidizing alkylpyridines to pyridine aldehydes.

As mentioned already (p. 265), oxidation with acidic permanganate disrupts the pyridine ring. In the absence of acid, however, permanganate has frequently been used to produce pyridine-carboxylic acids. The picolines behave satisfactorily, and under the conditions the products are not decarboxylated[267]. Lutidines give either dicarboxylic or methylpyridine-carboxylic acids[267, 269-70]. Higher alkyl groups are in some cases more rapidly oxidized than methyl groups, but not always so[271]. Thus, 3-ethyl-5-[272] and 3-ethyl-6-methylpyridine[273-4] give the methylpyridine-carboxylic acids, but with 4-ethyl-3,5-[275] and 4-ethyl-2,6-lutidine[276] the methyl groups are oxidized. There are indications from the results of oxidizing polymethyl-pyridines that the 3-methyl group is the most reactive[277]. Manganese dioxide in sulphuric acid has been used to oxidize alkylpyridines to the carboxylic acids[278a, 279a].

Benzylpyridines can usefully be oxidized to benzoylpyridines by permanganate[29a, 32, 132g, 280-1a, 282-3].

Under the conditions of the Kuhn–Roth oxidation, the picolines give only small amounts of acetic acid; presumably the methyl groups are oxidized more rapidly than the ring[279b]. Chromic acid has been used to oxidize a benzylpyridine to a benzoylpyridine[281b].

Nitric acid is a useful, cheaper oxidizing agent for the preparation of pyridine-carboxylic acids but has to be used at high temperatures and pressures[284]. It has been employed to oxidize a number of alkylpyridines[89a, 278b, 284], especially for preparing nicotinic acid. It also provides a useful way to isonicotinic acid[285].

Selenium dioxide has frequently been used to oxidize alkylpyridines to pyridine-carboxylic acids. Reactions have been effected in ethyl acetate[286], ethylene glycol[278c], diphenyl ether[287] and xylene[288], and without a solvent[22, 289-90]. 2- and 4-Methyl groups are oxidized to carboxyl groups[22, 286-90] (though yields are not always good), and sometimes a trace of aldehyde is also formed[286, 288-9]. 3-Nitro-4-picoline gives the aldehyde in high yield[291]. Although 3-picoline has been reported to give nicotinic acid[288], it did not react under conditions which worked well with 2- and 4-picoline[290]. In a sealed tube at 160°–170°, only 5 per cent of nicotinic acid was formed[290], and selenium dioxide oxidation of mixtures of alkylpyridines has been used to separate the 3-alkylated compounds[278c]. Selenium dioxide oxidation of 2- and 4-methyl groups in picolines and lutidines proceeds efficiently without a solvent[290]. Under these conditions, 2- and 4-picoline 1-oxide gave picolinic and isonicotinic acids mixed with their oxides. Pyridine bases form complexes with selenium dioxide of the form $B(SeO_2)_2$; that from 3-picolines gives 14 per cent of nicotinic acid[209] when heated at 250°.

A comparison has been made of the reactions of picolines and lutidines, picoline 1-oxides and picoline quaternary salts with selenium dioxide in

pyridine, dioxan, 3-picoline and isoquinoline[292]. In the pyridine homologues, 2- and 4-methyl groups were oxidized to carboxyl groups, but 3-
methyl groups were not affected. As in other conditions, the order of reactivity
was 4-Me > 2-Me ≫ 3-Me. An interesting case was the preparation of
pyridine-pentacarboxylic acid from 2,4,6-trimethylpyridine-3,5-dicarboxylic
acid. The behaviour of the oxides depended on the solvent; 2-picoline
1-oxide in pyridine gave pyridine-2-aldehyde 1-oxide hydrate, but in dioxan,
picolinic acid and pyridine-2-aldehyde 1-oxide hydrate were formed. In
contrast, 4-picoline 1-oxide gave isonicotinic acid 1-oxide in pyridine and
isonicotinic acid in dioxan. 2,6-Lutidine 1-oxide gave pyridine-2,6-
dicarboxylic acid 1-oxide, and the 2,4-compound formed pyridine-2-
aldehyde-4-carboxylic acid 1-oxide. The different behaviours of 2- and
4-methyl groups in the oxides were attributed to protection of the aldehyde
group formed from the 2-methyl group in a hydrogen-bonded hydrate
(55). 1-Butyl-2- and -4-methylpyridinium chlorides gave the corresponding
quaternary derivatives of the carboxylic acids. In each of the three series,
the order of reactivity was 4-Me > 2-Me, and between the series, quaternary
salt > picoline > oxide. 2- and 4-Ethylpyridine but not 3-ethylpyridine
gave pyridine-carboxylic acids.

(55)

A cheaper version of oxidation by selenium dioxide uses concentrated
sulphuric acid and a small quantity of selenium which is repeatedly oxidized
and reduced during the reaction. By this means, picolines and lutidines have
been oxidized, and 3-picoline gives nicotinic acid[88-9a, 293].

The introduction of vapour-phase oxidation of alkylpyridines over vanadium and molybdenum oxides in 1951 made pyridine aldehydes and dialde-
hydes technically available[13d, 44a, 278d]. With a deficiency of air, pyridoins
result[44a, 278e, 294], with a sufficiency, aldehydes result (3-methyl groups are
oxidized, but pyridine-3-aldehyde does not appear to have been made by
this method[44a, 278e, 295]), and with an excess, the carboxylic acids, including
nicotinic acid, and pyridines are formed by decarboxylation[44a, 296-7].
Vapour-phase oxidation in presence of ammonia gives nitriles and pyridine-
carboxylic acid amides[278f, 296].

None of the three picolines is affected by potassium t-butoxide in
t-butanol in presence of oxygen, but change of solvent to dimethylformamide
causes oxidation to the corresponding acids in good yields. Picolyl anions
are probably involved[298].

Electro-oxidation has been used with the picolines, especially as a way
to nicotinic acid[293d, 299]. Small yields of aldehydes have also been obtained
in this way[300]. Photo-oxidation produces some picolinic acid from 2-
picoline[301] and some 2-benzoylpyridine from the benzyl compound[302].

Oxidations of alkylpyridines to alcohols are rare. Oparina[303] obtained
the diol by oxidizing 3,5-di-isopropylpyridine with permanganate, and 2-
benzylpyridine gave the alcohol with mercuric acetate[304].

2- and 4-Picoline undergo the Willgerodt reaction[305a, c-8], giving moderate yields of picolinic and isonicotinic acid thioamides. When aromatic amines are used, diaryl pyridine-amidines and pyridylbenzthiazoles are also formed. The reaction has been used with 4-ethyl- and 4-n-propyl-pyridine[306].

Heated with sulphur, 2- and 4-picoline undergo side-chain dehydrogenation, giving dipyridylethanes, dipyridylethylenes, tetrapyridylthiophen and other products[181b, 278g, 305b]. 4-Nitro-3-picoline 1-oxide with sodium ethoxide forms 1,2-di-(4-nitro-3-pyridyl)ethane dioxide, and under some conditions the ethylene. External oxidizing agents are unnecessary, for the oxide function acts in that capacity. Similar products arise from 4-nitro-3-picoline when treated with oxygen in 30 per cent potash solution[309].

More important dehydrogenations can be carried out catalytically on the higher pyridine homologues. The availability of ethylpyridines, especially 5-ethyl-2-methylpyridine, makes this an important route to vinylpyridines (p. 351). The method is applicable to 2-, 3- and 4-ethyl groups and generally involves reaction in the vapour phase over mixed metal oxide catalysts supported on alumina or silica. In the early forms of the reaction, conversions were low and re-cycling was used[310]. The dehydrogenation of isopropylpyridine has been carried out, and later work on the reaction, which is sometimes accompanied by dealkylation, has produced improvements by the use of diluents such as superheated steam and by depolymerizing some of the product[311].

(c) C- and N-Alkenyl and Alkynyl Groups (cf. Table 5.1)

It has been mentioned that *cis-trans* isomerism has been observed in the stilbazoles (pp. 122, 334, 336). Examples are fairly common[312-3]. Hydrogenation over Lindlar's catalyst of aryl-pyridyl- or dipyridyl-acetylenes gives the *cis*-stilbazoles or *cis*-1,2-dipyridylethylenes[314-5]. *cis*-Stilbazoles readily form the *trans*-isomers when heated or irradiated with iodine, or heated with palladized charcoal[197a, 313, 315].

Irradiation of *trans*-2-styrylpyridine produces the *cis*-isomer. *trans*-2-Styrylpyridine hydrochloride, irradiated in water, gives not only its *cis*-isomer but also a dimer hydrochloride. Irradiation of *trans*-2-styrylpyridine methiodide in benzene suspension yields the bis-methiodide of a dimer, whereas in water it gives the *cis*-isomer and subsequently the dimer. *cis*-2-Styrylpyridine methiodide heated at its melting point produces the *trans*-methiodide. Two stereoisomers of the dimer are known:

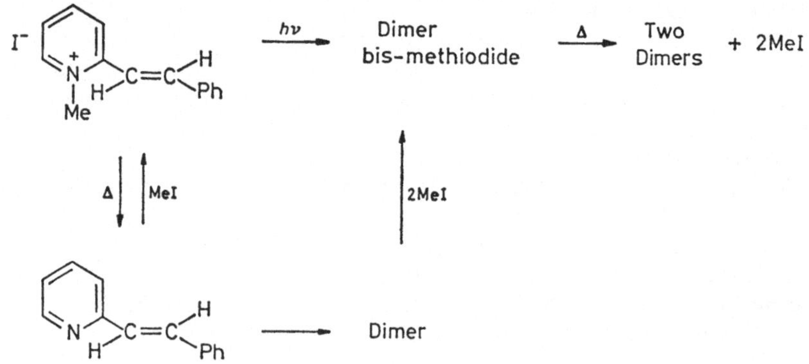

348

Conversion of *trans*-2-styrylpyridine methiodide into the *cis*-isomer by irradiation of the aqueous solution is very rapid compared with the dimerization[197a, 316].

Several examples of the dimerization of styrylpyridines and their salts are now known[317-8]. The degree of dimerization obtained under given conditions depends on the substituents present and, in the case of the salts, on the character of the anion. The dimers are derivatives of 1,3-diaryl-2,4-dipyridylcyclobutane.

Irradiation of stilbazoles in cyclohexane produces mixtures of the *cis*- and *trans*-compounds, and then causes cyclization. In this way, 3-stilbazoles give high yields (~60 per cent) of benz[f]isoquinolines, 2-stilbazoles yield moderate amounts (~30 per cent) of benzo[f]quinolines, and *trans*-4-stilbazole gives slowly a lower (22 per cent) yield of benz[h]isoquinoline[319]. Dipyridylethylenes similarly produce phenanthrolines[320].

The reactions of the unsaturated linkages in alkenyl- and alkynyl-pyridines are conveniently classified as involving attack by electrophiles, nucleophiles or radicals. As regards the first type, lack of information introduces a slight uncertainty, for reagents such as the halogens which most commonly are electrophilic might conceivably be acting as nucleophiles. This possibility has to be admitted for two reasons: first, the alkylpyridines and their salts formally resemble α,β-unsaturated carbonyl compounds [see (56)], acid-catalysed nucleophilic addition of halogen to which is well known, and secondly, many examples of nucleophilic addition are known in this series (see below). Information about the effect of substituents on the rates of what are probably electrophilic additions is needed before their character can be regarded as settled.

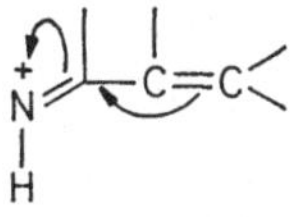

(56) 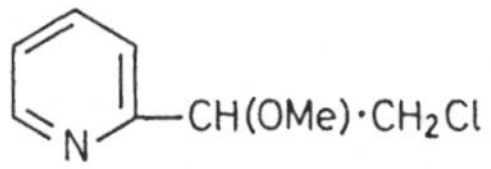   (57)

Electrophilic additions—Chlorine adds readily to 2-styrylpyridines in nonpolar solvents[183b, 321-3]. 2-Vinylpyridines give, with chlorine in methanolic caustic soda, compounds[324a] of the type (57) which are subsequently converted into 2-acetylpyridines.

α,β-Addition of bromine to 2- and 4-styryl- and distyryl-pyridines has usually been effected in chloroform, carbon tetrachloride or carbon disulphide[35, 42b, 43, 178a,b, 181a, 183b, 186c, 315, 325-7], and once in dilute hydrochloric acid[328]. 2-Vinylpyridine gives an unstable dibromide[329].

The initial products formed from styrylpyridines and bromine have been reported to be N-perbromides[222, 314, 330]. These rearrange in hot solvents, giving either α,β-dibromostilbazoles or, by loss of hydrogen bromide, the α-(or β?)-bromostyryl compounds (the orientations are not known); thus, 2-*p*-chlorostyrylpyridine perbromide gives the α,β-dibromo but 4-styrylpyridine a monobromo compound[222]. The role of the perbromides

349

in these additions generally has not been established. Perbromide formation may be responsible for some confusion over the melting point of 2-styryl-pyridine a,β-dibromide[152a].

N-Vinylpyridinium salts do not add on bromine, a fact which has been attributed to the electrophilic character of such additions[113b].

Little is known about the addition of halogen hydracids to alkenylpyridines. In the ozonization of 2-styrylpyridine in concentrated hydrochloric acid (see below), some of the addition product, 2-β-chlorophenethylpyridine, is formed (the orientation is unproved but probable)[181a]. With pyridine and pyridine hydrochloride, 2-vinylpyridine gives β-2-pyridylethylpyridinium chloride[324b]. An attempt to add hydrogen bromide to 2-styrylpyridine failed[331], but 3,5-dimethyl-4-vinylpyridine gave 4-β-bromethyl-3,5-dimethylpyri-dine[259b]. In this case, the orientation was proved and explained in terms of the resonance represented below (the possibility of nucleophilic addition [see above] does not appear to be suggested):

In connection with the failure to add hydrobromic acid to stilbazole, mentioned above, it is noteworthy that when 2-stilbazole dibromide reacts with a reagent such as phenol in amount equivalent to one 'positive' bromine atom, the expected 2-β-bromophenethylpyridine is not obtained. Instead, stilbazole is formed[321].

The hydration of alkynylpyridines has been of practical value, but it is less useful since the discovery of methods for acylating organometallic derivatives of the picolines (p. 379). 2-Pyridylacetylene is hydrated in presence of sulphuric acid to 2-acetylpyridine[329]. 2-Tolazole[35, 42b, 43, 327], 4-tolazole[326] and 2,6-diphenylethynylpyridine[42b] as well as various homologues and substituted analogues[222, 324c, 327b] have been hydrated in the same way to give phenacylpyridines in good yields. The products of hydration in the opposite direction, arylacetylpyridines, have never been detected. The result with 2-pyridylacetylene is surprising, but the orientation of hydration is the same as with o-nitrophenylacetylene[332a]. With the tolazoles the orientation is what would be expected, but there seems to be no reason why examination of examples containing electron-attracting substituents in the phenyl ring should not reveal some degree of hydration in the opposite direction.

Internal electrophilic substitution by a diazonium group can occur in alkenylpyridines of the types (58) and (59). In the case of (58; Ar = p-MeO . C_6H_4), the pyridyl group cannot remove electrons from the α-carbon atom sufficiently to affect the reaction, and the high yield of cinnoline is independent of pH, but with (58; Ar = Ph) cinnoline formation (in poor yield) occurs only in dilute acid. With (59), dilute acid must also be used, and cinnoline formation proceeds better with the 3- than with the

2- and 4-pyridyl compounds[283, 332b]. *o*-Amino-2-tolazole does not give a cinnoline when diazotized[323].

(58)

(59)

Oxidations—4-Vinylpyridine can be oxidized to the glycol by cold permanganate solution[333]. When the acetic anhydride reaction (p. 340) is applied to 2-styryl- and 2-propenyl-pyridine 1-oxides, some of the glycol is produced in each case[334]. Permanganate has more frequently been used to oxidize 2- and 4-styrylpyridines to pyridine-carboxylic acids[168, 180, 186h, 335].

Ozonization of 2- and 4-styrylpyridines provides a practicable way to pyridine-aldehydes[11, 181a, 335-6]. Under different conditions the pyridine-carboxylic acids are produced in good yields[337a, b].

Selenium dioxide oxidizes 2- and 4-styrylpyridine to the diketones and also gives rise[337c] to products of type (60).

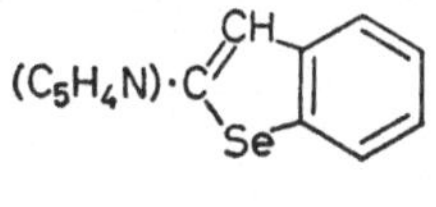

(60) 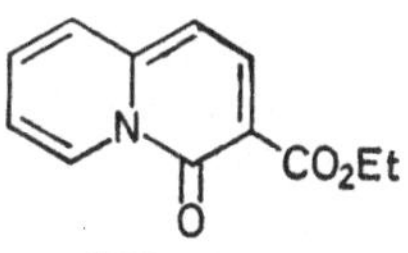   (61)

In the Willgerodt reaction, 2- and 4-vinylpyridine give pyridine-acetamides[337d, e].

Nucleophilic additions—We now come to a number of reactions, of great utility and some theoretical interest, which are clearly nucleophilic additions to the unsaturated linkages of alkenyl- and alkynyl-pyridines. Almost all the important types of reaction in this group found their first exemplification in the work of Doering and Weil[337f] who were the first to realize the importance of the electrophilic character of the β-carbon atoms of vinylpyridines [cf. (56), without the proton on nitrogen]. These reactions, 'pyridyl-ethylations', should be subject to acid catalysis (56), and in many cases strong bases should also assist by converting the nucleophile into the more reactive anion.

Ammonia[337g], hydrazines[338], hydroxylamine[339], primary amines[324d, 337g, 340-2] and secondary amines[155c, 156a, 324d,e,337f,g, 342-3] react satisfactorily with 2- and 4-vinylpyridines. Acid catalysis was useful but not necessary and was generally used for primary amines. In some cases, such as hydroxyl-amine, dipyridylethylation occurred. Among amines of similar basic strengths, steric factors may influence yield. Among the secondary amines used, the imidazoles, benzimidazoles and benzotriazole are noteworthy[343]. Pyrrole reacted only as its sodium salt[156a], whilst indole, which reacted at its nitrogen atom in presence of sodium ethoxide, gave with 4-vinylpyridine in acetic acid, 3-(β-4-pyridylethyl)indole[344a]. The case of quaternary amination was mentioned in connection with the addition of halogen hydracids to vinylpyridines (p. 350). Little has been reported on the reactions of 3-vinylpyridines with amines, but 2-methyl-5-vinylpyridine adds primary and secondary amines in presence of sodium[337g].

The addition of amides and imides to 2- and 4-vinylpyridines[337g, 339, 344b, 345-7a] occurs readily, usually with base catalysis.

Especially important are the additions to 2- and 4-vinylpyridines of compounds containing activated methyl, methylene or methine groups, and including sodium derivatives of malonic and acetoacetic ester types[163c, 337f, 348a, c, 349-50] and ketones[347c, 348d, 350-2], esters[352b-3], nitriles[337g, 347d, 348e, 352b], picolines[350, 354] and indene[344] in the presence of strong bases. Sometimes dipyridylethylation has been observed, as with malonic ester[348b] and some ketones[350]. The addition of acetoacetic ester is satisfactorily catalysed by hydrogen chloride[348a].

2-Isopropenylpyridine adds ketones in the same sense as 2-vinylpyridine[355a], but addition of sodio-malonic ester to ethyl 4-pyridylacrylate goes in the opposite sense[355b].

The product formed by adding sodio-malonic ester to 2-pyridylacetylene cyclizes[329] to (61).

The reactions between picolines and vinylpyridines in presence of sodium require further comment. Recent work[356], summarized below (the starting alkylpyridines were used in excess), requires the revision of earlier results[350]:

$$2-Me \cdot C_5H_4N + 4-CH_2{:}CH \cdot C_5H_4N \longrightarrow 2-C_5H_4N \cdot (CH_2)_3 \cdot 4-C_5H_4N$$
$$4-Me \cdot C_5H_4N + 2- CH_2{:}CH \cdot C_5H_4N$$
$$\longrightarrow 4-C_5H_4N \cdot (CH_2)_3 \cdot 4-C_5H_4N$$

$$2-Me \cdot C_5H_4N + 2CH_2{:}CH-(6-Me \cdot C_5H_3N) \longrightarrow 2-(6-Me \cdot C_5H_3N) \cdot (CH_2)_3 \cdot 2-C_5H_4N$$
$$2,6-Me_2 \cdot C_5H_3N + 2CH_2{:}CH \cdot C_5H_4N$$
$$\longrightarrow 2-(6-Me \cdot C_5H_3N) \cdot (CH_2)_3 \cdot 2-(6-MeC_5H_3N)$$

$$2,6-Me_2 \cdot C_5H_3N + 2CH_2{:}CH-(6-Me \cdot C_5H_3N)$$
$$2,6-Me_2 \cdot C_5H_3N + 4CH_2{:}CH \cdot C_5H_4N$$

$$4-Me \cdot C_5H_4N + 2CH_2{:}CH-(6-Me \cdot C_5H_3N) \longrightarrow 2-(6-Me \cdot C_5H_3N) \cdot (CH_2)_3 \cdot 4-C_5H_4N$$

From all the reactions, 1,3,5-tripyridylpentanes were secondary products. Clearly we have a situation in which the initial dipyridylpropane $A(CH_2)_3B$ gives rise to a new vinylpyridine $A.CH{:}CH_2$ and the anion of a new picoline $B\bar{C}H_2$:

$$A \cdot CH_2-CH_2-CH \cdot B \longrightarrow A \cdot \bar{C}H_2 + B \cdot CH{:}CH_2$$
$$A \cdot \bar{C}H-CH_2-CH_2 B \longrightarrow A \cdot CH{:}CH_2 + B \cdot \bar{C}H_2$$

Consideration of these equilibria and the products formed suggests the order of nucleophilicity

$$2\text{-}(6\text{-Me-C}_5\text{H}_3\text{N}).\bar{\text{C}}\text{H}_2 > 4\text{-C}_5\text{H}_4\text{N}.\bar{\text{C}}\text{H}_2 > 2\text{-C}_5\text{H}_4\text{N}.\bar{\text{C}}\text{H}_2$$

Recently the reaction of 2- and 4-vinylpyridine with the pyrrolidine enamines of cycloalkanones was found to be a good route to 2-pyridylethyl-cycloalkanones[357].

Additions of aryl residues to vinylpyridines have not been observed, except in the case of indole mentioned above.

Hydrogen cyanide adds to 2- and 4-vinylpyridine, giving pyridyl-propionitriles in high yields[337f, 348e].

Addition of sodium alkoxides produces ethers[337f, 347b] and occurs also with 2-pyridylacetylene[329].

The expected sulphonic acids are formed by addition of sulphurous acid to 2- and 4- but not 3-vinylpyridine[337f, 347e], and C–S bonds have also been formed by additon of hydrogen sulphide[358], thiols[358–9a, 360], sulphinic acids[360], thioacids[359a, b] and thiourea[361] (with which 2-methyl-5-vinylpyridine did not react).

Other substances with which addition occurs are hydrazoic acid[362], nitrous acid[359c] and dialkyl phosphites[363].

Reductions—Cases of the reduction of pyridine nuclei in which side chains are also saturated have been mentioned. It is possible, however, to saturate the double bonds of alkylpyridines without modifying the nucleus. For this purpose have been used hydriodic acid[364a], also with phosphorus[186h, 364b], sodium amalgam[365a], Raney alloy and alkali[365a], and various forms of catalytic hydrogenation[365–7]. Benzyl alcoholic potash can evidently reduce stilbazoles (p. 336).

Tolazoles, as mentioned above, can be partially reduced to stilbazoles.

Catalytic reduction of 2-stilbazole 1-oxide can be arranged to give 2-stilbazole, 2-phenethylpyridine 1-oxide or 2-phenethylpyridine[368–9].

Radical and miscellaneous reactions—The photochemical reactions of alkenylpyridines discussed involve radical intermediates. The same may be true of the conversion by irradiation or reaction with nitrosobenzene of 2′-nitrotolazole into an isatogen[322]

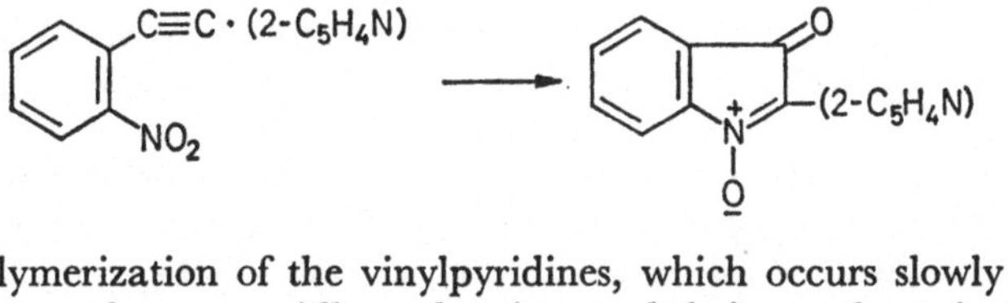

The polymerization of the vinylpyridines, which occurs slowly at room temperature and more rapidly on heating, and their copolymerization with other 'monomers' is of technical importance in the production of synthetic fibres, synthetic rubbers, polyelectrolytes and other valuable substances. The presence of the nitrogen atom makes possible quaternizations to give products with properties different from those arising from purely carbocyclic residues[370]. The polymerization of 1-vinylpyridinium salts has also been studied[113b].

2-Vinylpyridine gives in the Meerwein reaction with arene diazonium chlorides, 2-(α-chloro-β-arylethyl)pyridines, some of which readily lose hydrogen chloride, yielding stilbazoles[371a]. 2- and 3-Vinylpyridines have been used in Diels–Alder reactions with dienes[371b-2].

(d) Amino, Hydrazino, Nitramino, Nitrosamino, Diazonium, Diazo, Azo and Azoxy Groups

The dipole moments (p. 123), spectra (pp. 137, 143), ionization constants (p. 153) and tautomerism (p. 154) of amino-, acylamino-, arenesulphonyl-amino- and nitramino-pyridines have already been discussed.

The general fact that aminopyridines react at the exocyclic nitrogen atom with the common acylating and sulphonylating agents has been noted (p. 196). Like other aromatic amines, the aminopyridines form ureas by reaction with potassium cyanate[373], alkyl and aryl isocyanates[374], phosgene[375] and urea[375-6]. One report claims that with potassium cyanate, 2-aminopyridine gives the biuret[375]. Thioureas are formed analogously[374, 376-7b] and by reaction with carbon disulphide[373b, 377a, c]. From 2-aminopyridine and carbon disulphide in the presence of potash the dithiocarbamate results[378a].

In the same way, pyridylhydrazines give semicarbazides[379] and thiosemi-carbazides[379a, 380a].

With ethyl chloroformate, the aminopyridines form urethanes[377a, 378b].

Phosgene and 2-aminopyridine 1-oxide give the bicyclic compound (62), which reacts with ammonia and alcohols[380b, 381]:

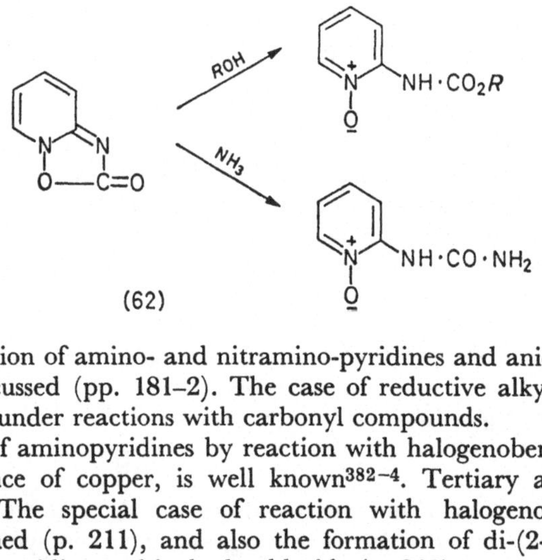

(62)

The alkylation of amino- and nitramino-pyridines and anions from them has been discussed (pp. 181–2). The case of reductive alkylation is mentioned below under reactions with carbonyl compounds.

Arylation of aminopyridines by reaction with halogenobenzenes, usually in the presence of copper, is well known[382-4]. Tertiary amines can be prepared[385]. The special case of reaction with halogenopyridines has been mentioned (p. 211), and also the formation of di-(2-pyridyl)amine from 2-aminopyridine and its hydrochloride (p. 210).

Whilst 3- and 4-aminopyridine give quaternary salts with a substance having an activated halogen atom, such as 2,4-dinitrochlorobenzene[386] (p. 184), 2-aminopyridine reacts with this reagent and with picryl chloride at the exocyclic nitrogen atom[387a], and with alkyl and aryl picryl ethers gives 2-N-alkyl- and -aryl-picrylaminopyridines[387b]. 2-Picrylaminopyridine cyclizes

Table 6.8. Some Aminopyridines and Aminopyridine 1-Oxides
(see also Tables 6.1 and 6.2)

Substituent	M.p., °C	B.p., °C	Derivatives (m.p., °C)
2-NH$_2$[390a]	56	204	acetyl (71)[377a]
			benzenesulphonyl
			(171–2)[391a]
			benzoyl (165)[390a]
			methiodide (149–150)[390b]
			picrate (216–7)[390a]
2-:NH-1-Me[390b]		108/16 mm	picrate (201)[390b]
3-NH$_2$[392a, 393a, 394a]	65	250	acetyl (133–4)[395a]
		115/12 mm	benzoyl (119)[394b]
		131–2/12 mm	
4-NH$_2$[396a]	159		acetyl (150)[377a]
			benzoyl (202)[393b]
			methiodide (187–8)[390c]
			picrate (215–6)[393b]
4-:NH-1-Me[390c]	150–1	180–1/10 mm	picrate (188–9)
2-NH$_2$-3-Me[390e, 1]	26–26·4;	221·5	acetyl (64)
	33·5		
		95/8 mm	picrate (229)
2-NH$_2$-4-Me[390g]	98		acetyl (102–3)
			picrate [227(dec.)]
2-NH$_2$-5-Me[390e, 395b]	76·5–79·5	226·9	
2-NH$_2$-6-Me[391b, 397a,c, 398a]	36·5; 40	208–9	acetyl (90)
		124–5/20 mm	picrate (202)
2-NH$_2$-4,6-Me$_2$[262]	69–70		picrate [205–7(dec.)]
3-NH$_2$-2-Me[391b]	114–6		picrate [236(dec.)]
3-NH$_2$-4-Me[93b]	106	254/735	acetyl (84)
			picrate (179–180)
3-NH$_2$-5-Me[395c]	57–9	153/21 mm	
3-NH$_2$-6-Me[391b, 392b, 399]	95–96·5		picrate [201 (dec.)]
4-NH$_2$-2-Me[396e, 400a]	95·5–96		picrate (193)
4-NH$_2$-3-Me[391c, 401a]	108–9		picrate (219–220)
4-NH$_2$-3-Et[401a, b]	73–4		picrate (196–7)
4-NH$_2$-2,6-Me$_2$[390a, 400d]	186–8	246	acetyl (113)[390a]
			picrate (194–5)[390a]
4-NH$_2$-3,5-Me$_2$[401a, b]	83–4		picrate (226–7)
4-NH$_2$-3-i-Pr[401a, b]	52–3		picrate (156–7)
4-NH$_2$-2,3,5,6-Me$_4$[401a, b]	197–8		picrate [225–6 (dec.)]
2-NHMe[390d, 397a]	15	90/9 mm	picrate (190)
1-Me-2-:NMe[398b]		138/38 mm	
3-NHMe[399b]		118–120/12	
		mm	acetyl (64)
4-NHMe[401a]	124–5		
3-Me-2-NHMe[391d]	~21	113/21 mm	benzoyl (92–93·5)
6-Me-2-NHMe[402]		209–210	acetyl (b.p. 148/14mm)
			picrate (192)
3-Me-4-NHMe[401a, b]	125–6		picrate (199–200)
3,5-Me$_2$-4-NHMe[401a, b]	120–1		picrate (194·5–195·5)
3-Et-4-NHMe[401a, b]	117–8	120–2/0·5 mm	picrate (182–3)
4-NHMe-3-i-Pr[401a, b]	95–6		picrate (159–160)
2,3,5,6-Me$_4$-4-NHMe[401a, b]	118–9		picrate (160–1)
2-NMe$_2$[390b, 397a]		196	picrate (182)
3-NMe$_2$[393a]		108–110/	
		12 mm	dihydrochloride (143)
4-NMe$_2$[393c]	114		methiodide (140)[402b]
			picrate (204)[393c]
2-NMe$_2$-6-Me[397a]		198–200	picrate (153)
4-NMe$_2$-3-Me[401a, b]		68–9/0·9 mm	picrate (172–3)
4-NMe$_2$-3-5-Me$_2$[401a, b]		50–2/0·1 mm	picrate (172–3)
4-NMe$_2$-3-Et[401a, b]		82–4/0·8 mm	picrate (118–9)
4-NMe$_2$-3-i-Pr[401a, b]		53–4/0·1 mm	picrate (138–9)

Table 6.8 (continued)

Substituent	M.p., °C	B.p., °C	Derivatives (m.p., °C)
2-NHPh[385a, 390i, 397a]	108		methiodide [176–9 (dec.)] picrate (219)
1-Me-2-:NPh[398b]	69–70		
2-NH.(2-C5H4N)[403a]	94–5		picrate (225)
2-NH.(3-C5H4N)[396b]	143·8–144·8		
2-NH.(4-C5H4N)[396b]	183–4		
3-NHPh[393d]	142		
3-NH.(3-C5H4N)[396b]	128·6–129·6		
3-NH.(4-C5H4N)[396b]	154–5		
4-NHPh[401c]	175–6		acetyl (112–3)
4-NH.(4-C5H4N)[388]	273–5		picrate [235 (dec.)]
2-N.(2-C5H4N)2[385a]	130		methiodide [204–6 (dec.)] picrate (150–1)
Aminohalogenopyridines			
2-NH2-3-Cl[399d]	134–5		
2-NH2-4-Cl[396d]	130–1		
2-NH2-5-Cl[397b]	134–5		acetyl (171)[399e]
2-NH2-6-Cl[396c]	75		
2-NH2-3-Br[396f]	64·5–65·5		picrate (231·5–232·5)
2-NH2-4-Br[396f]	143–144·5		picrate (261–3)
2-NH2-5-Br[396g]	137		acetyl (175)[399c]
2-NH2-6-Br[396h]	90		
3-NH2-2-Cl[393e]	79–80		acetyl (90–1) diacetyl (67–8)
3-NH2-5-Cl[392c]	82		
3-NH2-6-Cl[393a]	83		
3-NH2-2-Br[401d]	79		
3-NH2-5-Br[403b]	66–7		acetyl (127–8) picrate (212–3)
3-NH2-6-Br[393a]	77		
4-NH2-2,3,5,6-F4[401e]	85–6		
4-NH2-2-Cl[396d]	91–91·5		
4-NH2-2-Br[396f]	97·5–98·5		picrate (129–130)
4-NH2-3-Br[396f]	69·5–70·5		picrate (235–6)
3-Br-4-NHMe[401a]	92·5–93·5		
3-Br-4-NMe2[401a, b]		78–80/0·1 mm	picrate (182–3)
Aminohydroxypyridines			
2-NH2-3-OH[391e, 404]	169–170		N-benzoyl (95–6) tribenzoyl (169–170) picrate (257)
2-NH2-5-OH[391e]	116–7		N-benzoyl (180–1) tribenzoyl (182–3) picrate [225–7 (dec.)]
2-NH2-6-OH[390h, 404]	198–200; 214		acetyl (213) diacetyl (162–3) picrate (191)
3-NH2-2-OH[396i]			picrate (214)
3-NH2-4-OH[392d, 401f]			acetyl (213–4) dihydrochloride (228–230)
3-NH2-5-OH[392a]	158–160		
3-NH2-6-OH[405a]			N-acetyl (232–3)
4-NH2-2-OH[406a]	219–221		

Substituent	M.p., °C	B.p., °C	Derivatives (m.p., °C)
Aminonitropyridines			
2-NH$_2$-3-NO$_2$[405b]	163–4		
2-NH$_2$-5-NO$_2$[405b]	188		
3-NH$_2$-2-NO$_2$[406b]	203–4		
3-NH$_2$-6-NO$_2$[406b]	234–5 (dec.)		
4-NH$_2$-3-NO$_2$[400e]	200		
2-NH$_2$-4-Me-3-NO$_2$[405b]	134–6		
2-NH$_2$-5-Me-3-NO$_2$[405b]	190		
2-NH$_2$-6-Me-3-NO$_2$[405b]	141		
2-NH$_2$-3-Me-5-NO$_2$[405b]	255		
2-NH$_2$-4-Me-5-NO$_2$[405b]	220		
2-NH$_2$-6-Me-5-NO$_2$[405b]	187		
Di- and triaminopyridines			
2,3-(NH$_2$)$_2$[401g]	118·5–119·5		picrate [258 (dec.)]
2,4-(NH$_2$)$_2$[405c]	106–7		
2,5-(NH$_2$)$_2$[220b, 402c]	109–110		dibenzoyl (229–230)
2,6-(NH$_2$)$_2$[397a, 402d]	122		dibenzoyl (176)
3,4-(NH$_2$)$_2$[393g]	218–9		picrate (235–7)
3,5-(NH$_2$)$_2$[393f, 400b, 402d]	110–1		diacetyl (247–8)
			dibenzoyl (211–2)
			diethoxycarbonyl (195–6)
2,3,4-(NH$_2$)$_3$[401h]			dihydrochloride (246–8)
2,3,6-(NH$_2$)$_3$[373a, 398d, 405d, 406d]			2,6-diacetyl (171–5)[398c, e]
			triacetyl (251–2)[401g, 406d]
			dihydrochloride (230)
2,4,6-(NH$_2$)$_3$[402e]	185		
Pyridine 1-oxides			
2-NH$_2$[83d, 405e, 406c]	164–5		acetyl (130–1)
			diacetyl (159–160)
			hydrochloride (153–6)
3-NH$_2$[395a, 405f]	124–5		acetyl (215·5–216·5)[405f]
4-NH$_2$[395d, 400c, 401i]	235–6		hydrochloride (180–1)
			picrate [199–200 (dec.)]
			picrolonate [241–4 (dec.)]
2-NH$_2$-3-Me[405g]	128–9		
2-NH$_2$-4-Me[405g]	130–2		
2-NH$_2$-5-Me[405g]	150–1		
2-NH$_2$-6-Me[405g]	153–4		acetyl (123–4)[405e]
2-NH$_2$-4,6-Me$_2$[405g]	149–150		
4-NH$_2$-2-Me[400a]			hydrochloride (189–191)
4-NH$_2$-3,5-Me$_2$[401b]	227–9		picrate (221–3)
3-Me-4-NHMe[401b]	106–7		picrate (184–5)
3,5-Me$_2$-4-NHMe[401b]	94·5–95·5		picrate (172–3)
4-NHMe-3-i-Pr[401b]			picrate (164–5)
2,3,5,6-Me$_2$-4-NHMe[401b]			picrate (140–1)
4-NMe$_2$-3-Me[401b]		142–4/0·15 mm	picrate (130–1)
4-NMe$_2$-3,5-Me$_2$[401b]	83–4		picrate (115–6)
4-NMe$_2$-3-Et[401b]		178–180/1 m	picrate (139–140)
4-NMe$_2$-3-i-Pr[401b]			picrate (151–2)
3-Br-4-NHMe[401b]			picrate [189–191 (dec.)]
3-Br-4-NMe$_2$[401b]			picrate (160–1)
3-NH$_2$-4-NO$_2$[392e]	237 (dec.)		
2,6-(NH$_2$)$_2$[405e]	206–7		diacetyl (212–3)

readily[387a], giving (63), and an independent proof of the structure of the analogous product from 2-aminopyridine and 1-chloro-2,4-dinitronaphthalene confirms that the initial reactions occur at the exocyclic nitrogen atom[380c]. [The formation of a phenol containing the same tricyclic nucleus as that in (63) from 2-aminopyridine and benzoquinone has been mentioned (p. 194)].

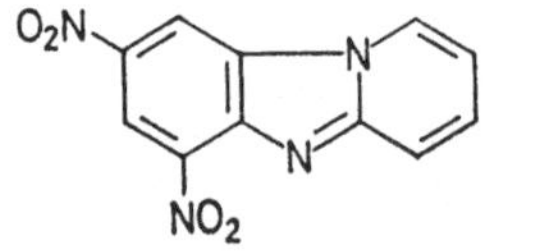

(63) 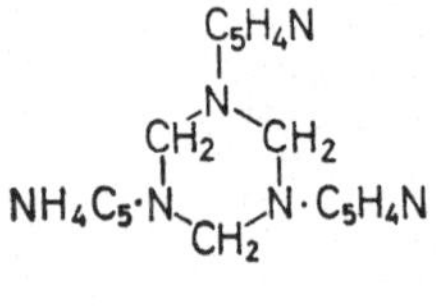   (64)

2-Aminopyridine and the naphthols, in presence of iodine, give 2-naphthylaminopyridines[387c].

4-Aminopyridine is converted by phosphorus trichloride into di-(4-pyridyl)amine[388].

Whilst 5-amino-2-chloropyridine reacted readily with nitrosobenzene in acetic acid, giving 2-chloro-5-phenylazopyridine[387d], 2- and 4-aminopyridine did not react with p-nitrosodimethylaniline under these conditions[389]. The failure has been attributed to protonation of both the aminopyridines and the nitroso compound[407], and indeed 2-, 3- and 4-aminopyridines[407] and their 1-oxides[408] react readily with aromatic nitroso compounds in presence of alkali. The amino-nitropyridines and 2,6-diaminopyridine do not react[407].

Azo compounds can also be prepared satisfactorily from p-nitrosodimethylaniline and the sodium derivatives of 2- and 4-aminopyridine, or from the aminopyridines and disodium derivatives of aromatic nitro compounds[389].

Of the reactions of aminopyridines with carbonyl compounds, those with aromatic aldehydes are the best defined. Numerous cases of Schiff-base formation with 2- and 4-amino-[373b, 403a, 409-13], 3-amino-[413-6a] and diaminopyridines[412] have been reported. Benzaldehyde and 2-aminopyridine at room temperature give the benzylidene derivative, $Ph.CH(NHC_5H_4N)_2$, which gives the Schiff base when heated[403a, 414]. The latter is converted by water into the former, and other examples are known[373b, 414].

With formalin and alkali, 2-aminopyridine is said to give bis-(2-pyridylamino)methane[417a], whilst distillation of the product from formalin and 2-aminopyridine[418] gives (64). In the presence of formic acid, different products are formed (see below). 2-Amino-3- and -5-nitropyridine give the bis-(nitro-2-pyridylamino)methanes with formaldehyde[419-20], whilst 3-aminopyridine evidently produces a polymer of 3-methyleneiminopyridine[393a]. Acetaldehyde and propionaldehyde give products of the type $RCH(NHC_5H_4N)_2$ with 2-aminopyridine[417b, 421], and a compound of the same type arises from 3-aminopyridine and ethyl glyoxalate[422]. Chloral has been reported to react with 2-aminopyridine in the same sense[403a, 410], but the product has also been described[421] as the carbinol $Cl_3C.CH(OH).NHC_5H_4N$.

Although acetophenone would not react with 2-aminopyridine, aceto-phenone diethyl acetal gave the Schiff base[423a]. With acetonylacetone, 2,5-dimethyl-1-(2-pyridyl)pyrrole was formed[423b]. The reactions of amino-pyridines with β-diketones and β-ketoesters are sources of bicyclic com-pounds and have been discussed already (p. 198). Internal cyclization between a 2-amino group and a carbonyl group in a quaternizing residue gives imidazo-[1,2-a]pyridines (p. 182). 3-Aminopyridines form glycosides with glucose[416b].

The reaction of aminopyridines with aldehydes in reducing conditions is useful. Thus, 3-aminopyridine and formaldehyde with zinc and sulphuric acid give 3-dimethylaminopyridine[393a]. The benzylidene derivatives and Schiff bases from aminopyridines and aromatic aldehydes are reduced to arylaminopyridines by formic acid[411, 418, 419b]. The compound from 3-aminopyridine and ethyl glyoxalate, $EtO_2C.CH(3-NHC_5H_4N)_2$ can be catalytically hydrogenolysed to N-3-pyridylglycine[422].

The reaction between 2-aminopyridine and formaldehyde in the presence of formic acid is exceptional, giving bis-(2-dimethylamino-5-pyridyl)-methane[419a]. 2-Dimethylaminopyridine and 2-dimethylamino-5-pyridyl carbinol have also been isolated from the reaction[417a]. The nuclear substitu-tion must precede N-methylation, for 2-dimethylaminopyridine does not react with formaldehyde[419a] (with benzaldehyde and zinc chloride, it gives phenyl-bis-(2-dimethylamino-5-pyridyl)methane[419b]).

Comments on the diazotization of aminopyridines and their 1-oxides have been made in connection with nucleophilic substitutions leading to cyano- (p. 226), halogeno- (p. 229) and hydroxy-pyridines (p. 239). The forma-tion of a diazotate from 2-aminopyridine (p. 229), and of diazo oxides from 3-amino-2- and -4-hydroxypyridine (p. 270) has also been noted. 2,3-[390a, 399d, 405d, 424b] and 3,4-diaminopyridines[424c] give pyridotriazoles when diazotized.

With nitrous acid, alkylamino-[390d, 399b, 425b] and arylamino-pyrid-ines[373b, 426] give nitrosamines. Whilst it has not proved possible to nitrosate 2-acetamidopyridine[428b], a number of 3-acylaminopyridines have been converted into N-nitroso compounds by reaction with nitrosyl chloride in acetic acid[427-8a, c]. The N-nitration of the aminopyridines has been dis-cussed (p. 172).

In addition to their conversion into 1-oxides, there are two main modes of oxidation of aminopyridines, giving the azo (or azoxy) and the nitro compounds. Alkaline hypochlorite oxidizes 2-aminopyridines[407, 414, 429-30a], 3- and 4-aminopyridine[407] to the azo compounds. As well as 2,2'-azopyri-dine, 2-aminopyridine gives some 5-chloro-2,2'-azopyridine which is also produced by oxidation of a mixture of 2-amino- and 2-amino-5-chloro-pyridine[407, 429b]. 2-Amino-5-nitropyridine can give the dichloramine as well as the azo compound[407, 420, 430a]. Addition of persulphuric acid to 3-aminopyridine in sulphuric acid (rather than the other way round, see below) gives 3,3'-azoxypyridine[430b-2]. Alkaline persulphate oxidizes 4-aminopyridine to 4,4'-azoxypyridine[433a].

The oxidation to nitropyridines has been effected with persulphate[434], usually in the form of a solution of perhydrol in 15 per cent oleum. By this means 2-aminopyridines[420, 429a, 432, 435-6a], 2-aminopyridine 1-oxides[405g],

3-aminopyridines[393e, 437] (note the consequence of reverse addition with 3-aminopyridine, see above) and 4-aminopyridines[429a, 436b] give nitro compounds in fair to good yields. 3-Aminopyridine is converted into the nitro compound and its 1-oxide by peroxytrifluoracetic acid[431]. These oxidations still provide the best routes to 2- and 3-nitropyridines, but 4-nitropyridines are more conveniently obtained by nitrating pyridine 1-oxides (p. 174).

The Dimroth rearrangement (p. 269) and the nuclear reduction (p. 263) of aminopyridines have been noted.

The pyridylhydrazines require little discussion. With carbonyl compounds they react normally, and removal of 2- and 4-hydrazino groups can be effected conveniently by oxidation with copper sulphate[377c, 433b, 435b, 436b, c]. Silver acetate has also been used as the oxidizing reagent[430b, 452b]. Hydrazine analogues of the pyridone imines have been obtained[438a] as

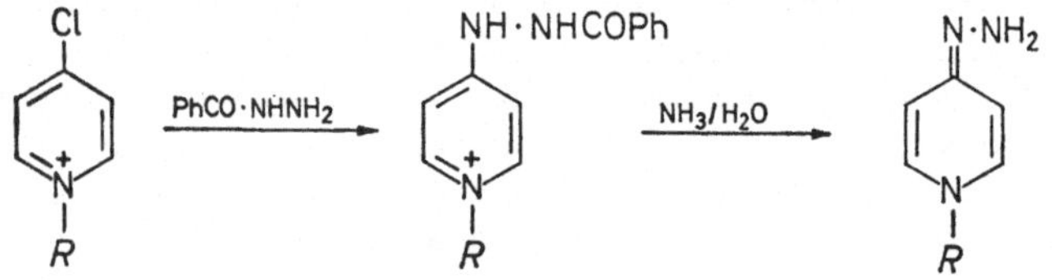

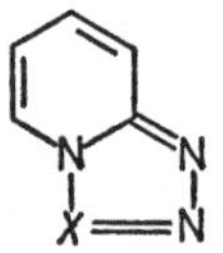

(65)

They have useful coupling properties (p. 361).

A number of pyridotetrazoles (65; $X = N$) have been prepared from 2-pyridylhydrazines and nitrous acid[431, 438c, 40a], and with formaldehyde[433c] and carbon disulphide or thiophosgene[440b], 2-pyridylhydrazine gives (65; $X = CH$) and (65; $X = C.SH$), respectively. The behaviour of pyridylhydrazones in the Fischer synthesis has been mentioned (p. 199).

Arylhydrazinopyridines do not undergo the benzidine rearrangement[441]. They are easily oxidized to azo compounds, even by air. Preparatively the oxidation has been effected by alkali[440c-1], mercuric oxide[379a], and by nitrous acid in acetic acid[441-2a, c, d]. Surprisingly, 3,3',5,5'-tetranitro-2,2'-hydrazopyridine is oxidized by silver acetate to 3,5-dinitropyridine[452b]. 2-Arylhydrazinopyridine 1-oxides give 2-arylazopyridine 1-oxides in alkali, but boiled in alcohol or acetic acid they form 2-arylazopyridines[442b]. Recrystallization of 4,4'-hydrazopyridine 1,1'-dioxide picrate causes oxidation to the azo compound[400c]. Arylhydrazinopyridines are reduced to the two primary amines by zinc and hydrochloric acid[442c, d].

The alkylation of nitraminopyridines (p. 182) and their acid-catalysed rearrangements to amino-nitropyridines (p. 172) have been described. It was noted that 3-nitraminopyridines do not rearrange (p. 172). The behaviour of

3-nitraminopicolinic and 3-nitramino-isonicotinic acid in concentrated sulphuric acid is interesting: they give 3,3'-azoxypyridine-2,2'- and -4,4'-dicarboxylic acid, respectively[442e]. Similarly, the chloro-3-nitraminopyridines produce dichloro-3,3'-azopyridines[443]. The reduction of nitramino- and N-nitroso-alkylamino-pyridines is a useful source of pyridylhydrazines and alkyl-pyridylhydrazines. The nitramines have been reduced with zinc and alkali[424a, 444a], and the nitrosamines with zinc and acetic acid[390d, 399b]. However, with zinc and alkali, 3-nitraminopyridine gives mainly 3-amino-pyridine[425a], and 2-nitramino-5-nitropyridine is said (improbably) to give 5-nitro-2-nitrosaminopyridine which with zinc and acid forms 2,5-diamino-pyridine[220b]. N-Nitroso-N-3-pyridylglycine is converted by acetic anhydride into 3-pyridylsydnone which with hydrochloric acid gives 3-pyridylhydrazine[422].

N-Nitroso-acylaminopyridines serve as sources of pyridyl radicals, as do pyridine-diazonium salts, sodium 2-pyridinediazotate and 3,3-dimethyl-1-(3-pyridyl)triazine (see p. 395).

There is no difficulty in carrying out coupling between pyridine-3-diazonium salts and phenols[377c, 445–6a, 447a, c], and aromatic amines[377c, 408b, 446a, 447a, c] including 2,6-diaminopyridine[447b, d, e]. The diazoamino compound has been prepared from 3-aminopyridine[445], and pyridine-3-diazonium cation gives the triazine with dimethylamine[427].

The diazonium cation prepared from 4-aminopyridine by Witt's method couples with phenols[393b, 448] and with primary aromatic amines[393b], but the value of this method in the coupling with dimethylaniline[393b] is doubtful[389]. 2-Aminopyridine fails also, and indeed no 2-pyridyl azo compounds have been obtained directly by diazotization and coupling. Successful coupling of dimethylaniline to pyridine-4-diazonium cation prepared in a mixture of phosphoric and nitric acids has been reported[406c].

Azo compounds are formed from sodium pyridine-diazotate and phenols or primary aromatic amines in ethanol in presence of carbon dioxide[444b], but coupling did not occur with dimethylaniline[389, 444b].

2-, 3- and 4-Aminopyridine 1-oxide can all be diazotized and coupled successfully[406c, 449].

The hydrazine analogues of pyridine imines (see above), oxidized by ferricyanide in the presence of phenols, give the coupling products[438a] shown

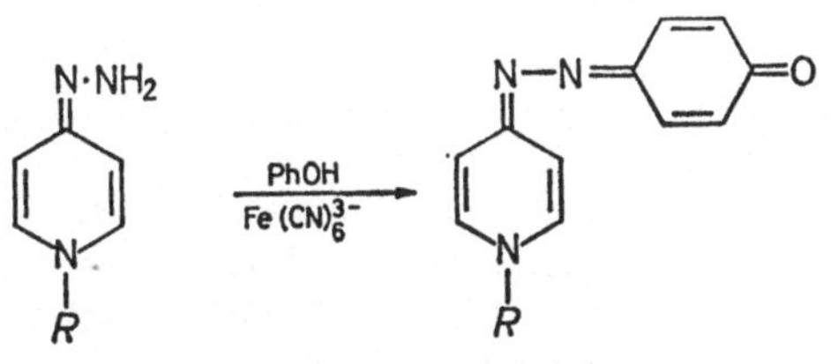

A 4-diazonium group has been removed by reduction with ethanol[448] and a 3-diazonium group by means of hypophosphorous acid[450]. Several pyridine-3-diazonium cations have been reduced to hydrazines by stannous chloride[377c, 446b, 451].

Table 6.9. *Hydrazino- and Hydrazoderivatives of Pyridine and Pyridine 1-Oxide*

Substituent	M.p., °C	Derivative (m.p., °C)
2-NHNH$_2$[433c, 444a]	46–8 (B.p. 140°/20 mm)	picrate (160–1 dec.)
		PhCH: (150–1)
3-NHNH$_2$[377c, 423]	53–5	dioxalate (226–227·5)
		Ph(Me)C: (156)
4-NHNH$_2$[379b]	(B.p. 185–7°/18 mm)	hydrochloride (238)
		PhCH: (195)
5-Cl-2-NHNH$_2$[452a]	135	
2-NHNH$_2$-5-NO$_2$[446c, 453]	205 (dec.)	PhCH: (93–5)
3-Cl-2-NHNH$_2$-6-NO$_2$[452a]	180–2 (dec.)	
2-NHNH$_2$-5-Me-3NO$_2$[436b]	167–8	
2-NHNH$_2$-6-Me-3NO$_2$[436c]	138–9	PhCH: (177–8)
2-NHNH$_2$-6-Me-5-NO$_2$[436c]	119–121	PhCH: (176–8)
2-NHNH$_2$-3,5-(NO$_2$)$_2$[452b]	173	Me$_2$C: (157)
5-Br-3-NHNH$_2$[451]	109–110	
2,3,5,6-F$_4$-4-NHNH$_2$[401e]	56–7	PhCH: (201·5–202·5)
4-NHNH$_2$-3-NO$_2$[454a]	200	
4-NHNH$_2$-3,5-(NO$_2$)$_2$[430b]	161–161·5	Me$_2$C: (170–1)
2-NMeNH$_2$[390d]	(B.p. 105°/10 mm)	picrate (153–5)
		PhCH: (67–8)
3-NMeNH$_2$[399b]	(B.p. 191°/11 mm)	
2-(4-ClC$_6$H$_4$NHNH)[442c]	110–1	hydrochloride (226 dec.)
2-(2,4-Cl$_2$C$_6$H$_3$NHNH)[442c]	113–5	hydrochloride (219–220 dec.)
2-(4-NO$_2$.C$_6$H$_4$NHNH)[442a]	158–9	hydrobromide(227–8 dec.)
2-(2-C$_5$H$_4$N.NHNH)[455a]	168	hydrochloride (238–240)
		hydrobromide (305)
6-Cl-2-NHNHPh[442d]	126–8	
3-NHNHPh[455b]	134–5	
3-(3-C$_5$H$_4$N.NHNH)[454b]	202	
4-NHNHPh[441]	171–2	hydrochloride (255 darkening)
4-(4-ClC$_6$H$_4$NHNH)[441]	106–7	hydrochloride (243 darkening)
4-(2,4-Cl$_2$C$_6$H$_3$NHNH)[441]	(120) 135	
4-(4-C$_5$H$_4$N.NHNH)[369]	240 (dec.)	
2,6-Me$_2$-4-(PhNHNH)[379a]	(160) 172–180	
1-Oxides		
4-NHNH$_2$[456a]	183–5 (dec.)	picrate (192–3 dec.)
2-NH.NHPh[457]	136 (dec.)	
2-(4-BrC$_6$H$_4$NHNH)[442d]	139 (dec.)	hydrobromide (177–8)
2-(4-ClC$_6$H$_4$NHNH)[442d]	132(dec.)	
4(4-C$_5$H$_4$NO.NHNH)[400c]		picrate 2H$_2$O (> 320)

The protonation (p. 148) and N-oxidation (p. 197) of azopyridines have been discussed.

4-Phenylazopyridine gives, with hydrochloric acid, 4-(4-chlorophenyl-hydrazino)pyridine. By oxidizing this product to the azo-compound the reaction can be repeated, giving 4-(2,4-dichlorophenylhydrazino)pyridine[441], and hydrobromic acid produces analogous products. 2-Phenylazopyridine[442c, d] and its 1-oxide[442d] react similarly, though 3-phenylazopyridine does not[455b], and 2,2'-azopyridine gives 2,2'-hydrazopyridine[455a]. Addition reactions occur with Grignard reagents and diazoalkanes. Thus, 2-phenylazopyridine gives with the former[455a, 460a] products of the type (66), and

Table 6.10. *Some Nitramino- and Nitrosamino-pyridines*

Substituent	M.p.,°C	Substituent	M.p.,°C
2-NHNO₂[399c, 405b, 444a]	185–9 (dec.)	2-NMeNO[309a]	(b.p. 123–4°/ 30 mm)*
3-NHNO₂[425]	170–5 (dec.)	3-NMeNO[399b]	(b.p. 135°/ 10 mm)
4-NHNO₂[393b, 458]	243–4	2-NPhNO[373b, 426]	102
2-NHNO₂-3-NO₂[459]	137 (dec.)	2-NMeNO-3-NO₂[425b]	102–3
2-NHNO₂-5-NO₂[444a]	—	2-NMeNO-5-NO₂[425b]	112–3
2-NMeNO₂[425b]	30–1	3-N(NO).CH₂CO₂H[423]	158–9 (dec.)
2-NMeNO₂-3-NO₂[425b]	89†		
2-NMeNO₂-5-NO₂[425b]	59–60	3-N(NO).COMe[427]	—
3-NMeNO₂[399b]	54–5	3-N(NO).COCHMe₂[427]	78–9
		3-N(NO).COMe-2-Cl[428c]	45–6 (dec.)
		3-N(NO).COMe-6-Cl[428c]	69–70 (dec.)
		3-N(NO).COMe-2-OMe[428a]	oil
		3-N(NO).COMe-4-OMe[428a]	84 (dec.)
		3-N(NO).COMe-6-OMe[428a]	51·5–52·5 (dec.)

* Picrate[???d], m.p. 186–7°.
† Giving a turbid liquid which cleared at 105°.

with the latter (R_2CN_2), azopyridines form[460b, c] products (67). 3-Phenyl-azopyridine is reduced slowly by phenyl magnesium bromide without phenylation[455b].

$$2\text{-}(C_5H_4N)\cdot NH\cdot N\,(R)\,Ph \qquad\qquad (C_5H_4N)\cdot N\!-\!N\cdot Ar$$

(66) (67)

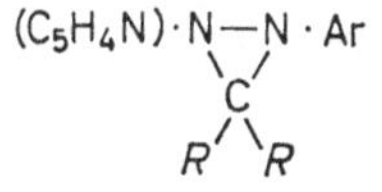

Table 6.11. *Some Azo- and Azoxyderivatives of Pyridine and Pyridine 1-Oxide*

Substituent	M.p., °C
2-PhN₂[407]*	32–4
3-PhN₂[407]†	52–3
4-PhN₂[407]‡	98
2-(4-ClC₆H₄N₂)[389, 442c]	115–8; 123–4
2-(2,4-Cl₂C₆H₃N₂)[442c]	147–8
2-(2,4-[HO]₂C₆H₃N₂)[444b]	186–8 (dec.)
2-(4-Me₂NC₆H₄N₂)[389]	cis 108–9; trans 111–2
2-(4-MeOC₆H₄N₂)[389]	50–2
2-(4-MeC₆H₄N₂)[389]	72–4
3-(2-ClC₆H₄N₂)[407]	60
3-(2,4-[HO]₂C₆H₃N₂)[445]	218 (dec.)
3-(4-Me₂NC₆H₄N₂)[408b]	123
4-(4-ClC₆H₃N₂)[441]	99–100
4-(2,4-Cl₂C₆H₄N₂)[441]	110–1
4-(2,4-[HO]₂C₆H₃N₂)[393b]	—
4-(4-Me₂NC₆H₄N₂)[389]	207–9
2,6-(NH₂)₂-3-PhN₂[463]	137
2-(2-C₅H₄N.N₂)[407,414]	cis 87; trans 83, 87
3-(3-C₅H₄N.N₂)[407]	cis 82; trans 140
4-(4-C₅H₄N.N₂)[407]	108–9
4-(4-C₅H₄N.NO:N)[433a,461a]	126–7; 131–2

Table 6.11 (continued)

Substituent	M.p., °C
1-*Oxides*	
2-PhN$_2$[408a]	110
3-PhN$_2$[408a]	78–9
4-PhN$_2$[408a]	150
2-(4-BrC$_6$H$_4$N$_2$)[442b]	204–6
2-(4-ClC$_6$H$_4$N$_2$)[442b]	200–1
2-(4-Me$_2$NC$_6$H$_4$N$_2$)[406c]	181–2
2-(4-NO$_2$C$_6$H$_4$N$_2$)[442a]	214–5
3-(4-Me$_2$NC$_6$H$_4$N$_2$)[406c]	189–191
4-(4-Me$_2$NC$_6$H$_4$N$_2$)[406c, 408b]	210–1, 212
4-(4-C$_5$H$_4$N.N$_2$)[461b]	160–1
4-(4-C$_5$H$_4$NO.N$_2$)[400c, 461b]	243 (dec.); 253–4
2-(C$_6$H$_5$NO:N)[457]	138
3-(C$_6$H$_5$NO:N)[455b]	89–90
4-(C$_5$H$_4$N.N:NO)[461b]	160–1
4-(C$_5$H$_4$NO.N:NO)[400c, 461b]	223–4; 236–7

* Picrate[442b], m.p. 141°. The picrate (m.p. 135–7°) and methiodide (m.p. 162–3°) of [383] appear to be those of 2-phenylazopyridine rather than 2-(4-dimethylaminophenylazo)pyridine to which they are referred.
† Polymorphous; the m.p. is that of the stable form[407].
‡ Methiodide[442], m.p. 188°.

The reduction of azo- to hydrazo-pyridines has been carried out with zinc and alkali[454b], sodium sulphide[442e–3] and stannous chloride[414, 443]. Azoxy compounds have been reduced to azo compounds by zinc and alkali[454b, 461a] and by arsenite[420], and to hydrazo compounds by stannous chloride[442d, 457]. With azoxypyridine 1-oxides the possibilities are more complicated, as illustrated[369, 457]:

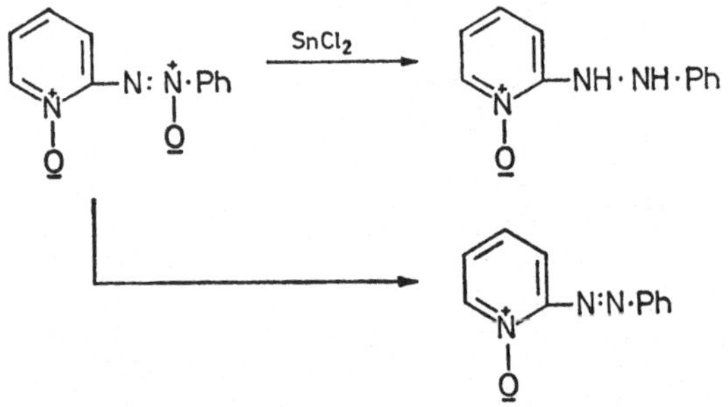

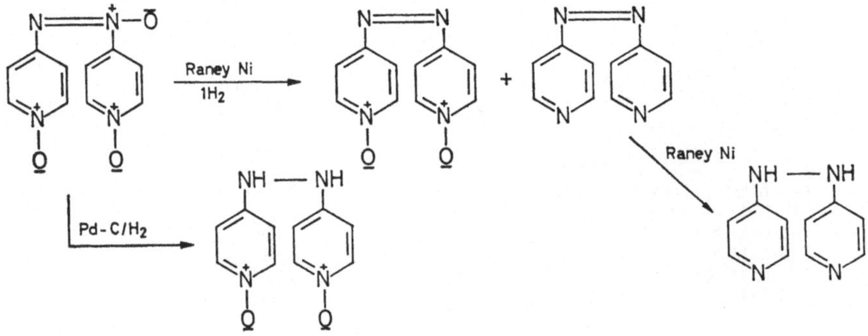

Zinc and acetic acid convert 4,4'-azoxypyridine 1,1'-dioxide into 4,4'-azopyridine 1,1'-dioxide[400c], and hydrazine in presence of copper reduces it to 4-aminopyridine 1-oxide[464].

The stable forms of the azopyridines are *trans*-isomers. In the case of *trans*-4,4'-azopyridine 1,1'-dioxide, the configuration has been demonstrated by x-ray studies (p. 120). Under preparative conditions *trans*-isomers are isolated, but one report[389] claims the isolation of both forms of 2-(4-dimethyl-aminophenylazo)pyridine. The dipole moment of 2,2'-azopyridine shows it to be the *trans*-compound, and illumination of solutions containing it causes partial isomerization to the *cis*-form[465]. Similar isomerization occurs with 4,4'-azopyridine 1,1'-dioxide[466]. Chromatography of solutions of 2,2'- and 3,3'-azopyridine after irradiation made possible the separation of *cis*- and *trans*-forms[407].

(e) Aryl Groups

The phenylpyridines and polypyridyls are included in *Table 5.1*. The dipole moments (p. 127), spectra (p. 136) and ionization constants (p. 147) of these compounds and their 1-oxides, and the ability of the polypyridyls to form complexes with metals (p. 159) have been noted.

It was pointed out that the ultra-violet spectra of 2-arylpyridine quaternary salts and 1-oxides indicate the presence of steric inhibition of resonance, and the same is true of those of 4-aryl-3,5-diethoxycarbonyl-2,6-lutidines[467]. These examples recall, of course, the hindered diphenyls, and in pursuing the similarity a number of attempts have been made to achieve optical resolutions of suitably substituted arylpyridines[468]. Success was achieved[469] with the 3,3'-bipyridyl (68), and salts of (69) with different rotations were obtained[470].

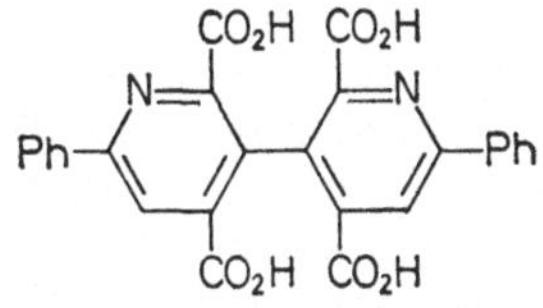

(68) 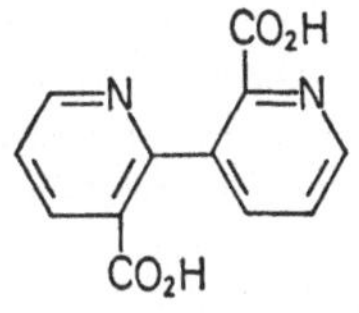   (69)

The catalytic hydrogenation (p. 263), their behaviour on oxidation (p. 265) and the nitration of the phenylpyridines and their derivatives (p. 175) have been discussed. The interesting intramolecular cyclizations (70)–(71) are brought about by hydrobromic acid[471].

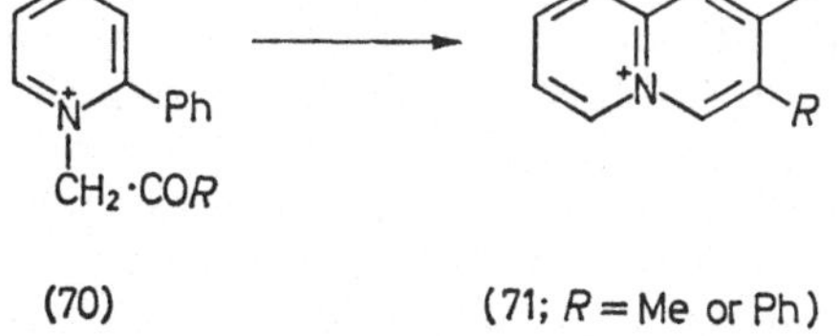

(70)   (71; R = Me or Ph)

365

(*f*) *Cyano Groups*

Table 6.12 lists the cyanopyridines and a selection of substituted cyanopyridines. The dipole moments (p. 125), spectra (p. 130) and ionization constants (p. 148) of the cyanopyridines have been mentioned.

In a number of reactions, such as those leading to amidines and the formation of ketones by reaction with organometallic reagents, the cyanopyridines behave normally, and these call for no comment. The simultaneous attack of propyl magnesium bromide on the cyano group and $C_{(4)}$ in nicotinonitrile has been mentioned (p. 200).

Table 6.12. Some Cyanopyridines[16b]

Substituent	*M.p.* (*b.p.*), °C	*Derivative* (*m.p.*, °C)
2-CN[472a, b, 473–4a]	29 (118–120°/25 mm) (212–5°)	methiodide (183–4)
3-CN[472c]	50–1 (204–8°)	
4-CN[472d, 475a]	80–2	picrate (200–3); methiodide (200–2 dec.)
2-CN-3-Me[472d]	87–90	methiodide (93–5)
2-CN-4-Me[472d]	89–91	methiodide (100–2)
2-CN-5-Me[476]	75–7 (134–6°/20 mm)	
2-CN-6-Me[472d, 476]	71–72·5 (112–4°/15 mm)	methiodide (105–8)
3-CN-2-Me[477]	58	methiodide (170)
3-CN-4-Me[15, 472e]	43–4 (64°/1–2 mm)	methiodide (185–186·5)
3-CN-5-Me[478–9]	84–5	hydrochloride (210)
3-CN-6-Me[475b, 479]	84–5	
4-CN-2-Me[472d]	46–8	
4-CN-3-Me[472d]	51–52·5	
2-CN-4,6-Me_2[472d]	55–6	hydrochloride (100–2)
4-CN-2,6-Me_2[472d]	77–81	hydrochloride (178–181)

Since ring-synthetic methods are available which produce cyanopyridines, the hydrolysis of these compounds is of some importance. Its ease varies greatly with the kind and degree of substitution of the ring. Monosubstituted 3-[479] and 4-cyano-pyridines[480] are hydrolysed to the acids by alkali. Nicotinonitrile and 4-methylnicotinonitrile are hydrolysed to the amides in high yield by the hydroxyl form of an anionic exchange resin[15, 481a], and the second compound gave ethyl 4-methylnicotinate with alcohol and sulphuric acid[15]. Nicotinonitrile and 2-amino-5-cyanopyridine give high yields of the amides when hydrolysed with aqueous ammonia[481b–2]. 2-Cyanopyridines form the amides on acid hydrolysis[401d, 483], though 2-cyano-5-nitropyridine can give the acid as well[483]. Alkaline hydrolysis of 3-cyano-2-hydroxy-6-propylpyridine readily yields the acid[484], but the presence of a 4-substituent in similar compounds complicates the situation. Thus, 3-cyano-2-hydroxy-4,6-dimethylpyridine with either fused alkali or concentrated acid gave only the amide[485], 3-cyano-2,4-dimethylpyridine with alkali produced the amide[486], and hindered nitrile groups generally either cannot be hydrolysed or give the amide[475b]. From 3-cyano-6-ethoxymethyl-2-hydroxy-4-methylpyridine, fuming sulphuric acid produces the amide, whilst alkali forms the acid or, with longer treatment, the decarboxylated product[487a]. From 3-cyano-4-ethoxycarbonyl-2-hydroxypyridines, boiling with concentrated hydrochloric acid yields 2-hydroxy-isonicotinic acids[488-9], a conversion representing an inversion of the usual order of ease of decarboxylation (p. 316). The cyano group in cyanopyridine quaternary salts suffers nucleophilic displacement by alkali (p. 239).

The hydrolysis of nicotinonitrile to nicotinamide (see above) is of commercial importance and is the subject of several patents[490].

The modification of cyano groups during the N-oxidation of cyanopyridines (p. 198) produces different results according to the position of the cyano group. Thus, with hydrogen peroxide at pH 7·5–8, picolinonitrile gives picolinamide 1-oxide (70 per cent), nicotinonitrile forms nicotinamide 1-oxide (44 per cent) and isonicotinonitrile yields isonicotinamide (45 per cent) with only 4 per cent of the oxide. The first result is ascribed to intramolecular oxidation by the peroxycarboximidic acid, the second to similar intermolecular oxidation. The small proportion of 1-oxide from the 4-isomer may be due merely to deactivation by the 4-substituent[487b].

The reduction of cyanopyridines to aminomethylpyridines is a reaction of some importance, particularly in its application to polysubstituted compounds related to pyridoxin. It has been effected by catalytic hydrogenation under a variety of conditions. Raney nickel works efficiently with 3- and 4-cyanopyridines, sometimes giving in the first case small proportions of secondary amine[121, 491-2]. Palladium catalysts cause simultaneous hydrogenolysis of chloro-substituents[484, 493-5a, 496]. Palladized barium carbonate has been used to reduce a nitro group and hydrogenolyse a chloro-substituent, and then Raney nickel to reduce the cyano group[475b, 497]. Similarly a platinum catalyst has served to reduce a nitro group before hydrogenolysis of halogen and reduction of a cyano group by mixed platinum and palladium catalysts[495c, 498-500]. A platinum catalyst has occasionally been used to reduce the cyano group[495b].

Chromous acetate reduces 2-, 3-, and 4-cyanopyridines to the aminomethyl compounds and has the virtue of not removing halogen substituents, though excess of this reagent cleaves the aminomethyl groups[501-2]. Electrolytic reduction of a 3-cyanopyridine has been described[503].

Lithium aluminium hydride reduces 3-cyanopyridines to the aminomethyl compounds[504]. It has been seen (p. 258) that lithium aluminium hydride can attack the pyridine nucleus; in compounds such as 3,5-dicyano- and 3,5-diethoxycarbonyl pyridine, it is reduced (to the dihydro stage) and the substituents remain intact[505]. In the borohydride reduction of quaternary salts (p. 261) the nucleus is reduced and a 3-cyano group left untouched. Controlled reduction with lithium aluminium hydride and with sodium triethoxyaluminium hydride failed to produce more than a trace of aldehyde from 3-cyano-4-methylpyridine[15], but the second reagent worked well with nicotino- and isonicotino-nitrile[102b]. Nicotinonitrile gives good yields of pyridine-3-aldehyde with lithium triethoxyaluminium hydride[103c].

Nicotinonitrile gives a high yield (70 per cent) of the aldehyde when reduced with Raney nickel and sodium hypophosphite. Raney nickel with formic acid is less efficient[506].

The Stephen reduction works well with nicotinonitrile[99a] but not generally with cyanopyridines[15, 289]. It converts 4-cyano-2,6-dihalogenopyridines into the aminomethyl compounds[507].

Polarographically the cyanopyridines are significantly easier to reduce than benzene and aliphatic nitriles[508a]. Such reduction of isonicotinonitrile in acid solution gives 4-aminomethylpyridine, whilst in alkaline solution the C–CN linkage is split, giving cyanide[508b].

(g) Halogen Substituents

The dipole moments (p. 125), spectra (pp. 131, 145) and ionization constants of the halogenopyridines have been discussed. Their simple properties are recorded in *Table 6.13*.

Table 6.13. Some Halogenopyridines and Halogenopyridine 1-Oxides
(see also Tables 6.2, 6.8 and 6.18)*

Substituent	M.p., °C	B.p., °C	Derivatives (m.p., °C)
2-F[509a, 510a, 511a]		124·8–125·4/ 755 mm	
3-F[377c, 509a, b 511a]		106·0–106·6/ 734 mm	
4-F[512]		24–5/20 mm	picrolonate (195–7)
2-Cl[509a]		167·3–167·5/ 744 mm	2HgCl₂ (177–8)[513a]
3-Cl[377c, 509a]		145·2–146·2/ 722 mm	hydrochloride (60)[377c]
4-Cl[514a]	−42·5	147–8[515]	hydrochloride (141·5–142·5) HgCl₂ [250–260 (dec.)][513a] picrate (146)[511b]
2-Br[373b, 377c]	−40·1†[514c]	192–4 86·8–87·5/ 21·5 mm	HgCl₂ (184–5)[514c] picrate (105–6)
3-Br[377c, 509a]	−29·5†[514c]	172–3/752 mm 73·5–73·8/23·4 mm	HgCl₂ (201·5–202·5)[514c] picrate (154)
4-Br[514a]	8·5–9·5	25·5–26·5/0·3 mm	picrate [223(dec.)][458, 516]
2-I[509a, 510a]		93–6/12·2 mm	picrate (119–120)[510a]
3-I[509a, 517a]	52·3–53·0	198–202	picrate (154)[517a] dichloride (128–130)[377c]
4-I[518–9a]	100(dec.); 106–8		picrate (156–7)
2-F-3-Me[509c, d]		144–6/737 mm 150·5–151/ 757 mm	
2-F-4-Me[509e]		157	
2-F-5-Me[509d]		155–6/752 mm	
2-F-6-Me[509e]		142	
3-F-4-Me[520a]		135·6/748 mm	picrate (129–130)
3-F-5-Me		139[395a].; 139/ 700 mm[521a]	
4-F-2-Me[522]		128–132	methiodide (191) picrate (152)
2-Cl-3-Me[390e]		192–3/751 mm	
2-Cl-4-Me[280b, 390g, 562]		194–5	
2-Cl-6-Me[397c, 511c, 523a]		183·5–184/ 749 mm 60–60·5/8 mm 75–80/10 mm	
3-Cl-4-Me[520b, 521b]		178 101/74 mm	
3-Cl-6-Me[273]	19	163	

Substituent	M.p., °C	B.p., °C	Derivatives (m.p., °C)
4-Cl-2-Me[511d]		162·5–163·5	picrate (176–7)[519b]
			picrate (203)[517c]
4-Cl-3-Me[519a]		68–70/18 mm	picrate (152–3)
2-Br-3-Me[523b, 524a]		218–9	
		67–9/6 mm	
2-Br-4-Me[524a]		223–4	
2-Br-5-Me[524a]	49–50		
2-Br-6-Me		198–201[514b]	picrate (115–6)[514b]
		205·5–207/	
		772 mm[514b]	picrate (163)[511c]
		80–90/10 mm[511c]	
3-Br-4-Me[520b]		47–8/0·6	
3-Br-6-Me[273]	32		
4-Br-2-Me		180–1[517c]	methiodide [214(dec.)]
			[522]
			picrate (184)[517c]
		74–6/14 mm[522]	picrate (186)[522]
3-I-2-Me[513b]	36–7		picrate [168(dec.)]
3-I-4-Me	160[517d]	90–5/5 mm[523c]	
3-I-6-Me	48–9[273]	205–215[392b]	hydriodide (235–8)[273]
		105–6/18 mm[392b]	picrate (150)[392b]
4-I-2-Me	43–5[519a]	84–6/14 mm[522]	methiodide [238 (dec.)][522]
			picrate [209(dec.)][522]
4-I-3-Me[519a]	46–8		picrate (157–8)
3-Cl-2,6-Me₂[525]			picrate (150–1)
4-Cl-2,6-Me₂		178[513c]	nitrate (207·5)[400d]
			picrate (166–7)[519c]
Halogeno-hydroxypyridines			
2-Cl-3-OH	169–170[514d]		PhCO (53–5)[153]
2-Cl-5-OH			hydrochloride [208 (dec.)][393f, 620c, d]
2-Cl-6-OH[514d, 520c]	128·5–129		
3-Cl-2-OH[520c]	180–1		
3-Cl-6-OH[523d]	163–4		PhCO (95·0–95·8)
4-Cl-2-OH[526a]	184		
2-Br-3-OH[435a]	186·5–187		hydrobromide [360 (dec.)]
2-Br-5-OH[514e]	135·5–136·5		
2-Br-6-OH[396h]	123		
3-Br-5-OH[514e]	166·5–167·5		
3-Br-6-OH[510b]	177–8		
4-Br-3-OH[514e]	123·5–124		
3-OH-2-I	190–2 (dec.)[153]; 195–6[523e]		PhCO(90–1)[153]
4-OH-3-I[527]	303		
6-OH-3-I[513d, 519d, 528]	192		
2-OH-4-I[526]	195		
Halogeno-nitropyridines			
2-F-5-NO₂[518]	19–21	79–81/5–6 mm	
2-Cl-3-NO₂[401d]	101		
2-Cl-5-NO₂	106[511e]; 119–121[437]		
3-Cl-6-NO₂[420]	120·5–121		
4-Cl-3-NO₂[523f]	45	95/5 mm	hydrochloride (156)[93b]
2-Br-3-NO₂[401d, 524b]	125		
2-Br-5-NO₂	137–8[524b]; 151–2[437]		
3-Br-4-NO₂[401b]		66–8/0·1 mm	picrate (165–7)
3-Br-6-NO₂[420, 435a]	149·5–150		
2-I-5-NO₂[524a]	165–6		

Table 6.13 (continued)

Substituent	M.p., °C	B.p., °C	Derivatives (m.p., °C)
*Polyhalogenopyridines**			
2,3,4,5,6-F$_5$[511f]	−41·5	83·3	
2,3-Cl$_2$	66·5–67	202·5–203·5	
2,4-Cl$_2$	−1–0	189–190	
2,5-Cl$_2$	59–60	192·5—193·5	
2,6-Cl$_2$	87–8	211–2	
3,4-Cl$_2$	23–23·5	182–3	
3,5-Cl$_2$	66–7	178–9	
2,3,4-Cl$_3$	45·5–46	227–8	
2,3,5-Cl$_3$	48–9	219–220	
2,3,6-Cl$_3$	67–8	232·5–233·5	
2,4,5-Cl$_3$	8·5–9	221·5–222·5	
2,4,6-Cl$_3$	32·5–33	217·5–218·5	
3,4,5-Cl$_3$	71·5–72·5	213·5–214·5	
2,3,4,5-Cl$_4$	21–2	250·5–251·5	
2,3,4,6-Cl$_4$	37·5–38	248·5–249·5	
2,3,5,6-Cl$_4$	90·5–91·5	251–2	
2,3,4,5,6-Cl$_5$	125–6	279–280	
2,3-Br$_2$	58·5–59·5	249–249·5	
2,4-Br$_2$	38–38·5	237–237·5	
2,5-Br$_2$	93–4	238–238·5	
2,6-Br$_2$	118·5–119	255–255·5	
3,4-Br$_2$	71–2		
3,5-Br$_2$	111·5–112	227–227·5	
2,5-I$_2$[523g]	154		
2,6-I$_2$[517b]	183		
Pyridine 1-oxides			
2-Cl[405g]			hydrochloride (138–142)
4-Cl	152·5–153·5 (dec.)[514f]		picrate (147·5–149)[456]
	169·5 (dec.)[395d]		picrolonate [167–8 (dec.)][456]
	179·5–180 (dec.)[456]		
	187–9 (dec.)[449]		
2-Br[524c]			hydrochloride (135–6)
3-Br		143–8/4 mm[395a]	hydrochloride (181–2)[514g, 519e]
			picrate (139–141)[519e]
4-Br	142–3(dec.) [514f]		picrate (142–142·5)[514f]
	163[395d, 449]		picrate (223–4)[519f]
4-I[519a]	171		
2-Cl-3-Me[405g]			hydrochloride (146–150)
2-Cl-4-Me[405g]			hydrochloride (121–5)
2-Cl-5-Me[405g]			hydrochloride (125–8)
2-Cl-6-Me[405g]			hydrochloride (155–160)
2-Br-3-Me[524c]			hydrochloride (179–180)
2-Br-4-Me[524c]			hydrochloride (147–8)
2-Br-5-Me[524c]			hydrochloride (141–2)
2-Br-6-Me[524c]			hydrochloride (185–6)
4-Br-2-Me[522]		178–182/14 mm	
2-Cl-4-NO$_2$[405g]	152–3		
4-Cl-3-NO$_2$[519g]	147–9		
2-Br-4-NO$_2$[396e]	145·5–146		
3-Br-4-NO$_2$[401b]	156–7		

Substituent	M.p., °C	B.p., °C	Derivatives (m.p., °C)
Side-chain halogenopyridines			
2-CH₂Cl[218]		43–7/1·5 mm	picrate (152–3)
		73–6/10 mm	
3-CH₂Cl[524d]			hydrochloride (142–5)
			picrate (130·5–132)
4-CH₂Cl[524d]			hydrochloride (170–5,
			solidifying 190)
			picrate (146–7)
2-CH₂Br[529]			hydrobromide (146)
4-CH₂Br[524e, 529]			hydrobromide (149–150)
			hydrobromide (145–150)
2-CHCl₂[218]		62–4/1·2 mm	picrate (117–8)
		90–2/15–16 mm	
4-CHCl₂[217a]		78–80/18 mm	
2-CCl₃[218]	−10	76–80/1·2 mm	
		112–5/15 mm	
		125–6/25 mm	
4-CCl₃[217a]		105–7/18 mm	
2-(CHCl)₂Ph[321]	153–4		dichloride (180–1)
2-(CHBr)₂Ph[321]	172		
2,6-[(CHCl)₂Ph]₂[183b]	213		
2,6-[(CHBr)₂Ph]₂[183b]	179 (dec.)		

* For a survey of the chloropyridines and bromopyridines see [396f, 396b] from which, unless otherwise indicated , the data on polyhalogenopyridines are taken.
† Solidification points.

The production from halogenopyridines of pyridyl lithiums is a reaction which generally proceeds smoothly and in high yield (p. 170). The formation of pyridyl Grignard reagents is more difficult, and indeed has not been achieved by direct interaction of a halogenopyridine with magnesium[396h, 530a]. The first successful preparation of a pyridyl Grignard reagent was achieved by submitting 2-bromopyridine to the entrainment method in presence of ethyl magnesium bromide[530a]. The method has frequently been used with 2-bromopyridine[530b–4], 2-iodopyridine[533], 2-bromopicolines[535], 3-bromo-[133a, 531, 536] and 4-chloropyridine[34a]. The yields of Grignard reagents produced by this method are said to be 40–55 per cent, and the best procedure for obtaining the 2-pyridyl Grignard reagent to be the conversion of 2-bromopyridine into 2-pyridyl lithium and then reaction with magnesium bromide or iodide[537].

Numerous examples of the removal of halogen substituents during the saturation of the pyridine nucleus (p. 264) and the reduction of substituents attached to it (pp. 267, 284) have been recorded. In nucleophilic halogen exchange, reductive removal of a halogen substituent is sometimes caused by the use of hydriodic acid (p. 230). Hydriodic acid and phosphorus[538] replace 2- and 4-halogen substituents selectively[397b]. Hydriodic acid and zinc with acid give similar results[539].

The polarographic reduction of halogenopyridines other than the iodo compounds involves the cations and proceeds irreversibly to give pyridine and halide ions. The ease of reduction[540] is $I > Br > Cl$ and $4 > 2 > 3$.

The interesting halogen migration which can occur with derivatives of 2,4-dihydroxypyridine has been mentioned (p. 169). In the example

o

quoted, 3-bromo-2,4-dihydroxypyridine was converted during chlorination with hydrochloric acid and hydrogen peroxide into 5-bromo-3-chloro-2,4-dihydroxypyridine. Many similar cases have been described[396d,f, 435b, 514d, 541], some of which are shown (the tautomeric possibilities are ignored in the structures):

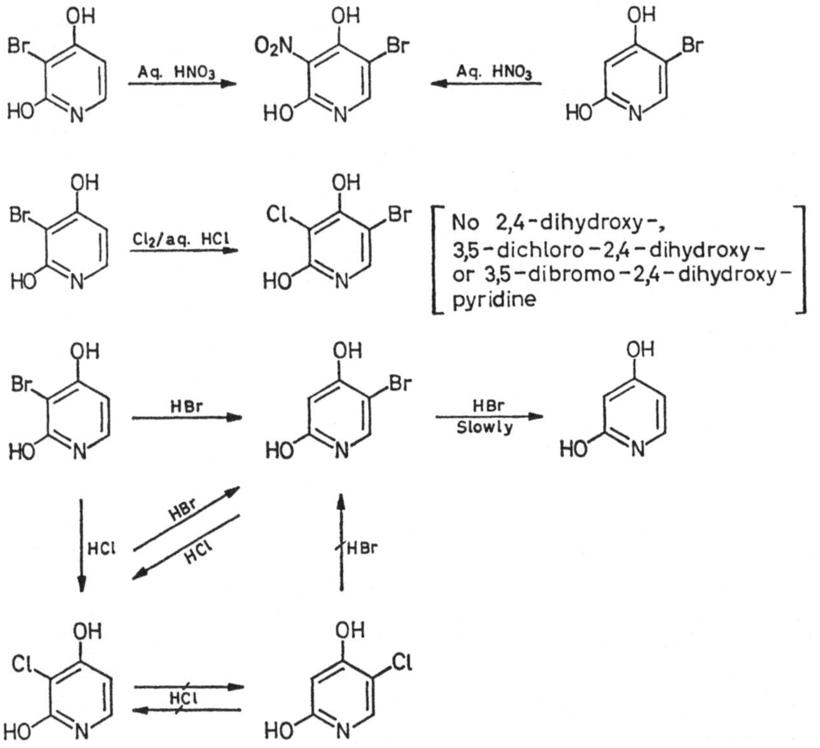

Characteristically a 3-bromo substituent in a 2,4-dihydroxypyridine migrates to $C_{(5)}$ under the influence of acid, and subsequently a slow reaction can occur in which the bromine atom is removed completely or, if an electrophilic substituting reagent is present, substitution can occur at $C_{(3)}$. Presumably the migration and subsequent removal of bromine are electrophilic protonations, the second being slower than the first. The interesting question is whether the migration of bromine occurs intra- or intermolecularly. Some of the evidence suggests the change to be intramolecular but cannot be regarded as quite conclusive.

The mechanism of the Ullmann diaryl synthesis is still uncertain[542], and for this reason the reaction is mentioned here rather than at other possible points. 2,2'-Bipyridyls have been obtained from 2-bromo- and 2-iodo-pyridine[543-4], 2-bromopicolines[524a, 544b] and 2-iodo-5-nitropyridine[524a] by heating with copper. 2,5-Dihalogenopyridines give 5,5'-dihalogeno-2,2'-bipyridyls[524a], and the method has been used to produce polypyridyls[543].

Like other iodo aromatics, 3-iodopyridines form dichlorides (*Table 6.13*).

2-Iodopyridine does not give such a derivative[377c, 545]. With alkali the 3-iodopyridine dichlorides form iodosopyridines[545-6].

The reactions of halogenomethylpyridines (*Table 6.13*) with nucleophilic reagents are normal and require no comment[218]. The familiar reaction of dehydrohalogenation to produce acetylene derivatives is of considerable value in its application to side-chain halogen derivatives of the pyridine series. Its value arises from the possibility of carrying out the sequence of reactions, 2(or 4)-picoline → 2(or 4)-styrylpyridine (p. 334) → 2(or 4)-styrylpyridine dihalide (p. 349) → 2-(or 4)-tolazole. The dehydrohalogenation is brought about by base, and most commonly the reaction has been applied to stilbazole dibromides[35, 42b, 222, 315, 326-7], although dichlorides have been used[322-3]. The dihalogen compounds can, of course, under appropriate conditions, give monohalogeno-vinyl compounds[321-2], and the latter can themselves be further transformed into acetylenes[222, 322, 329-30, 547a]. The unstable dibromide of 2-vinylpyridine gives 2-pyridyl-acetylene[329]. Both 3- and 4-monohalogenoethylpyridines have been converted into vinylpyridines[169, 547b,c], and 3-α-chlorovinylpyridine into 3-pyridylacetylene[547a].

(h) Hydroxyl and Substituted Hydroxyl Groups

X-Ray structural data (p. 120), dipole moments (p. 123), spectra (pp. 139, 144), ionization constants, the tautomerism of compounds in these groups (pp. 154-5) and the behaviour of hydroxypyridines towards alkylating (p. 183) and acylating reagents (p. 196) have been discussed. Some of these compounds are described in *Table 6.14*.

The colours given by a number of hydroxypyridines with ferric chloride and with the Folin–Denis phenol reagent are reported in *Table 6.15*. The reagent is seen not to be specific for 3-hydroxypyridines[548], though a negative result in this test might indicate the absence of a 3-hydroxy group.

The oxidation of some hydroxypyridines gives azaquinones. Examples are 2,3-dihydroxypyridine and the amino-hydroxypyridines[566f] illustrated:

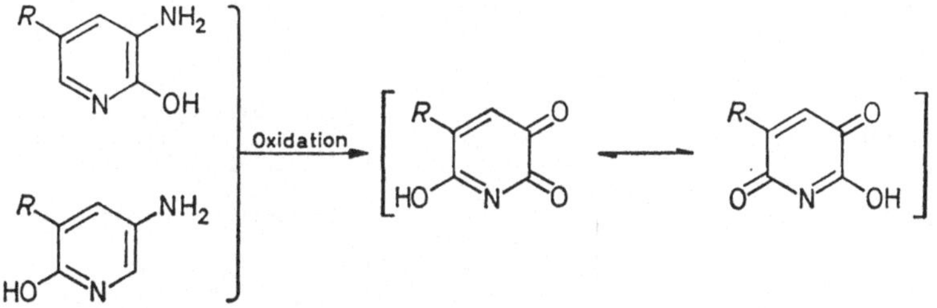

(R = H or Me)

Among the products formed when 2-amino-5-hydroxy-4-methyl-3-phenyl-pyridine is treated with nitrous acid is a compound $C_{12}H_9O_3N$. This is also formed when another of the products, 2,5-dihydroxy-4-methyl-6-nitroso-3-phenylpyridine, is hydrolysed, and it is clearly 2-aza-3-hydroxy-5-methyl-6-phenyl-1,4-benzoquinone (or a tautomer of this structure)[391e].

The behaviour of hydroxypyridines with reducing agents has been discussed (pp. 257-8, 264). The hydroxyl groups have been removed from several

Table 6.14. Some Hydroxypyridines and Hydroxypyridine 1-Oxides
(cf. Tables 6.1, 6.2, 6.8 and 6.13)

Substituents	M.p., °C	B.p., °C	Derivatives (m.p., °C)
2-OH	106–7[549a]	280–1[550]	O-acetyl (b.p. 110–2°/10 mm)[551a] O-benzoyl (47)[552a] (b.p. 183–6°/ 30 mm)[551b] O-benzyl (b.p. 134–5°/2 mm)[551c] N-methyl (30)[553a] (b.p. 122–4°/ 11 mm)[554] N-methyl picrate (145)[409] O-methyl (b.p. 142·4)[555] O-phenyl (b.p. 134–5/11 mm)[556a] picrate (170–1)[551a] O-p-toluenesulphonyl (53)[552b]
3-OH	124·5–125[523e] 129[557a]		O-acetyl (b.p. 210°)[549c] O-benzoyl (51)[552] O-benzyl (b.p. 114°/0·6 mm)[558a] methiodide (115–7)[559a] O-methyl (b.p. 70–1°/12 mm)[559a] O-methyl picrate (136–9)[559a] O-phenyl (b.p. 147–9°/17 mm)[556a] picrate (200—1·5)[523e] O-p-toluenesulphonyl (80)[552b]
4-OH	148·5[515] 66–7 (mono- hydrate)[515]	200/3 mm[560d]	O-benzoyl (79)[552] O-benzyl (55–6) (b.p. 155–160°/ 4 mm)[561a, b] N-methyl (92–4) (b.p. 230–3°/ 13 mm)[390c] O-methyl (b.p. 95·6°/31 mm)[556a] O-phenyl (44–6)[549d] (b.p. 157– 8°/21 mm[556a] picrate (238–9)[560d] O-p-toluenesulphonyl (dec. > 70°)[552b]
2-OH-3-Me	142–3[58]	288–290/752 mm[549b]	N-ethyl (b.p. 92°/2·2 mm)[561c] N-methyl (b.p. 86–8°/1·45 mm)[561c]
2-OH-4-Me	130[192] 65 (mono- hydrate)[280b, 390g, 562]	186–7/12 mm[192]	O-benzyl (b.p. 142–5°/5 mm)[556b] N-methyl (59) (b.p. 110°/1 mm)[192]
2-OH-5-Me	185[58]		N-ethyl (b.p. 104–6°/1·7 mm)[561c] N-methyl (36·8) (b.p. 109–111°/ 1·7 mm)[561c]
2-OH-6-Me	158–9[192] 163–5[228a]	282[192] 202–4/40 mm[514b]	O-acetyl (b.p. 118°/0·4 mm)[228a] O-benzyl (51–2)[556c] N-methyl (56·5–58) (b.p. 110°/ 2 mm)[192] picrate (149·5–150)[563a]
3-OH-2-Me	170–1[556d]		hydrochloride (223–4)[564a] picrate (206–7)[556d]
3-OH-4-Me	120–1[565a]		O-acetyl (b.p. 97–8°/4·5 mm)[566a] picrate (203–5)[565a]
3-OH-5-Me	140–1[567]	153/5 mm[551d]	picrate (189–190)[565a]
3-OH-6-Me	167–9[566b]		O-acetyl picrate (147)[557b] O-methyl (b.p. 43–5°/1 mm)[566b] O-methyl picrate (138–140)[557b] picrate (203)[568a]

Substituents	M.p., °C	B.p., °C	Derivatives (m.p., °C)
4-OH-2-Me	174–6[519b]	180–190/ 0·003 mm[519b]	nitrate (161·5–162)[565b] picrate (178·5–179·5)[565b] (199–200)[560b]
4-OH-3-Me	97–8[558b]	190–2/2 mm[560c]	picrate (205–6)[560c]
Aryl-hydroxypyridines 2-OH-5-Ph			N-methyl (73–4)[560a] N-methyl picrate (131–3)[560a]
2-OH-6-Ph	197[563b, c]		hydrochloride (104)[563b]
3-OH-2-Ph	206–207·5[564c]	320[563d]	O-methyl (b.p. 110–2°/ 0·34 mm)[564c] O-methyl picrate (153·5–155)[564c] picrate (201·0–201·4)[551e]
4-OH-2-Ph	155 (hemi- hydrate)[553b]		
Hydroxy-nitropyridines 2-OH-3-NO₂	224–225·5[431]		hydrochloride (221–2)[425b] N-methyl (175–6)[459] O-methyl (57–9)[564d]
2-OH-5-NO₂	190–1[425b]		N-methyl (175)[553c] O-methyl (110)[569a]
3-OH-2-NO₂	69–70[566c, 568b]		
4-OH-3-NO₂	269–270 (dec.)[454a] 279[454a]		N-methyl (220)[454a] (233)[523f] O-methyl (75)[454a]
Hydroxypyridine-sulphonic acids 2-OH-5-SO₃H			Na salt[570] N-methyl[570]
3-OH-2-SO₃H	302[568c, d]		
4-OH-3-SO₃H	265[448]		
Polyhydroxypyridines 2,3-(OH)₂	247–8*[559b] 250–2 (dec.)[564b]		monoacetyl (155)[393e] 2,3-O-diethyl (b.p. 215–7°)[393f] 2-O-methyl (b.p. 82°/12 mm) (68–9)[393e]
2,4-(OH)₂	260[393e] 271·5–272·5[559b]		2,4-O-diethyl picrate (139·5– 140)[559b] 2,4-O-diethyl picrolonate(158)[559b] 2-O-methyl (133·5–134)[559c] 4-O-methyl (165–6)[559c] N,2-O-dimethyl (141–3)[559c] N,4-O-dimethyl (113–4)[559c] 2,4-O-dimethyl picrate (163–4)[559c]
2,5-(OH)₂	225–235 (dec.)*[559b] 240–250 (dec.)*[566d] 248[393f]		monobenzoyl (187–9)[391e] 2,5-O-diethyl (b.p. 215–7°)[393f] 2,5-O-diethyl picrate (118–9)[559b] 2,5-O-diethyl picrolonate (139)[559b] 2-O-methyl (81)[566d]
2,6-(OH)₂	184–5[559b] 203·5–204 (hydrate)[571]		diacetyl (69)[571] 2,6-O-diethyl (21·5)[396h]

Table 6.14 (continued)

Substituents	M.p., °C	B.p., °C	Derivatives (m.p., °C)
3,4-(OH)₂	239·5–240 (dec.)[566e] 240–1 (dec.)*[559b]		monoacetyl (145·5–146·5)[566e] diacetyl (138·5–140)[566e] 3,4-O-diethyl picrate (169–170)[559b] 3,4-O-diethyl picrolonate (212–3)[559b] 3-O-methyl (180·5–181·5)[566e] 3,4-O-dimethyl picrate (173–4)[572a]
3,5-(OH)₂	252–3 (dec.)*[559b]		3,5-O-diethyl picrate (124·5–125·5)[559b] 3,5-O-diethyl picrolonate (153)[559b]
2,3,4-(OH)₃[573a] 2,3,5-(OH)₃ 2,3,6-(OH)₃ 2,4,6-(OH)₃	220–230(dec.) [569b, 573b]		O-triethyl (30·5–31·5)[559d] triacetyl (159)[566f] diacetyl (247)[573b] O-trimethyl (48·5–49·5)[574a] O-triphenyl (79–79·5)[574b]
Hydroxypyridine 1-oxides†			
2-OH	149–150[561a]		N-allyl[575] O-allyl[575] O-benzyl (103–6)[561a] Ox-benzyl (85–6)[561a] Cu salt (283–4)[524c] O-ethyl (71–3; hydrate, 49–52)[558c] Ox-ethyl[575] O-methyl (78–9; hydrate, 66·5–67·5)[401i] O-methyl picrate (145–146·5)[574c] Ox-methyl[401i, 575]
3-OH	189–191[559a,561a]		O-ethyl phthalate (83·5–84·5)[396e] O-methyl (101–2)[559a] O-methyl picrate (162–4)[559a]
4-OH	243–4[561a] 246–7(dec.)[572b]		O-benzyl (178–9)[561a] O-benzyl picrate (123–4)[561a] Ox-benzyl (112–3)[561a] O-ethyl (33)[456, 560e] O-ethyl picrate (125–6)[456, 560e] O-methyl (81·5–82·5)[560e] O-methyl picrate (143–4)[560e] Ox-methyl (b.p. 166°/0·01 mm)[449] (60–5)[401i] Ox-methyl picrate (200–2)[401i] O-phenyl (130–1)[560e] O-phenyl picrate (138–9)[560e] picrate (166–7)[572b]
2-OH-3-Me	138–9[561b]		
2-OH-4-Me	131–3[561b]		O-benzyl (81–2)[556b] O-ethyl (hydrate, 101–2)[561b] O-ethyl picrate (139)[561b]
2-OH-6-Me	141–2[556c]		O-benzyl (99–100)[556c] O-ethyl (102–3)[561b] O-ethyl picrate (88–9)[561b]
4-OH-2-Me[568e]	192–3		O-benzyl (160–1) hydrochloride (116) O-methyl (78–80)
4-OH-3-Me[391c, 629–30]	224		O-acetyl (220 dec.) O-benzyl (hydrate, 181) O-benzyl picrate (185) hydrochloride (148·5)

Substituents	*M.p.*, °C	*B.p.*, °C	*Derivatives* (*m.p.*, °C)
4-OH-3-Cl	258–259·5 (dec.)[572b]		
2-OH-3-Br	208–9[556b]		
2-OH-5-Br[556b]	137–9		O-benzyl (127–8)
3-OH-2-Br			O-ethyl hydrochloride (159–160)[524c]
3-OH-5-Br			O-methyl (200–1)[574d]
4-OH-3-I	304–7[560f]		
2-OH-4-NO_2			O-ethyl (132·5–133·5)[396e] O-methyl (154·5–158·5)[574c]
2-OH-5-NO_2	198–9[556b, 574c]		
3-OH-4-NO_2			O-ethyl (134–5)[396e] O-methyl (134–5)[574c]
4-OH-3-NO_2[519g]	219–221		O-acetyl (191–3)
3,5-$(OH)_2$			O-diethyl (115·5–116·5)[574e] O-diethyl picrate (114·5–115·5)[574e] O-dimethyl (91–3)[574d] O-dimethyl picrate (131–2)[574d]

* With darkening at lower temperatures.
† The prefix 'Ox' indicates that the substituent is attached to the oxide oxygen atom.

Table 6.15. Colour Reactions of Hydroxypyridines

Substituents	$FeCl_3$	*Folin–Denis reagent**
2-OH	red[576]†	*n*[548, 577]
3-OH	red[576]†	blue[548, 577]
4-OH	yellow[576]	*n*[548]
6-NH_2-2-OH		*p*[548]
6-Br-2-OH		*n*[548]
3,5-Br_2-2-OH		*n*[548]
5-CO_2H-2-OH		*n*[577]
3-CN-2-OH-4,6-Me_2		*n*[548]
2-OH-6-MeO		*p*[548]
2-OH-4,6-Me_2		*n*[577]
2-OH-4,5,6-Me_3		*n*[548]
5-CO_2H-3-OH		*p*[577]
3-OH-2-MeO		*p*[548]
3-OH-6-Me		*p*[577]
3-OH-1-Me-2-:O		*p*[548]
3-OH-1-Me-4-:O		*p*[548]
4-HO-3-MeO		*n*[548]
4-OH-2,6-Me_2		*n*[577]
2,3-$(OH)_2$	blue[559b]	blue-violet[559b]
2,4-$(OH)_2$	brown red[559b]	*n*[559b]
2,5-$(OH)_2$	pink red[559b]	dark blue[559b]
2,6-$(OH)_2$	sepia ppte[559b]	dark blue[559b]
3,4-$(OH)_2$	violet[559b]	dark blue[559b]
3,5-$(OH)_2$	brown red[559b]	green blue[559b]
3-Et-2,4-$(OH)_2$-6-Me		*n*[577]
2,4-$(OH)_2$-5,6-Me_2		*n*[548]
4-CO_2H-2,6-$(OH)_2$		*p*[548]

* *n* = negative, *p* = positive.
† The colours are described as 'identical' reds[576].

2-[539a, 578-9], 3-[580a] and 4-hydroxypyridines[580b,c], and from 3,3',5,5'-tetra-hydroxy-2,2'-bipyridyl[581] by heating with zinc dust. Red phosphorus and iodine produce the same result with 6-hydroxy-2,3,4- and -3,4,5-trimethyl-pyridine[433b].

2- and 4-Methoxypyridine are isomerized by heating to 1-methyl-2- and -4-pyridone, respectively[474b, 580d]. With the 4-methyl and 4-ethyl homologues of 2-methoxypyridine the isomerization is complete[582] after 14 h at 200°, and the use of these compounds labelled in the methoxyl group with ^{13}C showed the reaction to be intermolecular. Since it was catalysed by benzoyl peroxide, it may be radical in character. 2-Phenoxy-[583] and 2-benzyloxy-pyridine[584] are similarly isomerized, and the reaction has been used to prepare a number of 1-(2-dialkylamino-ethyl)-2-pyridones[585]. Some surprising variations have been reported; 2-methoxy-3-[564d] and -5-nitropyridine[553c] could not be isomerized, whilst 4-methoxy-3-nitropyridine rearranged[586] at 170°–180°. The tetra-acetyl-glucosyl residue migrates from oxygen to nitrogen in the same way as simple alkyl groups[587]. Heating 2-alkoxypyridine-carboxylic acids produces 2-hydroxypyridine-carboxylic acid esters, or simply decarboxylation, rather than O → N migration[474b].

Isomerization of a 3-alkoxypyridine has not been observed, and an attempt to cause N → O migration in a 1-benzyl-3-hydroxypyridine betaine failed[558a]. The conversion of 2-methoxypyridines into 1-methyl-2-pyridones by heating with methyl iodide (p. 183) is probably a reaction of different character from the thermal isomerizations just discussed; presumably a preliminary quaternization occurs in these cases.

In view of the above reactions it is not surprising that Claisen rearrangement of 2-allyloxypyridine produces 1-allyloxy-2-pyridone as well as 3-allyl-2-hydroxypyridine[588]. This rearrangement is conveniently carried out in dimethylaniline as solvent: 2-α- and 2-γ-methylallyloxy-pyridine give products of rearrangement to N and to $C_{(3)}$ in which the linkages to these atoms are made through the γ-carbon atom of the migrating group in typical Claisen fashion. The same is true of the rearrangement of the 2-α-compound when heated without a solvent, but under these circumstances 2-γ-methyl-allyloxy-pyridine gives the following products:

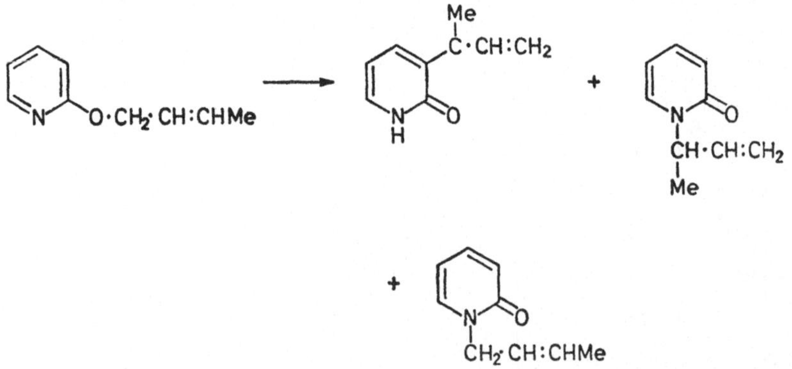

The 'abnormal' product does not arise from the 'normal' one. The mechanistic significance of this result has not been elucidated, but in view of the possible superimposition of a reaction of the kind involved in the $O \rightarrow N$ rearrangement of alkyl ethers (see above) upon the normal Claisen process a complex result is not surprising.

The 2-alkoxypyridine 1-oxides undergo thermal rearrangement more easily than the corresponding pyridines, giving 1-alkoxy-2-pyridones. The ease of this rearrangement is such that it may occur during an attempt to prepare a 2-alkoxypyridine 1-oxide; thus, 2-chloropyridine 1-oxide and sodium benzyloxide give 1-benzyloxy-2-pyridone, 2-benzyloxypyridine 1-oxide being isomerized at 100°. 2-Allyloxypyridine 1-oxide rearranges similarly. It is thought that these reactions do not involve free radicals, for they are unaffected by the addition of a scavenger[575].

A noteworthy reaction undergone by 1-methyl-4-pyridone has been mentioned briefly (p. 164). This is the deuteration which it undergoes; heating with 0·2 N sodium deuteroxide at 100° produces $C_5H_5D_2NO$ (61 per cent) and C_5H_6DNO (32 per cent) in 40 h. The deuteration occurs at $C_{(2)}$ and $C_{(6)}$, is reversible and does not occur in neutral or acidic conditions. The reaction is similar to that which occurs with quaternary pyridinium salts and probably goes through an ylid[589]:

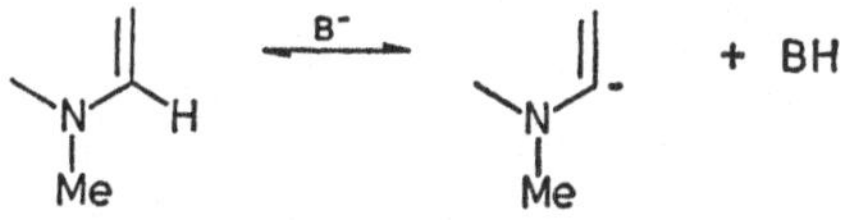

The esters of hydroxypyridines with carboxylic acids are all easily hydrolysed, but the 3-pyridyl esters are more stable than the other isomers. The esters of 4-hydroxypyridine must be prepared under anhydrous conditions[552a]. The catalytic hydrogenolysis of pyridyl benzoates and p-toluenesulphonates has been described (p. 264).

2- and 3-Pyridyl acetates will acetylate amines, alcohols and phenols; 2-pyridyl acetate is more effective than the 3-isomer[590].

The 2-acetyloxy- and 2-benzoyloxy-1-methylpyridinium cations, formed *in situ* from 2-iodopyridine methiodide and acetic or benzoic acid in presence of triethylamine, acylate amines[636].

(i) Metallic Derivatives

The reactions of pyridyl and picolyl metallic derivatives with organic reagents are of great synthetic value. Since the courses taken are, more often than not, normal, no attempt will be made to do more than substantially exemplify these uses and to note some of the infrequent irregularities. It can be said that almost always the yields realized in reactions using the Grignard reagents are inferior to those obtained with lithium, potassium or sodium compounds.

The acylation of these reagents to produce ketones is a particularly valuable reaction. The 2-picolyl Grignard reagent gave with acetyl chloride a mixure of ketone and carbinol[130]; it is interesting to note that in this reaction and others to be mentioned below, this showed none of the anomalies met with benzyl Grignard reagents. The 2-pyridyl Grignard reagent gives, with nitriles, ketones, with esters, mixtures of ketones and carbinols[533].

379

The 3-pyridyl Grignard reagent has been acetylated with acetic anhydride, and with ethyl *ortho*formate it gave pyridine-3-aldehyde diethyl acetal[536a]. The 4-pyridyl Grignard reagent reacts similarly with the latter reagent, and with benzamide gives 4-benzoylpyridine[34a].

Ketones have also been prepared by reaction of 2- or 3-pyridyllithium with carboxylic acid anhydrides[34b], esters[34b] and nitriles[34b, 537, 591]. 4-Pyridyl lithium gives, with nitriles, ketones and with ethyl, pyridinecarboxylates, but with ethyl benzoate the carbinol is formed[34a]. 2- and 4-Picolyl lithium have been acylated with acid halides[41, 119, 120c] and esters[36b, 38, 40-1, 592b], the former also with anhydrides[120c] and nitriles[122a, 592a]. With 2-picolyl lithium, carbinols are occasionally formed as well as ketones[120c, 594a], and sometimes diacylation occurs[120c]. Lithium derivatives of 2-benzyl, 2-ethyl- and 2-isobutyl-pyridines have been acylated with esters[594], and of 2-n-amylpyridine with acetonitrile[122a]. The difficulty arising from addition to the azomethine linkage during the formation of 4-picolyl lithium has been mentioned (p. 327).

Sodium or potassium derivatives of 2-[594a, 595-6], 3-[37b, 594c] and 4-picoline[36b, 594c] and of 2-benzylpyridine[594a] have also been acylated with esters.

Whilst 2-picoline failed to react with ethyl oxalate in the presence of sodium ethoxide[192], with potassamide or sodamide (72) was formed[597].

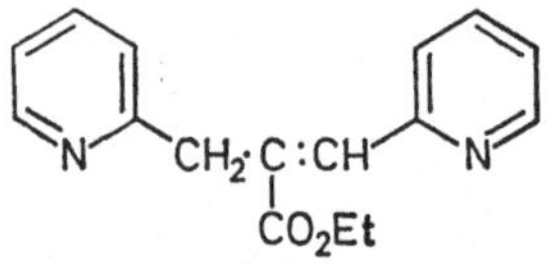

(72)

With the more reactive ethyl 6-methylnicotinate, the normal acylation occurred[125], as it did also with 1,2- and 1,4-dimethylpyrid-2-one[192].

2-Picoline 1-oxide can be satisfactorily acylated by esters in the presence of sodamide and is more reactive than 2-picoline[598]. In line with this, 2- and 4-picoline 1-oxide react satisfactorily with oxalic ester in presence of potassium ethoxide[556c].

Di-(2-picolyl) cadmium gave a low yield of ketone with ethyl oxalate, and the 4-isomer did not react. On the other hand, di-(2-picolyl) mercury gave some ethyl pyridine-4-pyruvate[122c].

The details of the mechanism of acylation of picolyl metallic derivatives have attracted attention. To account for the formation of ketones and carbinols a sequence of the following kind was postulated:

$$C_5H_4N \cdot CH_2R \ + \ BM \longrightarrow C_5H_4N \cdot \bar{C}HRM^+ \ + \ BH$$

$$C_5H_4N \cdot \bar{C}HRM^+ \ + \ R'CO_2Et \longrightarrow C_5H_4N \cdot CHR(COR') \ + \ EtOM$$

$$C_5H_4N \cdot CHR(COR') \ + \ C_5H_4N \cdot \bar{C}HRM^+ \longrightarrow \begin{cases} C_5H_4N \cdot \bar{C}R(COR')M^+ \\ \quad + \ C_5H_4N \cdot CH_2R \\ (C_5H_4N \cdot CHR)_2 \cdot C(OM)R' \end{cases}$$

Such a mechanism implies that reaction between the picoline and the ester should produce results similar to that between the ketone and the picolyl salt, which is not observed. Furthermore, it does not account for the fact that 3- and 4-picolines give only ketones, whilst 2-picolines give ketones and carbinols. The mechanism for 2-picolines was therefore re-written

$$C_5H_4N \cdot \bar{C}HR \; M^+ + R' \cdot CO_2Et \longrightarrow$$

The ketone/carbinol ratio would then depend on competition between the elimination and substitution reactions, and since a bulky R' group should favour elimination, the ratio should increase as R' changes from ethyl to isopropyl to isobutyl, as in fact it does. The mechanism also accounts for the difference between 2-picoline and 3- and 4-picoline[594a].

Pyridine-carboxylic acids have been prepared by the reaction of the 3-pyridyl Grignard reagent[536a] and of 2-[599] and 3-pyridyl lithium[600a] with carbon dioxide. Nicotinic acid isotopically labelled in the carboxyl group has been obtained in this way[600b]. The reaction is especially useful for preparing pyridine-acetic acids and has been applied to 2-[213, 601] and 4-picolyl lithium[41]. Under conditions where phenyl lithium adds to the azomethine linkage of 4-picoline, 2-phenylpyridine-4-acetic acid[121] is produced. 2,6-Lutidine gave 6-methylpyridine-2-acetic acid, but the reaction failed with 2-methyl-6-phenoxypyridine[511c].

The 3-pyridyl Grignard reagent has been alkenylated by reaction with allyl halides[536] and the 2-picolyl Grignard reagent homologated with alkyl halides and ω-diethoxyalkyl halides[131]. Of much greater importance has been the use of picolyl lithium, potassium and sodium compounds in reactions with alkyl and aralkyl halides.

Ziegler and Zeiser[118] were the first to use 2-picolyl lithium in this way, condensing it with benzyl chloride, and 3-[602] and 4-picolyl lithium[120a] have since been employed in alkylations. 2-Picolyl lithium has also been used in reactions with α,ω-dihalogenoalkanes[354b], and this and 4-picolyl lithium with ω-diethoxyalkyl halides[41, 131]. 5-Alkoxy-[566b, 557c] and 5-nitro-2-picolyl lithium[557c] have likewise been alkylated, and secondary alkylpyridines were obtained from the higher primary 2- and 4-alkylpyridines[120a, 129].

381

Tschitschibabin first used sodium derivatives, formed from 2- and 4-methyl- and -ethylpyridine and sodamide, to produce alkylpyridines by reaction with alkyl halides; 3-ethylpyridine failed to react under the conditions used[124]. The potassium derivatives of some 2-picolines were reported to fail in the reaction[126a], but this report was corrected, and the reaction assumed much greater importance after the work of Brown and Murphey[127d]. They examined in detail the reaction between 2- and 4-picolyl sodium and methyl halides and found conditions for introducing one, two or three methyl groups into the side chains. Equally important was their discovery that 3-picoline could also be alkylated by this method. The formation of secondary and tertiary alkylpyridines could be carried out in one reaction or stepwise. Whilst 2-picoline is monoalkylated using sodamide in xylene at 140°, in liquid ammonia low temperatures ($< 15°$) are necessary to prevent dialkylation. 4-Picoline reacts more vigorously and optimum monoalkylation is achieved at 0° but even then is accompanied by some dialkylation. 3-Picoline, being the least reactive, can be satisfactorily monoalkylated even in ammonia[127f].

A number of alkylations of potassium or sodium derivatives of 2-[113a, 126b], 3-[602-3] and 4-picoline[113a, 126b] have been described, and 11-bromo-N,N-dimethylundecamide has been used in these reactions, giving products[604] $C_5H_4N \cdot (CH_2)_{11}CONH_2$. Dipyridylalkanes have been prepared from 2-, 3- and 4-picoline and a,ω-dihalogenoalkanes[354b, 568e, 605] and from 4-ethylpyridine[605], though 4-picolyl potassium is said to give 1,2-di-(4-pyridyl)ethane when the attempt is made to condense it with ethylene dibromide[568e]. 2,3- and 2,5-Lutidine are alkylated at the 2-position; 2,4-lutidine gives a mixture in which the product of reaction at the 4-methyl group predominates, and 2,6-lutidine forms the monoalkylated product[603]. Some higher 2-alkylpyridines have been alkylated[129], as have the three diphenylmethyl-pyridines[606], 2-benzyloxy-3-, -4-, -5- and -6-methyl-pyridine[604], 2,6-dibenzyloxy-4-[605] and 5-methoxy-2-methylpyridine[566b, 607], 5-methylpyridine-2-carboxylic acid (the 6-methyl isomer failed)[263] and 6-methylpyridine-2-carboxylic acid diethylamide[604]. 5,6,7,8-Tetrahydro-iso-quinoline and 2-phenyl-5,6,7,8-tetrahydroquinoline have also been alkylated through their sodium derivatives[604]. Those of 2- and 4-picoline 1-oxide have been used in reactions with a,ω-dihalogenoalkanes[604-5].

As would be expected, 6-cyanomethyl- and 6-ethoxycarbonylmethyl-1-methyl-2-pyridone are readily alkylated[192].

The side-chain ethylation and propylation of picolines with olefins under pressure, in the presence of sodium, has been carried out[113a, 608]. These reactions are evidently nucleophilic attacks by the picolyl anions upon the olefins. With increasing substitution of the side chain only the more acidic 4-alkyl compounds will undergo ethylation. 3-Picoline gave low yields of products, but this was attributed to the occurrence of side reactions rather than to its inability to form an anion.

Pyridyl and picolyl metallic derivatives have frequently been used to prepare carbinols by reaction with aldehydes and ketones. Those of aldehydes have been recorded with the 2-[530a, 535, 609], 3-[536] and 4-pyridyl[34a] Grignard reagents, with the double reagent from 2,6-dibromopyridine[530b] and with lithium derivatives of 2,4-di- and 2,4,6-trimethylpyridine (reaction occurring at the 2-methyl group)[610]. With ketones, reactions have been carried out on the 2-[530b, 532], 3-[133a, 536] and 4-pyridyl[34a] Grignard reagents, with 2-[129, 599], 3-[591] and 4-pyridyl lithium[34a], with 2-picolyl lithium[120b, 130, 611a] and with the lithium derivative of 2,4-dimethylpyridine (reaction also at the 2-methyl group)[120b]. Whilst the sodium derivatives of 2- and 3-picoline reacted normally in liquid ammonia with 2-amino-4'-methoxybenzophenone, 4-picolyl sodium would not take part in the reaction and only the ketimine was isolated[611b].

The sodium derivative of methyl pyridine-3-acetate gave the corresponding stilbazole with benzaldehyde[593], and 2- and 4-picoline 1-oxide behaved similarly in presence of potassium methoxide[191, 196b].

3-Pyridyl lithium gives a small yield of the Michael addition product with benzalacetophenone[591], as does 4-picolyl sodium with methyl vinyl ketone[612a]. In contrast, 2-picolyl sodium forms the carbinol, perhaps because of initial reaction at the nitrogen atom

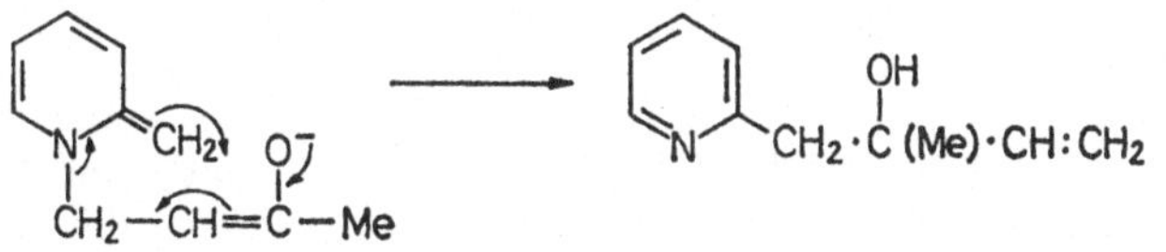

Carbinols have also been prepared from 2-picolyl lithium[130] and its nuclear and side-chain homologues by reaction with ethylene oxide and trimethylene oxide[612b].

The sodium derivatives of 2- and 4-picoline and of their 1-oxides react with alkyl nitrites to give good yields of the pyridine aldoximes[247a, 613], but the 3-isomers fail or give traces of products. The oxides are more reactive than the parent bases. 4-Ethylpyridine performed satisfactorily in the reaction, 2,6-lutidine gave a good yield of the monoxime, and 2,4-lutidine reacted at the 4-methyl group[247a].

In the above discussion some of the reactions of 6-methyl-2-picolyl lithium have been mentioned. These are normal, but discrepancies are to be found in the literature regarding the possibility of generating and using a dilithium derivative of 2,6-lutidine. The product obtained from 2,6-lutidine and benzyl chloride with two equivalents of phenyl lithium was not 2,6-diphenethylpyridine[119] but 2-dibenzylmethyl-6-methylpyridine[614]. Other products supposed to arise by the symmetrical dialkylation with ethyl bromide, and condensation with benzaldehyde and 4-ethoxypropiophenone of the dilithium derivative[122b] must be regarded as of doubtful character. Experiments which might have produced 2,6-diphenacylpyridine give 6-methyl-2-phenacylpyridine and its enol benzoate[614-5], and with benzonitrile a complex product is formed[614]. On the other hand, that from 2,6-lutidine and phenyl lithium gave with bromine a trace of 2,6-di-(bromomethyl)pyridine[123].

The 2-pyridyl Grignard reagent has been used in the preparation of phosphines[531, 616], arsines[534] and lead compounds[537], 2-pyridyl lithium in preparing arsines[537] and the 3-pyridyl Grignard reagent as a route to phosphines[531, 534].

In the presence of strong bases (sodium or potassium ethoxide, or sodium anilide), 2-methyl-3-[558b, 617] 3-methyl-2-[618], 3-methyl-4-[558b, 619a] and 4-methyl-3-acylaminopyridines[93b, 619b] form aza-indoles. Although these applications of the Madelung synthesis usually give poor yields, they are useful in providing compounds which are otherwise difficult to obtain.

(j) Nitro-, Nitroso- and Hydroxylamino Groups

The dipole moments (p. 126), spectra (pp.132, 142), and ionization constants (p. 159) of the nitropyridines have been mentioned. *Table 6.16* records some nitro-, nitroso- and hydroxylamino pyridines.

The reduction of nitro- to aminopyridines is a reaction of some importance. Stannous chloride has been used with 2-[461a] and 3-nitro-pyridine[454b], their homologues[557b, 620a] and alkoxy-derivatives[620c-1a]. With 4-nitropyridines, iron and acetic acid[401b], hydrazine hydrate with Raney nickel[401b] and catalytic hydrogenation[391c, 621b-2] have proved useful. When halogen substituents are present, special problems arise; reduction of halogeno-3-nitropyridines to the amines with elimination of halogen has often been observed when catalytic methods (particularly those using palladium) have been used[391b, 393a, 395c, 399a, 406b, 521a, 623-4a], and sometimes with metals and acids[393a, 624b]. Reduction with the retention of the halogen substituent has been achieved with stannous chloride[391b, 621a, 625], iron and acetic acid[393a, 406b, 541a, 626], electrochemical reduction[393a] and by catalytic methods[391b, 393a]. In some catalytic reductions of halogeno-3-nitropyridines in alcohols, alkoxy-3-aminopyridines are formed[393a].

Diaminopyridines have been prepared by reducing amino- or alkylamino-2-nitropyridines with stannous chloride[399b], catalytic methods again causing loss of a halogen substituent[406b]. More common are examples of diamine formation from amino-3-nitropyridines, for which purpose sulphides[390a, 393g, 396i], iron and acids[393g, 396i, 400b, 401g, 632-3a], stannous chloride[390a, 633b-4] and catalytic hydrogenation[396i, 632, 635] have been used. In some of these cases halogen substituents were retained in reductions using iron and acids[401g], stannous chloride[633b] and catalytic procedures[632], and eliminated when reduction was effected with Raney alloy and alkali[632]. 3,5-Diaminopyridine is produced by the catalytic hydrogenation of 4-chloro-3,5-dinitropyridine[400b] and 3,5-diamino-4-ethoxypyridine from the dinitro compound and stannous chloride[620c]. Diamino-nitro-[398c] and -nitroso-pyridines[401g] have been reduced to triamines.

The reduction of some amino-nitropyridines with stannous chloride and hydrochloric acid is accompanied by nuclear chlorination (p. 227). This difficulty can be avoided by the use of sodium hydrosulphite, a method which also has the feature of not removing halogen substituents[405d, 424b, c, 438b].

3-Amino-2- and -6-methylpyridine have been prepared by catalytic hydrogenation of the corresponding 4-chloro-3-nitropicoline 1-oxides[399a], and 2-amino-3,5-diethoxypyridine from the nitro-oxide by reduction with iron and acetic acid[574e] but, of course, the use of pyridine 1-oxides as a source of aminopyridines has become of importance only since the discovery that nitration occurs readily at $C_{(4)}$ (p. 174). 4-Nitropyridine 1-oxides have been reduced to 4-aminopyridine 1-oxides by ammonium sulphide[395d, 400c] and by catalytic hydrogenation[400a, 401i, 637], and to 4-aminopyridines by

Table 6.16. Some Nitro-, Nitroso- and Hydroxylaminopyridines and
-pyridine 1-Oxides (cf. Tables 6.2, 8–10, 13, 14, 18)

Substituent	M.p., °C	Derivatives (m.p., °C)
2-NO$_2$[434]	71 (b.p. 256°)	
3-NO$_2$[454b, 461a]	41 (b.p. 216°)	hydrochloride (154)
		nitrate (151)
4-NO$_2$[429a]	50	
3-Me-2-NO$_2$[432]	43–4	
4-Me-2-NO$_2$[432]	61–2	
5-Me-2-NO$_2$[432]	94–5	
6-Me-2-NO$_2$[432]	113–4	
4,6-Me$_2$-2-NO$_2$[405g]	59–60	
2-Me-3-NO$_2$[436c, 627]	(b.p. 99–100°/8 mm)	hydrochloride (165–7)
4-Me-3-NO$_2$[93b, 586, 627]	(b.p. 85°/3 mm)	hydrochloride (176–7 dec.)
		picrate (118)
5-Me-3-NO$_2$[405g]	90–1	
6-Me-3-NO$_2$[436c, 620a, 627]	112	picrate (132)
2,6-Me$_2$-3-NO$_2$[620a]	37 (b.p. 227°/738 mm)	picrate (143)
2,4,6-Me$_3$-3-NO$_2$[620a, b]	38	picrate (176)
		methiodide (210–1)
2-Me-4-NO$_2$[405g, 622]	32–4, 42–5	
3-Me-4-NO$_2$[405g]	28·9 (b.p. 67–9°/1·5 mm)	picrate (128–9)
3-Et-4-NO$_2$[401b]	(b.p. 56–8°/0·25 mm)	
4-NO$_2$-3-i-Pr[401b]	(b.p. 82–4°/0·85 mm)	picrate (106–7)
3,5-Me$_2$-4-NO$_2$[401b]	38–9	picrate (169–170)
2,3,5,6-Me$_4$-4-NO$_2$[401b]	198–200*	picrate (174–6)
3,5-(NO$_2$)$_2$[452b]	106	
1-Oxides		
2-NO$_2$[405g]	85–6	
3-NO$_2$[628]	169–169·5	
4-NO$_2$[395d, 514f]	159	picrate (161·5–162)
3-Me-2-NO$_2$[405g]	110–1	
4-Me-2-NO$_2$[405g]	118–9	
5-Me-2-NO$_2$[405g]	112–3	
6-Me-2-NO$_2$[405g]	120–1	
4,6-Me$_2$-2-NO$_2$[405g]	108–9	
2,6-Me$_2$-3-NO$_2$[395d]	105	
2-Me-4-NO$_2$[395d]	153–4	
3-Me-4-NO$_2$[629–30]	137·5	
3-Et-4-NO$_2$[401b]	68–9	
4-NO$_2$-3-i-Pr[401b]	138–9	
2,6-Me$_2$-4-NO$_2$[395d]	163	
3,5-Me$_2$-4-NO$_2$[401b]	174–5	picrate (137·5–138·5)
2,3,5,6-Me$_4$-4-NO$_2$[401b]	115–6	picrate (160–1)
Nitrosopyridines		
3-NO[631]	94†	
Hydroxylaminopyridines		
2-NHOH[83d]	83–5	
3-NHOH[631]	109	

* Dihydrate.
† A green melt is formed, and this colourless compound also gives green solutions.

iron and acetic acid[396e, 400d, 401b, 514g, 629, 637–8] (bromine being retained[514g, 574e]) and catalytically[391c, 393c, 638–40].

Products other than amines can be obtained by reducing nitropyridines. With amalgamated aluminium and wet ether at 0°, 3-nitropyridine gives 3-hydroxylamino-pyridine (some 3-nitrosopyridine may be formed, and it can be prepared by oxidizing the hydroxylamine)[631]. Hydroxylamines have been prepared from 2-nitropyridine[83d] and ethyl 2-methyl-5-nitronico-tinate[450] by hydrogenation over platinum in ethanol. Sodium arsenite reduces 2-nitropyridine to the azo compound[429a]. 3-Nitropyridines give the azoxy compounds with zinc and caustic soda[631] and sodium arsenite[454b], and sometimes on catalytic hydrogenation[393a]. The azo compound is formed from 4-nitropyridine and sodium stannite[514f], and the azoxy compound with sodium arsenite[461b].

The possibilities with 4-nitropyridine 1-oxides are obviously complicated. Azoxy-dioxides have been produced by reduction with zinc and acetic acid[400a, c] and by catalytic hydrogenation[638], azo-dioxides by reduction with ammonium sulphide[400a, c], warm caustic soda[400c], alcoholic ammonia or benzylamine[560e] and caustic soda with sodium nitrite[400a]. The azopyridine, its monoxide and dioxide are all formed in the reduction with sodium stannite[461b]. Stannous chloride produces the hydrazopyridine dioxide, and zinc and alkali the hydrazo-pyridine[400c]. Warm 15 per cent caustic soda converts 4-nitropyridine 1-oxide into 4,4'-azopyridine[514f].

(k) Oxides

The dipole moments (p. 126), spectra (p. 132) ionization constants (p. 149) and ability to complex with metals (p. 160) have been discussed. Some pyridine 1-oxides are recorded in *Table 6.17*.

Since the latter are useful sources of pyridine derivatives, methods for their deoxygenation have been much studied. These compounds lack the oxidizing properties found with some amine oxides. Polarographic studies show that in acid media pyridine 1-oxide has a much lower reduction potential than oxides of tertiary amines with exocyclic nitrogen atoms[646-7]. The half-wave potentials are not useful guides to the abilities of the N–O bonds in pyridine oxides to take part in oxygen transfer reactions, and in 4-nitro-pyridine 1-oxide the nitro group is reduced first[648].

The reduction of nitropyridine 1-oxides to aminopyridine 1-oxides or aminopyridines has been discussed (p. 384). Iron and acetic acid, which produce the second result, are useful in deoxygenating oxides lacking nitro groups and leave halogen substituents intact[514f, 574d, 645]. Zinc and acetic acid have also been used[641]. It has been noticed (p. 174) that in the nitration of some pyridine 1-oxides, nitropyridines are formed as well as their oxides, and conditions have been described[649] which convert pyridine 1-oxide into 4-nitropyridine in 56 per cent yield.

Catalytic reduction of nitropyridine 1-oxides can give either aminopyri-dine 1-oxides or aminopyridines (p. 384). Raney nickel can be used to deoxygenate 4-hydroxy- and 4-alkoxy-pyridine 1-oxides[560d]. With Raney nickel, 4-benzyloxypyridine 1-oxide gives 4-benzyloxypyridine, with palladized charcoal it forms 4-hydroxypyridine 1-oxide, and with a mixture of these catalysts, 4-hydroxypyridine[639]. A palladium catalyst has been used to deoxygenate a number of oxides; double bonds and halogen atoms could be reduced before the oxide group[368]. Whilst Raney nickel, used to catalyse hydrogenations in acetic acid–acetic anhydride, deoxygenated pyridine and picoline oxides and left halogen substituents intact, it failed with the oxides of 2,6-lutidine, 2-hydroxypyridine and picolinic acid[650].

Table 6.17. Some Pyridine 1-Oxides*

Substituents	M.p., °C	B.p., °C	Derivatives (m.p., °C)
—[227b, 641]	66–8	122–4/5 mm	hydrochloride (180–1) picrate (179·5)
2-Me[227]	—	123–4/15 mm	picrate (125–126·5)
3-Me[227a, 391c]	33–6	146–9/15 mm	picrate (141–3)
4-Me[86, 227]	186–8	151–3/11 mm	picrate (158·7–159·7)
2-Et[227b]	—	109–113/4 mm	—
3-Et[401b]	—	123–5/12 mm	picrate (95)
2,3-Me$_2$[8]	85–93	—	—
2,4-Me$_2$[642]	—	—	hydrochloride (178) picrate (140)
2,6-Me$_2$[227a, 642]	—	115–9/18 mm	hydrochloride (219·5) picrate (127·5–129)
3,5-Me$_2$[401b]	—	116–8/0·1 mm	picrate (135–6)
3-i-Pr[401b]	—	120–2/0·8 mm	picrate (125–6)
5-Et-2-Me[227b]	—	147/11 mm	—
2,4,6-Me$_3$[642]	—	—	hydrochloride (246) picrate (166–7)
2-t-Bu[401b]	—	132–4/1 mm	picrate (143–4)
2,3,5,6-Me$_4$[401b]	139–140	—	picrate (144–5)
2-CH$_2$Ph[643]	100·5	—	picrate (119)
4-CH$_2$Ph[643]	151	—	picrate (94)
2-Ph[643–4]	157	—	picrate (150–2, 130–1)
3-Ph[643]	119	—	picrate (161)
4-Ph[643]	152–152·5	—	picrate (171–3)
2,6-Ph$_2$[644]	125–6	—	picrate (161–2)
2,4,6-Ph$_3$[641]	184	—	hydrochloride (171–2) picrate (189)
2-CH:CHPh[196b]	162	—	—
4-CH:CHPh[196b]	169	—	—
4-C:CPh[191]	184·5–185·5	—	picrate (147·5–148·5)

* For 1-oxides of pyridines containing functional groups other than alkyl, aryl, benzyl, styryl and phenyl-thynyl, see the Table appropriate to the functional group.

The most widely used reagent for deoxygenating pyridine 1-oxides has been phosphorus trichloride in chloroform. Sometimes it causes the replacement of a 4-nitro group by chlorine to a small extent (p. 233), but can nevertheless be used in the preparation of 4-[391c, 395d, 401b, 621b, 649, 651] and 2-nitro-pyridines[405g]. Phosphorus tribromide can also cause both deoxygenation and replacement of a nitro group by bromine (p. 233), but used in ethyl acetate with 4-nitropyridine 1-oxide it gives 4-nitropyridine[652a]. Phosphorus trichloride in chloroform can convert 4-hydroxypyridine 1-oxides into 4-hydroxypyridines[621b, 630, 652b] and 4-aminopyridine 1-oxide into 4-amino-pyridine[652b]. In the latter reaction, yields are poor, and it is better to use 4-acetamidopyridine 1-oxide[652b]. In the absence of complicating functional groups, phosphorus trichloride[621b] and phosphorus tribromide[652a] cause smooth deoxygenation. With the latter compound, 4-hydroxy-2-methyl-pyridine 1-oxide gives 4-bromo-2-methylpyridine[651b].

The kinetics of the deoxygenation of some pyridine 1-oxides by phosphorus trichloride in chloroform have been found to be complex[653a]. A 4-nitro group slows down the reaction, as does protonation of the oxide group, both facts being consistent with the idea that the reaction involves nucleophilic attack upon phosphorus by the oxide oxygen atom. The much higher temperatures required for deoxygenation by triethyl phosphite also fit this picture, though this reaction probably involves radicals.

Triphenyl phosphite deoxygenates pyridine 1-oxide but not 4-nitropyridine 1-oxide[652c]. Triethyl phosphite deoxygenates pyridine 1-oxide in the cold in diethylene glycol diethyl ether. 4-Methoxypyridine 1-oxide behaved similarly, but the 4-nitro compound reacted much more slowly[653b]. These reactions require oxygen and peroxides present in the solvent, and a free radical chain mechanism has been suggested.

From time to time, other reagents have been used for deoxygenating pyridine 1-oxides. These include ferrous oxalate (which may be suitable for amino but not for nitro compounds)[654], sulphur, mercaptans and hydrazobenzene[655a]. Dichlorocarbene[656a] and fluorene carbene also deoxygenate pyridine and picoline oxides[656b]. Photolysis of pyridine 1-oxide gives pyridine in a reaction whose characteristics depend on the frequency of radiation used. 2-Picoline 1-oxide photolysed at 2,537 Å gives picoline, but at 3,261 Å 2-hydroxymethylpyridine is formed[657].

The value of the quaternary salts derived from pyridine 1-oxides in certain nucleophilic substitutions has been discussed (e.g. p. 225). These salts react with alkali giving an aldehyde and the deoxygenated pyridine[456, 658]. Good yields of some aromatic aldehydes have been obtained in this way[659]:

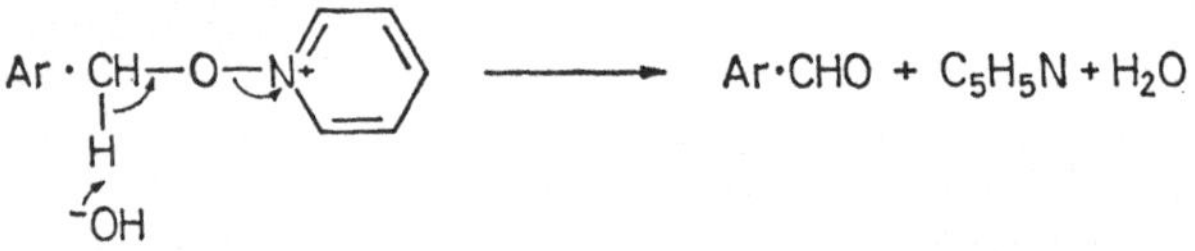

$$Ar \cdot CH_2\text{—}O\text{—}N^+ \longrightarrow Ar \cdot CHO + C_5H_5N + H_2O$$

The scope of this reaction is increased by the fact that the Ortoleva–King method can be used for quaternizing pyridine 1-oxides as well as pyridines (p. 179); thus, 2,4-dinitrotoluene with iodine and pyridine 1-oxide gives the quaternary salt, whence 2,4-dinitrobenzaldehyde can be obtained, and similarly desoxybenzoin forms benzil[212d].

Some cyclic quaternary derivatives behave rather differently[660]:

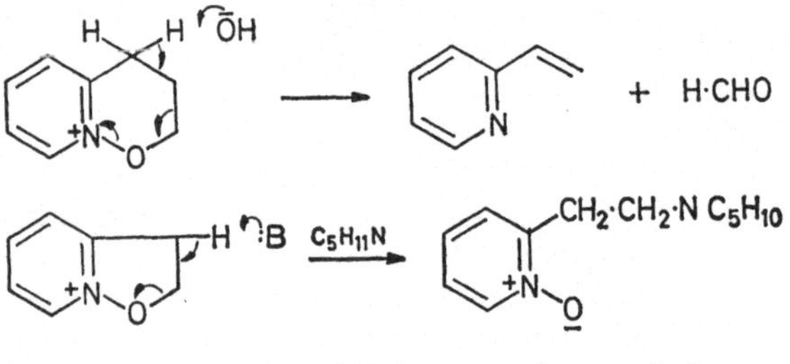

$$\longrightarrow \quad + \quad H \cdot CHO$$

$$\xrightarrow{\text{—H ···B } C_5H_{11}N}$$

With some bases, 1-methoxypyridinium p-toluenesulphonate acts as an alkylating reagent; with aniline, pyridine 1-oxide and methylaniline[661] are formed (p. 209).

(l) Quaternary Salts

When alkyl- and aralkyl-pyridinium salts are heated, some loss of alkyl or aralkyl halide and some decomposition usually occurs, but also the quaternizing group migrates into the pyridine nucleus. Ladenburg, after whom this reaction is named, first observed it with ethylpyridinium iodide which gave mainly 2-ethyl-, some 4-ethyl- and a small proportion of 2,4-diethyl-pyridine, as well as ethane and ethylbenzene. In the reaction with n-propylpyridinium iodide, the n-propyl group became an i-propyl group[75d, 662]. 2- and 4-Picoline ethiodide gave 4-ethyl-2- and 2-ethyl-4-methylpyridine, respectively[663], and benzylpyridinium salts formed 2- and 4-benzylpyridine[664a] with some toluene and stilbene[664b]. The reaction assumed real practical value when it was demonstrated that catalysts such as copper powder permitted it to be carried out in open vessels[665]. In this form it provides a good method for preparing 2- and 4-benzylpyridine[29c, 282, 665] which are accompanied by some 2,4- and 2,6-dibenzylpyridine[666], and it has been used on the large scale[667]. In its most efficient form, in which pyridine hydrochloride serves as solvent, the reaction gives 75 per cent of mixed 2- and 4-benzylpyridine[668]. 2-Picoline[669a] and 4-propylpyridine[280b] have been similarly benzylated, and α-naphthylmethyl-[669a] and 2-tolylmethyl-pyridines[669b] have been prepared. The formation of small proportions of 3-alkylpyridines in some of these rearrangements has been reported[665, 667].

It has been pointed out (p. 265) that nucleophilic attack upon a quaternary salt may have various consequences: the nucleophile may replace substituents from the pyridine nucleus, it may attack and open the ring and it may displace the pyridine ring from the quaternizing group. Numerous examples of the first two kinds of reaction (pp. 200–51, 265), and of the third the important case of the formation of 4-substituted pyridines from 4-pyridyl-pyridinium chloride (pp. 210, 239, 251) have been quoted. This last kind is a particular case of arylation of nucleophiles by arylpyridinium salts, and the general case of this reaction and analogous alkylations will now be considered. These are presumably S_N2 reactions (but see p. 269).

Substituted alkylation of primary and secondary alcohols has been carried out with alkoxymethyl-pyridinium salts[670], and these give esters with sodium salts of carboxylic acids[671]. Diethylaminoethylpyridinium p-toluenesulphonate alkylates the amino group of 8-amino-6-methoxyquinoline[672]. Although benzylpyridinium chloride gives benzyl sulphide with sodium sulphide[673a], it did not benzylate Grignard reagents[673b]. The reaction in which an alkylpyridinium salt is converted into a pyridine by being heated with triethanolamine presumably belongs in the present class, but the second product, possibly the alternative quarternary salt, has not been identified[193].

Arylations are commoner than alkylations. Ullmann and Nadai's preparation of dinitrochlorobenzenes, and its extension to that of diphenyl ethers—reactions in which the nucleophiles are chloride ions and phenols, respectively—have been mentioned already (p. 268). Applied to 1-(2,4-dinitronaphthyl)pyridinium chloride, the reaction gives 1-chloro-2,4-dinitronaphthalene, and the same salt reacts with water to give the naphthol, with hydrogen sulphide to yield the thionaphthol, and with carboxylic acids to form dinitronaphthyl esters[674]. Amines[675], alcohols[675], mercaptans and thiophenols[676] have been dinitrophenylated with pyridinium salts. Styrylpyridinium salts give N-styrylpiperidine with piperidine[677].

As would be expected, these arylations involve aryl groups which are highly activated to nucleophilic attack. The following kind, in which indole nuclei are involved, are surprising[678]:

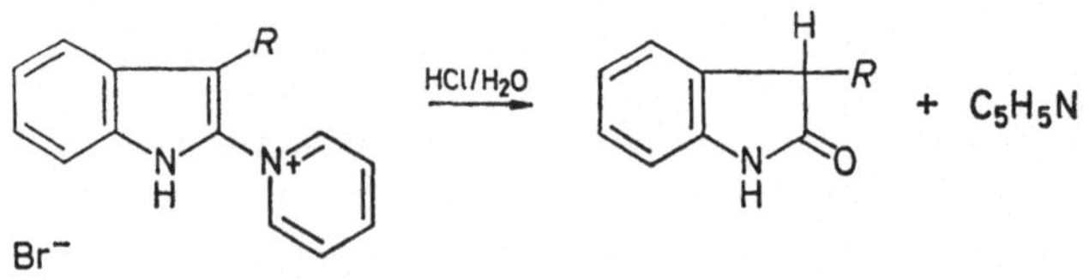

The relationship between phenacylpyridinium cations and 1,3-dicarbony compounds has been pointed out (p. 330). In view of this similarity it is not surprising that these cations are split by alkali to give carboxylic acids:

$$\overset{+}{C_5H_5N}\ CH_2COAr + \overset{-}{OH} \rightarrow \overset{+}{C_5H_5N}\ Me + ArCO_2^-$$

Since methyl ketones are readily available from a number of syntheses and can easily be converted into the quaternary pyridinium derivatives (p. 179), this reaction is a useful method of synthesizing carboxylic acids[212d]. The kinetics of this reaction have been studied[679], as have those of the reactions between phenacylpyridinium ions and ethoxide ions[680]. The latter have been written as

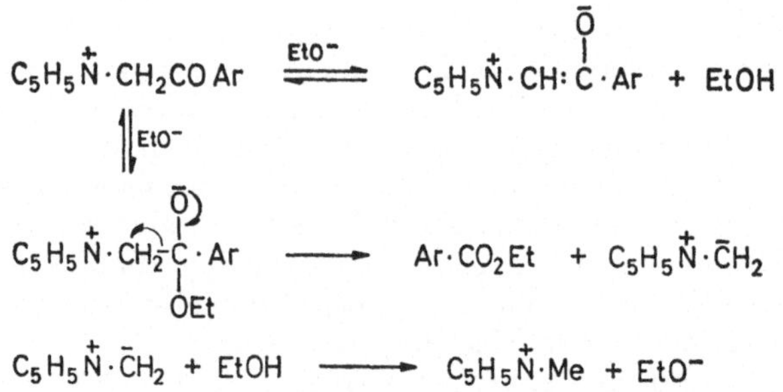

For the compounds in which Ar = Ph, p-Br.C_6H_4. and p-MeO.C_6H_4., the rates of this reaction were all about the same. In contrast, in the alkaline fission the p-bromophenacyl cation reacted twice as fast as the phenacyl cation, which in turn reacted about 1·5 times as fast as the p-methoxy-phenacyl cation. In the alcoholysis, the first step is favoured by the electron-attracting substituents and the second step hindered by them. In the hydro-lysis, two hydroxyl anions are involved, and besides addition to the carbonyl group and displacement of the $\overset{+}{C_5H_5N}.\overset{-}{C}H_2$ group, there is also loss of proton to the second hydroxyl anion.

(m) Sulphur-containing Groups

The x-ray structural examination of 2-pyridthione (p. 120) and the dipole moments (p. 125), spectra (p. 139), ionization constants (p. 153) and tautomerism of this and other compounds of this class have been mentioned, as have the spectra (p. 132) and ionization constants (p. 148) of the pyridine-sulphonic acids. The yellow thiones are among the simplest of coloured pyridine derivatives. *Table 6.18* records some thiones, sulphides, disulphides, sulphoxides, sulphones, sulphonic acids and thiocyanates of the pyridine series; the pyridthiones are for convenience recorded as thiols, with no implication for their tautomeric composition (p. 154).

The alkylation of pyridthiones has been discussed (p. 183).

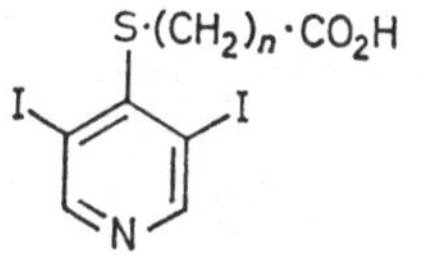

$$S \cdot (CH_2)_n \cdot CO_2H$$

(73)

Pyridthiones are useful sources of other sulphur-containing pyridine derivatives which can be obtained from them by oxidation. Thus, pyridthiones have been oxidized to disulphides by iodine and alkali[681-4], bromine[511b], hydrogen peroxide[683], ferric chloride[621a, 685], ferricyanide[687] and nitric acid[682, 683e]. Atmospheric oxidation can also effect this change[686], as can air and alkali[687]. Aqueous solutions of the sodium salts of the acids (73; $n = 1$ and 2) deposit the disulphide, the carbon side chains being lost; the analogous compounds without iodine substituents are stable under these conditions[688].

2- and 4-Pyridthiones can also be oxidized to sulphonic acids. For this purpose have been used nitric acid[681a, 683a, e, 689-90], hydrogen peroxide[405a, 511b, 690] and potassium permanganate[683e, 691-2]. The substance originally described as pyridine-4-sulphonic acid[681b], obtained by oxidizing 4-pyridthione with nitric acid, was a mixture of the sulphonic acid and the disulphide dinitrate[684, 689]. The pure sulphonic acid was obtained by carrying out the oxidation with hydrogen peroxide[684, 690]. With alkaline hydrogen peroxide, 3-cyano-4,6-dimethyl-2-pyridthione can give either the sulphinic or the sulphonic acid[682].

From 3,5-dihalogeno-4-pyridthiones, potassium permanganate produced the sulphonic acids, but hydrogen peroxide gave the dihalogenopyridines and nitric acid the dihalogenopyridones[691] (p. 249).

1-Substituted 4-pyridthiones can be oxidized to the 1-substituted 'sulphobetaines'[699, 711a].

2- and 4-Pyridthiones have been oxidized to sulphonyl chlorides by chlorine and hydrochloric acid[405a, 711b, c-2].

Useful products are also obtained by oxidizing alkyl and aryl pyridyl sulphides. With hydrogen peroxide, 2-methylthiopyridine[703] and 6-methylthionicotinamide[696] give the sulphoxides, and perbenzoic acid produces the sulphoxide from 2-benzylthiopyridine[524c]. Hydrolysis of the dibromide from 2-pyridyl sulphide gives the sulphoxide[703]. More commonly the oxidation of

Table 6.18. Some Sulphur-containing Derivatives of Pyridine

Substituents	M.p., °C	Derivatives (m.p., °C)
Thiols and sulphides		
2-SH[693]	130–2	1-oxide (65–7)[524c, 694]
		S-methyl (b.p.100–4°/ 33 mm)
		N-methyl (90)
3-SH[693]	81	S-methyl (b.p. 102°/17 mm)
4-SH[511b, 693]	179–189 (dec.)	1-oxide [140 (dec.)][694]
		S-methyl (47)
		N-methyl (168·5–170)
3-CO₂H-2-SH[683e, 695a]	270, 260–1 (dec.)	
5-CO₂H-2-SH[683d]	272 (dec.)	
5-NH₂-2-SH[683d]	170–1	S-methyl (71–2)[696]
5-CN-2-SH[696]	255	S-methyl (78)
5-Cl-2-SH[683d]	198	
5-Br-2-SH[683d]	203–4	
5-I-2-SH[683d]	210–1	
3,5-Br₂-2-SH[691]	148	
3,5-I₂-2-SH[695b]	206–206·5	
3-NO₂-2-SH[697]	174–5	
5-NO₂-2-SH[405a]	188–191	S-methyl (115)[696]
4-Et-3-SH[698]	122–5	
2-CO₂H-3-SH[683e]	183·5	
3-Me-4-SH[402b]	159–160	
2,6-Me₂-4-SH[400d, 681a, 699]	224	S-methyl (51)
		N-methyl (267–8)
3-CO₂H-4-SH[702]	236–8	Me ester (170–1)
3-NH₂-4-SH[700b, c]	213 (dec.)	S-methyl picrate (165–6)
3,5-Cl₂-4-SH[691]	188	
3,5-Br₂-4-SH[691]	222	
3,5-I₂-4-SH[691]	206 (dec.)	
3-NO₂-4-SH[400e, 700c]	190 (dec.)	S-methyl (133–4)
Disulphides		
2,2′-Dipyridyl[683d]	58	
3,3′-Dipyridyl[701a]	—	picrate (185)
		dihydrochloride (183)
4,4′-Dipyridyl[681b]	155	
Pyridyl sulphoxides		
2-SOMe[703]	(b.p. 122°/5 mm)	
5-CONH₂-2-SOMe[696]	224–6	
2-SOEt[703]	(b.p. 123°/4 mm)	
2-SO(2-C₅H₄N)[703]	(b.p. 178°/6 mm)	
2-SOCH₂Ph[524c]	87–8	
Pyridyl sulphones		
2-SO₂Me[681a, 703]	(b.p. 325°; 157°/5 mm)	
4-SO₂Me[511b]	81	
5-NH₂-2-SO₂Me[696]	171–3	
5-CONH₂-2-SO₂Me[696]	210	
5-CN-2-SO₂Me[696]	133	
5-NO₂-2-SO₂Me[696]	115	
2,6-Me₂-4-SO₂Me[681a]	—	picrate (221)
3-NH₂-4-SO₂Me[700c]	109–110	acetyl (159–160)
3-NO₂-4-SO₂Me[700c]	123	
Pyridinesulphonic acids		
2-SO₃H[690]	251–2	amide (133)[704]
		anilide (168)[704]

Substituents	M.p., °C	Derivatives (m.p., °C)
3-SO$_3$H[690, 705]	358–360	amide (110–1)[705]
		anilide (145)[706]
		chloride hydrochloride
		(141–4)[706]
4-SO$_3$H[690]	317–8	amide (168–9)[704]
2,6-Me$_2$-3-SO$_3$H[690]	>350	
2,6-Me$_2$-4-SO$_3$H[690]	>350	
6-Cl-3-SO$_3$H[683d]	265 (darkening)	amide (159)
Pyridyl thiocyanates		
2-SCN		1-oxide (158–160)[707]
3-SCN[708]	32 (b.p. 124°/12 mm)	
4-SCN		1-oxide (190 dec.)[707]
3-NO$_2$-2-SCN[709]	120	
4-NO$_2$-2-SCN		1-oxide (180–1)[707]
5-NO$_2$-2-SCN[709]	129–130	
3,5-(NO$_2$)$_2$-2-SCN[710]	145–6	
3-NO$_2$-4-SCN[700a]	139	

sulphides yields sulphones, and for this purpose have been used permanganate[681a, 686–7, 696, 713–4], chromic acid[715–6] and hydrogen peroxide[687] which also converts the sulphoxide from 6-methylthionicotinamide into the sulphone[696].

With benzoyl chloride in alkali, 4-pyridthione gives 4-benzoylthiopyridine, whilst acetic anhydride causes N-acetylation[717]. Some interesting reactions occur in the ethoxycarbonylation of 4-pyridthione. With ethyl chloroformate in ethanol, reaction occurs at the sulphur atom, whilst in sodium bicarbonate solution the nitrogen atom is attacked[717]:

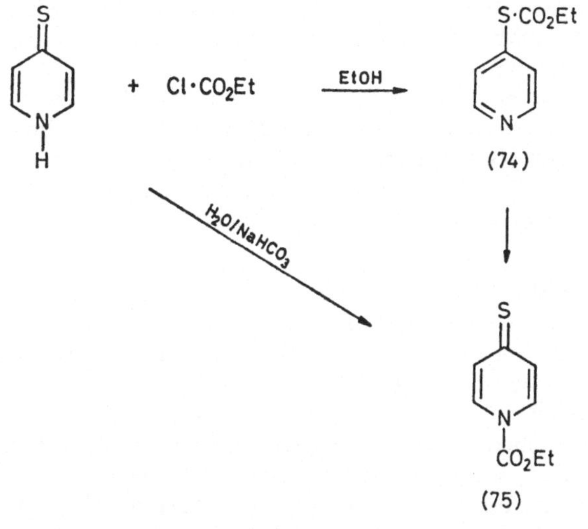

(74) is unstable in the absence of acid, isomerizing to (75). When kept at

room temperature or irradiated with ultra-violet light, (75) gives di-4-pyridyl sulphide:

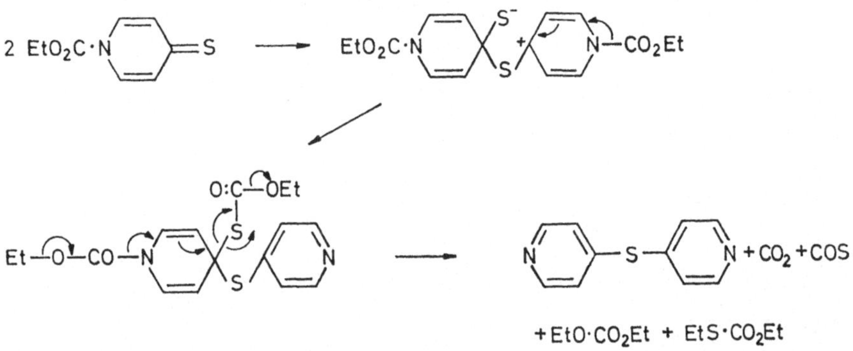

$$2\ EtO_2C\cdot N\!\!\bigcirc\!\!=S \longrightarrow EtO_2C\cdot N\!\!\bigcirc\!\!\underset{S}{\overset{S^-}{\diagdown}}\!\!\bigcirc\!\!N\!\!-\!CO_2Et$$

$$Et\!-\!O\!-\!CO\!-\!N\!\!\bigcirc\!\!\underset{S}{\overset{O:C\!-\!OEt}{\diagdown}}\!\!\bigcirc\!\!N \longrightarrow N\!\!\bigcirc\!\!-S\!-\!\bigcirc\!\!N + CO_2 + COS$$

$$+ EtO\cdot CO_2Et\ +\ EtS\cdot CO_2Et$$

When heated, 1-alkoxycarbonyl-4-pyridthiones give 4-alkylthiopyridines and carbon dioxide[717].

3-Cyano-4,6-dimethyl-2-methylthiopyridine is reduced by zinc and acetic acid to 4,6-dimethylnicotinonitrile[682].

The pyridinesulphonic acids are normal in the derivatives which they form. Sodium and ammonium 3-cyano-4,6-dimethyl-2-sulphinate heated with copper and acetic anhydride, respectively, lost the sulphinate group[682].

Pyridine-2- and -4-sulphonyl chlorides are unstable compounds, which are said to decompose above 0° to give sulphur dioxide and the chloropyridines[704]. Pyridine-3-sulphonyl chloride is stable as its hydrochloride[706]. All three isomers readily give amides.

In 50 per cent aqueous acetone at 30°, the ratios of hydrolysis rates[718] for p-nitrobenzenesulphonyl, m-nitrobenzenesulphonyl, pyridine-3-sulphonyl and benzenesulphonyl chloride are 5·15 : 2·52 : 2·28 : 1, whilst in aqueous dioxan the 3-pyridyl group was found to be more activating than either m- or p-nitrophenyl[719].

Pyridine-3-sulphonyl chloride is reduced to the thiol by stannous chloride[701b] and to the disulphide by sulphur dioxide and hydriodic acid[701a]. With hydrazine hydrate in diethylene glycol, the pyridinesulphonyl chlorides give the disulphides[712], and with zinc and water, pyridine-2-sulphonyl chlorides yield the sulphinic acids[711c].

(n) Pyridyl and Related Radicals

The reactions of pyridines with radicals have been discussed (p. 251). The properties and reactions of pyridyl and related radicals formed from pyridines will now be considered.

The formation of the radical (76) by reduction of the quaternary salt has been mentioned (p. 259). This remarkable substance is a dark-green, distillable oil of which the spectra have been recorded[720]. Its unusual stability has been ascribed to the steric hindrance which its structure offers to dimerization and to the resonance stabilization, augmented by the presence of the hetero-atoms. The rates of halogen abstraction from halocarbons by the radical have been measured; these are atom transfers and give products[721] of the kind shown

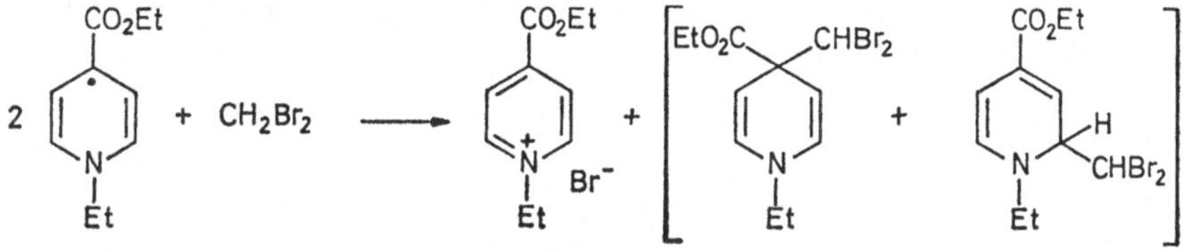

(76)

2-Pyridyl radicals from sodium 2-pyridine-isodiazotate effect o- and p-substitution in phenol, and some 2-phenoxypyridine is also formed[722]. Radicals from diazotized 3-aminopyridines cause 3-pyridylation of benzene[428c] and of ferrocene[723]. In the useful variation of the Gomberg–Hey reaction in which an aromatic amine, pentyl nitrite, and an aromatic hydrocarbon react homogeneously in the absence of acid, benzene and 3-aminopyridine give 52 per cent 3-phenylpyridine[724]. Several 3-N-nitroso-acylaminopyridines have been used in pyridylations of benzene[428c, 725–6] and of ethyl pyrrole-1-carboxylate[727]. 3,3-Dimethyl-1-(3-pyridyl)triazine is too stable to be used in phenylating benzene[427], and the similar triazine from 5-amino-1-methyl-2-pyridone gave a poor yield of 1-methyl-5-phenyl-2-pyridone with benzene[428a].

The reactions of pyridines with certain metals are important sources of 2,2'-bipyridyls and probably involve radical intermediates formed from the metal and the pyridine. These reactions may be homogeneous or heterogeneous. In the homogeneous form they involve the alkali metals.

Anderson first observed the reaction of pyridine with sodium[728], and bipyridyls and dihydrobipyridyls were recognized as products[729]. Bipyridyls are also formed as byproducts in the Tschitschibabin amination reaction (p. 206), particularly when hydrocarbon solvents are employed; and its course depends markedly on the quality of the sodamide used[730–1]. The reaction of pyridine with sodium became of practical importance when it was shown that passage of air or oxygen through the solution gave a mixture of bipyridyls in useful amount; in addition to the previously observed 4,4'-compound, the 2,2'-, 3,3'- and 3,4'-isomers were also obtained[732]. Later, the 2,4'-isomer was also said to be formed[733]. When Emmert's 'sodium dipyridine' (see below) was dispersed in pyridine at 90° and oxidized with air or oxygen, the chief product was 4,4'-bipyridyl. If the solution was kept for several hours at 90°–115° before oxidation, increasing amounts of the other isomers were formed[732].

There is some confusion about the character of the initial reaction between sodium and pyridine, which gives a dark-green solution. Emmert[729] isolated a compound formulated as $(C_5H_5N)_2Na$, ('sodium dipyridine'), and this gave $C_5H_5N.Na$ when heated. In this reaction and similar ones with potassium and lithium, hydrogen was said not to be evolved[729, 732]. Setton[734] claimed that hydrogen was evolved and sodium pyridyl, $C_5H_4N.Na$, formed.

When the attempt was made to prepare the pyridine negative ion, by reaction with potassium at low temperatures in 1,2-dimethoxyethane, a purple solution was formed which became yellow and paramagnetic in 5–10 min. Electron-spin resonance showed that 4,4'-bipyridyl was formed and then its negative ion[735].

The formation of bipyridyls in these reactions is best formulated as

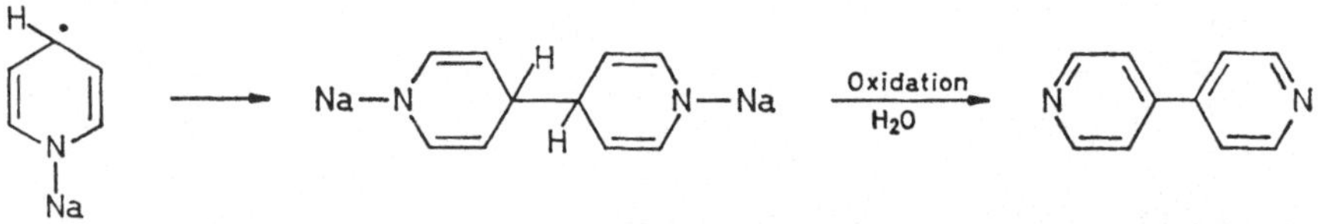

The formation of 2,2'-bipyridyl can obviously be accounted for in the same way, and it has been suggested that linking through the 3-position occurs because at higher temperatures the sodium becomes attached to $C_{(2)}$ rather than to nitrogen, giving an odd electron at the β-position[736].

With ammonium chloride and sodium in liquid ammonia, pyridine 1-oxide forms 2,2'-bipyridyl and a small amount of pyridine[737].

A number of heterogeneous catalysts have been used to produce 2,2'-bipyridyl from pyridine, but only palladized charcoal and Raney nickels are active at ordinary pressures[738]. When pyridine is boiled with 5 per cent palladized charcoal, a small conversion to 2,2'-bipyridyl occurs, and 2,4-lutidine gives some 4,4',6,6'-tetramethyl-2,2'-bipyridyl[739]. Better conversions are obtained with Raney nickel[740], and this method is now of great practical value for producing 2,2'-bipyridyl. 3- and 4-Alkyl groups facilitate the formation of bipyridyls, whilst electron-attracting groups in the same positions and 2-methyl groups have the opposite effect. 2-Methyl groups can be eliminated, as in the case of 2,5-lutidine which gives 3,3',5,5'-tetramethyl- and 5,5'-dimethyl-2,2'-bipyridyl. 3-Substituted pyridines give 5,5'-disubstituted 2,2'-bipyridyls[740b–1a].

The effects of substituents, the fact that only 2,2'-bipyridyls are formed and the efficacy of Raney nickels of low hydrogen content are believed to indicate that hybrid radicals (as illustrated) are involved, that the linking occurs when the two organic residues are bonded to one nickel atom and that linking is followed by dehydrogenation and desorption[740b, 741b]:

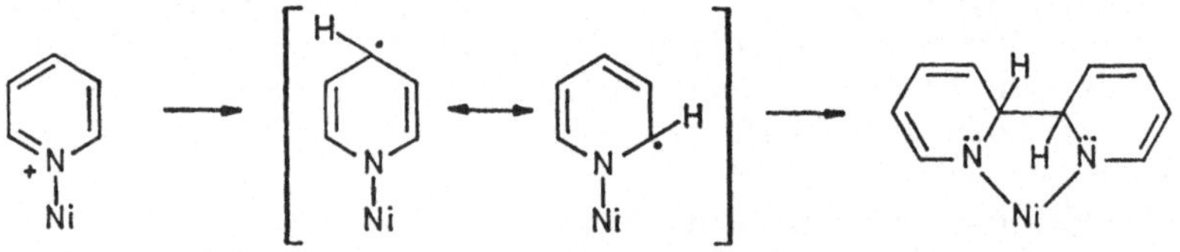

The conversion of pyridine into 2-picoline in high yield when it is boiled with a primary alcohol and Raney nickel may be closely related to this method of producing 2,2'-bipyridyl. It has been suggested that the alcohol is a source of carbon monoxide or some related, partially reduced species, which attacks the 2-position of the adsorbed pyridine[742].

Addendum chiefly on work reported during 1965

(*To a*) The ionization constants of pyridine-2,6-dialdehyde dioxime, and the effect upon these of coordination with ferrous ions have been reported[743]. Studies of the cupric complex of 2-benzoylpyridine oxime have been published[744] (the *anti*-pyridyl oxime, but not its stereoisomer, gives a complex).

Quaternary salts of O-acetyl-oximes of formylpyridines give with alkali the corresponding nitriles as well as the parent oximes[745].

2-Formyl-1-methylpyridinium cation monomethyl acetal (and also 2-hydroxymethyl-1-methylpyridinium) give 1-methylpyridinium ions with alkali[746] by way of the ylid(77).

(*To b*) 2- and 4-Methylpyridines, but not 3-methylpyridines are oxidized to the carboxylic acids by manganese dioxide[747].

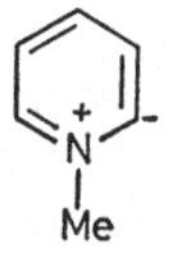

(77) 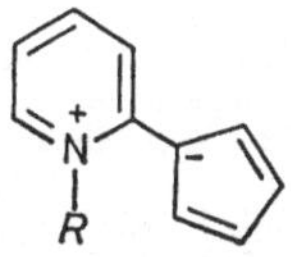   (78)

Kinetic studies of base-catalysed deuterium exchange in the methyl groups of picolines give the sequence 4>2>3, in the oxides 2>4>3 and in the quaternary salts 2>4, and quaternary salt > oxide > picoline. The reactions resemble nucleophilic substitutions in their ρ values, and delocalization energies and π-electron densities at nuclear carbon atoms carrying the methyl groups are indices of reactivity in the methyl groups[748].

Methides of the type (78) have pK_a values of 8·5–12. A theoretical discussion of their ground and excited state properties has been given[749].

In contrast to 2-picoline, 2- and 4-picoline 1-oxide and their O-ethyl quaternary salts do not couple with *p*-nitrobenzenediazonium ions[750].

(*To c*) Further examples of the photo-isomerization and dimerization of 2-styrylpyridinium salts have been reported[751].

(*To d*) 4-Aminotetrafluoropyridine is a very weak base. Through the diazonium compound it gives 4-bromotetrafluoropyridine and 4-(*p*-dimethylaminophenylazo)tetrafluoropyridine. It has been oxidized to the nitro compound[752].

(*To h*) A full report of the base-catalysed deuteration of 1-methyl-4-pyridones has appeared[753a].

With carboxylic acid chlorides, 2-ethoxypyridine 1-oxide gives ethyl chloride and 1-acyloxy-2-pyridones. The latter are acylating agents, and the corresponding sulphonyl compounds, formed similarly, sulphonylate amines[753b].

(*To i*) In the presence of catalytic amounts of alkali, 2- and 3-picoline are alkylated in the side chain by styrene and butadiene[754].

4- and 6-Methyl groups in 3-cyano-2-pyridones can be alkylated by alkyl halides when the compounds are converted into dianions by potassamide[755].

(*To k*) Arylacetic acid anhydrides react distinctively with pyridine and picoline 1-oxides. The main products are the corresponding pyridine or picoline, the arylacetic acid, the arylaldehyde and carbon dioxide[756]. The mechanism of this type of reaction is not yet clear.

(*To l*) Reactions of the relatively stable acylpyridinium salt, 1-dimethyl-aminocarbonyl pyridinium chloride, mentioned above, with nucleophiles have been examined[757]. In non-hydroxylic solvents, chloride anion removes the acyl group. In aqueous solution the reactions are not S_N1 in character.

(*To n*) Ultra-violet and electron-spin resonance spectra of some stable pyridinyl radicals have been reported[758].

REFERENCES

[1] Harries, C. and Lenart, G. H. *Justus Liebigs Annln. Chem.* **410** (1915) 1

[2] Dyson, P. and Hammick, D. L. *J. chem. Soc.* (1939) 781

[3] Mathes, W. and Sauermilch, W. *Chem. Ber.* **84** (1951) 648; *U.S.Pat.* 2,620,342 (1952); *Br. Pat.* 710,201 (1954)

[4] Panizzon, L. *Helv. chim. Acta* **24** (1941) 24E

[5] Felder, E. and Pitre, D. *Gazz. chim. ital.* **86** (1956) 386; *Br. Pat.* 711,307 (1954)

[6] Metzler, D. E. and Snell, E. E. *J. Am. chem. Soc.* **77** (1955) 2431; Mason, S. F. *J. chem. Soc.* (1959) 1253

[7] Nakamoto, K. and Martell, A. E. *J. Am. chem. Soc.* **81** (1959) 5857

[8] Ginsburg, S. and Wilson, I. B. *J. Am. chem. Soc.* **79** (1957) 481

[9] Klein, T. *Chem. Ber.* **86** (1953) 584

[10] Angyal, S. J., Barlin, G. B. and Wailes, P. C. *J. chem. Soc.* (1953) 1740

[11] Wibaut, J. P., Kooyman, E. C. and Boer, H. *Recl Trav. chim. Pays-Bas Belg.* **64** (1945) 30

[12] Poziomek, E. J., Kramer, D. N., Mosher, W. O. and Michel, H. O. *J. Amer. chem. Soc.* **83** (1961) 3916

[13] (*a*) Mathes, W. and Sauermilch, W. *Chem. Ber.* (*a*) **85** (1952) 1008; (*b*) **90** (1957) 758; (*c*) with Klein, T. **86** (1953) 584; (*d*) *Chemikerzeitung* **80** (1956) 475; (*e*) Volke, J. *Chem. Listy* **55** (1961) 26

[14] Dornow, A. and Bormann, H. *Chem. Ber.* **82** (1949) 216

[15] Bobbitt, J. M. and Scola, D. A. *J. org. Chem.* **25** (1960) 560

[16] For an exhaustive tabulation see *Pyridine and its Derivatives*, ed. Klingsberg, E. (*a*) Chapter 14; (*b*) Chapter 10 and 11, New York (Interscience) 1964

[17] Queguiner, G. and Pastour, P. *C.r.hebd. Séanc. Acad Sci., Paris* **258** (1964) 5903

[18] Furukawa, S. *Chem. Abstr.* **53** (1959) 3219

[19] Mathes, W. and Sauermilch, W. *Justus Liebigs Annln Chem.* **618** (1958) 152

[20] Kolloff, H. G. and Hunter, J. H. *J. Am. chem. Soc.* **63** (1941) 490

[21] Clemo, G. R., Holmes, T. and Leitch, G. C. *J. chem. Soc.* (1938) 753

[22] Henze, H. R. and Knowles, M. B. *J. org. Chem.* **19** (1954) 1127

[23] Clemo, G. R. and Holmes, T. *J. chem. Soc.* (1934) 1739

[24] Katritzky, A. R. *J. chem. Soc.* (1955) 2586

[25] Warner, C. R., Walsh, E. J. and Smith, R. F. *J. chem. Soc.* (1962) 1232

[26] Emmert, B. and Wolpert, A. *Ber. dtsch. chem. Ges.* **74** (1941) 1015

[27] Shivers, J. C., Dibbon, M. L. and Hauser, C. R. *J. Am. chem. Soc.* **69** (1947) 119

[28] Engler, C. *Ber. dtsch. chem. Ges.* **24** (1891) 2539

[29] (*a*) Teague, P. C. *J. Am. chem. Soc.* **69** (1947) 714; (*b*) Huntress, E. H. and Walter, H. C. **70** (1948) 3702; (*c*) LaForge, F. B. **50** (1928) 2484

[30] Villani, F. J., King, M. S. and Papa, D. *J. org. Chem.* **17** (1952) 249

[31] Wolffenstein, R. and Hartwich, F. *Ber. dtsch. chem. Ges.* **48** (1915) 2043

[32] Tschitschibabin, A. E. *Zh. russk. fiz.-Khim. Obshch.* **33** (1901) 700

[33] Müller, A. and Dorfman, M. *Mh. Chem.* **65** (1935) 411

[34] (*a*) Wibaut, J. P., with Heeringa, L. G. *Recl Trav. chim. Pays-Bas Belg.* **74** (1955) 1003; (*b*) with de Jonge, A. P. *et al.* **70** (1951) 1054

[35] Nakashima, T. *Chem. Abstr.* **52** (1958) 6345

[36] (*a*) Case, F. H. and Butte, W. A. *J. org. Chem.* **26** (1961) 4415; (*b*) Osuch, C. and Levine, R. **22** (1957) 939

[37] (*a*) Burger, A. and Walter, C. R. *J. Am. chem. Soc.* **72** (1950) 1988; (*b*) Miller, A. D., Osuch, C., Goldberg, N. N. and Levine, R. **78** (1956) 674

[38] (*a*) Goldberg, N. N. and Levine, R. *J. Am. chem. Soc.* **74** (1952) 5217; (*b*) with Barkley, L. B. **73** (1951) 4301

[39] Oda, D. *Chem. Abstr.* **56** (1962), 10089

REFERENCES

40 Beckett, A. H., Kerridge, K. A., Clark, P. and Smith W. G. *J. Pharm. Pharmac.* **7** (1955) 717
41 Hey, J. W. and Wibaut, J. P. *Recl Trav. chim. Pays-Bas Belg.* **72** (1953) 522
42 (a) Kröhnke, F. and Ellegast, K. *Justus Liebigs Annln Chem.* **600**, (1956) 198;
   (b) Schewing, G. and Winterhalter, L. **473** (1929) 126
43 *Germ. Pat.* 594,849 (1934)
44 Mathes, W., Sauermilch, W. and Klein, T., *Chem. Ber.* (a) **84** (1951) 452; (b) **87**
   (1954) 1870; (c) Kramer, F. and Krum, W. **86** (1953) 1586
45 Ladenburg, A. and Kroener, E. *Ber. dtsch. chem. Ges.* **36** (1903) 119
46 Katritzky, A.R. and Lagowski, J. M. in *Advances in Heterocyclic Chemistry*, Vol. I, ed.
   Katritzky, A. R., New York (Academic Press) 1963
47 Volke, J. and Valenta, P. *Colln. Czech. chem. Commun.* **25** (1960) 1580
48 Mathes, W. and Sauermilch, W. *Chem. Ber.* **89** (1956) 1515
49 (a) Buehler, C. A., Addleburg, J. W. and Glenn, D. M. *J. org. Chem.* **20** (1955) 1350;
   (b) Marvel, C. S. and Stille, J. K. **21** (1956) 1313
50 Hensel, H. R., *Angew. Chem.* **65** (1953) 491
51 Graf, R. and Weinberg, A. *J. prakt. Chem.* **134** (1932) 177; Graf, R. and László, P. **138**
   (1933) 231
52 Williams, J. L. R., Adel, R. E. *et al.*, *J. org. Chem.* **28** (1963) 387
53 Capuano, L. and Jamaigne, F. *Chem. Ber.* **96** (1963) 798
54 Liu, C. H. and C. F. *J. Am. chem. Soc.* **83** (1961) 4167, 4169
55 Hanania, G. I. H. and Irvine, D. H. *J. chem. Soc.* (1962) 2745
56 Hackley, B. E., Poziomek, E. J., Steinberg, G. M. and Mosher, W. A. *J. org. Chem.* **27**
   (1962) 4220
57 Szmant, H. H. and Harmuth, C. M. *J. Am. chem. Soc.* **86** (1964) 2909
58 Bain, B. M. and Saxton, J. E. *J. chem. Soc.* (1961) 5216
59 Nienburg, H. *Ber. dtsch. chem. Ges.* (a) **67** (1934) 874; (b) Gighi, E. **75** (1942) 1316
60 Oparina, M. P. *Chem. Abstr.* **29** (1935) 1820
61 Galinowsky, F. and Kainz, G. *Mh. Chem.* **77** (1947) 137
62 van der Meer, S., Kofman, H. and Veldstra, H. *Recl Trav. chim. Pays-Bas Belg.* **72** (1953) 236
63 Branch, R. F., Beckett, A. H. and Cowell, D. B. *Tetrahedron* **19** (1963) (a) 401; (b) 413
64 Beckett, A. H. and Kerridge, K. A. *J. chem. Soc.* (1954) 2948
65 Eistert, B. *Bull. Soc. chim. Fr.* (1955) 288; with Munder, H. *Chem. Ber.* **88** (1955) 215,
   226; with Schade, W. **91** (1958) 1404
66 Reid, W. and Keil, G. *Justus Liebigs Annln Chem.* **616** (1958) 96
67 Lüttke, W. and Marsen, H. *Z. Elektrochem.* **57** (1953) 680
68 (a) Strubell, W. and Baumgärtel, H. *J. prakt. Chem.* **6** (1958/4) 115; (b) Klosa, J. **10** (1960/
   43) 335
69 Mathes, W. and Sauermilch, W. *Chem. Ber.* **86** (1953) 109
70 Hirokawa, S. and Ashida, T. *Acta crystallogr.* **14** (1961) 774
71 Kuick, L. F. and Adkins, H. *J. Am. chem. Soc.* **57** (1935) 143
72 Ley, H., Schwarte, C. and Münnich, O. *Ber. dtsch. chem. Ges.* **57** (1924) 349; Dubský, J. V.
   and Okáč, A. *Colln. Czech. chem. Commun.* **3** (1931) 465; Lukeš, R. and Jureček, M.
   31 (1948) 131; Cox, E. G., Wardlaw, W. and Webster, K. C. *J. chem. Soc.* (1936) 775;
   Gorvin, J. H. (1944) 25; Holmes, F. and Crimmin, W. R. C. (1955) 1175, 3467;
   G. Anderegg, *Helv. chim. Acta* **43** (1960) 414, 1530
73 Skraup, Z. H. *Mh. Chem.* **7** (1886) 210; cf. Wolff, L. *Justus Liebigs Annln Chem.* **322** (1902)
   351
74 (a) Acheson, R. M. and Taylor, G. A. *J. chem. Soc.* (1959) 4140; (b) Lever, A. B. P.,
   Lewis, J. and Nyholm, R. S. *J. chem. Soc.* (1962) 5262
75 (a) Skraup, Z. H. with Cobenzl, A. *Mh. Chem.* **4** (1883) 455; (b) with Vortmann, G.
   ibid. 594; (c) Hoogewerff, S. and van Dorp, W. A. *Recl Trav. chim. Pays-Bas Belg.* **4**
   (1885) 285; (d) Ladenburg, A. *Justus Liebigs Annln Chem.* **247** (1888) 1
76 Hopff, H. and Krieger, A. *Helv. chim. Acta* **44** (1961) 1058
77 Hoogewerff, S. and van Dorp, W. A. *Ber. dtsch. chem. Ges.* **13** (1880) 61
78 (a) Cartwell, N. H. and Brown, E. V. *J. Am. chem. Soc.* **74** (1952) 5967; **75** (1953) 1489,
   4466; (b) Dox, A. W. **37** (1915) 1948; (c) Doering, W. von E. and Pasternak, V. Z.
   72 (1550) 143
79 Goldschmiedt, G. and Strache, H. *Mh. Chem.* **10** (1889) 156; Kirpal, K. **20** (1889) 766;
   23 (1902) 929; **28** (1907) 439; Ternajgo, L. **21** (1900) 446; Kaas, K. **23** (1902) 250, 681
80 (a) Pinner, A. *Ber. dtsch. chem. Ges.* **34** (1901) 4234; (b) Sucharda, E. **58** (1925) 1727
81 (a) Bylicki, A. *Chem. Abstr.* **54** (1960) 17026; (b) Blanchard, K. C., Dearborn, E. H.,
   Lasagna, L. C. and Buhle, E. L. **47** (1953) 10535
82 Weber, J. *Justus Liebigs Annln Chem.* **241** (1887) 1
83 (a) Kenyon, J. and Thaker, K. *J. chem. Soc.* (1957) 2531; (b) Falkner, P. R. and Harrison,
   D. (1960) 1171; (1962) 2148; (c) Clemo, G. R. and Koenig, H. (1949) S231; (d) Newbold, G. T. and Spring, F. S. (1949) S133
84 Strache, H. *Mh. Chem.* **11** (1890) 133

399

[85] Dorn, H. and Hilgetag, G. *Chem. Ber.* **97** (1964) 695

[86] Jerchel, D. and Jacobs, W. *Angew. Chem.* **66** (1954) 298

[87] *U.S. Pat.* 2,475,969 (1949); 2,505,568 (1950); 2,513,099 (1950); 2,586,555 (1952)

[88] Jordan, T. E. *Ind. Engng Chem. analyt. Edn* **44** (1952) 332

[89] (a) Murahashi, T. and Otsuka, S. *Chem. Abstr.* **45** (1951) 9054; **47** (1953) 4919; (b) Tanabe, K. and Arai, K. **49** (1955) 2523

[90] Schenkel, H. and Klein, A. *Helv. chim. Acta* **28** (1945) 1211; Schenkel, H. and Rudin, M. S. **31** (1948) 924

[91] Clark, L. W. *J. phys. Chem., Ithaka* **66** (1962) 125

[92] Panizzon, L. *Helv. chim. Acta* **27** (1944) 1748

[93] (a) *Germ. Pat.* 638,596 (1936); 644,193 (1937); (b) Koenigs, E., with Fulde, A., *Ber. dtsch. chem. Ges.* **60** (1927) 2106; (c) with Jaeschke, W. **54** (1921) 1351

[94] Singh, D. *J. chem. Soc.* (1925) 2445; Oparina, M. P. *Chem. Abstr* **30** (1936) 1789

[95] Miescher, K. and Kägi, H. *Helv. chim. Acta* **24** (1941) 1471

[96] Dimitrijević, D. M., Tadić, Z. D. and Muskatirović, M. D. *Glas. hem. Društ., Beogr.* **27** (1962) 397, 407; **28** (1963) 82

[97] Behrmann, A. and Hofmann, A. W. *Ber. dtsch. chem. Ges.* **17** (1884) 2681

[98] Rabe, P., Spreckelsen, O., Wilhelm, L. and Höter, H. *J. prakt. Chem.* **151** (1938/2) 65

[99] (a) Mosettig, E. *Org. React.* **8** (1954) 218; (b) Supniewski, J., Bany, T. and Krupinska, J. *Chem. Abstr.* **50** (1956) 7800

[100] Weygand, F., Eberhardt, G. et. al., *Angew. Chem.* **65** (1953) 525; Brown, H. C. and Tsukamoto, A. *J. Am. chem. Soc.* **81** (1959) 502; **86** (1964) 1089

[101] Burrus, H. O. and Powell, G. *J. Am. chem. Soc.* **67** (1945) 1468; Bloom, M. S., Breslow, D. S. and Hauser, C. R. *ibid.* 2206; Gilman, H. and Broadbent, H. S. **70** (1948) 2755

[102] (a) Winterfeld, K. and Müller, E. *Justus Liebigs Annln Chem.* **581** (1953) 77; (b) Hesse, G. and Schrödel, R. **607** (1957) 24

[103] (a) Levine, R. and Sneed, J. K. *J. Am. chem. Soc.* **73** (1951) 5614; (b) Brown, H. C. with Subba Rao, B. C. **80** (1958) 5377; (c) with Garg, C. P. **86** (1964) 1085

[104] Rojahn, C. A. and Schulten, J. *Arch. Pharm., Berl.* **264** (1926) 348

[105] (a) Yabuta, T. *J. chem. Soc.* **125** (1924) 575; (b) Mann, F. G. and Reid, J. A. (1952) 2057

[106] Bachmann, G. B. and Barker, R. S. *J. org. Chem.* **14** (1949) 97

[107] Kirpal, A. *Mh. Chem.* **29** (1908) 227

[108] Gabriel, S. and Colman, J. *Ber. dtsch. chem. Ges.* **35** (1902) 2831

[109] McLean, A. C. and Spring, F. S. *J. chem. Soc.* (1949) 2582

[110] *U.S. Pat.* 2,349,896 (1944)

[111] Pozdeeva, A. P. and Gepshteïn, E. M. *Chem. Abstr.* **47** (1953) 9325; Mairanovskii, S. G. **57** (1962) 12252

[112] Mason, S. F. and Reynolds, C. F., unpublished work; Reynolds, C. F. *Ph.D. Thesis*, University of Exeter, 1963

[113] (a) Pines, H. and Notari, B. *J. Am. chem. Soc.* **82** (1960) 2209; Notari, B. and Pines, H. *ibid.* 2945; (b) Duling, I. L. and Price, C. C. **84** (1962) 578

[114] Kursanov, D. N., Setkina, V. N. and Bykova, E. V. *Chem. Abstr.* **49** (1955) 10946

[115] Shatensteïn, A. I. *Dokl. Akad. Nauk SSSR* **60** (1950) 1029; with Zvyagintseva, E. N. **117** (1957) 852; Abramovich, T. I., Gragerov, I. P. and Perekalin, V. V. *Zh. obshch. Khim.* **31** (1961) 1962

[116] Zatsepina, N. N., Tupitsyn, I. F. and Efros, L. S. *Zh. obshch. Khim.* **33** (1963) 2705; *Dokl. Akad. Nauk SSSR* **154** (1964) 148

[117] Streitwieser, A. and van Sickle, D. E. *J. Am. chem. Soc.* (a) **84** (1962) 249; (b) with Reif, L. *ibid.* 258

[118] Ziegler, K. and Zeiser, H. *Justus Liebigs Annln Chem.* **485** (1931) 174

[119] Bergmann, E. and Rosenthal, W. *J. prakt. Chem.* **135** (1932/2) 267

[120] (a) Wibaut, P. J. and Hey, J. W. *Recl Trav. chim. Pays-Bas Belg.* **72** (1953) 513; (b) Arens, J. F., van Dorp, D. A. and van Dijk, G. M. **69** (1950) 287; (c) Kloppenburg, C. C. and Wibaut, J. P. **65** (1946) 393

[121] Prijs, B., Lutz, A. H. and Erlenmeyer, H., *Helv. chim. Acta* **31** (1948) 571

[122] (a) Burger, A. and Ullyot, G. E. *J. org. Chem.* **12** (1947) 342; (b) Bergmann, E. D. and Pinchas, S. **15** (1950) 1184; (c) Amstutz, F. D. and Besso, M. M. **25** (1960) 1688

[123] Barnes, R. A. and Fales, H. M. *J. Am. chem. Soc.* **75** (1953) 3830

[124] Tschitschibabin, A. E. *Bull. Soc. chim. Fr.* **3** (1936/5) 1607

[125] Morris, I. G. and Pinder, A. R. *J. chem. Soc.* (1963) 1841

[126] (a) Bergstrom, F. W. *J. Am. chem. Soc.* **53** (1931) 4065; (b) with Norton, T. R. and Seibert, R. A. *J. org. Chem.* **10** (1945) 452

[127] (a) Coulson, E. A. and Jones, J. I. *J. Soc. chem. Ind., Lond.* **65** (1946) 169; (b) Jones, J. I. **69** (1950) 99; (c) Coulson, E. A., Hales, J. L., Holt, E. C. and Ditcham, J. B. *J. appl. Chem.* **2** (1952) 71; (d) Brown, H. C. with Murphey, W. A. *J. Am. chem. Soc.* **73** (1951) 3308; (e) with Kanner, B. *U.S. Pat.* 2,780,626 (1957); (f) Kyte, C. T., Jeffery, G. H. and Vogel, A. I. *J. chem. Soc.* (1960) 4454

400

REFERENCES

[128] Sidgwick, N. V., *The Organic Chemistry of Nitrogen*, revised Taylor, T. W. J. and Baker, W., p. 553 *et seq.*, London (Oxford University Press) 1942

[129] Sperber, N., Papa, D., Schwenk, E., Sherlock, M. and Fricano, R. *J. Am. chem. Soc.* **73** (1951) 5752

[130] Gilman, H. and Towle, J. L. *Recl Trav. chim. Pays-Bas Belg.* **69** (1950) 428

[131] Profft, E. with Schneider, F. *J. prakt. Chem.* **2** (1955) 316; with Stumpf, R. **19** (1963) 266

[132] (a) Decker, H. *Ber. dtsch. chem. Ges.* **38** (1905) 2493; (b) Schneider, W., Gaertner, K. and Jordan, A. **57** (1924) 522; (c) Mumm, O. and Hingst, G. **56** (1923) 2301; (d) Hantzsch, A. and Kalb, M. **32** (1899) 3109 (cf. Aston, J. G. and Lasselle, P. A. *J. Am. chem. Soc.* **56** (1934) 426); (e) Koenigs, E., Köhler, K. and Blindow, K. **58** (1925) 933; (f) Tschitschibabin, A. E. and Benewolenskaja, S. W. **61** (1928) 547; (g) with Kuindshi, B. M. **58** (1925) 1580

[133] (a) Anderson, L. C. and Seegers, N. V. *J. Am. chem. Soc.* **71** (1949) 343; (b) Traynelis, V. J. and Pacini, P. L. **86** (1964) 4917; (c) Berson, J. A., Evleth, E. M. and Hamlet, Z. **82** (1960) 3793

[134] Kursanov, D. N. and Baranetskaya, N. K. *Izv. Akad. Nauk SSSR Otd. khim. nauk* (1961) 1703; with Setkina, V. N. *Dokl. Akad. Nauk SSSR* **113** (1957) 116

[135] (a) Kumler, W .D. *J. org. Chem.* **28** (1963) 1731; (b) Jones, R. A. and Katritzky, A. R. *Aust. J. Chem.* **17** (1964) 455

[136] (a) Kröhnke, F., Ellegast, K. and Bertram E. *Justus Liebigs Annln Chem.* **600** (1956) 176; (b) Mumm, O. **443** (1925) 272; (c) Berson, J. A. and Evleth, E. M. *Chemy Ind.* (1961) 1362; (d) Boyd, G. V. and Jackman, L. M. *J. chem. Soc.* (1963) 548

[137] (a) Pinck, L. A. and Hilbert, G. E. *J. Am. chem. Soc.* **68** (1964) 2011; (b) Kröhnke, F. *Chem. Ber.* **83** (1950) 253

[138] Lloyd, D. and Sneezum, J. S. *Chemy Ind.* (1955) 1221

[139] Kosower, E. M. and Ramsey, B. G. *J. Am. chem. Soc.* **81** (1959) 856

[140] (a) Krollpfeiffer, F. and Müller, A. *Ber dtsch. chem. Ges.* **66** (1933) 739; (b) Kröhnke, F. **70** (1937) 543

[141] Stafford, W. H. *J. chem. Soc.* (1952) 580

[142] Bohlmann, F. and Kröhnke, F. *Naturwissenschaften* **39** (1953) 43

[143] Brown, D. A. and Dewar, M. J. S. *J. chem. Soc.* (1953) 2406

[144] Arnold, Z. *Colln. Czech. chem. Commun.* **28** (1963) 863

[145] Jacobsen, E. and Reimer, C. L. *Ber. dtsch. chem. Ges.* **16** (1883) 2602

[146] Kuhn, R. and Bär, F. *Justus Liebigs Annln Chem.* **516** (1935) 155

[147] Manly, D. G., Richardson, A., Stock, A. M., Tilford, C. H. and Amstutz, E. D. *J. org. Chem.* **23** (1958) 373

[148] Taurinš, A. *J. prakt. Chem.* **153** (1939) 177, 189

[149] Barton, A. W. H. and McClelland, E. W. *J. chem. Soc.* (1947) 1574

[150] Scholtz, M. *Ber. dtsch. chem. Ges.* **45** (1912) 734; Tschitschibabin, A. E. and Stepanow, F. N. **62** (1929) 1068

[151] Mosby, W. L. *Heterocyclic Systems with Bridgehead Nitrogen Atoms*, I, 239, New York (Interscience) 1961

[152] (a) Agawa, T. and Miller, S. I. *J. Am. chem. Soc.* **83** (1961) 449; (b) Brooker, L. G. S. and Keyes, G. H. **57** (1935) 2488

[153] Baker, B. R. and McEvoy, F. J. *J. org. Chem.* **20** (1955) 118

[154] (a) Kröhnke, F. and Kübler, H. *Ber. dtsch. chem. Ges.* **70** (1937) 539; (b) Kröhnke, F. **70** (1937) 1114; (c) **68** (1935) 1177; (d) with Gerlach, K. *Chem. Ber.* **95** (1962) 1108

[155] (a) Héou-Féo, T. *C.r. hebd. Séanc. Acad. Sci., Paris* **192** (1931) 1242; *Bull. Soc. chim. Fr.* **2** (1935/5) 103; (b) Monti, L. and Felici, L. *Chem. Abstr.* **35** (1941) 3241; (c) Sommers, A. H., Freifelder, M., Wright, H. B. and Weston, A. W. *J. Am. chem. Soc.* **75** (1953) 57; (d) Matuszko, A. J. and Taurins, A. *Can. J. Chem.* **32** (1954) 538

[156] (a) Reich, H. E. and Levine, R. *J. Am. chem. Soc.* **77** (1955) 4913; (b) van Tamelen, E. E., Hughes, D. L. and Taylor, C. W. **78** (1956) 4625

[157] Lathewood, E. and Suschitzky, H. *J. chem. Soc.* (1964) 2477

[158] Adamcik, J. A. and Flores, R. D. *J. org. Chem.* **29** (1964) 572

[159] Strell, M., Braunbruck, W. B. and Reithmayr, L. *Justus Liebigs Annln Chem.* **587** (1954) 195

[160] Kröhnke, F., Schmeiss, H. and Gottstein, W. *Chem. Ber.* **84** (1951) 131

[161] Thesing, J., with Ramloch, H., Willersinn, C.-H. and Funck, F. *Angew. Chem.* **68** (1956) 387; with Willersinn, C.-H. *Chem. Ber.* **89** (1956) 2896; with Festag, W. *Experientia* **15** (1959) 127

[162] Kröhnke, F., Schmidt, E. and Zecher, W. *Chem. Ber.* **97** (1964) 1163

[163] Ladenburg, A. (a) *Justus Liebigs Annln Chem.* **301** (1898) 117; (b) 1477; Meisenheimer, J. and Mahler, E. **462** (1928) 301; (c) Winterfeld, K. and Heinen, C. **573** (1951) 85; (d) Bohlmann, F., Ottawa, N. and Keller, R. **587** (1954) 162

[164] Ladenburg, A. *Ber. dtsch. chem. Ges.* **22** (1889) 2583

[165] (a) Tullock, C. W. and McElvain, S. M. *J. Am. chem. Soc.* **61** (1939) 961; (b) McElvain, S. M. and Johnson, H. G. **63** (1941) 2213

[166] Koton, M. M. *Chem. Abstr.* **55** (1961) 16546; *USSR Pat.* 135,488 (1961) [*ibid.* 16571]; *Jap. Pat.* 18285 (1960) [*Chem. Abstr.* **56** (1962) 2429]; Gluzman, L. D., Tsin, R. M. and Rok, A. A. *Chem. Abstr.* **57** (1962) 8537; *Br. Pat.* 906,469 (1962)

[167] (a) Koenigs, W. and Happe, G. *Ber dtsch. chem. Ges.* **35** (1902) 1343; **36** (1903) 2904; (b) Koenigs, W. **35** (1902) 1349; (c) Lipp, A. and Richard, J. **37** (1904) 737

[168] Melichar, F. *Chem. Ber.* **88** (1955) 1208

[169] Meisenheimer, J. *Justus Liebigs Annln Chem.* **420** (1919) 190

[170] Winterfeld, K. *Chem. Abstr.* **24** (1930) 3790

[171] Michalski, J. and Studniarski, K. *Roczn. Chem.* **29** (1955) 1141

[172] Löffler, K. (a) with Grosse, A. *Ber. dtsch. chem. Ges.* **40** (1907) 1325; (b) with Plöcker, P. *ibid.* 1310

[173] Rubtsov, M. V., Nikitskaya, E. S. and Usovskaya, V. S. *Chem. Abstr.* **50** (1956) 9401

[174] (a) Tschitschibabin, A. E. *J. prakt. Chem.* **69** (1904) 310; (b) Stoehr, C. **42** (1890/2) 420

[175] Koller, G. *Mh. Chem.* **47** (1926) 393

[176] (a) Kleiman, M. and Weinhouse, S. *J. org. Chem.* **10** (1945) 562; (b) Rabe, P., Huntenburg, W., Schultze, A. and Volger, G. *Ber. dtsch. chem. Ges.* **64** (1931) 2487

[177] Kakimoto, S., Nishie, J. and Yamamoto, K. *Chem. Abstr.* **54** (1960) 21079

[178] (a) Baurath, H. *Ber. dtsch. chem. Ges.* **20** (1887) 2719; (b) Friedländer, C. **38** (1905) 159; (c) Schwarz, P. **24** (1891) 1676

[179] Katsumoto, T. *Bull. chem. Soc. Japan* **32** (1959) 1019; **33** (1960) 242

[180] Kögl, F., van der Want, G. M. and Salemink, C. A. *Recl Trav. chim. Pays-Bas Belg.* **67** (1948) 29

[181] (a) Lénart, G. H. *Justus Liebigs Annln Chem.* **410** (1915) 95; (b) Thayer, H. I. and Corson, B. B. *J. Am. chem. Soc.* **70** (1948) 2330; (c) Fields, E. K. **71** (1949) 1495; (d) Drefahl, G., with Gerlach, E. *J. prakt. Chem.* **6** (1958/4) 72; (e) with Plötner, G. and Buchner, G. *Chem. Ber.* **94** (1961) 1824

[182] (a) Langer, G. *Ber. dtsch. chem. Ges.* **38** (1905) 3704; (b) Bacher, F. **21** (1888) 3071; (c) Proske, H. **42** (1909) 1450; (d) Schuster, F. **25** (1892) 2398

[183] (a) Clemo, G. R. and Gourlay, W. M. *J. chem. Soc.* (1938) 478; (b) Shaw, B. D. **125** (1924) 2363

[184] Koenigs, W. and Bentheim, A. v. *Ber. dtsch. chem. Ges.* **38** (1905) 3907

[185] Khromov-Borisov, N. V., Karlinskaya, R. S. and Ageeva, L. N. *Chem. Abstr.* **50** (1956) 9429

[186] (a) Feist, K. *Ber. dtsch. chem. Ges.* **34** (1901) 464; (b) Späth, E., Kubiczek, G. and Dubensky, E. **74** (1941) 873; (c) Knick, R. **35** (1902) 2790; (d) Roth, E. **33** (1900) 3476; (e) Bach, H. **34** (1901) 2223; (f) Räth, C. and Lehmann, E. **58** (1925) 342; (g) Mustafa, A. and Hilmy, M. K. *J. chem. Soc.* (1947) 1698; (h) Shaw, B. D. and Wagstaff, E. A. (1933) 79; cf. Wagstaff, E. A. (1934) 276

[187] Akkerman, A. M. and Veldstra, H. *Recl Trav. chim. Pays-Bas Belg.* **73** (1954) 629

[188] Hebký, J. and Staněk, J. *Chem. Abstr.* **47** (1953) 9971

[189] Bennett, G. M. and Pratt, W. L. C. *J. chem. Soc.* (1929) 1465

[190] Blout, E. K. and Eager, V. W. *J. Am. chem. Soc.* **67** (1945) 1315

[191] Katritzky, A. R., Short, D. J. and Boulton, A. J. *J. chem. Soc.* (1960) 1516

[192] Adams, R. and Schrecker, W. *J. Am. chem. Soc.* **71** (1949) 1186

[193] Jerchel, D. and Heck, H. E. *Justus Liebigs Annln. Chem.* **613** (1958) 171

[194] Binns, F. and Swan, G. A. *J. chem. Soc.* (1962) 2831

[195] Staněk, J. and Horák, M. *Colln. Czech. chem. Commun.* **15** (1951) 1037

[196] (a) Bragg, D. R. and Wibberly, D. G. *J. chem. Soc.* (1961) 5074; (b) Katritzky, A. R. and Monro, A. M. (1958) 150

[197] (a) Williams, J. L. R., Webster, S. K. and van Allan, J. A. *J. org. Chem.* **26** (1961) 4893; (b) Horwitz, L. **21** (1956) 1039

[198] Avramoff, M. and Sprinzak, Y. *J. Am. chem. Soc.* **78** (1956) 4090

[199] Parker, E. D. and Furst, A. *J. org. Chem.* **23** (1958) 201

[200] Mills, W. H. and Pope, W. J. *J. chem. Soc.* **121** (1922) 946

[201] Cocker, W. and Turner, D. G. *J. chem. Soc.* (1940) 57; Hirshberg, Y., Knott, E. B. and Fischer, E. (1955) 3313; Koelsch, C. F. *J. Am. chem. Soc.* **66** (1944) 2126; Doja, M. Q. and Lal, A. B. *Chem. Abstr.* **41** (1947) 3102

[202] Doja, M. Q. and Prasad, K. B. *J. Indian chem. Soc.* (a) **19** (1942) 125; (b) **24** (1947) 301

[203] Phillips, A. P. *J. org. Chem.* (a) **12** (1947) 333; (b) **13** (1948) 622; (c) **14** (1949) 302

[204] Staněk, J. and Zekja, Z. *Chem. Abstr.* **49** (1955) 314

[205] Clemo, G. R. and Swan, G. A. *J. chem. Soc.* (1938) 1454

[206] (a) Brooker, L. G. S., Sklar, A. L. *et al.*, *J. Am. chem. Soc.* **67** (1945) 1875; (b) Phillips, A. P. **74** (1952) 3296

[207] Wizinger, R. and Wagner, K. *Helv. chim. Acta* **34** (1951) 2290

[208] (a) Doja, M. Q. and Prasad, K. B. *Chem. Abstr.* **43** (1949) 7024; (b) Crippa, G. B. and Maffei, S. **42** (1948) 2973; (c) Staněk, J. Hebký, J., and Zvěřina, V. **47** (1953) 12378

[209] Tschitschibabin, A. E. *Ber. dtsch. chem. Ges.* **60** (1927) 1607

[210] Bragg, D. R. and Wibberley, D. G. *J. chem. Soc.* (1963) 3277

REFERENCES

211 Mills, W. H. and Raper, R. *J. chem. Soc.* **127** (1925) 2466
212 Kröhnke, F. *Chem. Ber.* (a) **84** (1951) 388; (b) with Meyer-Delius, M. *ibid.* 411; (c) with Wolff, J. and Jentzsch, G. *ibid.* 399;(d) *Angew. Chem.* **65** (1953) 605; (e) *Ber. dtsch. chem. Ges.* **68** (1935) 1351; (f) **67** (1934) 656
213 (a) Westphal, O., Jann, K. and Heffe, W. *Arch. Pharm., Berl.* **294** (1961) 37; (b) Frank, R. L. and Phillips, R. R. *J. Am. chem. Soc.* **71** (1949) 2804
214 Kharkharov, A. A. *Zh. obshch. Khim.* **23** (1953) 1175
215 Krollpfeiffer, F. and Braun, E. *Ber. dtsch. chem. Ges.* **70** (1937) 89
216 McBee, E. T., Hass, H. B. and Hodnett, E. M. *Ind. Engng Chem. analyt. Edn* **93** (1947) 389; cf. Sell, W. J. *J. chem. Soc.* **93** (1908) 1993; Seyfferth, E. *J. prakt. Chem.* **34** (1886) 241
217 Brown, B. R., Hammick, D. Ll. and Thewlis, B. H. *J. chem. Soc.* (a) (1951) 1145; (b) with Walbridge, D. J. (1953) 1369
218 Mathes, W. and Schüly, H. *Angew. Chem. (int. Edn)* **2** (1963) 144
219 *U.S.Pat.* (a) 2,670,352 (1954); (b) 2,748,141 (1956); (c) 2,743,277 (1956)
220 Knudsen, P. *Ber. dtsch. chem. Ges.* **25** (1892) 2985; Dehnel, E. **33** (1900) 3498; (b) Räth, C. and Prange, G. **58** (1925) 1208
221 Linnell, W. H. and Vyas, A. F. *Qu. Jl Pharm. Pharmac.* **20** (1947) 119
222 Smith, J. M., Stewart, H. W., Roth, B. and Northey, E. H. *J. Am. chem. Soc.* **70** (1948) 3997
223 Edwards, O. E., Chaput, M., Clarke, F. H. and Singh, T. *Can. J. Chem.* **32** (1954) 785
224 (a) Bauer, L. and Gardella, L. A. *J. org. Chem.* **28** (1963) 1323; (b) Vozza, J. F. **27** (1962) 3856
225 Matsumura, E. *J. chem. Soc. Japan* **74** (1953) 363; with Hirooka, T. and Imagawa, K. *Chem. Abstr.* **57** (1962) 12466
226 (a) Murakami, M. and Matsumura, E. *J. chem. Soc. Japan* **70** (1949) 393; (b) Matsumura, E. **74** (1953) 446; (c) Villiers, P. A. de and den Hertog, H. J. *Recl Trav. chim. Pays-Bas Belg.* **75** (1956) 1303; **76** (1957) 747; (d) with Buurman, D. J. **80** (1961) 325;(e) Oae, S., Kitao, T. and Kitaoka, Y. *Tetrahedron* **19** (1963) 827
227 (a) Boekelheide, V. and Linn, W. J. *J. Am. chem. Soc.* **76** (1954) 1286; (b) Bullitt, O. H. and Maynard, J. T. *ibid.* 1370
228 (a) Kobayashi, G. and Furukawa, S. *Pharm. Bull., Tokyo* **1** (1953) 347; Furukawa, S. *Yakugaku Zasshi* **59** (1959) 487; (b) idem, *Chem. Abstr.* **53** (1959) 18029
229 Berson, J. A. and Cohen, T. *J. Am. chem. Soc.* **77** (1955) 1281
230 Hardegger, E. and Nikles, E. *Helv. chim. Acta* **40** (1957) 1016
231 Boekelheide, V. and Feely, W. J. *org. Chem.* **22** (1957) 589
232 Endo, M. and Nakashima, T. *Chem. Abstr.* **54** (1960) 24705
233 Biland, H., Lohse, F. and Hardegger, E. *Helv. chim. Acta* **43** (1960) 1436
234 Farley, C. P. and Eliel, E. L. *J. Am. chem. Soc.* **78** (1956) 3477
235 Sato, Y., Iwashige, T. and Mujalera, T. *Chem. Abstr.* **55** (1961) 12401
236 Rimek, H.-J. *Justus Liebigs Annln Chem.* **670** (1963) 69
237 (a) Hamana, M. and Yamazaki, M. *J. pharm. Soc. Japan* **81** (1961) 574; (b) Kobayashi, G., Furukawa, S. and Kawada, Y. **74** (1954) 790; (c) Kato, T. **75** (1955) 1228
238 Yoshida, M. and Kumagae, H. *Chem. Abstr.* **55** (1961) 6477; (b) Nishimoto, N. and Nakashima, T. *ibid.* 13420
239 Furukawa, S. *Pharm. Bull., Tokyo* **3** (1955) 413
240 Biemann, K., Büchi, G. and Walker, B. H. *J. Am. chem. Soc.* **79** (1957) 5558
241 Oae, S., Kitaoka, Y. and Kitao, T. (a) *Tetrahedron* **20** (1964) 2677; (b) *J. Am. chem. Soc.* **84** (1962) 3362; (c) *ibid.* 3359; (d) *Tetrahedron* **20** (1964) 2685
242 Boekelheide, V. and Harrington, D. L. *Chemy Ind.* (1955) 1423
243 (a) Traynelis, V. J. and Martello, R. F. *J. Am. chem. Soc.* **80** (1958) 6590; (b) with Gallagher, A. I. *J. org. Chem.* **26** (1961) 4365
244 Muth, C. W., Darlak, R. S., English, W. H. and Hamner, A. T. *Analyt. Chem.* **34** (1962) 1163
245 Kato, T., Goto, Y. and Yamamoto, Y. *Chem. Abstr.* **59** (1963) 2765
246 (a) Tschitschibabin, A. E. *Zh. russk. fiz.-khim. Obshch.* **26** (1894) 16; (b) Oparina, M. P. **57** (1925) 319
247 (a) Forman, S. E. *J. org. Chem.* **29** (1964) 3323; (b) Poziomek, E. J. and Melvin, A. R. **26** (1961) 3769; with Poirier, R. H. *et al.* **29** (1964) 217
248 Kaufmann, A. and Valette, L. G. *Ber. dtsch. chem. Ges.* (a) **46** (1913) 49; (b) **45** (1912) 1736
249 de Waal, H. L. and Brink, C. v. d. M. *Chem. Ber.* **89** (1956) 636
250 (a) Kröhnke, F., with Leister, H. and Vogt, I. *Chem. Ber.* **90** (1957) 2792; (b) *Angew. Chem. (int. Edn)* **2** (1963) 380
251 Schulze, W. *J. prakt. Chem.* **17** (1962) 24
252 Lal, A. B. and Petrow, V. *J. chem. Soc.* (1949) S115
253 Földi, Z. *Acta chim. hung.* **19** (1959) 205
254 Bridger, R. F. and Russell, G. A. *J. Am. chem. Soc.* **85** (1963) 3754
255 Johnston, K. M. and Williams, G. H. *J. chem. Soc.* (1960) 1446
256 Graf, R. and Zettl, F. *J. prakt. Chem.* **147** (1936) 188

403

P

257 Buu-Hoï, N. G. *Justus Liebigs Annln Chem.* **556** (1944) 1
258 Hasegawa, M. *Pharm. Bull., Tokyo* **1** (1953) 293
259 (a) Kutney, J. P. and Tabata, T., with Cretney, W. and Frank, M. *Can. J. Chem.* **42** (1964) 698; (b) **41** (1963) 695
260 (a) Ramirez, F. and Paul, A. P. *J. org. Chem.* **19** (1954) 183; (b) Walker, B. H. **25** (1960) 1047
261 Hurst, J. and Wibberley, D. G. *J. chem. Soc.* (1962) 119
262 Mariella, R. P. and Belcher, E. P. *J. Am. chem. Soc.* **74** (1952) 1916
263 Steiner, K., Graf, U. and Hardegger, E. *Helv. chim. Acta* **46** (1963) 690
264 (a) Clark, W. G. and Lothian, G. F. *Trans. Faraday Soc.* **54** (1958) 1790; (b) Hardwick, R., Mosher, H. S. and Passailaigue, P. **56** (1960) 44
265 Mosher, H. S., Souers, C. and Hardwick, R. *J. chem. Phys.* **32** (1960) 1888
266 Wettermark, G. *J. Am. chem. Soc.* **84** (1962) 3658; Sousa, J. A. and Weinstein, J. *J. org. Chem.* **27** (1962) 3155; with Bluhm, A. L. **28** (1963) 1989; Hardwick, R. and Mosher, H. S. *J. chem. Phys.* **36** (1962) 1402; Margerum, J. D., Miller, L. J., Mosher, H. S., Hardwick, R. *et al.*, *J. phys. Chem., Ithaka* **66** (1962) 2423; Kortum, G., Kortum-Seiler, M. and Bailey, S. D. *ibid.* 2439; Wettermark, G. and Sousa, J. **67** (1963) 874
267 Black, G., Depp, E. and Corson, B. B. *J. org. Chem.* **14** (1949) 14 (with bibliography of early work)
268 Kucharczyk, N. *Chem. Listy* **10** (1961) 1199
269 Heyns, K. and Vogelsang, G. *Chem. Ber.* **87** (1954) 13
270 Efimovsky, O. and Rumpf, P. *Bull. Soc. chim. Fr.* (1954) 648
271 Vaculík, P. and Kuthan, J. *Colln. Czech. chem. Commun.* **25** (1960) 1591
272 Goutarel, R., Janot, M.-M., Mathys, F. and Prelog, V. *C.r. hebd. Séanc. Acad. Sci., Paris* **237** (1953) 1718
273 Graf, R. *J. prakt. Chem.* **133** (1932/2) 19
274 Plattner, P. A., Keller, W. and Boller, A. *Helv. chim. Acta* **37** (1954) 1379
275 Tschitschibabin, A. E. and Oparina, M. P. *Zh. russk. fiz.-khim. Obshch.* **54** (1922) 428
276 Altar, S. *Justus Liebigs Annln Chem.* **237** (1886) 182
277 Oparina, M. P. *Chem. Abstr.* **24** (1930) 3790; **44** (1950) 1108; *Ber. dtsch. chem. Ges.* **64** (1931) 562
278 *U.S. Pat.* (a) 2,109,954 (1938); (b) 2,524,957 (1950); 2,884,415 (1959); *Swiss Pat.* 335,519 (1959); *Germ. Pat.* 1,161,563 (1964); (c) *Germ. Pat.* 697,759 (1940); (d) 837,535 (1952); 936,871 (1955); (e) *Br. Pat.* 710,192 (1954); (f) *U.S. Pat.* 2,510,605 (1950); (g) 2,496,319; 2,515,233 (1950)
279 (a) Tyabji, A. N. *Chem. Ber.* **92** (1959) 2677; (b) Franck, B. and Knoke, J. **95** (1962) 579
280 (a) Bryans, F. and Pyman, F. L. *J. chem. Soc.* (1929) 549; (b) Solomon, W. (1946) 934
281 (a) Koenigs, E., Mensching, H. and Kirsch, P. *Ber. dtsch. chem. Ges.* **59** (1926) 1717; (b) Benary, E. and Psille, H. **57** (1924) 828
282 Crook, K. E. and McElvain, S. M. *J. Am. chem. Soc.* **52** (1930) 4006
283 Schofield, K. *J. chem. Soc.* (1949) 2408; with Nunn, A. J. (1952) 583
284 Plažek, E. and Kozdrojowna, H. *Roczn. Chem.* **25** (1951) 509
285 Bengtsson, E. B. *Acta chem. scand.* **9** (1955) 832
286 Borsche, W. and Hartmann, H. *Ber. dtsch. chem. Ges.* **73** (1940) 839
287 Cook, D. J. and Yunghans, R. S. *J. Am. chem. Soc.* **74** (1952) 5515
288 Henze, M. *Ber. dtsch. chem. Ges.* **67** (1934) 750
289 Fox, H. H. *J. org. Chem.* **17** (1952) 555
290 Jerchel, D., Bauer, E. and Hippchen, H. *Chem. Ber.* **88** (1955) 156
291 Baumgarten, H. E. and Krieger, A. I. *J. Am. chem. Soc.* **77** (1955) 2438
292 Jerchel, D., Heider, J. and Wagner, H. *Justus Liebigs Annln Chem.* **613** (1958) 153
293 (a) Woodward, C. F., Badgett, C. O. and Kaufman, J. G. *Ind. Engng Chem. analyt. Edn* **36** (1944) 544; (b) Kaufman, J. G. *J. Am. chem. Soc.* **67** (1945) 497; (c) *U.S. Pat.* 2,436,660 (1948); (d) Ochiai, E. and Okuda, S. *J. pharm. Soc. Japan* **70** (1950) 156
294 Ishiguro, T. and Utsumi, I. *J. pharm. Soc. Japan* **72** (1952) 861
295 Mathes, W. and Sauermilch, W. *Chem. Ber.* **88** (1955) 1276; **93** (1960) 286
296 Mayurnik, G., Moschetto, A. F., Bloch, H. S. and Scudi, J. V. *Ind. Engng Chem. analyt. Edn* **44** (1952) 1630
297 Roy, S., Banerjee, P. K. and Basu, A. N. *Chem. Abstr.* **55** (1961) 16967
298 Bartok, W., Rosenfeld, D. D. and Schriesheim, A. *J. org. Chem.* **28** (1963) 410
299 Yokoyama, M. *Bull. chem. Soc. Japan* **7** (1932) 69; Kulka, M. *J. Am. chem. Soc.* **68** (1946) 2472; Kruglikov, S. S. and Khomyakov, V. G. *Chem. Abstr.* **57** (1962) 16542
300 Ferguson, L. N. and Levant, A. J. *Nature, Lond.* **167** (1951) 817
301 John, H. and Behmel, G. *Ber. dtsch. chem. Ges.* **66** (1933) 426
302 Müller, A. and Dorfman, M. J. *J. Am. chem. Soc.* **56** (1934) 2787
303 Oparina, M. P. *Zh. russk. fiz.-khim. Obshch.* **57** (1925) 319
304 Anker, R. M., Cook, A. H. and Heilbron, I. M. *J. chem. Soc.* (1945) 917
305 Emmert, B.; with Groll, M. *Chem. Ber.* **86** (1953) (a) 208; (b) 205; (c) with Holz, A. **87** (1954) 676

REFERENCES

306 Porter, H. D. *J. Am. chem. Soc.* **76** (1954) 127
307 Miller, P. E., Oliver, G. L., Dann, J. R. and Gates, J. W. *J. org. Chem.* **22** (1957) 664
308 Saikachi, H. and Hisano, T. *Chem. pharm. Bull., Tokyo* **7** (1959) 716
309 Taylor, E. C. and Driscoll, J. S. *J. org. Chem.* **26** (1961) 3796
310 *U.S. Pat.* 2,611,769 (1952); 2,716,118; 2,716,119; 2,728,770 (1955); 2,732,376 (1956); Folz, J. M., Mahan, J. E. and White, D. H. *Petrol. Process.* **7** (1952) 1802
311 *U.S. Pat.* 2,677,688 (1954); 2,769,811 (1956); 2,996,509 (1959); 2,962,498 (1960); 2,980,684 (1961); *East Germ. Pat.* 13,099 (1957); Farberov, M. I., Ustavshchikov, B. F., Kut'in, A. M. *et al., Chem. Abstr.* **53** (1959) 11364; **55** (1961) 17630; Yoshioka, S., Ohmae, T. and Hasegawa, N. **59** (1963) 2762; Oga, T. **58** (1963) 5631; Gechele, G. B. and Pietra, S. *J. org. Chem.* **26** (1961) 4412
312 Sperber, N., Papa, D. and Sherlock, M. *J. Am. chem. Soc.* **78** (1956) 4489
313 Clarke, F. H., Felock, G. A., Silverman, G. B. and Watnick, C. M. *J. org. Chem.* **27** (1962) 533
314 Jerchel, D. and Melloh, W. *Justus Liebigs Annln Chem.* **622** (1959) 53
315 Katsumoto, T. and Honda, A. *Chem. Abstr.* **59** (1963) 15254
316 Williams, J. L. R. *J. org. Chem.* **25** (1960) 1839; *J. Am. chem. Soc.* **84** (1962) 1323
317 Koller, G. *Ber. dtsch. chem. Ges.* **60** (1927) 1920
318 Williams, J. L. R., Carlson, J. M., Reynolds, G. A. and Adel, R. E. *J. org. Chem.* **28** (1963) 1317
319 Loader, C. E., Sargent, M. V. and Timmons, C. J. *Chem. Commun.* (1965) 127
320 Perkampus, H.-H. and Senger, P. *Ber. Bunsenges. phys. Chem.* **67** (1963) 876
321 Blood, J. W. and Shaw, B. D. *J. chem. Soc.* (1930) 504
322 Ruggli, P. and Cuenin, H. *Helv. chim. Acta* **27** (1944) 649
323 Schofield, K. and Swain, T. *J. chem. Soc.* (1949) 2393
324 *U.S. Pat.* (a) 2,624,750 (1953); (b) 2,512,789 (1950); (c) 2,482,521 (1949); (d) 2,615,892 (1952); (e) 2,792,403 (1957)
325 Dierig, W. *Ber. dtsch. chem. Ges.* **35** (1902) 2774
326 Lukeš, R. and Ernest, I. *Colln. Czech. chem. Commun.* **14** (1949) 679
327 (a) Beyerman, H. C., Eveleens, W. and Muller, Y. M. F. *Recl Trav. chim. Pays-Bas Belg.* **75** (1956) 63; (b) Clemo, G. R., Fletcher, N., Fulton, G. R. and Raper, R. *J. chem. Soc.* (1950) 1140
328 Butter, F. *Ber. dtsch. chem. Ges.* **23** (1890) 2697
329 Leaver, D., Gibson, W. K. and Vass, J. D. R. *J. chem. Soc.* (1963) 6053
330 Drefahl, G., Plötner, G. and Hertzer, H. *Chem. Ber.* **94** (1961) 1833
331 Räth, C. *Ber. dtsch. chem. Ges.* **57** (1924) 840
332 (a) Schofield, K. and Simpson, J. C. E. *J. chem. Soc.* (1945) 512; (b) Nunn, A. J. and Schofield, K. (1953) 3700
333 Aaron, H. S., Owens, O. O. *et al., J. org. Chem.* **30** (1965) 1331
334 Furukawa, S. *Chem. Abstr.* **53** (1959) 18028
335 Hart, E. P. *J. chem. Soc.* (1952) 4540
336 Callighan, R. H. and Wilt, M. H. *J. org. Chem.* **26** (1961) 4912
337 (a) Kaslow, C. E. and Stayner, R. D. *J. Am. chem. Soc.* **67** (1945) 1716; (b) Lochte, H. L., Barton, A. D., Roberts, S. M. and Bailey, J. R. **72** (1950) 3007; (c) Buehler, C. A., Harris, J. O. and Arendale, W. F. *ibid.* 4953; (d) Pattison, D. B. and Carmack, M. **68** (1946) 2033; (e) Noller, C. R. and Wunderlich, E. A. **74** (1952) 3835; (f) Doering, W. E. and Weil, R. A. N. **69** (1947) 2461; (g) Magnus, G. and Levine, R. **78** (1956) 4127
338 Kost, A. N., Suminov, S. I., Vinogradova, E. V. and Kozler, V. *Zh. obshch. Khim.* **33** (1963) 3606
339 Bauer, L., Shoeb, A. and Agwada, V. C. *J. org. Chem.* **27** (1962) 3153
340 Reich, H. E. and Levine, R. *J. Am. chem. Soc.* **77** (1955) 5434
341 Profft, E. *J. prakt. Chem.* **4** (1956/4) 19
342 Phillips, A. P. *J. Am. chem. Soc.* **78** (1956) 4441
343 Profft, E. and Georgi, W. *Justus Liebigs Annln Chem.* **643** (1961) 136
344 Gray, A. P. and Archer, W. L. *J. Am. chem. Soc.* (a) **79** (1957) 3554; (b) with Spinner, E. E. and Cavallito, C. J. *ibid.* 3805
345 Kirchner, F. K., McCormick, J. R., Cavallito, C. J. and Miller, L. C. *J. org. Chem.* **14** (1949) 388
346 Shapiro, S. L., Rose, I. M. and Freedman, L. *J. Am. chem. Soc.* **79** (1957) 2811
347 *Br. Pat.* (a) 667,457 (1952); (b) 889,748 (1962); *U.S. Pat.* (c) 2,579,419 (1951); (d) 2,713,050 (1955); (e) 2,508,904 (1950)
348 Boekelheide, V. with Rothchild, S. *J. Am. chem. Soc.* (a) **71** (1949) 879; (b) **69** (1947) 3149; (c) with Agnello, E. J. **72** (1950) 5005; (d) with Mason, J. H. **73** (1951) 2356; (e) with Linn, W. J. *et al.* **75** (1953) 3243
349 Burckhalter, J. H. and Stephens, V. C. *J. Am. chem. Soc.* **73** (1951) 4460
350 Magnus, G. and Levine, R. *J. org. Chem.* **22** (1957) 270
351 Albertson, N. F. *J. Am. chem. Soc.* **72** (1950) 2594

[352] Levine, R. and Wilt, M. H. *J. Am. chem. Soc.* (a) **74** (1952) 342; (b) **75** (1953) 1368

[353] Winterfield, K., Wald, G. and Rink, M. *Justus Liebigs Annln Chem.* **588** (1954) 125

[354] (a) Leonard, N. J. and Boyer, J. H. *J. Am. chem. Soc.* **72** (1950) 4818; (b) Jampolsky, L. M., Baum, M., Kaiser, S., Sternbach, L. H. and Goldberg, M. W. **74** (1952) 5222

[355] (a) Shono, T., Yasumura, T. and Oda, R. *Chem. Abstr.* **50** (1956) 999; (b) Rubtsov, M. V. and Mikhlina, E. E. **48** (1954) 3978

[356] Michalski, J. and Zajac, H. *J. chem. Soc.* (1963) 593

[357] Singerman, G. and Danishefsky, S. *Tetrahedron Lett.* (1964) 2249

[358] Thompson, R. B., Chenicek, J. A. and Symon, T. *Ind. Engng Chem. analyt. Edn.* **44** (1952) 1659

[359] *U.S. Pat.* (a) 2,607,775; 2,607,776; (b) 2,606,190 (1952); (c) 2,490,672 (1949)

[360] Achmatowicz, O., Maruszewska-Wieczorkowska, E. and Michalski, J. *Roczn. Chem.* **29** (1955) 1029

[361] Bauer, L. and Gardella, L. A. *J. org. Chem.* **26** (1961) 82

[362] Boyer, J. H. *J. Am. chem. Soc.* **73** (1951) 5248; Westland, R. D. and McEwen, W. E. **74** (1952) 6141

[363] Maruszewska-Wieczorkowska, E. and Michalski, J. *J. org. Chem.* **23** (1958) 1886

[364] (a) Baurath, H. *Ber. dtsch. chem. Ges.* **21** (1888) 818; (b) Friedländer, K. **38** (1905) 2837

[365] (a) Papa, D., Schwenk, E. and Klingsberg, E. *J. Am. chem. Soc.* **73** (1951) 253; (b) Lochte, H. L., Thomas, E. D. and Fruitt, P. **66** (1944) 550; (c) Gregg, E. C. and Craig, D. **70** (1948) 3138; (d) Yamin, M. and Fuoss, R. M. **75** (1953) 4860; (e) Phillips, A. P. **72** (1950) 1850

[366] (a) Beyerman, H. C. and Bontekoe, J. S. *Recl Trav. chim. Pays-Bas Belg.* **74** (1955) 1395; (b) Späth, E. and Galinovsky, F. *Ber. dtsch. chem. Ges.* **71** (1938) 721

[367] Chiang, M.-C. and Hartung, W. H. *J. org. Chem.* **10** (1945) 21

[368] Katritzky, A. R. and Monro, A. M. *J. chem. Soc.* (1958) 1263

[369] Hayashi, E. Yamanaka, H., Iijima, C. and Matsushita, S. *Chem. pharm. Bull., Tokyo* **8** (1960) 649

[370] Wallsgrove, E. R. *Mfg Chem.* **30** (1959) 206

[371] (a) Dale, W. J. and Ise, C. M. *J. Am. chem. Soc.* **76** (1954) 2259; (b) Doering, W. v. E. and Rhoads, S. J. **75** (1953) 4738; (c) Meek, J. S., Merrow, R. T. and Cristol, S. J. **74** (1952) 2667

[372] Petrov, A. A. and Ludwig, W. *J. gen. Chem. U.S.S.R.* **25** (1955) 703

[373] (a) Korte, F. *Chem. Ber.* **85** (1952) 1012; (b) Fischer, O. *Ber. dtsch. chem. Ges.* **32** (1899) 1297; (c) Fox, H. H. *J. org. Chem.* **17** (1952) 542

[374] Buu-Hoï, Ng. Ph., Xuong, Ng. D. and Suu, V. T. *J. chem. Soc.* (1958) 2815

[375] Gerchuk, M. P. and Taïts, S. Z. *Zh. obshch. Khim.* **20** (1950) 910

[376] Schmidt, L. and Becker, B. *Mh. Chem.* **46** (1926) 671

[377] (a) Camps, R. *Arch. Pharm., Berl.* **240** (1902) 345; (b) Roy, A. C. and Guha, P. C. *J. sci. ind. Res.* **9B** (1950) 262; (c) Räth, C. *Justus Liebigs Annln Chem.* **486** (1931) 95

[378] (a) *Germ. Pat.* 832,891 (1952); (b) *Swiss Pat.* 324,439 (1957); *Chem. Abstr.* **52** (1958) 18475

[379] (a) Marckwald, W. and Rudzik, K. *Ber. dtsch. chem. Ges.* **36** (1903) 1111; (b) Koenigs, E., Weiss, W. and Zscharn, A. **59** (1926) 316

[380] (a) Gardner, T. S., Smith, F. A., Wenis, E. and Lee, J. *J. Am. chem. Soc.* **74** (1952) 2106; (b) Boyer, J. H., Borgers, R. and Wolford, L. T. **79** (1957) 678; (c) Adams, R. and Pomerantz, S. H. **76** (1954) 702

[381] Hoegerle, K. *Helv. chim. Acta* **41** (1958) 548

[382] Tschitschibabin, A. E. *Zh. russk. fiz.-khim. Obshch.* **50** (1918) 497

[383] Gibson, C. S., Johnson, J. D. A. and Vining, D. C. *Recl Trav. chim. Pays-Bas Belg.* **49** (1930) 1006

[384] Price, C. C. and Roberts, R. M. *J. org. Chem.* **11** (1946) 463

[385] (a) Wibaut, J. P., with La Bastide, G. L. C. *Recl Trav. chim. Pays-Bas Belg.* **52** (1933) 493; (b) with Tilman, G. *ibid.* 987

[386] Turitsyna, N. F. and Vompe, A. F. *Dokl. Akad. Nauk SSSR* **74** (1950) 509

[387] (a) Morgan, G. and Stewart, J. *J. chem. Soc.* (1938) 1292; (b) Berg, S. S. and Petrow, V. (1952) 784; (c) Buu-Hoï, Ng. Ph. *ibid.* 4346; (d) Mills, W. H. and Widdows, S. T. **94** (1908) 1372

[388] Koenigs, E. and Jung, G. *J. prakt. Chem.* **137** (1933) 141

[389] Faessinger, R. W. and Brown, E. V. *J. Am. chem. Soc.* **73** (1951) 4606

[390] (a) Marckwald, W. *Ber. dtsch. chem. Ges.* **27** (1894) 1317; (b) Tschitschibabin, A. E., with Konowalowa, R. A. and A. A. **54** (1921) 814; (c) with Ossetrowa, E. D. **58** (1925) 1708; (d) with Knunjanz, I. L. **61** (1928) 2215; (e) Seide, O. **57** (1924) 1802; *U.S. Pat.* 2,456,379 (1948); *Chem. Abstr.* **43** (1949) 7050; (f) Seide, O. *Ber. dtsch. chem. Ges.* **58** (1925) 1733; (g) **57** (1924) 791; (h) with Titow, A. I. **69** (1936) 1884; (i) Fischer, O. **32** (1899) 1297

REFERENCES

391 (a) English, J. P., Chappell, D., Bell, P. H. and Roblin, R. O. *J. Am. chem. Soc.* **64** (1942) 2516; (b) Parker, E. D. and Shive, W. **69** (1947) 63; (c) Herz, W. and Tsai, L. **76** (1954) 4184; (d) Robison, M. M. and B. L. **77** (1955) 6554; (e) Moore, J. A. and Marascia, F. J. **81** (1959) 6049

392 (a) Marcinkow, A. and Plažek, E. *Roczn. Chem.* **16** (1936) 136 (cf. p. 298681b–d); (b) Plažek, E., with Rodewald, Z. **21** (1947) 150; (c) with Sorokowska, A. and Tolopka, D. **18** (1938) 210; (d) Bojarska-Dahlig, H. **29** (1955) 119; (e) Talik, T. **36** (1962) 1465

393 (a) Binz, A. and Schikh, O. v. *Ber. dtsch. chem. Ges.* **68** (1935) 315; (b) Koenigs, E., with Kinne, G. and Weiss, W., **57** (1924) 1172; (c) with Friedrich, H. and Jurany, H. **58** (1925) 2571; (d) Späth, E. and Eiter, K. **73** (1940) 719; (e) Schickh, O., Binz, A. and Schultz, A. **69** (1936) 2593; (f) Maier-Bode, H., *ibid.* 1534; (g) Koenigs, E., Bueren, H. and Jung, G. *ibid.* 2690

394 Philips, A. *Justus Liebigs Annln Chem.* **288** (1895) 253; (b) Binz, A. and Räth, C. **486** (1931) 95

395 (a) Murray, J. G. and Hauser, C. R. *J. org. Chem.* **19** (1954) 2008; (b) Leonard, N. J. and Ryder, B. L. **18** (1953) 598; (c) Hawkins, G. F. and Roe, A. **14** (1949) 328; (d) Ochiai, E. **18** (1953) 534

396 (a) Wibaut, J. P., Herzberg, S. and Schlatmann, J. *Recl Trav. chim. Pays-Bas Belg.* **73** (1954) 140; (b) Zwart, C. and Wibaut, J. P. **74** (1955) 1081; (c) Wibaut, J. P. and Nicolai, J. R. **58** (1939) 709 ;(d) Kolder, C. R. and den Hertog, H. J. **72** (1953) 285; (e) den Hertog, H. J., with Kolder, C. R. and Combé, W. P. **70** (1951) 591; (f) **64** (1945) 85; (g) **65** (1946) 129; (h) with Wibaut, J. P. **55** (1936) 122; (i) Salemink, C. A and van der Want, G. M. **68** (1949) 1013

397 (a) Tschitschibabin, A. E., with Seide, O. A. *Zh. russk. fiz.-khim. Obshch.* **46** (1914) 1216; (b) with Egorov, A. F. **60** (1928) 683; (c) Seide, O. A. **50** (1920) 534

398 (a) *Germ. Pat.* 374,921 (1923) [*Chem. Abstr.* **18** (1924) 2176]; (b) 595,361 (1934) [*Chem. Abstr.* **28** (1934) 4069]; (c) 628,605; 670,920 [*Chem. Abstr.* **30** (1936) 6137; **33** (1939) 6351]; (d) *U.S. Pat.* 2,637,731 (1954) [*Chem. Abstr.* **48** (1954) 5228]; (e) *Swiss Pat.* 212,197 (1941) [*Chem. Abstr.* **36** (1942) 3632]

399 (a) Suzuki, I. *Chem. Abstr.* **46** (1952) 4004; (b) Plažek, E., with Marcinikow, A. and Stammer, C. **30** (1936) 1377; (c) with Sucharda, E. **22** (1928) 1813; (d) Tschitschibabin, A. E. and Kirsanov, A. W. **22** (1928) 2563; (e) Kucherova, N. F., Kucherov, V. F. and Kocheskov, K. A. **41** (1947) 6243

400 (a) Ochiai, E., with Suzuki, I. *J. pharm. Soc. Japan* **67** (1947) 158 [*Chem. Abstr.* **45** (1951) 9541]; (b) with Futaki, K. **72** (1952) 274; (c) with Katada, M. **63** (1943) 186; **67** (1947) 30; (d) with Fujimoto, M. *Pharm. Bull., Tokyo* **2** (1954) 131; (e) Takahashi, T. and Ueda, K. *ibid.* 78

401 Essery, J. M. and Schofield, K. *J. chem. Soc.* (a) (1961) 3939; (b) (1960) 4953; (c) Petrov, V. A. (1945) 927; (d) Berrie, A. H., Newbold, G. T. and Spring, F. S. (1952) 2042; (e) Chambers, R. D., Hutchinson, J. and Musgrave, W. K. R. (1964) 3736; (f) Fraser, J. and Tittensor, E. (1956) 1781; (g) Petrow, V. and Saper, J. (1948) 1389; Leese, C. L. and Rydon, H. N. (1954) 4039; (h) Albert, A. and Hampton, A. (1952) 4985; (i) Gardner, J. N. and Katritzky, A. R. (1957) 4375

402 (a) Feist, K., Awe, W. and Kuklinski, M. *Arch. Pharm., Berl.* **274** (1936) 418; (b) Jerchel, D., Fischer, H. and Thomas, K. *Chem. Ber.* **89** (1956) 2921; (c) Meyer, H., with Staffen, F. *Mh. Chem.* **34** (1913) 517; (d) with Tropsch, H. **35** (1914) 207; (e) with Beck, F. R. v. **36** (1915) 731

403 (a) Steinhäuser, E. and Diepolder, E. *J. prakt. Chem.* **93** (1916/2) 387; (b) Graf, R., Lederer-Ponzer, E. *et al.* **138** (1933) 244

404 Bray, H. G., Neale, F. C. and Thorpe, W. V. *Biochem. Z.* **46** (1950) 506

405 (a) Caldwell, W. T. and Kornfield, E. C. *J. Am. chem. Soc.* **64** (1942) 1695; (b) Pino, L. N. and Zehrung, W. S. **77** (1955) 3154; (c) Bernstein, J., Stearns, B., Dexter, M. and Lott, W. A. **69** (1947) 1147; (d) Vaughan, J. R., Krapcho, J. and English, J. P. **71** (1949) 1885; (e) Adams, R. and Miyano, S. **76** (1954) 2785; (f) Jaffé, H. H. and Doak, G. O. **77** (1955) 4441; (g) Brown, E. V. **79** (1957) 3565

406 (a) Bäumler, J., Sorkin, E. and Erlenmeyer, H. *Helv. chim. Acta* **34** (1951) 496; (b) Talik, Z. *Bull. Acad. pol. Sci. Sér. Sci. tech.* **9** (1961) No. 9,567; (c) Faessinger, R. W. and Brown, E. V. *Trans. Ky Acad. Sci.* **24** (1963) 106; (d) Tschitschibabin, A. and Hoffmann, C. *C. r. hebd. Séanc. Acad. Sci., Paris* **205** (1937) 153

407 Campbell, N., Henderson, A. W. and Taylor, D. *J. chem. Soc.* (1953) 1281

408 (a) Pentimalli, L. *Gazz. chim. ital.* **90** (1960) 1203; (b) **89** (1959) 1843; *Tetrahedron* **9** (1960) 194

409 Fischer, O. and Chur, M. *J. prakt. Chem.* **93** (1916/2) 363

410 Feist, K., Awe, W., Schultz, J. and Klatt, F. *Arch. Pharm., Berl.* **272** (1934) 100

411 Kaye, I. A. and Kogon, I. C. *Recl Trav. chim. Pays-Bas Belg.* **71** (1952) 309

412 Savich, I. A., Pikaev, A. K., Lebedev, I. A. and Spitsyn, V. I. *Chem. Abstr.* **53** (1959) 1334

413 Saikachi, H. and Hoshida, H. *J. pharm. Soc. Japan* **71** (1951) 982

[414] Kirpal, A. and Reiter, E. *Ber. dtsch. chem. Ges.* **60** (1927) 664
[415] Takahashi, T., Saikachi, H., Yoshina, S. and Mizuno, C. *J. pharm. Soc. Japan* **69** (1949) 284
[416] (a) *Br. Pat.* 625,327 (1949); (b) *U.S. Pat.* 2,426,012 (1947)
[417] (a) Titov, A. I. and Baryshnikova, A. N. *Chem. Abstr.* **48** (1954) 2704; (b) Zalukajevs, L. *ibid.* 10024
[418] Kahn, H. J. and Petrow, V. A. *J. chem. Soc.* (1945) 858
[419] (a) Tschitschibabin, A. E. and Knunjanz, I. L. *Ber. dtsch. chem. Ges.* **62** (1929) 3048; (b) **64** (1931) 2839
[420] Bystritskaya, M. G. and Kirsanov, A. V. *Chem. Abstr.* **35** (1941) 4023
[421] Schmid, L. and Becker, B. *Mh. Chem.* **46** (1926) 675
[422] Tien, J. M. and Hunsberger, I. M. *J. Am. chem. Soc.* **77** (1955) 6604
[423] (a) Schmid, L. and Bangler, B. *Ber. dtsch. chem. Ges.* **58** (1925) 1971; **59** (1926) 1360; (b) Gilman, H., Stuckwisch, C. G. and Nobis, J. F. *J. Am. chem. Soc.* **68** (1946) 326
[424] (a) Koenigs, E., Mields, M. and Gurlt, H. *Ber. dtsch. chem. Ges.* **57** (1924) 1179; (b) Bremer, O. *Justus Liebigs Annln Chem.* **518** (1935) 274; (c) **514** (1934) 279
[425] Tschitschibabin, A. E. and Kirsanov, A. W. *Ber. dtsch. chem. Ges.* (a) **60** (1927) 2433; (b) **61** (1928) 1223
[426] Danyushevskiĭ, Y. L. and Gol'dfarb, Y. L. *Chem. Abstr.* **44** (1950) 9446
[427] Rapoport, H., Look, M. and Kelly, G. J. *J. Am. chem. Soc.* **74** (1952) 6293
[428] (a) Ahmad, Y. and Hey, D. H. *J. chem. Soc.* (1954) 4516; (b) Haworth, J. W., Heilbron, I. M. and Hey, D. H. (1940) 372; (c) Adams, W. J., Hey, D. H., Mamalis, P. and Parker, R. E. (1949) 3181
[429] Kirpal, A. and Böhm, W. *Acta chim. hung.* (a) **65** (1932) 680; (b) Kirpal, A. **67** (1934) 70
[430] (a) Bystritskaya, M. G. and Kirssanov, A. V. *Chem. Abstr.* **35** (1941) 4380; (b) Kimura, M. and Takano, Y. **53** (1959) 18030
[431] Taylor, E. C. and Driscoll, J. S. *J. org. Chem.* **25** (1960) 1716
[432] Wiley, R. H. and Hartman, J. L. *J. Am. chem. Soc.* **73** (1951) 494
[433] (a) Boyland, E. and Sims, P. *J. chem. Soc.* (1958) 4198; (b) Bailey, A. S. and Brunskill, J. S. A. (1959) 2554; (c) Fargher, R. G. and Furness, R. **107** (1915) 688
[434] Kirpal, A. and Böhm, W. *Ber. dtsch. chem. Ges.* **64** (1931) 767
[435] (a) den Hertog, H. J., Jouwersma, C., van der Wal, A. A. and Willebrands-Schogt, E. C. C. *Recl Trav. chim. Pays Belg.* **68** (1949) 275; (b) *U.S. Pat.* 2,129,294 (1938); den Hertog, H. J., with Schogt, J. C. M. *et al.*, *Recl Trav. chim. Pays-Bas Belg.* **69** (1950) 673; with Jouwersma, C. **72** (1953) 44
[436] (a) Bambas, L. L. *J. Am. chem. Soc.* **67** (1945) 668; (b) Brown, E. V. **76** (1954) 3167; (c) Baumgarten, H. E. and Chien-Fan Su, H. **74** (1952) 3828
[437] Shibasaki, J. *Chem. Abstr.* **47** (1953) 6403
[438] (a) Hünig, S. and Köbrich, G. *Justus Liebigs Annln Chem.* **617** (1958) 181, 203, 216; (b) Bremer, O. **539** (1939) 276
[439] (a) Mangini, A. and Colonna, M. *Gazz. chim. ital.* **75** (1943) 313; (b) Graf, R. *J. prakt. Chem.* **138** (1933/2) 244
[440] (a) Boyer, J. M., McCane, D. L., McCarville, W. J. and Tweedie, A. T. *J. Am. chem. Soc.* **75** (1953) 5298; (b) Tarbell, D. S., Todd, C. W. *et al.* **70** (1948) 1381; (c) Tschitschibabin, A. E. and Ossetrowa, E. D. **56** (1934) 1711
[441] Koenigs, E., Freigang, W., Lobmayer, G. and Zscharn, A. *Ber. dtsch. Chem. Ges.* **59** (1926) 321
[442] (a) Pentimalli, L. *Gazz. chim. ital.* **93** (1963) 404; (b) Colonna, M. **86** (1956) 705; (c) with Risaliti, A. *ibid.* 288; (d) and Serra, R. **85** (1955) 1508; (e) Carboni, S. and Berti, G. **84** (1954) 683
[443] Czuba, W. *Roczn. Chem.* **34** (1960) 905, 1639, 1647
[444] (a) Tschitschibabin, A. E., with Rasorenov, B. A. *Zh. russk. fiz.-khim. Obshch.* **47** (1915) 1286; (b) **50** (1918) 512; Ostromislensky, I. *U.S. Pat.* 1,680,109, 1,680,111 (1928); 1,724,305 (1929); 1,809,352, 1,820,483 (1931); Tisza, E. T. and Joos, B. *U.S. Pat.* 1,856,602 (1932); 2,029,315 (1936); Renshaw, R. R. and Tisza, E. T. 2,135,293 (1938); cf. p. 287[142c]
[445] Mohr, E. *Ber. dtsch. chem. Ges.* **31** (1898) 2495
[446] *U.S. Pat.* (a) 2,294,380 (1934); (b) 1,671,257 (1928); (c) *French Pat.* 641,422 (1926) [*Chem. Abstr.* **23** (1929) 1139]
[447] (a) Saikachi, H. *J. pharm. Soc. Japan* **63** (1943) 328; (b) Takahashi, T. and Saikachi, H. **62** (1942) 143; (c) with Suzuki, K. **66** (1946) 28; (d) with Akai, H. **63** (1943) 153; (e) Takahashi, T. and Senda, S. **66** (1946) 26
[448] Koenigs, E. and Jungfer, O. *Ber. dtsch. chem. Ges.* **57** (1924) 2080
[449] Ochiai, E., Teshigawara, T., Oda, K. and Naito, T. *Chem. Abstr.* **45** (1951) 8527
[450] Fanta, P. E. *J. Am. chem. Soc.* **75** (1953) 737
[451] Zwart, C. and Wibaut, J. P. *Recl Trav. chim. Pays-Bas Belg.* **74** (1955) 1062
[452] (a) Czuba, W. and Plažek, E. *Recl Trav. chim. Pays-Bas Belg.* **77** (1958) 92; (b) Plažek, E. **72** (1953) 569

REFERENCES

453 Colonna, M. and Dal Monte Casoni, D. *Chem. Abstr.* **44** (1950) 3494
454 (a) Koenigs, E. and Freter, K. *Ber. dtsch. chem. Ges.* **57** (1924) 1187; (b) Friedl, F. **45** (1912) 428; *Mh. Chem.* **34** (1913) 759
455 (a) Colonna, M. and Pentimalli, L. *Annali Chim.* **48** (1958) 1403; (b) Pentimalli, L. and Risaliti, A. **46** (1956) 1037
456 (a) Katritzky, A. R. *J. chem. Soc.* (1956) 2404; (b) Ochiai, E., Itai, T. and Yoshmo, K. *Proc. imp. Acad. Japan* **20** (1944) 141
457 Colonna, M. and Risaliti, A. *Gazz. chim. ital.* **85** (1955) 1148
458 (a) Kalthod, G. G. and Linnell, W. H. *Qu. Jl Pharm. Pharmac.* **21** (1948) 63; (b) Wibaut, J. P., Overhoff, J. and Geldof, H. *Recl Trav. chim. Pays-Bas Belg.* **54** (1935) 807
459 Tschitschibabin, A. E. and Konowalowa, R. A. *Ber. dtsch. chem. Ges.* **58** (1925) 1712
460 Colonna, M. and Risaliti, A. *Gazz. chim. ital.* (a) **86** (1956) 698; (b) **87** (1957) 923; (c) **89** (1959) 2493
461 (a) den Hertog, H. J., with Overhoff, J. *Recl Trav. chim. Pays-Bas Belg.* **49** (1930) 552; (b) with Henkers, C. H. and van Roon, J. M. **71** (1952) 1145
462 Hantzsch, A. and Burawoy, A. *Ber. dtsch. chem. Ges.* **63** (1930) 1775
463 Tschtschibabin, A. E. and Zeide, C. A. *Zh. russk. fiz.-khim. Obshch.* **50** (1918) 522
464 Kubota, S. and Akita, T. *Chem. Abstr.* **52** (1958) 11834
465 Le Fèvre, R. J. W. and Worth, C. V. *J. chem. Soc.* (1951) 1814
466 Eichhorn, E. L. *Acta crystallogr.* **12** (1959) 746
467 Phillips, A. P. and Graham, P. L. *J. Am. chem. Soc.* **74** (1952) 1552
468 Steele, C. S. and Adams, R. *J. Am. chem. Soc.* **52** (1930) 4528; Lions, F. **53** (1931) 1176; Chalmers, A. J., Lions, F. and Robson, A. O. *J. R. Soc. N.S. Wales* **64** (1930) 320
469 Woodruff, E. H. and Adams, R. *J. Am. chem. Soc.* **54** (1932) 1977
470 Brydowna, W. *Roczn. Chem.* **14** (1934) 304
471 Bradsher, C. K. and Beavers, L. E. *J. Am. chem. Soc.* **77** (1955) 453
472 (a) Craig, L. C. *J. Am. chem. Soc.* **56** (1934) 231; (b) Ellin, R. I. **80** (1958) 6588; (c) Teague, P. C., Ballentine, A. R. and Rushton, G. L. **75** (1953) 3429; (d) Feely, W. E. and Beavers, E. M. **81** (1959) 4004; (e) Webb, J. L. and Corwin, A. H. **66** (1944) 1456
473 Camps, R. *Arch. Pharm., Berl.* **240** (1902) 366
474 Meyer, H. *Mh. Chem.* (a) **23** (1902) 437; (b) **28** (1907) 47
475 (a) Poziomek, E. J. *J. org. Chem.* **28** (1963) 590; (b) Reider, M. J. and Elderfield, R. C. **7** (1942) 286
476 Moynehan, T. M., Schofield, K., Jones, R. A. Y. and Katritzky, A. R. *J. chem. Soc.* (1962) 2637
477 Baumgarten, P. and Dornow, A. *Ber. dtsch. chem. Ges.* **72** (1939) 563
478 Räth, C. and Schiffmann, F. *Justus Liebigs Annln Chem.* **487** (1931) 127
479 McElvain, S. M. and Goese, M. A. *J. Am. chem. Soc.* **65** (1943) 2233
480 Suzuki, Y. *J. pharm. Soc. Japan* **81** (1961) 1204
481 (a) Galat, A. *J. Am. chem. Soc.* **70** (1948) 3945; (b) Krewson, C. F. and Couch, J. F. **65** (1943) 2256
482 Vejdělek, Z. *J. Chem. Abstr.* **47** (1953) 11191
483 Schmidt-Thome, J. and Goebel, H. *Hoppe-Seyler's Z. physiol. Chem.* **288** (1951) 237
484 Gruber, W. and Schlögl, K. *Mh. Chem.* **81** (1950) 83
485 Moir, J. *J. (Trans.) chem. Soc.* **81** (1902) 100
486 Meyer, J. *J. prakt. Chem.* **78** (1908) 519
487 (a) Wenner, W. and Plati, J. T. *J. org. Chem.* **11** (1946) 751; (b) Payne, G. B. **26** (1961) 668
488 Bardhan, J. C. *J. chem. Soc.* (1929) 2223
489 Isler, O., Gutmann, H. *et al., Helv. chim. Acta* **38** (1955) 1033
490 *U.S. Pat.* 2,446,957 (1948); 2,471,518 (1949); 2,685,586 (1954)
491 (a) Huber, W. *J. Am. chem. Soc.* **66** (1944) 876; (b) Adkins, H., Wolff, I. A., Pavlic, A. and Hutchinson, E. *ibid.* 1293; (c) Bruce, W. F. and Coover, H. W. *ibid.* 2092
492 Vejdělek, Z. J. and Protiva, M. *Colln. Czech. chem. Commun.* **16** (1951) 344
493 Mowat, J. H., Pilgrim, F. J. and Carlson, G. H. *J. Am. chem. Soc.* **65** (1943) 954; Mariella, R. P., with Leech, J. L. **71** (1949) 331; with Belcher, E. P. **74** (1952) 4049; Heyl, D., Luz, E., Harris, S. A. and Folkers, K. **75** (1953) 4079
494 Henecka, H. *Chem. Ber.* **82** (1949) 36
495 (a) *U.S. Pat.* 2,615,896 (1952); (b) 2,516,673 (1950); (c) *Swiss Pat.* 224,314 (1943) [*Chem. Abstr.* **43** (1949) 1811]
496 Tsuda, K., Ikekawa, N. *et al., Pharm. Bull., Tokyo* **1** (1953) 122
497 Velluz, L. and Amard, G. *Bull. Soc. chim. Fr.* (1947) 136
498 Harris, S. A. and Folkers, K. *J. Am. chem. Soc.* **61** (1939) 1245, 3307
499 Wagtendonk, H. M. and Wibaut, J. P. *Recl Trav. chim. Pays-Bas Belg.* **61** (1942) 728
500 Martin, G. J., Avakian, S. and Moss, J. *J. biol. Chem.* **174** (1948) 495
501 Graf, R. *J. prakt. Chem.* **140** (1934) 39; with Perathoner, G. and Tatzel, M. **146** (1936) 88
502 Erlenmeyer, H. and Epprecht, A. *Helv. chim. Acta* **20** (1937) 690
503 Matsukawa, T. and Matsuno, T. *J. pharm. Soc. Japan* **64** (1944) 145

409

[504] Vanderhorst, P. J. and Hamilton, C. S. *J. Am. chem. Soc.* **75** (1953) 656; Sculley, J. D. and Hamilton, C. S. *ibid.* 3400

[505] Bohlmann, F. and M. *Chem. Ber.* **86** (1953) 1419

[506] Backeberg, O. G. and Staskun, B. *J. chem. Soc.* (1962) 3961; (1964) 5880

[507] Wibaut, J. P. and Overhoff, J. *Recl Trav. chim. Pays-Bas Belg.* **52** (1933) 55

[508] (a) Volke, J., with Kubíček, R. and Šantavý, F. *Colln. Czech. chem. Commun.* **25** (1960) 1510; (b) with Holubek, J. **28** (1963) 1597

[509] (a) Brown, H. C. and McDaniel, D. H. *J. Am. chem. Soc.* **75** (1955) 3752; (b) Roe, A. and Hawkins, G. F. **69** (1947) 2443; (c) Ferm, F. L. and Vander Werf, C. A. **72** (1950) 4809; (d) Minor, J. T., Hawkins, G. F., VanderWerf, C. A. and Roe, A. **71** (1949) 1125; (e) Roe, A., Cheek, P. H. and Hawkins, G. F. *ibid.* 4152

[510] (a) Tschitschibabin, A. E., with Rjasanjew, M. D. *Zh. russk. fiz.-khim. Obshch.* **47** (1915) 1571; (b) with Tyazhelova, V. S. **50** (1920) 483, 492

[511] (a) Beaty, R. D. and Musgrave, W. K. R. *J. chem. Soc.* (1952) 875; (b) King, H. and Ware, L. L. (1939) 873; (c) Clemo, G. R., Fox, B. W. and Raper, R. (1954) 2693; (d) Sedgwick, A. P. and Collie, N. **67** (1895) 399; (e) Phillips, M. A. (1941) 9; (f) Banks, R. E., Ginsberg, A. E. and Haszeldine, R. N. (1961) 1740

[512] Wibaut, J. P. and Holmes-Kamminga, W. J. *Bull. Soc. chim. Fr.* (1958) 424

[513] (a) Bobrański, B., Kochańska, L. and Kowalewska, A. *Ber. dtsch. chem. Ges.* **71** (1938) 2385; (b) Dornow, A. **73** (1940) 78; (c) Conrad, M. and Epstein, W. **20** (1887) 162; (d) Magidson, O. and Menschikoff, G. **58** (1925) 113

[514] (a) Wibaut, J. P. and Broekman, F. W. *Recl Trav. chim. Pays-Bas Belg.* **58** (1939) 885; (b) Willink, H. D. T. and Wibaut, J. P. **53** (1934) 417; (c) den Hertog, H. J., with Wibaut, J. P. **51** (1932) 381, 940; (d) with de Bruyn, J. **70** (1951) 182; (e) with Schepman, F. R., de Bruyn, J. and Thysse, G. J. E. **69** (1950) 1281; (f) with Combé, W. P. **70** (1951) 581; (g) with Overhoff, J. **69** (1950) 468

[515] Haitinger, L. and Lieben, A. *Mh. Chem.* **6** (1885) 279

[516] *Germ. Pat.* 510,432 (1930)

[517] (a) Rodewald, Z. and Plažek, E. *Roczn. Chem.* **16** (1936) 444; (b) *ibid.* 130; (c) Talik, T. and Plažek, E. **33** (1959) 387; (d) Plažek, E., Dohaniuk, K. and Grzyb, Z. **26** (1952) 106

[518] Gruber, W. *Can. J. Chem.* **31** (1953) 1020

[519] (a) Suzuki, Y. *J. pharm. Soc. Japan* **81** (1961) 1206; (b) Kato, T. **75** (1955) 1239; (c) with Ohta, M. **71** (1951) 217; (d) Sugii, Y., Shimoya, I. and Shindo, H. **50** (1930) 727; (e) Ochiai, E., with Ishikawa, M. and Sai, Zai-Ren **65** (1945) 72; (f) with Ito, T. and Okuda, S. **71** (1951) 591; (g) Hayashi, E. **70** (1950) 142

[520] Roe, A. and Seligman, R. B. *J. org. Chem.* **20** (1955) 1729; (b) Pearson, D. E., Hargrove, W. W., Chow, J. K. T. and Suthers, B. R. **26** (1961) 789; (c) Cava, M. P. and Bhattacharyya, N. K. **23** (1958) 1287

[521] (a) *U.S. Pat.* 2,516,830 (1950); (b) *British Pat.* 726,378 (1955)

[522] Profft, E. and Richter, H. *J. prakt. Chem.* **9** (1959/4) 164

[523] (a) Kochetkov, N. K. *Chem. Abstr.* **47** (1953) 3309; (b) Nakashima, T. **52** (1958) 18399; (c) Palát, K., Celadník, M. Nováček, L. and Urbančík, R. **54** (1960) 11016; (d) Kabachnik, M. I. **32** (1938) 560; (e) Bojarska-Dahlig, H. and Urbański, R. **48** (1954) 1337; (f) Reitmann, J. **29** (1935) 4359; Wibaut, J. P. and la Bastide, G. **21** (1927) 3619

[524] (a) Case, F. H. *J. Am. chem. Soc.* **68** (1946) 2574; (b) Brown, E. V. and Burke, H. T. **77** (1955) 6053; (c) Shaw, E., Bernstein, J., Losee, K. and Lott, W. A. **72** (1950) 4362; (d) Mosher, H. S. and Tessieri, J. E. **73** (1951) 4925; (e) Niemann, C., Lewis, R. N. and Hays, J. T. **64** (1942) 1678

[525] Bocchi, O. *Gazz. chim. ital.* **30** (1900) i, 89

[526] Graf, R. (a) *Ber. dtsch. chem. Ges.* **64** (1931) 21; (b) *J. prakt. Chem.* **133** (1932/2) 36

[527] Binz, A. and Maier-Bode, H. *Angew. Chem.* **49** (1936) 486

[528] Binz, A., Räth, C. and Maier-Bode, H. *Justus Liebigs Annln Chem.* **478** (1930) 22

[529] Šorm, F. and Šedivý, L. *Colln. Czech. chem. Commun.* **13** (1948) 289

[530] (a) Overhoff, J. and Proost, W. *Recl Trav. chim. Pays-Bas Belg.* **57** (1938) 179; (b) Proost, W. and Wibaut, J. P. **59** (1940) 971

[531] Davies, W. C. and Mann, F. G. *J. chem. Soc.* (1944) 276

[532] Tilford, C. H., Shelton, R. S. and van Campen, M. G. *J. Am. chem. Soc.* **70** (1948) 4001

[533] de Jonge, A. P., den Hertog, H. J. and Wibaut, J. P. *Recl Trav. chim. Pays-Bas Belg.* **70** (1951) 989

[534] Mann, F. G. and Watson, J. *J. org. Chem.* **13** (1948) 502

[535] Knott, R. F., Ciric, J. and Breckenridge, J. G. *Can. J. Chem.* **31** (1953) 615

[536] Wibaut, J. P. and van der Voort, H. G. P. *Recl Trav. chim. Pays-Bas Belg.* (a) **71** (1952) 798; (b) with Markus, R. **69** (1950) 1048

[537] Gilman, H., Gregory, W. A. and Spatz, S. M. *J. org. Chem.* **16** (1951) 1788

[538] Ruzicka, L. and Fornasir, V. *Helv. chim. Acta* **2** (1919) 338

[539] (a) Oparina, M. P. *Ber. dtsch. chem. Ges.* **64** (1931) 569; (b) Schroeter, C., Seidler, C., Sulzbacher, M. and Kanitz, R. **65** (1932) 432; (c) Schroeter, G. and Finck, E. **71** (1938) 671

REFERENCES

[540] Holubek, J. and Volke, J. *Colln. Czech. chem. Commun.* **27** (1962) 680

[541] Kolder, C. R. and den Hertog, H. J. *Recl Trav. chim. Pays-Bas Belg.* (a) **72** (1953) 853; (b) with Combé, W. P. **73** (1954) 704; (c) den Hertog, H. J. and Schogt, J. C. M. **70** (1951) 353

[542] Fanta, P. E. *Chem. Rev.* **38** (1946) 139; Nursten, H. E. *J. chem. Soc.* (1955) 3081; Forrest, J. (1960) 594; Nilsson, M. *Acta chem. scand.* **12** (1958) 537

[543] Burstall, F. H. *J. chem. Soc.* (1938) 1662

[544] Wibaut, J. P. (a) with Overhoff, J. *Recl Trav. chim. Pays-Bas Belg.* **47** (1928) 761; (b) with Willink, H. D. T. **54** (1935) 275

[545] Magidson, O. G. *Ber. dtsch. chem. Ges.* **67** (1954) 1329

[546] *French Pat.* 621,989 (1927)

[547] (a) Alberts, A. A. and Bachman, G. B. *J. Am. chem. Soc.* **57** (1935) 1284; (b) Iddes, H. A., Lang, E. H. and Gregg, D. C. **59** (1937) 1945; (c) Stevens, J. R., Beutel, R. H. and Chamberlin, E. **64** (1942) 1093

[548] Kleipool, R. J. C. and Wibaut, J. P. *Recl Trav. chim. Pays-Bas Belg.* **69** (1950) 59

[549] (a) Späth, E. and Galinovsky, F. *Ber. dtsch. chem. Ges.* **69** (1936) 2059; (b) Tschitschibabin, A. E. **57** (1924) 1802; (c) Fischer, O. and Renouf, E. **17** (1884) 1896; (d) Koenigs, E. and Greiner, H. **64** (1931) 1049

[550] Pechmann, H. v. and Baltzer, O. *Ber. dtsch. chem. Ges.* **24** (1891) 3144

[551] (a) Tschitschibabin, A. E., with Szokow, P. G. *Ber. dtsch. chem. Ges.* **58** (1925) 2650; (b) with Oparina, O. P. *Zh. russk. fiz.-khim. Obshch.* **56** (1925) 153; (c) *U.S. Pat.* 2,540,218 (1951); (d) 2,557,076 (1952) [*Chem. Abstr.* **46** (1952) 145]; (e) 2,636,882 (1953)

[552] (a) Cavallito, C. J. and Haskell, T. H. *J. Am. chem. Soc.* **66** (1944) 1166; (b) *ibid.* 1927

[553] (a) Binz, A. and Räth, C. *Justus Liebigs Annln Chem.* **489** (1931) 107; (b) Borsche, W. and Peter, W. **453** (1927) 148; (c) Räth, C. **484** (1930) 52

[554] *Org. Synth., Coll. Vol.* **2** (1943) 419

[555] Grave, T. B. *J. Am. chem. Soc.* **46** (1924) 1460

[556] (a) Renshaw, R. R. and Conn, R. C. *J. Am. chem. Soc.* **59** (1937) 297; (b) Lott, W. A. and Shaw, E. **71** (1949) 70; (c) Adams, R. and Miyano, S. **76** (1954) 3168; (d) Rapaport, H. and Volcheck, E. J. **78** (1956) 2451

[557] (a) Weidel, H. and Murmann, E. *Mh. Chem.* **16** (1895) 749; (b) Gruber, W. and Schlögl, K. **80** (1949) 499; (c) **81** (1950) 473

[558] (a) Cohen, A., Haworth, J. W. and Hughes, E. G. *J. chem. Soc.* (1952) 4374; (b) Clemo, G. R. and Swan, G. A. (1948) 198; (c) Newbold, G. T. and Spring, F. S. (1948) 1864

[559] (a) Prins, D. A. *Recl Trav. chim. Pays-Bas Belg.* **76** (1957) 58; (b) den Hertog, H. J., with Wibaut, J. P. *et al.*, **69** (1950) 700; (c) with Buurman, D. J. **75** (1956) 257; (d) with Mulder, B. **67** (1948) 957

[560] (a) Sugasawa, S. and Kirisawa, M. *Pharm. Bull., Tokyo* **3** (1955) 187; (b) Ochiai, E., Fujimoto, M. and Ichimura, S. **2** (1954) 137; (c) Suzuki, Y. **5** (1957) 78; (d) Ishii, T. *J. pharm. Soc. Japan* **71** (1951) 1092 [*Chem. Abstr.* **46** (1952) 5046]; (e) Ochiai, E. and Katada, M. **63** (1943) 265; Itai, T. **65** (1945, No. 9/10A) 8; Takeda, K. and Tokuyama, M. **75** (1955) 286, 620; (f) Hayashi, E. **71** (1951) 213

[561] (a) Shaw, E. *J. Am. chem. Soc.* **71** (1949) 67; (b) Cunningham, K. G., Newbold, G. T., Spring, F. S. and Stark, J. *J. chem. Soc.* (1949) 2091; (c) Bradlow, H. L. and Vander Werf, C. A. *J. org. Chem.* **16** (1951) 73

[562] Tschitschibabin, A. E., with Vidonova, M. S. *Zh. russk. fiz.-khim. Obshch.* **53** (1921) 238; with Kirssanow, A. W. *Ber. dtsch. chem. Ges.* **57** (1924) 1163; Bergstrom, F. W. and Fernelius, W. C. *Chem. Rev.* **20** (1937) 413; Levine, R. and Fernelius, W. C. *ibid.* **54** (1954) 537; Lecocq, J. *Bull. Soc. chim. Fr.* (1950) 188; Roe, A. M. *J. chem. Soc.* (1963) 2195

[563] (a) Ochiai, E. and Ito, Y. *Ber. dtsch. chem. Ges.* **74** (1941) 1111; (b) Leben, J. A. **29** (1896) 1673; (c) Dornow, A. and Neuse, E. **84** (1951) 296; (d) Leditschke, H. *Chem. Ber.* **85** (1952) 202

[564] (a) Clauson-Kaas, N., with Elming, N. and Tyle, Z. *Acta chem. scand.* **9** (1955) 1; (b) with Nedenskov, N. *ibid.* 14; (c) Abramovitch, R. A. and Notation, A. D. *Can. J. Chem.* **38** (1960) 1445; (d) Gruber, W. **31** (1953) 1181

[565] (a) Bojarska-Dahlig, H. *Roczn. Chem.* **30** (1956) 475; (b) with Guda, I. **31** (1957) 1147

[566] (a) Traynelis, V. J. and Martello, R. F. *J. Am. chem. Soc.* **82** (1960) 2744 ; (b) Marion, L. and Cockburn, W. F. **71** (1949) 3402; (c) Wuest, H. M. and Sakal, E. H. **73** (1951) 1210; (d) Adams, R. and Govindachari, T. R. **69** (1947) 1806; (e) Bickel, A. F. *ibid.* 1805; (f) Boyer, J. H. and Kruger, S. **79** (1957) 3552

[567] (a) Duesel, B. F. and Scudi, J. V. *J. Am. chem. Soc.* **71** (1949) 1866; (b) Marion, L. and Cockburn, W. F. *ibid.* 3402; (c) Jacobs, W. A. and Sato, Y. *J. biol. Chem.* **191** (1951) 71

[568] (a) Aso, K. *Chem. Abstr.* **34** (1940) 6278; (b) Plažek, E., with Rodewald, Z. **31** (1937) 3918; (c) *ibid.* 4669; (d) *ibid.* 1808; (e) Ishikawa, M. **45** (1951) 8529; Suzuki, I. *J. pharm. Soc. Japan* **68** (1948) 126

[569] (a) Magidson, O. and Menschikoff, G. *Ber. dtsch. chem. Ges.* **58** (1925) 113; (b) Stokes, H. N. and Pechmann, H. v. **19** (1886) 2694

411

[570] Tschitschibabin, A. E. and Tjashelowa, V. S. *Zh. russk. fiz.-khim. Obshch.* **50** (1918) 495; Naegeli, C., Kündig, W. and Brandenburger, H. *Helv. chim. Acta* **21** (1938) 1746; Skrowaczewska, Z. *Chem. Abstr.* **48** (1954) 7568

[571] Titov, A. I. and Levin, B. B. *J. gen. Chem. U.S.S.R.* **11** (1941) 9

[572] (a) Kleipool, R. J. C. and Wibaut, J. P. *Recl Trav. chim. Pays-Bas Belg.* **69** (1950) 37; (b) den Hertog, H. J. and Combé, W. P. **71** (1952) 745

[573] (a) Ost, H. *J. prakt. Chem.* **19** (1879) 203; **23** (1881) 441; **27** (1883) 257; (b) Niementowski, S. v. and Sucharda, E. **94** (1916) 193

[574] (a) den Hertog, H. J., with Farenhorst, E. *Recl Trav. chim. Pays-Bas Belg.* **67** (1948) 381; (b) with de Jonge, A. P. *ibid.* 385; (c) with van Ammers, M. **74** (1955) 1160; (d) and Schukking, S. *ibid.* 1171; (e) with Henkens, C. H. and Dilz, K. **72** (1953) 296

[575] Dinan, F. J. and Tieckelmann, H. *J. org. Chem.* **29** (1964) 1650

[576] Albert, A. *Heterocyclic Chemistry. An Introduction*, p. 55, London (Athlone Press) 1959

[577] Kuhn, R. and Wendt, G. *Ber. dtsch. chem. Ges.* **72** (1939) 305

[578] Joshi, S. N., Kaushal, R. and Deshapande, S. S. *J. Indian chem. Soc.* **18** (1941) 479

[579] Fried, J. and Elderfield, R. C. *J. org. Chem.* **6** (1941) 566

[580] (a) Fischer, O. and Renouf, E. *Ber. dtsch. chem. Ges.* **17** (1884) 755; (b) Lieben, A. and Haitinger, L. **16** (1882) 1259; (c) Benary, E. and Bitter, G. A. **61** (1928) 1057; (d) Haitinger, L. and Lieben, A. **18** (1885) 929

[581] den Hertog, H. J. and Mulder, B. *Recl Trav. chim. Pays-Bas Belg.* **68** (1949) 433

[582] Wiberg, K. B., Shryne, T. M. and Kintner, R. R. *J. Am. chem. Soc.* **79** (1957) 3160

[583] Tschitschibabin, A. E. and Jeletzky, N. P. *Ber. dtsch. chem. Ges.* **57** (1924) 1158

[584] Hanson, M. P., Nelsen, D. C. and Kintner, R. R. *Chem. Abstr.* **54** (1960) 19670

[585] *Br. Pat.* 355,017 (1930) [*Chem. Abstr.* **26** (1932) 5574]

[586] Bremer, O. *Justus Liebigs Annln Chem.* **529** (1937) 290

[587] Wagner, G. and Pischel, H. *Naturwissenschaften* **48** (1961) 454

[588] (a) Moffett, R. B. *J. org. Chem.* **28** (1963) 2885; (b) Dinan, F. J. and Tieckelmann, H. **29** (1964) 892

[589] Beak, P. and Bonham, J. *Tetrahedron Lett.* (1964) 3083

[590] Ueno, Y., Takaya, T. and Imoto, E. *Bull. chem. Soc. Japan* **37** (1964) 864

[591] French, H. E. and Sears, K. *J. Am. chem. Soc.* **73** (1951) 469

[592] (a) Wibaut, J. P. and de Jong, J. I. *Recl Trav. chim. Pays-Bas Belg.* **68** (1949) 485; (b) Beets, M. G. J. **63** (1944) 120

[593] Hartmann, M. and Bosshard, W. *Helv. chim. Acta* **24** (1941) 28E

[594] (a) Levine, R. and Raynolds, S. *J. org. Chem.* **25** (1960) 530; (b) Osuch, C. and Levine, R. **21** (1956) 1099; (c) Raynolds, S. and Levine, R. *J. Am. chem. Soc.* **82** (1960) 472

[595] Weiss, M. J. and Hauser, C. R. *J. Am. chem. Soc.* **71** (1949) 2023

[596] Dornow, A. and Bruncken, K. *Chem. Ber.* **85** (1950) 189; with Machens, H. **84** (1951) 147

[597] Hauser, C. R. and Humphlett, W. H. *J. Am. chem. Soc.* **72** (1950) 3805

[598] Osborne, D. R. and Levine, R. *J. heterocycl. Chem.* **1** (1964) 138

[599] Adamson, D. W. and Billinghurst, J. W. *J. chem. Soc.* (1950) 1039

[600] (a) Gilman, H. and Spatz, S. M., *J. Am. chem. Soc.* **62** (1940) 446; (b) Murray, A., Foreman, W. W. and Langham, W. **70** (1948) 1037

[601] Woodward, R. B. and Kornfeld, E. C. *Org. Synth.* **29** (1949) 44

[602] Miller, A. D. and Levine, R. *J. org. Chem.* **22** (1957) 168

[603] Lochte, H. L. and Cheavens, T. M. *J. Am. chem. Soc.* **79** (1957) 1667

[604] Ames, D. E. and Warren, B. T. *J. chem. Soc.* (1964) 5518

[605] Ames, D. E. and Archibald, J. L. *J. chem. Soc.* (1962) 1475

[606] Martensson, O. and Nilsson, E. *Acta chem. scand.* **15** (1961) 1021

[607] Govindachari, T. R., Narasimhan, N. S. and Rajadurai, S. *J. chem. Soc.* (1957) 560

[608] Profft, E. and Schneider, F. *Arch. Pharm., Berl.* **289** (1956) 99

[609] Sperber, N., Papa, D., Schwenk, E. and Sherlock, M. *J. Am. chem. Soc.* **71** (1949) 887

[610] Cale, A. D., McGinnis, R. W. and Teague, P. C. *J. org. Chem.* **25** (1960) 1507

[611] Nunn, A. J. and Schofield, K. *J. chem. Soc.* (a) (1952) 589; (b) (1953) 716

[612] Ross, N. C. and Levine, R. *J. org. Chem.* **29** (1964) 2346; (b) Reinecke, M. G. and Kray, L. R. *ibid.* 1738

[613] Kato, T. and Goto, Y. *Chem. pharm. Bull., Tokyo* **11** (1963) 461

[614] de Jong, J. I. and Wibaut, J. P. *Recl Trav. chim. Pays-Bas Belg.* **70** (1951) 962

[615] Goldberg, N. N. and Levine, R. *J. Am. chem. Soc.* **77** (1955) 4926

[616] Petrow, V., Rewald, E. L. and Sturgeon, B. *J. chem. Soc.* (1947) 1407

[617] Adler, T. K. and Albert, A. *J. chem. Soc.* (1960) 1794

[618] Clemo, G. R. and Swan, G. A. *J. chem. Soc.* (1945) 603; Albert, A. and Willette, R. E. (1964) 4063; Robison, M. M. and B. L. *J. Am. chem. Soc.* **77** (1955) 457

[619] (a) Okuda, S. and Robison, M. M. *J. org. Chem.* **24** (1959) 1008; (b) Herz, W. and Murty, D. R. K. **25** (1960) 2242

REFERENCES

620 (a) Plažek, E. *Ber. dtsch. chem. Ges.* **62** (1929) 577; (b) van Rijn, P. J. *Recl Trav. chim. Pays-Bas Belg.* **45** (1926) 267; (c) Blau, F. *Mh. Chem.* **10** (1889) 372; (d) Koenigs, E., Gerdes, H. C. and Sirot, A. *Ber. dtsch. chem. Ges.* **61** (1928) 1022

621 (a) Yamamoto, Y. *Chem. Abstr.* **46** (1952) 8109; (b) Hamana, M. *ibid.* 4542

622 Ochiai, E. and Suzuki, I. *Pharm. Bull., Tokyo* **2** (1954) 247

623 Clemo, G. R. and Holt, R. J. W. *J. chem. Soc.* (1953) 1313

624 *Germ. Pat.* 622,345 (1935); (b) 653,200 (1938) [*Chem. Abstr.* **30** (1936) 1396 and **32** (1938) 4180]

625 Plažek, E., Sorokowska, A. and Tolopka, D. *Chem. Abstr.* **33** (1939) 3379

626 Cragoe, E. J. and Hamilton, C. S. *J. Am. chem. Soc.* **67** (1945) 536

627 Baumgarten, H. E., Chien-fan Su, H. and Krieger, A. L. *J. Am. chem. Soc.* **76** (1954) 596

628 Katritzky, A. R., Beard, J. A. T. and Coats, N. A. *J. chem. Soc.* (1959) 3680

629 Taylor, E. C. and Crovetti, A. J. *J. org. Chem.* **19** (1954) 1633

630 Itai, T. and Ogura, H. *J. pharm. Soc. Japan* **75** (1955) 292

631 Kirpal, A. and Reiter, E. *Ber. dtsch. chem. Ges.* **58** (1925) 699

632 Ziegler, J. B. *J. Am. chem. Soc.* **71** (1949) 1891

633 (a) Wedenhagen, R. and Train, G. *Ber. dtsch. chem. Ges.* **75** (1942) 1936; (b) Tschitschibabin, A. E., with Kirsanow, A. W. **60** (1927) 766; (c) with Knunianz, I. L. **61** (1928) 427

634 Räth, C. and Prange, G. *Justus Liebigs Annln Chem.* **467** (1928) 1

635 Lappin, G. R. and Slezak, F. B. *J. Am. chem. Soc.* **72** (1950) 2806

636 Sutherland, J. K. and Widdowson, D. A. *J. chem. Soc.* (1964) 4650

637 Berson, J. A. and Cohen, T. *J. org. Chem.* **20** (1955) 1461

638 Ferrier, B. M. and Campbell, N. *Chemy Ind.* (1958) 1089

639 Hayashi, E., Yamanaka, H. and Shimizu, K. *Chem. pharm. Bull., Tokyo* **7** (1959) 141

640 (a) Ochiai, E., with Arima, K. and Ishikawa, M. *J. pharm. Soc. Japan* **63** (1943) 79; (b) with Katoh, T. **71** (1951) 156; [*Chem. Abstr.* **45** (1951) 9542]

641 Meisenheimer, J. *Ber. dtsch. chem. Ges.* **59** (1926) 1848

642 Ishikawa, M. and Sai, Zai-Ren *J. pharm. Soc. Japan* **63** (1943) 78

643 Hands, A. R. and Katritzky, A. R. *J. chem. Soc.* (1958) 1754

644 Gilman, H. and Edward, J. T. *Can. J. Chem.* **31** (1953) 457

645 den Hertog, H. J., Maas, J., Kolder, C. R. and Combé, W. P. *Recl Trav. chim. Pays-Bas Belg.* **74** (1955) 59

646 Ochiai, E. *Proc. imp. Acad. Japan* **19** (1943) 307 [*Chem. Abstr.* **41** (1947) 5880]

647 Kubota, T. and Miyazaki, H. *Bull. chem. Soc. Japan* **35** (1962) 1549

648 Emerson, T. R. and Rees, C. W. *J. chem. Soc.* (1962) 1923

649 Kröhnke, F. and Schäfer, H. *Chem. Ber.* **95** (1962) 1098

650 Jerchel, D. and Melloh, W. *Justus Liebigs Annln Chem.* **613** (1958) 144

651 (a) Hamana, M. and Yoshimura, H. *J. pharm. Soc. Japan* **72** (1952) 1051; (b) Ochiai, E. and Suzuki, I. *Chem. pharm. Bull., Tokyo* **2** (1954) 147 [*Chem. Abstr.* **50** (1956) 1015]

652 (a) Hamana, M. *J. pharm. Soc. Japan* **75** (1955) 135; (b) *ibid.* 130; (c) *ibid.* 139 [*Chem. Abstr.* **50** (1956) 1818]

653 (a) Emerson, T. R. and Rees, C. W. *J. chem. Soc.* (1964) 2319; (b) (1962) 1917

654 Abramovitch, R. A. and Adams, K. A. H. *Can. J. Chem.* **39** (1961) 2134

655 Relyea, D. I., Tawney, P. O. and Williams, A. R. *J. org. Chem.* **27** (1962) 477

656 (a) Schweizer, E. E. and O'Neill, G. J. *J. org. Chem.* **28** (1963) 2460; (b) with Wemple, J. N. **29** (1964) 1744

657 Hata, N. and Tanaka, I. *J. chem. Phys.* **36** (1962) 2072

658 Ochiai, E., Katada, M. and Naito, T. *J. pharm. Soc. Japan* **64** (1944) 210 [*Chem. Abstr.* **45** (1951) 5154]

659 Feely, W., Lehn, W. L. and Boekelheide, V. *J. org. Chem.* **22** (1957) 1135

660 Boekelheide, V. and Feely, W. *J. Am. chem. Soc.* **80** (1958) 2217

661 Coats, N. A. and Katritzky, A. R. *J. org. Chem.* **24** (1959) 1836

662 Ladenburg, A. *Ber. dtsch. Chem. Ges* **16** (1883) 1410, 2059; **18** (1885) 2961

663 Eckert, A. and Loria, S. *Mh. Chem.* **38** (1917) 225

664 (a) Tschitschibabin, A. E. *Zh. russk. fiz.-khim. Obshch.* **33** (1901) 249; **34** (1901) 133; (b) *ibid.* 130

665 Tschitschibabin, A. E. and Rjumschin, P. F. *Zh. russk. fiz.-khim. Obshch.* **47** (1915) 1297 [*Chem. Abstr.* **9** (1915) 3057]

666 Braun, J. v. and Pinkernelle, W. *Ber. dtsch. Chem. Ges.* **64** (1931) 1871

667 Hands, C. H. G. and Whitt, F. R. *J. Soc. chem. Ind., Lond.* **66** (1947) 407

668 Crook, K. E. *J. Am. chem. Soc.* **70** (1948) 416

669 Braun, J. v. and Nelles, J. *Ber. dtsch. chem. Ges.* **70** (1937) (a) 1767; (b) 1760

670 Setkina, V. N., Baranetskaya, N. K. and Kursanov, D. N. *Bull. Acad. Sci. USSR, Div. chem. Sci.* (1955) 667

671 Setkina, V. N. and Kursanov, D. N. *Chem. Abstr.* **43** (1949) 6161

672 Tatsuoka, S. and Inui, Y. *J. pharm Soc. Japan* **69** (1949) 535

[673] (a) Snyder, H. R. with Speck, J. C. *J. Am. chem. Soc.* **61** (1939) 2895; (b) with Eliel, E. L. and Carnahan, R. E. **73** (1951) 970

[674] Zincke, T. and Krollpfeiffer, F. *Justus Liebigs Annln Chem.* **408** (1915) 293

[675] Zincke, T. *J. prakt. Chem.* **82** (1910/2) 17

[676] Biedig, H.-J. and Reidies, A. *Chem. Ber.* **89** (1956) 550

[677] Kröhnke, F. and Vogt, I. *Justus Liebigs Annln Chem.* **589** (1954) 52

[678] Kobayashi, T. and Inokuchi, N. *Tetrahedron* **20** (1964) 2055

[679] Kröhnke, F. and Heffe, W. *Ber. dtsch. chem. Ges.* **70** (1937) 864

[680] Pearson, R. G. and Sandy, A. C. *J. Am. chem. Soc.* **73** (1951) 931

[681] (a) Marckwald, W., Klemm, W. and Trabert, H. *Ber. dtsch. chem. Ges.* **33** (1900) 1556; (b) Koenigs, E. and Kinne, G. **54** (1921) 1357

[682] Schmidt, U. and Giesselmann, G. *Chem. Ber.* **93** (1960) 1591

[683] (a) Bittner, K. *Ber. dtsch. chem. Ges.* **35** (1902) 2933; (b) Koenigs, E. and Kantrowitz, H. **60** (1927) 2097; (c) *U.S. Pat.* 1,753,658 (1930); (d) Räth, C. *Justus Liebigs Annln Chem.* **487** (1931) 105; (e) Sucharda, E. and Troskiewicz, C. *Roczn. Chem.* **12** (1932) 493; (f) Thirtle, J. R. *J. Am. chem. Soc.* **68** (1946) 342

[684] Comrie, A. M. and Stenlake, J. B. *J. chem. Soc.* (1958) 1853

[685] Yamamoto, Y. *J. pharm. Soc. Japan* **71** (1951) 1436

[686] Takahashi, T. and Ueda, K. *Chem. Abstr.* **48** (1954) 5187

[687] Profft, E. and Rolle, W. *Chem. Abstr.* **55** (1961) 1609; *J. prakt. Chem.* **11** (1960) 22

[688] Klingsberg, E. and Papa, D. *J. Am. chem. Soc.* **71** (1949) 2373

[689] den Hertog, H. J., van der Plas, H. C. and Buurman, D. J. *Recl Trav. chim. Pays-Bas Belg.* **77** (1958) 963

[690] Evans, R. F. and Brown, H. C. *J. org. Chem.* **27** (1962) 1329

[691] Dohrn, M. and Diedrich, P. *Justus Liebigs Annln Chem.* **494** (1932) 284

[692] Mangini, A. and Colonna, M. *Chem. Abstr.* **41** (1947) 1224

[693] (a) Jones, R. A. and Katritzky, A. R. *J. chem. Soc.* (1958) 3610; (b) Albert, A. and Barlin, G. B. (1959) 2384

[694] Jones, R. A. and Katritzky, A. R. *J. chem. Soc.* (1960) 2937

[695] (a) Fibel, L. R. and Spoerri, P. E. *J. Am. chem. Soc.* **70** (1948) 3908; (b) Klingsberg, E. and Papa, D. **73** (1951) 4988

[696] Forrest, H. S. and Walker, J. *J. chem. Soc.* (1948) 1939

[697] Saikachi, H. *J. pharm. Soc. Japan* **64** (1944) 201

[698] Hansch, C., Carpenter, W. and Todd, J. *J. org. Chem.* **23** (1958) 1924

[699] Michaelis, A. and Hölken, A. *Justus Liebigs Annln Chem.* **331** (1904) 245

[700] (a) Takahashi, T. and Ueda, K. *Pharm. Bull., Tokyo* **2** (1954) 34; (b) with Ichimoto, T. *ibid.* 196; (c) **3** (1955) 356

[701] (a) *Br. Pat.* 582,638 (1946); (b) 637,130 (1950); *U.S. Pat.* 2,544,904 (1951)

[702] Katz, L., Schroeder, W. and Cohen, M. *J. org. Chem.* **19** (1954) 711

[703] Courtot, C. and Zwilling, J. P. *Congrès chim. ind.* **18II** (1938) 796; *Chem. ZentBl.* **II** (1939) 3412

[704] Talik, Z. and Plažek, E. *Chem. Abstr.* **51** (1957) 17911

[705] Machek, G. *Mh. Chem.* **72** (1938) 77

[706] Reinhart, F. E. *J. Franklin Inst.* **236** (1943) 316

[707] Leonard, F. and Wajngart, A. *J. org. Chem.* **21** (1956) 1077

[708] Panouse, J.-J. *C.r. hebd. Séanc. Acad. Sci., Paris* **230** (1950) 846

[709] Takahashi, T. and Goto, H. *Chem. Abstr.* **45** (1951) 4716

[710] Takahashi, T. and Yamamoto, Y. *J. pharm. Soc. Japan* **70** (1950) 187

[711] (a) Elkaschef, M. A.-F. and Nosseir, M. H. *J. Am. chem. Soc.* **82** (1960) 4344; (b) Caldwell, W. T., Tyson, F. T. and Lauer, L. **66** (1944) 1479; Hurd, C. D. and Morissey, C. J. **77** (1955) 4658

[712] Kubota, S. and Akita, T. *Chem. Abstr* **55** (1961) 19926

[713] Takahashi, T., with Saikachi, H. and Itoh, S. *Chem. Abstr.* **45** (1951) 4716; with Shibasaki, J. and Uchibayashi, M. **50** (1956) 336

[714] Reinhart, F. E. *J. Franklin Inst.* **249** (1950) 248

[715] Surrey, A. R. and Lindwall, H. G. *J. Am. chem. Soc.* **62** (1940) 173

[716] Colonna, M. *Chem. Abstr.* **34** (1940) 4737; **37** (1943) 4399

[717] Comrie, A. M. *J. chem. Soc.* (1963) 688

[718] Linetskay, Z. G. and Sapozhnikova, N. V. *Dokl. Akad. Nauk SSSR* **86** (1952) 753

[719] Martin, E. C. *Dissert. Abstr.* **20** (1960) 3533

[720] Kosower, E. M. and Poziomek, E. J. *J. Am. chem. Soc.* **85** (1963) 2035; **86** (1964) 5515

[721] Kosower, E. M. and Schwager, I. *J. Am. chem. Soc.* **86** (1964) 5528

[722] Tschitschibabin, A. E. *Zh. russk. fiz.-khim. Obshch.* **50** (1918) 502

[723] Schlögl, K. and Fried, M. *Mh. Chem.* **94** (1963) 536

[724] Cadogan, J. I. G. *J. chem. Soc.* (1962) 4257

[725] Dirstine, P. H. and Bergstrom, F. W. *J. org. Chem.* **11** (1946) 55

[726] Williams, G. H. *Homolytic Aromatic Substitution*, London (Pergamon) 1960

REFERENCES

[727] Rapoport, H. and Look, M. *J. Am. chem. Soc.* **75** (1953) 4606

[728] Anderson, T. *Justus Liebigs Annln Chem.* **154** (1870) 270

[729] Emmert, B. *Ber. dtsch. chem. Ges.* **47** (1914) 2598; **49** (1916) 1060; **50** (1917) 31; with Buchert, R. **54** (1921) 204

[730] Wibaut, J. P. and Dingemanse, E. *Recl Trav. chim. Pays-Bas Belg.* **42** (1923) 240

[731] Leffler, M. T. *Org. React.* **1** (1942) 91

[732] Smith, C. R. *J. Am. chem. Soc.* **46** (1924) 414

[733] Zahlan, A. B. and Linnell, R. H. *J. Am. chem. Soc.* **77** (1955) 6207

[734] Setton, R. *C.r. hebd. Séanc. Acad. Sci., Paris* **244** (1957) 1205

[735] Ward, R. L. *J. Am. chem. Soc.* **83** (1961) 3623

[736] Barnes, R. A. in *Pyridine and Its Derivatives*, Part 1, ed. Klingsberg, E., New York (Interscience) 1960

[737] Tokuyama, M. *Chem. Abstr.* **49** (1955) 15895

[738] Jackson, G. D. F., Sasse, W. H. F. and Whittle, C. P. *Aust. J. Chem.* **16** (1963) 1126

[739] Rapoport, H., Iwamoto, R. and Tretter, J. R. *J. org. Chem.* **25** (1960) 372

[740] Badger, G. M. and Sasse, W. H. F. (*a*) *J. chem. Soc.* (1956) 616; (*b*) *Advances in Heterocyclic Chemistry*, Vol. 2, edit. Katritzky, A. R., New York (Academic Press) 1963

[741] Sasse, W. H. F. and Whittle, C. P. (*a*) *J. chem. Soc.* (1961) 1347; (*b*) *Aust. J. Chem.* **16** (1963) 14

[742] Reinecke, M. G. and Kray, L. R. *J. Am. chem. Soc.* **86** (1964) 5355

[743] Hanania, G. I. H., Irvine, D. H. and Shurayh, F. *J. chem. Soc.* (1965) 1149

[744] Meek, T. L. and Cheney, G. E. *Can. J. Chem.* **43** (1965) 64

[745] Blanch, J. H. and Onsager, O. T. *J. chem. Soc.* (1965) 3729, 3734

[746] Golding, S. and Katritzky, A. R. *Can. J. Chem.* **43** (1965) 1250

[747] Hughes, B. and Suschitzky, H. *J. chem. Soc.* (1965) 875

[748] Zatsepina, N. N., Tupitsyn, I. F. and Efros, L. S. *J. gen. Chem. U.S.S.R.* **34** (1964) 4124, 4130

[749] Berson, J. A. and Evleth, E. M., with Hamlet, Z. *J. Am. chem. Soc.* **87** (1965) 2887; with Manatt, S. L. *ibid.* 2901, 2908

[750] Miskina, I. M. and Efros, L. S. *J. gen. Chem. U.S.S.R.* **32** (1962) 2185

[751] Williams, J. L. R., Carlson, J. M., Adel, R. E. and Reynolds, G. A. *Can. J. Chem.* **43** (1965) 1345

[752] Chambers, R. D., Hutchinson, J. and Musgrave, W. K. R. *J. chem. Soc.* (1965) 5040

[753] (*a*) Beak, P. and Bonham, J. *J. Am. chem. Soc.* **87** (1965) 3365; (*b*) Paquette, L. A. *ibid.* 5186

[754] Wegler, R. and Pieper, G. *Chem. Ber.* **83** (1950) 6; Chumakov, Y. I. and Ledovskikh, V. M. *Tetrahedron* **31** (1965) 937

[755] Boatman, S., Harris, T. M. and Hauser, C. R. *J. org. Chem.* **30** (1965) 3593

[756] Ruchardt, C., Eichler, S. and Kräty, O. *Tetrahedron Lett.* (1965) 233; Cohen, T., Song, I. H. and Fager, J. H. *ibid.* 237; Koenig, T. *ibid.* 3127; Cohen, T. and Fager, J. H. *J. Am. chem. Soc.* **87** (1965) 5701

[757] Johnson, S. L. and Rumon, K. A. *J. phys. Chem., Ithaka* **68** (1964) 3149; *J. Am. chem. Soc.* **87** (1965) 4782

[758] Itoh, M. and Nagakura, S. *Tetrahedron Lett.* (1965) 417

INDEX: TABLE OF SUBSTITUTION REACTIONS OF PYRIDINES

Substrates	Leaving group	Entering group	PAGE

Alkylation and substituted alkylation—continued

Substrates	Leaving group	Entering group	PAGE
Pyridines	halogen	alkyl	201, 203
		.CH(CN)Ph, .CEt(CN)Ph, .C(CN)Ar$_2$, .C(Me$_2$).CO$_2$Et, .CEt.(CO$_2$Et)$_2$, .CHAr.SO$_2$R, .CH(CO$_2$Et)$_2$, .CEt(CO$_2$Et)$_2$	203
Pyridine (via pyridyne)	Br	.CH$_2$COPh	204
Pyridine 1-oxides	Br	.CHAc.CO$_2$Et, .CHPh.CO$_2$Et	204
Pyridinium ions	halogen	alkyl	203
		.CH(COPh).COCH$_2$R, .CH$_2$.C$_5$H$_4$$\overset{+}{N}$Et	205
Pyridines	.OMe	.CH(CO$_2$Et)$_2$	203
Pyridones	:O	:CHR	201
Pyridinium ions	.OR	alkyl	203

Amination and substituted amination

Substrates	Leaving group	Entering group	PAGE
Pyridines	H	.NH$_2$, .NHPh, .NHR, .NHNHR	206–8
Pyridinium ions		:$\overset{+}{\underset{\shortmid}{N}}$.	186, 209–10
Pyridines	.NH$_2$	.NH.C$_5$H$_4$N	210
Pyridinium ions	:$\overset{+}{\underset{\shortmid}{N}}$.	.NH$_2$, .NHPh	210
	.CN	.NH$_2$	210
Pyridines	halogen	.NH$_2$, .NHR, .NR$_2$	191, 210–1, 214, 216
		.NHAr, .NHNHR	211, 216
		:$\overset{+}{\underset{\shortmid}{N}}$.	191, 212, 216
Pyridines (via pyridynes)		.NH$_2$, .NR$_2$	214
Pyridine 1-oxides		.NH$_2$, .NR$_2$, .NHNH$_2$	213, 215–6
Pyridine 1-oxides (via pyridynes)		.NH$_2$	215
Pyridinium salts		.NH$_2$, .NHR, .NHAr	213
		.NHNHCOPh	213
		:$\overset{+}{\underset{\shortmid}{N}}$. (intramolecular)	
Pyridines	.OH, .OAr .O.COMe,	.NH$_2$, .NHPh	218–9
	.O.SO$_2$C$_7$H$_7$	:$\overset{+}{\underset{\shortmid}{N}}$.	219
Pyridine 1-oxides	.OAr	.NHR	219
Pyridinium ions	.OR, .OAr	.NHR	219
Pyridines	.NO$_2$	.NH$_2$, .NHR, .NR$_2$	212–3, 219
	.SO$_3$H	.NH$_2$, .NHR	219–20
Pyridine 1-oxides	.SR	.NHR, .NHAr	219
Pyridinium ions	.SR, .SAr	.NHR	219

Arylation

Substrates	Leaving group	Entering group	PAGE
Pyridines	H	.Ar	220–3
Pyridine 1-oxides		.Ar	220–3
Pyridinium ions		.Ar	220, 223
Pyridines	Br	.Ph	223
	Cl	.Ar(intramolecular)	223–4

Modified carbonyl reactions

Substrates	Leaving group	Entering group	PAGE
Pyridines	H	.CR$_2$OH	224–5

* When the entering group is in brackets the substitution is not completed but stops at the stage of addition.

SUBJECT INDEX

(Page numbers in italics refer to Tables.)